Foundations For College Mathematics 2e

Edward D. Laughbaum
The Ohio State University
Columbus, Ohio

RED BANK PUBLISHING
849 WEDGEWOOD DRIVE
MARYSVILLE, OH 43040
www.redbankpublishing.com
SAN: 856-4531

ISBN-13: 978-0-9817536-2-1, and ISBN-10: 0-9817536-2-0

TABLE OF CONTENTS

CHAPTER ONE
NUMBERS

CHAPTER TWO
REPRESENTATION AND BEHAVIOR OF FUNCTIONS

CHAPTER THREE
COMMON BEHAVIORS OF FUNCTIONS

CHAPTER FOUR
FUNCTIONS: NOTATION AND OPERATIONS

CHAPTER FIVE
ADVANCED ANALYSIS OF THE LINEAR FUNCTION

CHAPTER SIX
EQUATIONS AND INEQUALITIES

CHAPTER SEVEN
ADVANCED ANALYSIS OF THE EXPONENTIAL FUNCTION

CHAPTER EIGHT
ANALYSIS OF THE RATIONAL FUNCTION

CHAPTER NINE
ADVANCED ANALYSIS OF THE SQUARE ROOT FUNCTION

CHAPTER TEN
ADVANCED ANALYSIS OF THE QUADRATIC FUNCTION

CHAPTER ELEVEN
BASIC GEOMETRY

CHAPTER TWELVE
BASIC TRIGONOMETRY

CHAPTER THIRTEEN
SYSTEMS OF EQUATIONS AND INEQUALITIES

CHAPTER FOURTEEN
INTRODUCTION TO THE ANALYSIS OF THE LOGARITHMIC FUNCTION

Preface to Students and Instructors

This text contains terminology, content, and algorithms that may not be found in a traditional textbook because it is the author's intention to break from tradition and prepare students for the mathematics needed in a modern society. Further, as learning progresses, terminology may change to reflect new understandings. The author also recognizes that mathematics is learned by understanding, not by memorization. Developing mathematical ideas in the context of a situation helps students understand mathematics; at the same time, students understand that mathematics relates directly to the world.

The use of this text requires a graphing calculator with function notation; the calculator will be used as a tool to help us understand mathematical concepts and perform mathematical algorithms. Technology, which offers students and teachers a variety of methods for solving problems, is used in this text to explore mathematical ideas, and it changes what is considered important. Many traditional mathematical topics have diminished importance. This text offers content that is important to students directly entering the work force and to students continuing the study of mathematics and science. Many of the exercise sets offer questions that are engaging and demand higher level thinking.

Students have the responsibility to read and study the text in order to learn. Students have the responsibility to view the exercise sets as questions that are to stimulate thinking, not questions that are designed to encourage memorization of facts. Students have the responsibility to recognize that the exercise sets contain questions to help them formulate their own ideas about mathematical relationships.

There are limited references to calculator keystrokes; when they are included, they are for theTI-83 or TI-84. With slight modification, the keystrokes will also work on other calculators.

The teaching features of this text include:

- Technology integrated throughout to enhance the learning and teaching of mathematical concepts.
- Technology integrated throughout to provide options for performing mathematical algorithms.
- Mathematical concepts introduced in the context of real-world situations.
- Thorough analysis of applications.
- Guided discovery exercises.
- Reduced emphasis on the use of symbol manipulation and increased emphasis on the use of function as a central theme.
- Three methods (numeric, graphic, and algebraic) of representing a function.

- Distributed learning has been incorporated in this text. For example, the idea of function is introduced in Section 2.1 as a data relationship. The intuitive idea of function is further developed in Chapter Three by looking at algebraic expressions that model data relationships introduced in Chapter Two. Chapter Four continues with formal function notation, and functions are used in equation and inequality solving in Chapter Six. Chapters Seven, Eight, Nine, Ten, Twelve, and Fourteen offer analysis of individual function types.
- A variety of methods for solving problems are encouraged. Students are encouraged to explore on their own.
- Higher level thinking skills are encouraged through projects, open ended questions, and concept questions.
- Sample problems are checked numerically or graphically and students are encouraged to do likewise.
- Exercise sets contain low-level difficulty problems that prepare students for future topics.
- Exercise sets contain high-level difficulty problems that review topics previously taught.
- Exercise sets contain writing questions.
- Exercise sets contain concept questions.
- Exercise sets contain exploration problems. Many of the explorations can be used for group work. Many can be used as portfolio exercises.
- Exercise sets contain open ended questions.
- Mathematical words are defined by bold text.
- Extended laboratory projects on modeling are included with the text.

Neuro and cognitive science research has provided considerable information about how the brain functions. This textbook capitalizes on this research through the implementation of the cognitive processes of associations, pattern recognition, attention, visualizations, priming, meaning, and the enriched teaching/learning environment.

- We remember algebra longer and have better memory by using associations – made through function permeating the content. That is, we are more likely to remember the mathematics taught because we capitalize on associations made through using a function approach.
- Learning is made simpler, faster, and more understandable by using pattern building as a teaching tool. In the function approach used in this text, almost all of the pencil and paper activities, e-teaching activities, and class discussions use pattern building to reach a generalization about a concept or skill.
- We cannot learn if we are not paying attention. The graphing calculator is used to draw attention to the mathematics through its basic functionalities **including,** various app software.
- Without visualizations, we do not understand or remember the mathematics as well. In the function approach visualizations are used <u>first</u> before any symbolic development. This greatly increases the likelihood that we will remember the mathematical concept being taught.

- Considerable brain processing takes place in the subconscious side of the brain, including a learning module. To make this processing possible for our purposes, the brain must be primed. The function implementation module (Chapters Two and Three) and early e-learning activities prime the brain for all the algebra that follows.
- The enriched teaching/learning environment promotes correct memory of math content. The wide variety of teaching activities facilitated by the function approach provides the enriched environment.
- Contextual situations (often represented as relationships) provide meaning to the algebra learned. Algebra taught without meaning creates memories without meaning that are quickly forgotten.

Algebra approached through function, as in this textbook, has reordered the content and capitalized on function concepts to develop understanding, long-term memory, and skills.

This text has been published as a preliminary edition, as a revised preliminary edition, a first edition test, and now as a second edition. It has been class tested over a ten year period and has been revised and edited no less than twelve times. It has been reviewed by over twenty-five reviewers.

Edward D. Laughbaum
The Ohio State University
Columbus, OH 43210

CHAPTER ONE
NUMBERS

1.0 Introduction and Numbers

Before algebra can be studied with any great detail, the properties of real numbers -- the properties upon which most mathematical topics are based -- must be analyzed. It is also important for you to know how to perform the arithmetic operations of addition, subtraction, multiplication, division, powers, absolute values, and square roots of numbers before you continue the study of algebra. Nearly *all* work in mathematics and applied mathematics assumes a working knowledge of arithmetic of real numbers. Number properties are integral to work in arithmetic and algebra.

Pictured below is a graph of all real numbers with a few sample real numbers labeled. You should recognize the picture as the **number line**; it is the graphical representation of all real numbers – as implied by the arrows at the ends of the number line.

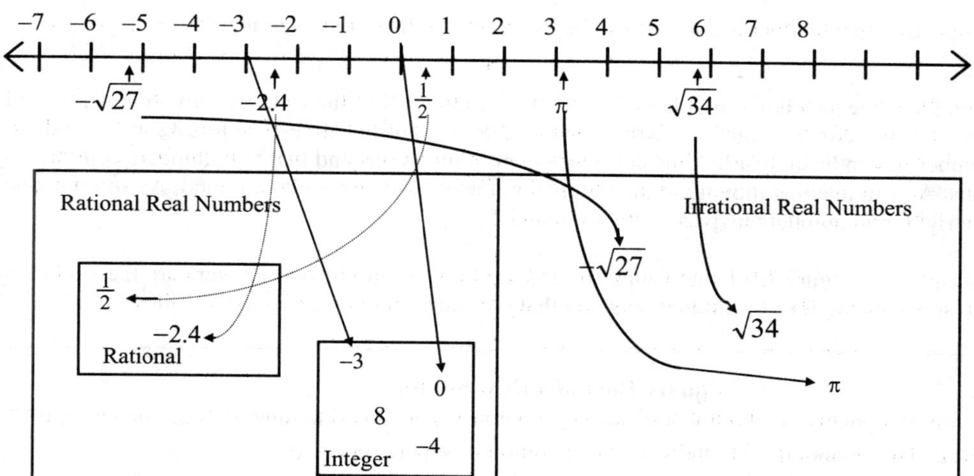

Figure 1.0.1 Real Numbers

A few of the number sets you should be familiar with are the **counting** or **natural numbers** $\{1, 2, 3 \ldots\}$; **integers** $\{\ldots -3, -2, -1, 0, 1, 2 \ldots\}$; **rational numbers** $\{p/q$, where p and q are integers, but q is not zero$\}$; for example $\left\{\ldots -\frac{25}{3}, -\frac{9}{2}, -\frac{4}{2}, -\frac{1}{8}, \frac{2}{3}, \frac{7}{9}, \frac{63}{17}, \ldots\right\}$ and finally, **irrational numbers,** such as $\{\ldots -\sqrt{6}, -\sqrt{2}, \sqrt{3}, \sqrt{5}, \pi \ldots\}$. Visualize the number line. A real number (excluding 0) that is the negative of another number is the same distance from zero but on the opposite side of zero.

Opposite Numbers

Visually (on a number line), two numbers are **opposites** when they are on opposite sides of the number zero and the same distance from zero. The **opposite** of zero is zero. Symbolically, a and $-a$ are opposites. The sum of opposites is zero; that is, $a + (-a) = 0$.

Real numbers are either rational or irrational. This is to say that any real number can be written as a ratio (fraction} of two integers – a rational number, or it cannot be written as a ratio of two integers – an irrational number. The integers are labeled on the top of the number line in Figure 1.0.1 and a few examples of other real numbers are labeled below the number line. The numbers -2.4, $\frac{1}{2}$, and the integers are examples of rational; $-\sqrt{27}, \sqrt{34}$, and π are examples of irrational numbers.

Other real numbers that you have seen in grade school are repeating decimals like the number 0.6666… They are rational numbers because they can be expressed as ratios of two integers, like the rational number $\frac{2}{3}$.

Repeat it until you drop.

Another example of a real number is 2.8%. It can be expressed as the rational number 0.028 or $\frac{28}{1000}$.

Imagine the number line as a horizontal line. When you do, numbers to the right of zero are positive and numbers to the left of zero are negative. Zero is neither positive nor negative. The arrows at the ends of the above number line indicate that the line continues in both directions and that real numbers continue in both directions. As you imagine moving to the left, real numbers decrease without bound. As you imagine moving to the right, real numbers increase without bound.

The numbers shown in Figure 1.0.1 that cannot be expressed as a ratio of two integers are the irrational numbers. The above examples of irrational numbers that are square roots are defined as follows:

Square Root of a Real Number

The **square root** of a positive real number or zero a, is a non-negative real number that can be multiplied by itself to get a. The symbol used to indicate the operation of square root is $\sqrt{}$.

The square root of 36 is 6 because when 6 is multiplied by itself, the product is 36. This operation is normally written as $\sqrt{36}$. The square root of -25 is not a real number because there is no real number that

can be squared (multiplied by itself) to get -25. Square roots are needed in geometry, for example, given the sides of a square or rectangle, the calculation of the diagonal quite often is an irrational number. If the graphing calculator is used to calculate a square root like $\sqrt{15}$, it will likely not return the exact root. It may return a rational number that is approximately equivalent to $\sqrt{15}$. If the graphing calculator is used to approximate the square root of a sum, difference, product, quotient, etc. of numbers, the calculator's grouping symbols must enclose the operation. For example, $\sqrt{13-5}$ must be entered as $\sqrt{}(13-5)$.

Another arithmetic operation you will use is the absolute value of a real number. Absolute values will be used in several chapters when you encounter problems with constant rates of change – like federal taxes. If you imagine a number on the number line, the distance to zero is called the **absolute value** of that number.

Absolute Value of a Real Number

The **absolute value** of a real number x is the distance from x to zero.

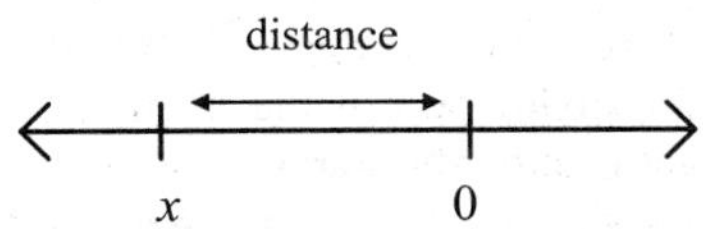

Distances are positive numbers or 0.

The notation used for absolute value is a pair of vertical bars around the number. For example, $|4|$ means the absolute value of 4; the absolute value of 4 is 4 because the distance from 0 to 4 is 4 units. The $|-6|$ means the distance from 0 to –6, which is 6. The $|0|$ is 0 because the distance from 0 to 0 is 0. The graphing calculator can calculate absolute values; however, it does not use vertical bars. Typically, it uses the notation *abs* with parentheses as the grouping notation. This is commonly used by computers and graphing calculators. For example, $|3-5|$ looks like *abs* (3 – 5) on some calculators and computers.

Example 1: Find $|-4|$, $\sqrt{49}$, and $\sqrt{8}$.

Solution: $|-4|$ is 4 because the distance from –4 to 0 is 4. $\sqrt{49}$ is 7 because 7·7 is 49. $\sqrt{8}$ is approximately 2.828 because (2.828)·(2.828) is nearly 8. The exact value of $\sqrt{8}$ cannot be written in decimal form; however, it can be symbolized as $\sqrt{8}$!

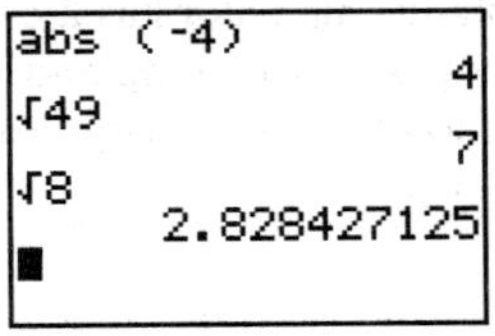

Figure 1.0.2

⁂

Example 2: Find: $|7-(-6)|, \sqrt{9+4}$, and $\sqrt{9 \cdot 4}$

Solution: $|7-(-6)|$ is 13 because 7–(–6) is 13 and the absolute value of 13 is 13. $\sqrt{9+4}$ is approximately 3.605551275 because 9 + 4 is 13 and the square root of 13 is approximately 3.605551275. $\sqrt{9 \cdot 4}$ is 6 because 9 × 4 is 36 and the square root of 36 is 6.

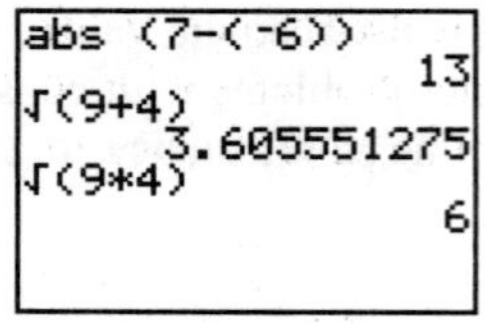

Figure 1.0.3

The calculator *approximates* irrational numbers. If you enter the keystrokes: $\sqrt{}$ 7) ENTER your calculator will approximate the exact irrational number $\sqrt{7}$. If you enter the symbol π on the home screen of the calculator, you will see an approximation of the irrational number π on the screen. That is, the calculator will approximate any exact irrational number you enter. It will use the approximation in place of the exact value when entered on the screen – except in CAS-based calculators.

Number lines do not always have to be horizontal. In mathematics, they are usually either horizontal or vertical. To the right is another number line. The arrow pointing up implies that the numbers continue increasing forever in a positive direction. The arrow pointing down means that the numbers continue decreasing in the negative direction forever. The symbol ∞ means increasing without bound. The symbol −∞ means decreasing without bound.

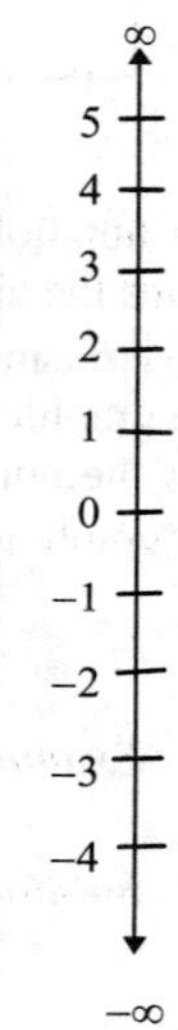

Figure 1.0.4

When analyzing graphical relationships between two sets of numbers, it will be necessary to use two number lines at the same time - one number line in which to show numbers from the first set and another to show numbers from the second set (see Section 2.1). The study of relationships between two sets of numbers is a major focus in this text.

While all of the arithmetic operations on numbers can be done without a calculator, this text utilizes the graphing calculator. The graphing calculator does more than just display graphs; it

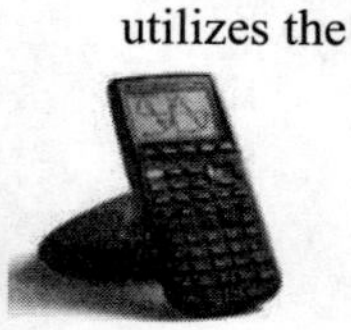

is excellent at performing the operations reviewed in this chapter. Unlike scientific calculators, the graphing calculator has a computer-like screen. You can see both the expression you are evaluating and the value of the expression on the same screen! Arrow keys allow you to move the cursor around the expression, which allows you to edit with the insert and delete keys.

1.1 Properties of Numbers, Equality, and Inequality

The Addition Problem

What do you suppose would be a good way of adding the following quiz scores {3, 6, 7, 4, 9, 5, 5, 1, 7}? Well, if you are in a hurry or want an accurate method, you would probably do what most people who know a little mathematics would do; you would mentally re-group the data like this: {3, 7, 6, 4, 9, 1, 5, 5, 7}. Then you would add 3 + 7, 6 + 4, 9 + 1, and 5 + 5 in your head to give 40 and then add 7. Then sum is 47. You have just used basic number properties to help add the data. The method is faster and it is more accurate than adding from left to right.

Symbols are used in our daily lives; for example, the symbol *chair* is used to represent all objects that fit your definition of a chair. In mathematics, symbols are commonly used to represent numbers. While the symbol 3 is used to represent a collection of three objects or being third in a sequence, a symbol like x can be used to represent *any* real number. In the properties below, think of the symbols a, b, and c as representing any real number. The properties are simply listed here as a reference. It is assumed that you have seen and used these properties before.

Properties of Real Numbers. Did you ever notice when you perform some arithmetic operations the order in which you perform them does not affect the answer? For example, suppose you add –4 + 9 and then add 9 + (–4) as shown in Figure 1.1.1. On the other hand, what happens if you change the order when you subtract? As you can see in Figure 1.1.2, the answer depends on the order in which you subtract. Thus, subtraction does not have the same property as does addition. This property is called the commutative property of addition. This basic property of real numbers does not apply to subtraction.

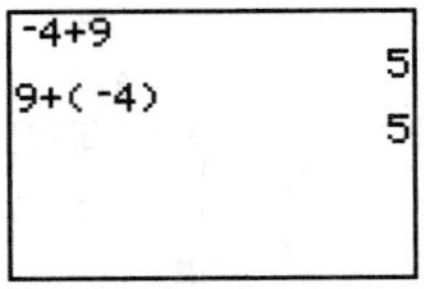

Figure 1.1.1

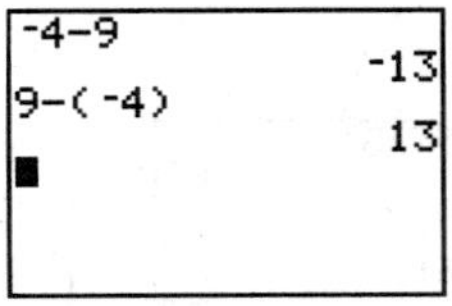

Figure 1.1.2

Commutative Property of Addition $a + b = b + a$

The Commutative Property of Addition shows that the *order* in which addition is done makes no difference; that is, $a + b$ is the same number as $b + a$.

Confirmation: $-5 + 7 = 7 + (-5)$ where each side simplifies to 2

$2 = 2$ therefore, $-5 + 7 = 7 + (-5)$ ⁂

If you investigate further, you will discover this property applies to another operation as well. Note the results of multiplying two numbers. Change the order and multiply again as shown below in Figure 1.1.3.

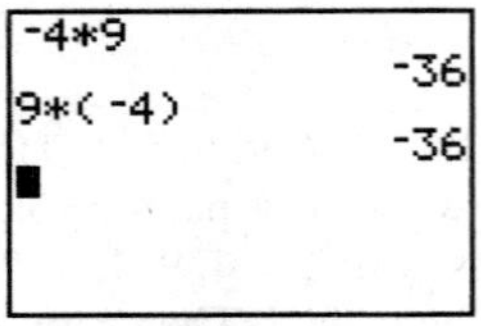

Figure 1.1.3

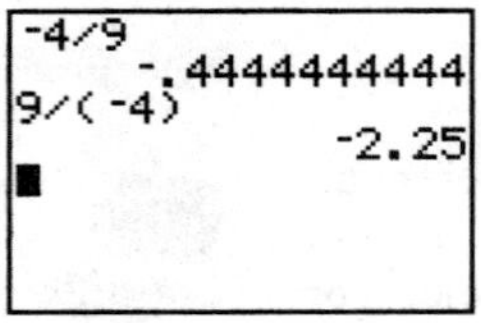

Figure 1.1.4

If you try this same procedure with division, you will quickly see that division does not exhibit the property of order - the commutative property.

Commutative Property of Multiplication

$ab = ba$

The commutative property of multiplication suggests that the *order* in which multiplication is done makes no difference; that is, ab is the same number as ba.

Confirmation: $(-4) \cdot (-7) = (-7) \cdot (-4)$ where each side simplifies to 28;

therefore, $(-4) \cdot (-7) = (-7) \cdot (-4)$

⁂

Do grouping symbols have an effect on the outcome of addition, subtraction, multiplication, or division? The calculator can help you explore this notion. Consider the following screens and then try changing grouping symbols with numbers of your own. What can you conclude about the use of grouping symbols?

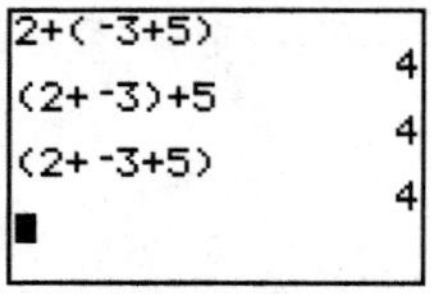

Figure 1.1.5a

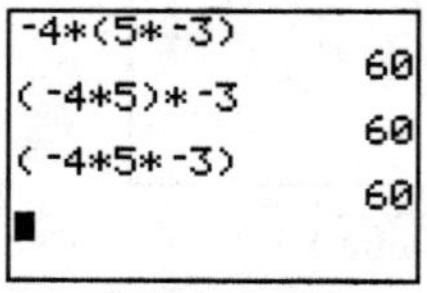

Figure 1.1.5b

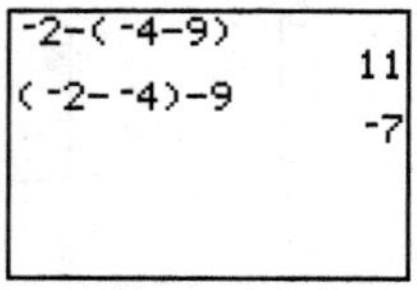

Figure 1.1.5c

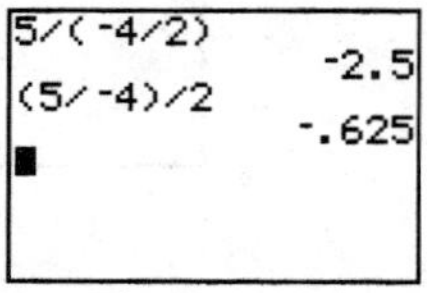

Figure 1.1.5d

Figures 1.1.5a and b demonstrate a property called the **associative property**. Hopefully, it is clear from the Figures 1.1.5c and d that the only operations that show this characteristic are addition and multiplication.

Associative Property of Addition

$a + (b + c) = (a + b) + c$

The associative property of addition indicates that when adding, the grouping may be done in any fashion. The same sum is obtained no matter how the numbers are grouped; $a + (b + c)$ is the same number as $(a + b) + c$. Recall that the operation inside grouping symbols is done first; thus, in the expression $a + (b + c)$, b is added to c first and a is added to this sum. In the expression $(a + b) + c$, a and b are added first and this sum is added to c.

Confirmation: $\{4 + (-3)\} + (-8) = 4 + \{-3 + (-8)\}$

$$1 \quad + (-8) = 4 + \quad (-11)$$

$$-7 = -7$$

Caution: Do not use { } on the calculator for this problem. Use ( ) for all grouping symbols.

⁂

Associative Property of Multiplication

$a(bc) = (ab)c$

The associative property of multiplication indicates that when multiplying, the groupings may be done in any fashion. The same product is obtained no matter how the numbers are grouped; $a(bc)$ is the same number as $(ab)c$. Recall that the operation inside grouping symbols is done first; thus, in the expression $a(bc)$, b is multiplied by c and this product is multiplied by a. In the expression $(ab)c$, a and b are multiplied first and this product is multiplied by c.

Confirmation: $-3\{(-7) \cdot 4\} = \{-3 \cdot (-7)\} \cdot 4$

$$-3 \cdot (-28) = 21 \cdot 4$$

$$84 = 84$$

Caution: Do not use { } on the calculator for this problem. Use ( ) for all grouping symbols.

⁂

All of the previous properties apply to a single operation. The question you may have in mind is "What if two operations are involved; are there any properties that apply?" The answer is yes. One property relates multiplication with addition or subtraction. How do you calculate the product of –2 and the sum of 5 and –8; that is, how do you calculate $-2(5 + (-8))$? Of course, add 5 and –8 and multiply this sum by –2 is the way most would calculate. The answer is 6. Think about another method for calculating as shown in Figure 1.1.6. Instead of finding the sum first, the correct answer can be found by multiplying the –2 times 5 and then –8 followed by adding the results. This second method has significant impact on simplification of symbols in algebra.

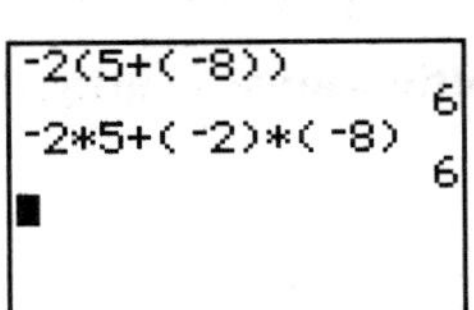

Figure 1.1.6

Distributive Property

$$a(b + c) = ab + ac$$

$$a(b - c) = ab - ac$$

The distributive property relates two operations; it is the connection between addition and multiplication. A real number a times a sum of real numbers $b + c$ has the same value as the number a times the number b, added to the number a times the number c. This same reasoning applies to multiplying a number times a difference.

Confirmation: $\mathbf{-4}(6 + (-5)) = \mathbf{-4} \cdot 6 + \mathbf{(-4)} \cdot (-5)$

$-4 \cdot 1 = -24 + 20$ simplify

$-4 = -4$ simplify

⁂

Another interpretation of the distributive property is found if it is read from right to left. A sum or difference can be written as a product. The sum $ab + ac$ is written on the left side as the product $a \cdot (b + c)$. The process of writing a sum or difference as a product is called **factoring**. Factoring is discussed in Chapter Four.

Consider the results of multiplying the number 1 times a variety of other numbers as shown in Figure 1.1.7.

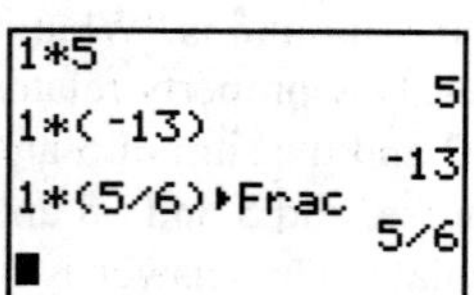

Figure 1.1.7

No matter what the real number 1 is multiplied by, the product is always the number it is being multiplied by! A number that behaves in this fashion is called the multiplicative identity. One is called the **multiplicative identity**.

Multiplicative Identity

$$\mathbf{1 \cdot a = a}$$

One times any real number is the same as the real number.

Confirmation: $1 \cdot (-8) = -8$

$1 \cdot \pi = \pi$

$1 \cdot 0 = 0$

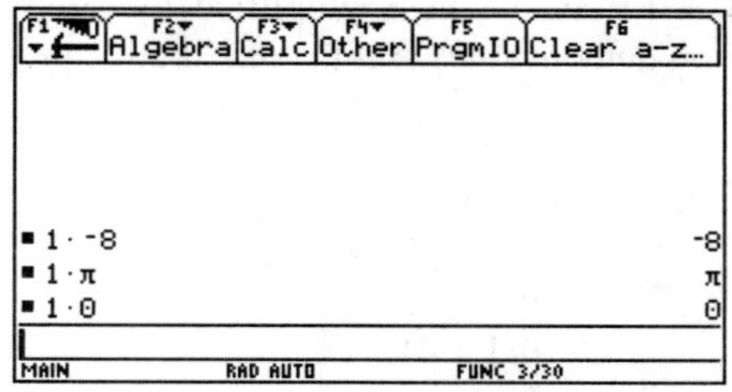

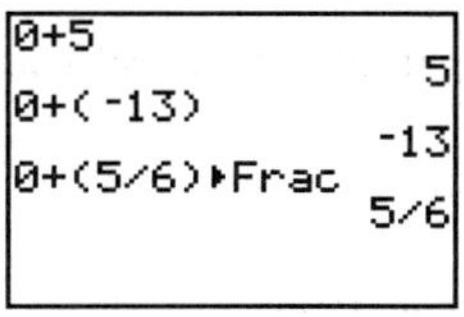

Figure 1.1.8

There is another number that behaves in a similar fashion when used in addition. It is the number 0. Any real number added to 0 has a sum of that number! The number that behaves in this fashion is called the additive identity (See Figure 1.1.8.) Zero is the additive identity.

Additive Identity

$0 + a = a$

The additive identity property suggests that zero plus any real number is the same as the real number.

Try these on your calculator.

Confirmation: $0 + (-34) = -34$

$0 + \pi = \pi$

$0 + \sqrt{7} = \sqrt{7}$

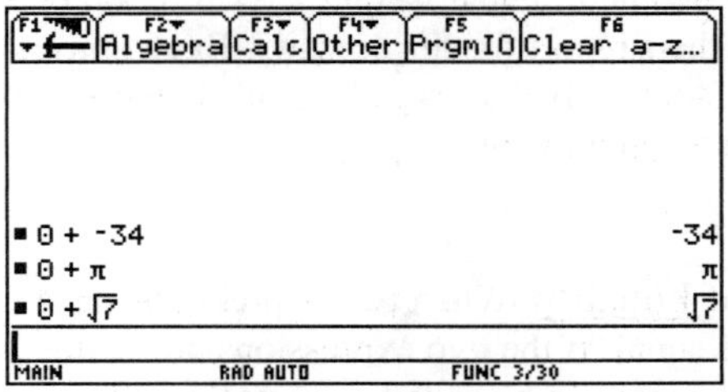

You should recognize when a real number property is being used. Further, you must be able to use the properties when they are applied in the chapters containing algebra, geometry, or trigonometry. In the arithmetic examples below, the expressions are being simplified to demonstrate examples of the properties being used. Some simplifications can be worked faster by other methods.

Confirmation:	$7(4+5)$	given (a numerical value of 63)
	$7 \cdot 4 + 7 \cdot 5$	distributive property
	$28 + 35$	multiplication
	63	addition

While 7(4 + 5) can be best simplified by simplifying inside the grouping symbols first, the above simplification demonstrates that the distributive property can also be used.

Confirmation:		
	$9 + 9 \cdot 4$	given (a numerical value of 45)
	$1 \cdot 9 + 9 \cdot 4$	multiplicative identity
	$9 \cdot 1 + 9 \cdot 4$	commutative property for multiplication
	$9(1 + 4)$	distributive property
	$9 \cdot 5$	addition
	45	multiplication

⁂

The expression $9 + 9 \cdot 4$ cannot be simplified by doing the addition first. The above simplification demonstrates that the properties allow for the correct simplification.

Confirmation:		
	$6 + (-2) + 3 - (-2) + 1$	given (a numerical value of 10)
	$6 + (-2) + 3 + 2 + 1$	subtraction
	$6 + 3 + 1 + 2 + (-2)$	commutative property of addition
	$(6 + 3 + 1) + (2 + (-2))$	associative property of addition
	$10 + 0$	addition
	10	additive identity

⁂

The Seesaw Problem
Suppose two people of equal weight are sitting on opposite ends of a seesaw and the same distance from the fulcrum (the support for the seesaw). Experience suggests that the seesaw is balanced. What happens when one person is given a bag of sugar? Is the seesaw balanced? How can you make it balanced again without one person moving?

Equal or not?

Properties of Equality When two expressions have the same numerical value, they are said to be equal. If the two expressions are written with an equal symbol between them, the statement can be called an **equation**. Other uses of the = symbol do not call for the use of the word equation. In the above simplifications, some people will use the = sign after each line of the simplification with no intention of thinking of the idea of an equation. The statement $2(3 + 5) = 2 \cdot 3 + 2 \cdot 5$ can be thought of as an equation. You may recall from earlier work in mathematics that many equations contain symbols such as an x. Eventually, in Chapter Five, this text will have equations with x in them as well. For this section, the equations are numerical equations. There is nothing to solve for; they are simply statements of equality.

Consider the equation:			
	$2 + 3 =$	$12 - 7$	a given true statement
	$2 + 3 + \mathbf{9} \neq$	$12 - 7$	a false statement
	$2 + 3 - \mathbf{1} \neq$	$12 - 7$	a false statement
	$\mathbf{6}(2 + 3) \neq$	$12 - 7$	a false statement

$(2+3)/\mathbf{10}$	$\neq$	$12-7$	a false statement

The point of the examples is that you *cannot* perform an operation on just *one side* of an equation; if you do, the resulting equation normally becomes a false statement. That is, the left side is not equal in value to the right side. On the other hand, if the same operation is done to *both* sides of an equation, the resulting equation remains a true statement as shown below -- just like the balanced seesaw.

$4 \cdot 3$	$=$	$10+2$	a given true statement	
$4 \cdot 3 + \mathbf{5}$	$=$	$10+2+\mathbf{5}$	a true statement	Verify these.
$4 \cdot 3 - \mathbf{6}$	$=$	$10+2-\mathbf{6}$	a true statement	
$4 \cdot 3 \cdot \mathbf{7}$	$=$	$(10+2) \cdot \mathbf{7}$	a true statement	
$4 \cdot 3/\mathbf{6}$	$=$	$(10+2)/\mathbf{6}$	a true statement	

Several properties -- as demonstrated above -- are needed to solve equations by the analytical method. The **analytical** method for solving equations means that you will use algebra or logical reasoning to find the solution. Equations are solved in Chapters Five, Six, Seven, Eight, Nine, Thirteen, and Fourteen. Solving equations will be introduced in Chapter Five, but the equality properties will be used earlier.

In the properties below, the symbols a, b, and c represent real numbers.

Properties of Equality

Addition Property of Equality	If $a = b$, then $a + c = b + c$
Subtraction Property of Equality	If $a = b$, then $a - c = b - c$
Multiplication Property of Equality	If $a = b$, then $ac = bc$, where $c \neq 0$
Division Property of Equality	If $a = b$, then $\frac{a}{c} = \frac{b}{c}$, where $c \neq 0$

The properties of equalities allow you to manipulate an equation by adding, subtracting, multiplying, and dividing *both* sides of the equation by the same number -- exceptions are noted above. This manipulation process transforms an equation into another true statement. Try using any of the properties of equalities on any equation of your choice, you will find that the transformed equation will always be a true statement, provided you start with a true statement.

Example 1: Why can't both sides of an equation be divided by 0?

Solution: Division by 0 is undefined! Thus, the transformed equation would not be a true statement. It is undefined = undefined, a nonsense statement.

Example 2: Why can't both sides of the equation $15 = (-3)(-5)$ be multiplied by 0?

Solution: While multiplying by 0 results in a true statement $0 = 0$; *any* equation would result in the same statement $0 = 0$. This would not be helpful. Also, if you inadvertently had a false statement and multiplied both sides by zero, it would become a TRUE statement: For example:

$$\begin{aligned} 3+4 &= 9 && \text{a false statement} \\ 0(3+4) &= 9\cdot 0 && \text{multiplication by 0} \\ 0 &= 0 && \text{a true statement!} \end{aligned}$$

No matter how you look at the problem, multiplication by 0 should not be done.

Properties of Inequality How do you symbolize the statement "My mother earns more than $100,000 per year." using only the equal sign? How do you symbolize the statement "It is less than 800 miles to the Outer Banks of North Carolina." using only the equal sign? While you may think it is sufficient to use English, it is impossible, for example, to use English in accounting software that must determine if your earned income is more than $22,100 but not more than $53,500.

$= \quad \neq \quad <$

If two numbers are different in value, they are said to be not equal. If two numbers are not equal, then the first must be larger than the second or the first must be smaller than the second. If the numbers are visualized on the number line, the number on the right is larger; the number on the left is smaller. Comparing numbers gives rise for the need to symbolize these statements of inequality.

$>$ means is greater than

$<$ means is less than

$\geq$ means is greater than *or* equal to

$\leq$ means is less than *or* equal to

The statement $4 < 13$ is read 4 is less than 13.

The statement $3 > -15$ is read 3 is greater than −15.

The statement $-6 \leq -6$ is read −6 is less than *or* equal to −6.

The statement $1 \geq -9$ is read 1 is greater than *or* equal to −9.

The inequality statements have properties similar to the properties of equality. One of the properties allows for both sides of an inequality to be multiplied by a negative, but unlike the multiplication property of equality, a change must be made in the inequality to maintain a true statement. Consider the inequality:

$3 > -15$ is a true statement

If both sides are multiplied by any negative number, the statement is false.

$(-2) \cdot 3 > (-2)(-15)$

$-6 > 30$ is a false statement

However, if the sense of the inequality is reversed, it remains a true statement.

$-6 < 30$ is a true statement

The same thing happens when both sides of an inequality are divided by a negative number. The sense of the inequality must be reversed to maintain a true statement. Consider the inequality:

$3 > -15$ is a true statement

If both sides are divided by any negative number, the statement is false.

$$\frac{3}{-3} > \frac{-15}{-3}$$

$-1 > 5$ is a false statement

However, if the sense of the inequality is reversed, it remains a true statement.

$-1 < 5$ is a true statement

Changing direction is good.

Properties of Inequality

Addition Property of Inequality

If $a < b$, then $a + c < b + c$

Subtraction Property of Inequality

If $a <, b$ then $a - c < b - c$

Multiplication Property of Inequality

If $a < b$, then $ac < bc$, **when $c > 0$**

Multiplication Property of Inequality

If $a < b$, then $ac > bc$, **when $c < 0$**

Division Property of Inequality

If $a < b$, then $\frac{a}{c} < \frac{b}{c}$, **when $c > 0$**

Division Property of Inequality

If $a < b$, then $\frac{a}{c} > \frac{b}{c}$, **when $c < 0$**

You will notice how similar the properties of inequality are to the properties of equality. However, there is a major difference. When multiplying or dividing both sides of an inequality by a negative ($c < 0$) number, the sense of the inequality changes to its opposite. The less than becomes a greater than; the greater than becomes a less than.

Example 3: If $-5 < 3$, is $-5 + 7 < 3 + 7$?

Solution: Simplify each side of the inequality: $-5 + 7 < 3 + 7$

$2 < 10$ yes.

⁂

Example 4: Show why multiplying both sides of an inequality by a negative number causes the inequality symbol to change to the opposite.

Solution: Consider the statement $6 \geq -13$. If the left side is multiplied by a negative, it becomes a negative number. If the right side is multiplied by a negative, it becomes a positive number. The inequality becomes a negative $\geq$ a positive. This is always a false statement, which leads you to believe that when you multiply by a negative, the inequality symbol must be changed to the opposite symbol $\leq$. When $\leq$ is used, the statement is true.

⁂

Example 5: Why can't both sides of an inequality be multiplied by 0?

Solution: If both sides are multiplied by 0, each side simplifies to 0. If each side is 0, then the $>$ and the $<$ inequalities are false statements and the $\leq$ and $\geq$ inequalities are not of any interest.

⁂

Example 6: Does the inequality $4 > -2$ imply that $6 > 0$?

Solution: Yes. If 2 is added to each side of $4 > -2$, the inequality simplifies to $6 > 0$. Sometimes starting with a false statement (assumption) will lead you to a false conclusion!

⁂

1.1 EXERCISES

Please read the text before you do the exercises. Read the **next assigned section** after you finish the assignment below.

A Quick Look To The Future

-4. What does the word “average” mean?

-3. What does the word “mean” mean?

-2. Describe the data $\{2, 3, 4, 5, 6\}$.

-1. Is 9.5×10^{-4} a big number or little number? Explain.

Arithmetic Revisited

1. Evaluate $-3|4-(-2)|+7$
2. Evaluate $\dfrac{-4-(-1)}{5-(-4)}$
3. Evaluate $-2(-1)^2 + 3(-1) - 8$
4. Add $-11 + 4$
5. Subtract $-13 - (-15)$
6. Divide $\dfrac{12}{15} \div \dfrac{4}{5}$

New Topic Exercises

Give a reason why each statement is true.

7. $-7 + (-2 + 5) = (-7 + (-2)) + 5$
8. $-7 + (-2 + 5) = -7 + (5 + (-2))$
9. $2 + (3 + 4) = (2 + 3) + 4$
10. $2(3 + 4) = 2 \cdot 3 + 2 \cdot 4$
11. $1 \cdot 6 = 6$
12. $3 \cdot 5 \cdot 7 = 3(5 \cdot 7)$

Give a reason why each line is true.

13.
$7 + 7 \cdot 5$ given
$1 \cdot 7 + 7 \cdot 5$
$7 \cdot 1 + 7 \cdot 5$
$7(1 + 5)$

14.
$2(4 - 7)$ given
$2(4 + (-7))$
$2 \cdot 4 + 2 \cdot (-7)$
$8 + 2 \cdot (-7)$
$8 + (-14)$
-6

Give reasons why each statement is true in Exercises 15 - 20.

15. Since $-4 + 9 = 5$ then $-4 + 9 - 9 = 5 - 9$.

16. If $3 \cdot 11 = 27 + 6$ then $\frac{3 \cdot 11}{3} = \frac{27+6}{3}$.

17. If $-5 > -13$ then $-5 - (-20) > -13 - (-20)$.

18. Since $-4 + 9 < 5 + 9$ then $-4 < 5$.

19. If $2 \geq -5$ then $(-3)(2) \leq (-5)(-3)$.

20. If $0 > -3$ then $\frac{0}{-2} < \frac{-3}{-2}$.

21. Simplify the expression $-2(4 + 6) - 3$. Give a reason for each step taken.

Using the properties of equalities, isolate the number 7 on one side of the = symbol.

22. $7 + 13 = 42 - 22$

23. $7 \cdot (-3) = (-1)(4 \cdot 5 + 1)$

24. $\frac{42}{7} = 5 + 1$

25. $0 = 2 + 7 - 9$

26. $5 = \frac{4 \cdot 7 \cdot 5}{28}$

Using the properties of inequalities, isolate the number 7 on the left side of the ≤ symbol.

27. $7 + 13 \leq 42 - 12$

28. $7 \cdot (-3) \leq (-1)(4 \cdot 5 + 1) + 6$

29. $\frac{42}{7} \leq 5 + 4$

30. $0 \leq 2 + 7 - 4$

31. $5 \leq \frac{4 \cdot 7 \cdot 9}{28}$

In the next four exercises, consider a, b, c, and d as representing real numbers.

32. If $a = b$, for what value of c does $a = b + c$?

33. If $a = b$, under what conditions does $a = bc$?

34. If $a > b$, under what condition on c and d is $a + c > b + d$?

35. If $\frac{a}{c} < \frac{b}{c}$, what values of c allows $a < b$?

From Mathematics to English

36. Do the properties of inequality apply to the inequality ≠ (not equal to)? Why or why not?

37. Explain the difference between the distributive property and the associative property of addition using paragraph format.

38. Explain why you do not have to simplify inside the parentheses first, in the expression $-2(7-(-4))$.

39. Describe the addition property of equality.

40. What is the meaning of "If $a = b$, then $a - c = b - c$."

41. If you have a balance with the right side containing more weight than the left and if you add equal weights to both sides, describe how this will affect the balance.

42. Give your definition of:
    a. a real number
    b. a rational number
    c. an irrational number

    Do not use examples in your definition.

43. Is there a number larger than 999,999,999. . . ?

44. Describe how you might most efficiently add the set of numbers { 2, 6, 5, 4, 5, 3, 1, 6 }

45. After reading this section, make a list of questions that you want to ask your instructor.

46. Start a daily journal and make an entry. In addition to your normal entry on thoughts about the mathematics in this section, list at least two positive comments about what you have learned about this topic.

47. In paragraph format, summarize the material in this section of the text in your daily journal.

48. Describe how your classroom instructor made this topic more understandable and clear.

49. After reading the text and listening to your instructor, what do you not understand about this topic?

From English to Mathematics

Rewrite the English statements as mathematical statements. For example, "six is less than forty" is written as "$6 < 40$" using mathematical symbols.

50. five is larger than negative two

51. two is smaller than or equal to the number x

52. the wind is blowing at least thirty miles per hour

53. the car is traveling forty miles per hour or under

54. the number c is negative

55. the number c is at most zero

56. the number c is positive

Explorations

57. Give a counter-example to verify that the following properties are *not true and valid* properties. A counter-example is **a** single case or circumstance showing the statement to be *false*. For example, the statement $a(b \cdot c) = ab \cdot ac$ is a *false statement;* however, many beginning students use the property as being a true and valid statement. A counter-example is simply one case that verifies that the statement is *NOT always true.* The counter-example is: suppose $a = 5$, $b = 2$, and $c = 3$; then,

$a(b{\cdot}c) = ab{\cdot}ac$	given
$5(2{\cdot}3) = 5{\cdot}2 \cdot 5{\cdot}3$	substitution
$5{\cdot}6 = 10 \cdot 15$	simplification
$30 = 150$	a false statement

This verifies that the statement $a(b{\cdot}c) = ab{\cdot}ac$ is **not** always a true statement. Actually, it is only true under the **very** limited conditions!

Give arguments showing the following statements are *not true* in general.

Commutative property of subtraction →	$a - b = b - a$
Associative property of subtraction →	$a - (b - c) = (a - b) - c$
Commutative property of division →	$a/b = b/a$
Distributive property with respect to multiplication →	$a(b{\cdot}c) = ab{\cdot}ac$
Subtractive identity →	$0 - a = a$
Square root property →	$\sqrt{a^2 + b^2} = \sqrt{a^2} + \sqrt{b^2}$
Absolute value property →	$\lvert a + b \rvert = \lvert a \rvert + \lvert b \rvert$

58. Develop numeric reasoning why the following property *is true*:

    Distributive property with respect to subtraction → $a(b-c) = ab - ac$

59. The graphing calculator, in many cases, can identify if a statement is true or false. In the following screen copy, 1 implies a true statement and 0 implies a false statement.

    Statements: $\frac{3}{4}+\frac{3}{4}=\frac{6}{8}$, $-1 > -142(3+4) = 2 \bullet 3 + 4$

```
3/4+3/4=6/8
                 0
-1>-14
                 1
2(3+4)=2*3+4
                 0
█
```

Verify that the following statements are all false.

a. $\frac{1}{2}+\frac{3}{5}=\frac{4}{7}$ b. $\sqrt{2^2+3^2}=2+3$

c. $-10000 > 4$ d. $\pi = 3.14$

e. $\frac{1}{10} < \frac{1}{100}$ f. $\frac{100}{100} > 1$

1.2 Data Analysis

The world abounds with numeric data. From the numeric scores you receive on quizzes, explorations, lab projects, and tests, to the numeric representations of the ever increasing ever changing number of people on the planet Earth, to the radio signals representing numerical values sent to Earth from remote spaceships. Everywhere you look you will find numeric data. In this section, you will look at individual data much like the raw scores received on tests and quizzes, and you will also start the study of data that comes as pairs of numbers.

Consider the following 1992 rounded data on the average per hour dollar earnings of a variety of different types of jobs. (Source: U. S. Bureau of Labor Statistics.)

Table 1.2.1

Manufacturing	Mining	Construction	Transportation	Wholesale Trade	Retail Trade	Finance, Ins.	Service
11	15	14	13	11	7	11	11

The hourly earnings from one group of workers are not related to another group. One group may go up and another go down or stay the same. This data is like a snapshot of earning conditions in 1992. While none of the numbers are related to any other of the numbers, it may still prove interesting to analyze the data as a set of numbers. To get mathematics involved, call the set of data L_1, that is $L_1 = \{11, 15, 14, 13, 11, 7, 11, 11\}$. This is the numeric representation of the data. Because the data set is small, it is easy to scan the data and pick out useful information. In other words, as a student, you may be interested to know that the highest paid workers are in mining. The maximum value in L_1 is 15. Likewise, you may be interested to know that the lowest paid workers are in the retail business. The minimum value in L_1 is 7. The difference between the maximum and minimum tells you what range of earnings to expect. In this case, it is \$8. This number, called the **range**, tells you that there is a fairly wide gap between the minimum and maximum of L_1. That is, it helps describe the data. The range is information you can use in decision making. For example, if the range is \$2, it would be easier to decide what field to pursue because all jobs would pay within \$2 of one another (nearly the same.) On the other hand, if the range were \$30, you may want to give serious consideration to your employment choice.

Show me the money!

Since the data set is quite small, looking at the graphical representation of the earnings may not be very helpful. However, when the amount of data is large, a picture of the data can be very telling. Below is the graphical representation of the data set.

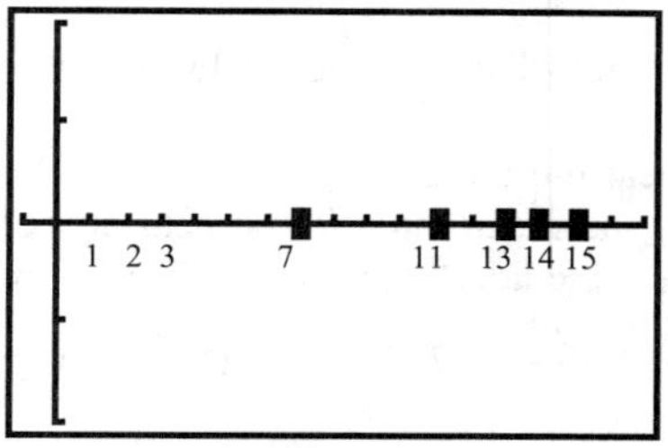

Figure 1.2.1

You can see that four numbers are grouped together and one of the numbers is out by itself. The picture (graphical representation) is somewhat misleading because there are actually four eleven's. Another graphical representation below in Figure 1.2.2 shows that there are four eleven's.

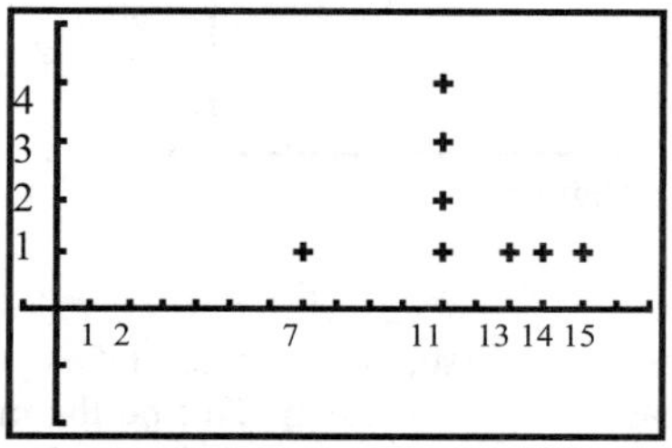

Figure 1.2.2.a

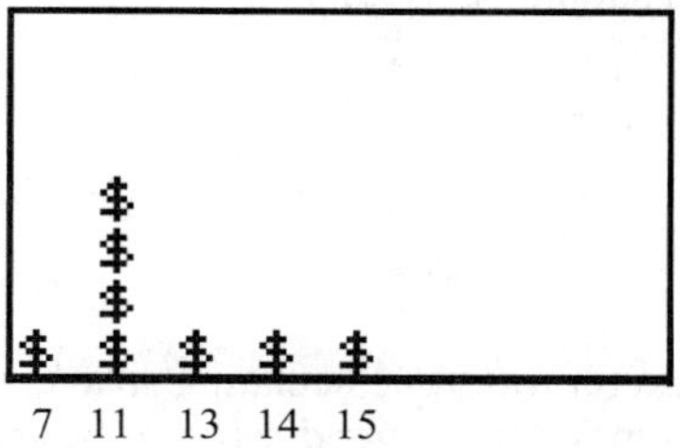

Figure 1.2.2.b

This graphical representation is much like a bar graph that is used in business. The height of the bar is a measure of the number of times each datum is in the set.

Another number that proves to be of value in decision making is a measure of "average". The idea behind the concept of average is that it is a number that can be used to best describe the entire set of numbers. A number that the other numbers in the set have a tendency be close to. Perhaps the measure of average you are most familiar with is the arithmetic mean. The **mean** is the sum of all of the data in the set divided by the number of numbers in the set. If this idea is converted from English to a symbolic mathematical form, it looks like $\overline{x} = \frac{x_1 + x_2 + x_3 + ... + x_n}{n}$, where $\overline{x}$ is the mean, $x_1 + x_2 + x_3 + ... + x_n$, is the sum of n numbers.

The mean of the data in set L_1 is $\overline{x} = \frac{11+15+14+13+11+7+11+11}{8} = 11.6$. The graphical representation of the data and the mean will show the data grouped around the mean.

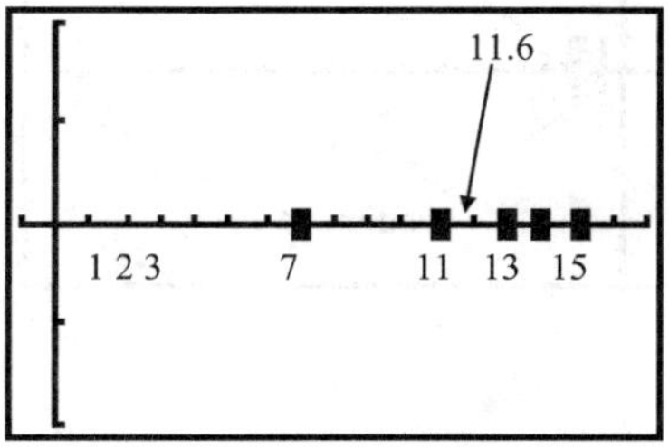

Figure 1.2.3

The "average" hourly earnings for the jobs listed above is \$11.60 per hour.

In the case of the mean, it may be near the middle of the data as in the example above, or it may be near one end or the other. In some cases the mean may not be near any of the data. What if lawyers earned \$150 per hour and they are included in the above groupings? The mean becomes $\bar{x} = \dfrac{11+15+14+13+11+7+11+11+150}{9} = 27$. None of the data is grouped around the mean as can be seen in Figure 1.2.4.

The point of the discussion is that sometimes the mean is not a very good measure of "average". If it isn't, what should be used?

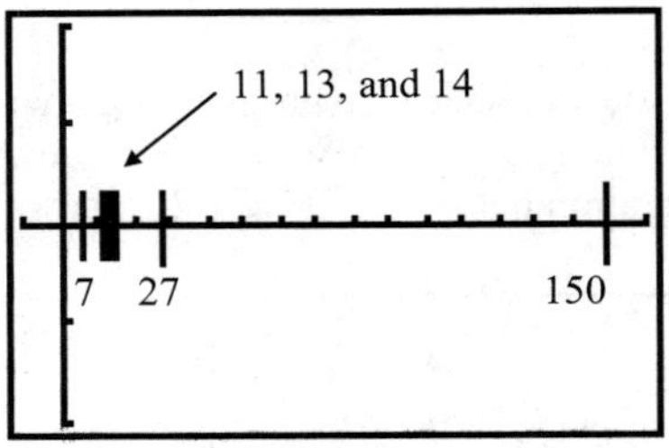

Figure 1.2.4

There are several ways of calculating an average. The mean is just one measure of average and as you can see, it sometimes is not a good measure. A second measure of average is the median. To find the median, the data must be sorted either in ascending or descending order. A graphing calculator will put the data in order as shown in Figure 1.2.5.

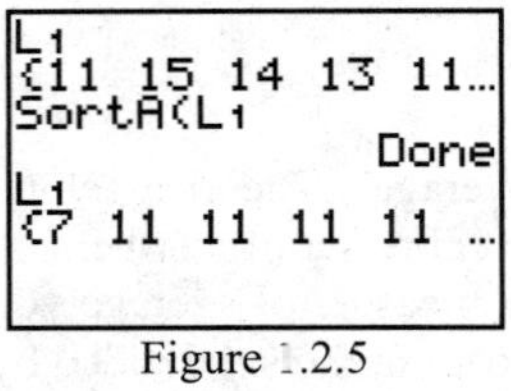

Figure 1.2.5

The set L_1 is {11, 15, 14, 13, 11, 7, 11, 11} and the sorted set L_1 is {7, 11, 11, 11, 11, 13, 14, 15}. The median of the set of data is the number in the middle. In the case of L_1, which has an even number of numbers in the set, the median is the mean of the two numbers in the middle. Since there are two eleven's in the middle, the median is $\dfrac{11+11}{2} = 11$. If the median is used as the measure of average in the set of data that contains the lawyers hourly earnings ({7, 11, 11, 11, 11, 13, 14, 15, 150}), the median remains 11. It is a much better measure of average than the mean of 27 because most of the data is grouped around the 11, not 27.

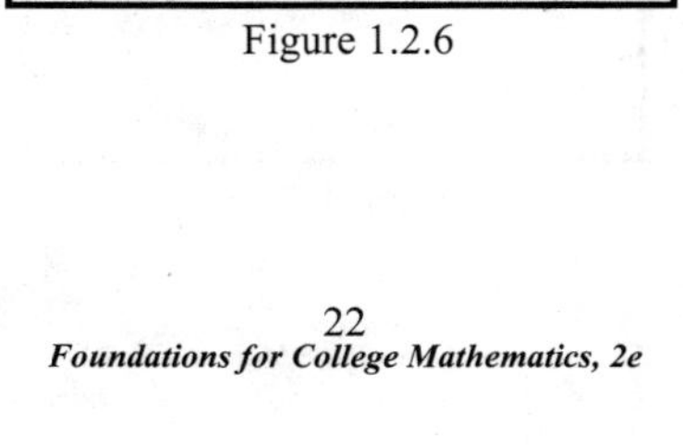

Figure 1.2.6

Example 1: Find the mean, median, maximum, minimum, and range of the ages of the following youngest and oldest presidents upon taking office: T. Roosevelt (42), Harrison (68), Grant (46), Pierce (48) Bush (64), Garfield (49), Reagan (69), Kennedy (43), Cleveland (47), Taylor (64), Buchanan (65), Polk (49), and Clinton (46).

Solution: Call the data set L_1. L_1 is {42, 68, 46, 48, 64, 49, 69, 43, 47, 64, 65, 49, 46}. The sorted set L_1 is {42, 43, 46, 46, 47, 48, 49, 49, 64, 64, 65, 68, 69}. The range is the maximum minus the minimum, 69 – 42 = 27. This means there is a fairly wide spread in the ages of people becoming president. The number in the middle is the seventh number from the smallest or the largest; it is 49. The median age of the above presidents upon becoming president is 49. The mean of the data is found by summing all the data and then dividing by 13, the number of president listed.

$$\overline{x} = \frac{42+43+46+46+47+48+49+49+64+64+65+68+69}{13} = 53.8$$

The mean age of the youngest and oldest presidents is 53.8. ⁂

This ends the analysis of "single variable data". There is much more to be developed in this analysis and you will find your graphics calculator has many other built in statistics used to analyze data. You may want to read the manual of your calculator to find out more about descriptive statistics.

1.2 EXERCISES

Please read the text before you do the exercises. Read the **next assigned section** after you finish the assignment below.

A Quick Look To The Future

-4. Create a method for describing the set of all real numbers between −2 and 5.

-3. What is a concise (short-cut) way of writing 2,000,000,000?

-2. Can your gross weekly wages be represented with mathematical symbols? Explain.

-1. If you throw a ball straight up, use English to describe how the height of the ball changes.

Previous Topics Revisited

1. What property is being demonstrated in the statement $4(3 + 7) = 4 \times 3 + 4 \times 7$?

2. Write a statement that shows the use of the associative property for multiplication.

3. Isolate the number 7 using the properties of equality. $7(4 - 3) + 4 = 5 + 6$.

4. Evaluate $-4+5^2+3(2-7)$.

5. Evaluate $\sqrt{3^2+4^2}$.

6. Evaluate $\dfrac{-6-5}{13-(-4)}$.

New Topic Exercises
Calculate the mean, median, maximum, minimum, and range of the following data in Exercises 7 - 10.

7. Number of U.S. battle deaths: (rounded to hundreds)

Revolutionary War	4,400
War of 1812	2,300
Mexican War	1,700
Civil War	140,400
Spanish-American War	400
World War I	53,400
World War II	291,600
Korean Conflict	33,700
Vietnam Conflict	47,400

For this data, which measure of average is better, mean or median? Why?

8. Highest number of reported cases of aids as of 2003: (rounded to thousands)

South Africa	5,300,000
India	4,900,000
Nigeria	3,600,000
Zimbabwe	1,800,000
Tanzania	1,600,000
Ethiopia	1,500,000
Mozambique	1,300,000
Kenya	1,200,000
Zambia	920,000
China	840,000

Source: World Health Organization

For this data, which measure of average is better, mean or median? Why?

9. Gravitational attraction (acceleration) relative to the Earth at 32.1 ft/sec^2:

Mercury	38	In % of the Earth's.
Venus	88	
Earth	100	
Mars	38	
Jupiter	234	
Saturn	92	
Uranus	79	
Neptune	112	
Pluto	16	

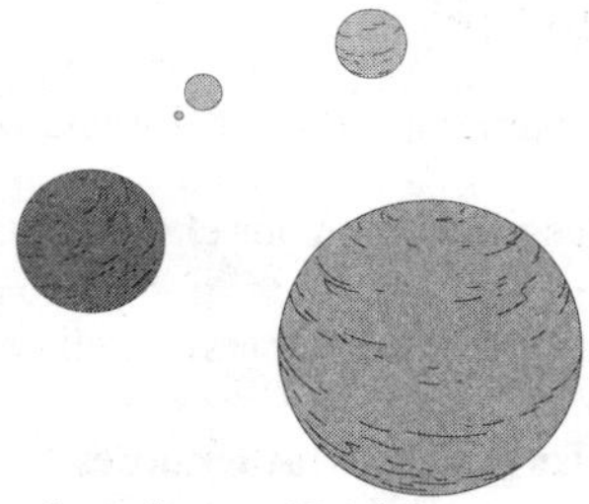

On an "average" planet, is the gravitational attraction weaker or stronger than that on Earth?

10. Deaths due to earthquakes:

Year	Location	Deaths
526	Turkey	250,000
1556	China	830,000
1731	China	100,000
1737	India	300,000
1828	Japan	30,000
1868	Peru	40,000
1899	Alaska	0
1906	San Francisco	3,000
1908	Sicily	75,000
1927	China	200,00
1970	Peru	66,800
1976	China	750,000
1978	Iran	25,000
1991	Los Angeles	2

Are the number of deaths due to earthquakes in the United States above or below the World average?

11. The ages of students in an algebra class are as follows {18, 18, 18, 18, 18, 18, 18, 18, 18}. Find the mean, median, mode, max, and min.

12. Create any data set with 8 elements, and a range of 0.

13. In the data set {3, 2, 4, 2, 3, 2, 129, 2}, which measure of "average" is the best choice when describing the data?

14. Create a data set where the mode is the best measure of average.

From Mathematics to English

15. Describe how you find the median of the data set {4, 3, 7, 2, 9, 1, 13, 4, 5}. Do not actually find the median.

16. After reading this section, make a list of questions that you want to ask your instructor.

17. Start a daily journal and make an entry. In addition to your normal entry on thoughts about the mathematics in this section, list at least two positive comments about what you have learned about this topic.

18. In paragraph format, summarize the material in this section of the text in your daily journal.

19. Describe how your classroom instructor made this topic more understandable and clear.

20. After reading the text and listening to your instructor, what do you not understand about this topic?

From English to Mathematics
Rewrite each English statement as a mathematical statement. For example, the English statement "the mean of 12 and 14 is 13" is written in mathematical symbols as $\frac{12+14}{2}=13$.

21. The range of the data 2, 5, 1, 13, 28, 3, 13 is 27.

22. The mean of 24, 18, and 15 is 19.

Explorations

23. Write a paragraph describing the reason why people use average and range to describe a data set.

1.3 Describing Sets of Numbers with Interval Notation

The Ball Doesn't Know My Watch Is Only Marked in Seconds

A ball is thrown straight upward with an initial velocity of 32 feet per second from a point 20 feet above ground level. Below is data showing the height of the ball at *t* seconds.

Table 1.3.1

t	0	0.2	0.4	0.6	0.8	1	1.2	1.4	1.6	1.8	2	2.2	2.4	2.5
Height	20	25.76	30.24	33.44	35.36	36	35.36	33.44	30.24	25.76	20	12.96	4.64	0

The set of numbers {0, 0.2, 0.4, 0.6, 0.8, 1, 1.2, 1.4, 1.6, 1.8, 2, 2.2, 2.4, 2.5} represents the times at which the height was measured? The question is: "Does time jump from 0 second to 0.2 second?" Does it never include 0.1 second, or 0.001 second? Likewise, does the clock not include any time between 1.6 and 1.8 seconds? Etc.

The ball is increasing in height from 0 second to 1 second. It moves from 20 feet above the ground to 36 feet above ground. The ball is increasing in height when time is 0, 0.2, 0.4, 0.6, 0.8, and 1 second. Does this mean it is not increasing at 0.23336 second, or 0.3275 second? The difficulty with collected data from a real-world occurrence is that data may not be collected continuously. The fact is that the ball is increasing in height for ALL real numbers between 0 and 1 second, and the height is ALL real numbers from 0 to 36.

As in the above situation, there are many occasions when it is necessary to describe sets of real numbers, as you will see in the remaining chapters. When solving an inequality, it may be necessary to describe all real numbers between two known real numbers; such as all real numbers between –2 and 5. You might argue that these numbers have just been described, in English, as all real numbers between –2 and 5, so why is another method needed? While it is certainly true that the numbers have been described, there is a more concise method for accomplishing this called **interval notation**.

It is helpful to visualize the real number line when using interval notation. Below is the real number line with a left parenthesis at –2 and a right parenthesis at 5.

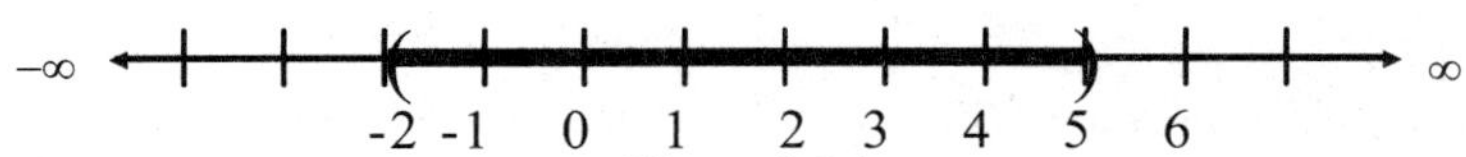

Figure 1.3.1

Notice how it looks like a group of numbers, just as you may have used parentheses as grouping symbols in earlier work in mathematics. The real numbers between –2 and 5 form a set of numbers. A set is a collection of things; in mathematics, it is usually a collection of numbers. This set of numbers is symbolized using interval notation as (–2, 5). This notation means the ***set*** of all real numbers between –2 and 5. This set <u>does not</u> contain the number –2 nor the number 5. The interval <u>does</u> include, for example, 4, 4.9, 4.99, 4.999, 4.99993, 4.99999927, 4.999999869, etc. and –1, –1.8, –1.9, –1.95, –1.99, –1.998, –1.9999999704, etc. In interval notation such as (–2, 5), the smaller of the two numbers is *always* on the left while the larger number is *always* on the right -- *just like on the number line, the smaller number is on* *the left and the larger on the right.* The parentheses indicate that the beginning and ending numbers do *not* belong to the interval. To show that the end numbers belong to the set, study Figure 1.3.2.

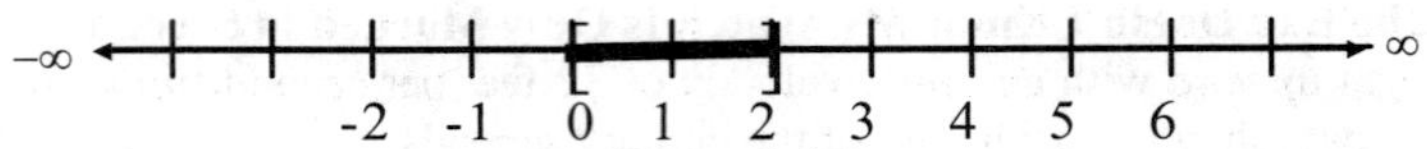

Figure 1.3.2

The group of numbers displayed is all real numbers between and including 0 and 2. Remember, real numbers are more than just integers. The use of a bracket at an endpoint indicates that the endpoint is *included in the set* of numbers. The interval notation to describe the set of numbers is [0, 2]. Again, the smaller number is on the left and the larger on the right. The endpoints 0 and 2 are elements of the set; that is, both 0 and 2 belong to the set of numbers [0, 2].

In some sets of numbers, instead of both endpoints being in the set or both endpoints not being in the set, the set may have just one endpoint. For example, consider the graphical representation of the temperature readings from 1 am to 6 am on a cold winter morning.

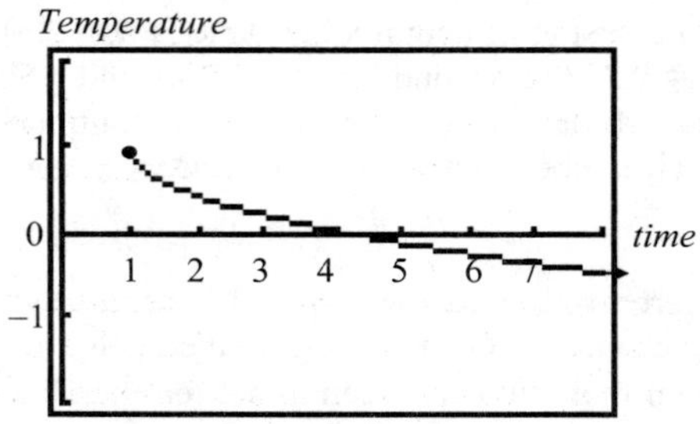

Figure 1.3.3

Use interval notation to describe when the temperature is above 0°. Starting at 1 am the temperature is above 0° and it remains so until 4 am. At 4 am the temperature is 0°.The temperature is above zero degrees from 1 to 4, including 1 and not including 4. The interval notation is [1, 4). This notation describes all real numbers from 1 to 4, including 1. In the notation [1, 4), the smallest number is on the left and the largest is on the right. The bracket on the left means to include the left endpoint and the parenthesis on the right means to not include the right endpoint in the set.

Example 1: Use interval notation to describe all real numbers between 3 and 2.

Solution: Since neither 2 nor 3 belongs to the set, parentheses will be used. Because 2 is the smallest endpoint, it is put on the left. The interval notation is (2, 3).

⁂

Example 2: Use interval notation to describe the set of numbers shown on the number line below:

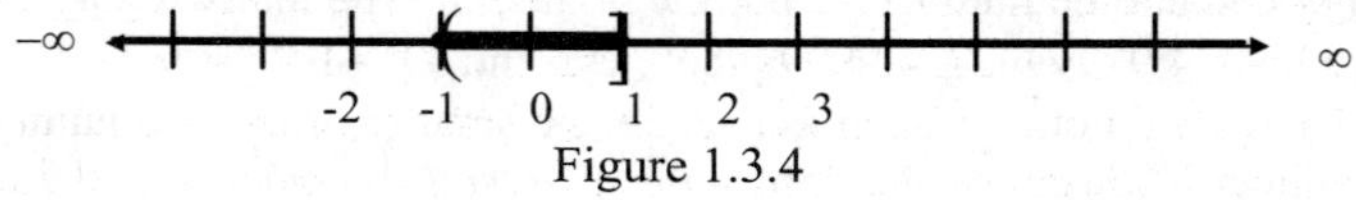

Figure 1.3.4

Solution: The smallest endpoint is −1 and the largest is 1. The number 1 is included in the set and the number −1 is not. The notation is (−1, 1].

⁂

Example 3: Describe all real numbers larger than −2 *and* smaller than 4 using interval notation.

Solution: If numbers are larger than −2 and at the same time smaller than 4, they must be between −2 and 4. Since the smallest number is −2, it belongs on the left. Parentheses should be used because neither −2 nor 4 belongs to the set of numbers. The notation is (−2, 4).

⁂

Interval notation is also used to describe all real numbers larger than a given number or smaller than a given number. For example, suppose the number 1 symbolizes that a species of animal exists and the number −1 symbolizes non-existence of the species. Using the graphical representation of this data, describe when the animal will be non-existent.

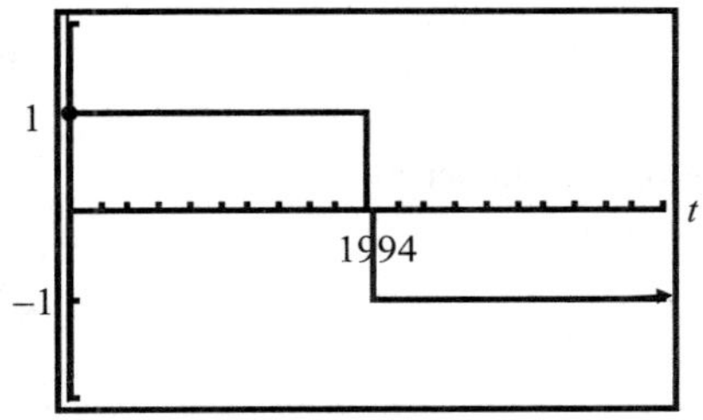

Figure 1.3.5

The data shows that the animal no longer exists from and including 1994, to forever. Mathematically representing this time span is done with the interval [1994, ∞). Consider the graph of the set of all real numbers larger than 3 in Figure 1.3.6.

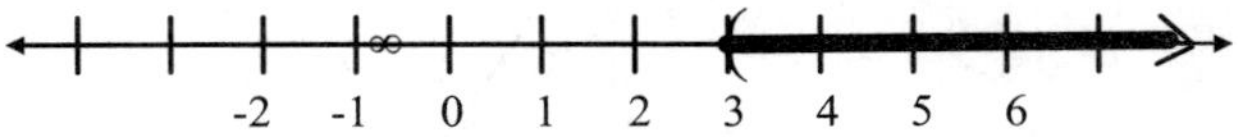

Figure 1.3.6

As you move to the right of 3, the numbers become larger and larger and there is no biggest number. The numbers increase without limit; that is, there is no numeric upper endpoint. The symbol used to indicate no upper numeric endpoint or, increasing without limit, is ∞. So, the set of numbers larger than 3 is (3, ∞). The right parenthesis is used because interval notation always starts and ends with a parenthesis or bracket. The symbol on the right end of the notation cannot be a bracket because a bracket means that the largest number is included in the set; however, there is no largest number in this set.

The set of numbers smaller than 12 can be thought of as starting at 12 (but it does not include 12) and then getting smaller and smaller. In this example, the numbers decrease without limit -- there is no numeric lower endpoint. The symbol indicating decreasing without bound or getting smaller and smaller is −∞. The interval notation for the set of numbers smaller than 12 is (−∞, 12). The number 12 is the largest, so it is on the right.

Example 4: Use interval notation to describe all real numbers less than or equal to 5.

Solution: The number 5 is the largest number in the set and it is to be included in the set, so it is placed on the right with a bracket. The other numbers in the set are all smaller than 5 and they get smaller and smaller (decrease without bound). The notation is (−∞, 5]. The left parenthesis is used because there is no smallest number; thus, there is no smallest number to be included in the set.

Example 5: Using interval notation, describe all numbers larger than or equal to −3.

Solution: The number −3 is the smallest in the set, so it is placed on the left and a bracket is used because it is included in the set. There is no largest number. On the right is the symbol meaning increasing without bound. The correct notation is $[-3, \infty)$.

To describe the set of numbers that belongs to $(3, 5]$ OR to the interval $[-4, 2]$, the symbol $\cup$ is used. The set of numbers $(3, 5] \cup [-4, 2]$ is the set of all real numbers belonging to the set $(3, 5]$ *or* belonging to the set of numbers $[-4, 2]$. The symbol $\cup$ is called the **union symbol**. The set of numbers $(3, 5] \cup [-4, 2]$ is called the union of $(3, 5]$ and $[-4, 2]$. A visualization of the number set is below.

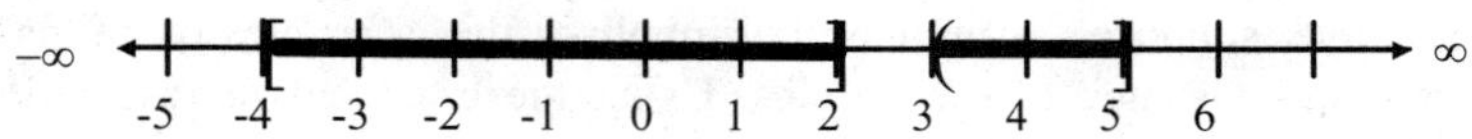

Figure 1.3.7

The numbers shaded belong to $[-4, 2]$ OR $(3, 5]$, that is, $[-4, 2] \cup (3, 5]$.

Example 6: Using interval notation, describe when the V is above the number line.

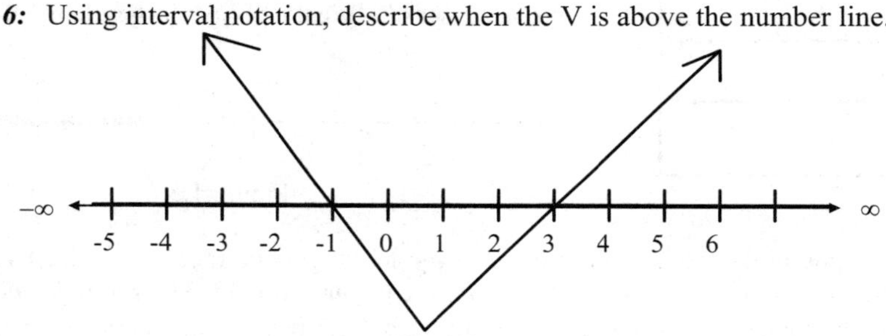

Figure 1.3.8

Solution: The V is above the number line to the left of −1 or to the right of 3. The V is above the number line on the interval $(-\infty, -1)$ *or* on the interval $(3, \infty)$. Using the union symbol notation, $(-\infty, -1) \cup (3, \infty)$.

To describe the set of numbers that belong to $(-1, 3)$ <u>and</u> $[1, 5)$, the symbol $\cap$ is used. The set of numbers $(-1, 3) \cap [1, 5)$ is the set of all real numbers belonging to the set $(-1, 3)$ AND at the same time also belonging to the set $[1, 5)$. The symbol $\cap$ is called the **intersection symbol**. The set of numbers $(-1, 3) \cap [1, 5)$ is called the intersection of $(-1, 3)$ and $[1, 5)$. A visualization of the number set is below.

(−1, 3)
[1, 5)

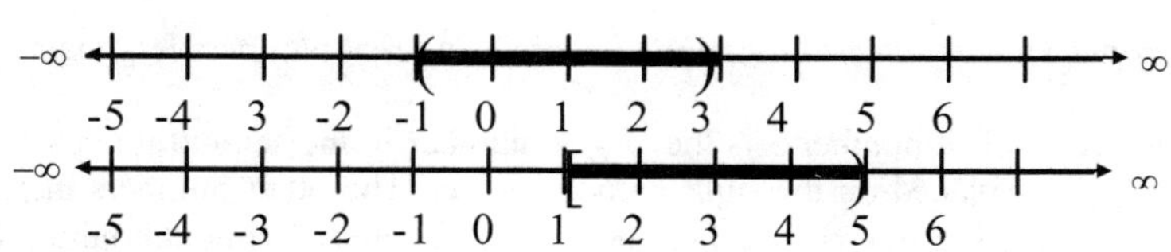

$(-1, 3) \cap [1, 5)$

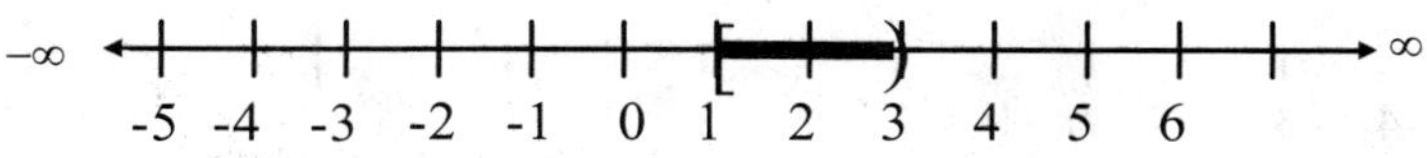

Figure 1.3.9

The number 1 is an element of both sets $(-1, 3)$ AND $[1, 5)$; thus, it belongs to $(-1, 3) \cap [1, 5)$. The number 3 does not belong to both sets $(-1, 3)$ AND $[1, 5)$, so it does not belong to $(-1, 3) \cap [1, 5)$. The intersection must therefore contain all real numbers from 1 to 3, not including 3. The intersection is the interval $[1, 3)$.

Example 7: Using interval notation, describe when the Z is above AND below the number line.

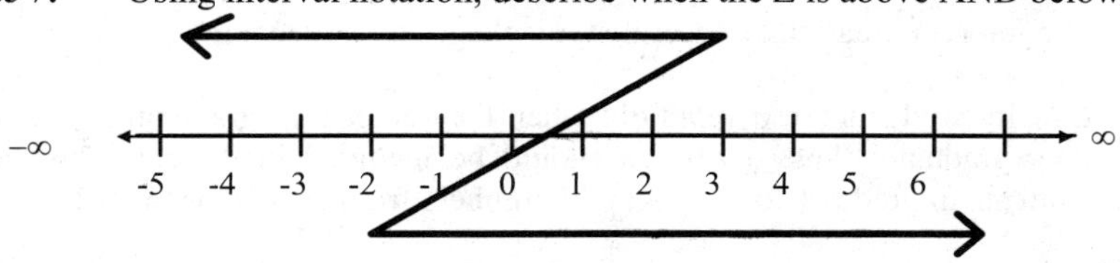

Figure 1.3.10

Solution: The Z is above the number line from $-\infty$ to 3 or on the interval $(-\infty, 3]$. The Z is below the number line from -2 to ∞ or on the interval $[-2, \infty)$. Therefore, it is above AND below the number line from -2 to 3 or on the interval $[-2, 3]$. The intersection $(-\infty, 3] \cap [-2, \infty)$ is $[-2, 3]$.

⁂

Example 8: Describe all real numbers in the set $(-\infty, 0) \cup (0, \infty)$.

Solution: The mathematical notation indicates to describe all numbers that belong to the first set OR the second set. The first set is all real numbers less than 0 and the second set is all real numbers larger than 0. All numbers less than 0 OR bigger than 0 must be all real numbers except 0.

⁂

Example 9: Draw the visual representation (graph) of numbers in the interval $[-3.5, 2]$.

Solution: Draw a picture of the numbers on the number line.

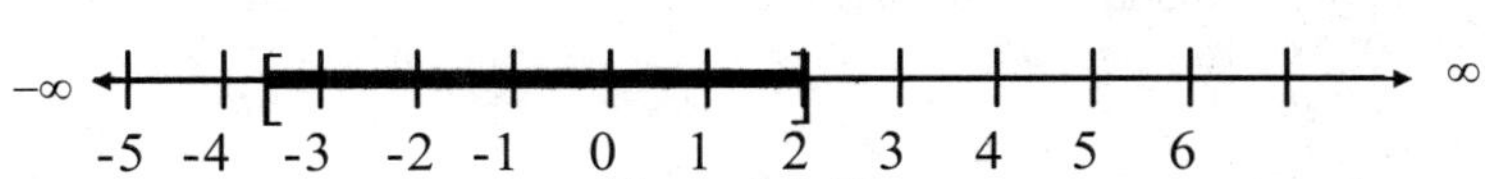

Figure 1.3.11

⁂

Example 10: Describe the set of numbers $(2, 5) \cap [7, 12]$.

Solution: There are no real numbers that are in the set $(2, 5)$ AND in the set $[7, 12]$. The sets are separate and have no common elements.

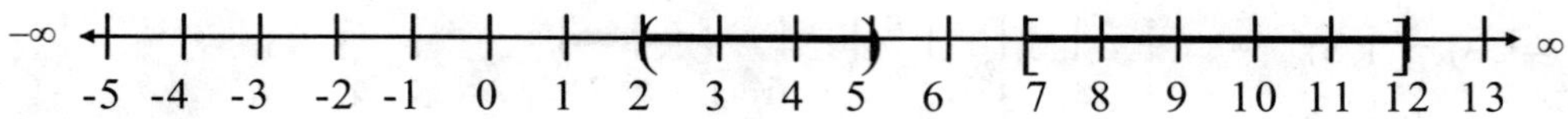

Figure 1.3.12

A notation used to describe the set that contains no elements is ∅. It is called the **empty set**.

⁂

Example 11: Use interval notation to describe typical ages of students in grades1 through 12.

Solution: Usually students are 6 years old when they enter first grade and 18 years old when they graduate; thus, 6 and 18 should be included in the set. The set of ages of students in grades 1 to 12 is all real numbers from 6 to 18 or [6, 18].

⁂

Example 12: Describe legal driving ages using interval notation.

Solution: Assuming you can start driving legally at age 16 and you die at age 78, interval notation describing these numbers is [16, 78].

⁂

Example 13: Describe when the parabola is below the number line.

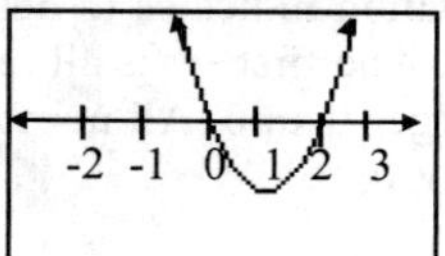

Figure 1.3.13

Solution: It appears to be <u>on</u> the number line at 0 and 2; thus, it is below the number line between 0 and 2. To describe all real numbers between 0 and 2, interval notation is (0, 2).

⁂

Example 14: Describe when the inverted V is below the number line. The geometric shape of V will be discussed further in Chapters Two, Three, and Five.

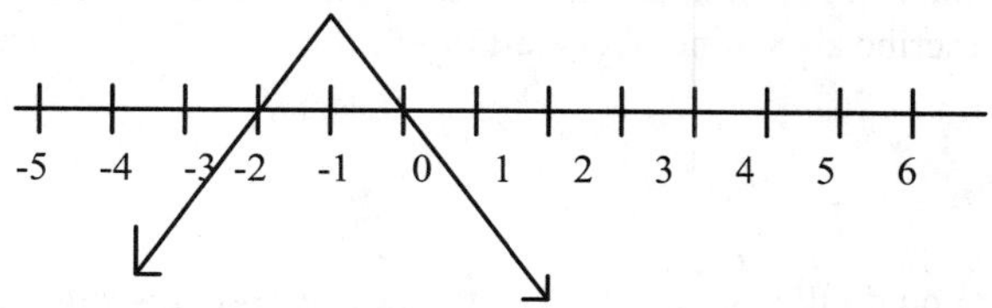

Figure 1.3.14

Solution: The V is on the number line at –2 and it is below the number line to the left of –2. To describe numbers left of –2, interval notation is (–∞, –2). The V is on the number line at 0 and below the number line to the right of 0. To describe all numbers to the right of 0, the interval notation is (0, ∞). So, the V is below the number line on the interval (–∞, –2) *or* on the interval (0, ∞). This is commonly known as the **union**; the V is below the number line on the set of numbers
(–∞, –2) ∪ (0, ∞).

⁂

Example 15: Describe when the geometric shape is on or below the number line. The shape will be described further in Chapters Two and Nine.

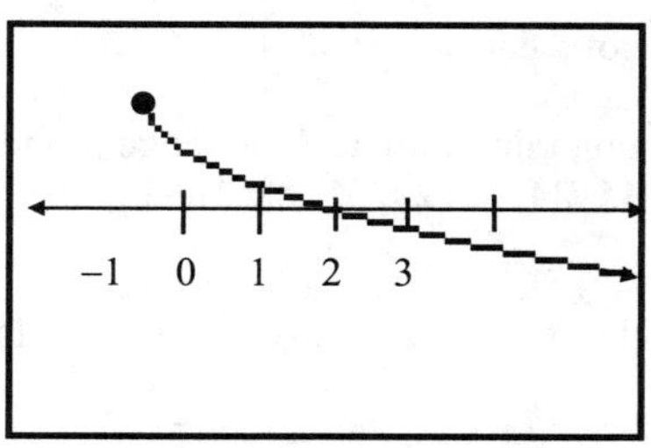

Figure 1.3.15

Solution: The graph is on the number line at 2. It is below the number line to the right of 2. Thus, the graph is on or below the number line for the numbers [2, ∞).

⁂

Example 16: Describe when the line is above the number line. Lines are discussed further in Chapters Three, Five, and Six.

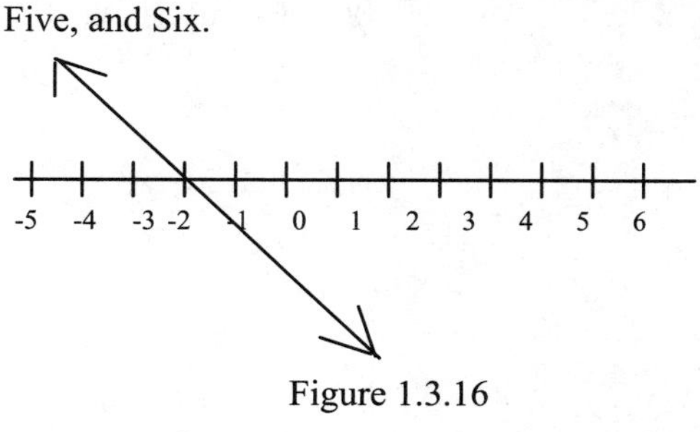

Figure 1.3.16

Solution: The line is on the number line at –2 and it is above the number line to the left of –2. To describe all real numbers left of –2, use the interval notation $(-\infty, -2)$.

⁂

1.3 EXERCISES

Please read the text before you do the exercises. Read the **next assigned section** after you finish the assignment below.

A Quick Look To The Future

-4. Write the current national debt.

-3. Write, symbolically, the area of a rectangle with a length of l and a width of w.

-2. Is the number of people on earth increasing or decreasing?

-1. Can the relationship between time in the air and the height of a ball tossed straight up be represented graphically?

Previous Topics Revisited

1. Write a "formula" for the mean of a data set.

2. Write a "formula" for the range of a data set.

3. Find the mean, median, maximum, minimum, and the range of the following lab project scores for a class of 23 students: {23, 38, 15, 34, 32, 25, 31, 28, 18, 36, 28, 37, 31, 30, 35, 29, 27, 38, 33, 35, 24, 32, 35}.

4. Identify the property being used: If $6 \cdot 7 = 42$, then $6 \cdot 7 - 9 = 42 - 9$

5. Identify the property being used: $3(4 + 7) = 3 \cdot 4 + 3 \cdot 7$

6. Simplify: $-5|-2-3|+25$

New Topic Exercises

Shade in the following sets of numbers on a number line.

7. $(-5, 0)$

8. $[-5, 0]$

9. $(-3, \infty)$

10. $(-\infty, 2]$

11. $(-1, 3]$

12. $[-1, 3)$

13. $[0, \infty)$

In Exercises 14 through 24, use interval notation to describe the following sets of numbers.

14. All real numbers between –3 and 259.

15. All real numbers less than 2000.

16. All real numbers larger than or equal to –5.03.

17. Positive real numbers.

18. Negative real numbers.

19. Positive real numbers and 0.

20. All real numbers between 6 and –3, including 6.

21. All real numbers between –4 and 3 OR between 5 and 7.

22. $[-2, 5] \cap (3, 9]$

23. $(-1, 4) \cap [4, 6)$

24. All real numbers between –1 and 5, including 5 OR between 7 and 9.

In Exercises 25 to 33, use interval notation to answer the questions.

25. Describe when the line is above the number line.

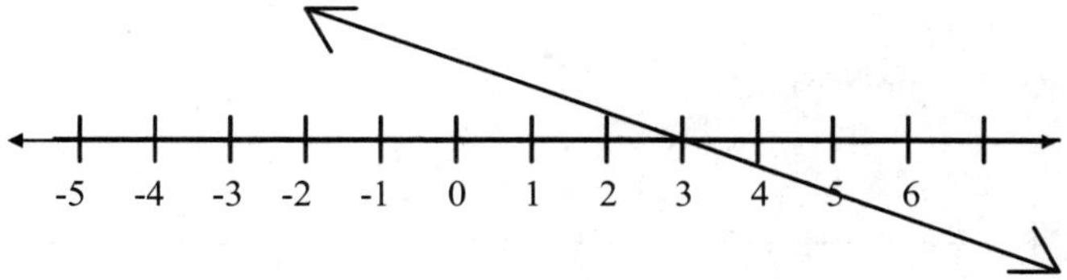

26. Describe when the parabola is below the number line.

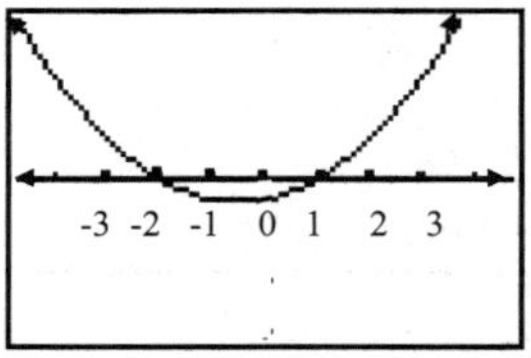

27. Describe when the curve is below the number line.

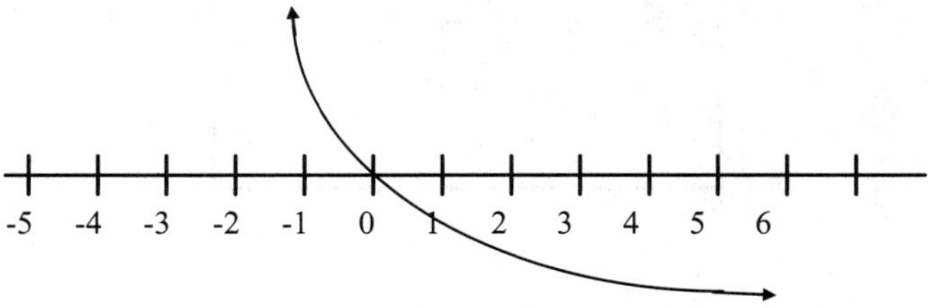

28. Describe when the inverted V is above the number line.

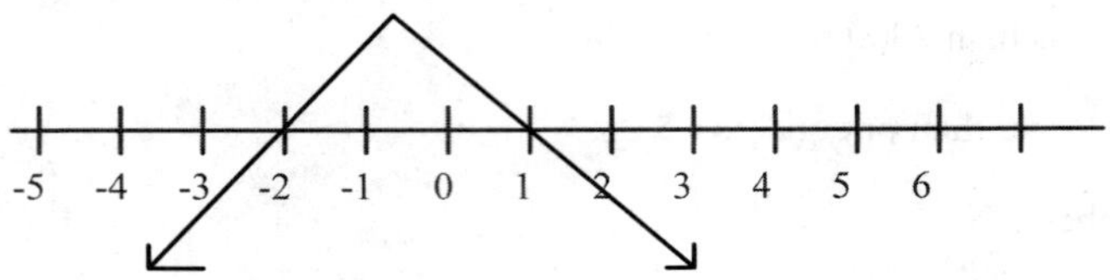

29. When is the parabola below the number line?

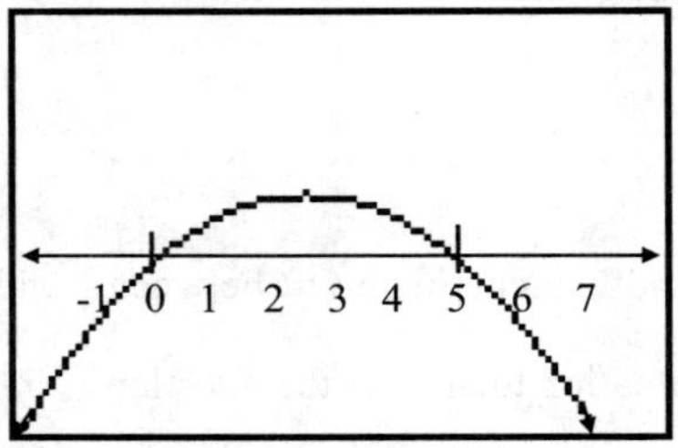

30. When is the V above the number line?

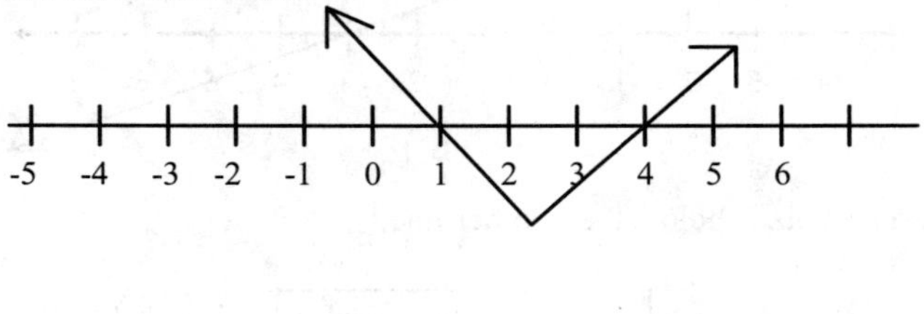

31. When is the curve above the number line?

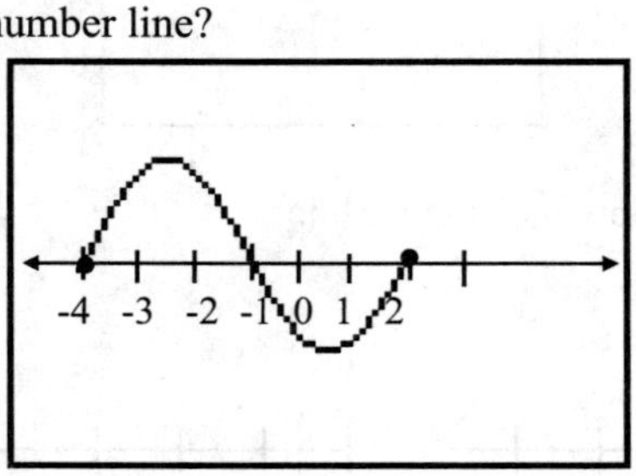

32. When is the curve above and below the number line?

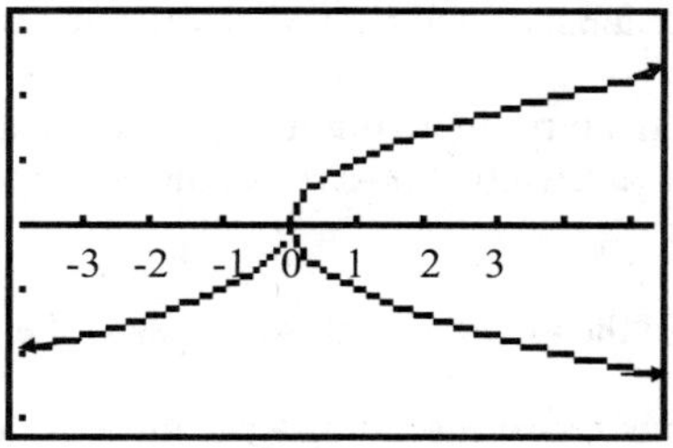

33. When is the curve above and below the number line?

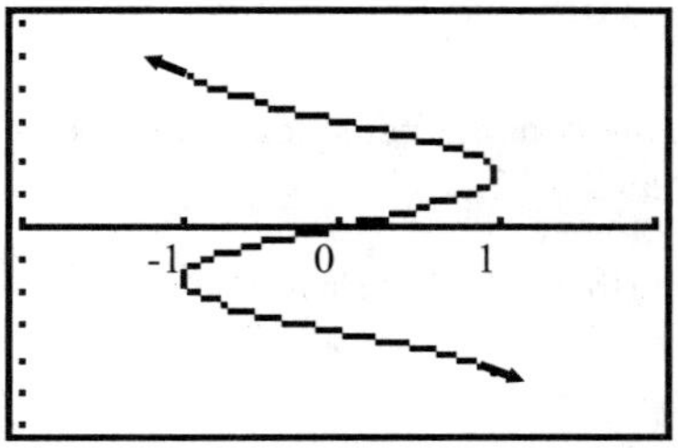

34. Draw any curve (shape) that is above the number line for all numbers in the set $(-\infty, 3)$.

35. Draw any curve (shape) that is below the number line for all numbers in the set $[-2, 3]$.

36. Draw any curve (shape) that is on or above the number line for all numbers in the set $[-3, 1)$.

37. Draw any curve (shape) that is above the number line on $(-\infty, -2)$ and below the number line on $(-2, \infty)$.

38. Draw any curve (shape) that is above *and* below the number line on the interval $[-4, 3]$.

39. Draw any curve (shape) that is above *or* below the number line on the interval $(-\infty, \infty)$.

40. Draw any curve (shape) that is on the number line on the interval $(-2, 3]$.

From Mathematics to English

41. Describe, as best you can, the largest number in the interval $[-1, 3)$.

42. Make up a definition for the symbol ∞.

43. List the advantages of using interval notation to describe sets of real numbers.

44. Does the interval $[-6, -6]$ contain any numbers?

45. What is incorrect with the way the interval $[6, -2)$ is written?

46. Describe the numbers in the interval (4, 4).

47. After reading this section, make a list of questions that you want to ask your instructor.

48. Start a daily journal and make an entry. In addition to your normal entry on thoughts about the mathematics in this section, list at least two positive comments about what you have learned about this topic.

49. In paragraph format, summarize the material in this section of the text in your daily journal.

50. Describe how your classroom instructor made this topic more understandable and clear.

51. After reading the text and listening to your instructor, what do you not understand about this topic?

From English to Mathematics

Write the English statement as an equivalent mathematical statement. For example, the mathematical statement for real numbers bigger than 2 is $(2, \infty)$.

52. real numbers larger than one fourth and less than one half

53. real numbers larger than five

54. real numbers at least five

Explorations

55. Think of the symbol x as representing all elements in each of the following sets. Describe each set using x and one of the following inequality symbols: $<, >, \leq, \geq$.

a. $(6, \infty)$ b. $(-\infty, 4)$ c. $[2, \infty)$ d. $(-\infty, -3]$ e. $(-1, 5)$

56. Develop several ways of describing all real numbers between –4 and 6, including 6, OR larger than 10.

1.4 Notation for Big and Little Numbers

How big is big? If you are a child of six months in age, the house you live in is big. If you are 40 years old, the same house may not seem so big. A gift of \$1,000 is big if you are six years old. A gift of \$1,000 is not big if you earn five million dollars (\$5,000,000) per year. President Franklin D. Roosevelt thought a national debt of \$2,000,000 was too big. Do you think it is big, considering that the national debt today is around \$8,360,000,000,000? Is \$8,360,000,000,000 really much bigger than \$2,000,000? Is 45,000 a big number? There were approximately 45,000 United States soldiers killed in the 15-year Vietnam War. Around 45,000 people are killed on American roads in one year. Close to 45,000 soldiers were killed in the 3-day battle of Gettysburg in the Civil War. Is 45,000 a big number? You decide.

How little is little? Is your little finger little at 1.5 centimeters (cm) in diameter? This is 0.015 meter (*m*). What about the diameter of a typical hair at 0.0005 meter (*m*), or a bacterium with a length of 0.000002 meter? Is the length of a virus a little number? It has a length of about 0.00000002 meter. Or is the width of a proton or neutron at 0.00000000000001 meter a little number? The time it takes you to read this page is about 45 seconds. Is 45 seconds a little number? You can open a window in about 1 second. A neutrino requires about 0.04 second to pass through the entire Earth. The time it takes light to bounce off this text and into your eyes is around 0.000000002 second.

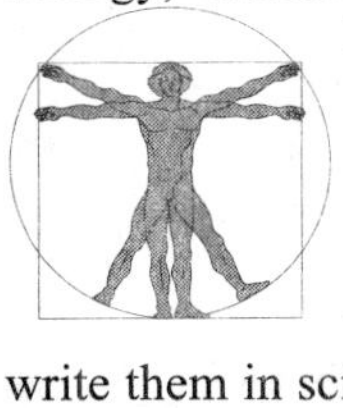

One point of this discussion is that unlike a traditional mathematics class where most of the numbers you study are nice simple numbers like 2 or –5, numbers outside the classroom can range from rather small numbers like the rate of pay on your job to the very large numbers like the distance to the next galaxy. You will encounter a variety of numbers in this text, just as you do when you read a newspaper or study biology, astronomy, chemistry, and physics. The problem with numbers like 0.0000000000012 or 15,000,000,000,000 is that they are unwieldy. Try entering the distance light travels in a year (30,970,874,880,000,000 feet) on a scientific calculator. Is it possible? Try it on your graphing calculator. Can you do it? Even if a calculator or computer accepts the number, it is still a major task to enter it and make certain you have entered it correctly. Entering it correctly may still not be sufficient; your calculator may only use 10 to 14 of the digits you entered. To make big and little numbers manageable, you must learn to write them in scientific notation -- an invention by people who work with these numbers on a daily basis. Not only must you be able to write numbers in scientific notation, but you must also be able to calculate with numbers in scientific notation. Your calculator has scientific notation built in. It looks a bit different; however, you will discover that it really is nothing but scientific notation.

To understand scientific notation, please consider the following reasoning:

34,000. is the same as 3,400. $\times$ 10 or 3,400. $\times 10^1$

3,400. $\times 10^1$ is the same as 340. $\times$ 10 $\times$ 10 or 340. $\times 10^1 \times 10^1$

340. $\times 10^1 \times 10^1$ is the same as 34. $\times$ 10 $\times$ 10 $\times$ 10 or 34. $\times 10^1 \times 10^1 \times 10^1$

34. $\times 10^1 \times 10^1 \times 10^1$ is the same as $3.4 \times 10^1 \times 10^1 \times 10^1 \times 10^1$

$3.4 \times 10^1 \times 10^1 \times 10^1 \times 10^1$ in scientific notation is 3.4×10^4

Thus, 34,000 is the same number as 3.4×10^4. Notice that each time the decimal point is moved one place to the left, a factor of 10 is introduced. When writing a number in **scientific notation**, the goal is to multiply a number between 1 and 10, times 10 to an exponent. Thus, 3.4×10^4 is scientific notation for the number 34,000.

Example 1: The distance to the star Alpha Centauri is 23,000,000,000,000 miles. Write it in scientific notation.

Solution: Move the decimal point 13 places to the left and make the exponent on 10, 13. The distance to the star Alpha Centauri is 2.3×10^{13} miles.

Example 2: In the year 2004, 4.12×10^6 humans were born in American. Write this number in standard notation.

Solution: The exponent of 6 indicates that the decimal point has been moved 6 places to the left to get the scientific notation; therefore, move the decimal point RIGHT 6 places to get the standard notation of $4,120,000. This is often read as 4.12 million.

What if a number is a little number, how do you write it in scientific notation? Please consider the reasoning below. Remember, dividing by 10 is the same thing as multiplying times $\frac{1}{10}$ and, $\frac{1}{10}$ is the same thing as 10^{-1}.

$$0.00058 = \frac{0.0058}{10} = 0.0058 \times \frac{1}{10} =$$

$$0.0058 \times 10^{-1} = \frac{0.058}{10} \times 10^{-1} = 0.058 \times \frac{1}{10} \times 10^{-1} =$$

$$0.058 \times 10^{-1} \times 10^{-1} = \frac{0.58}{10} \times 10^{-1} \times 10^{-1} = 0.58 \times \frac{1}{10} \times 10^{-1} \times 10^{-1} =$$

$$0.58 \times 10^{-1} \times 10^{-1} \times 10^{-1} = \frac{5.8}{10} \times 10^{-1} \times 10^{-1} \times 10^{-1} = 5.8 \times \frac{1}{10} \times 10^{-1} \times 10^{-1} \times 10^{-1} =$$

$$5.8 \times 10^{-1} \times 10^{-1} \times 10^{-1} \times 10^{-1} = 5.8 \times 10^{-4}$$

From the above chain of equalities you can see that $0.00058 = 5.8 \times 10^{-4}$. Thus, 5.8×10^{-4} is scientific notation for 0.00058. Notice that each time the decimal point is moved one place to the right, a factor of 10^{-1} is introduced. When writing a number in scientific notation, the goal is to multiply a number between 1 and 10, times 10 to an exponent. The number 5.8×10^{-4} is a number between 1 and 10 (5.8) times 10 to an exponent of (–4).

Example 3: A typical virus is around 0.000 000 090 meter. Write this number in scientific notation.

Solution: Move the decimal point 8 places to the RIGHT and multiply by 10^{-8}. A virus is about 9.0×10^{-8} meters long.

Example 4: The mass of a hydrogen atom is 0.000 000 000 000 000 000 000 001 673 gram. Convert this to scientific notation.

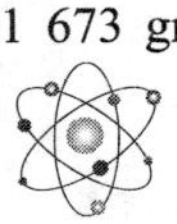

Solution: Move the decimal point 24 places to the right to get 1.673; then, multiply this times 10^{-24}. The mass of a hydrogen atom is 1.673×10^{-24} gram. (As a reference, a box of cereal has a mass of around 350 to 450 grams.)

Example 5: A garden snail can reach a top speed of about 3.0×10^{-2} miles per hour (mph). Convert this number to standard notation.

Solution: The decimal point has been moved 2 places to the RIGHT to get scientific notation; thus, move it LEFT 2 places to get standard notation. The top speed of a snail is 0.03 mph.

Example 6: Human hair grows at an average rate of 1.0×10^{-8} mph. Convert this to standard notation.

Solution: The decimal place has been moved 8 places to the RIGHT to get scientific notation; thus, move it back to the LEFT 8 places to get standard notation. The average rate of growth of a hair is 0.000 000 01 mph.

How does the calculator handle scientific notation, and how do you use it? The graphing calculator makes it especially easy to use scientific notation. Your graphing calculator will probably have a key labeled EE or EXP that is used to enter numbers in scientific notation. If your calculator does not have this key, read the calculator instruction book or ask your instructor to help you find the correct key. All you need to do is enter the number between 1 and 10 (the coefficient), type the EE key and finish by entering the exponent. The base of 10 is not entered and it does not show on the display of the calculator. To enter the number 5.6×10^9, enter 5.6 EE 9. Your calculator display will probably look like 5.6E9. Of course your calculator can perform any of the arithmetic operations with numbers in scientific notation that it does with numbers in standard notation. The answer you get will either be in scientific or standard notation – depending on the mode settings. Although you are expected to do arithmetic on your calculator, below are a few examples done without the calculator.

Example 7: With a 2005 annual national defense budget of around $\$4.659\times10^{11}$ and 2.95×10^8 people in the country, what was the yearly share for each person?

Solution: This is a division problem because the desired number is dollars per person. *Per* is usually an English word for division.

$$\frac{4.659\times10^{11}}{2.95\times10^8}=\frac{4.659}{2.95}\times\frac{10^{11}}{10^8}=1.579\times10^3=\$1,579.\text{ per person per year.}$$

(To divide numbers in scientific notation without a calculator, divide the coefficients and divide the powers of 10. Then multiply these numbers together. This idea is developed further in Chapter Seven.)

The graphing calculator keystrokes are: 4.659 EE 11 ÷ 2.95 EE 8 ENTER

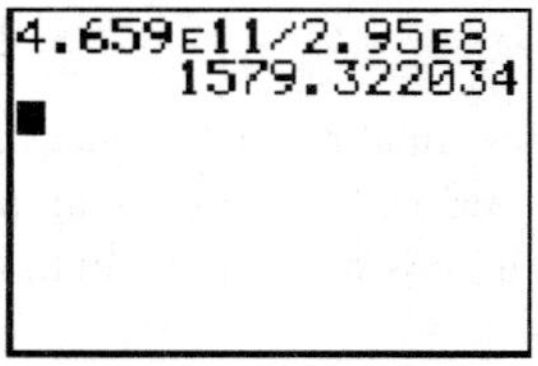

Example 8: In 1995, there were enough stored nuclear weapons on the planet Earth to give **every** person 1×10^4 equivalent pounds of TNT. If the Earth had 5.6×10^9 people, how much total nuclear power was available in a TNT equivalent measure? (Note: a few pounds of TNT can destroy your home, and you have 10,000 pounds stored.)

Solution: This is a multiplication problem, using the amount each person has times the total number of people. To multiply numbers in scientific notation, multiply the coefficients and multiply the powers of 10. Then multiply these numbers together. $(1\times10^4)\times(5.6\times10^9)=(1\times5.6)\times(10^4\times10^9)=5.6\times10^{13}$. In standard notation, this is 56,000,000,000,000 or 56 trillion pounds of TNT stored for future use. The calculator keystrokes are: 1 EE 4 × 5.6 EE 9 ENTER

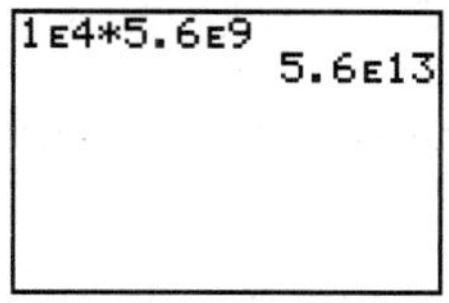

⁂

Example 9: The Mojave Desert is 1.5×10^4 square miles, and the Painted Desert is 7×10^3 square miles. What is the total area of the two deserts?

Solution: This is an addition problem. Add the areas. $1.5\times10^4+7\times10^3=?$ The calculator keystrokes are: 1.5 EE 4 + 7 EE 3 ENTER

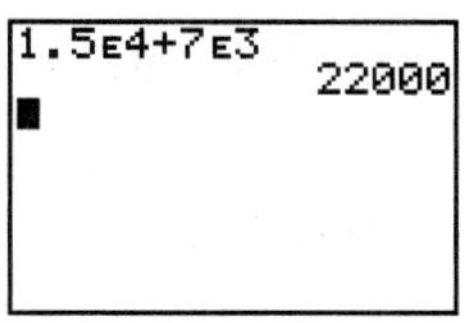

The area of both deserts is 22,000 sq. mi.

⁂

1.4 EXERCISES

Please read the text before you do the exercises. Read the **next assigned section** after you finish the assignment below.

A Quick Look To The Future

-4. Can the relationship between the hours you work and the amount of your gross pay be represented with symbols? Why or why not?

-3. As you work longer at your job, does your salary increase or decrease?

-2. Can the relationship between time and the national debt be represented as a graph?

-1. What does the physics teacher mean when he/she says that the relationship between the force applied to a spring and the amount of stretch in the spring is linear?

Previous Topics Revisited

1. Draw the graph of all numbers in the set $[-3, 2)$.

2. Describe the numbers in the intersection using interval notation: $(-\infty, 6]\cap(-3, 9]$.

3. Use interval notation to describe when the semi-circle is above the number line.

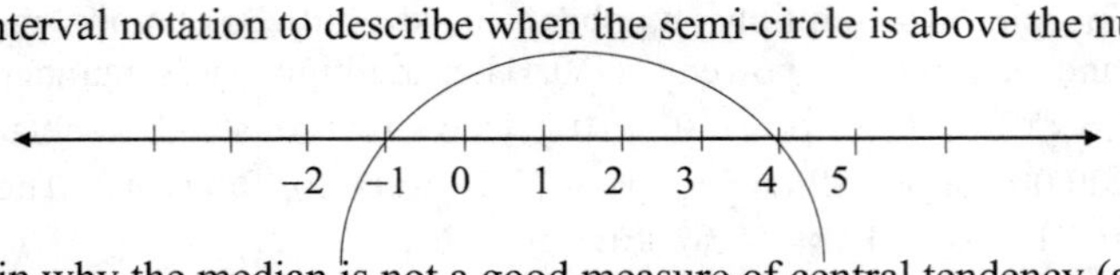

4. Explain why the median is not a good measure of central tendency (average) for the data {2, 2, 2, 2, 2, 83, 120, 427}.

5. Explain why the mean is not a good measure of central tendency (average) for the data {2, 4, 6, 1, 4, 7, 5, 83}.

6. Using the properties of equality, isolate the number 7 on the right side on the equality $3 + 4(2 - 7) = 0$.

New Topic Exercises

Convert the following numbers to scientific notation:

7. The velocity of light in a vacuum: 29,979,250,000 *cm*/sec.

8. The 2006 human population of the planet Earth: 6,500,000,000.

9. The approximate annual number of males in the 45 to 64 age range that have a heart attack in the United States: 424,000.

10. The mass of the Sun: 1,983,000,000,000,000,000,000,000,000,000,000 grams.

11. The rest mass of an electron: 0.000 000 000 000 000 000 000 000 000 911 gram.

12. Assuming an interest rate of 4% paid on the national debt in 2006, it is $334,400,000,000 a year.

13. The maximum amount of mercury allowed in our drinking water in the United States: 0.000 002 gram per liter.

14. The maximum amount of coli-forms from human and animal fecal matter allowed in our drinking water: 0.01 *ml* per liter.

15. The 1975 concentration of methane in the atmosphere: 0.00017%.

Convert the following numbers to standard notation.

16. The distance from the Sun to the Earth: 9.2×10^{7} miles.

17. Printing paper is around 4×10^{-3} inches thick.

18. An angstrom (used to measure wave lengths of light) is 4×10^{-9} inch.

19. The amount of garbage generated by each person in Los Angeles is 6.4×10^{0} pounds per day.

20. The core of the Earth is currently around 1.1×10^{4} °F.

Do the calculations in Exercises 21 to 31:

21. In 2002, each of the 2.87×10^8 people in the US generated 4.3 pounds of garbage per day. How much garbage is generated in one year by everyone in the United States?

22. If the Earth, Mars, and the Sun are collinear, and the Earth is 1.496×10^8 kilometers from the Sun, and Mars is another 7.83×10^7 kilometers from Earth. How far is Mars from the Sun?

23. In 1989, the worldwide military spent $\$1.9 \times 10^6$ every minute! How much did they spend throughout the year?

24. About 3×10^4 people die from earthquakes each year. How many have died in the last 50 years?

25. There are about 1.24×10^9 non-religious or atheists people on the earth. The total 2006 population of the earth was 6.5×10^9 people; what percent are non-religious or atheists?

26. If a piece of paper 1×10^{-2} inch thick is cut in half and the pieces are stacked so as to double the thickness, and this is done 25 times, what is the thickness of the final paper?

27. The sun and the earth are orbiting the center of our Milky Way Galaxy. It takes 2.4×10^8 years to complete one orbit. What percent of the way around the galaxy do we travel in one year?

28. The sun has a mass (something like weight) of 1.989×10^{30} kg. The mass of the Milky Way Galaxy is 2×10^{11} times as massive as the sun. What is the mass of the galaxy in kilograms?

29. At the current rate of growth, 2.65×10^5 additional people are added to the human population every day. How many additional humans will call earth home this year?

30. The number of paid civilians working for the federal government in 2000 was 11.484×10^6. If they were paid an average salary of \$45,000 per year, how much was needed to pay the employees for the year?

31. The mass of the earth is 5.976×10^{24} kg. How many times more massive is the sun?

32. Find any two numbers, expressed in scientific notation, whose sum is 2.812×10^6.

33. Find any two numbers, expressed in scientific notation, whose difference is 4.5×10^3.

34. Find any two numbers, expressed in scientific notation, whose product is 8.4×10^{-2}.

35. Find any two numbers, expressed in scientific notation, whose quotient is 7.16×10^5.

The number 75,209 written in polynomial form is $\mathbf{7} \times 10^4 + \mathbf{5} \times 10^3 + \mathbf{2} \times 10^2 + \mathbf{0} \times 10^1 + \mathbf{9} \times 10^0$.
The zero place holder ($\mathbf{0} \times 10^1$) may be omitted leaving $\mathbf{7} \times 10^4 + \mathbf{5} \times 10^3 + \mathbf{2} \times 10^2 + \mathbf{9} \times 10^0$.

36. Write the following numbers in polynomial form.

one thousand	1,000
one million	1,000,000
one billion	1,000,000,000
one trillion	1,000,000,000,000

37. Can you see a virus that has a length of 0.000 000 02 meter if you observed it under a microscope that magnifies an object 1,000,000 times? Why or why not?

38. People who use big or little numbers on a regular basis do not always use a coefficient that is between 1 and 10. Write 5,430,000 in an exponential notation where the coefficient is:
    a. between 0 and 1
    b. between 1 and 10
    c. between 10 and 100
    d. between 100 and 1000
    e. between 1000 and 10000.

39. Write 0.00038 in scientific notation where the coefficient is:
    a. between 0.0001 and 0.001
    b. between 0.001 and 0.01
    c. between 0.01 and 0.1
    d. between 0.1 and 1
    e. between 1 and 10.

From Mathematics to English

40. Why is scientific notation used?

41. Describe any situation where you have used scientific notation.

42. Do you think $9.999... \times 10^{999999999...}$ is the biggest real number? Explain.

43. After reading this section, make a list of questions that you want to ask your instructor.

44. Start a daily journal and make an entry. In addition to your normal entry on thoughts about the mathematics in this section, list at least two positive comments about what you have learned about this topic.

45. In paragraph format, summarize the material in this section of the text in your daily journal.

46. Describe how your classroom instructor made this topic more understandable and clear.

47. After reading the text and listening to your instructor, what do you not understand about this topic?

From English to Mathematics

Rewrite each English statement using a mathematical notation.

48. All real numbers larger than 2.3×10^{-4}.

49. -5.99×10^4 is smaller than 9.34×10^3.

50. 259.4×10^3 equals 2.594×10^5.

Explorations

51. When you do a calculation on your calculator in FLOAT mode, the answer is expressed either in standard or scientific notation. By doing the series of multiplications below, determine how big a number must be before your calculator expresses it in scientific notation. Determine how small it must be for the calculator to express it in scientific notation by using a series of divisions by 10, 100, etc.
    a. 5×10
    b. 5×100
    c. 5×1000
    d. etc.

CHAPTER ONE TEST

1. List an example of the use of the distributive property.

2. What property is used in the statement $-2 \times (5 \times 7) = (-2 \times 5) \times 7$.

3. Isolate the 7 on the left side of the equation $-1(3 + 4) = -7$. Do not simplify the right side.

4. Isolate the 3 on the left side of the equation $-4(3 + 2) = -20$. Do not simplify the right side.

5. Isolate the 5 on the left side of the inequality $2(4 - 1) < 13 - 5$. Do not simplify the right side.

6. Find the mean, median, and range of the monthly precipitation (in inches) in Houston, Texas.

J	F	M	A	M	J	J	A	S	O	N	D
3.2	3.3	2.7	4.2	4.7	4.1	3.3	3.7	4.9	3.7	3.4	3.7

Source: PC USA

7. If the mean salary of all 21 people employed at the XYZ Publishing Company is \$28,000, are there exactly 10 employees below and 10 employees above the mean?

8. Find the maximum and minimum for the data set $\{4, 4, 4, 4, 4, 4, 4\}$.

9. What is the range of the data set in Exercise 8?

10. How many numbers are in the set of numbers (3, 3)?

11. Describe the set of numbers (–2, 5] in English.

12. Describe the set of numbers $(-\infty, 4)$ using a number line.

Use interval notation to describe the set of numbers in Exercises 13 - 15.

13. When is the V above the number line?

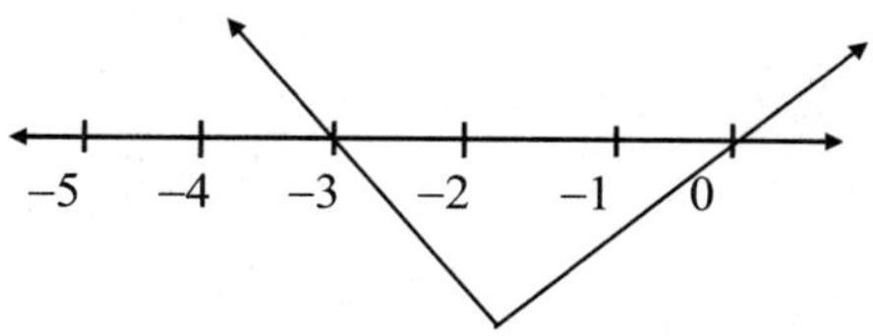

14. When is the Z below the number line?

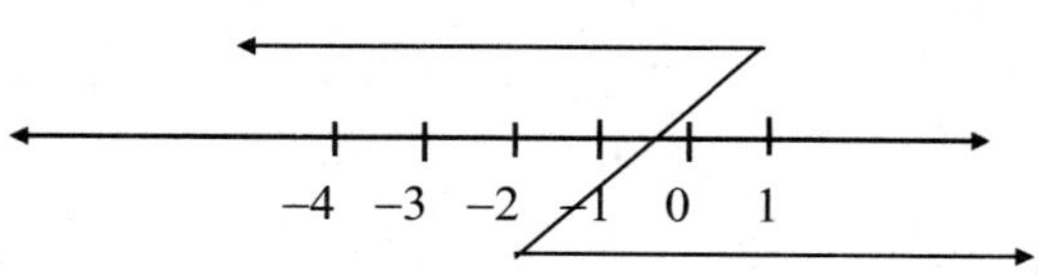

15. When is the graph above the number line?

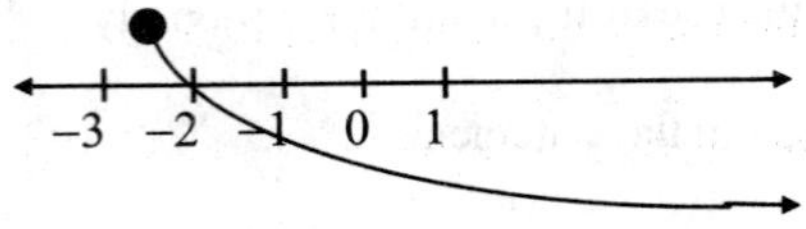

16. The total amount of blood in all humans on Earth is about 5×10^9 gallons. Write this in standard notation.

17. The odds of being in a commercial airline crash is 0.0000022 to 1. Write this number (0.0000022) in scientific notation. (Saying this another way, there are on average, 2.2 crashes for every 1,000,000 flights.)

18. If the average weight of all humans on earth is 105 pounds, how much do all 6.512×10^9 humans weigh? (2006 data)

19. If the sun is 5×10^9 years old and it is expected to live to 1.5×10^{10} years, how many years are left in the life of the sun?

20. Approximately 55 square feet of rain forests have been cut down for every fast food hamburger that comes from cattle raised on this land. How many square feet of rain forest have been destroyed to produce just 1×10^9 of the hamburgers already eaten?

CHAPTER TWO
REPRESENTATION AND BEHAVIOR OF FUNCTIONS

2.0 Introduction

The biggest technological and mathematical advantage of the graphing calculator is that you can 'see' visualizations of relationships between numbers. This provides you with valuable information about the behavior of the data, and will allow you to further analyze the relationship between the data (numbers). For example, consider the following data:

Table 2.0.1

Number of calculators sold (N)	0	500	1000	1500	2000	2500	3000	3500	4000
Profit made (P)	0	12500	25000	37500	50000	62500	75000	87500	100000

Reading this numeric representation of the relationship between the number of graphing calculators sold and the profit earned by the calculator company tells you that as more calculators are sold, the profit increases. You can also see that if they sell exactly 2,000 calculators, they will earn $50,000 in profit. This type of analysis of the relationship between the set of numbers N and the set of numbers P is developed throughout this text. This kind of information is useful, for example, to an accountant for this company, she may be asked to estimate the profit when 26,250 of the calculators are sold. Based on the numerical data above, she may be asked to estimate the calculator sales needed to bring in $1,000,000 in profit. To help answer these and other questions and to explain the answers to these questions, a visualization of this data can be useful. The visualizations (pictures) may not be similar to those studied in arithmetic. You may have seen graphs like in Figure 2.0.1.a. From now on, you will not use this type. You will use graphs like shown in Figure 2.0.1.b. If the number of calculators sold is represented by the appropriate numbers on a horizontal number line and the profit from the number of calculators sold is displayed above the number sold, you will see a visualization of the data.

Profit number line

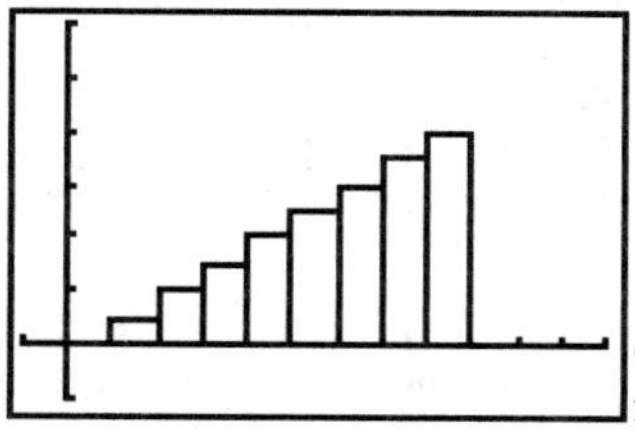

Calculators sold number line

Figure 2.0.1.a

Profit number line

25000

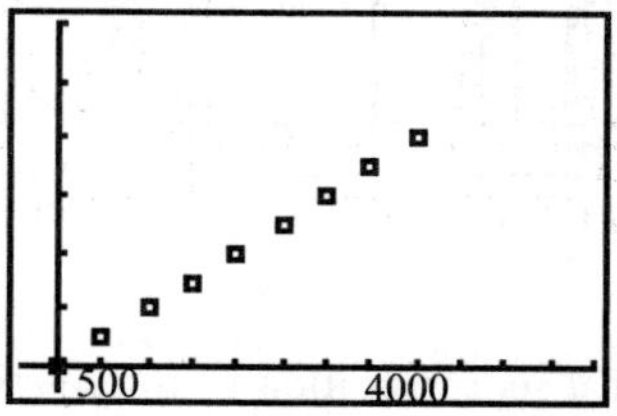

Calculators sold number line

Figure 2.0.1.b

The visualization or **graphical representation** shows that the data is following a linear growth pattern. The data lie in a straight line, as do the tops of the bars. The linear growth pattern indicates a constant growth rate. It is this idea of a pattern that will help you analyze relationships at a higher mathematical level than just looking at the height of each bar as in Figure 2.0.1.a.

This same kind of analysis, although in much more detail, will be developed in this text. Not all data relationships are linear. As you progress through this text, you will analyze several types of relationships. You will look at behaviors in data relationships such as data that sometimes increases and sometimes decreases, or has a maximum or minimum value(s). Knowing when data is positive, negative, or zero gives useful information to the person analyzing the data. Relationships can be quadratic, for example, many profit relationships are quadratic rather than linear. Relationships can be exponential, for example, the relationship between time and the number of humans on earth is exponential. Throughout this text, you will investigate seven elementary relationships and several other combinations of relationships.

Have you ever wondered where things like $\frac{1}{2}x-12$ or $-16t^2+48t+100$ come from? Or, why do mathematicians study such things as $2|x-3|+5$ or $\sqrt{x+4}-2$? Hopefully these and other questions will be answered in Chapter Two and in the remainder of the text. The process begins with the analysis of the data relationship in Table 2.0.2.

Consider the numeric relationship between time (t) and the average amount of garbage (g) generated each day by each person in the United States.

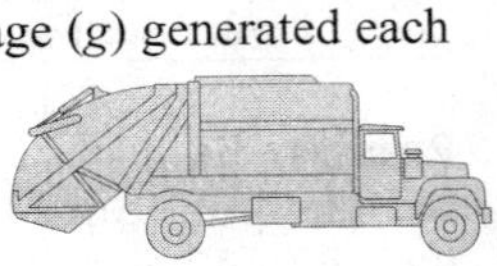

Table 2.0.2

t	1960	1970	1980	1988	1993	1998	2004	2006
g	2.7	3.2	3.6	4.0	4.1	4.2	4.3	4.4

(g is in pounds)

It has been projected that in the year 2010, the average amount of daily garbage generated by each person in the United States will be 4.6 pounds. How is it possible to make such a prediction? Perhaps if the shape of the relationship is known, you can think of a way of answering the question. The four steps necessary to get the graphical representation of the relationship are shown again in Figure 2.0.2.

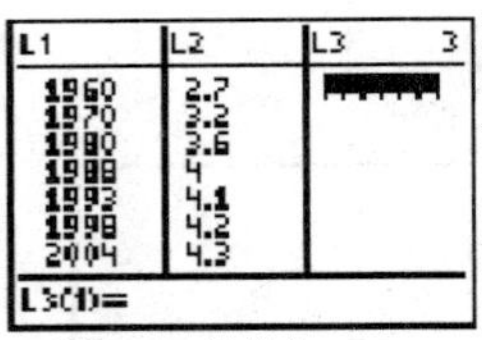

Figure 2.0.2.a

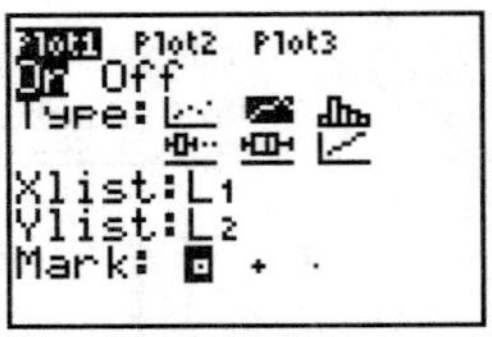

Figure 2.0.2.b

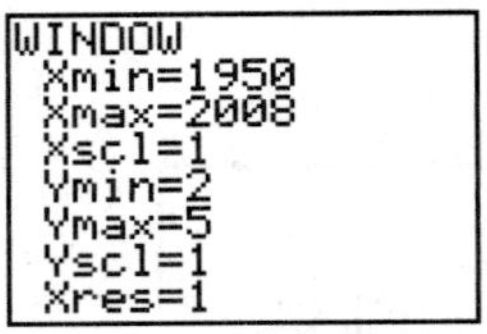

Figure 2.0.2.c

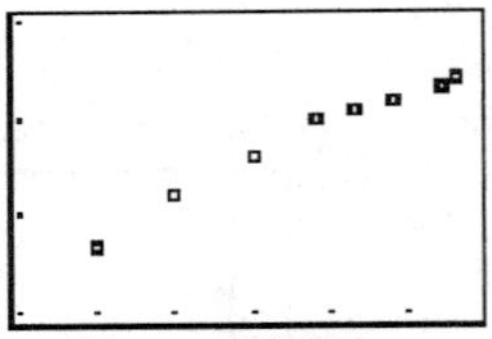
Figure 2.0.2.d

The graphical representation of the relationship suggests that the data has been nearly linear since 1960. If, in fact, it is a linear relationship, then a line drawn through the data points and extending the domain to include the year 2010 should reveal a data point at 4.6. Should you be curious if this method works, try it using a correctly scaled drawing and a good ruler. Of course, you can do the same thing on the calculator and not waste your time and paper. While this method is acceptable in the classroom, mathematicians, engineers, business people, and scientists may predict the garbage generation in the year 2010 by the method described below.

The numeric and graphic representations of the garbage problem have already helped to estimate what the per person per day garbage will be in the year 2010 by drawing a line. However, there is one more very important and useful idea that will allow you to not only predict for the year 2010, but for any year –

within reason. There is one more way of representing the data relationship; it can be represented symbolically. Many data relationships can be represented symbolically. As this chapter unfolds, you will discover that symbols like $\frac{1}{2}x-12$ can be used to represent data relationships! To find a "good" symbolic representation requires that you first know the shape of the data relationship. For more complicated relationships, you must learn more about their behavior. That is, thus far, you know how to find the shape of a relationship. Chapters Two and Three will help you learn about other behaviors of relationships. This, in turn, will help you learn how to find the symbolic representation of a relationship.

One suggested symbolic representation of the garbage problem is $0.035t - 65.69$ where t is time in calendar years and $0.035t - 65.69$ is the amount of garbage created per person per day. The symbolic representation can also be used to obtain the numeric or graphic representations. For example, Table 2.0.2 is created by using the symbolic representation for the garbage with t being replaced with members of the domain of your choosing!

Table 2.0.3

T	1960	1970	1980	1988	1990	1993	1995	1998	2000	2010
$0.035t-65.69$	2.9	3.3	3.6	3.9	3.96	4.1	4.14	4.24	4.3	4.66

The symbolic representation provides a very simple method for predicting any year you want to put in the problem domain.

You should not yet concern yourself with where the symbolic representation of the garbage problem comes from. You will learn where it comes from and how it is developed in Section 2.2. The point for now is that the relationship "can" be expressed in symbolic form. If data relationships are expressed in symbolic form, it may become easier to analyze certain features of the relationship.

This discussion may cause you to wonder if all of the "algebra-like" symbols are nothing but symbolic representations of data relationships. Of course it is possible; however, it is not always necessary or desirable to relate all symbolic representations to actual data relationships. To help you learn mathematics, it is desirable to develop new ideas in the context of a data relationship. After you have learned a mathematical idea, it may no longer be necessary to relate mathematical symbols to real-world data relationships. It is true that "algebra-like" symbols such as $3.5T+80$, $-800|t-30|+24000$, $-16t^2+48t+25$, and $10000\cdot 2^{\frac{t}{20}}$ do have graphical representations that behave like the graphical representations of the data relationships presented in Section 2.1. The types of data relationships found in Section 2.1 are linear, absolute value, quadratic, and exponential.

$3.5t + 80$

Linear

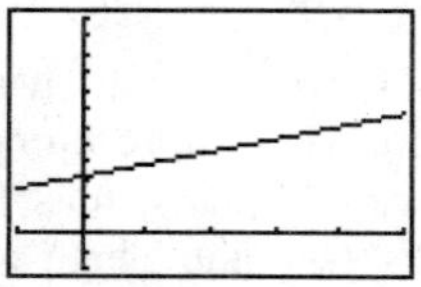

Figure 2.0.3a

$-800|t - 20| + 24000$

Absolute Value

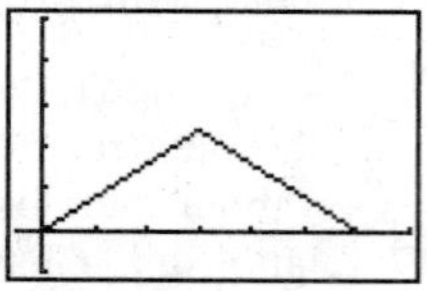

Figure 2.0.3b

$-16t^2 + 48t + 25$

Quadratic

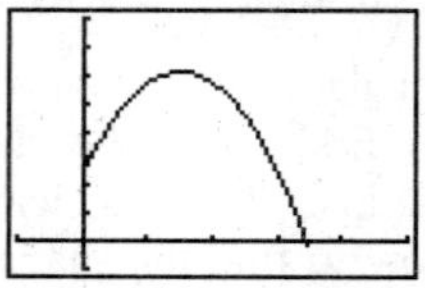

Figure 2.0.3c

$10000 \cdot 2^{\frac{t}{20}}$

Exponential

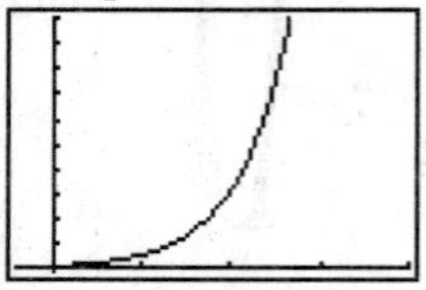

Figure 2.0.3d

These relationships will be studied numerically, graphically, and symbolically in Chapter Two.

The analysis of data relationships studied in Chapter Two will also introduce the concept of a variable and of a function. By symbolizing the real numbers in data relationships, you can make generalizations about the characteristics (behaviors) of relationships. Once some of the behaviors of relationships are known, you can solve problems encountered in the everyday world of business people, teachers, engineers, scientists, and others. As students, you will use the mathematics you learn to solve a variety of real-world problems often encountered by these professional people and others.

Chapter Two serves a two-fold purpose. You will learn about variables, behaviors of relationships, and three different ways of representing relationships - numeric, symbolic, and graphic. Second, Chapter Two gives you an opportunity to *experiment* with and learn how to use the graphing calculator to help you analyze and solve problems involving relationships. Using the three representations in conjunction with the calculator will enhance and enrich the teaching and learning of mathematics in this and all remaining chapters.

2.1 Data Relationships Represented Numerically and Graphically

Related Data The study of related data extends throughout this text. In this section, you will study the geometric shapes of the data relationships. Yes, if you look at a picture (**graphical representation**) of the related data, many times you will see a very clear and definite shape to the data. This is not to say that all data has a clear and distinct shape; it doesn't. In addition to identifying shapes of data relationships, you will also be asked to identify if the relationship is increasing or decreasing. This behavior of the relationship becomes important in later studies. The data relationships you will analyze in this text come as pairs of numbers. If they are related, you will likely see a recognizable geometric shape. To get the graphical representation of the related data, the graphing calculator will be used. It will use the first number in the pair to determine the horizontal position of the data point, just like in Figure 2.1.3. The second number in the pair will determine how far above or below the horizontal number line to put a mark representing the graph of the pair of numbers. Below in Table 2.1.1 is the numeric representation of seven pairs of numbers. The first number in each pair is time and the second number is the percent of the Unites States Gross National Product spent on health care.

Table 2.1.1

Time	1960	1965	1970	1975	1980	1985	1990	1995	2002	2003
% of GNP	5.3	5.9	7.3	8.3	9.2	10.5	12.2	14	14.9	15.3

Source: U.S. Dept. of Health and Human Services

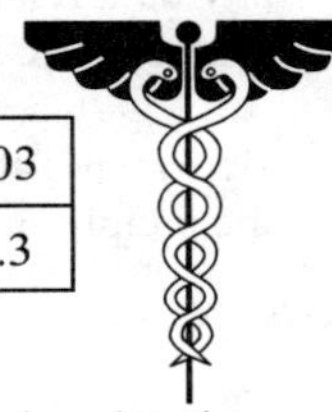

Hopefully, you can see that the amount of money spent on health care in the United States is related to time. As time passes, money spent on health care increases. It does not fluctuate wildly. If the numbers were not related, you would see no pattern in the percent of GNP. While the data is useful in numeric form and will be further analyzed, the purpose here is to look at the graphical representation of the data and give a name (if possible) to the shape. The graphical representation can be obtained by using your graphing calculator. Below are screens from a TI-84 graphics calculator outlining the steps needed to graph the data relationship.

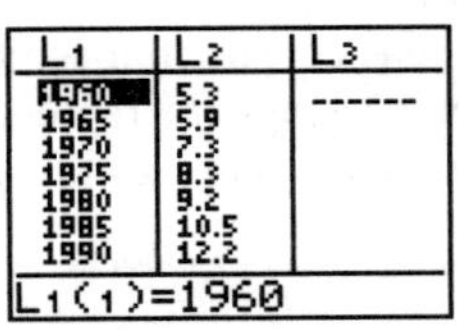

Figure 2.1.2a

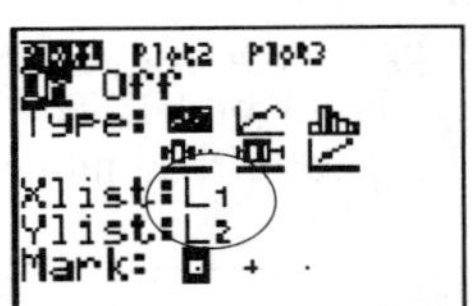

Figure 2.1.2b

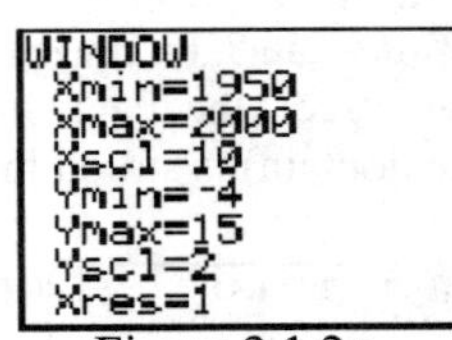

Figure 2.1.2c

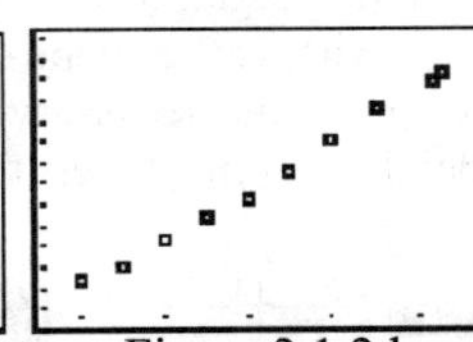
Figure 2.1.2d

The data is entered in L_1 and L_2 as shown in Figure 2.1.2a. The set of all numbers in L_1 is called the **domain** (Xlist) of the relationship and the set of all data in L_2 is called the **range** (Ylist) of the relationship. Figure 2.1.2b tells the calculator that the first number in each pair is stored in L_1 and the corresponding numbers are in L_2. This figure also shows that plot mode is on, the calculator is to plot points only, and to use the □ sign as the mark where each data point is to be plotted. Figure 2.1.2c shows the calculator window to be used. It tells the calculator to start time in 1950, stop in 2000, and put a tic mark every 10 years. It also tells the calculator to show any health care money from –4% to 15% and to put a tic mark every 2%. The final figure shows the graphical representation of the data. This is also shown below in Figure 2.1.3.

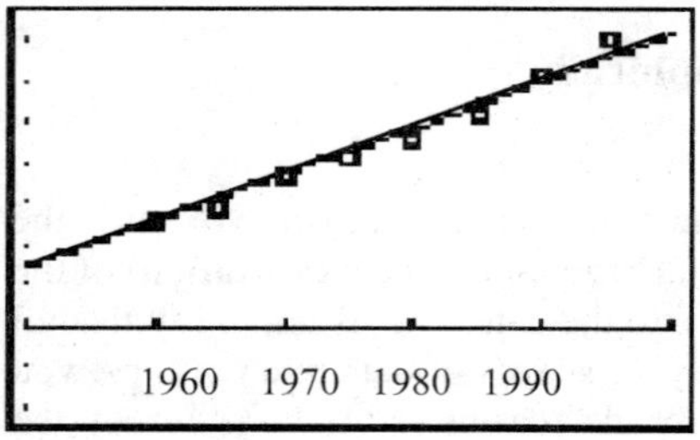

Figure 2.1.3

The issue here is simple. What is the shape of the data? Of course, it is nearly linear. The growth of money spent on health care as a percent of the Gross National Product follows a linear growth pattern. The growth of money spent on health care as a percentage of the Gross National Product seems to depend on time. On many occasions, real data will not produce a "perfect" shape when it is graphed. The graphical representation of the data shown in Figure 2.1.3 indicates that the data in not exactly in a straight line. The solid line drawn on top of the data is a good approximation to the actual data. Because the graphical representation of the data is a straight line, the relationship is called **linear**.

As you think about linear data relationships, you quickly realize that the graph (being a straight line) either rises as time passes, falls as time moves forward, or is constant (a horizontal graph). In the example above, we know the GNP is rising as time passes, so we say it is increasing. We will extend this idea for any data relationship to say that if the graph is rising as we trace to the right, it will be an increasing data relationship. If it falls as we read it from left to right, we will call it decreasing.

Below in Table 2.1.2 is data showing the relationship between the Centigrade (Celsius) temperature and the weight of one gallon of water in pounds. If you look closely at the weight of the water,

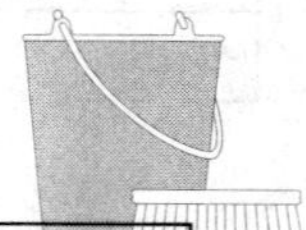

Table 2.1.2.

T	0°	1°	2°	3°	4°	5°	6°	7°	8°
W	8.33461	8.33513	8.33551	8.33575	8.33585	8.33582	8.33565	8.33536	8.33495

Source: *CRC Handbook of Chemistry and Physics College Edition.* 1965-1966. The Chemical Rubber Company. Cleveland, OH

you will notice that as the temperature changes from 0° to 8°, the weight of the water increases part of the time and then it decreases in weight. Are the data related? If so, how are they related? Is it a linear relationship? Can the available data in Table 2.1.2 be used to predict the weight at 20°, or at –30° Centigrade? Again, these and other questions will be answered later in the text. For now, the question remains what is the shape of this data? To find out, proceed as in the first data set. The four screens below show the steps necessary to "see" the data on a TI-83 or 84. The set of numbers representing the temperature of the water (the domain) is stored in L_1. The corresponding set of numbers for the weight

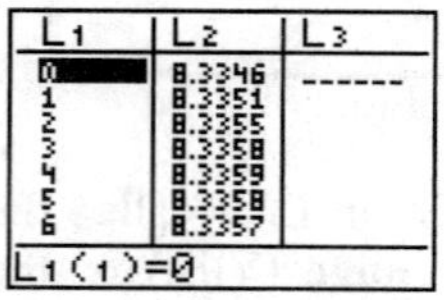

Figure 2.1.4a

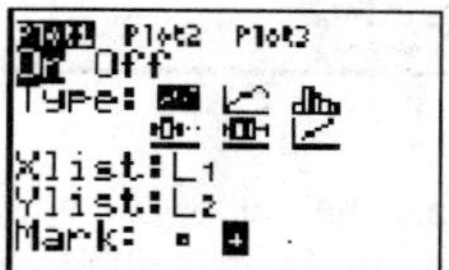

Figure 2.1.4b

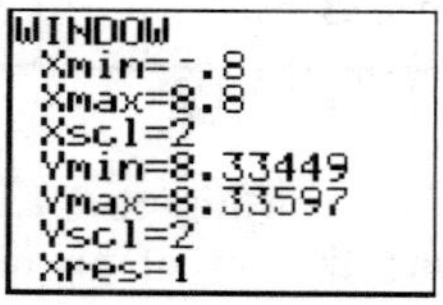

Figure 2.1.4c

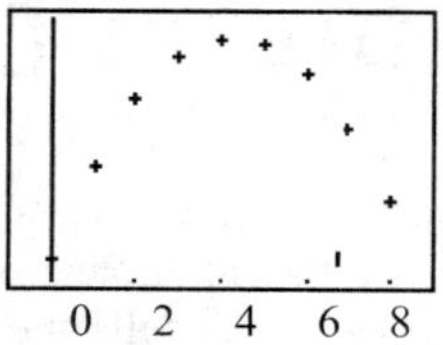

Figure 2.1.4d

of the water (the range) is stored in L_2. As with the first set of related data, Figure 2.1.4b shows that the data plotter is turned on, what type of graph to create, where the domain (Xlist) and range (Ylist) are stored, and what type of mark to use at each data point (+). The window shown in Figure 2.1.4c can be set manually or use ZOOM - ZoomStat on the TI calculators. You may recognize the shape as something like the path of a ball tossed lightly from one person to another. It is similar to the shape of water coming out of a drinking fountain. It looks like a parabola. The relationship between the temperature of a gallon of water and the weight of the water is parabolic in nature. This relationship is called a **quadratic**

relationship. You will see why later in Chapter Three. The weight of the gallon of water depends on the temperature of the water in a quadratic-like fashion. Looking a Figure 2.1.4d or Table 2.1.2 we see that the weight (or the temperature-weight relationship) is increasing until the temperature is 4°, and then the relationship is decreasing when the temperature is above 4° Centigrade.

The third example is the relationship between the time and the national debt. As time moves forward, the national debt just keeps rising; that is, it appears that the national debt depends (is related to) on time. Table 2.1.3 shows the relationship between time and the national debt.

Table 2.1.3

Year	1940	1950	1960	1970	1980	1990	1994	2002	2006
National Debt	51	257	291	381	909	3113	4500	6000	8400

(Debt is in billions.)

Is the national debt growing in a linear fashion? Is it following a quadratic relationship? Or does it have still another shape? To see the shape of the data, do as before and enter the data in the calculator's memory and then graph the data. As with the previous examples, the domain is stored in L_1 and the range in L_2. The graph type is data points connected and the mark to be used is the □ mark. Figure 2.1.5d shows the shape of the data.

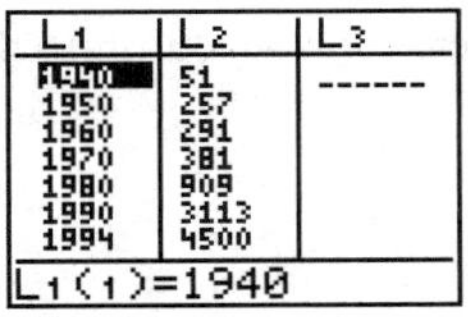

Figure 2.1.5a

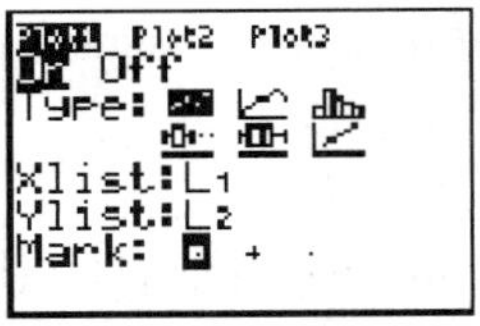

Figure 2.1.5b

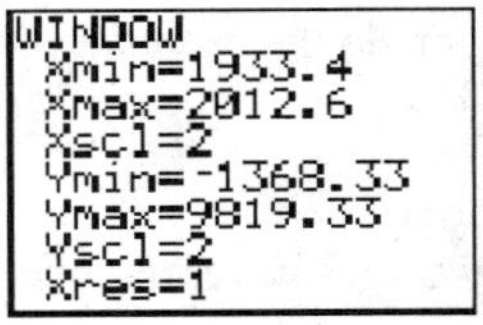

Figure 2.1.5c

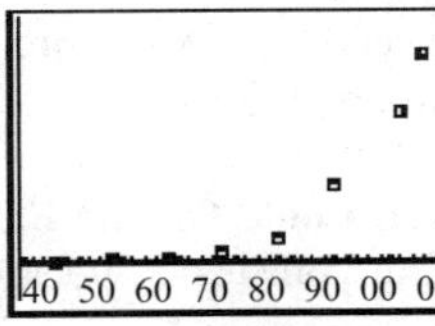

Figure 2.1.5d

Hopefully it is obvious that the data is not linear, not quadratic, but still another shape. This shape stays fairly flat for a while and then increases very sharply. This relationship is called **exponential**. Currently, human population is growing exponentially. Money invested in a compound interest account grows exponentially. Many bacteria grow in numbers (population) exponentially. In each of these examples, the population depends on the length of time spent growing. In the example exponential relationship above, it is increasing. That is, the relationship between time and the USA national debt is increasing.

The few samples above demonstrate that real-world data relationships do behave according to very definite mathematical patterns. You will encounter several more shapes throughout the remainder of the text. The last example may convince you that not all data fits into a recognizable shape.

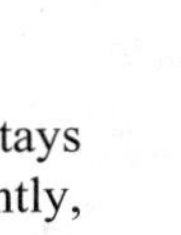

Table 2.1.4 shows the relationship between IQ and weight of 10 randomly selected students.

Table 2.1.4

Weight	110	115	118	127	135	138	142	150	155	170
IQ	120	115	130	103	95	137	120	112	128	116

Using the same methods as before, the steps used to graph the data are shown in Figure 2.1.6 with the numbers that make-up the domain stored in L_1 and the range is stored in L_2.

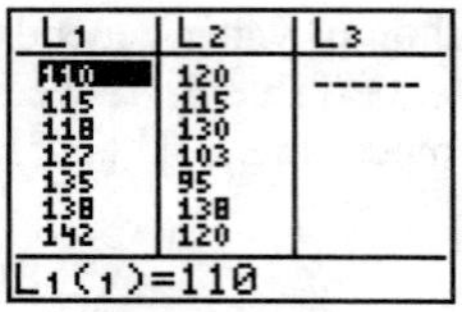

Figure 2.1.6a

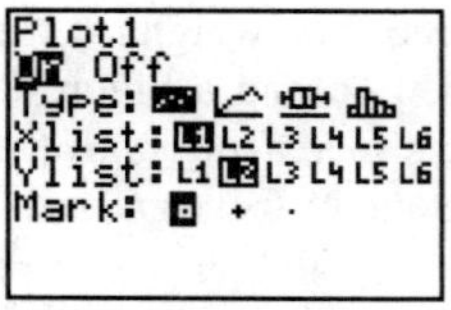

Figure 2.1.6b

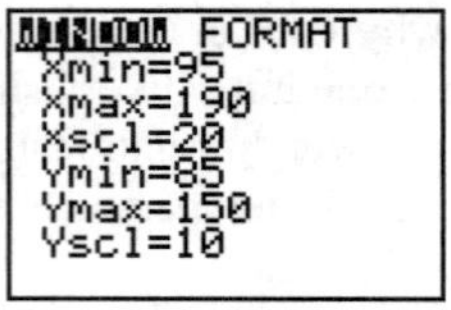

Figure 2.1.6c

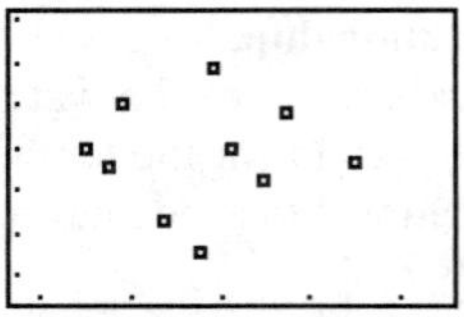
Figure 2.1.6d

As you can see in Figure 2.1.6d, the data represents no clear recognizable shape. The conclusion you may draw from this is that your IQ is not related to weight. That is, IQ does not depend on a person's weight. In looking at the visualization in Figure 2.1.6d, hopefully it is clear that the concept of increasing or decreasing is meaningless in this relationship.

Data relationships are found in your work life as well as your daily life. For example, the amount of gasoline in the gas tank of your car depends on the distance your car is driven. The time it takes you to get to work depends on how fast you drive. Your weekly gross wages may depend on the number of hours worked. The length of time it takes to paint a room depends on the surface area of the room. If you have a good understanding of mathematical relationships, you will have the power to control some of these relationships in your life.

2.1 EXERCISES

Please read the text before you do the exercises. Read the **next assigned section** after you finish the assignment below.

A Quick Look To The Future

-4. Can your gross weekly wages be represented with mathematical symbols? Explain.

-3. If you throw a ball straight up, use English to describe how the height of the ball changes.

-2. What name might you give to the largest height the ball reaches (see above)?

-1. What is another word that describes the velocity (or speed) of the ball?

Previous Topics Revisited

1. What property is being demonstrated in the statement $4(3 + 7) = 4 \times 3 + 4 \times 7$?

2. Write a statement that shows the use of the associative property for multiplication.

3. Isolate the number 7 using the properties of equality. $7(4 - 3) + 4 = 5 + 6$.

4. Evaluate $-4 + 5^2 + 3(2 - 7)$.

5. Evaluate $\sqrt{3^2 + 4^2}$.

6. Write 0.0000000001 in scientific notation.

New Topic Exercises

In Exercises 7 – 29, identify the shape of the data relationships, when the relationship is increasing and when it is decreasing. If a new shape occurs -- give it a name of your choice.

7. The number of paid civilians in the U.S. government is increasing as time passes. The number of employees is in thousands.

Year (*Time*)	1901	1910	1920	1930	1940	1950	1960	1970	1980	1990	2000
Employees	239	389	655	601	1042	1961	2399	2985	2876	3128	11484

Source: U.S. Bureau of the Census, *Statistical History of the U. S.*

8. The average cost (tuition and fees) of attending a public four-year college is increasing as time passes. Time is in academic years starting in 1980 - 81.

Time	1980	1981	1982	1983	1984	1985	1986	1987	1988	1989	1990	1991	1992	2002
Cost	804	909	1031	1148	1228	1318	1414	1537	1646	1781	1908	2137	2315	3754

Source: The College Board

9. The poverty threshold for a family of four has been on the rise as indicated in the data below.

Time	1960	1965	1970	1975	1980	1985	1990	1991	1992	2000
Income	3022	3223	3968	5500	8414	10989	13359	13942	14463	17524

Source: U. S. Bureau of the Census, *Current Population Reports.*

10. A ball is thrown straight upward with an initial speed of 32 feet per second from a point 20 feet above ground level. Below is the data that shows the height of the ball above the ground at t seconds.

t	0	0.2	0.4	0.6	0.8	1	1.2	1.4	1.6	1.8	2	2.2	2.4	2.5
Height	20	25.76	30.24	33.44	35.36	36	35.36	33.44	30.24	25.76	20	12.96	4.64	0

11. Below is the data showing the number of inmates in state and federal prisons for recent years.

(inmates are in thousands)

Time	1980	1981	1982	1983	1984	1985	1986	1987	1988	1989	1990	1991	1992	1997	2003
Inmates	330	370	414	437	462	503	545	585	632	713	773	824	884	1198	1470

12. As a commercial airliner flies from Columbus to Pittsburgh (approximately 200 miles), its height (in miles) is related to the ground distance from Columbus. Below is the data showing the ground distance from Columbus and the related height of the plane.

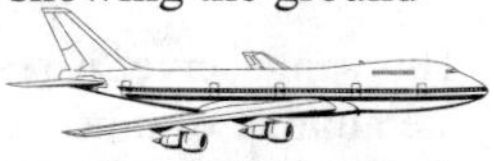

Ground Distance	0	25	50	75	100	125	150	175	200
Height	0	.853	1.705	2.558	3.410	2.558	1.705	.853	0

13. The amount of land being farmed is shown below as a percentage of the total land in the United States. The data shows that the amount increased for a while and recently has been decreasing.

Time	1910	1930	1940	1950	1960	1970	1980	1990	2000
% Farmed	38.8	43.6	46.8	51.1	49.5	47	44.8	42.7	41.6

14. If you invest $1000 at 6% compounded annually, the data below shows that the amount of money in your account depends on how long it is in the account. (time is in years)

Time	0	5	10	15	20	25	30	35	40	45	50
Amount	1000	1338	1791	2397	3207	4292	5744	7686	10286	13765	18420

15. The data below shows a checking account balance for the month of March 1994. The time is in the day of the month and the balance is rounded to the nearest dollar.

Time	1	3	5	7	9	11	13	15	17	19	21	23	25	27
Balance	3245	3202	958	885	623	623	623	623	623	1847	551	464	198	43

Source: author's checkbook

16. The data below shows that the number of hospitals in the United States has recently been decreasing; earlier, the number of hospitals was increasing.

Time	1950	1955	1960	1965	1970	1975	1980	1985	1990	2000	2002	2003
Number	6788	6956	6876	7123	7123	7156	6965	6872	6649	6116	5794	5764

Source: American Hospital Association, *Hospital Statistics* (annual).

17. The human population of the earth has been steadily increasing as time passes. Below is the data showing the relationship between time and population. The population is in billions. For example, 2.8 means 2.8 times 1,000,000,000. In 1950 the earth had 2,800,000,000 people. All numbers are rounded.

Time	1750	1800	1850	1900	1950	1994	1999	2001	2006
Population	0.8	1.0	1.2	1.7	2.8	5.9	6	6.2	6.6

18. Below are the possible wages for a server working at the Blue Point Cafe. He is paid a salary plus he earns an average of $3.50 in tips per person served. His wages depend on how many people he serves on during the week.

Persons	10	15	20	25	30	35	40	45	50	55
Wages	75.00	92.50	110.00	127.50	145.00	162.50	180.00	197.50	215.00	232.50

19. This same server is responsible for filling the ice chest in the cafe. He kept the following records on the number of times per day he had to fill the chest. Time is expressed in the day of the month.

Day	1	2	3	4	5	6	7	8	9	10	11	12	13	14
Number	6	6	8	4	12	9	3	4	6	5	8	11	6	5

20. Below is the data showing the number of people enrolled in Medicare from 1975 to 1993. The numbers have been rounded to the nearest 100.

Time	1975	1980	1985	1990	1991	1992	1993
Number	23,800	27,500	30,100	33,100	33,800	34,400	35,100

21. Below is the volume of water as it changes with temperature changes. The original volume of water in this experiment was 1 milliliter at 4° C.

T	−10	−5	0	2	4	6	10	20
V	1.00186	1.00070	1.00013	1.00003	1.00000	1.00003	1.00027	1.00177

Source: *CRC Handbook of Chemistry and Physics College Edition.* 1965-1966. The Chemical Rubber Company. Cleveland, OH

22. The internal revenue service has been collecting taxes at a per-person per year rate as shown in the data below that has been rounded to the nearest dollar.

Year	1960	1965	1970	1975	1980	1985	1986	1987	1988	1989	1990	1991	2004
Collected	508	589	955	1386	2276	3099	3233	3627	3792	4063	4222	4344	6874

23. An author can write his book according to the schedule below.

Time (in days)	1	2	3	4	5
Pages Written	2.5	5	7.5	10	12.5

24. Below is data showing how charges from the electric company depend on the number of kilowatt-hours of electricity used.

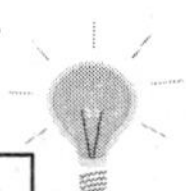

kWh	200	400	600	800	1000	1200	1400	1600	1800	2000
Charge	23.82	38.14	52.47	66.79	81.11	95.43	109.75	124.08	138.4	152.72

25. Here is data showing the relationship between the number of E. coli bacteria (*N*) on an uncooked piece of hamburger and time (*t*) in minutes.

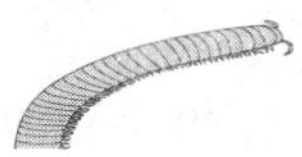

t	0	30	60	90	120	150	180
N	5000	14142	40000	113137	320000	905097	2560000

26. Below is the data for the relationship between the selling price (*s*) of a graphing calculator and the daily profit earned (*P*) from the sale of the calculators.

s	50	55	60	65	70	75	80	85	90
P	15890	17540	18690	19340	19490	19140	18290	16940	15090

27. Below is data for the relationship between time and the number of immigrants (in thousands) entering the United States, for the last several decade? (30 means the decade of the 1930's)

Decade	30	40	50	60	70	80	90
Number	528	1035	2515	3322	4493	7338	9095

Source: U.S. Bureau of Census.

28. The table below is data for the relationship between public education expenditures (in billions) in the United States and time?

Time	1940	1950	1960	1970	1980	1990
Expenses	3.3	8.9	23.9	68.5	165.6	377.5

Source: U.S Department of Education, National Center for Educational Statistics, *Digest of Education Statistics*, 1992

29. Below is data for the relationship between blood velocity (v) in an artery (that has a radius of 0.5 *cm*) vs. the distance (d in *cm*) the blood is from the center of the artery. Velocity is in *cm*/sec.

d	0	0.1	0.2	0.3	0.4	0.49
v	25	24	21	16	9	0.99

30. Using the graphical representation of the number of prisoners in state and federal prisons you made in Exercise 11, predict how many prisoners there will be in 2012. Explain how you arrive at your answer.

31. Find or make-up one example of a relationship that is linear. Identify the domain and range and describe why you think the relationship is linear.

From Mathematics to English

32. After reading this section, make a list of questions you want to ask your instructor.

33 Continue in your daily journal and make an entry. In addition to your normal entry on thoughts about the mathematics in this section, list at least two positive comments about what you have learned about this topic.

34. In paragraph format, summarize the material in this section of the text in your daily journal.

35. Describe how your classroom instructor made this topic more understandable and clear.

36. After reading the text and listening to your instructor, what do you not understand about this topic?

37. In this section, you have used data relationships in numeric and graphic forms. Express you opinion as to whether these relationships could be represented in symbolic form.

From English to Mathematics

In the next three exercises, write the English statement as a mathematical statement. For example, "five times the difference of x and 4", would be $5(x - 4)$.

38. \$3.50 plus x

39. x squared minus 3, times 43

40. the length x times the width $(x - 1)$

Explorations

41. Create (make-up) a data relationship with 6 data points that is linear and increasing.

42. Create (make-up) a data relationship with 6 data points that is linear and decreasing.

2.2 Data Relationships Represented Symbolically

Thus far in Chapter Two, you have analyzed many data relationships using the numeric and graphical representations. In Section 2.2 we will learn how to represent data relationships in symbolic form. In Section 2.0 we suggested that $0.035t - 65.69$ can represent the garbage problem. What do we mean? Simply stated when t is replaced with any year, the symbols $0.035t - 65.69$ approximate the amount of garbage generated per person per day. So in 2004 ($t = 2004$), the symbols $0.035t - 65.69$ become 4.45. The actual data for 2004 is 4.3. Or for example in 1993, the symbols $0.035t - 65.69$ become 4.07, and the actual data is 4.1. The symbolic representation $0.035t - 65.69$, nearly produces the "real" data. The symbols $0.035t - 65.69$ appear to "model" the actual data relationship.

From Numeric to Symbolic Many times mathematical relationships are available in numeric form or are described using words. It is extremely important that you be able to convert this information to symbolic mathematical form. Below is a method that is based on you recognizing a pattern as the data relationship is developed from a situation.

House Painter Measurements on a potential customer's house shows a surface area of 1792 square feet. Let t be time in hours and A the area that remains to be painted. Since you paint at an average rate of 64 square feet per hour, and the initial conditions are known, we can develop the algebraic model.

The initial conditions are known. At time 0, the area remaining to be painted is 1792 square feet.

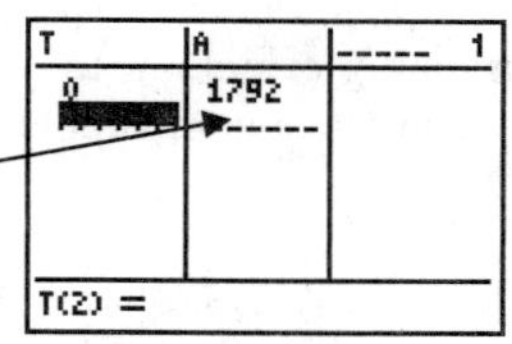

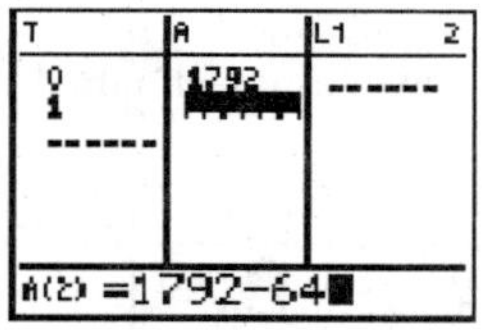

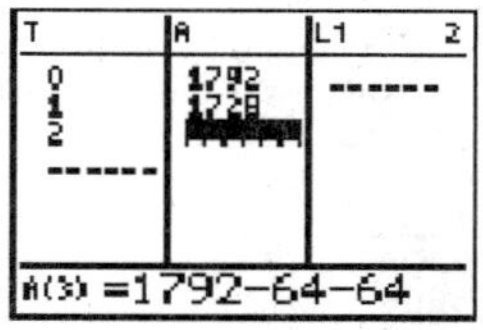

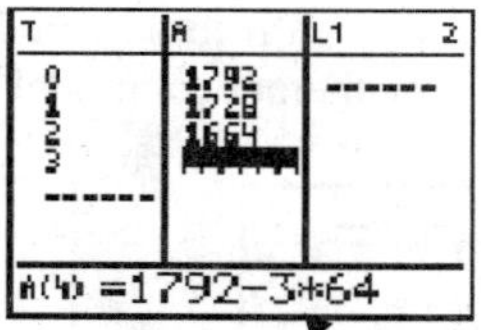

Since the rate is known, you also know that each hour worked is another 64 square feet less to paint. The mathematical process is subtraction of 64. Thus, we can create a numeric representation from the verbal description.

But repeated subtraction of 64 can best be accomplished by subtracting the product of 64 and the number of times it has been subtracted – as shown in the edit line.

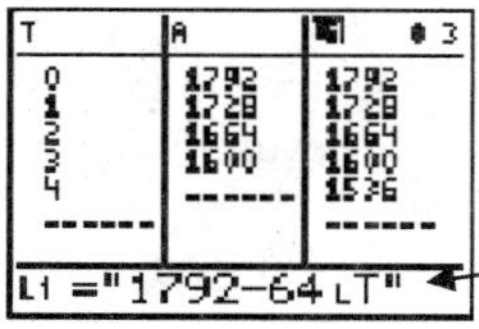

From a pattern of arithmetic operations, you can generalize to algebraic symbols such as $1792 - 64T$, or $1792 - 64x$.
(LT represents all numbers stored in T.)

The symbols 1792 – 64ʟT represent the house painter data and so they are called the symbolic representation of the situation. That is, as seen above, the symbols automatically created the area. If we enter new values for time, the symbols will create the corresponding areas – see below.

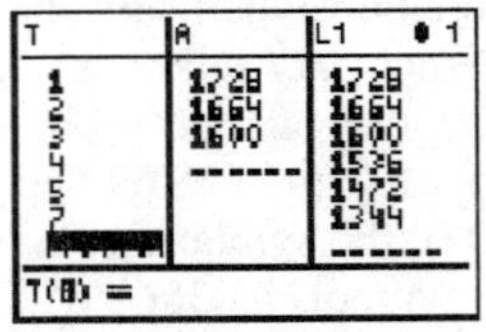
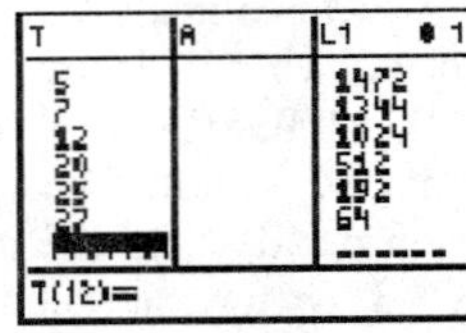

Typically, you may not find common or standard use of the symbols such as 1792 – 64ʟT to represent a situation. In mathematics, we normally us an "x" to represent a list of number such as is used above. So, we will likely use $1792 - 64x$. When we graph the data relationship and the model $1792 - 64x$, or $-64x + 1792$, we see the graph of the data points are on the graph of the model.

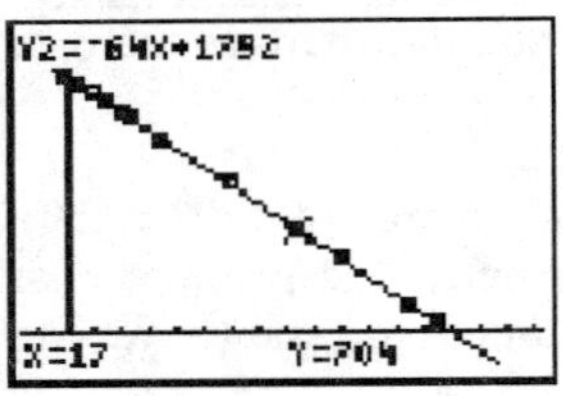

On the figure to the left, we see the graphical representation of the data relationship between time and the area remaining to be painted, and also displayed is the graphical representation of the symbols Y2 = –64X + 1792 with the data point (17, 704) displayed through the use of trace on the calculator. Through trace, we can see three representations of the house painter situation on the screen at the same time – with only one data point displayed at a time.

When using a graphing calculator to graph the symbolic representation you will note the use of another symbol, Y2. It or another comparable symbol is used to represent the symbolic form of the situation $-64x + 1792$.

Driving Home On the 720-mile return trip home from Kitty Hawk, North Carolina to Columbus, Ohio, Professor Ed drove at an average rate of 60 mph throughout the night. Create a mathematical model that will find the distance D left to travel after T hours driven.

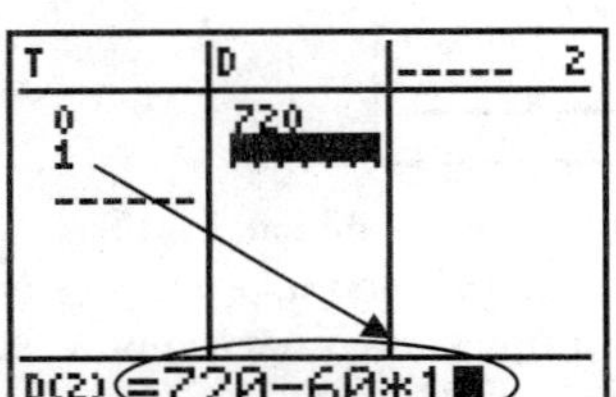

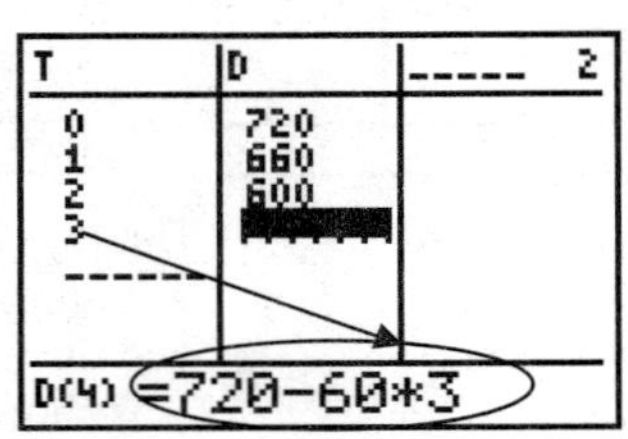

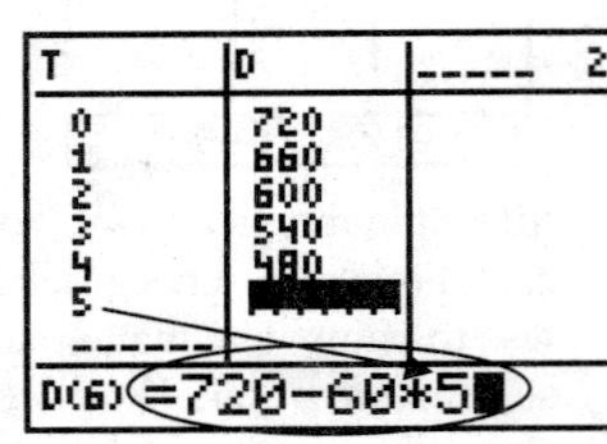

After each hour traveled, the distance left to travel changes (decreases) by 60 miles. The pattern shown above is very useful in developing the symbols $720 - 60T$, where T is the time traveled. We may find it useful to write the symbolic representation of the driving situation as $D = 720 - 60T$, to be consistent with the mathematics community. Since $720 - 60T$ represents the distance remaining in the trip, we might use $D = 720 - 60T$ as the symbolic representation of the data relationship. However, if we want the graphical representation displayed on the graphing calculator, we will simply enter $720 - 60x$ in Y1, Y2, etc. So the technology may force you into writing the symbolic representation as $Y1 = 720 - 60x$.

The Blue Point Café At many restaurants, servers are paid a set weekly salary plus tips. At the Blue Point Café in Duck, North Carolina, servers are paid \$80 per week plus tips. A serious job seeker calls the local Restaurant Association and discovers that servers average \$3.50 in tips per person served for the dinner meal.

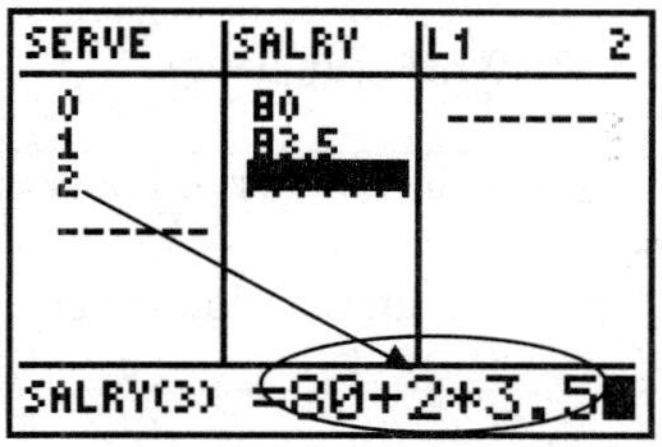

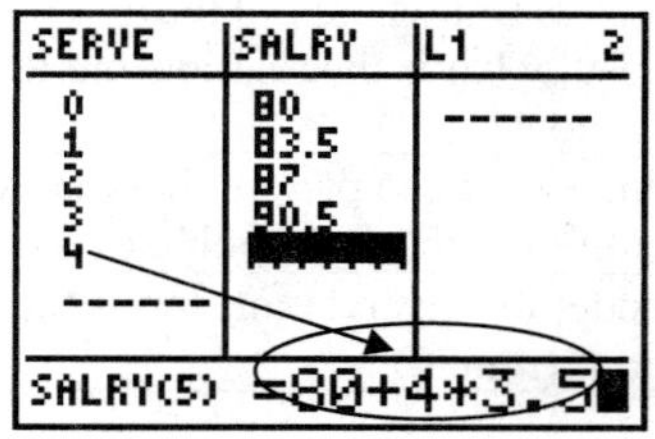

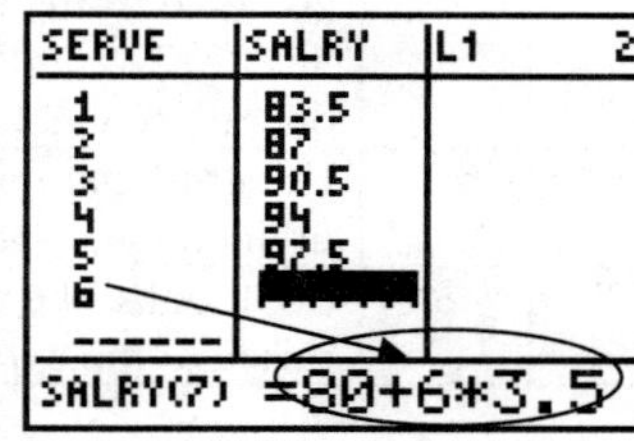

The pattern established in the series of screens above suggests that the relationship is 80 + 3.5*SERVE*. Perhaps a more common way of writing the symbols is $80 + 3.5S$, or even $3.5S + 80$. Using a more common notation, we may even write *SALRY* = 3.5*SERVE* + 80. Like before, if we choose to find the graphical representation of the relationship using technology, we would likely use $y = 80 + 3.5x$, or $Y1 = 80 + 3.5x$.

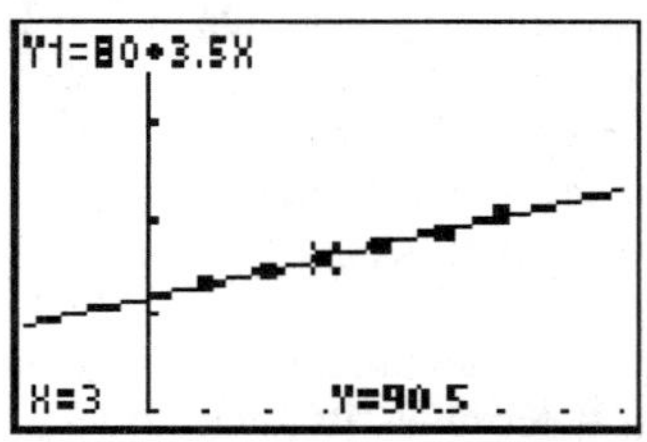

TB Bacteria Below is the data relationship between time and the population of tuberculosis bacteria growing under unrestricted conditions. This is something like what might happen in a petri dish in a medical laboratory. Time is in hours and the bacteria population is in thousands.

Tuberculosis Data

t	0	6	12	18	24	30	36	42	48
B	5	10	20	40	80	160	320	640	1280

The initial conditions are shown and the bacteria population is doubling every 6 hours.

T	*B*	Thoughts
0	5	At time 0-hours, the number of bacteria is obvious – the initial condition.
6	10	6 hours later, the number of bacteria is what? Sure, it is 5 doubled, or 10. How did you calculate it? OK, 5×2, but what is the relationship to "6 hours later?" Would it help to think of 5×2 as 5×2^1? But what is the relationship to "6 hours later?" What arithmetic operation on 6 will yield 1?
12	20	6 hours later, the number of bacteria is what? Sure, it is doubled, or 20. How did you calculate it? OK, 10×2, but where did the 10 come from? OK, 5×2. So we now have $5\times2\times2$, or 5×2^2. But what is the relationship to "12 hours?" What arithmetic operation on 12 will yield 2?

18	40	6 hours later, the number of bacteria is what? Sure, it is doubled, or 40. How did you calculate it? OK, 20 × 2, but where did the 20 come from? OK, 10×2. But where did the 10 come from? OK, 5×2. So we now have 5×2×2×2, or 5×2^3. But what is the relationship to "18 hours?" What arithmetic operation on 18 will yield 3?
24	80	6 hours later, the number of bacteria is what? Sure, it is doubled, or 80. How did you calculate it? OK, 40 × 2, but where did the 40 come from? OK, 20×2. But where did the 20 come from? OK, 10×2. But where did the 10 come from? OK 5×2. So we now have 5×2×2×2×2, or 5×2^4. But what is the relationship to "24 hours?" What arithmetic operation on 24 will yield 4?
…	…	Hopefully, at this point we see the pattern that the exponent on 2 is the number of hours divided by 6 – the doubling rate (they double every 6 hours).
T	B	So it looks like the model for this relationship is the initial condition times 2 (doubling) raised to the exponent of (time/6).

Finally, the generalized symbolic form is $5 \times 2^{\frac{T}{6}}$, or $B = 5\times2^{\frac{T}{6}}$

In every data relationship, there are two sets of numbers involved. One set contains those numbers that depend on input numbers selected by you. For example, in the tuberculosis bacteria situation above, time is determined by you, and it is independent of anything else. Once you select a time, you find that this time value determines the number of bacteria in the dish. All of the values you select are called the domain of the relationship, and the variable that represents these numbers is called the independent variable. All of the values determined from the input numbers you select are called the range of the relationship, and the variable that represents these numbers is called the dependent variable.

The data relationships used in Section 2.1 all had domains that were determined by the problem or the available information. A real-world problem determines its own domain. That is, the garbage problem presented in Section 2.0 has the domain [1960, 2004] because the data only exists for those years. However, when the symbolic representation is used to describe the garbage situation, there is no mathematical reason why the domain of $0.035t - 65.69$ cannot be $(-\infty, \infty)$. Thus, you will quite frequently use a different domain when the symbolic representation is used.

> A **variable** is a symbol that represents any element in the **domain** of a relationship. The *largest* set of real numbers that the variable in a relationship *can* represent is the **normal domain** of the relationship. The set of real numbers used for the variable when applied to a problem is the **problem domain**.

Since the domain is a set of real numbers, interval notation may be the best method of describing the domain.

Example 1: If the variable x represents all real numbers except 4, how can it be written?

Solution: It is sometimes written using set notation: $x \in \{\text{all real numbers}, x \neq 4\}$. The symbol $\in$ means belongs to, or is a member of. Another method is interval notation, $x \in (-\infty, 4) \cup (4, \infty)$.

⁂

Example 2: If the variable z represents all real numbers, how can the domain be described?

Solution: One method is to use set notation with English, $z \in \{\text{all real numbers}\}$, or use interval notation $z \in (-\infty, \infty)$.

⁂

The domain can be selected from the set of real numbers. If the domain of a relationship contains just one real number, the variable in the symbolic form of the relationship is called a **constant**. For example, the symbol π is used to represent the one number, 3.14159...; thus, π is a constant. There are other symbols you will encounter that also represent only one real number; they are all constants.

Recalling the data relationships discussed in Section 2.1, you may have noted that the ranges of data relationships were also real numbers. Throughout most all of this text you will find that the range must be a set of real numbers.

> When a relationship is represented symbolically, the set of real numbers that it represents is called the **range**.

Since the range must be a set of real numbers, you must check the symbolic from of the relationship for arithmetic operations that can create non-real numbers. You should recall two occurrences in your previous mathematical education that have given rise to non-real numbers, division by zero and the square root of a negative real number. Thus, if the symbolic representation of a relationship contains division or square root, you must exclude the values from the normal or problem domain that cause the relationship to be non-real.

The symbolic representation of relationships will display a variety of different arithmetic operations. For example, the garbage problem ($0.035t - 65.69$) contains multiplication and subtraction. Many symbolic representations of relationships are terms and polynomials.

> A **term**, or **monomial**, is a product of real numbers and/or variables.

Examples of terms: $5x^2, -3x, 14.3x^0$.

Recall that $x^0 = 1$, when x is not zero.

Also recall that $x^2 = x \cdot x$.

> A **polynomial** is a sum or difference of terms.

For example, $3x + 7$ and $-3.6x^5 - 17x^3$ are polynomials. The following representations are not polynomials: $\frac{2}{x}$, $\sqrt{x} - 15$, and $\frac{x-5}{x^2 - 3x + 4}$.

<u>Function Defined</u> Finally, you may have noticed that all of the data relationships in this text have had exactly one element of the range (represented by the dependent variable) for each element of the domain (represented by the independent variable). Each relationship has *not* been like the following examples. Table 2.2.1 shows the relationship between the numbers (*N*) and letters (*L*) on your phone. For mathematical reasons think of the letters of the alphabet as integers 1 through 26.

Table 2.2.1

N	2	3	4	5	6	7	8	9
L	1,2,3	4,5,6	7,8,9	10,11,12	13,14,15	16,18,19	20,21,22	23,24,25

In this relationship, each number in the domain is related to three numbers from the range. Another example of this type of relationship comes from a family of five seated at a restaurant table designed for four as shown in Table 2.2.2.

Table 2.2.2

Seat	1	2	3	4
Person	1	2	3	4,5

In this example, the number 4 from the domain is related to both 4 and 5 from the range. Both of these examples demonstrate an important and basic difference between relationships. Nearly all relationships you will study will be of the first type -- where each element of the domain is related to only one element in the range. For the purpose of clarification, each type of relationship will be given a name. Instead of using the word relationship, mathematicians use the word **mathematical relation**, or just **relation**, to describe both types of relationships. The first type is also given the name function.

> A **function** is a relation where every real number in the domain is related to exactly one real number in the range.

Example 3: Is the relation $-2\sqrt{x+3}$ a function?

Solution: Yes, because any number from the domain will cause $-2\sqrt{x+3}$ to be exactly one real number.

⁂

Example 4: Is the relation below a function?

Table 2.2.3

t	−2	−1	0	1	2
$3t \pm 6$	−12, 0	−9, 3	−6, 6	−3, 9	−12, 0

Solution: No. For a number like −1 from the domain, there are two related numbers from the range, −9 and 3.

⁂

Example 5: Is the relation in Figure 2.2.1 a function?

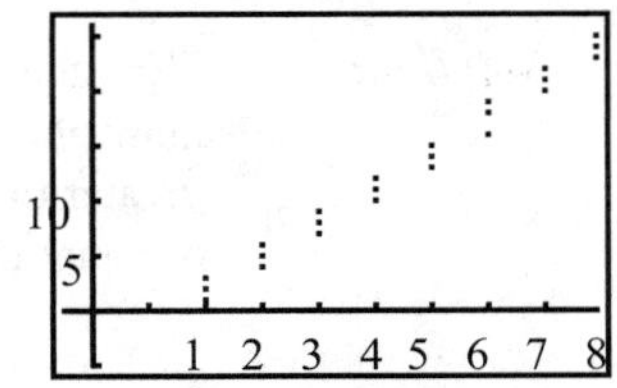

Figure 2.2.1

Solution: No. For example, the number 7 from the domain is related to three numbers from the range, 16, 18, and 19. (Note: this is the graphical representation of the phone relationship.)

⁂

Domain The variable in a function must have a domain that causes the function to be a real number. Suppose some collected data suggests the function $\dfrac{x^2 - x - 2}{x-2}$ as its symbolic representation; the domain of the function $\dfrac{x^2 - x - 2}{x-2}$ cannot contain the number 2. If x has a value of 2, the function is undefined, not a real number. The domain is $(-\infty, 2) \cup (2, \infty)$.

Table 2.2.4

x	. . .	0	1	2	3	4	. . .
$\dfrac{x^2 - x - 2}{x-2}$	. . .	1	2	und.	4	5	. . .

The function $\sqrt{x+2}$ does not represent a real number if the domain contains numbers smaller than −2. The square root of a negative number is not a real number, and if x is less than −2, the function is the square root of a negative number. Therefore, the normal domain must be real numbers −2 and larger, or $[-2, \infty)$.

Table 2.2.5

x	. . .	−4	−3	−2	−1	0	1	2.5	. . .
$\sqrt{x+2}$	. . .	not real	not real	0	1	1.41	1.73	2.12	. . .

Example 6: Suppose the function $0.00153\sqrt{80000 - m}$ models the thickness of a brake pad on a car driven m miles. What is the problem domain? What is the normal domain?

Solution: If m is larger than 80000, the function becomes a non-real number because $80000 - m$ is negative. The number of miles (m) driven cannot be less than 0. This means that the problem domain must be [0, 80000]. When the function is considered out of the context of the problem, it is mathematically acceptable for m to be negative (try it). Thus, the normal domain is $(-\infty, 80000]$.

Example 7: Find the normal domain of the function described by $\sqrt{x-5}$.

Solution: The square roots of negative numbers are not real numbers, thus the domain must be numbers that cause $(x - 5)$ to be on the interval $[0, \infty)$. If $x \in [5, \infty)$, then $x - 5$ is greater than or equal to zero. Thus the normal domain is all real numbers larger than or equal to 5. In interval notation, the normal domain is $[5, \infty)$.

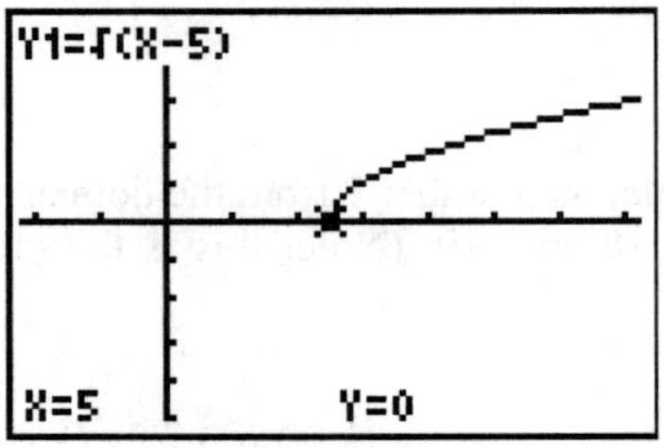

Example 8: Find the normal domain of the function described by $\dfrac{x-17}{x-17}$.

Solution: A real number divided by zero is not a real number; thus, the domain cannot contain 17. The normal domain is all real numbers except 17. In interval notation this is $(-\infty, 17) \cup (17, \infty)$.

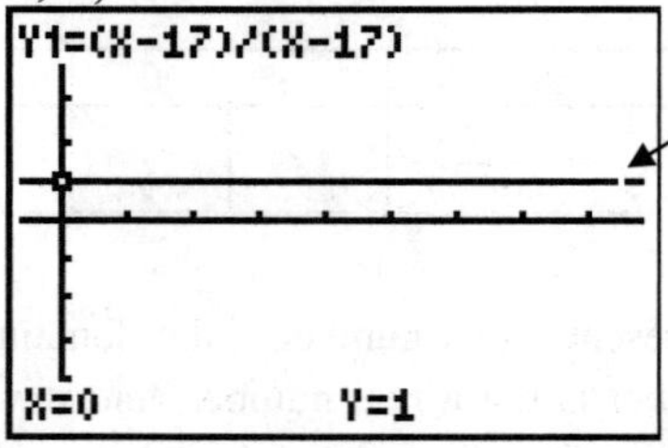

Notice the hole in the graph when x is 17.

Example 9: Find the normal domain of the function described by $\dfrac{2}{(x-1)(x+2)}$.

Solution: Division by zero is undefined. Thus the quotient is not a real number, if x has a value of 1, $(x - 1)$ is zero. The function contains division by $(x - 1)$. So $x \neq 1$. If x has a value of -2, $(x + 2)$ is zero. The function contains division by $(x + 2)$. So $x \neq -2$. The normal domain is all real numbers except 1 and -2. Interval notation is a little messy; it is $(-\infty, -2) \cup (-2, 1) \cup (1, \infty)$.

⁂

Example 10: Find the normal domains of $\frac{1}{x+1}$, $\frac{1}{x+2}$, and $\frac{-5}{x+3}$.

Solution: The only operation that can cause the functions to not be real numbers is division by 0. Thus to avoid division by zero, eliminate –1 from the domain of $\frac{1}{x+1}$, eliminate –2 from the domain of $\frac{1}{x+2}$, and –3 from the domain of $\frac{-5}{x+3}$. The respective normal domains are $(-\infty, -1) \cup (-1, \infty)$, $(-\infty, -2) \cup (-2, \infty)$, and $(-\infty, -3) \cup (-3, \infty)$.

⁂

Example 11: Find the normal domain of $\sqrt{x-1}$, $\sqrt{x-2}$, and $\sqrt{x-3}$.

Solution: The operation that causes non-real numbers is the square root of negative numbers. To avoid the square root of negative numbers, $x - 1$ must be zero or positive. This can only happen if x is 1 or larger. Thus the normal domain of $\sqrt{x-1}$ is $[1, \infty)$. By the same reasoning, the normal domain of $\sqrt{x-2}$ is $[2, \infty)$, and the normal domain of $\sqrt{x-3}$ is $[3, \infty)$.

⁂

2.2 EXERCISES

Many times you can work the exercises by algebraic or numeric methods. You can learn about mathematics no matter which method you use. Part of the learning process is for you to decide which method is best for you to use. Please read the text before you do the exercises. Read the **next assigned section** after you finish the assignment below.

A Quick Look To The Future

-4. If you throw a ball straight up, does it have a maximum height?

-3. If you throw a ball straight up, does it have a zero height?

-2. Do you think the linear data below is increasing or decreasing? Why?

t	1	2	3	4	5
Salary	\$6.50	\$13	\$19.50	\$26	\$34.50

-1. When you get a drink at a drinking fountain, what do you think is the shape of the water path?

Previous Topics Revisited

1. Fifty-four percent of the current contribution to global warming comes from carbon dioxide. It is 355 parts per million of the current atmosphere. Write this number in scientific notation.

2. In the United States, 14×10^9 pounds of trash is dumped into the ocean each and every hour. Write this number in standard notation.

3. Assuming that every person, on average, eats 1.5 pounds of food per day, how much food is consumed by all 6.5×10^9 people on earth in one year? (2006 data)

4. Use interval notation to describe all numbers greater than 4.02.

5. Describe the set of numbers $[-4, 3)$.

6. Is the function below linear, quadratic, exponential, or none of these?

Yearly Medical Charges	50	80	130	200	225	290	350	480	600
After Insurance Charge	50	80	130	200	205	218	230	256	280

New Topic Exercises

Find the symbolic representation of the relationships in Exercises 7 – 12.

7. A telemarketer makes \$100 per week plus \$1.00 per call made. Express salary in terms of calls made.

8. A lawn spreader (A device that puts fertilizer on your lawn.) is loaded with 50 pounds of fertilizer and is set to release approximately 0.01 pounds per square foot of lawn. Express fertilizer left in the spreader in terms of the area covered.

9. The electric company charges residential customers \$9.50 a month as a meter fee plus they charge \$0.11 per kilowatt hour of electricity used in the monthly billing cycle. Express monthly charges in terms of electricity used.

10. A 4 meg file can be down-loaded from the Internet at the rate of 0.25 meg per second. Express the amount remaining to be down-loaded in terms of time (in seconds).

11. 4000 quarters are dropped on the table. Approximately ½ of them are heads for each toss and the heads are removed after each toss. Express the number of heads up in terms of the number of tosses x.

12. Control mice showed an estimated mean tumor volume doubling rate of 9.0 days. The size of the tumor upon first observation was 4 ml^3. Time is in days. Express the size of the tumor in terms of time passed (in days).

time	0	9	18	27	36	45
population	4	8	16	32	64	128

Are the relations in Exercises 13 – 18 functions?

13.

D	1	2	3	4	5	6	7	9
R	12	19	6	32	47	2	26	12

14.

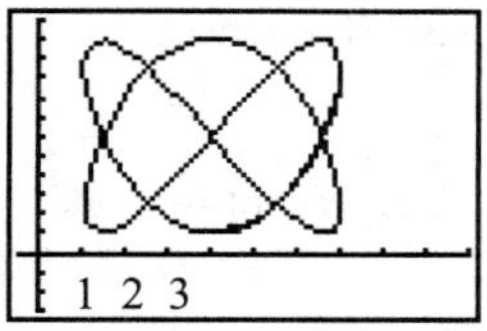

15. $\sqrt{-x}$

16.

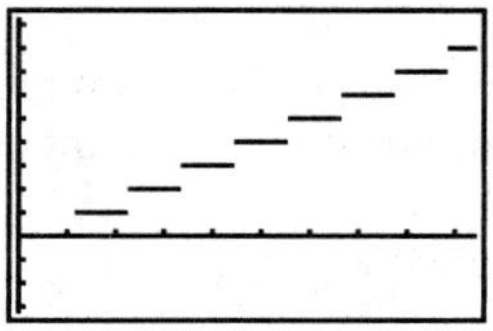

17. x^3+1

18.

D	1	2	3	4	5	4	3	2	1
R	1	3	5	7	9	8	6	4	2

Specify the normal domain of the functions in Exercises 19 – 40 .

19. $\frac{1}{2}x-5$

20. $\frac{x-5}{3}$

21. $\frac{3}{4}x^2-\frac{5}{8}x+\frac{2}{3}$

22. $\frac{37}{x+5}$

23. $\frac{-58}{x+5}$

24. $\frac{9}{x+5}$

25. $\frac{(x-2)(x+3)(x-17)}{x+5}$

26. $\frac{x+5}{x+5}$

27. $\sqrt{x+3}$

28. $\sqrt{x+4}$

29. $\sqrt{x+5}$

30. $\sqrt{x-4}$

31. $\sqrt{x-5}$

32. $\sqrt{x-5}+8$

33. $-4\sqrt{x-5}+6$

34. $\sqrt{x+4}-\sqrt{x-2}$

35. $-\frac{\sqrt{x+4}}{x-2}$

36. $\sqrt{2-x}$

37. $\frac{x+5}{(x-1)(x+2)(x+5)}$

38. $\frac{1}{\sqrt{-x}}$

39. $\frac{-4}{\sqrt{x+3}}$

40. $\frac{\frac{1}{x}}{\sqrt{x+5}}$

Identify a problem domain that makes sense with respect to the situation described in Exercises 41 – 48.

41. $0.055x$ — 5.5% sales tax function, with x representing the selling price

42. $0.27t + 0.50$ — long distance telephone charge of $0.27 per minute and a $0.50 connect charge, with t representing time in minutes

43. $1.10m + 1.25$ — taxi fare at $1.10 per mile plus a $1.25 user charge, with m representing the distance traveled in miles

44. $0x$ — financial return on investing x dollars in the state lottery

45. $-0.105|d-60|+6.2$ — the height of an airplane above ground level on a trip from Columbus to Cincinnati, with d representing the ground distance from Columbus

46. $-16t^2 + 45t$ — the height of a rock thrown upward with t being time in seconds.

47. $\frac{R}{1-R}$ — the rate of markup based on the cost price, where R is the rate of markup based on the selling price

48. $\frac{100}{1+R}$ — the cost price of an item selling for $100, where R is the markup rate based on the cost price

In Exercises 49 – 54, create any function that has a domain of:

49. $(-\infty, 6) \cup (6, \infty)$

50. $(-\infty, -4) \cup (-4, 9) \cup (9, \infty)$

51. $(-\infty, 3]$

52. $(-\infty, -2]$

53. $[4, \infty)$

54. $(4, \infty)$

In the next three exercises, find the symbolic form of the given relationship.

55. Elizabeth the nurse set the drip rate on a 1,000 *ml* IV bottle at 2 *ml* per minute. Find the symbolic form of the *TIME-IV* relationship. (Where *IV* is the amount left in the bottle at time *TIME*.)

56. The North Carolina Utility Commission sent a letter to all North Carolina Power customers with the following proposed usage charges. Where k is the number of *kWh* used and E_c is the monthly electric bill.

k	200	500	700	1000	1500	2000	3000
E_c	21.37	42.69	56.90	78.21	113.74	149.26	220.31

57. Farmer Tom bought a 100 bushel grain dryer and it came with the following "moisture chart" for corn initially at 24% moisture. Find the symbolic form of the *time-moisture* relationship.

Drying Time	0	15	30	45	60	75	90	105
% Moisture	24	22.8	21.6	20.4	19.2	18	16.8	15.6

From Mathematics to English

58. After reading this section, make a list of questions you want to ask your instructor.

59 Continue in your daily journal and make an entry. In addition to your normal entry on thoughts about the mathematics in this section, list at least two positive comments about what you have learned about this topic.

60. In paragraph format, summarize the material in this section of the text in your daily journal.

61. Describe how your classroom instructor made this topic more understandable and clear.

62. After reading the text and listening to your instructor, what do you not understand about this topic?

63. Explain why the normal domain of $\sqrt{x^2}$ is all real numbers.

64. Why can the domain of $x^2 + x + 1$ contain $\sqrt{7}$?

65. Can the domain of $x - 1$ contain 0.3333...? Why?

66. Describe the normal domain of the function $\dfrac{1}{(x-1)(x+1)(x-2)(x-3)(x+2)(x-5)(x+7)}$.

67. What is a function?

68. Is the relationship between the length of time a copier is running and the number of copies made a function relationship? Explain.

69. If the symbolic representation of the relationship between time and the number of hotel rooms cleaned by the housekeeping staff is $3.6t$, what are the differences between the normal domain and the problem domain? (t is time in hours.)

70. Why does the normal domain of $\dfrac{-7}{\sqrt{x+3}}$ not contain -3?

From English to Mathematics

In the next three exercises, write the English statement as a mathematical statement. For example, "seven times the sum of x and 4", would be $7(x + 4)$.

71. seventeen plus x

72. the difference of x minus 3, times 43

73. the length x times the width $(x + 4)$

Explorations

74. In a function, for every x there must be only one related value of the function. List the symbolic representations of two functions where not only is this true but it is also true that for every value of the function there is only one corresponding value for x.

2.3 Geometric Behaviors of Data Relationships

The Short Trip

Suppose you are driving your car between New York and Boston. You know that the longer you drive the farther you go; that is, the number used to represent time and the number used to represent distance are related by a function. For example, when time is 1 hour, the distance traveled may be 60 miles. And when time is 2 hours, the distance may be 120 miles. To analyze the functional relationship between time and distance, it is very useful to express it numerically in a table. Table 2.3.1 shows you the relationship between time and distance traveled.

Table 2.3.1

TIME	0	1	2	3	4	5	6	7
DISTANCE	0	60	120	180	240	300	360	420

For any given value of *time*, the related distance value has been calculated. For example, when *time* is 2, the distance function, has a value of 60×2 or 120. In every pair of numbers in the numeric relationship, the *distance* is related to the *time.* It is *time* multiplied by 60. The symbolic representation of the distance function is $60t$. Or we might use $D = 60t$.

A major emphasis of this chapter and the next is the study of the behavior of functions. You may wonder what kind of behavior will be studied. While you may look at the *DISTANCE* function above and think that the distances are just a bunch of unrelated numbers, you are incorrect in your thinking. In fact, there are many types of functions and each one displays its own unique behavior.

You may have noticed that the distance function values are increasing in value. Not all functions do that. Others may increase for awhile and then decrease. Some types of functions increase faster than others. The rate of increasing or decreasing can and does vary from function to function. Some relationships are constant; they never increase or decrease in value. You will study relationships that have a maximum value; others have a minimum value. Some functions are always positive, some are always negative, and some vary between positive and negative. Many functions have a value of 0 for certain x-values; others do not. Some functions have a value of 0 more than once. A close look at the numeric representation of the distance function shows that *distance* is increasing at a constant rate of 60 miles per hour, it has a zero value, and there is no maximum or minimum on the normal domain.

So why are these things important to study? Well, engineers, businesspersons, scientists, or any professional person can solve problems by knowing the behavior of a particular function. For example, a physicist or astronomer may need to know the maximum height of a projectile. If he or she has a function that relates the time in flight to the height above ground, the maximum can be found. When the function is zero, the scientist knows that the projectile is back on the ground. Do you think a business person would like to know when profits are increasing or decreasing, or know when to expect maximum profits, is worried about zero or negative profits? Like the scientist, if the businessperson has a function that relates profit to time or some other variable, he or she will know the answers to these questions.

The numeric representation of the *distance* function above is commonly written in symbolic form as $60t$. Since the word *time* begins with the letter t, it makes sense to write the function as $60t$ instead of $60x$. When you investigate the function, your graphing calculator may require that you enter it as $60x$, not $60t$.

You would no doubt agree that the functions $-\frac{1}{2}x+1$ and $-|x-1|+2$ look totally different. Since the symbolic representations of the above functions look different, wouldn't you also expect the numeric representation of the two functions will look different as well? If you do, you are correct. Likewise, wouldn't you think that the functions $2x^2-x-3$ and $-\sqrt{x+4}+2$ have vastly different characteristics when you look at the numeric representation? Of course, the functions look different in the symbolic representation as well. On the other hand, if the symbolic representations of the functions look similar, hopefully you would expect their numeric representation to have similar behaviors or characteristics. In this section, you will study the behavior of functions and be expected to identify the following characteristics/behaviors of functions when represented in numeric form (a table):

Behaviors

- when the function is increasing or decreasing in value
- the maximum or minimum value of the function
- the rate of change of the function
- when the function is positive or negative
- when the function is zero
- the domain of the function
- the range of the function

In the numeric relationships that will be analyzed, only a small sample of x-values from the problem or normal domain will be used. It normally is an impossible task to look at a complete numeric relationship using every possible value for x in the problem or normal domain. The following numeric representations include mostly numbers from the problem domain. Each table (numeric representation) in this text and each table you make must be a complete table. **A complete table** should indicate established trends at both the left and right ends of the table. All behaviors should be determinable from a complete table. In this section, you are introduced to a numeric analysis of functions. Later sections analyze for behavior from the graphical representation rather than the numeric representation. When the analysis is done on the graphical representation of a function, many more values of x can and will be used.

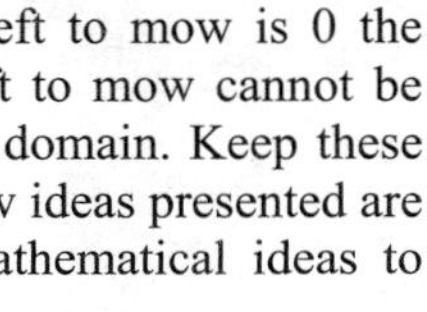

Several "problems" follow in this section. The reason for studying these data relationships found outside the mathematics classroom is that scientists, engineers, business persons, teachers, etc. actually must be familiar with the functions that model these relationships. The models include the linear, quadratic, absolute value, and square root functions. A second and probably more important reason for starting each section with problems is that you can learn mathematics by relating a situation you understand to the mathematics being presented. For example, you know that the amount of lawn left to mow (see below) is decreasing; therefore, you also know that the model of the lawn-mowing problem is also decreasing. You know when the amount of lawn left to mow is 0 the function that models the area left to mow is zero. You know the area of lawn left to mow cannot be negative; thus, you know the model is never negative (0 or larger) on the problem domain. Keep these ideas in mind as you study the behaviors of the linear function below. Most of the new ideas presented are developed from a problem situation; thus, you should be able to relate these mathematical ideas to something that makes sense from the real world.

The Lawn Mower

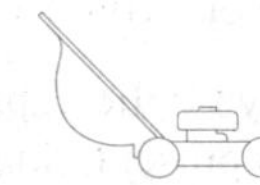

Suppose you are planning on mowing grass as your summer job. Estimating that you can push a 24 inch mower at a little under 2 miles per hour, you can mow at a rate of 300 square feet per minute. Can you confirm this? Further, suppose a prospective client has a 12,000 square foot lawn. The function that models the amount of lawn left to mow at time t minutes is $-300t+12000$. Why? (See Section 2.2)

Analysis of $-300t+12000$

Please study the following numeric representation of the function. The numeric representation in Table 2.3.2 shows the problem domain. All of the behaviors you might be interested in are in the table.

Table 2.3.2

t	0	5	10	15	20	25	30	35	40
$-300t+12000$	12000	10500	9000	7500	6000	4500	3000	1500	0

You were taught to read English from left to right. To read the numeric representation, you will also read from left to right. In this way, the numeric values for ***x* will always change from low to high.** The table feature of the calculator creates the numeric representation of the function in a vertical fashion with values for x changing from low to high just as you read from the top to the bottom. You can make the calculator scroll forward (making x larger) or backward (making x smaller) through the domain. Thus, it appears that the calculator is using the normal domain of the function. While you read the x values from low to high, you will notice that the function values will have a variety of behaviors depending on the type of function.

You can generate the numeric representation of a function using your graphing calculator. If you think of your calculator as a computer (which it is), you may know that numbers can be stored in memory. To store a number or a function, you must assign it an address. For example, the TI-84 keystrokes for storing the number 7 in address A are: " 7 STO▶ ALPHA A ENTER ." You use a slightly different process to store a function. Functions must be stored in addresses Y_1, Y_2, through Y_0, while numbers are stored in addresses A through Z. The three TI - 84 screens below summarize how the function $-300t+12000$ is stored in address Y1, the numeric representation is setup, and the table (numeric representation) displayed.

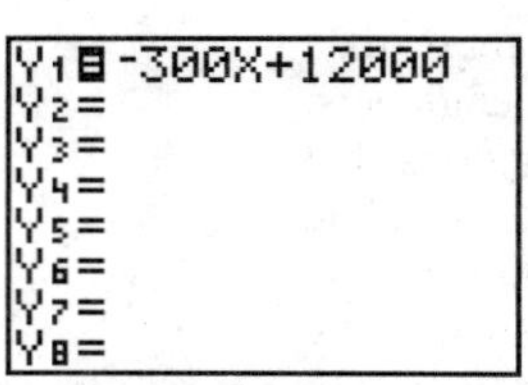

Figure 2.3.1a

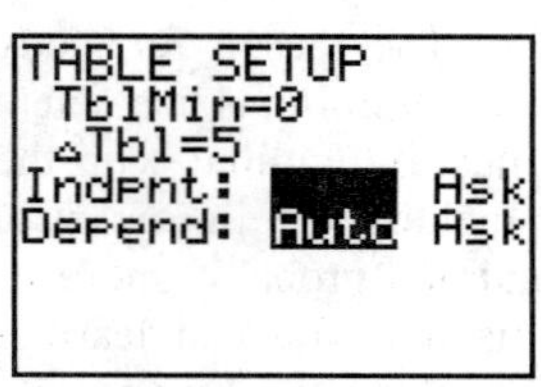

Figure 2.3.1b

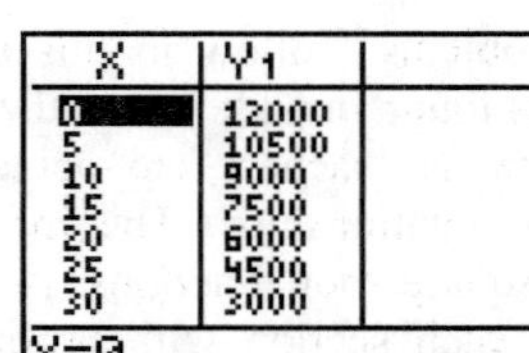

Figure 2.3.1c

Figure 2.3.1a shows that the calculator must use an x in place of the function variable t.

It is very helpful to think of the TABLE SETUP as *xmin* and Δx, instead of TblMin and ΔTbl.

That is, since the function contains the variable x and you control the domain (the set of all x values), you must tell the calculator the smallest value of x to use (*xmin*) and how much to change x by in making the numeric representation. Mathematicians have long used the letter Δ as meaning change. Thus, Δx is the amount x is to be changed by each time the calculator evaluates another data pair of the numeric representation, as shown in Figure 2.3.1c. The purpose of creating the numeric representation of the function is to analyze it for the behaviors described below.

Table 2.3.2

t	0	5	10	15	20	25	30	35	40
$-300t + 12000$	12000	10500	9000	7500	6000	4500	3000	1500	0

Domain The real numbers from 0 to 40 form the problem domain. Time before 0 means time before the job is started and time past 40 minutes is after the job has been completed. All of the action takes place when $t \in [0, 40]$. The normal domain is the set of all real numbers because the function $-300t + 12000$ has no operation that will cause it to be non-real when t is real.

Range When the normal domain is used, the range is all real numbers. Of course, the problem domain creates the range shown in Table 2.3.2 as being all real numbers on the interval [0, 12000].

Increasing/decreasing The most obvious characteristic (behavior) of the function is that it is decreasing in value. As you read t values from low to high, the function values are getting smaller and smaller. Hopefully, it is clear that the amount of yard left to mow as modeled by the function $-300t + 12000$ cannot increase nor remain constant -- it must decrease. The mowing function is decreasing on any domain.

Maximum/minimum If you restrict the domain to the problem domain, the maximum yard left to mow is 12,000 square feet. This is called the **maximum** of the function $-300t + 12000$. The least amount of yard left to mow is 0. This is called the **minimum** of the function. If the domain is the normal domain, there is no maximum nor is there a minimum. Try scrolling toward small values in the normal domain on your calculator. Does it appear that there is a maximum value to $-300t + 12000$? Try scrolling toward the larger numbers in the normal domain, does there appear to be minimum value to the function? There is no minimum value on the normal domain, nor is there a maximum value.

The faster I mow, the quicker I get to zero.

Rate of change How fast is $-300t + 12000$ changing You know it is changing by the fact that it is decreasing in value. But, what is the rate of this change? How fast is it decreasing? If you study the numeric representation, you can see that each time x gets bigger by 5 minutes, the function decreases by 1500 square feet. The quotient of 1500 divided by 5 is called the **rate of change** of the function. Since it is getting smaller and smaller by 1500 units as x increases by 5 units, the rate of change is $-\frac{1500}{5}$ or -300 square feet per minute. If the function were changing by getting bigger by 300 units, the rate of change would be a positive 300. *To find the rate of change, simply find how much the function changes and divide this by the corresponding change in x.* Using the notation that Δ means change, the calculation for rate of change is $\frac{\Delta(-300t + 12000)}{\Delta t}$. If the function is decreasing, the rate of change will be negative because the change in the function is negative while the variable in the function increases. If the function is increasing, the rate of change will be positive.

RATE OF CHANGE -- THE MEASURE OF HOW FAST A FUNCTION IS CHANGING

$$\frac{\Delta \text{ in the function}}{\Delta \text{ in } t}, \text{ where } t \text{ is the function variable.}$$

Positive/negative If you consider the numeric values of the function when t is on the normal domain, it is clear that $-300t + 12000$ has both positive and negative values. The function is positive when t is less than 40. In Section 1.3, you learned how to describe numbers less than 40. In interval notation, the function is positive when $t \in (-\infty, 40)$. The function is negative when t is larger than 40, or in interval notation, the function is negative when t is on the interval $(40, \infty)$. As you can see from Table 2.3.2, which shows the problem domain, the function is positive when t is between 0 and 40, including 0. The interval notation for when $-300t + 12000$ is positive is $t \in [0, 40)$. The function is never negative on the problem domain.

Zeros While being positive or negative may not seem important yet, what it leads to is a very important mathematical idea. This idea is that a function can have a value of zero between when the function is negative and when it is positive. **Caution:** This is only true for a select group of functions. However, in this text, every function you study except the rational function in Chapter Eight will have the characteristic that between when the function is negative and when it is positive, it will be zero. As you can see in the Table 2.3.2, the function $-300t + 12000$ is zero when t has a value of 40. The number 40 is called a **zero** of the function. That is, a **zero** is a value of the variable in the function that causes the function to be 0. In the case of the mowing problem, the zero is significant to the person mowing because it means she is finished.

Analysis of $-1100|t-20|+22000$ - **The Airplane Flight**

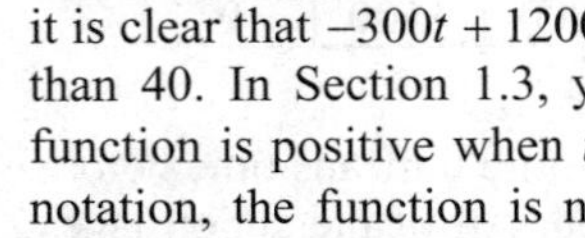

The behaviors of the function $-1100|t-20|+22000$ are uncovered in the numerical analysis below. You may wonder why use a function with such large numbers? The reason is that this function models the height of an airplane flight plan from Columbus, Ohio to Pittsburgh, Pennsylvania. It is a 40-minute flight and the pilot uses the flight plan shown in the numeric representation below in Table 2.3.3. The variable in the function is t and is measured in minutes. The function $-1100|t-20|+22000$ is the on-board computer's plan for the height of the plane at t minutes into the flight and is measured in feet.

Table 2.3.3

t	0	5	10	15	20	25	30	35	40
$-1100\|t-20\|+22000$	0	5500	11000	16500	22000	16500	11000	5500	0

Domain Table 2.3.3 shows the problem domain of [0, 40]. This is the problem domain because the flight starts at 0 minutes and it is over at 40 minutes because you can see that the plane is back on the ground (a height of 0). The normal domain of the function $-1100|t-20|+22000$ is all real numbers, $t \in (-\infty, \infty)$. There are no operations that cause the function to be non-real. The operations shown in the symbolic

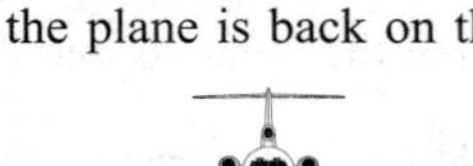

representation of the function are multiplication, absolute value, subtraction, and addition. None of these operations on real numbers will cause the answer to be non-real.

Range When the problem domain is used, the range is all real numbers from 0 to 22,000 because the function models the intended height of the airplane and it changes from 0 height to 22,000 feet. Using interval notation, the range of $-1100|t-20|+22000$ is [0, 22 000]. If the normal domain is used, the function values drop below 0. When you scroll to t values larger than 40 or to t values smaller than 0, you will see negative numbers for the function on your calculator display. The largest function value of 22,000 is not changed, it still remains 22,000 regardless of the two domains. The range is the set of real numbers from $-\infty$ to 22,000. When the normal domain is used, the interval notation for the range of $-1100|t-20|+22000$ is $(-\infty$, 22 000].

Increasing/decreasing You want to know **when** the function is getting larger and when it is getting smaller. By using a function that models a real-world situation, you have the ability to visualize the airplane as it increases in height for the first 20 minutes and then decreases in height the last 20 minutes. If the problem domain is used, the function (the height) increases when $t \in (0, 20)$ and decreases when $t \in (20, 40)$. When the normal domain is used and you no longer visualize the airplane, the function still increases until t is 20, but the difference is that it increases for all t up to 20. The function is increasing when $t \in (-\infty, 20)$. Likewise, when the normal domain is used, the function does not stop decreasing at 40 but continues decreasing forever. The function is decreasing when $t \in (20, \infty)$. Remember, you are describing **when** the function is increasing or decreasing. You cannot describe this by using function numbers. For example, would you describe **when** you were 4 feet tall by saying you were 4 feet tall when you were 4 feet tall? NO! You would say you were 4 feet tall when you were about 9 years old. Would you describe when you were growing in height by saying you were growing higher when you were growing higher? NO! You would say you were growing up when you were ages 0 to 18. Would you say the airplane is increasing in height when its height is 0 to 22,000 feet? No! The plane is both rising and falling when its height value is 0 to 22,000 feet.

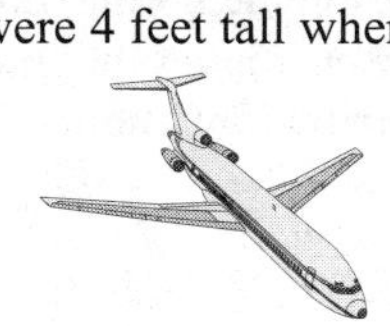

Maximum/minimum Using either the problem or normal domain, the airplane and the function modeling the height of the airplane have a maximum value of 22,000. The airplane reaches a maximum height of 22,000 feet. On the problem domain, the function has a minimum height of 0 feet; however, since the function drops to $-\infty$ on the normal domain, there is no minimum. Remember that the range is the set of all numbers that the function represents. The **maximum of a function** is a number in the range that is larger than or equal to any other number in the range. The **minimum of a function** is a number in the range that is smaller than or equal to any other number in the range.

Rate of change The idea of rate of change is how fast something is happening. This function models the height of the airplane, so the rate of change will represent how fast the height of the plane is changing. The calculation for rate of change is change in the function divided by the change in variable in the function, $\frac{\Delta\ (-1100|t-20|+22000)}{\Delta t}$. You know the airplane is increasing in height during the first 20 minutes of flight; the question, however, is how fast is it increasing in height? The plane changed from 0 feet high to 5,500 high while time changed by 5 minutes. The rate of change during the first half of the flight is $\frac{5500}{5}$, or 1100 feet per minute. This calculation can be done for any time (t) on the interval $(-\infty$, 20]. The rate of change calculated for values of t on the interval $[20, \infty)$ will be different. Using the numeric representation of the height function in Table 2.3.3, you can see that from 30 minutes to 40 minutes ($\Delta t = 10$) the height decreased from 11,000 feet to 0 feet. This is a change of –11,000 feet (Δ in $-1100|t - 20| + 22000 = -11000$). The rate of change is $\frac{-11000}{10}$, or –1100 feet per minute.

You should start to see a pattern developing between the some of the behaviors and the numbers (constants) in the symbolic representation of these functions. Did you notice the rate of change of –1100 is the same as one of the numbers in the function? As you continue reading, please take note of connections between the numbers in the symbolic representation of the function and some of the behavior of the function. The numbers (constants) in the symbolic representation of the function are commonly called the **parameters** of the function or function parameters.

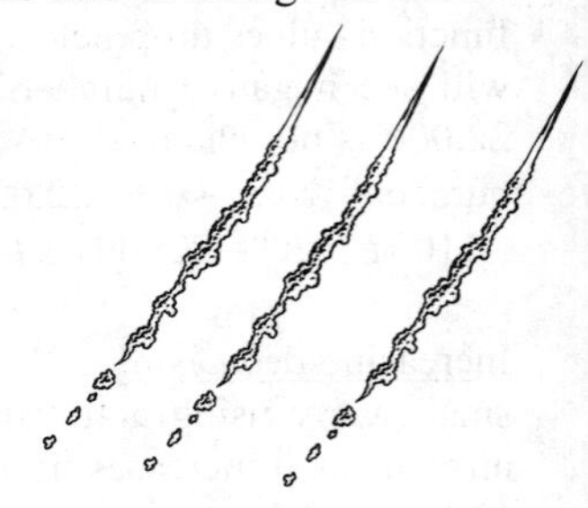

Positive/negative If you think of ground level as being 0 height, then above ground is a positive height. The numeric representation in Table 2.3.3 shows you that the height of the plane is positive when t is on the interval (0, 40). When the normal domain is used, the height function becomes negative to the left of 0 or to the right of 40. You are encouraged to use your calculator to look at the numeric representation for these values. The height function is negative when $t \in (-\infty, 0) \cup (40, \infty)$.

Zeros Functions represent real numbers and real numbers must either be negative, zero, or positive. If you have found when the function is positive and when it is negative, all that is left is when it is zero. Using the normal domain, the height function is positive when t is on the interval (0, 40) and it is negative when t is on the interval $(-\infty, 0) \cup (40, \infty)$. Therefore, the function must be zero at 0 and 40 as further verified by Table 2.2.3.

Analysis of: $9\sqrt{t} + 35$ - **The Slow Car**

The third example of a numerical analysis of a function is $9\sqrt{t} + 35$. When you are first introduced to a new function, you usually can't identify the behavior of a function by looking at the symbolic representation. This function is no different than the first two; the symbolic representation doesn't tell you much about the behavior of the function -- yet. However, as you progress through this text you will soon learn what the symbols suggest about the behavior of the function.

The function $9\sqrt{t} + 35$ typically models the velocity of a car on the ramp entering the freeway. The variable in the function (t), measures time in seconds. The function models the velocity of the car in miles per hour. The numeric representation will help you analyze the function for its behavior.

Table 2.3.4

t	0	2	4	6	8	10	12	14	16	18
$9\sqrt{t} + 35$	35	48	53	57	60	63	66	69	71	73

(Function values have been rounded to the nearest whole number.)

The numeric representation shows that the entrance ramp velocity is 35 mph and after 4 seconds the car is traveling at 53 mph. Or, for example, it takes 10 seconds for the car to reach 63 mph. As you think about the mathematical model for the velocity of the car entering the freeway, visualize the car and relate the visualization to some of the behaviors below.

Domain The numeric representation of the velocity function in Table 2.3.4 shows a reasonable problem domain; although you may want to use the domain of [0, 12) and assume the driver does not break the speed limit. All of the behavior can be determined from the table. The normal domain also starts at 0 because of the square root operation. If t is on the interval $(-\infty, 0)$, the velocity function is not a real number. Confirm this using the table feature of your calculator. The normal domain continues to the right of 0 to ∞.

Range Since the velocity function is not real if t is less than zero, the range starts at 35. As shown in Table 2.2.4, the function values continue upwards with no end, assuming a normal domain. The range of the velocity function using a normal domain is $[35, \infty)$.

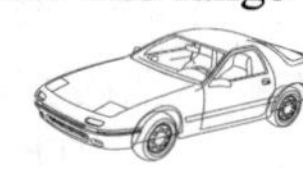

Increasing/decreasing A look at the function values indicates that they are increasing when you read them while the domain values change from low to high – time can only pass from low to high. This tells you the function is increasing. If you relate this mathematical behavior of increasing to the action happening to the car, you know the car is increasing in velocity. Think: "the car has increasing velocity -- the velocity function is increasing." The velocity function has real values when t is on the interval $[0, \infty)$; thus, you would say that the velocity function is increasing when t is an element of the interval $[0, \infty)$. In mathematical notation, $9\sqrt{t} + 35$ is increasing when $t \in [0, \infty)$.

Maximum/minimum The smallest value of the function you see is 35 and it increases from there. The minimum is 35 and there is no maximum. However, using a problem domain of [0, 12], the car has a minimum velocity of 35 and a maximum velocity 66 as shown in the numeric representation.

Rate of change The functions $-300t + 12000$ and $-1100|t - 20| + 22000$ analyzed in the first two examples have the unique behavior of having a constant rate of change. Calculating the rate of change of the velocity function $9\sqrt{t} + 35$ will give a different value for each change in time. For example, the change in the function as t changes from 0 to 2 seconds is 13; thus, the rate of change is $\frac{13}{2}$ or 6.5. The rate of change as t changes from 2 to 4 seconds is $\frac{5}{2}$ or 2.5. The rate of change as t changes from 4 to 6 seconds is $\frac{4}{2}$ or 2. There is no constant rate of change for the square root function. A change in meaning is called for.

> If the rate of change is not constant, $\frac{\Delta \text{ in the function}}{\Delta \text{ in the variable}}$ represents an average rate of change.

For example, if you travel from Washington, DC to Richmond, VA, the distance traveled is about 100 miles and the time it might take is 2 hours. The average rate of change is $\frac{100}{2}$ or 50 miles per hour. The change in distance divided by the change in time gives an average, not an instantaneous rate of change. You probably rarely travel at 50 mph, yet the calculation of velocity is 50 -- the correct average velocity. The best you can do with functions that have a changing rate of change is to calculate an average between two given values of the variable in the function. For functions that do not have a constant rate of change, the instantaneous rate of change analysis will be left for mathematics textbooks at a higher level.

Positive/negative As you can see from the numeric representation of the function, when t is on the problem domain of [0, 12), the function is positive. When t is on the interval $[0, \infty)$, the function remains positive. Of course, the function is not real when $t \in (-\infty, 0)$. It is not positive. It is not negative. It is not zero. It is not real. Since the velocity function starts at 35 and increases, it is never negative.

Zeros A zero is a value for t that makes the function 0. The function never changes from positive to negative -- an indication of when it might be zero. Further, the function starts at 35 and is increasing which means there cannot be a zero.

Analysis of $-16(t-1)^2+20$ - **The Flying Ball**

The fourth and last numerical analysis of a function is $-16(t-1)^2+20$. You may recognize the function as a quadratic function. When simplified, it looks like $-16t^2+32t+4$. You will study quadratic functions in standard form $(-16(t-1)^2+20)$ and in general form $(-16t^2+32t+4)$ throughout this text. This particular quadratic function models the height of a ball when you throw it straight upwards with an initial velocity of 32 feet per second and it leaves your hand at 4 feet above the ground – assuming no air resistance. Using a function to model the height of the ball means that $-16(t-1)^2+20$ is the height and t is the time into the flight in seconds. Just as a reminder of what your calculator does when it makes the numeric representation, consider the following: at 1.5 seconds into the flight of the ball, the height of the ball is at 16 feet above the ground. Please note this pair of numbers (1.5, 16) in the numeric representation below. Each pair of related numbers is created in a similar fashion by using the table feature of the graphing calculator. You may certainly do a similar calculation for each pair of numbers in the traveling ball relationship. Of course, if you are not using a calculator with a table feature, you may have to create the numeric representation by using this calculation process. Even before you look at the numeric representation of the traveling ball function, you know from experience that the height of the ball (the function) will rise from 4 feet to a maximum height and then fall to a zero height. Below is the numeric representation of the height function $-16(t-1)^2+20$.

$$
\begin{aligned}
&-16(1.5-1)^2+20\\
&-16(0.5)^2+20\\
&-16(0.25)+20\\
&-4+20\\
&16
\end{aligned}
$$

Table 2.3.5

t	0	0.2	0.5	0.6	0.8	1	1.3	1.5	1.7	2	2.1	2.2
$-16(t-1)^2+20$	4	9.76	16	17.4	19.4	20	18.6	16	12.2	4	0.6	–3.0

(function values have been rounded to tenths)

Domain Table 2.3.5 gives a reasonable problem domain of [0, 2.2). Sometime between 2.1 and 2.2 seconds the ball hits the ground; thus, the problem domain must stop at this number. If you are interested in finding the exact time when the ball strikes the ground, it is fully developed in Chapter Ten. For now, approximate values are perfectly acceptable. The normal domain has to be all real numbers. There is no mathematical operation in the function that causes division by zero nor the square root of a negative number.

Range Under the problem domain of the height function, the ball has a height of 0 feet (when it hits the ground) and it appears to reach a high point of 20 feet during the flight. You should conclude from this that the height function $-16(t-1)^2+20$ has a range of [0, 20]. Using the normal domain of all real numbers, the range is different. If you now ignore the restrictions put on the function of modeling the height of the ball and let the domain be all real numbers, the function has values all the way to negative infinity. The high point of 20 feet does not change. You should confirm this idea on your calculator by scrolling to where the domain is below 0 or above 2.1. You will see the function values tend to go to $-\infty$. The range of the height function with a normal domain is $(-\infty, 20]$.

Increasing/decreasing Reading the numeric representation from low to high t values, the function increases until t is 1; at this point the function starts to decrease. The function is increasing when t is smaller than 1. This is true not only for the problem domain, it is true for the normal domain. Visualizing the ball decreasing in height may help you accept the fact that the function is decreasing when t is larger than 1. The height function is decreasing when $t \in (1, \infty)$ on the normal domain and approximately (1, 2.2) on the problem domain.

Maximum/minimum The largest value of the function is 20 when t is 1. It is smaller than 20 both for smaller and larger values of t. The function $-16(t-1)^2 + 20$ has a maximum of 20. When the problem domain is used, the ball has a minimum height of 0. However, if t is allowed to be an element of $(-\infty, \infty)$ the normal domain, there is no minimum value.

Rate of change Try throwing a ball straight upwards and pay particular attention to how it changes velocity. It will rise quickly and when it is near the maximum it will slow down, stop, and then it will start to fall slowly and very quickly pick up speed until it hits the ground. In terms of this situation, what you are observing is the rate of change of the height function. The height function models this behavior; thus, when you calculate an average rate of change, you should expect a larger number near the beginning of the flight than near the high point of the flight. You must also expect negative values for the rate of change when the ball is falling back the ground because height is decreasing. For example, the average

X	Y1	
0	4	
.1	7.04	
.2	9.76	
.3	12.16	
.4	14.24	
.5	16	
.6	17.44	

X=0

Figure 2.3.2

rate of change between 0 and 0.1 second is about 30 feet per second. As you can see in the height changes by about three feet (from 4 to 7.04) while time has changed by 0.1 second. The average rate of change is $\frac{\Delta(-16(t-1)^2+20)}{\Delta t} = \frac{3.04}{0.1} = 30.4$. Remember, it is thrown from your hand at 32 feet per second.

To confirm that the rate of change is small and negative when it is falling, notice the height function table in Figure 2.3.3. The height changes from 20 feet to 19.84, a decrease of 0.16 feet, while time has increased from 1 second to 1.1 seconds, or by 0.1 second. Thus the average rate of change is $\frac{\Delta(-16(t-1)^2+20)}{\Delta t} = \frac{-0.16}{0.1} = -1.6$. Not only is it negative, it is much smaller than when it first leaves your hand. Like the square root function, the quadratic function does not have a constant rate of change.

X	Y1	
1	20	
1.1	19.84	
1.2	19.36	
1.3	18.56	
1.4	17.44	
1.5	16	
1.6	14.24	

X=1

Figure 2.3.3

Positive/negative Hopefully, it is obvious from looking at the numeric representation on the problem domain that the function is positive when t is between 0 and about 2.11. That is, when $t \in [0, 2.11)$. Considering the normal domain; however, the function is also positive from around -0.1 second to 2.11 seconds. You should make a table on your calculator and confirm that this statement is correct. If the function is positive when $t \in (-0.1, 2.11)$, you can also see from your table that it is negative when t is smaller than approximately –0.1 and when t is larger than approximately 2.11. The height function is negative when $t \in (-\infty, -0.1) \cup (2.11, \infty)$.

Zeros The ball is at zero height when it is on the ground. The zero heights of the ball (on the normal domain) are at approximately –0.1 and 2.11 seconds.

Below are a few sample exercises regarding the topics of increasing, decreasing, maximum, minimum, rate of change, positive and negative, zeros, domain and range.

Example 1: Does the function $\frac{2}{5}x - 3$ have a constant rate of change? If so, what is it?

Solution: Make the numeric representation of the function and check.

Table 2.3.6

x	–3	–2	–1	0	1	2	3
$\frac{2}{5}x - 3$	$\frac{-21}{5}$	$\frac{-19}{5}$	$\frac{-17}{5}$	-3 or $\frac{-15}{5}$	$\frac{-13}{5}$	$\frac{-11}{5}$	$\frac{-9}{5}$

Each time x increases by one unit, the function changes by $\frac{2}{5}$. Yes the function has a constant rate of change; it is a positive $\frac{2}{5}$.

⁂

Example 2: Is the function $2\sqrt{x+3} - 4$ increasing or decreasing?

Solution: Make the numeric representation and read the function values from left to right.

Table 2.3.7

x	–4	–3	–2	–1	0	1	2
$2\sqrt{x+3}-4$	not real	–4	–2	–1.172	–0.536	0	0.472

From looking at the function values you can tell that they are getting larger and larger as you read x values from low to high. The function is increasing when x is on the interval $(-3, \infty)$.

⁂

Example 3: Find the zeros of the function $-3|x+2|-1$.

Solution: A zero is a value of x that makes the function zero. Make a table and scroll through the domain until you see the function become zero. Caution: the exact zero may not show on your table because of your selection of Δx. Look for approximate zeros.

Table 2.3.8

x	–4	–3	–2	–1	0	1	2
$-3\|x+2\|-1$	–7	–4	–1	–4	–7	–10	–13

The function is never zero! The biggest it gets is –1. There is no zero of the function $-3|x+2|-1$. No value of x will cause the function to be zero.

⁂

Example 4: The weight of a constant volume of water changes as the temperature of the water changes. For example, the function that models the weight of one gallon of water as the temperature (in Centigrade) changes is $-0.00007(T-4.2)^2+8.33585$. Does the weight of the water have a maximum or minimum? What is it?

Solution: The numeric representation of the weight function should help in answering the question. It should show the behavior of either maximum or minimum.

Table 2.3.9

T	0°	1°	2°	3°	4°	5°	6°	7°	8°
W	8.33461	8.33513	8.33551	8.33575	8.33585	8.33582	8.33565	8.33536	8.33495

Source: *CRC Handbook of Chemistry and Physics College Edition.* 1965-1966. The Chemical Rubber Company. Cleveland, OH

The numeric representation of the temperature-weight function shows a maximum weight of around 8.33585 pounds.

⁂

STATEMENT OF PHILOSOPHY

After reading this section and doing Exercises 2.3, you should have a general idea of what it means for a function to be increasing or decreasing. You should be able to specify a reasonable problem domain, normal domain, and a range of any function. You should have an idea of how to find a maximum or minimum value of a function and you should begin to have an understanding of the idea of maximum or minimum. You should be able to find an average rate of change of a function and you may have an inkling of the idea behind the concept of average rate of change. You should know what it means for a function to be positive, negative, or zero. These behaviors have just been introduced; therefore, you may not fully understand all of the ideas behind the behaviors. The philosophy used in this text is to study these same important ideas several times, at different levels and using different methods to develop understanding. Using this approach recognizes that many mathematical ideas may not be understood the first time you see and hear about them. By the end of the text, however, you should have a full and thorough understanding of the material included in this text.

2.3 EXERCISES

Many times you can work the exercises by algebraic, numeric, or graphical methods. You can learn about mathematics no matter which method you use. Part of the learning process is for you to decide which method is best for you to use. Please read the text before you do the exercises. Read the **next assigned section** after you finish the assignment below.

A Quick Look To The Future

-4. Why do many business reports contain graphs of data?

-3. What is the initial value of your weekly salary at the start of the week?

-2. What name have you seen given to symbols that "look" like x^2+5x-3?

-1. You know what $|-3|$ and $|4|$ are. What do you think $|x|$ is?

Previous Topics Revisited

1. Does the curve in Exercise 6 represent a function?

2. What is the normal domain of the function $\sqrt{\dfrac{4}{x+3}}$?

3. What is the normal domain of the function $\sqrt{-x}$?

4. In 1985, US citizens generated 300,000,000,000 pounds of waste. Write this number in scientific notation.

5. In 1989 the US used an average of 1.724×10^7 barrels of oil per day! Write this number in standard notation.

6. Use interval notation to describe when the curve is below the number line.

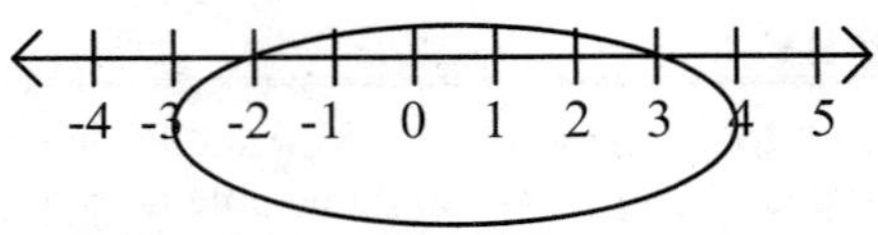

New Topic Exercises

7. Describe the normal domain, the range, when the function is increasing, when decreasing, the maximum and the minimum value of the function on the normal domain, the average rate of change when x changes from 1 to 2, when the function is positive, when negative, and the zeros of the function.
   a. $-\frac{1}{2}x+5$ b. $|x+2|-4$ c. $\sqrt{x-1}-3$ d. $-x^2+4$.

8. Make a numeric representation (table) for the function $\dfrac{1}{x-2}$. Use the following values of x in the table: 2.00001, 2.0001, 2.001, 2.01, 2.1, 2.2, 2.5, 3, 5, 10, 15, 30, 50, 200, and 5000.

9. Is the function $\dfrac{1}{x-2}$ increasing or decreasing?

10. Does the function $\dfrac{1}{x-2}$ have a zero? If yes, what is the zero?

11. The function $\dfrac{1}{2}x-3$ has a constant rate of change; it is the same no matter where you calculate it. What is the rate of change of the function? Is the function increasing or decreasing?

12. The function $\dfrac{3}{2}x-4$ has a constant rate of change; it is the same no matter where you calculate it. What is the rate of change of the function? Is the function increasing or decreasing?

13. The function $\dfrac{4}{3}x+2$ has a constant rate of change; it is the same no matter where you calculate it. What is the rate of change of the function? Is the function increasing or decreasing?

14. The function $\frac{5}{4}x+1$ has a constant rate of change; it is the same no matter where you calculate it. What is the rate of change of the function? Is the function increasing or decreasing?

15. The function $\frac{7}{2}x-4$ has a constant rate of change; it is the same no matter where you calculate it. What is the rate of change of the function? Is the function increasing or decreasing?

16. Based on the previous five exercises, answer the following questions without doing any written work. What is the rate of change of the function $\frac{3}{5}x+1$? Is it increasing or decreasing?

17. Just by looking at the symbolic representation of the function $\frac{3}{5}x+2$, what is the rate of change of the function? Is it increasing or decreasing?

18. What is the rate of change of the function $-\frac{1}{2}x+4$? Is it increasing or decreasing?

19. What is the rate of change of the function $-\frac{3}{5}x-2$? Is it increasing or decreasing?

20. What is the rate of change of the function $-\frac{4}{5}x-3$? Is it increasing or decreasing?

From Mathematics to English

21. After reading this section, make a list of questions that you want to ask your instructor.

22. Continue in your daily journal and make an entry. In addition to your normal entry on thoughts about the mathematics in this section, list at least two positive comments about what you have learned about this topic.

23. In paragraph format, summarize the material in this section of the text in your daily journal.

24. Describe how your classroom instructor made this topic more understandable and clear.

25. After reading the text and listening to your instructor, what do you not understand about this topic?

26. Describe how you can tell whether a function of the form $dx + e$ is increasing or decreasing.

27. Describe how you know the rate of change of a function in the form of $dx + e$.

28. Describe what a rate of change of 20 means.

29. Describe what it means for a function to be increasing.

30. List three functions that have a maximum value on the normal domain.

31. List three functions that have a minimum value on the normal domain.

From English to Mathematics
Write each English statement in a symbolic mathematical form.

32. Starting at 22 inches, the human height function increases at a rate of 3 inches per year.

33. The range of the human height function is 22 to 72 inches.

34. A human at age 2 is 28 inches tall, at age 5 is 37 inches tall, at age 8 is 46 inches tall, and at age 15 is 67 inches tall.

Explorations
35. Circle the functions that have a constant rate of change over the normal domain of the function.

$-2\sqrt{x-1}$	$2x^2+3x$	$-3\lvert x+1\rvert-2$	$\frac{1}{4}x$
$4x-3$	$\frac{1}{3x}$	$2^{x-1}+1$	x^3+1

36. Each of the following functions is either always increasing or always decreasing on its domain. Find a method for identifying which it is by only looking at the symbolic representation. Write your answer in sentence form.

$-2\sqrt{x-1}$	$3\sqrt{x+1}-2$	$\sqrt{x+3}-4$	$-5\sqrt{x-9}$
$-\frac{1}{2}\sqrt{x+3}-1$	$-\sqrt{x}$	$6\sqrt{x+3}-1$	$5500\sqrt{x+4}-130$

37. Find the exact zero of the function $2\sqrt{x-1}-3$. Explain in detail how you found it.

38 For each of the following square root functions, identify the ***first number*** in the domain, the maximum or minimum, and whether the function is increasing or decreasing. You may want to make a chart like the chart below.

		# in Domain	Max/Min	Inc/Dec
a.	$2\sqrt{x-1}+3$	__________	__________	__________
b.	$3\sqrt{x+4}-5$	__________	__________	__________
c.	$1\sqrt{x-2}+4$	__________	__________	__________
d.	$-\sqrt{x+1}-2$	__________	__________	__________
e.	$-2\sqrt{x+3}-6$	__________	__________	__________
f.	$-3\sqrt{x-2}+4$	__________	__________	__________

39. Given the square root function of the form $d\sqrt{x+e}+f$, where *d*, *e*, and *f* are real numbers, answer the following questions:

   a. If *d* is positive, is the function increasing or decreasing?
   b. If *d* is negative, is the function increasing or decreasing?
   c. What is the first number in the normal domain of the function?
   d. What is the maximum or minimum value?
   e. What is the first number in the range of the function?

40. For each of the following functions, find the maximum or minimum. Specify the range. You may want to make a chart like the one below.

	Maximum	Minimum	Range
$3\lvert x+2\rvert-5$	______	______	__________
$5\lvert x-3\rvert+7$	______	______	__________
$2\lvert x+4\rvert+3$	______	______	__________
$-2\lvert x-3\rvert+6$	______	______	__________
$-5\lvert x+1\rvert+4$	______	______	__________
$-2.6\lvert x-5\rvert-7$	______	______	__________

41. Given the absolute value function of the form $d\lvert x+e\rvert+f$, where *d*, *e*, and *f* are real numbers, answer the following questions:

   a. What is the maximum or minimum value of the function?
   b. What is the first number in the range of the function?
   c. What number, *d*, *e*, or *f*, help you decide if the there is a maximum or a minimum?

2.4 Functions Represented Graphically

Real numbers are normally plotted or graphed as points on the real number line as demonstrated in Section 1.2. In this chapter, however, you have seen that the numeric representation of a function consists of many pairs of real numbers, with the first number in the pair representing the variable in the function and the second related number representing the function. Whereas real numbers are graphed on a number line, a pair of number lines is required to graph a function. One number line is for the elements of the domain and the other number line is used to graph the related values of the function (the range). As shown in the graph below, the values for x (elements of the domain) will be used to determine the horizontal positions of a point to be graphed. At the same time, the related value of the function (elements of the range) will be used for the vertical position of the point. So, like numbers being plotted on the real number line, now the ***value of the function will be plotted in a vertical direction above or below the x-value from the horizontal position.***

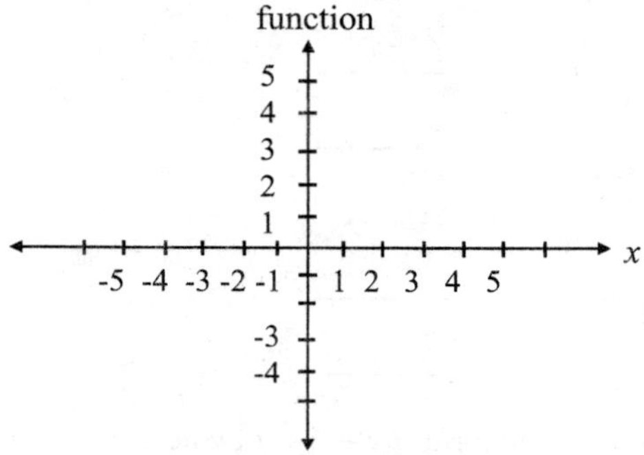

Figure 2.4.1

From the numeric representation of a function, the graphical representation of the function is developed by following the procedure outlined below.

1. Start at the **origin** (the intersection of the number lines) and move in the horizontal direction an amount equal to the value of the variable in the pair of numbers (left for negative values, right for positive values).

2. Next, move in the function direction (up or down) in the amount of the value of the function (up for positive values of the function, down for negative values of the function).

3. Plot the point (a dot).

4. Continue this process until all points from the numeric representation of the function have been plotted.

Figure 2.4.2 shows the plot (graph) of each pair of numbers from each pair of numbers from the numeric representation of $2x - 3$.

Table 2.4.1

x	–4	–3	–2	–1	0	1	2	3	4
$2x - 3$	–11	–9	–7	–5	–3	–1	1	3	5

$2x - 3$

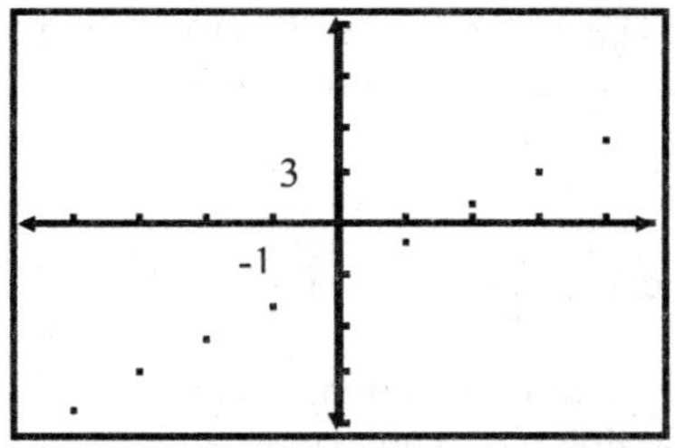

Figure 2.4.2

From looking at the graphical representation in Figure 2.4.2, you see that the points of the graph form a geometric pattern. They lie in a straight line. If more values from the normal domain were chosen, they would also be on the same straight line. If a very large number of elements from the domain were chosen and then graphed, the line would be so packed full of points that it would look like the graph below; a solid straight line.

$2x - 3$

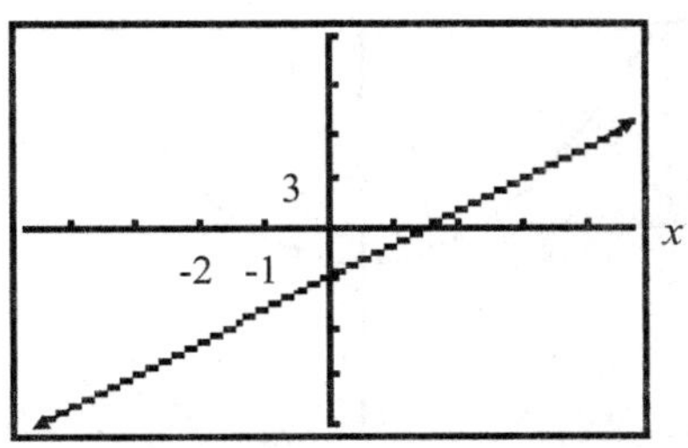

Figure 2.4.3

GUIDELINES FOR GRAPHING WITHOUT THE AID OF TECHNOLOGY

Usually, the problem or normal domain of a function contains an infinite number of elements; thus, it is an impossible task to find corresponding values of the function for all elements of the domain. Therefore, graph enough points from the numeric representation to give the general shape of the graphical representation; then, if enough of the proper points are graphed, the complete graph can be sketched.

The arrowheads at the ends of the graph are used to indicate that the graph trends continue with nothing else unusual happening. Although the geometric shape of the function $2x - 3$ is a straight line, you might expect that functions that have a different symbolic form may not have graphical representations that are straight lines. The quadratic function $x^2 - 3$ is certainly not like the linear function $2x - 3$. Does this mean the graphical representation has a different shape? As you progress through this text, the answer to this question will become obvious.

The points that make up the graphical representation of the function consist of pairs of numbers called **coordinates**. The first is called the ***x*-coordinate** (or whatever variable is in the function) and the second is called the **function-coordinate**. The horizontal and vertical number lines are called the ***x*-axis** (or whatever the variable in the function is) and the **function-axis**.

Example 1: Find the graphical representation for $x^2 - 3$.

Solution: From the numeric representation, a graph can be drawn. Choose enough elements from the normal domain to get a clear picture of the graph. If the function is applied to a real-world data relationship, the problem domain should be acceptable.

Table 2.4.2

x	–4	–3	–2	–1	0	1	2	3	4
$x^2 - 3$	13	6	1	–2	–3	–2	1	6	13

$x^2 - 3$

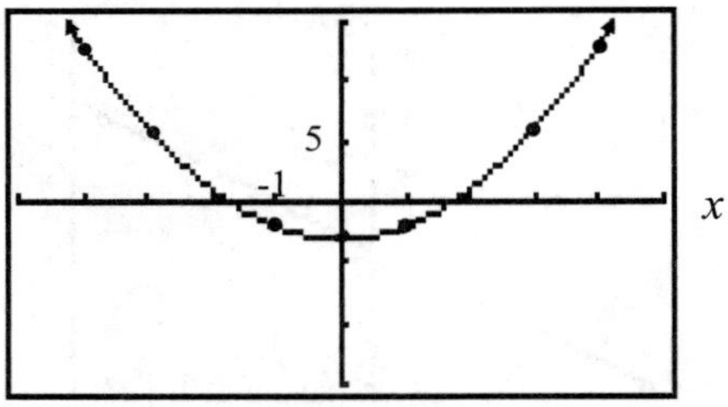

Figure 2.4.4

⁂

The geometric shape of the function is called a **parabola**, you will see more parabolas throughout this text and you saw them before in the data relationships of Section 2.1. Now that you have a picture of the function, it might be easier to "see" the minimum value, or to "see" when the function is increasing and when decreasing, or to "see" most of the behaviors you have found by looking at the numeric representation. These ideas behind "seeing" the behaviors are developed in Chapter 3.

If the elements *chosen* from the normal domain of the numeric representation contained numbers from 4 to 10, the graph will not be a complete graphical representation of the function. It has interesting behaviors that do not show in the window. The graph of $x^2 - 3$ would look like the graph in Figure 2.4.5.

More information about complete graphs will be presented in Chapter Three.

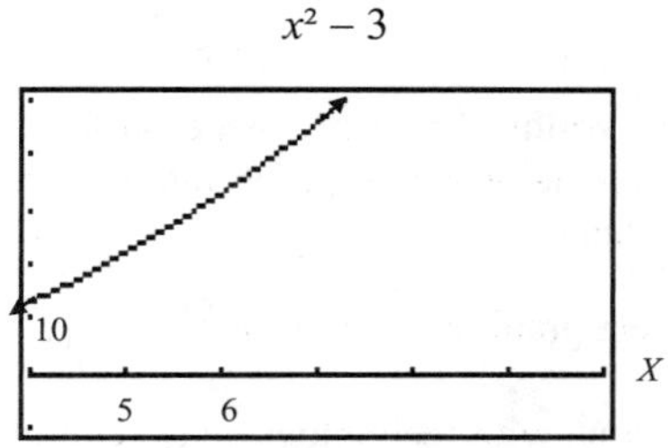

Figure 2.4.5

UNDERSTANDING OF A COMPLETE GRAPHICAL REPRESENTATION

A complete graph must show all of the interesting behavior of the function and the behavior trends on the left and right ends of the viewing window must continue.

Using Technology A graphing calculator will find the graphical representation of a function if it is entered in symbolic form. *Calculators graph functions by plotting points from the numeric representation.* You enter the symbolic representation of the function, followed by the domain and range to be used. The calculator can calculate and plot around one hundred points and it connects these points to form the graph. That is, even though most graphical representations are very large pictures, with domains extending to negative and positive infinity, calculators are capable of only displaying a small portion of the full graphical representation. The portion that you see on the display of the calculator is the **viewing window**. See the example below in Figure 2.4.6.

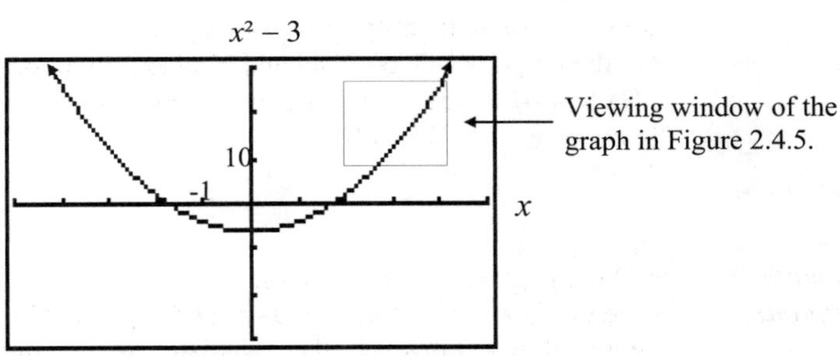

Figure 2.4.6

HOW CALCULATORS GRAPH

Calculators start with the left-most number in the domain you specify (x-coordinate) and find the related value of the function; next, it increments x (*increases the size of x by a small amount* $\triangle x$) and calculates the related value of the function for the new value of x. It plots these two points and draws a line segment connecting the points. The calculator increments x again, and the process continues until the points in the domain you specified have been graphed. The lines drawn between successive points can be so short that the graph "looks" like just points are being graphed.

In a book, you cannot be shown the action that takes place on the calculator; thus, these examples are done in a manner similar to how a calculator would graph them. You should try these on your calculator using a similar viewing domain and range.

Example 2: Find the complete graphical representation of $|x + 2| - 1$. What is its shape?

Solution: Graphing functions with technology may require a three-step process. The first step is to store the function. The second, is to select a window containing a domain that you think will show all of the interesting behavior (a complete graph). Selecting a good range for your window may be a little more complicated; however, you will get better with experience (The purpose of this section is to gain experience.) The third step is to push the graph key. These three steps are outlined below in Figure 2.4.7.

Store the function.

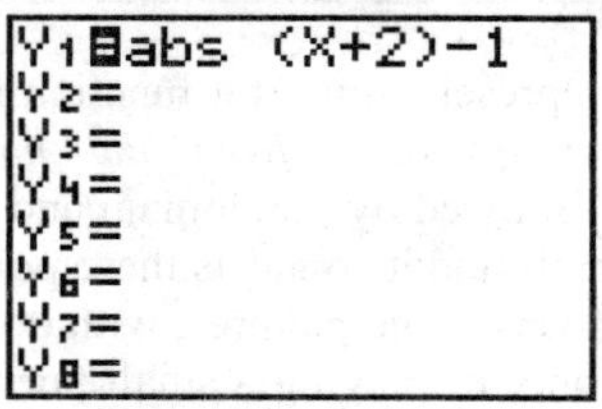

Figure 2.4.7a

Set the window.

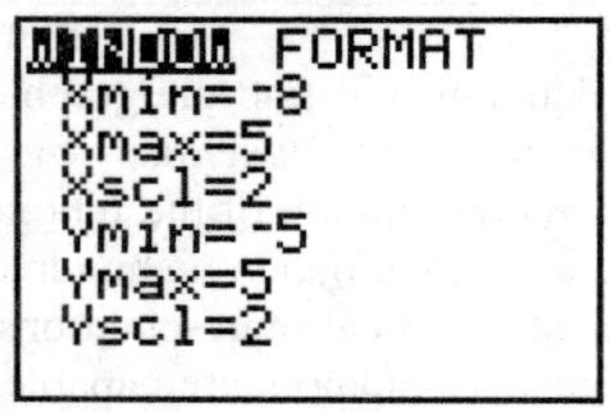

Figure 2.4.7b

Type the graph key.

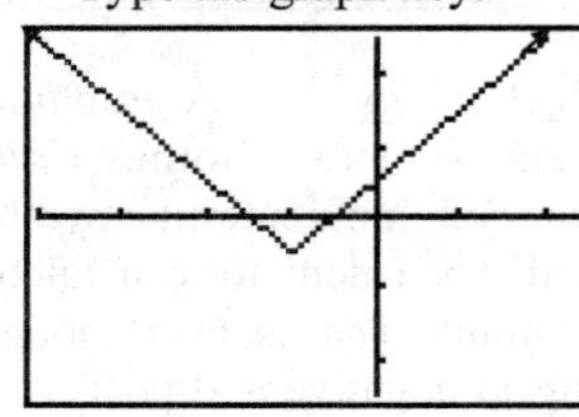

Figure 2.4.7c

Regardless of the variable in the function, the calculator may require the variable x. *Xmin* and *Xmax* identify the viewing domain as $[-8, 5]$, and *Ymin* and *Ymax* specify that the range is $[-5, 5]$. Xscl and Yscl are directions to put tic marks at multiples of 2 (except 0) on the x-axis and the *function*-axis. The graphical representation of the function $|x + 2| - 1$ is a V.

⁂

Example 3: Find the graphical representation of $\sqrt{t-5}$.

Solution: Store the function, set the window, and type the GRAPH key. The function contains the variable t; however, the calculator may require the variable x. The normal domain of the function is $[5, \infty)$ which means that you will see no graph before 5. You may still set a window that contains numbers less than 5.

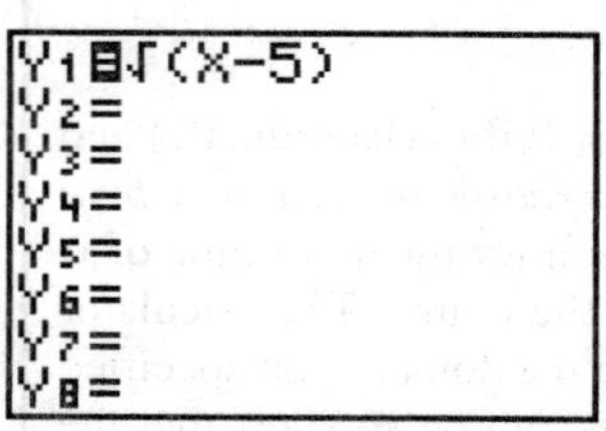

Figure 2.4.8a

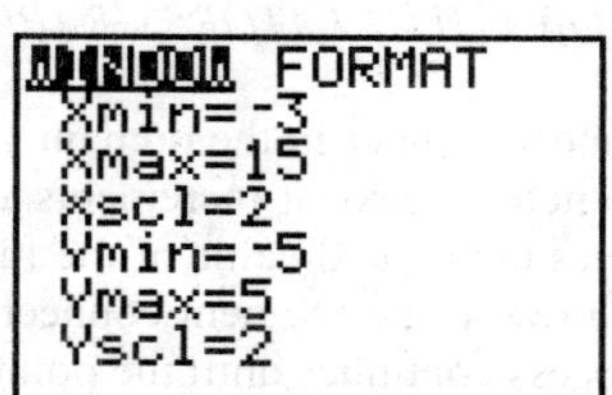

Figure 2.4.8b

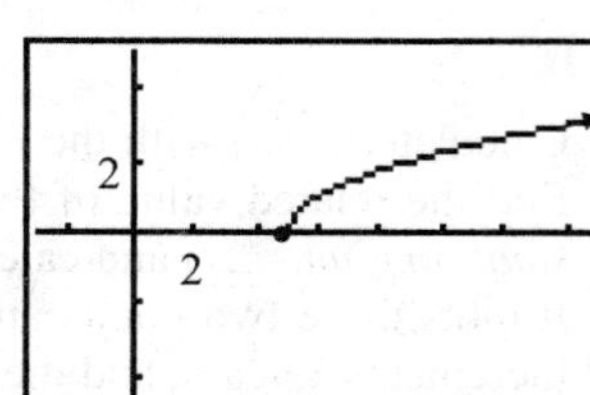

Figure 2.4.8c

⁂

2.4 EXERCISES

Many times you can work the exercises by algebraic, numeric, or graphical methods. You can learn about mathematics no matter which method you use. Part of the learning process is for you to decide which method is best for you to use. Please read the text carefully before you do the exercises. Read the **next assigned section** after you finish the assignment below.

A Quick Look To The Future

-4. Explain why you think expressions like $2x + 1$ are called linear.

-3. If you throw a ball straight up, describe when the height of the ball is zero.

-2. If a plane ascends at a constant rate and then descends at a constant rate, what is the shape of the path of the plane?

-1. What kind of number does $\sqrt{x}$ represent, if $x \in (0, \infty)$?

Previous Topics Revisited

1. Make a numeric representation for $-|x-3|$ and describe the behaviors listed.
   a. When is the function negative, zero, or positive?
   b. What is the rate of change of the function when $x > 3$?
   c. What is the rate of change of the function when $x < 3$?
   d. When is the function increasing? When is it decreasing?
   e. What is the maximum value of the function?
   f. What is the domain? What is the range?

2. Make a numeric representation for $\frac{1}{4}x + 1000$ and describe the behaviors listed.
   a. Is the function increasing or decreasing?
   b. What is the rate of change of the function?
   c. When is the function 0?
   d. When is the function positive?
   e. When is the function negative?
   f. What is the domain? What is the range?

3. Make a numeric representation for $\frac{x}{x^2+1}$ with a domain for x of $\{$all integers from -5 to $5\}$.

4. What is the normal domain of $\frac{x^2}{(x-1)(x+3)}$?

5. What is the normal domain of $\frac{\sqrt{x+3}}{(x-1)}$?
6. From 1976 to 1980, 2.2×10^7 acres of tropical forest were destroyed. From 1981 to 1990, 4.1×10^7 acres were destroyed. Find the total area destroyed from 1976 to 1990.

New Topic Exercises
Use your calculator to find graphical representations of the functions in Exercises 7 to 14. What is the general shape of each graph?

7. $x-3,\ x-2,\ x-1,\ x+0,\ x+1,\ x+2$, and $x+3$
8. $-3x-1,\ -2x-1,\ -x-1,\ 0x-1,\ 1x-1,\ 2x-1$, and $3x-1$
9. $|x+3|-1,\ |x+2|-1,\ |x+1|-1,\ |x+0|-1,\ |x-1|-1,\ |x-2|-1$, and $|x-3|-1$

   Reminder: $|x+4|-1$ is entered as abs(x + 4) – 1

10. $-2|x+3|-1,\ -2|x+2|-1,\ -2|x+1|-1,\ -2|x+0|-1,\ -2|x-1|-1,\ -2|x-2|-1$, and $-2|x-3|-1$
11. $\sqrt{x+3}-2,\ \sqrt{x+2}-2,\ \sqrt{x+1}-2,\ \sqrt{x+0}-2,\ \sqrt{x-1}-2,\ \sqrt{x-2}-2$, and $\sqrt{x-3}-2$

    Reminder: $5\sqrt{x+7}-14$ is entered as 5 $\sqrt{\ }$(x + 7) – 14 because the calculator does not use a bar as a grouping symbol -- only parentheses.

12. $1\sqrt{x+3}-2,\ 2\sqrt{x+2}-2,\ 3\sqrt{x+1}-2,\ -1\sqrt{x+0}-2,\ -2\sqrt{x-1}-2,\ -3\sqrt{x-2}-2$, and $-4\sqrt{x-3}-2$
13. $(t-3)^2+2,\ (t-3)^2+1,\ (t-3)^2,\ (t-3)^2-1,\ (t-3)^2-2$, and $(t+3)^2-2$
14. $(t-3)^2-2,\ (t-2)^2-2,\ (t-1)^2-2,\ (t-0)^2-2,\ (t+1)^2-2,\ (t+2)^2-2$, and $(t+3)^2-2$
15. How much above the x-axis is the function x^2-x+2 when x has values of –4, –1, 0, 2, and 5?
16. How much below the x-axis is the function $-2|x+1|-3$ when x has values of –2, –1, 0, 1, and 2?
17. What is the value of the function $-3x+12$ when x is 4?
18. What is the value of the function $-|x+2|+1$ when x is –1? When x is –3?
19. What is the value of the function $(x-3)^2-1$ when x is 2? When x is 4?
20. What is the value of the function $\sqrt{x+2}-3$ when x is 7?
21. If a data relationship (a function) contains the related pair of numbers (–2, 5), what is one point on the graphical representation of this relationship?
22. If a data relationship (a function) contains the related pair of numbers (4, 7), what is one point on the graphical representation of this relationship?

From Mathematics to English

23. After reading this section, make a list of questions that you want to ask your instructor.

24. Continue in your daily journal and make an entry. In addition to your normal entry on thoughts about the mathematics in this section, list at least two positive comments about what you have learned about this topic.

25. In paragraph format, summarize the material in this section of the text in your daily journal.

26. Describe how your classroom instructor made this topic more understandable and clear.

27. After reading the text and listening to your instructor, what do you not understand about this topic?

28. What is the most important idea presented in this section? Why?

29. For the function x^2, would a good viewing domain (values for x chosen by you) contain just positive numbers, $(0, \infty)$? Why?

30. Is it acceptable to use a viewing domain of $[-1000, 1000]$ for the function $2x + 3$? Why?

31. Can the *viewing domain* of $\dfrac{1}{x-2}$ contain the number 2? Why?

32. In the function in Exercise 31, is there a graph when $x = 2$? Explain.

33. Make up several functions of the form $d|x + e| + f$, where d, e, and f are numbers of your choice. What general shape is each graph?

34. Make up several functions of the form $d(x + e)^2 + f$, where d, e, and f are numbers of your choice. What general shape is each graph?

35. Make up several functions of the form $d\sqrt{x+e}+f$, where d, e, and f are numbers of your choice. What general shape is each graph?

36. Make up several functions of the form $dx + e$, where d, e, and f are numbers you supply. What general shape is each graph?

37. Describe how to graph the function $7x - 3$ by hand. Do not actually graph the function.

38. What is the graph of a function?

From English to Mathematics

In Exercises 39 - 41, convert the English statement to an equivalent mathematical statement.

39. four more than x squared

40. double the number x

41. a domain of all real numbers between -2 and 6 inclusive

Explorations

42. You have now briefly looked at the symbolic, numeric, and graphical representations of the following functions: $dx + e$, $dx^2 + ex + f$, $d|x+e| + f$, and $d\sqrt{x+e} + f$. Using sentence format, describe the behavior of : $-4|x+2| - 1 + \frac{1}{2}x^2$ and $|x-5| + 3\sqrt{x+4}$.
Include when increasing/decreasing, when positive/negative, when zero, the maximum/minimum values, the domain and range. Are the average rates of change constant?

43. Using the graphical representation of each function, identify the minimum or maximum and the range of the function.

a. $-3|x+2|+4$ b. $5|x-4|-8$ c. $-0.5|t+3|+12$

d. $\frac{2}{3}|t-1|-3$ e. $5+|t+6|$ f. $-4|x-5|-2$

Based on these examples, what is the maximum or minimum of the function $d|x+e|+f$, where *d*, *e*, and *f* are known numbers? Remember: *d*, *e*, and *f* are called function parameters.
What parameter tells you when there is a maximum or when there is a minimum?

CHAPTER TWO TEST

1. What is the normal domain of $\sqrt{x-4}$?

2. What is the normal domain of $\dfrac{x-2}{x+3}$?

3. Identify the characteristics of: $|x+1|+|x-2|-5$.

   When is the function increasing?
   When is the function decreasing?
   When is the function constant?
   When is the function positive?
   When is the function negative?
   When is the function zero?
   What is the minimum value of the function?

   What is the average rate of change of the function as x changes from –2 to –1?

4. What is the symbolic representation of the function describing the fare in a taxi with a $3.50 entrance fee and a rate of $2.45 per mile?

In Exercises 5 - 6, what is the geometric shape of the graphical representation?

5. $x^2 + 2x + 3$

6. $2|x-1|+3$

7. Find the zero of $-3x + 60$.

8. As x increases in value, does $-6x + 40$ rise, fall, or neither?

9. As x increases in value, does $0x + 2$ rise, fall, or neither?

10. What is the rate of change of the function $\dfrac{1}{4}x-17$?

11. What is the value of the function $-2.4x^2 + 5.2x - 1$ when x is –3?

12. Does $x^2 - 5x - 14$ have a minimum or does it have a maximum? Find its approximate value.

13. What is the range of $x^2 - 5x - 14$?

14. Find the zeros of $x^2 - 16$.

15. What is the maximum value of the function $-|x-2|+3$?

16. Find the zeros of $-2|x+1|+3$.

17. What is the range of $3|x+7|-5$? (Assume a normal domain.)

18. What is the rate of change of the function $-2|x+3|-8$ when x is left of –3?

19. Does the function $-\sqrt{x+2}-\dfrac{1}{2}$ have a maximum or a minimum? What is its value?

20. Is the function $\dfrac{1}{4}\sqrt{x-5}+12$ increasing or decreasing over the normal domain?

CHAPTER THREE
COMMON BEHAVIORS OF FUNCTIONS

3.0 Introduction to Geometric Behaviors of Functions

Chapter Two introduced you to behaviors of data relationships. You also saw how data relationships can be written in numeric, graphic, and symbolic forms. In this chapter, we will continue with the development of behaviors of functions, but here we will primarily use them in symbolic form. This is to say that using contextual situations like in Chapter Two, allows your brain to better remember the concepts presented. However, after you understand the mathematical concept, we no longer need the real-world context to further your understanding. At the same time, we will build on what you learned in Chapter Two. In doing so, we distribute the learning over time. This has the effect of increasing the likelihood you will remember the mathematics learned for a much longer period of time – months and years instead of days and weeks.

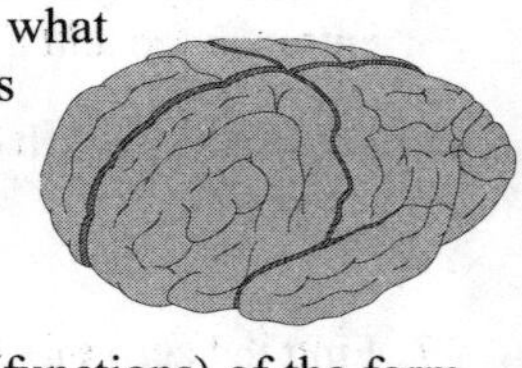

In this chapter, we will investigate the geometric behaviors of linear relationships (functions) of the form $dx + e$. Think back to Chapter Two and recall that many data relationships were linear (their graphs were straight lines and they had a constant rate of change). The world abounds with relationships that are linear. Earning salary that depends on the hours worked and the rate at which you work, is a simple yet prevalent method of earning wages.

We will analyze functions that are quadratic in nature and have the symbolic form $d(x+e)^2 + f$. You should recall relationships that are quadratic – their graphs are in the shape of a parabola. When you toss a ball straight up in the air, the time-height relationship follows a quadratic model (ignoring air resistance).

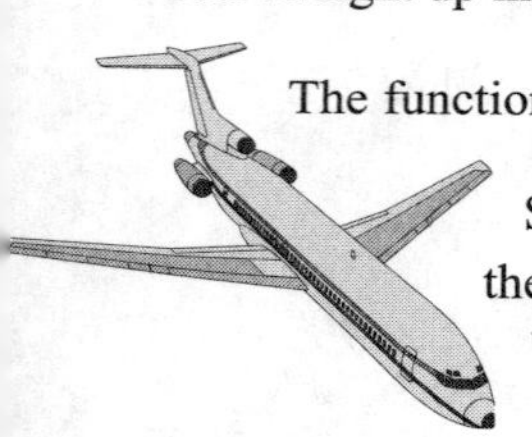

The function that models the data relationships that have graphs that look like V's are also found in nature. We will discover that the absolute value function has a graph that is a "V." Symbolically, the absolute value function looks like $d|x+e|+f$. If you collect data on the height of an airplane and the temperature outside the plane as the plane ascends and then immediately descends in a "normal" atmosphere, you will find the absolute function in the data.

Because of the mathematics studied in Chapter Nine, we will study the behaviors of the square root function $d\sqrt{x+e}+f$. As it turns out, the type of function makes no difference as we are only studying the behaviors of the function at this time. Velocity vs. stopping distances varies according to the square root function as does time-velocity of an object in free fall.

Finally, we will analyze behaviors of the exponential function. You have seen several data relationships that fit the exponential model $d \cdot b^{x+e} + f$. Exponential growth and decay occurs quite often in the world outside the classroom. You have experienced your food cooling exponentially. It is a good thing it does, because otherwise we would need to wait 4 to 5 times as long for hot food to cool.

The graphing calculator will be needed to graph these functions. Likewise, many of the features will be required to help you answer questions about behaviors of functions. This will be an excellent time to learn how the graphing calculator works and how to use it to do mathematics.

3.1 An Introduction to the Analysis of the Linear Function $dx + e$

The House Painter

Suppose you want to paint houses to earn extra money while attending college. Your experience at painting has been to paint a 20×16-foot side of your own house in 5 hours (including windows). This means you can paint at a rate of 320 square feet per 5 hours, or $\frac{320}{5} = \frac{64}{1} = 64$ square feet per hour. Measurements on your first potential customer's house show a surface area of 1792 square feet. Table 3.1.1 represents your analysis of the painting job where t is time in hours and A is the area that remains to be painted. Since you paint at a rate of 64 square feet per hour, for each hour you work, the area left to

Table 3.1.1

t	0	1	2	3	8	16	24	26	27	28
A	1792	1728	1664	1600	1280	768	256	128	64	0

paint decreases by 64 square feet. Your numerical analysis of the painting job shows that you can finish the painting in 28 hours. But what if the customer wants to know how much of the house will be left to paint on Tuesday? Or at noon on Wednesday? What if your experience changes the rate at which you paint; how may this change the time it takes to finish? What if your second job is 5000 square feet of painting; do you start the entire analysis over again? What if you want to supply the customer with a picture of how your work will progress, so as to make it as clear as possible what is going to happen during the painting job? Is it possible to use mathematics to answer all of these questions and many more? Yes! This is the reason for the analysis of functions, and you will find many of the behaviors you have already studied in Chapter Two can be determined from the graphical representation of the function also. Each representation of the function -- English, numeric, symbolic, and graphic -- is useful in determining behavior.

What is the symbolic representation of the painting problem? If you start with 1792 square feet to paint and every hour (t) you paint there are 64 square feet less to paint, then the symbolic representation is $1792 - 64t$. Or you may want to write it as $-64t + 1792$. Further, if you choose to give it a name, you may want to call it A for area left to paint. You are now set to analyze the function for behaviors as you did in Chapter Two.

What does the graphical representation of the function $-64t + 1792$ (or $-64x + 1792$ on the graphing calculator) look like? How is it different from $64t + 1792$? How is it different from $-64t - 1792$? What geometric shape does it have? When is $-64t + 1792$ zero? When is it positive? When is it negative? What about increasing and decreasing? Do all graphical representations of functions of the form $dx + e$ have the same geometric shape? These and other questions can be answered easily if a graphing calculator is used to generate the graphical representations of functions of the form $dx + e$.

<u>A Little Background Material</u>
The *normal domain of all functions in the form dx + e is the set of real numbers.* There are no real numbers for x that cause the function to be non-real because there is no division or square root. The range (with a normal domain) is all real numbers as well because the function has no maximum value and no minimum value. The problem domain is determined by you and the actual problem you are studying. You

may recognize the shape of the function from work done in Chapter Two as a straight line. Further, $dx + e$ is called a **linear function**, and ***d*** and ***e*** are called the **parameters** of the function.

The graphical representation of the function comes from doing a three-step procedure -- much like getting the numeric representation. The three steps are:

a. store the symbolic representation in the calculator's memory

b. set a viewing window

c. strike the graph key.

Figure 3.1.1a

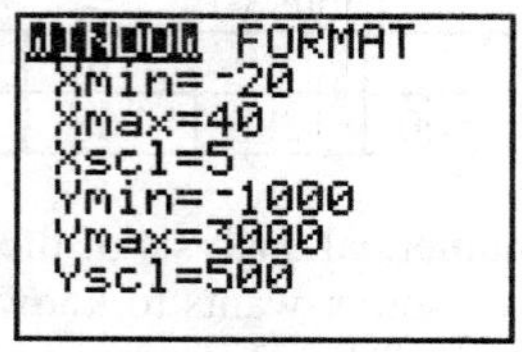

Figure 3.1.1b

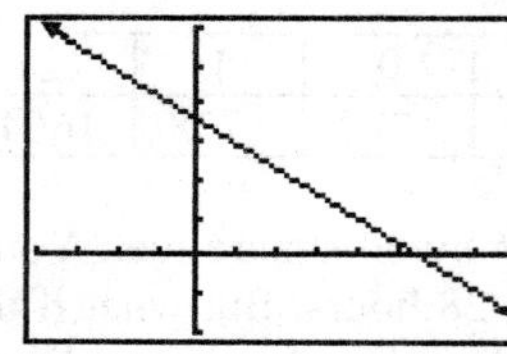

Figure 3.1.1c

The arrows on the ends of the graph indicate that the domain does not stop at –20 or at 40. They also tell you that the range does not stop at –1000 or at 3000. The above graph is complete because all of the interesting behavior is displayed and the trends in behavior continue both to the left and right of the viewing window. Remember, just like reading the numeric representation, you must read the graphical representation as the domain values change from low to high. Of course, this means you will read from left to right. You can "see" from the graphical representation that the function is decreasing. As you read the graph from low to high values of the domain, the values of the function are dropping. The trace key on your calculator will definitely show you that the function value is dropping. Remember to move the cursor from low to high values of the domain.

> The trace values for t and $-64t + 1792$ shown on the bottom of your calculator screen are the numbers the calculator uses in the numeric representation it makes before it graphs the function. Some calculators cannot use variables like t, nor can they use function names like A.

You should observe each entry of the numeric representation from trace and note the relationship to the graphical representation. For example, you should note that when the values of the function (from trace) are positive, the graph is *ABOVE* the t-axis (x-axis). When function values are negative, the graph is *BELOW* the t-axis (x-axis). When the function value is zero, the graph is *ON* the t-axis (x-axis). If the function is decreasing, the trace cursor drops as you move it to the right. The graphing calculator can help you determine most of the behaviors of the function. If the graph and trace don't help, use the numeric, symbolic, or English representations to help find the behaviors of the function.

To graph a function by making a numeric representation by hand and then a graphical representation by hand can be very time consuming. Just a reminder, the calculator starts with the smallest value in the viewing window you give it, then it:

- calculates the corresponding function value
- plots the point
- increments x (increases the value of x by a small amount)
- calculates the corresponding function value
- plots the point
- draws a line between these two points
- increments x
- calculates the corresponding function value
- etc.

until it gets to the biggest value in the viewing domain you entered. This is similar to the method you might use if the graph were done by hand.

> Understanding
>
> The variable in the function will be identified as x when using the calculator. The name of the function will be Y_1, Y_2, etc., so as to relate to the calculator's notation.

Several "problems" follow in this section. The reason for studying these data relationships found outside the mathematics classroom is that scientists, engineers, business persons, teachers, etc. actually must be familiar with the functions that model these relationships. The models are of the linear function. A second and probably more important reason for starting each section with real-world situations is that you can learn mathematics by relating a situation you understand to the mathematics being presented. For example, you know that the amount of painting left undone (see below) is decreasing; therefore, you also know that the model of the painting problem is decreasing. You know when the area of painting left to be painted is 0, the function that models the area left to paint is zero. You know the area left to paint cannot be negative; thus, you know the model is never negative (0 or larger) on the problem domain. Keep these ideas in mind as you study the behaviors of the linear functions below. Most of the new ideas presented are developed from a problem situation; thus, you should be able to relate these mathematical ideas to something that makes sense from the real world.

Analysis of $-64t + 1792$ - **The House Painter**

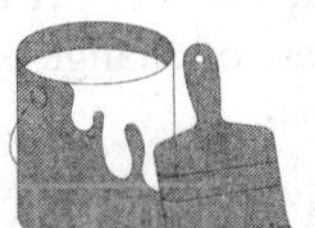

Domain Since there are no arithmetic operations in $-64t + 1792$ that cause it to be non-real, the normal domain is $t \in (-\infty, \infty)$. As developed in the setup of the painting problem, the values for t that make sense in the problem are [0, 28], this is the problem domain. The complete graphical representation of $-64t + 1792$ in Figure 3.1.1c shows that the domain continues to negative, and to positive infinity. The arrows on the ends of the graph are an indication that the normal domain is $(-\infty, \infty)$.

<u>Range</u> Assuming a normal domain, the range is $A \in (-\infty, \infty)$. As shown in Figure 3.1.1c, $-64t + 1792$ continues to decrease to $-\infty$ as t gets larger and as t gets smaller $-64t + 1792$ rises to ∞. Considering the problem domain of [0, 28], the function $-64t + 1792$ has a maximum of 1792 and it reaches 0 when x is 28. This makes the range [0, 1792] for the problem domain.

<u>Maximum/minimum</u> Because $-64t + 1792$ tends to negative or positive infinity, there is no maximum nor is there a minimum. The maximum and minimum on the problem domain are 1792 and 0 respectively.

<u>Increasing/decreasing</u> A look at the graphical representation of $-64t + 1792$ shows that it is decreasing for all t in the normal domain. This behavior can only be answered by reading the graph as t changes from low numbers to high numbers. This is why you must read the graph from left to right. Try trace to confirm it is decreasing.

<u>Positive/negative</u> Since $-64t + 1792$ is positive when it is above the t-axis and negative when it is below the t-axis, the graphical representation -- with help from trace or the numerical representation -- shows you that $-64t + 1792$ is positive when $t \in (-\infty, 28)$. By the same reasoning, $-64t + 1792$ is negative when $t \in (28, \infty)$.

$-64t + 1792$

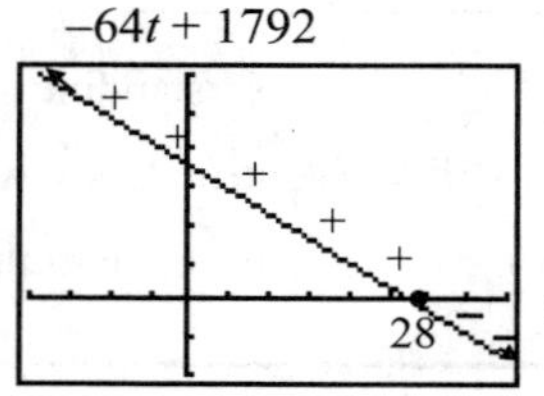

Figure 3.1.2

<u>Zeros</u> Since $-64t + 1792$ is positive if $t \in (-\infty, 28)$ and negative if $t \in (28, \infty)$, when t is 28 the function $-64t + 1792$ must either be non real or 0. Since it isn't nonreal, it must be 0. A look at the numeric representation confirms that when t is 28., the function $-64t + 1792$ is 0. The zero of the function $-64t + 1792$ is 28.

<u>Rate of change</u> The rate of change is not easily gotten from the graphical representation of the function. Of course the numeric representation will always help as it did in Chapter Two. However, the rate of change measures how fast a function is changing. The function $-64t + 1792$ models the area of the house that remains to be painted and the amount of area left to paint is decreasing at a rate of 64 square feet per hour -- as was given in the original English description of the function. The rate of change is -64 square feet per hour, or if you do not attach meaning to the function other than it being a real number, the rate of change is -64.

Analysis of $-2t + 2650$ - **The Beach Front**

Professor Ed has a house on the Outer Banks of North Carolina that sits 2650 feet from the ocean. The civic association published a newsletter in which the building inspector warned that the beach has been eroding at an annual rate of 2 feet per year. He said that in the last 32 years the beach has receded about 64 feet. If t is time in years, then the function that models the distance from Prof. Ed's house to the water is 2650 minus 2 feet per year for each year that passes. Since t represents the years that are passing, then the symbolic representation of the function is $2650 - 2t$. Another way of writing this is $-2t + 2650$.

<u>Domain</u> Because the function is a linear function, the normal domain is all real numbers. That is, $t \in (-\infty, \infty)$. The problem domain is the set of numbers for t that makes the distance positive or zero; further, time is usually thought of as positive number; thus t must be larger than or equal to 0. The graphical representation with trace shows that the function $-2t + 2650$ is positive or zero when $t \in [0, 1325]$. A reasonable problem domain is [0, 1325].

$-2t + 2650$

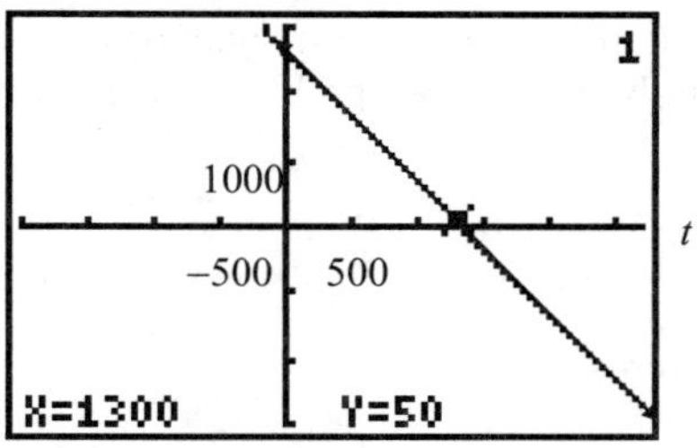

Figure 3.1.3a

$-2t + 2650$

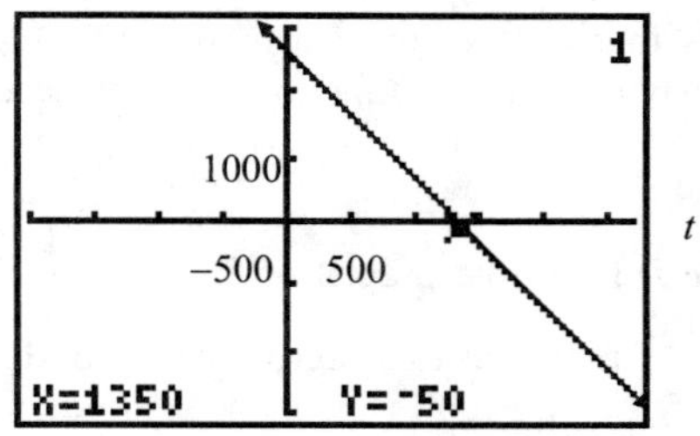

Figure 3.1.3b

<u>Range</u> With a normal domain, the range is all real numbers, as is the case for all linear functions. Using the problem domain, the range of the function $-2t + 2650$ is [0, 2650]. The function is zero when t is 1325, and the function is 2650 when $t = 0$. The reality of the situation suggests that the closest the house can be to the ocean is 0 feet, and the maximum is where it is initially – 2650 feet from the ocean. This makes the range [0, 2650] using the problem domain.

<u>Maximum/minimum</u> Since the function can represent all real numbers when t is in the normal domain, there is no maximum nor a minimum. However, if the problem domain of [0, 1325] is used, the maximum distance to the ocean is 2650 feet and the minimum distance is 0 feet. The actual situation also suggests the maximum and minimum. You may assume the house gets washed away by the ocean during the next storm after the distance becomes zero.

<u>Increasing/decreasing</u> If you think about what the function models (the distance from the house to the ocean), you know that the model of the beach front problem is decreasing. The distance between the house and the ocean is decreasing. The graphical representation also shows the function $-2t + 2650$ decreasing as time passes (It is going downhill.)

<u>Zero</u> Figure 3.1.3 shows that the function is zero someplace between 1300 and 1350 because at 1300 years the distance is 50 feet to the ocean and at 1350 years the distance is –50 feet to the ocean. A reasonable approximation to the zero is 1325. Trace will confirm this.

<u>Positive/negative</u> Since the graph of the beach front function is above the t-axis when t is less than 1325, then the function is positive when $t \in (-\infty, 1325)$. Since the graph of the beach front function is below the t-axis when t is greater than 1325, the function is negative when $t \in (1325, \infty)$.

$-2t + 2650$

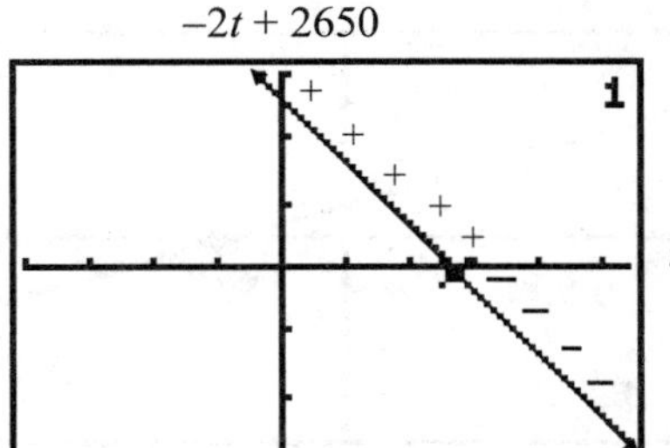

Figure 3.1.4

Rate of change While you may certainly use the numeric representation to find the rate of change as you did in Chapter Two, the English representation of the beach front problem gives the rate of change of the distance to the ocean as –2. It is negative 2 because the distance is decreasing; further, the definition confirms this notion: rate of change is $\frac{\Delta \text{ in the distance function}}{\Delta \text{ in time}} = \frac{-2}{1} = -2$.

The examples that follow are of a more traditional nature. Several linear functions are analyzed for a particular behavior, and the functions do not necessarily represent a problem taken from a real-world setting.

Example 1: From the graphs of $3x - 6$, $\frac{1}{4}x + 2$, and $-x + 3$, find a value for x that causes the functions to be zero. That is, find the zeros.

Solution: With a graphing calculator, graph the functions and use the trace key to find the x-coordinate of the graph when the function value is zero. The zeros may be approximate. Rules of acceptable approximate answers are developed later. The graphs are shown below, with the zero value indicated on the graph.

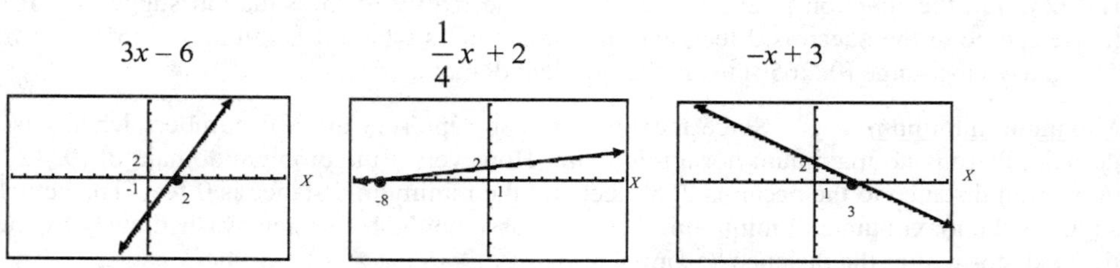

Figure 3.1.5a Figure 3.1.5b Figure 3.1.5c

The value of x that makes the function zero is called a zero of the function.

Example 2: Find the zero of the function $-\frac{3}{5}x + 2$.

Solution: Graph the function and trace to a point where the function has a value of zero. The approximate zero is marked on the graph below.

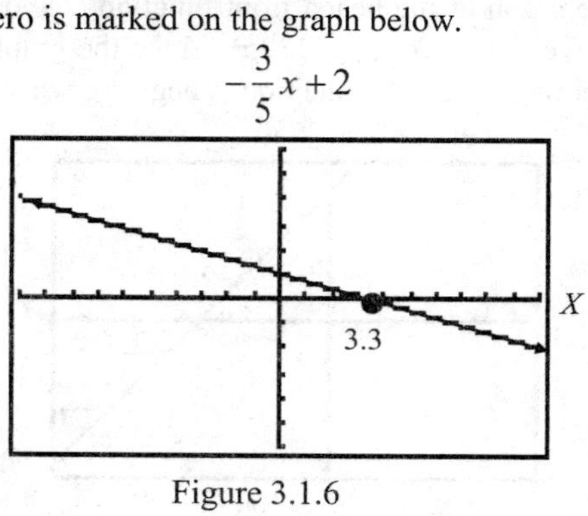

Figure 3.1.6

The zero can also be approximated using the numeric approach.

x	2	3	3.2	3.3	3.4	4	5
$-\frac{3}{5}x+2$	0.8	0.2	0.08	0.02	–0.04	–0.4	–1

As you can see from the numeric representation, there is a zero between 3.3 and 3.4. A better estimate is 3.35.

Example 3: Is the function $2x - 4$ increasing or decreasing?

Solution: There are several ways of answering this question. If you recall the work you did in Section 2.3, you know that with a coefficient of x that is positive, the function is increasing. You can watch your calculator graph the function and see if the graph is rising or falling as the calculator plots values from the numeric representation it makes. If you do this, you will see the graph rise as it is being plotted; it is increasing. A third method for deciding whether the function is increasing or decreasing is to use the trace key. As you move the cursor from small values of x to larger values of x, is it rising or falling? It is rising; therefore, the function is increasing. Finally, if the rate of change is a positive number, it is increasing.

Example 4: What is the rate of change of the function $-5x + 42$?

Solution: Graph the function and choose a window so that the values for x change one unit each time you move the cursor when using the trace key. As you move the cursor to the right, note how much the function changes. It changes by –5 units each time x increases by 1. The rate of change is –5. The numeric representation can also be used and you may have learned in Section 2.3 that the coefficient of x is the rate of change.

Example 5: If the zero of a linear function is 6 and the function is decreasing, when is the function positive? When is it negative?

Solution: Since the function is zero at 6 and it is decreasing, as x-values change to values larger than 6, the function drops in value because it is decreasing. Thus, it must be less than zero to the right of 6. The function is negative when x is in the interval $(6, \infty)$. If the function is negative to the right of 6 and it is zero at 6, the only possibility it has left is that it is positive to the left of 6. The function is positive when x is in the interval $(-\infty, 6)$.

A zero by any other name is still a zero.

Several examples.

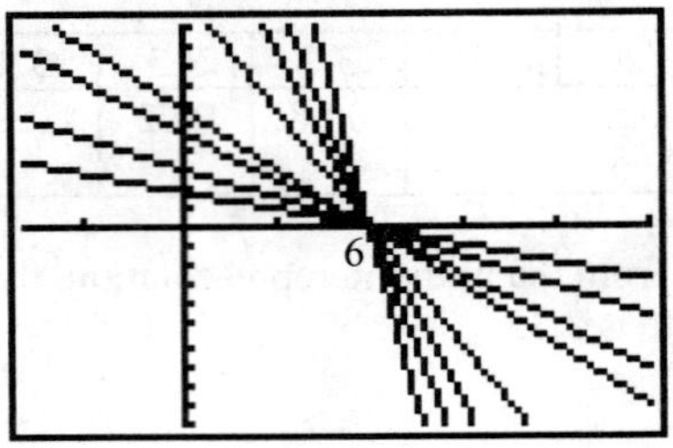

Figure 3.1.7

⁂

Having looked at the idea of increasing and decreasing from an intuitive perspective, a more formal statement follows.

> **Understanding**
>
> On the domain of the function $dx + e$ and as x changes from low to high, if the function increases in value, the function is increasing or it is an increasing function. On the domain of the function $dx + e$ and as x changes from low to high, if the function decreases in value, the function is decreasing or it is a decreasing function.

Example 6: Why do some graphs rise and some fall?

Solution: Below are graphs of $-\frac{3}{5}x+2$, $-4x + 0$ and $-\frac{2}{3}x-1$.

$-\frac{3}{5}x+2$

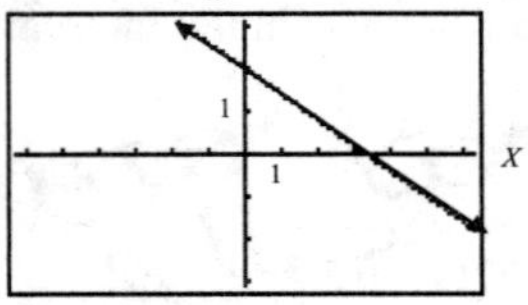

Figure 3.1.8a

$-4x + 0$

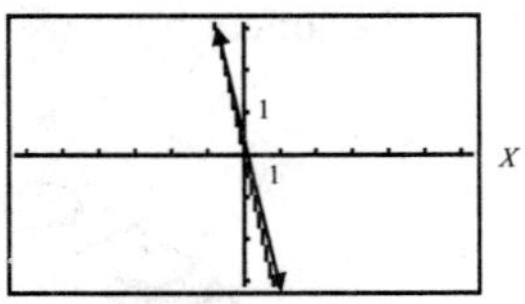

Figure 3.1.8b

$-\frac{2}{3}x-1$

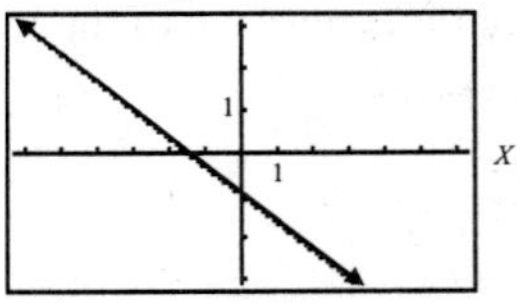

Figure 3.1.8c

All of the graphs fall **as x increases**; thus, the functions are decreasing. The only common characteristic of the three functions is that the coefficient of x (the number being multiplied by x) is negative. The graphs of $3x - 5$, $\frac{1}{2}x + 2$, and $\frac{1}{5}x + 0$ are below.

$3x - 5$

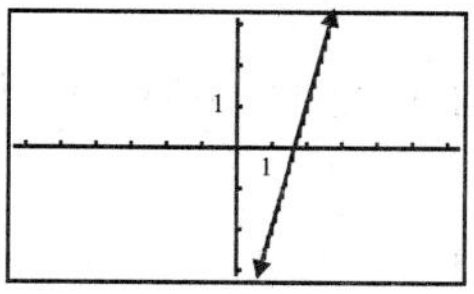

Figure 3.1.9a

$\frac{1}{2}x + 2$

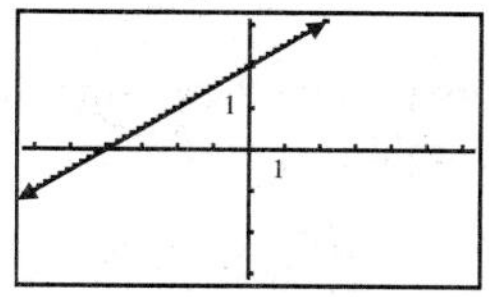

Figure 3.1.9b

$\frac{1}{5}x + 0$

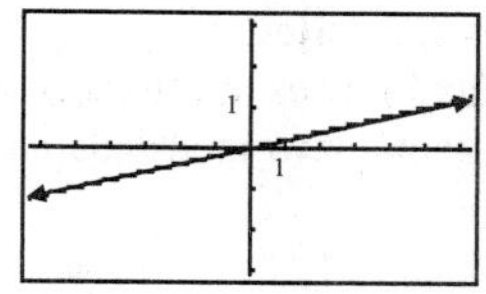

Figure 3.1.9c

All of the graphs rise **as x increases**; thus, the functions are increasing functions. The only common characteristic of the three functions is that the coefficient of x is positive. **As x increases**, graphs of functions in the form $dx + e$ rise (are increasing) if d is positive ($d > 0$) and they fall (are decreasing) if d is negative ($d < 0$).

⁂

Example 7: Does every graph of $dx + e$ increase or decrease as x increases?

Solution: Observe the graph of $0x + 2$ below, or graph it yourself on your calculator.

$0x + 2$

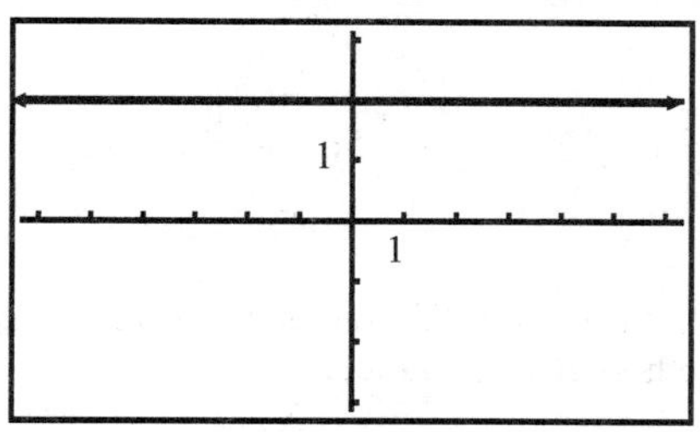

Figure 3.1.10

As x increases, this graph neither rises nor falls. The graph remains *constant*, no matter the value of x. Functions that remain constant over the entire domain are called constant functions.

⁂

3.1 EXERCISES

Many times you can work the exercises by algebraic, numeric, or graphical methods. You can learn about mathematics no matter which method you use. Part of the learning process is for you to decide which method is best for you to use. Please read the text carefully before you do the exercises. Read the **next assigned section** after you finish the assignment below.

A Quick Look To The Future

-4. If the relationship between the hours you work and the salary you earn is quadratic, would this be "good" for you? Explain.

-3. What kind of number is $-|x|$?

-2. Is $\sqrt{-x}$ a real number?

-1. Without using an example, describe a function.

Previous Topics Revisited
Using the graphical representations of the next 3 functions, find the approximate values for the maximum or minimum. (Exact answers are also acceptable.)

1. $x^2 - 1$
2. $|x-3| - 3$
3. $\sqrt{x-2}$
4. Is $\dfrac{2(x-2)}{(x-2)}$ increasing, decreasing, or neither?
5. When is $\sqrt{x+2}$ negative?
6. What is the normal domain of $\dfrac{x+2}{(x-3)(x+2)}$?

New Topic Exercises
In Exercises 7 - 10, find the rate of change of the functions. Use the method of your choice.

7. $4x-5,\ 5x-5,\ 6x-5,\ 7x-5,\ 8x-5,\ 9x-5,\ 10x-5,\ 11x-5$

8. $-x+20,\ -2x+20,\ -3x+20,\ -4x+20,\ -5x+20,\ -6x+20$

9. $\frac{1}{2}x-56,\ \frac{1}{3}x-56,\ \frac{1}{4}x-56,\ \frac{1}{5}x-56,\ \frac{1}{6}x-56$

10. $-\frac{1}{2}x-56,\ -\frac{1}{3}x-56,\ -\frac{1}{4}x-56,\ -\frac{1}{5}x-56,\ -\frac{1}{6}x-56$

In Exercises 11 - 17, find values of x that makes the function 0 (find the zeros). Approximate numbers are acceptable and you may use any method of your choice.

11. $x-1,\ x-2,\ x-3,\ x-4,\ x-5,\ x-6$

12. $x+1,\ x+2,\ x+3,\ x+4,\ x+5,\ x+6$

13. $-x+1,\ -x-1,\ -x+2,\ -x-2,\ -x+3,\ -x-3,\ -x+4,\ -x-4$

14. $-2x+1,\ -2x+2,\ -2x+3,\ -2x+4,\ -2x+5,\ -2x+6$

15. $3x-1,\ 3x-2,\ 3x-3,\ 3x-4,\ 3x-5,\ 3x-6,\ 3x-7,\ 3x-8,\ 3x-9$

16. $-6x+200,\ 7x+\pi,\ 4x+350,\ -5x+42,\ 19x-78$

17. $-3x,\ 5x,\ 42x,\ -560x,\ 39x,\ -162x$

18. For the functions in Exercises 11 and 13, find values of x that cause each function to be positive.

19. For the functions in Exercises 11 and 13, find values of x that cause each function to be negative.

Are the following functions increasing or decreasing?

20. $-x+2,\ -2x+2,\ -3x+2,\ -4x+2,\ -5x+2$

21. $x+2,\ 2x+2,\ 3x+2,\ 4x+2,\ 5x+2,\ 6x+2$

22. $-x,\ x,\ -2x,\ 2x,\ -3x,\ 3x,\ -4x,\ 4x$

23. $-\frac{1}{2}x+3,\ -\frac{1}{4}x-5,\ -\frac{3}{8}x+4,\ -\frac{5}{13}x-2$

24. $\frac{1}{3}x-4,\ \frac{2}{5}x+3,\ \frac{3}{4}t-1,\ \frac{5}{8}t+3,\ \frac{7}{3}t$

25. $0.01x-3,\ -0.001t+4,\ 0.0002x+16,\ -0.00001t-53$

26. $56t+3000,\ -23x-500,\ 47.8t-139,\ -95x+0.002$

27. $0x+7,\ 0t-13,\ -0x+5,\ 0t-23$

28. Ramon can read at a rate of 220 words per minute and is reading a 30,000-word book.

    a. What function models the number of words left to read at a time t?

    b. How long does it take Ramon to read half of the book? How many hours is this?

    c. How long does it take Ramon to read the entire book?

    d. How fast is Ramon reading?

    e. What mathematical term is used to describe how fast Ramon is reading?

    f. If Ramon can increase his reading speed to 300 words per minute, how long will it take to read the book?

    g. At this speed, how long will it take Ramon to read a book with 500,000 words?

29. Create a linear function that is increasing.
30. Create a linear function that is decreasing.
31. Create a linear function that has a zero of 1.
32. Create a linear function that is 5 when x is 0.
33. Create a linear function that is always constant.
34. Create a linear function that has a rate of change of –3.

35. Create a linear function that has a rate of change of $\frac{4}{5}$ and has a value of 6 when x is 0.

From Mathematics to English

36. Explain how to tell if a graph of $dx + e$ rises, falls, or remains constant.

37. Are there any functions of the form $dx + e$ that do not have a zero? If yes, give an example.

38. Does every graph of the function $dx + e$ cross the function-axis? Why?

39. Give your own definition of a linear function.

40. Describe what it means for a function to be increasing.

41. Write your own definition of the idea of rate of change.

42. Explain what the problem domain means to you.

43. Make up a problem situation similar to Exercise 28. Include a list of questions that you think should be answered about this situation.

44. Explain how the concepts presented in this section could be used in another course or on a job.

45. After reading this section, make a list of questions that you want to ask your instructor.

46. Continue in your daily journal and make an entry. In addition to your normal entry on thoughts about the mathematics in this section, list at least two positive comments about what you have learned about this topic.

47. In paragraph format, summarize the material in this section of the text in your daily journal.

48. Describe how your classroom instructor made this topic more understandable and clear.

49. After reading the text and listening to your instructor, what do you not understand about this topic?

50. What is the most important idea presented in this section? Why?

From English to Mathematics
In the next three exercises, convert the English statement to an equivalent mathematical statement.

51. negative four times a number x plus two hundred

52. three halves of a number x minus seventeen

53. the negative of a number x times thirty, plus forty-three

From English to Mathematics Again
Convert each of the English descriptions to a function. Describe in English what your function represents and what the variable in the function represents.

54. A Kansas City taxi service advertises on the side of each taxi that the fare is $5.40 plus $3.90 per mile.

55. Suppose the total number of brain cells is around 120,000,000,000, you normally lose 50,000 per day.

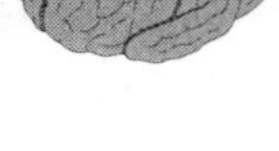

56. A waiter's weekly income is $284 in salary and an average of $5.50 per table served.

57. A 200-pound person maintains a diet that causes a 1.4 pound per week weight loss.

Explorations

58. Explain how to find a viewing window for the following graphical representations:

a. $6000x + 2$ b. $\frac{1}{3}x - 170$ c. $0.0005x + .001$

59. What parameter in the function $dx + e$ indicates that the function is increasing? What values of this parameter cause the function to increase?

What parameter in the function $dx + e$ indicates that the function is decreasing? What values of this parameter cause the function to decrease?

What parameter in the function $dx + e$ indicates that the function is constant? What value(s) of this parameter cause(s) the function to remain constant?

What parameter in the function $dx + e$ indicates the rate of change of the function?

What parameters in the function $dx + e$ help you find the zero of the function?

What parameters in the function $dx + e$ tell you where the graph crosses the function-axis?

60. In the function $dx + e$, what "visual effect" does the parameter d have on the graph? Graph each group of four functions on the same window as you attempt to answer the question.

a.

x	$2x$	$3x$	$7x$
x	$\frac{7}{8}x$	$\frac{3}{5}x$	$\frac{1}{3}x$
x	$-x$	$-2x$	$-5x$
x	$-\frac{7}{8}x$	$-\frac{3}{5}x$	$-\frac{1}{3}x$

In the function $dx + e$, what "visual effect" does the parameter e have on the graph? Graph each group of four functions on the same window as you attempt to answer the question.

b.

x	$x + 2$	$x + 5$	$x + 7$
$-x$	$-x - 2$	$-x - 5$	$-x - 7$
$2x$	$2x + 4$	$2x - 5$	$2x + 3$
$-\frac{1}{2}x$	$-\frac{1}{2}x - 5$	$-\frac{1}{2}x + 2$	$-\frac{1}{2}x - 7$

Summarize your investigation by identifying what effect the parameters d and e have on the graph of $dx + e$.

3.2 An Introduction to the Analysis of the Quadratic Function $d(x + e)^2 + f$

The Soda Can Explosion

What happens when you put a can of soda pop in the freezer component of your refrigerator to cool and you leave it in too long? Hopefully you do not know from first hand experience, but it may explode or simply puff out. This happens because the volume of the pop changes as the temperature changes. If you put a can on the stove and heat up the soda, the same event will happen.

A 12-ounce can of soda is equivalent to 354.8824 milliliters (*ml*). One milliliter is a little less than the size of your little finger from the base of the nail to the end of the finger. The volume of 354.8824 *ml* of water (soda is mostly water and will behave in a similar fashion) is measured at varying temperatures and the data is displayed in numeric form in Table 3.2.1. The problem starts with 354.8824 *ml* of water at 4°C (around 39°F) and the temperature is made lower and then raised.

Table 3.2.1

T	−10	−5	0	2	4	6	10	20
V	355.5425	355.1308	354.9285	354.8930	354.8824	354.8930	354.9782	355.5105

The data cannot be modeled with a linear function for two reasons; the volume both decreases (when $T \in [-10, 4)$) and increases (when $T \in (4, 20]$), and the rate of change is not constant. For example, from −10° to 0° the average rate of change is –0.614 and from 4° to 6° the average rate of change is 0.0053. To find what function can be used to model the volume of the water, find the graphical representation of the data. The shape of the data may give a clue to the type of function that can be used. Figure 3.2.1 shows the plot of the data and Figure 3.2.2 the graphical representation of the function $0.003(T-4)^2 + 354.8824$. Hopefully, it is clear that a quadratic function is an excellent model for the

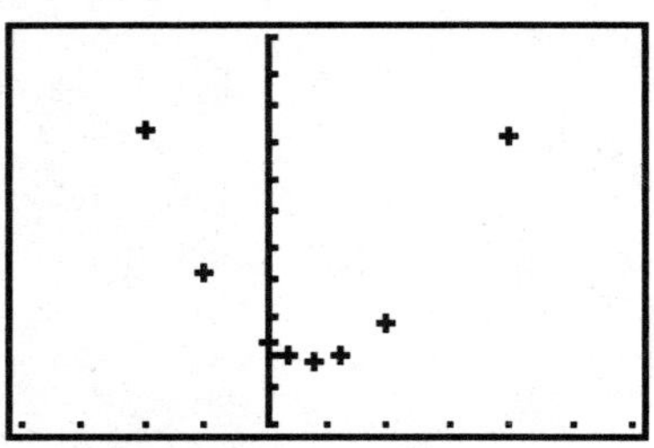

Figure 3.2.1

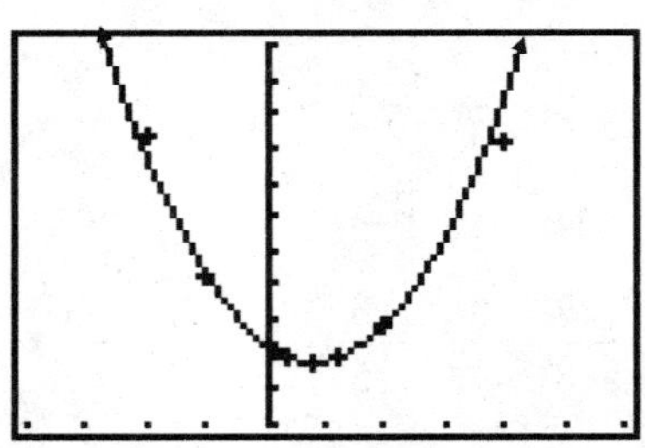

Figure 3.2.2

volume of the water. The volume of the 12 ounces of water has a minimum around 4°C and the volume gets larger as the temperature drops and as the temperature rises. As you can see from the graphical representation of the volume, if the temperature falls to –30°C or–40°C, or if it rises to 30°C to 40°C the volume will be quite large. This, of course, will cause the soda can to explode.

Recalling the symbolic representation of a linear function as $dx + e$, where d and e are the parameters of the function, the quadratic function in symbolic form is $d(x+e)^2 + f$. Please take note of the relationship between the parameters e and f and some of the behaviors in the "Pop Can Problem".

Analysis of $0.003(T-4)^2+354.8824$ - **The Soda Pop Can Explosion**

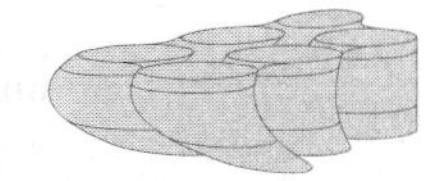

Domain The normal domain is $(-\infty, \infty)$ because the function contains no operations that cause division by zero or the square root of a negative number. Since the variable in the function represents temperature, a reasonable problem domain may be around $[-100, 100]$. Temperatures cannot vary from $-\infty$ to ∞.

Maximum/minimum The graphical representation shows a low point (the water has a minimum volume) and the numerical representation shows a smallest number for the function. This smallest number is the minimum value of the function $0.003(T-4)^2+354.8824$; it is 354.8824. Since the trends established by the function at the ends of the viewing window show it to be headed toward infinity, there must be no maximum value. Considering this statement in the context of the problem, the problem domain will not allow this to happen.

Range Knowing that the function has a minimum value and no maximum value establishes the range. The range of function $0.003(T-4)^2+354.8824$ is $[354.8824, \infty)$. If the problem domain is used, the minimum value remains the same; however, the maximum value occurs at either -100 or 100. Table 3.2.2 shows the maximum volume to be 387.3304 *ml*.

Table 3.2.2

T	-100	100
V	387.3304	382.5304

Increasing/decreasing The graphical representation of the volume function shows the volume decreasing (the graph is dropping) until the temperature rises to 4°. As the temperature continues to rise past 4° the volume starts to increase (the graph is rising) and continues increasing as the temperature rises. Once again, a reminder, you must read the graph or table as the temperature changes from low to high (left to right). The pop can function is decreasing when $T \in (-\infty, 4)$ and increasing when $T \in (4, \infty)$.

Positive/negative Since the minimum volume is 354.8824, $0.003(T-4)^2+354.8824$ is always positive. The volume of the soda cannot be a negative number.

Zeros The volume of the water never reaches zero. Thus, there is no zero.

Average Rate of Change The volume of the water changes as the temperature changes. The average rate of change (how fast the volume is changing) depends on where you check. The volume is changing only slightly near 4°, but is changing quite rapidly at temperatures different from 4°. The numeric representation below will confirm this statement.

Table 3.2.3

$\Delta T=2$ (from 4 to 6) $\Delta T=2$ (from 30 to 32)

T	4	6	30	32
V	354.8824	354.8944	356.9104	357.2344

$\Delta V=0.012$ $\Delta V=0.324$

The average rate of change between 4° and 6° is $\frac{0.012}{2}=0.006$, and between 30° and 32° it is $\frac{0.324}{2}=0.162$ - about 27 times faster.

The reason for studying data relationships found outside the mathematics classroom is that scientists, engineers, business persons, teachers, etc. must be familiar with the functions that model these relationships. The models include the quadratic function as well as the linear, absolute value, and square root functions already introduced. A second, and probably more important reason for starting each section with problems is that you can learn mathematics by relating a situation you understand to the mathematics being presented in this text. For example, you know that the volume of the soda can is increasing; therefore, you also know that the model of volume of the soda can is increasing. You know that the volume of the soda has a minimum; thus, you know the model has a minimum at the same temperature. Keep these ideas in mind as you study the behaviors of the quadratic function below.

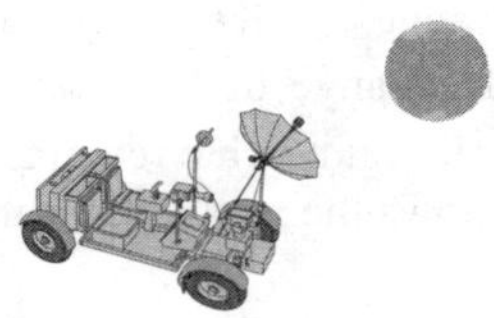

Analysis of $-2.735t^2+50t+5$ - **The Moon Toss**

On Earth, a ball thrown straight upward leaving your hand (Assume your hand is 5 feet above the ground when the ball is released.) at a speed of 50 feet per second will reach a maximum height of 44 feet (ignoring any air resistance). This is about 15-20 feet higher than a two-story house. The ball will hit the ground 3.22 seconds after it is released from your hand. Will the same thing happen on the moon? Hopefully your thinking leads you to the answer of no – not exactly. The gravitational attraction is much less on the moon and this will allow the ball to travel much higher. Certainly the astronauts who spent time on the moon knew that objects thrown would behave differently (but similar) than on earth.

The function that models the height of the ball on the moon is $-2.735t^2+50t+5$ where t is the time in seconds from when the ball leaves your hand. Please note that this function (in symbolic form) doesn't quite look like the function in the "Pop Can Problem". Both functions are quadratic because of the variable squared. The soda can function can be simplified to $0.003T^2-0.024T+354.9304$ (confirm this yourself). Both functions now match the pattern ax^2+bx+c, where x is the function variable and a, b, and c are the parameters of the function. Both forms of the quadratic are studied in this section as well as throughout Chapter Ten.

Domain Because the moon toss function contains no division or any square roots, the normal domain is all real numbers; that is, $t \in (-\infty, \infty)$. The problem domain should be only values for t while the ball is in flight. This flight time is found later.

Maximum/minimum Either with the normal or problem domain the ball reaches a maximum height. Since the function $-2.735t^2+50t+5$ models the height of the ball, it must also have a maximum. The graphical representation of the height function shows the maximum height to be 233.52 feet above the surface of the moon. With a normal domain, the function drops to negative infinity both to the left and right of the maximum. This tells you there is no minimum. If the problem domain is used, the smallest height of the ball is 0 feet.

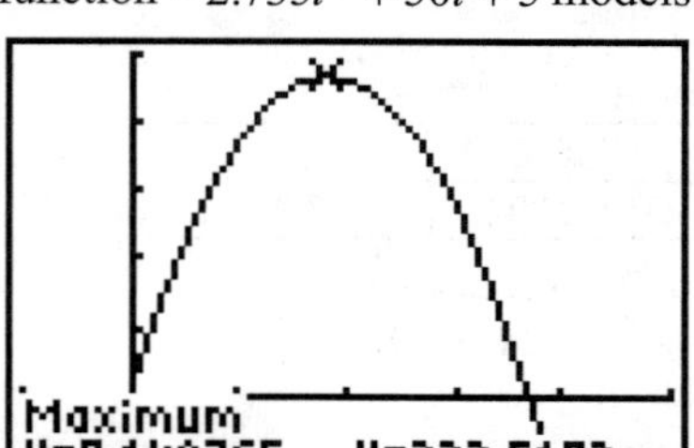

Figure 3.2.3

Range With a problem domain, the function $-2.735t^2 + 50t + 5$ has a maximum of 233.52 and a minimum of 0; this makes the range [0, 233.52]. Using the normal domain, the biggest value of the function is 233.52 and there is no minimum value; this makes the range (–∞, 233.52].

Increasing/decreasing As you visualize the ball, it increases in height -- reaches a maximum height -- then decreases in height. Figure 3.2.3 shows that at 9.14 seconds the ball has reached the maximum. Thus, the height function is increasing before 9.14 seconds and decreasing after 9.14 seconds. Using a normal domain, the moon toss height function is increasing when $t \in (-\infty, 9.14)$ and decreasing when $t \in (9.14, \infty)$.

Zeros Figure 3.2.4 shows one of the zeros to be approximately 18.38 seconds. This is when the ball is on the surface of the moon (the height is 0). It is the only zero on the problem domain because the flight of the ball starts at 5 feet above 0. If the normal domain is used, you can see another zero just to the left of when the ball is thrown. It is at about –0.1 second.

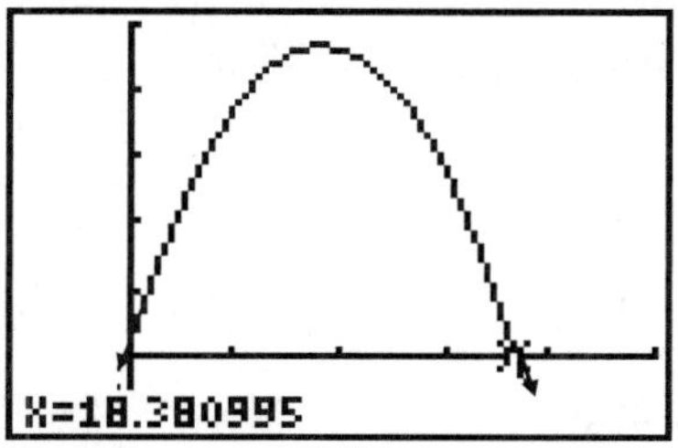

Figure 3.2.4

Positive/negative The height of the ball is positive when the ball is above the surface of the moon. Figure 3.2.4 shows the height function to be positive between the zeros. The function $-2.735t^2 + 50t + 5$ is positive when $t \in (-0.1, 18.38)$. If you ignore the problem domain and use the normal domain, you can also see in Figure 3.2.4 that the function is below zero (negative) when t is larger than 18.38 or when t is less than –0.1. The function $-2.735t^2 + 50t + 5$ is negative when $t \in (-\infty, -0.1) \cup (18.38, \infty)$.

Average rate of change Like the soda can function and like all other quadratic functions, the height function does not have a constant average rate of change. The best you can do is to calculate an average between two time values. Using the numeric representation may be the best way to calculate an average. Table 3.2.4 will help in finding the average rate of change between 0 and 1 second and between 8 and 9 seconds. Between 0 and 1 second the ball increases in height by 47.27 feet while time increases by 1

Table 3.2.4

t	0	1	8	9
height	5	52.27	229.96	233.47

second. The average rate of change is $\frac{\Delta \text{height}}{\Delta t} = \frac{47.27}{1} = 47.27$. Between 8 and 9 seconds the height increases by 3.51 feet while time increases by 1 second. This makes the average rate of change 3.51 feet per second. You may recognize these numbers as the speed or velocity of the ball.

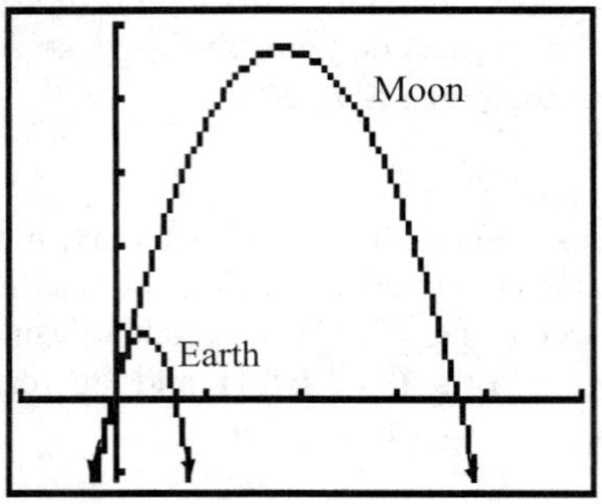

Figure 3.2.5

Figure 3.2.5 shows the height function on earth and on the moon. You can no doubt guess which graph is the height on the moon and which one is height on earth. They both share the point (0, 5) because they were both thrown upward from 5 feet above the surface. Remaining in this section are several examples of identifying behaviors of the quadratic function. The function $d(x+e)^2+f$, where d, e, and f are known real numbers (parameters) and $d \neq 0$, is called a **quadratic function**. In Section 2.3 you analyzed functions numerically. Below is a similar analysis of a quadratic function from a graphical perspective.

Analysis of $-(x+2)^2+4$

When you follow along with the analysis using your calculator, use a decimal or integer window.

$-(x+2)^2+4$

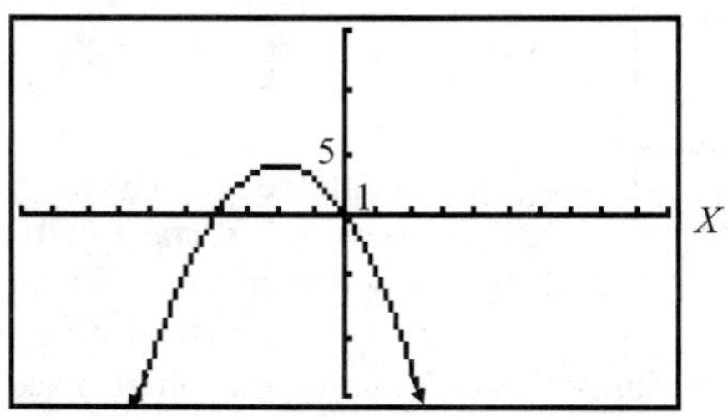

Figure 3.2.6

<u>Increasing/decreasing</u> Don't forget to read the graph from low to high domain values. As your eyes track along the graph, you see that it increases until x reaches −2. It then starts to decrease as you track to the right of −2. The function is increasing when x is on the interval $(-\infty, -2)$ and it is decreasing when x is on the interval $(-2, \infty)$.

<u>Maximum/minimum</u> The function has a maximum when it changes from increasing to decreasing. Visually, this is the peak of the graph. The maximum value of the function is 4, and it occurs when x is a −2. There is no minimum because the function tends toward $-\infty$; therefore, it has no smallest value.

<u>Positive/negative</u> Recall that the function is measured along the vertical axis and the vertical axis contains positive numbers above the x-axis and negative numbers below the x-axis. The visual effect is that when the function is positive, it is above the x-axis. When it is negative, it is below the x-axis. Using trace and a decimal window or integer window, you can see the function take on positive values when x is between −4 and 0. The function is negative when x is to the left of −4 or to the right of 0. The function is positive when x is on the interval $(-4, 0)$ and it is negative when x is on the interval $(-\infty, -4) \cup (0, \infty)$.

<u>Zeros</u> Visually, a function is zero when it is *on* the x-axis. Trace tells you the function is zero when x is −4 and 0.

<u>Average Rate of change</u> If an integer window is used, the rate of change is relatively easy to calculate because when you trace from low to high numbers in the domain, the x value increases by 1. This means that the average rate of change can be found by seeing how much the function changes as you trace from

point to point. Try tracing on your calculator and you will see that the average rate of change is changing from point to point; for example, when x changes from –8 to –7 the function changes by 11; from –7 to –6, the function changes by 9. This makes 9 the average rate of change between –7 and –6. You conclude that the rate of change is changing -- it is not a constant number.

Domain The normal domain is all real numbers because no value of x will cause the function to be non-real. Visually speaking, the trends set at the left and right edges of the viewing domain continue; thus, the x-values continue getting smaller to the left and continue getting bigger to the right of the viewing window.

From a graphical perspective, the domain is the set of all possible x-coordinates on the complete graph of the function.

Range Hopefully, you can "see" that the largest value of the function is the maximum value of 4! All other values of the function are less than 4. To describe all real numbers 4 and less, use interval notation. The range is $(-\infty, 4]$.

From a graphical perspective, the range is the set of all possible function-coordinates on the complete graph of the function.

A Different Direction

In the remainder of this section, the quadratic function will be given in simplified form. For example, the function $(x - 1)^2 + 2$ simplifies to the quadratic function $x^2 - 2x + 3$. Even if you have figured out how to identify some of the function behaviors without looking at the graph, you will find that the remaining quadratic functions in this section may require a graph and trace or the numeric representation to find these behaviors. In Chapter Ten, the relationship between the two forms of the quadratic function will be developed algebraically. The new form of the function is called **general form** and it looks like $ax^2 + bx + c$.

Recall that the set of values taken by the function is called the **range** of the function. The normal domain of the quadratic function is all real numbers because there is no division or square root. The range of a quadratic function is not so obvious. The graphing calculator will come to your aid here.

The graphs in Figure 3.2.7 demonstrate that the quadratic function is a parabola opening up or down.

$x^2 - 5$

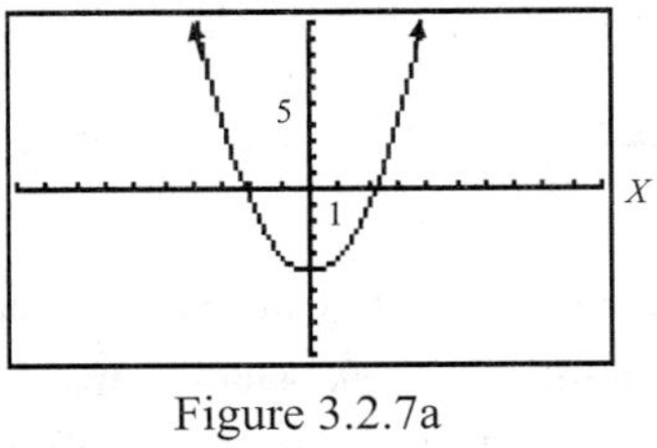

Figure 3.2.7a

$-x^2 + 2x - 1$

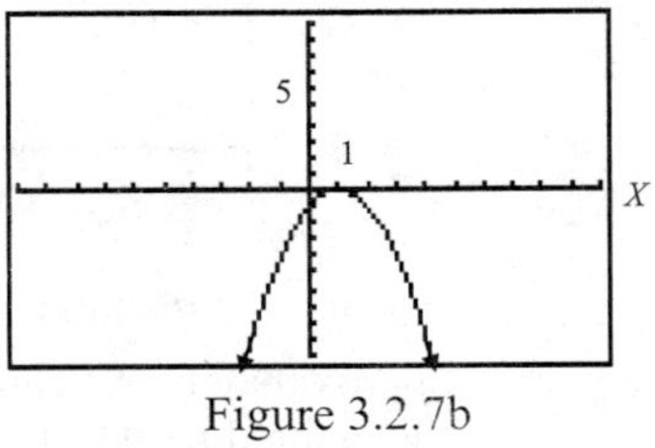

Figure 3.2.7b

$2x^2 + 4x - 5$

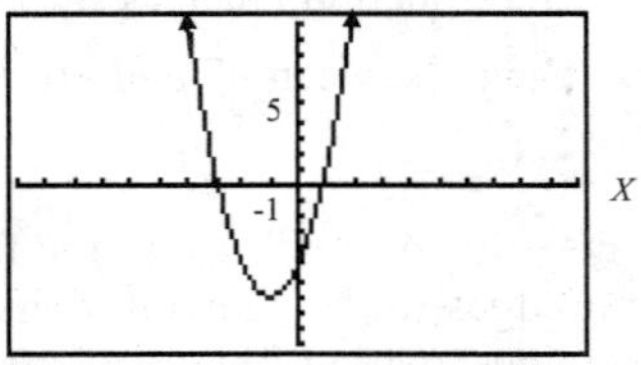

Figure 3.2.7c

$\frac{3}{4}x^2 - 2x + 0$

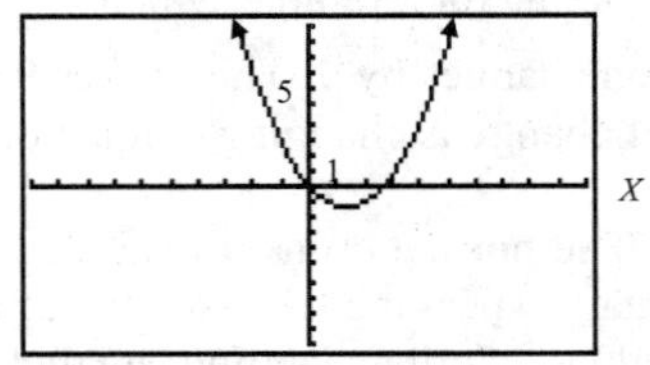

Figure 3.2.7d

If the graph opens up, it has a low point. At this point, the quadratic function reaches its lowest value. The **value** of the quadratic **function** at this point is called a **minimum.** The numbers in the range start at this value and only get bigger, because the range is the set of all values of the function and the function only rises in value from the minimum. If the graph opens down, it has a high point. At this point, the quadratic reaches its highest value. The **value** of the quadratic function at this point is called a **maximum.** The range numbers start at this value and only get smaller. In graphing quadratic functions, the viewing window should display this maximum or minimum.

If the window displays the maximum or minimum, where the graph crosses the x-axis, the function-axis, and left and right trends have been established; a complete graph is displayed.

If you study all of the graphs in Figure 3.2.7, you should notice that if the coefficient of x^2 is positive ($a > 0$), then the graph opens up and has a minimum. If the coefficient of x^2 is negative ($a < 0$), the graph opens down and has a maximum. Try graphing several quadratic functions with the calculator to confirm this statement.

Example 1: Find the range of $-x^2 + 6x - 15$.

Solution: Graph the function and locate the maximum point. Remember, the range is all values of the function at and below the maximum.

$-x^2 + 6x - 15$

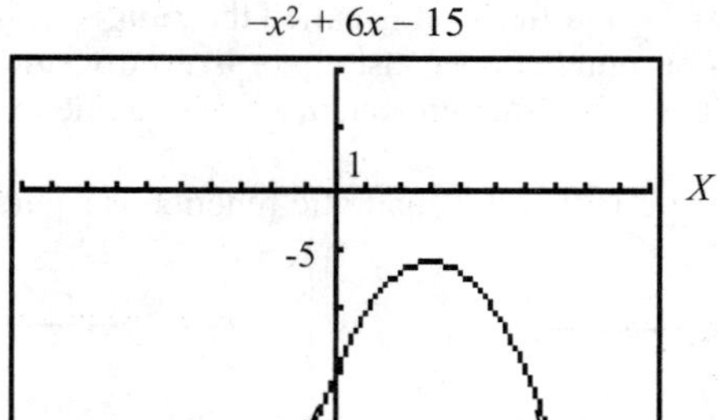

Figure 3.2.8

Because the maximum point is found using trace, it may be approximate. It is approximately –6. Thus, the range is all numbers –6 and smaller. Just a reminder, the graph does not stop at the bottom of the viewing range; it continues downward without limit. Remember, the mathematical symbol for decreasing without limit is

$-\infty$. The range is the interval $(-\infty, -6]$. Note: A decimal window will many times give you the exact value of the maximum.

⁂

Approximate answers are acceptable! You do not have to find a decimal window where trace gives the exact answer. At this time, no limitations are put on you as far as how accurate your answer must be. Error is discussed later. After the concept is addressed, any answer you give must meet the error specifications set.

Sometimes quadratic functions cross the x-axis. The value or values for x when the function crosses the x-axis are called **zeros of the function**. When a function crosses the x-axis, the function has a value of zero!

Example 2: Find the zeros of $x^2 - 3x - 2$.

Solution: Graph $x^2 - 3x - 2$ on the viewing domain of $[-10, 10]$ and a viewing range of $[-10, 10]$. The zeros are approximately -0.6 and 3.6, based on coordinates from trace.

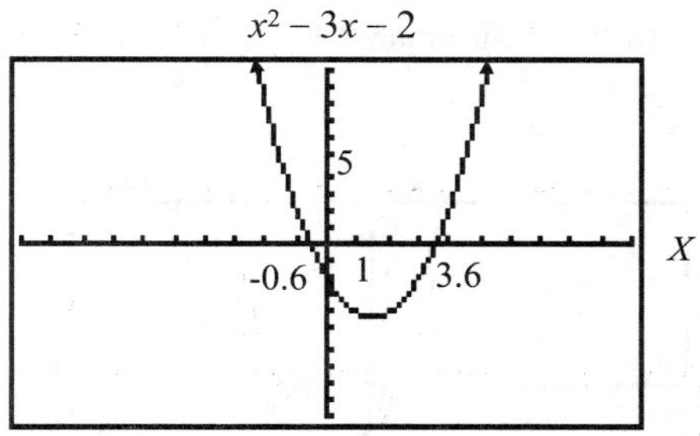

Figure 3.2.9

The zeros can also be approximated from the numeric representation of the function as shown below in Table 3.2.5. The function is approximately zero when x is -0.5 and 3.5.

Table 3.2.5

x	–2	–1	–0.5	0	1	2	3	3.5	4
$x^2 - 3x - 2$	8	2	–0.25	–2	–4	–4	–2	–0.25	2

⁂

Example 3: Find the maximum value of the quadratic function $-3x^2 + 5x - 1$.

Solution: Graph the function $-3x^2 + 5x - 1$ on the viewing domain $[-10, 10]$ and viewing range $[-10, 10]$. Use trace to find the maximum value of the quadratic function $-3x^2 + 5x - 1$. It is approximately 1.1.

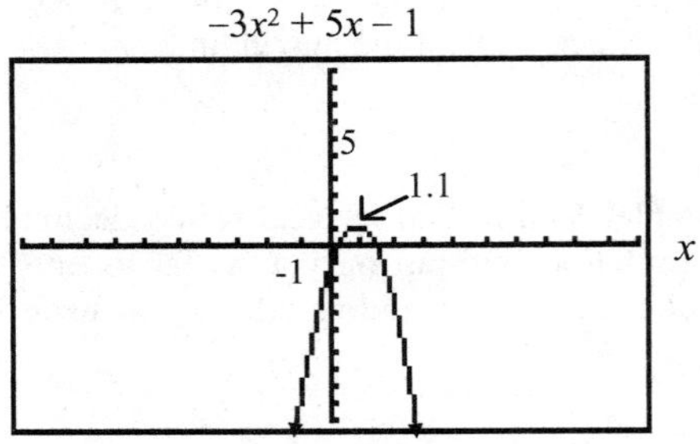

Figure 3.2.10
Graphing on a decimal window may give the exact maximum.

⁂

Example 4: Find all values of x where $x^2 - 3x - 2$ is increasing.

Solution: As you read the graph from low to high values of x, look for values of x where the function increases; that is, look where the graph rises. Graph $x^2 - 3x - 2$ and find the x-coordinate of the minimum. The function is increasing to the right of this minimum.

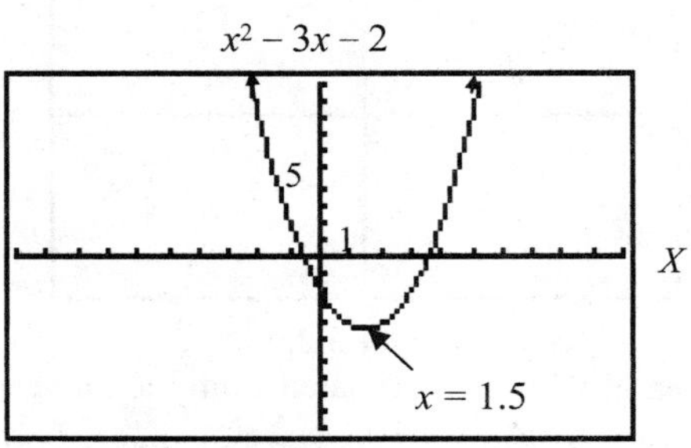

Figure 3.2.11

From the graph above, the position when the graph first starts to increase is at 1.5. The graph (function) keeps getting higher and higher as x gets larger. Thus, the function is increasing for all values of x from 1.5 to ∞. The function increases on the interval $(1.5, \infty)$. The numeric representation confirms the notion that the function is increasing on the interval $(1.5, \infty)$.

Table 3.2.6

x	−2	−1	0	1	**1.5**	**2**	**3**	**4**	**5**
$x^2 - 3x - 2$	8	2	−2	−4	−4.25	−4	−2	2	8

⁂

3.2 EXERCISES

Many times you can work the exercises by algebraic, numeric, or graphical methods. You can learn about mathematics no matter which method you use. Part of the learning process is for you to decide which method is best for you to use. Please read the text carefully before you do the exercises. Read the **next assigned section** after you finish the assignment below.

A Quick Look To The Future

-4. Is $|x+2|-1$ always positive?

-3. Is $-\sqrt{x-2}+1$ always negative?

-2. If you had the freedom to give a name to the symbols $\sqrt{x+1}-|2x|+x^2$, what name would you give them? Why?

-1. Describe how you might add $(3x^2+5x-2)$ and $(2x^2-4x+7)$.

Previous Topics Revisited

1. Does the graphical representation of $2x-6$ rise or fall as x increases?

2. Does the function $-\frac{1}{2}x+2$ increase or decrease as x increases?

3. What is the value of $\frac{2}{3}x-2$ when $x=0$?

4. Do $\sqrt{x}-5$ and $\sqrt{x-5}$ have the same graphical representation?

5. What one behavior do $-\frac{1}{2}x+4$ and $-\sqrt{x+3}$ have in common?

6. What is the minimum value of the function $|x+2|-4$?

New Topic Exercises

In Exercises 7 - 11, find the approximate zeros of each function. (Exact zeros are also acceptable.)

7. $(x)^2-9,\ \ (x+1)^2-9,\ \ (x+2)^2-9,\ \ (x+3)^2-9,\ \ (x+4)^2-9$

8. $-(x)^2+9,\ \ -(x+1)^2+9,\ \ -(x+2)^2+9,\ \ -(x+3)^2+9,\ \ -(x+4)^2+9$

9. $(x)^2-4,\ \ (x-1)^2-4,\ \ (x-2)^2-4,\ \ (x-3)^2-4,\ \ (x-4)^2-4$

10. $-(x)^2+4,\ \ -(x-1)^2+4,\ \ -(x-2)^2+4,\ \ -(x-3)^2+4,\ \ -(x-4)^2+4$

11. $-x^2+5x+19,\ \ 2x^2-15x+21,\ \ -x^2-3x+57,\ \ 7x^2-13x+43$

In Exercises 12 - 14, find approximate values for x that cause the functions to be negative. (Exact answers are also acceptable.)

12. $(x)^2-9,\ (x+1)^2-9,\ (x+2)^2-9,\ (x+3)^2-9,\ (x+4)^2-9$

13. $-(x)^2+4,\ -(x-1)^2+4,\ -(x-2)^2+4,\ -(x-3)^2+4,\ -(x-4)^2+4$

14. $2x^2-3x+5,\ -\frac{1}{2}x^2+3x+5,\ x^2+3x$ [illegible]

In Exercises 15 - 17, find when each function is d[illegible] acceptable.)

15. $-(x+4)^2-3,\ -(x+3)^2-3,\ -(x+2)$ [illegible]

16. $(x-5)^2+2,\ (x-4)^2+2,\ (x-3)^2+2,\ (x-$ [illegible]

17. $2x^2-5x+9,\ -3x^2-x+14,\ \frac{1}{2}x^2+$ [illegible]

In Exercises 18 - 20, find the maximum or mini[illegible] answers are acceptable.)

18. $-(x+2)^2+4,\ -(x+2)^2+6,\ -(x+$ [illegible]

19. $(x-1)^2+5,\ 3(x-1)^2+5,\ 7(x-1)^2$ [illegible] $(x-1)^2+5,\ (x+3)^2+5$

20. $-2x^2+4x+3,\ 2x^2-5x+3,\ \frac{1}{5}x^2+3x+3,\ -\frac{1}{3}x^2-6x+3,\ 0.2x^2-x+3$

Find the domain and range of the functions in Exercises 21 and 22.

21. $-2(x+6)^2+16,\ 2(x+6)^2+16,\ \frac{1}{2}(x-7)^2-23,\ -\frac{1}{4}(x-3)^2-15$

22. $x^2+x+1,\ -x^2+3x-5,\ \frac{1}{3}x^2-4x+1,\ 17x^2-12x+8$

In Exercises 23 - 24, find the average rate of change of each function between the given numbers.

23. $-(x+3)^2-5$, between 1 and 5, between 1 and 2, between 1 and 1.5, between 1 and 1.1

24. $-3x^2-5x-4$, between -5 and -3, between -5 and -4, between -5 and -4.5

25. The acceleration due to gravity on Mars is -12.9 feet per second squared. This makes the model for the height of a ball thrown straight upward from 5 feet above the surface of Mars at 50 feet per second $-6.45t^2+50t+5$. Based on this model, answer the following questions:

a. What is the maximum height of the ball?

b. When does the ball hit the surface of Mars?

c. When does the ball change from rising (increasing in height) to falling (decreasing in height)?

d. How high is the ball at 2 seconds?

e. Find the average speed of the ball between 0 and 1 second. Calculate it again for the last second of flight.

f. Approximate when the ball is traveling at the fastest speed (rate of change).

g. Does the ball ever have a speed of 0 feet per second (other than after it is on the ground)? If yes, when?

26. Create a quadratic function that opens down.

27. Create a quadratic function that opens down and has a maximum value of 7.

28. Create a quadratic function that opens up and has a minimum of 7.

29. Create a quadratic function that has a zero of 2.

30. Create a quadratic function that has no zero.

31. Create a quadratic function that is always negative.

32. Given the function $ax^2 + bx + c$, what conditions on the parameter a will cause it to have a maximum? A minimum?

33. Where does the function $ax^2 + bx + c$ cross the function-axis?

From Mathematics to English

34. Describe a quadratic relationship.

35. Describe the maximum value of a quadratic function.

36. What are the differences between a linear and quadratic function?

37. Throw a ball up and then describe the velocity (rate of change) of the ball as it moves.

38. Go to a water fountain and turn on the water. Describe any mathematics you see.

39. If the ball in Exercise 37 takes 1 second from when it leaves your hand until it reaches the high point, and then it takes 1.5 seconds to hit the floor, describe when the ball is increasing in height and when it is decreasing in height.

40. Describe all possible heights the ball in Exercise 39 is above the floor, if it reaches 10 feet in 1 second.

41. After reading this section, make a list of questions that you want to ask your instructor.

42. Continue in your daily journal and make an entry. In addition to your normal entry on thoughts about the mathematics in this section, list at least two positive comments about what you have learned about this topic.

43. In paragraph format, summarize the material in this section of the text in your daily journal.

44. Describe how your classroom instructor made this topic more understandable and clear.

45. After reading the text and listening to your instructor, what do you not understand about this topic?

46. What is the most important idea presented in this section? Why?

47. Write a paragraph about how you think you might use the quadratic function in another course.

From English to Mathematics

In the next three exercises, convert the English statement to an equivalent mathematical statement.

48. four times a number x squared plus three times the number x

49. the number x squared

50. ten times a number squared plus three times the number minus four

Explorations

51. List four different occurrences in the real world where you have seen the shape of a parabola. (A parabola is different than a semi-circle.)

52. Investigate many functions of the form $d(x+e)^2+f$ on your calculator by graphing the following functions: (Graph each group of four on the same screen before moving to the next group. Write what you think the numbers d, e, and f do to the graph of x^2.)

a.	x^2	$2x^2$	$5x^2$	$8x^2$
	x^2	$0.8x^2$	$0.6x^2$	$0.2x^2$
	x^2	$-3x^2$	$-0.4x^2$	$-x^2$
b.	x^2	$(x+2)^2$	$(x-5)^2$	$(x+6)^2$
	x^2	$(x-7)^2$	$(x+4)^2$	$(x-2)^2$
c.	x^2	x^2+4	x^2-6	x^2+1
	x^2	x^2-7	x^2+5	x^2-2
d.	x^2	$(x+3)^2+4$	$(x+5)^2+3$	$(x+1)^2+6$
	x^2	$(x-3)^2+4$	$(x-5)^2+3$	$(x-1)^2+6$
e.	x^2	$2(x-3)^2-5$	$-2(x-3)^2-5$	$\frac{1}{2}(x+6)^2+2$

Summarize your investigation by identifying what ***visual effect*** d, e, and f have on the graph of x^2.

3.3 An Introduction to the Analysis of the Absolute Value Function $d|x + e| + f$

The Alberta Clipper

When a fast moving "Alberta Clipper" cold front drops down from Alberta, Canada and sweeps across a relatively small path through the Northeast quarter of the US, the temperature drops quickly by as much as 25° and then recovers quickly. Below is the numeric representation of the temperature (T) in Columbus, Ohio during an "Alberta Clipper". The day the front arrived, the daily high had been 18° at 4 PM.

Table 3.3.1

Time (t) zero is midnight.

t	0	1	2	3	4	5	6	7	8	9	10
T	2	0	–2	–4	–6	–8	–6	–4	–2	0	2

If a meteorologist wants to model this data with a function, what function can be used? The numeric data suggests that the linear function cannot be used because the temperature (the function) is both decreasing and increasing -- not a behavior of the linear function. Can the data be modeled by a quadratic function? Quadratic functions both decrease and increase on the normal domain. However, the quadratic function may not be a good choice because the average rate of change of the quadratic is not constant. If you check the rate of change of the function in the numeric representation, you will find that from midnight to 5 am the temperature decreased by 2° per hour and after 5 am it increased by 2° per hour. This is not one of the behaviors of the quadratic function. If not linear and if not quadratic, then what? The graphical representation of the data may give a clue as to the type of function that can be used to model the temperature.

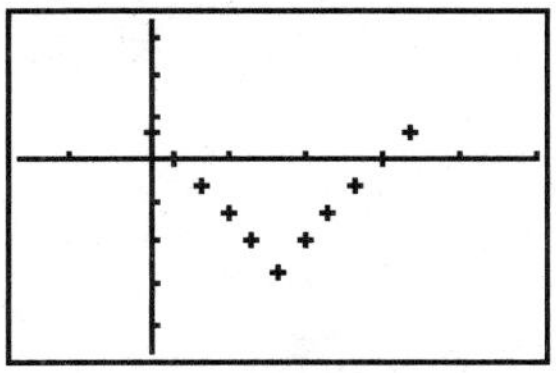

Figure 3.3.1a

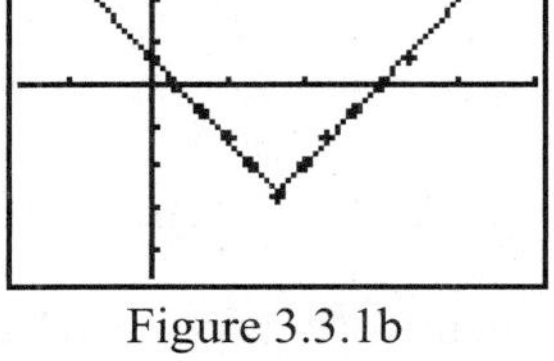

Figure 3.3.1b

The shape matches that of the absolute value function studied in Section 2.1. Figure 3.3.1b shows the data and the graph of the absolute value function $2|t - 5| - 8$ on the same coordinate system. The absolute value function $2|t - 5| - 8$ matches the data. The shape is a V and the graph of any absolute value function of the form $d|x + e| + f$ will always have this shape. You may want to confirm this conjecture by graphing several absolute value functions of the form $2|t-5|-8$. When an absolute value function looks like $d|x+e|+f$, it is in **standard form** and the function parameters are *d*, *e*, and *f*.

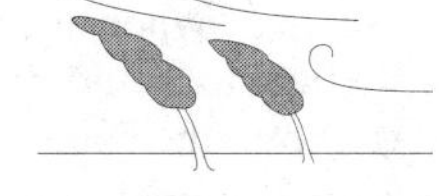

Analysis of $2|t-5|-8$ - **Alberta Clipper**

Domain The function $2|t-5|-8$ has no division or square root; thus, there is no operation that can cause the function to be a non-real number. The normal domain of $2|t-5|-8$ is $(-\infty, \infty)$. The problem domain should be about as long as the clipper lasts, around 24 hours. A reasonable problem domain for the "Alberta Clipper" is [0, 24], with the time the clipper arrives as 0 hours.

Maximum/minimum From a graphical perspective, a minimum value of a function is a low point on the graph. A high point is a maximum. The graphical representation of $2|t-5|-8$ shows a low value and no high value. The lowest value of the function is –8. This is the minimum -- the "low" temperature. With a normal domain, there is no maximum because the function rises to ∞ both to the left of 5 and the right of 5. Using a problem domain, the maximum value may be the temperature as the cold front arrives or as it leaves.

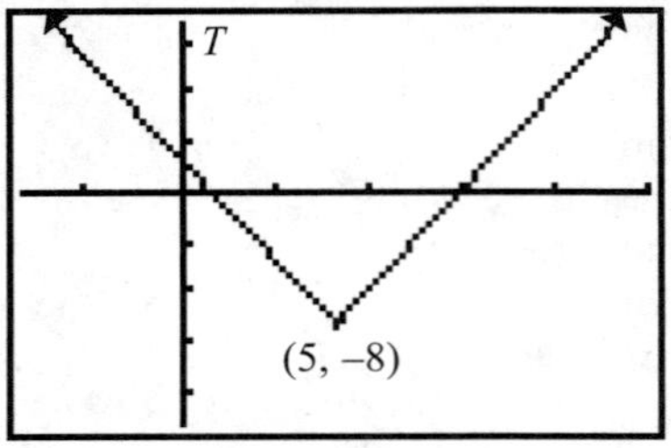

Figure 3.3.2

Range Since the minimum is –8 and the function tends toward ∞ both to the left and right of 5 hours, the range is [–8, ∞). In the "Alberta Clipper Problem", the front arrived at 4 PM with a temperature of 18°. After a 24 hour period the temperature is at 14°; thus, the range in this problem setting is [–8, 18]. Please confirm this by making a table on your calculator.

Increasing/decreasing The function $2|t-5|-8$ models the temperature as a cold front passes through an area. The temperature decreases until 5 am and then it increases. Before 5 am the temperature drops -- after 5 am the temperature rises. The function that models the temperature behaves the same as the temperature! The temperature function is decreasing when $t \in (-\infty, 5)$ and it is increasing when $t \in (5, \infty)$.

Zeros Either looking at the numeric representation of the data or the graphical representation of the function as shown in Figure 3.3.3, the temperature is zero on two occasions, at 1 am and 9 am.

$2|t-5|-8$

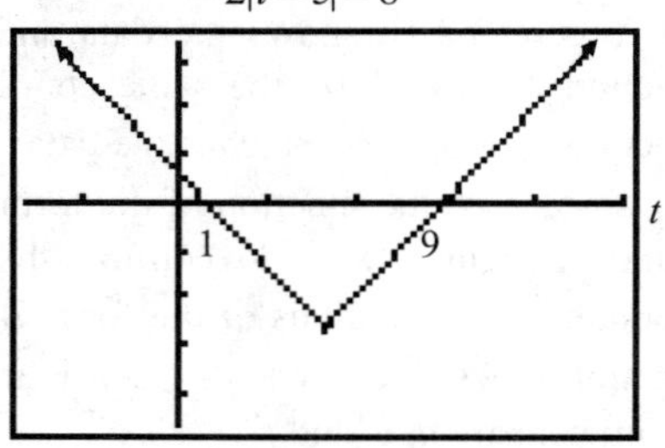

Figure 3.3.3

There are no other zeros because the function tends toward ∞ before the first zero and after the last zero.

Positive/negative The graphical representation of the temperature function (T) is very useful in "seeing" when the temperature (T) is positive or negative. Because the temperature is measured on the vertical axis and positive numbers on the vertical axis are above 0, the function is positive when the graph is above the t-axis. The graph in Figure 3.3.3 shows positive temperatures before 1 or after 9. Using the normal domain, $2|t-5|-8$ is positive when $t \in (-\infty, 1) \cup (9, \infty)$. The temperature function is below zero (negative) when the graphical representation is below the t-axis. Figure 3.3.3 shows the temperature is negative when $t \in (1, 9)$ -- between 1 am and 9 am.

Rate of change The opening discussion of the "Alberta Clipper Problem" indicates that the rate of change of the temperature is –2 before 5 am, and 2 after 5 am. You should calculate the rate of change to the left of 5 and then to the right of 5 to confirm this statement.

Several "problems" follow in this section. The reason for studying these data relationships found outside the mathematics classroom is that scientists, engineers, business persons, teachers, etc. must be familiar with the functions that model these relationships. The models include the linear, quadratic, absolute value, exponential, and square root functions. A second, and probably more important reason for starting each section with problems is that you can learn mathematics by relating a situation you understand to the mathematics being presented. For example, you know that the when the temperature is decreasing (see Alberta Clipper Problem), you also know that the model of the temperature is decreasing. You know when the temperature reaches a minimum, the function that models the temperature is at a minimum. Further, the concept of the minimum as being the smallest temperature leads directly to the concept of the minimum value of the function modeling the temperature. Keep these ideas in mind as you study the behaviors of the absolute value function below. Most of the new ideas presented are developed from a problem situation; thus, you should be able to relate these mathematical ideas to something that makes sense from the real world.

The family of functions $d|x+e|+f$ comes from the absolute value function $|x|$ where d, e, and f (the function parameters) are real numbers and $d \neq 0$. Below are the numeric and graphic representations of the function $|x|$. You know that the absolute value of a real number is always positive or zero. This comes

Table 3.3.2

x	–4	–3	–2	–1	0	1	2	3	4
$\lvert x \rvert$	4	3	2	1	0	1	2	3	4

$|x|$

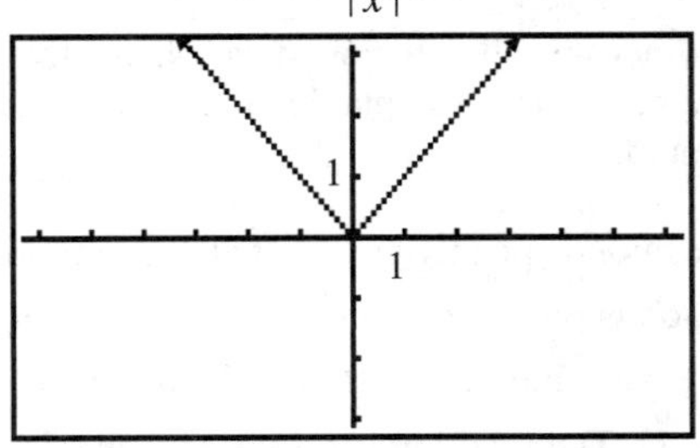

Figure 3.3.4

from the definition $|x| = \begin{cases} x, & \text{if } x \geq 0 \\ -x, & \text{if } x < 0 \end{cases}$. It leaves the value of x as x if it is larger than or equal to 0, and it replaces x with its opposite if x is negative. Further, you will note that on the domain $(-\infty, 0)$, the absolute value function becomes the function $-x$. Looking at the graphical representation in Figure 3.3.4, you see the graph of $-x$ when $x < 0$. On the domain $[0, \infty)$, the absolute value function becomes x. The graph above shows the function x, when $x \geq 0$. These two pieces together form the graph of the absolute value function (V). The parameters d, e, and f in the function $d|x+e|+f$ transform the graph in Figure 3.3.4 to a V in other positions.

The graph of the absolute value function is in the shape of the letter V, not a parabola as is the quadratic function. Just as the quadratic has a minimum or maximum, the absolute value function has a maximum or a minimum. A complete graph of the absolute value function must show this maximum or minimum as well as zeros.

Below is the analysis of the absolute value function $2|x+1|-4$. Please note all of the behaviors and try to make connections between the numbers in the symbolic representation of the function and behaviors described below. The analysis is from a graphical perspective; thus, the graph is needed first.

$$2|x+1|-4$$

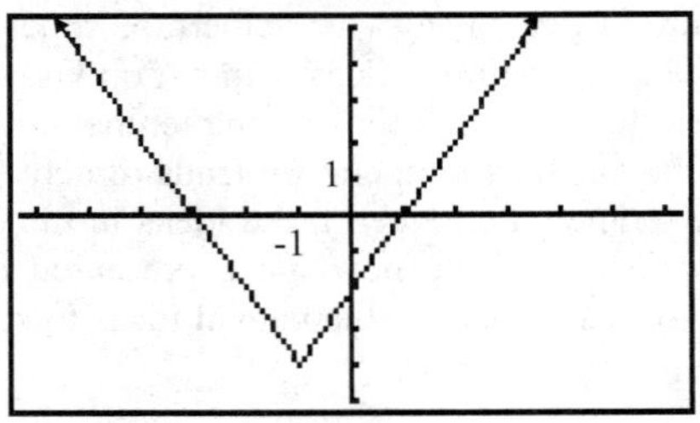

Figure 3.3.5

Analysis of $2|x+1|-4$

Increasing/decreasing If you move the trace cursor to the left side of the viewing window and then trace to the right, you will see the cursor dropping until the x-coordinate is –1. At this point the cursor starts to rise. Since the trace cursor moves along the graph, it is mimicking the behavior of the graphical representation. The function is decreasing when x is on the interval $(\infty, -1)$. It is increasing when x is on the interval $(-1, \infty)$. Remember to read the graph from low to high values of the domain.

Maximum/minimum Hopefully, you can "see" that the function has a minimum. Trace and a decimal window will tell you it is –4. At this time, an approximate value for the minimum is also acceptable; thus, you can use any window and trace to find the minimum.

Positive/negative Since any function is positive when it is above the x-axis, the trace key shows you that this function is positive when x is left of –3 or to the right of 1. As you trace between –3 and 1, you find the function to be negative; that is, the function is below the x-axis when x is on the interval $(-3, 1)$. The function is positive when x is on the interval $(-\infty, -3) \cup (1, \infty)$ and it is negative when x is on the interval $(-3, 1)$.

Zeros You now know when the function is negative and when it is positive. Since it is positive or negative everywhere except –3 and 1, it has to be zero at these values. Visually, you know the function is zero on the x-axis. The graphical representation of the function is on the x-axis at –3 and 1. The zeros are –3 and 1.

Domain/range The normal domain is all real numbers. To the left and to the right of the window used for the graph, the trends established continue. That is to say that the x-coordinates go to $-\infty$ left of the window and toward ∞ to the right of the window. Since the domain is all x-coordinates on the complete graph, the domain is $(-\infty, \infty)$. Looking at this analytically, the function $2|x+1|-4$ has no square roots or divisions; thus there is no operation that will cause it to be non-real. Since the function is always real, x can be all real numbers. The range is the set of all function-coordinates on the complete graph. Since the smallest function value is -4 and all other values of the function are larger, the range is $[-4, \infty)$.

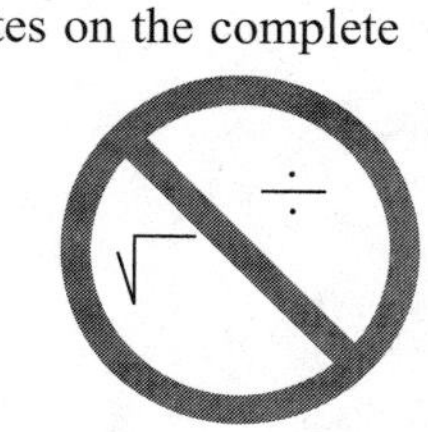

Rate of change Using an integer or decimal window, as you trace from left to right, you discover that the function changes by -1 as x changes by 1. The rate of change is $\dfrac{\text{change in } 2|x-1|+4}{\text{change in } x} = \dfrac{-1}{1} = -1$

This is true if you check when x is on the interval $(-\infty, -1)$. If you check the rate of change of the function when x is on the interval $(-1, \infty)$, you will find the function changes by 1 each time x changes by 1. The rate of change is now 1. The rate of change is -1 when x is on the interval $(-\infty, -1)$ and the rate of change is 1 when x is on the interval $(-\infty, -1)$.

Example 1: Find the approximate zeros of $\frac{1}{2}|x-2|-5$. Also find the value of the function when $x = 0$.

Solution: Graph $\frac{1}{2}|x-2|-5$ and locate the zeros with the trace feature of the calculator. The value of the function when x is 0 can also be found with the trace key or it can be found analytically as well.

$\frac{1}{2}|x-2|-5$

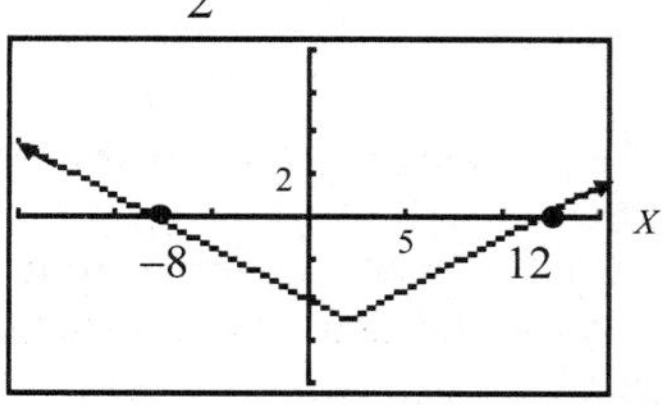

Figure 3.3.6

The zeros are at -8.0 and another at 12.0. The value of the function when $x = 0$ can be found from the function; let $x = 0$ and calculate.

$$\frac{1}{2}|x-2|-5$$

$$=\frac{1}{2}|0-2|-5$$

$$=\frac{1}{2}|-2|-5$$

$$= \frac{1}{2}(2) - 5$$

$$= 1 - 5$$

$$= -4$$

The value of the function is –4 when $x = 0$. The graph crosses the function-axis at –4.

⁂

Example 2: What is the approximate range of $-0.6|x - 3.5| + 5.9$?

Solution: Graph $-0.6|x - 3.5| + 5.9$ and find the maximum value. The range will be all numbers at and below the maximum.

$-0.6|x-3.5| + 5.9$

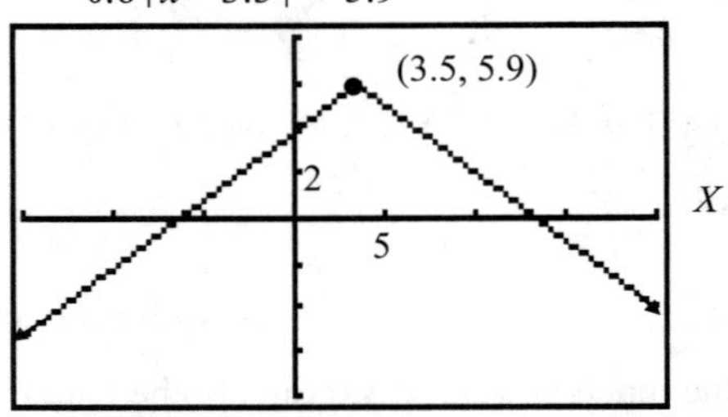

Figure 3.3.7

The maximum value is 5.9. Since the function only has values of 5.9 and lower, the range must be $(-\infty, 5.9]$.

⁂

Example 3: When is $-2|x + 3| + 5$ increasing and when is it decreasing?

Solution: As x gets larger, (that is, read the graph from left to right) look at the graph to see what values of x cause the graph to rise (increase) and what values of x cause the graph to drop (decrease). The graph of $-2|x + 3| + 5$ is shown below.

$-2|x+3| + 5$

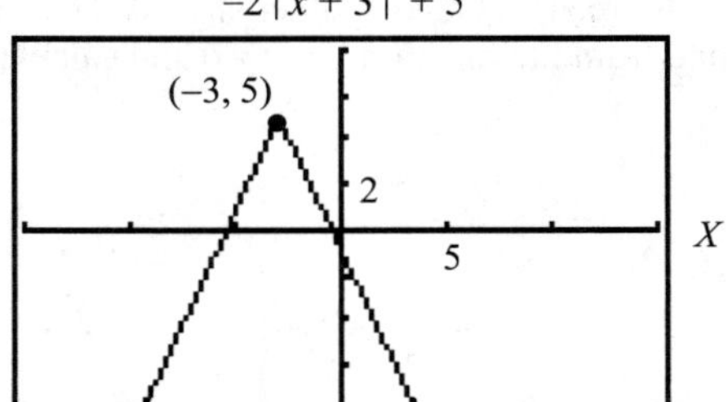

Figure 3.3.8

The graph reaches a maximum at the point (–3, 5), so, when x is left of –3, the graph is increasing. When x is to the right of –3, the graph gets lower and lower; thus, it is decreasing. The function is increasing on (–∞, –3) and it is decreasing on the interval (–3, ∞). This question can also be answered from looking at the numeric representation of the function. Notice that the function values rise until x is –3. The function values drop as x becomes larger than –3. Please note a connection between the function parameters e and f and the answers to the question posed in this example.

Table 3.3.3

x	–7	–6	–5	–4	–3	–2	–1	0	1	2
$-2\|x+3\|+5$	–3	–1	1	3	5	3	1	–1	–3	–5

⁂

Example 4: What is the rate of change of the function $-2|x+3|+5$ on the left branch as well as the right branch?

Solution: The simplest way to solve this problem may be to graph the function on an integer window where trace gives integer values for x. As you trace from left to right, the function increases by 2 as x changes by 1 until x has a value of –3. To the right of –3, the function decreases by 2 units as x increases by 1. The rate of change when x is left of –3 is 2, and when x is to the right of –3, the rate of change is –2. You are encouraged to also solve this problem using the numeric representation of the function.

⁂

If you can answer the above question by looking at the symbolic representation of the function, you are insightful. If you think the above solution is the only way of answering the question, concentrate a little more on connections between symbolic form and characteristics of functions. Usually there are several ways of answering a question. You have learned something about mathematics even if you only know one solution to a problem!

3.3 EXERCISES

Many times you can work the exercises by algebraic, numeric, or graphical methods. You can learn about mathematics no matter which method you use. Part of the learning process is for you to decide which method is best for you to use. Please read the text carefully before you do the exercises. Read the **next assigned section** after you finish the assignment below.

A Quick Look To The Future

-4. What kind of number is $\sqrt{x}$? Be specific.

-3. If the name of the function $2x^2 + 1$ is g, what is the variable in the function g?

-2. Can $3x^2 + 5x^2$ be written in a simpler form?

-1. Can $(3x^2)(5x^2)$ be written in a simpler form?

Previous Topics Revisited

1. Find the maximum value of $-(x + 6)^2 - 3$.
2. Find the range of $0.2x^2 + 4x - 9$.
3. Find the zeros of $0.1x^2 + 2x - 5$.
4. As x increases, does $\frac{4}{3}x - 9$ increase or decrease?
5. What is the value of $6x - 27$ when $x = 0$?
6. Find the range and zero of $\sqrt{1-x}$.

New Topic Exercises

7. The temperature in the "Alberta Clipper Problem" was 18° when it started and the temperature dropped to –8° 13 hours later, this is a drop of –26°. If a similar front moves through with the 4 PM beginning temperature of 33°, the model for the temperature becomes $2|t-5|+7$. For this clipper, find the minimum temperature, when the temperature reaches the minimum, when the temperature is dropping (decreasing), when the temperature is rising (increasing), how fast the temperature is decreasing and then increasing, when the temperature is zero, when it is negative, and when it is positive?

8. An airplane flies between two cities and the flight requires 48 minutes. It ascends for the first 24 minutes and descends for the last 24 minutes. The function that models the height of the airplane at any time t in the flight is $-1000|t-24|+24000$. What is the problem domain of this function? What is the maximum height of the plane? Using the problem domain, what is the range? How fast is the plane rising? How fast is it descending? What are the zeros of the height function? When is the height function positive? When is it negative?

In Exercises 9 - 10, find the maximum or minimum.

9. $3|x+2|-4,\ \ 3|x+2|-1,\ \ 3|x+2|+1,\ \ 3|x+2|+5,\ \ 3|x+2|+56,\ \ 5|x-2|-4$

10. $-\frac{1}{2}|x+3|+2,\ \ -\frac{1}{2}|x+3|+5,\ \ -\frac{1}{2}|x+3|-6,\ \ -\frac{1}{3}|x+3|+0,\ \ -5|x+3|+73$

In Exercises 11 - 13, find the zeros.

11. $1|t+4|-3,\ \ 1|t+3|-3,\ \ 1|t+2|-3,\ \ 1|t+1|-3,\ \ 1|t+0|-3,\ \ 1|t-1|-3,\ \ 1|t-1|-2$

12. $2|x-1|-6,\ \ 2|x-1|-4,\ \ 2|x-1|-2,\ \ 2|x-1|,\ \ 2|x-1|+2,\ \ 2|x-1|+4$

13. $-3|t+4|+24,\ \ 2|x-1|-15,\ \ \frac{1}{2}|t+3|+\frac{1}{3},\ \ -5|x-2|-1,\ \ 3-|t+1|,\ \ |x+4|-\pi$

In Exercises 14 - 16, find the range.

14. $-|x+3|+5,\ \ -|x+3|+2,\ \ -|x+3|+0,\ \ -|x+3|-2,\ \ -|x+3|-6$

15. $2|t-1|-7,\ 3|t+2|-7,\ 5|t-3|-7,\ -4\left|t-\frac{1}{2}\right|-7,\ -1|t+4|-7,\ -\frac{2}{5}|t-3|-7$

16. $2|x-1|+4,\ -\frac{3}{4}\left|x+\frac{1}{2}\right|,\ 7|x+5|-9,\ \frac{1}{10}|x-3|-2$

Find when each function is increasing and find when each is decreasing.

17. $3|x+2|-4,\ 3|x+2|-1,\ 3|x+2|+1,\ 3|x+2|+5,\ 3|x+2|+56,\ 5|x-2|-4$

18. $-\frac{1}{2}|x+3|+2,\ -\frac{1}{2}|x+3|+5,\ -\frac{1}{2}|x+3|-6,\ -\frac{1}{3}|x+3|+0,\ -5|x+3|+73$

Use interval notation to describe when each of the following functions is positive and when it is negative.

19. $1|t+4|-3,\ 1|t+3|-3,\ 1|t+2|-3,\ 1|t+1|-3,\ 1|t+0|-3,\ 1|t-1|-3,\ 1|t-1|-2$

20. $2|x-1|-6,\ 2|x-1|-4,\ 2|x-1|-2,\ 2|x-1|,\ 2|x-1|+2,\ 2|x-1|+4$

21. $-3|t+4|+24,\ 2|x-1|-15,\ \frac{1}{2}|t+3|+\frac{1}{3},\ -5|x-2|-1,\ 3-|t+1|,\ |x+4|-\pi$

Find the rate of change for both branches of each function.

22. $5|x-1|+4,\ 3|x-1|+4,\ 1|x-1|+4,\ \frac{1}{2}|x-1|+4,\ \frac{1}{5}|x-1|+4,\ \frac{1}{10}|x-1|+4$

23. $-\frac{1}{4}|t+1|-2,\ -\frac{1}{2}|t+1|-2,\ -1|t+1|-2,\ -2|t+1|-2,\ -5|t+1|-2$

24. $14|x-3|+8,\ -14|x-3|+8,\ 0.01|x-3|+8,\ -0.01|x-3|+8$

Create absolute value functions that meet the criteria described below in Exercises 25 - 33.

25. Opens up and has a minimum of 2.

26. Opens down and has a maximum of 12.

27. Has a range of $[-2, \infty)$.

28. Has only one zero at 4.

29. Changes from increasing to decreasing at 5.

30. Branches have rates of change of -2 and 2.

31. Give an example of an absolute value function that crosses the x-axis.

32. Give an example of an absolute value function that does not cross the x-axis.

33. Give an example of an absolute value function that intersects the x-axis only once.

From Mathematics to English

34. After reading this section, make a list of questions that you want to ask your instructor.

35. Continue in your daily journal and make an entry. In addition to your normal entry on thoughts about the mathematics in this section, list at least two positive comments about what you have learned about this topic.

36. In paragraph format, summarize the material in this section of the text in your daily journal.

37. Describe how your classroom instructor made this topic more understandable and clear.

38. After reading the text and listening to your instructor, what do you not understand about this topic?

39. Describe the range of a function.

40. How are the absolute value function and the quadratic function different? How are they the same?

41. Does the range of the function $-2|x+3|-4$ include -4? Explain.

42. Why is the domain an absolute value function $(-\infty, \infty)$?

43. Describe the relationship between the rates of change of the two branches of the absolute value function.

44. Describe conditions on d, e, and f that cause the absolute value function $d|x+e|+f$ to cross the x-axis?

45. Describe conditions on d, e, and f that cause $d|x+e|+f$ to have a maximum value. A minimum value.

46. Why does $d|x+e|+f$ always cross the function-axis?

47. When was the first time you learned about the absolute value functions? What are the differences between when you first learned about absolute values and the material in this section? How does the graph of the function help in understanding the behavior of absolute values?

From English to Mathematics

In the next three Exercises, convert the English statement to an equivalent mathematical statement.

48. seven times the absolute value of the sum of x and three

49. the absolute value of the difference of x and 7

50. the absolute value of a number squared

Explorations

51. Investigate many functions of the form $d|x+e|+f$ on your calculator by graphing the following functions. Graph each group of four on the same screen and then describe how the parameters in the function affect the parent graph $|x|$. The idea behind this exploration is that you start with a simple function like $|x|$ and perform arithmetic operations on it as shown in the functions below and observe what the operation does to the graph of the parent. What does d do? What does e do? What does f do? Be specific in your descriptions.

a.	$\|x\|$	$2\|x\|$	$5\|x\|$	$8\|x\|$
	$\|x\|$	$0.8\|x\|$	$0.6\|x\|$	$0.2\|x\|$
	$\|x\|$	$-3\|x\|$	$-0.4\|x\|$	$-\|x\|$
b.	$\|x\|$	$\|x+2\|$	$\|x-5\|$	$\|x+6\|$
	$\|x\|$	$\|x-7\|$	$\|x+4\|$	$\|x-2\|$
c.	$\|x\|$	$\|x\|+4$	$\|x\|-6$	$\|x\|+1$
	$\|x\|$	$\|x\|-7$	$\|x\|+5$	$\|x\|-2$
d.	$\|x\|$	$\|x+3\|+4$	$\|x+5\|+3$	$\|x+1\|+6$
	$\|x\|$	$\|x-3\|+4$	$\|x-5\|+3$	$\|x-1\|+6$
e.	$\|x\|$	$2\|x-3\|-5$	$-2\|x-3\|-5$	$\frac{1}{2}\|x+6\|+2$

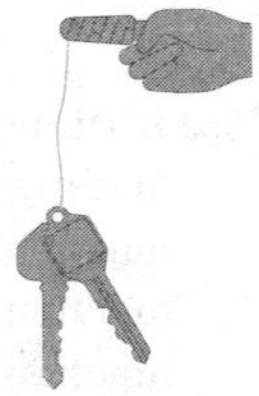

3.4 An Introduction to the Analysis of the Square Root Function $d\sqrt{x+e}+f$

The Pendulum

Try the following experiment. Take a small object about the size and weight of a set of keys. Tie a string to the object (Let's suppose you use keys.) and with a string length of 1 foot, hold the string and pull the keys to one side and then let go and let them swing freely. Measure how long it takes for the keys to travel from one side to the other and back again. This time is called the **period**, and the swinging keys is called a **pendulum**. If you measure carefully, the period should be just a little over one second. Try lengthening the string to two feet and measure the period of the pendulum. It should take just over 1.5 seconds. The period depends solely on the length of the pendulum. You should try another weight on the string and confirm that this is a true statement. Table 3.4.1 shows the length and

Table 3.4.1

l	0	1	2	3	4	5	6	7	8
t	0	1.1	1.6	1.9	2.2	2.5	2.7	2.9	3.1

period for several different lengths. The usual question remains what function can be used to model the period (t) of the pendulum? That is, given the length, what mathematics will tell us the period? Looking at the behaviors will give a clue as to what the function may be. It is clear that the function (t) is only increasing and this trend does not change. This information rules out the quadratic and absolute value functions as being models. The linear function cannot be used because the data is not increasing at a constant rate of change. Confirm this yourself by calculating the rate of change from data point to data point. A look at the graphical representation of the data may give a clue as to the function. While the data looks nearly linear (Figure 3.4.1a), once you look at the connected data it is clear that it is not linear.

Data (t)

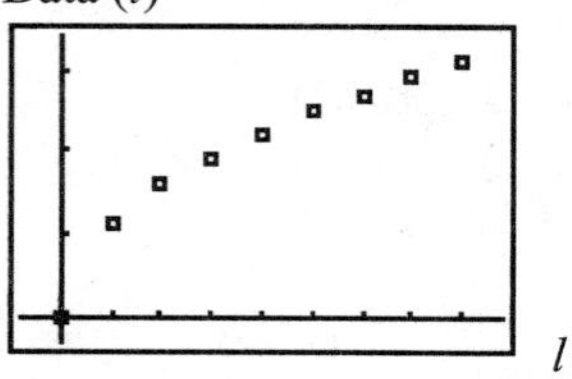

Figure 3.4.1a

Connected Data (t)

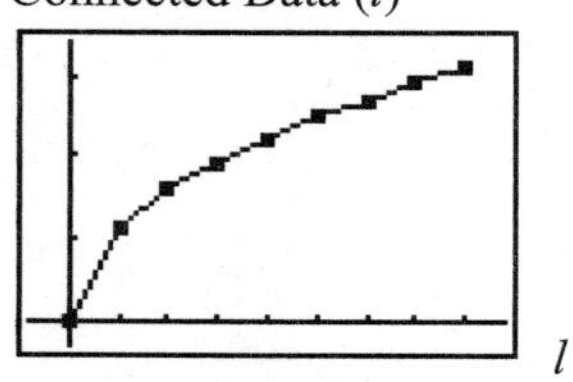

Figure 3.4.1b

Data & $1.1107\sqrt{l}$

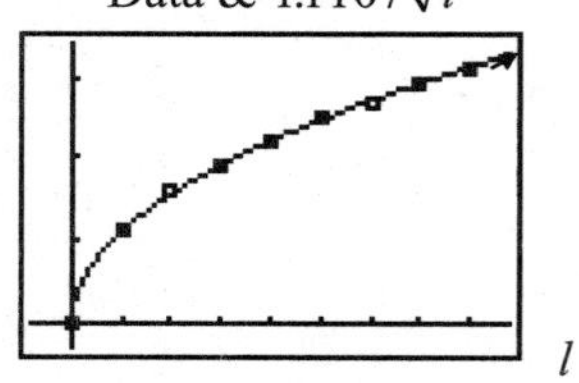

Figure 3.4.1c

Figure 3.4.1c shows the data and the square root function $1.1107\sqrt{l}$. The data relationship is a match with the square root function. The relationship between the length of a pendulum and the period is a square root relationship. You saw examples of this type of relationship in Section 2.3.

Each function studied in this chapter is useful in modeling real-world relationships. Modeling of relationships is a highly useful tool in making projections in many fields such as health, business, science, and engineering. The five functions analyzed in this chapter are five of the seven main functions that are studied in this text.

Analysis of $1.1107\sqrt{l}$ - **The Pendulum**

<u>Domain/range</u> Since l represents the length of a pendulum, the problem domain should be real numbers larger than 0; that is, $l \in (0, \infty)$. However in reality, the length of a pendulum may never be longer than several hundred feet when air resistance changes the relationship. The normal domain is

[0, ∞). The function variable l cannot be negative because the square root of a negative number is not real. If l were negative, the function would be non-real. Function values and the variable in the function must both be real numbers. The range is [0, ∞) because the function is a square root function and square roots are always 0 or positive.

<u>Maximum/minimum</u> On the normal domain, the function starts at 0 and becomes larger; thus, the function has a minimum of 0. Because the function grows larger from 0, there is no maximum.

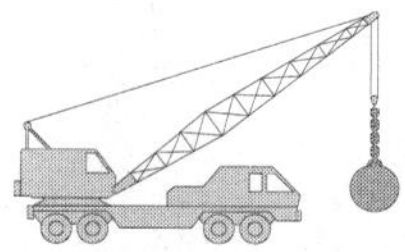

<u>Increasing/decreasing</u> The graphical representation clearly shows that the function is rising and does not ever drop. Thus the function $1.1107\sqrt{l}$ is increasing on the entire domain.

<u>Zeros</u> The data shows the function to be 0 when l is 0. There are no other zeros because the function is increasing which means it cannot become zero again.

<u>Positive/negative</u> Because the function is a basic square root function, it can never be negative. The function $1.1107\sqrt{l}$ is positive when $l \in (0, \infty)$ and never negative.

<u>Average Rate of change</u> Table 3.4.2 shows that the average rate of change is not constant and is different for each pair of numbers used to calculate the average rate of change.

Table 3.4.2

$\Delta l = 1$ (0 to 1) $\Delta l = 1$ (1 to 2) $\Delta l = 1$ (4 to 5) $\Delta l = 1$ (7 to 8)

l	0	1	2	3	4	5	6	7	8
t	0	1.1	1.6	1.9	2.2	2.5	2.7	2.9	3.1

$\Delta t = 1.1$ $\Delta t = 0.5$ $\Delta t = 0.3$ $\Delta t = 0.2$

Even with the function values rounded to tenths, the average rates of change shown in Table 3.4.2

are: $\frac{\Delta t}{\Delta l} = \frac{1.1}{1} \neq \frac{0.5}{1} \neq \frac{0.3}{1} \neq \frac{0.2}{1}$, with none of them being the same.

The square-root function $d\sqrt{x+e}+f$ where the function parameters d, e, and f are real and $d \neq 0$, does not have a normal domain of all real numbers. As discussed earlier, the square root of a negative number is not real. The domain must be specified so that the **radicand** or **argument** of the square root (the number under the radical sign) is always zero or positive. Remember that the idea behind a function requires that both the variable in the function and the function be real numbers. The linear, quadratic, and absolute value functions have normal domains of all real numbers; however, the square root function does not have a normal domain of all real numbers.

$d\sqrt{x+e}+f$

As with all of the previous functions studied, the major interest in the square root function is the characteristics of zeros, when it is increasing or decreasing, the range & domain, maximums or minimums, when it is positive or negative, and the rate of change. The viewing window used to show the complete graph should include the maximum or minimum, the zeros and any established trends near the left and right sides of the viewing window must continue with no other interesting behavior outside the viewing window. To show that a function starts at a specific point, put a dot at that point. To show that the graph continues, put an arrow on the end.

The family of square root functions exhibits unique behavior and is different than the three functions already studied. Below is an analysis of the square root function $-\sqrt{x+4}+3$. Before the analysis can take place, the graphical representation should be displayed.

$-\sqrt{x+4}+3$

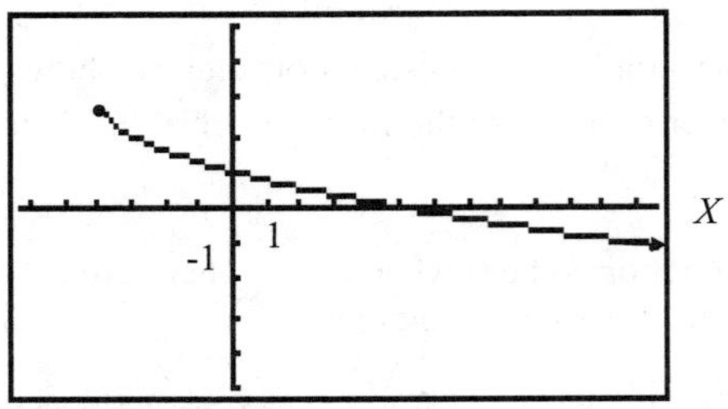

Figure 3.4.2

Analysis of $-\sqrt{x+4}+3$

Increasing/decreasing When the calculator graphs the function, it displays the graph using domain values from low to high, just as you are to read the graph. You should notice that the graph drops as it is displayed; thus, it is a decreasing function over the entire domain. Please pay close attention to the relationship between the function parameters and the behavior of the function. You may want to graph several square root functions and try to determine which parameter is the clue as to whether the function is increasing or decreasing.

Maximum/minimum Because the function is decreasing, there can be no minimum. However, use trace and a decimal window to trace to where the graph starts and you will find that the function reaches a maximum of 3. A decimal window is not necessary because approximate answers are also acceptable. That is, using any other window that shows the maximum value of the function, trace will give a number close to the actual maximum.

Zeros From looking at the above graphical representation, there appears to be a zero at about 5. Algebra can be used to see if your guess is correct. Replace x with 5 and simplify the expression. The expression simplifies to 0; therefore, 5 is a zero. As always, the zero can be found by trace. Remember, trace simply shows you the numerical representation the calculator uses to create the graphical representation of the function; it may not actually give the exact zero.

$$-\sqrt{x+4}+3$$
$$-\sqrt{5+4}+3$$
$$-\sqrt{9}+3$$
$$-3+3$$
$$0$$

Positive/negative The function is zero when x is 5; therefore, it must be positive or negative everywhere else. You can see from the graphical representation above that the function is positive to the left of 5 and negative to the right of 5. The function is positive when x is on the interval $[-4, 5)$ and it is negative when x is on the interval $(5, \infty)$. Remember that the function has positive values when it is above the x-axis and negative values when it is below the x-axis. You can easily verify this with the trace key.

Domain/range You can see in the graphical representation it appears that the x-coordinates (domain) start at –4. You can verify by using trace and a decimal window. Also, you know that $x + 4$ must be zero or positive, and if x is –4, $x + 4$ is zero. If x is larger than –4, then $x + 4$ is positive. The domain is $[-4, \infty)$.

Using the visual information in the graph, it appears the set of all function-coordinates starts at about 3 and decreases from there. Trace and a decimal window will verify this. The range is $(-\infty, 3]$. It may help you to remember domain and range by making a visual connection as shown in the Figures 3.4.3a and 3.4.3b.

Domain $[-4, \infty)$

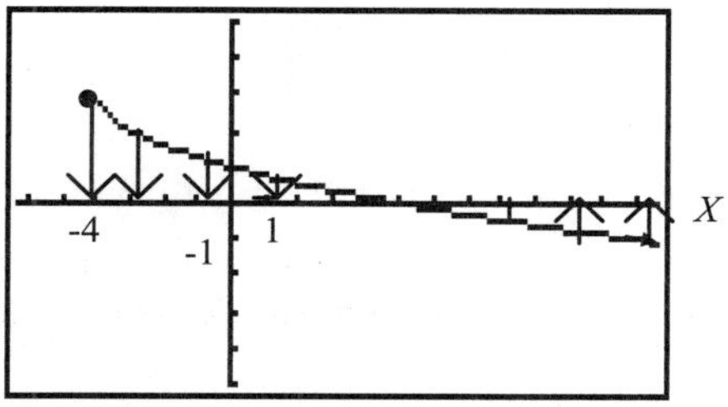

Figure 3.4.3a

Range $(-\infty, 3]$

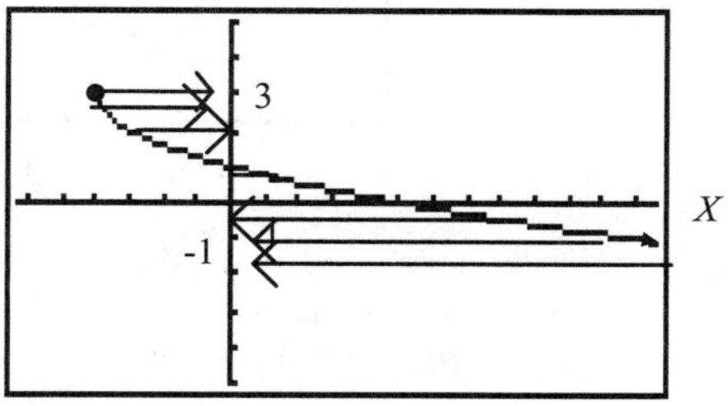

Figure 3.4.3b

Just a reminder: try to make connections between the numbers in the symbolic representation and the behavior of the function. Set a personal goal to try to determine as much of the behavior of the function as you can from the symbolic representation of the function.

Average Rate of change It might be best to analyze the average rate of change from an integer window. As you trace to the right, x changes by 1; thus, the change in the function is the average rate of change from point to point. Since trace cannot be duplicated in written form, a few trace values are shown below in numeric form.

Table 3.4.3

x	-4	-3	-2	-1	0	1	2	3
$-\sqrt{x+4}+3$	3	2	1.59	1.27	1	0.76	0.55	0.35

As x changes from -4 to -3, the function changes by -1. The average rate of change is -1. As x changes from -3 to -2, the function changes by -0.41. Now the average rate of change is -0.41. As you continue doing calculations the average rate of change is different each time. The function has no constant rate of change! The analysis of a function in which the average rate of change is changing from point to point will be left for a later course. That is, you might wonder if the rates of change can be expressed as a function.

Example 1: Find the normal domain and range of $\sqrt{x+2}-5$.

Solution: Graph $\sqrt{x+2}-5$ and find the minimum and leftmost point on the graph. If the minimum point on the graph is found, then the range must be all real numbers larger than or equal to the value of the function at the minimum. If the leftmost point on the graph is found, then the domain must be all real numbers larger than or equal to the x-coordinate of this point. The graphical representation is displayed below with approximate coordinates for the minimum point. If you try to find this point with trace, you must zoom-in several times to get similar results, or use a decimal window that gives integer values for x. Since -5 is the lowest value of the function, this value is the minimum. This point is also the leftmost point on the graph; thus, -2 is the beginning value of the normal domain.

$\sqrt{x+2}-5$

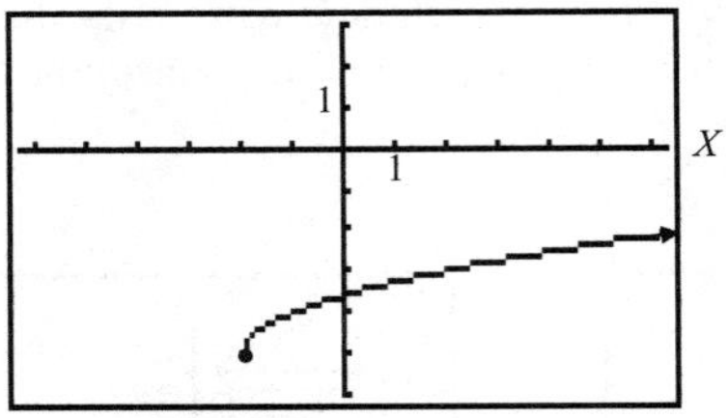

Figure 3.4.4

The domain is all real numbers in the interval [−2, ∞), and the range is all real numbers in the interval [−5, ∞). Please note how the function parameters can help you find the normal domain and range.

⁂

Example 2: Find the zero and when the function $-3\sqrt{x+2}-3.6$ is negative.

Solution: Graph $-3\sqrt{x+2}-3.6$ and use trace to find when it is zero and when it is negative.

$-3\sqrt{x+2}-3.6$

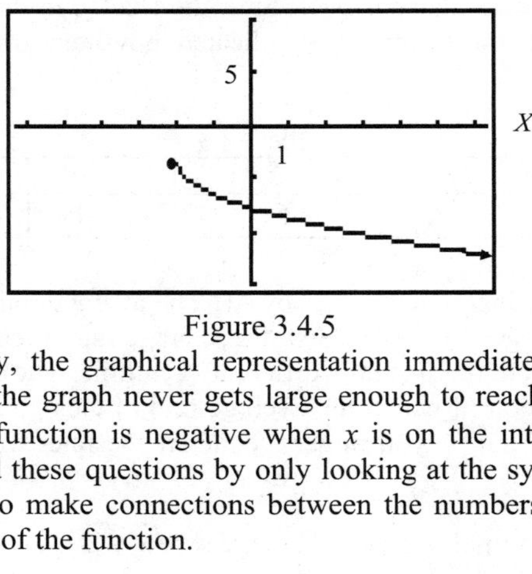

Figure 3.4.5

Hopefully, the graphical representation immediately shows you there is no zero because the graph never gets large enough to reach zero. It should also show you that the function is negative when x is on the interval [−2, ∞). Could you have answered these questions by only looking at the symbolic form of the function? If not, try to make connections between the numbers in the symbolic form and the behavior of the function.

⁂

Example 3: Find an interval when $-3\sqrt{x+2}-3.6$ is increasing and an interval when the function is decreasing.

Solution: The graphical representation of the function is shown in Figure 3.4.5. The function only gets smaller as x increases; this means the function only decreases. The function starts when x is −2, and as x gets larger in value the function decreases in value; thus, it is decreasing on (−2, ∞). A personal goal for you may be to see if you can answer similar questions without a graphical representation.

⁂

Example 4: Is the rate of change of the function $2\sqrt{x+3}-6$ a constant? If not, find the average rate of change between –3 and –2 and again between 2 and 3.

Solution: Make a numeric representation or use trace with a decimal window to calculate the rate of change. Since trace cannot be shown in written form, below is the numeric representation. Reading the function values from left to right, you can see that the increase in the function changes from x to x. The rate is not constant. As x changes from −3 to −2, the average rate of change is 2. As x changes from 2 to 3, the average rate of change is 0.43. Can you find *any* square root function that has a constant rate of change?

Table 3.4.4

x	–3	–2	–1	0	1	2	3	4	5
$2\sqrt{x+3}-6$	–6	–4	–3.17	–2.54	–2	–1.53	–1.10	–0.71	–0.34

⁂

If you can answer a few of the above questions by looking at the symbolic representation of the function, you are insightful. If you think the above solution is the only way of answering the questions, concentrate a little more on connections between symbolic form and characteristics of functions. Usually there are several ways of answering a question. You have learned something about mathematics even if you only know one method for solving a problem!

3.4 EXERCISES

Many times you can work the exercises by algebraic, numeric, or graphical methods. You can learn about mathematics no matter which method you use. Part of the learning process is for you to decide which method is best for you to use. Please read the text carefully before you do the exercises. Read the **next assigned section** after you finish the assignment below.

A Quick Look To The Future

-4. If the symbols $3x-2$ are called $f(x)$, what do you think $f(4)$ means?

-3. Can you add $|x|$ and $|2x|$? If yes, what is the sum?

-2. If the lengths of several rectangles are 2 feet more than the corresponding widths, explain why you may want to multiply w and $w+2$.

-1. Write 15 as a product.

Previous Topics Revisited

1. Find the maximum value of $-|x-3|+35$.
2. Find the zeros of $-6|x+2.1|$.
3. Find the range of $\frac{1}{2}|x-1|+6.2$.
4. Find the minimum value of x^2-5x-2.
5. Find the zeros of x^2-6x+9.
6. Does $0x+2$ increase or decrease?

New Topic Exercises

Find the domain of each function in Exercises 7 and 8.

7. $-2\sqrt{x+5}+1,\ -2\sqrt{x+3}+1,\ -2\sqrt{x+1}+1,\ -2\sqrt{x}+1,\ -2\sqrt{x-2}+1,\ -2\sqrt{x-4}+1$

8. $\frac{1}{2}\sqrt{t-2}+4,\ -\frac{3}{4}\sqrt{t-2}+6,\ 17\sqrt{t-2}-8,\ -6\sqrt{t-2}+5,\ 0.2\sqrt{t-2}-0,\ \frac{3}{8}\sqrt{t-2}+\frac{1}{2}$

Find the range of each function in Exercises 9 and 10.

9. $-\sqrt{x-1}+5,\ -\sqrt{x-1}+4,\ -\sqrt{x-1}+3,\ -\sqrt{x-1}+2,\ -\sqrt{x-1}+1,\ -\sqrt{x-1}$

10. $3\sqrt{x+2}+5,\ 3\sqrt{x+1}+5,\ 3\sqrt{x-2}+5,\ 2\sqrt{x-3}+5,\ 2\sqrt{x-3}+4,\ 2\sqrt{x-3}+3$

11. Find the minimum or maximum of each function in Exercise 9.

12. Find the minimum or maximum of each function in Exercise 10.

13. Identify whether each function in Exercises 9 and 10 are increasing or decreasing.

14. Do any of the functions in Exercises 7 - 10 have a constant rate of change? If yes, which ones.

Find the zero for each function in Exercises 15 - 17.

15. $3\sqrt{x+4},\ 2\sqrt{x+2},\ -3\sqrt{x+1},\ -\frac{1}{2}\sqrt{x-1},\ \frac{3}{8}\sqrt{x-3},\ 15\sqrt{x-6}$

16. $\sqrt{x+3}-1,\ \sqrt{x-2}-1,\ -3\sqrt{x+2}+9,\ 2\sqrt{x-3}-4,\ -\sqrt{x+5}+1,\ \frac{1}{2}\sqrt{x-1}-\frac{1}{2}$

17. $5\sqrt{x+4}-6,\ \frac{1}{8}\sqrt{x+2}-3,\ -4\sqrt{x-7}+19,\ -\frac{1}{3}\sqrt{x+4}+26$

18. Find when each function in Exercises 15 and 16 is positive and find when negative.

19. The velocity (in feet per second) of an object dropped and has fallen h feet is modeled by the function $8\sqrt{h}$, where air resistance is ignored. Based on this model, how fast is an object moving after falling 5 feet? After 12 feet?

20. How high must an object be dropped from to have a velocity of 100 feet per second when it hits the ground? 300 feet per second?

21. What is the average rate of change of the above velocity function between a fall of 0 and 1 foot? Between 6 and 7 feet?

22. Create a square root function that is increasing.

23. Create a square root function that is decreasing.

24. Create a square root function that has a maximum of 7.

25. Create a square root function that has a minimum of 2.

26. Create a square root function that has a domain of $[-3, \infty)$.

27. Create a square root function that has a range of $[-3, \infty)$.

28. Create a square root function that has a zero at 4.

29. Give an example of a square root function whose graph crosses the x-axis.

30. Give an example of a square root function whose graph does not cross the x-axis.

31. Give an example of a square root function that starts on the x-axis.

From Mathematics to English

32. After reading this section, make a list of questions that you want to ask your instructor.

33. Continue in your daily journal and make an entry. In addition to your normal entry on thoughts about the mathematics in this section, list at least two positive comments about what you have learned about this topic.

34. In paragraph format, summarize the material in this section of the text in your daily journal.

35. Describe how your classroom instructor made this topic more understandable and clear.

36. After reading the text and listening to your instructor, what do you not understand about this topic?

37. In the square root function $\sqrt{-x}$, explain why the domain is all real numbers less than or equal to zero.

38. Describe conditions on d, e, and f that make the square root function $d\sqrt{x+e}+f$, cross the x-axis.

39. Does $d\sqrt{x+e}+f$ always cross the function-axis? Explain.

40. Explain why the domain of $d\sqrt{x+e}+f$ contains $-e$.

41. Explain why the range of $d\sqrt{x+e}+f$ contains f.

42. Describe any unique behaviors of the square root function.

43. Describe what each function parameter of the square root function does.

44. What is your definition of the domain of a function?

45. Describe how a square root function can be negative.

46. Where is the graph of a square root function when the function is negative? Positive? Zero?

From English to Mathematics

In the next three exercises, convert the English statement to an equivalent mathematical statement.

47. the square root of the difference of a number x and four

48. the difference of the square root of a number x and four

49. seven times the square root of the sum of x and nine

Explorations

50. Investigate many functions of the form $d\sqrt{x+e}+f$ on your calculator by graphing the following group of four functions on the same window. Summarize your investigation by identifying what visual effect the parameters d, e, and f have on the graph of the parent function $\sqrt{x}$.

a.	$\sqrt{x}$	$2\sqrt{x}$	$5\sqrt{x}$	$8\sqrt{x}$
	$\sqrt{x}$	$.8\sqrt{x}$	$.6\sqrt{x}$	$.2\sqrt{x}$
	$\sqrt{x}$	$-3\sqrt{x}$	$-.4\sqrt{x}$	$-\sqrt{x}$
b	$\sqrt{x}$	$\sqrt{x+2}$	$\sqrt{x-5}$	$\sqrt{x+6}$
	$\sqrt{x}$	$\sqrt{x-7}$	$\sqrt{x+4}$	$\sqrt{x-2}$
c.	$\sqrt{x}$	$\sqrt{x+4}$	$\sqrt{x-6}$	$\sqrt{x+1}$
	$\sqrt{x}$	$\sqrt{x-7}$	$\sqrt{x+5}$	$\sqrt{x-2}$
d.	$\sqrt{x}$	$\sqrt{x+3}+4$	$\sqrt{x+5}+3$	$\sqrt{x+1}+6$
	$\sqrt{x}$	$\sqrt{x-3}+4$	$\sqrt{x-5}+3$	$\sqrt{x-1}+6$
e.	$\sqrt{x}$	$2\sqrt{x-3}-5$	$-2\sqrt{x-3}-5$	$\frac{1}{2}\sqrt{x+6}+2$

51. a. Is the graph of the function $-\frac{1}{2}x-2+8\sqrt{x}$ a straight line? What is the normal domain of the function? Make a sketch of the graph.

b. Is the graph of the function $-x^2+7x+10-20\sqrt{x}$ a parabola? What is the normal domain of the function? Make a sketch of the graphical representations.

c. Is the graph of the function $-2|x - 5| + 3 + 8\sqrt{x}$ a V? What is the normal domain of the function? Make a sketch of the graphical representation.

d. Is the graph of the function $-\frac{1}{2}x - 2 + 0\sqrt{x}$ a straight line?
What is the normal domain of the function? Make a sketch of the graphical representation.

e. Is the graph of the function $-x^2 + 7x + 10 + 0\sqrt{x}$ a parabola? What is the normal domain of the function? Make a sketch of the graphical representation.

f. Does the graph of the function $-2|x - 5| + 3 + 0\sqrt{x}$ look like the letter V? What is the normal domain of the function? Make a sketch of the graphical representation.

52. (Hint: Use trace and a decimal window to answer the following questions.) Draw sketches of the graphical representations in parts b, c, and d.

a. For what values of x is the function $-x^2 + 6x - 5$ positive? Use interval notation.

b. What is the normal domain of $\frac{3}{2}x - 1 + 0\sqrt{-x^2 + 6x - 5}$?

c. What is the normal domain of $\frac{1}{4}x^2 - 4 + 0\sqrt{-x^2 + 6x - 5}$?

d. What is the normal domain of $|x - 2| - 5 + 0\sqrt{-x^2 + 6x - 5}$?

53. (Hint: Use trace and a decimal window to answer the following questions.) Draw sketches of the graphical representations in parts b, c, and d.

a. For what values of x is the function $-x^2 - x + 6$ positive? Use interval notation.

b. What is the normal domain of $\frac{3}{2}x - 1 + 0\sqrt{-x^2 - x + 6}$?

c. What is the normal domain of $\frac{1}{4}x^2 - 4 + 0\sqrt{-x^2 - x + 6}$?

d. What is the normal domain of $|x - 1| - 5 + 0\sqrt{-x^2 - x + 6}$?

3.5 An Introduction to the Analysis of the Exponential Function $d \cdot b^{x+e} + f$

The National Debt

The amount of money the federal government owes is referred to as the national debt. The first time the country was in debt was in the 1930's. As of 2006 the national debt was $8,400,000,000,000. With a 2006 US population of 296,000,000 people; this averages $28,380 per person. So to pay off the debt in 2006, if each person had $28,380 added to their federal taxes, we would be out of debt. Or looking at debt another way, we (the tax payers) must pay the DAILY interest on this debt of around $1,000,000,000.

Table 3.5.1

Year (t)	1940	1950	1960	1970	1980	1990	1994	2002	2006
National Debt (n)	51	257	291	381	909	3113	4500	6000	8400

(Debt is in billions.)

You should wonder how it is growing; that it, has it grown in a linear fashion? Like a square root function? A quadratic function? Something new? If we knew how it was growing, we could predict what it will be in 2010, or 2012, etc. The numeric representation of this data relationship already tells us it is increasing, and in the last few years, it has been increasing faster than in the years before. A look at the graphical representation will help us decide how it is growing.

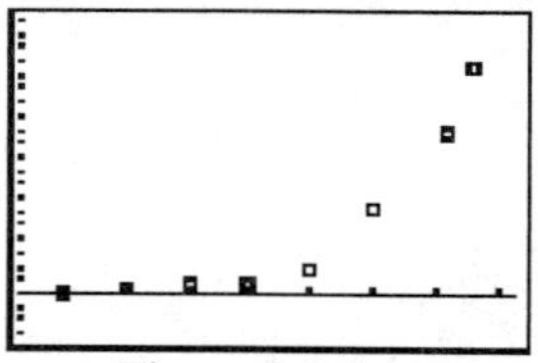

Figure 3.5.1a

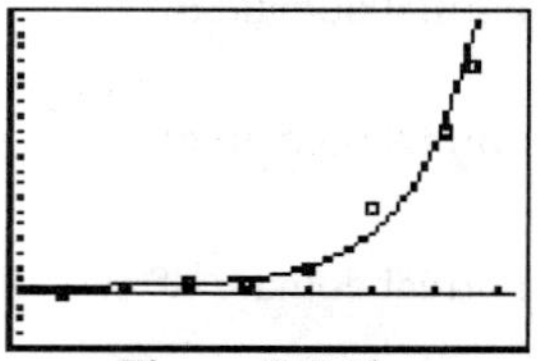

Figure 3.5.1b

You have seen the shape before, but you may not have seen symbols connected to it. The shape is called exponential because the symbols that model this kind of shape use the time variable as an exponent. In particular, one model of this data is $1.1^{x-1910} + 225$. The graph of this function is shown in Figure 3.5.1b. This function was created by using parameter-behavior connections discussed later in this textbook and so may not "look" like models found in related fields.

Analysis of $1.1^{x-1910} + 225$ - **The National Debt**

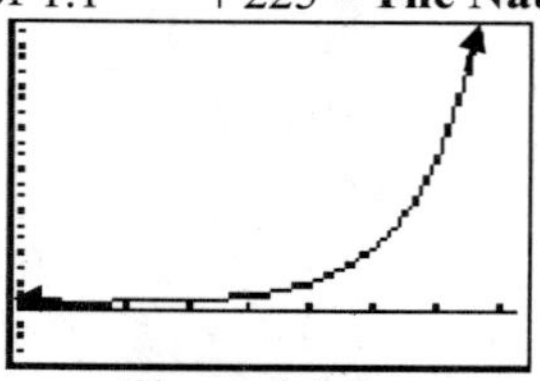

Figure 3.5.2

Domain/range The context of the situation tells us the problem domain is from about 1930 until the US nearly goes bankrupt. So, a guess may be something like [1930, 2020]. The symbols of the model do not show any division or square roots – the operations that normally cause difficulty. So, one would guess the normal domain to be $(-\infty, \infty)$. However, the graphical representation of the function shows something new about the range. The graph appears to level off as it heads to the left. When you look at the symbols, we know from arithmetic that 1.1 raised to any real exponent is always greater than zero. When we realize this, and see that 1.1^{x-1910} has 225 added to it, we strongly suspect the range is $(225, \infty)$. As for the range based on the contextual problem, we know the debt was 0 at one point in time and cannot exceed the wealth of the country – so maybe [0, 30 trillion] may be reasonable – on not.

<u>Maximum/minimum</u> Once we know the range, we immediately know the maximum and minimum. Mathematically, the function goes to ∞ and approaches 225 on the low end – but never gets to 225. So there is no maximum and no minimum. Considering the problem situation, we know there was a minimum of $0, and we do not know what the maximum will end up being.

<u>Increasing/decreasing</u> The complete graph is displayed in Figure 3.5.2, and it shows an increasing function on the entire domain.

<u>Zeros</u> Since the function approaches 225 to the left and goes to infinity to the right, there are no zeros.

<u>Positive/negative</u> The discussion under <u>zeros</u> suggests the function is always positive.

<u>Average Rate of change</u> This behavior is the most interesting for this type of function. Table 3.5.1 is

$\Delta t = 10$ (1950 to 1960) $\Delta t = 10$ (1980 to 1990) $\Delta t = 4$ (2002 to 2006)

Table 3.5.1

Year (t)	1940	1950	1960	1970	1980	1990	1994	2002	2006
National Debt (n)	51	257	291	381	909	3113	4500	6000	8400

(Debt is in billions.)

$\Delta n = 34$ $\Delta n = 2204$ $\Delta n = 2400$

shown again with several values for Δt's and Δn's. From 1950 to 1960 $\dfrac{\Delta t}{\Delta n}$ is <u>$3.4 billion per year</u>, but from 1980 to 1990 is it <u>$220.4 billion per year</u>, and from 2002 to 2006 it is <u>$600 billion per year</u>. So the average rate of change is increasing very rapidly. It is this behavior that makes the exponential function of interest. The population of the earth is growing exponentially as you saw in Section 2.1. Money invested at compound interest grows exponentially. Sometimes, relationships decay exponentially – like the value of your car as it levels off to the junk value.

Example 1: Find the normal domain and range of $3 \times 2^{x+1} - 4$.

Solution: Just like most exponential functions of the form $d \times b^{x+e} + f$, ($b > 0$ & $b \neq 1$) the domain is all real numbers because there are no operations that will cause it to be non-real. We avoid domain problems by making the base b larger than 0. To figure out the range, perhaps the graph will remind us.

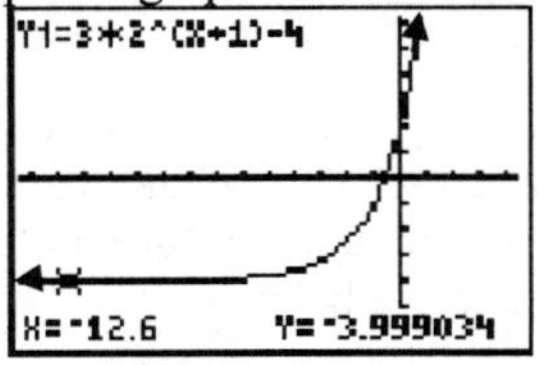

Figure 3.5.3

As x increases, we see $3 \times 2^{x+1} - 4$ headed to ∞. As we look to the left toward −∞, the function $3 \times 2^{x+1} - 4$ is approaching −4. The concept of approaching −4 (or any number) will be further discussed in Chapter Seven. So, the range appears to be (−4, ∞).

⁂

Example 2: Find the zero, and find whether $3 \times 2^{x+1} - 4$ is increasing or decreasing.

Solution: As x increases, the function $3 \times 2^{x+1} - 4$ is rising on the entire domain. You can convince yourself of this through trace. As for a zero, we see from Figure 3.5.3 above that the function crosses the x-axis. So we know it has a zero. Figure 3.5.4 below shows it to be about −0.6

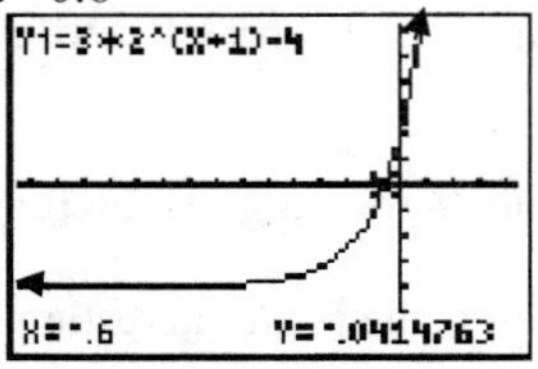

Figure 3.5.4

Trace shows the function is near zero when $x = -0.6$. You will learn how to find the exact zero in Chapter Seven.

⁂

Example 3: Find the zero and maximum/minimum of the exponential function $-3 \times 2^{x+2} + 4$.

Solution: In all the exponential functions we have seen so far, have not had a maximum or minimum. The graph shown in Figure 3.5.5 confirms this idea. Trace to the left and you will strongly suspect the function is approaching 4, thus no maximum. As you trace to the right it goes to −∞, so no minimum.

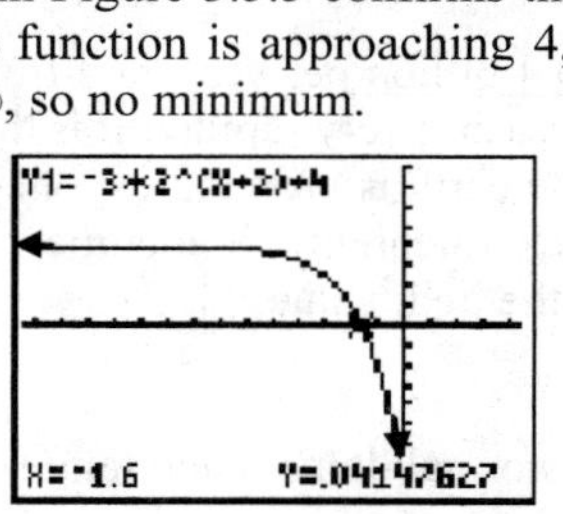

Figure 3.5.5

The trace cursor shows that when x is −1.6, the function is almost zero. So we can approximate the zero as −1.6.

⁂

3.5 EXERCISES

Many times you can work the exercises by algebraic, numeric, or graphical methods. You can learn about mathematics no matter which method you use. Part of the learning process is for you to decide which method is best for you to use. Please read the text carefully before you do the exercises. Read the **next assigned section** after you finish the assignment below.

A Quick Look To The Future

-4. Create your own definition of a function.

-3. Is it possible to add $(2x + 5)$ and $(3x - 17)$?

-2. Do you think $(x - 2)$ times $(x + 2)$ is $x^2 - 4$?

-1. True or false: $x^2 - 4 = (x + 2)(x - 2)$.

Previous Topics Revisited

1. Find the zero of $2\sqrt{x+1}+\frac{1}{2}$.

2. Is $-2\sqrt{x+1}+\frac{1}{2}$ increasing or decreasing?

3. What is the domain of $-2\sqrt{x+1}+\frac{1}{2}$?

4. What are the coordinates of the vertex of $-|x+2|-3$?

5. When is $-|x+2|-3$ increasing and when is it decreasing?

6. What are the coordinates of the vertex of $-(x+2)^2-3$?

New Topic Exercises

7. If 10,000 E. coli bacteria are put in a Petri dish (with all the food and space they need, & no predators), they double their numbers every 20 minutes. What model can be used to predict the number of bacteria after x minutes?

8. The depreciation on a \$5,000 piece of office equipment with an estimated life of 8 years can be modeled with the function $1250 \times 0.75^{x-1}$ where time in years is x ($x \in [1, 8]$). So for example in year 2, the depreciation is $1250 \times 0.75^{2-1} = 1250 \times 0.75^1 = \$937.50$. Is the depreciation function increasing or decreasing?

9. Suppose the human population doubled every 50 years. On April 21, 2006 at 8:50 PM, the world population was about 6.6 billion (6,600,000,000). The model for population growth under the 50 year doubling assumption is $6.6 \times 2^{\frac{x}{50}}$, where x is time in years measured from calendar year 2006 being year 0. When will there be 12 billion people?

10. Is the function $6.6 \times 2^{\frac{x}{50}}$ increasing or decreasing?

11. If the problem domain of the function $6.6 \times 2^{\frac{x}{50}}$ is [0, 50], what are the minimum and maximum values? What is the range?

12. Given the population growth function $6.6 \times 2^{\frac{x}{50}}$, what is the annual average rate of growth of humans from 2 years to 3 years (2008-2009)? What is it from the year 2020 to 2021 ($x = 14$ to 15)? What is it from the year 2040 to 2041 ($x = 34$ to 35)?

13. The exponential function $2^{x-1}+3$ approaches what horizontal line as x gets smaller and smaller toward $-\infty$? That is does $2^{x-1}+3 \rightarrow 2$? Does $2^{x-1}+3 \rightarrow -1$? Does $2^{x-1}+3 \rightarrow 3$? (Please note your graphing calculator will have rounding errors as x gets small.)

14. 3^x+1 crosses the vertical-axis at 2. Where does 3^x+2 cross the vertical-axis?

15. $-1\times 3^x+1$ crosses the vertical-axis at 0. Where does $-1\times 3^x+2$ cross the vertical-axis?

16. 2^x+1 crosses the vertical-axis at 2. Where does 2^x+2 cross the vertical-axis? Where does 2^x+3 cross the vertical-axis? Where does 2^x+4 cross the vertical axis?

17. Are the functions 2^{-x} and $\left(\frac{1}{2}\right)^x$ the same or different?

18. Is the function $3\times\left(\frac{1}{4}\right)^x$ increasing or decreasing?

19. Is the function $-3\times\left(\frac{1}{4}\right)^x$ increasing or decreasing?

20. Is the function $3\times\left(\frac{1}{4}\right)^{-x}$ increasing or decreasing?

21. Find the zero of $3\times\left(\frac{1}{4}\right)^{-x}$.

22. Is $4\times 1^{x-3}+2$ an exponential function?

From Mathematics to English

23. After reading this section, make a list of questions that you want to ask your instructor.

24. Continue in your daily journal and make an entry. In addition to your normal entry on thoughts about the mathematics in this section, list at least two positive comments about what you have learned about this topic.

25. In paragraph format, summarize the material in this section of the text in your daily journal.

26. Describe how your classroom instructor made this topic more understandable and clear.

27. After reading the text and listening to your instructor, what do you not understand about this topic?

28. In the exponential function 2^x, explain why the range isn't all real numbers.

29. In the exponential function 2^x, explain why the domain is all real numbers.

30. List three relationships you are aware of that are exponential.

31. Explain why you think the base b in the exponential function $d \times b^{x+e} + f$ cannot be negative.

32. Since the population of the Earth is growing exponentially, explain why this is a problem. Or, if you think it isn't a problem, explain why you think it isn't.

33. Would it be better or worse if the human population were growing in a linear fashion? Give reasons for your position.

From English to Mathematics

In the next three exercises, convert the English statement to an equivalent mathematical statement.

34. Two thousand E. coli bacteria double every 20 minutes.

35. The continuous cutting paper in half and stacking the haves doubles the thickness with every cut.

36. $\{1, 2, 4, 8, 16, 32, 64, 128, \ldots\}$

Explorations

37. After looking at the graphs of -2^x and 2^x. describe the relationship between the two graphs. Between -4^{2x} and 4^{2x}. Between 3^{x-1}, and -3^{x-1}. Between $\left(\frac{1}{2}\right)^{x-4}$ and $-\left(\frac{1}{2}\right)^{x-4}$.

38. Based on your experience with Exercise 37, make a conjecture of what you think the relationship will be between the graphs of $(2x - 3)$ and $-(2x - 3)$. Between x^2 and $-x^2$.

39. For each of the functions that follow, identify what horizontal line the functions are approaching as x gets smaller and smaller. $2^x + 3$, $2^x + 2$, $2^x + 1$, $2^x + 0$, $2^x - 1$, $2^x - 2$, $2^x - 3$.

40. For each of the functions that follow, identify what horizontal line the functions are approaching as x gets smaller and smaller. $5^x + 3$, $5^x + 2$, $5^x + 1$, $5^x + 0$, $5^x - 1$, $5^x - 2$, $5^x - 3$.

41. For each of the functions that follow, identify what horizontal line the functions are approaching as x gets larger and larger. $\left(\frac{1}{2}\right)^x - 3$, $\left(\frac{1}{2}\right)^x - 2$, $\left(\frac{1}{2}\right)^x - 1$, $\left(\frac{1}{2}\right)^x$, $\left(\frac{1}{2}\right)^x + 1$, $\left(\frac{1}{2}\right)^x + 2$, $\left(\frac{1}{2}\right)^x + 3$.

CHAPTER THREE TEST

1. What is the rate of change of the function $\frac{2}{3}x - \frac{5}{2}$?
2. Find the zero of the function $3x - 12$.
3. When is the function $3x - 12$ negative?
4. Is the function $3x - 12$ increasing or decreasing?
5. If the house-keeping person at a hotel can clean 1 room in 10 minutes, and she has 32 rooms to clean, what function models the number of rooms left to clean at time t (in minutes)?
6. Find the zeros of the function $-(x - 3)^2 + 4$.
7. Find values for x, that cause the function $-(x - 3)^2 + 4$ to be negative.
8. When is the function $-(x - 3)^2 + 4$ increasing, and when is it decreasing?
9. Find the minimum value of the function $3(x + 2)^2 - 1$.
10. If the lengths of the sides of a rectangle are x and $x + 3$, what is the model of the area of the rectangle?
11. A hot air balloon ascended to a height of 2,500 feet in 20 minutes, and then immediately descended at the same rate as going up. The function that models the air temperature during the flight is $2|t - 20| + 30$ (the function variable t is time in minutes). As you might figure the temperature started warm, dropped to a low temperature, and then rose back to warmer during this 40 minute trip.
    a. What is the minimum temperature during this flight?
    b. What is the problem domain of the temperature function?
    c. Using the normal domain, what is the range?
    d. How fast is the temperature changing on the trip up?
    e. At what time is the outside temperature 45°?
12. What is the missing number in this exponential function?

1	2	3	4	5
1	$\frac{1}{2}$	$\frac{1}{4}$	$\frac{1}{8}$	??

13. Is the function $-\left(\frac{3}{4}\right)^{x+1} - 5$ increasing, decreasing, or constant?
14. What is the average rate of change of the function $2 \times 5^{x-1} + 3$ as x changes from 1 to 2?
15. Find the zero of the function $2 \times 5^{x-1} - 3$.
16. Is $x^4 + 5$ an exponential function?

CHAPTER FOUR
FUNCTIONS: NOTATION AND OPERATIONS

4.0 Introduction

Functions are an important part of the study of mathematics. Actually, most of what you will study is the analysis of many different types of functions. Functions are defined again in Section 4.1 using a more formal approach.

Functions are typically represented in symbolic, numeric, graphical, and written forms. For example, in business, a revenue function is normally expressed in symbolic form; however, business people may also want to see the graphical representation of the function as well. The Internal Revenue Service expresses the tax function in numeric form (tax tables). Scientists may prefer to examine the nature of a temperature function in graphical form. Mathematics students may find it preferable to analyze functions in the graphical form, but it is necessary to study functions in each form because they are used and encountered outside the mathematics classroom in all four forms.

Chapter Four includes the commonly used *notation* to represent functions as well as addition, subtraction, multiplication, and factoring of polynomial functions. It ends with the analysis of sums and differences of some basic parent functions studied in Chapter Three, and how you can control the domain of any function by adding a square root function.

The Flight Plan

If you have ever taken a commercial airline flight over an hour in length, you may have noticed that the plane takes off and ascends to a cruising altitude -- usually taking around ½ hour, then it stays at the cruising altitude until the last ½ hour of the flight when it descends into the destination city. What you know from Chapter Three is that there is no elementary function that has this behavior. If you look ahead at new functions analyzed in Chapters Six, Seven, Eleven, Nine, and Fourteen, you still will not find any functions that can be used to model the flight of an airplane. The same is true if you study college algebra, precalculus, or calculus. So why bring up the topic if it is not found in almost all courses? The answer is simple; by adding two absolute value functions similar to those in Chapter Three, you can model the on-board computer flight plan of any commercial flight. However, because you are working in two dimensions, you cannot give a destination direction to the flight plan, only height. The model will tell you the rate of ascent and descent. It will tell you how long the flight is in the air, when it levels-off, when it starts the descent, and the cruising altitude. Consider the four-hour flight plan shown below and the typical questions that can be answered after learning the material in this chapter. The cruising height of the plane is 36,000 feet and it reaches this height in ½ hour. The graph also shows that the trip takes four hours. Typical questions you will be expected to answer if you are assigned a lab project similar to the flight plan lab project are:

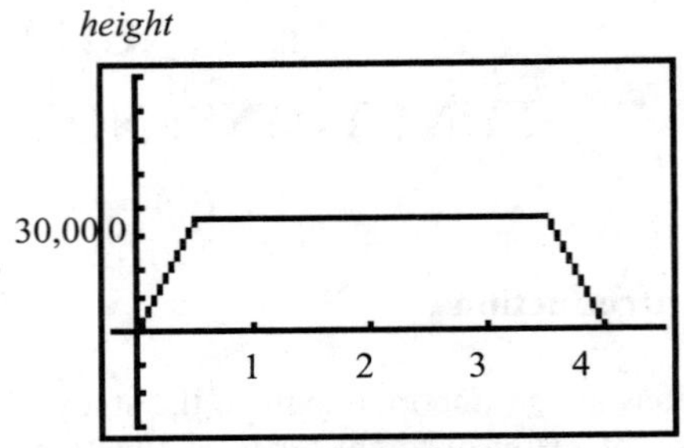

Figure 4.0.1

a. What is the problem domain?

b. What is the symbolic representation of the flight plan?

c. What are the limitations to the flight plan model?

d. Using the mathematical model, how high is the plane 20 minutes into the flight?

e. Using the mathematical model, how long has the plane been in-flight if it is 3 miles high?

f. What is the range of the model (not the airplane)?

g. When is the model increasing? When is the model decreasing?

h. Does the model have a maximum? Does the model have a minimum?

i. What is the average rate of change in the height of the plane as it is reaching cruising altitude?

j. Do you think anyone working at a professional job would need or could use the mathematical model to help solve a problem? If yes, what kind of professional? How would they use it?

4.1 Definition of a Function, Again

You should have noticed that all of the data relationships in this text have had exactly one element of the range for each element of the domain. Each relationship has *not* been like the following examples. Table 4.1.1 shows the relationship between the numbers (N) and letters (L) on your phone. For mathematical reasons think of the letters of the alphabet as integers 1 through 26.

Table 4.1.1

N	2	3	4	5	6	7	8	9
L	1,2,3	4,5,6	7,8,9	10,11,12	13,14,15	16,18,19	20,21,22	23,24,25

In this relationship, each number in the domain is related to three numbers from the range. Another example of this type of relationship comes from a family of five seated at a restaurant table designed for four as shown in Table 4.1.2.

Table 4.1.2

Seat	1	2	3	4
Person	1	2	3	4,5

In this example, the number 4 from the domain is related to both 4 and 5 from the range. Both of these examples demonstrate an important and basic difference between relationships. Nearly all relationships you will study will be of the nature of the relationships studied in Chapter Three -- where each element of the domain is related to only one element in the range. For the purpose of clarification, each type of relationship will be given a name. Instead of using the word data relationship as you may have been doing, mathematicians use the word mathematical relation, or just relation, to describe both types of relationships. The relationships studied in Chapters Two and Three are named functions as well as relations.

> A function is a relation(ship) where each real number in the domain is related to exactly one real number in the range.

Example 1: Is the relation $-2\sqrt{x+3}$ a function?

Solution: Yes, because any number from the domain will cause $-2\sqrt{x+3}$ to be exactly one real number. ⁂

Example 2: Is the relation in Table 4.1.3 a function?

Table 4.1.3

t	–2	–1	0	1	2
$3t \pm 6$	–12, 0	–9, 3	–6, 6	–3, 9	–12, 0

Solution: No. For a number like –1 from the domain, there are two related numbers from the range, –9 and 3. ⁂

Example 3: Is the relation in Figure 4.1.1 a function?

Solution: No. For example, the number 7 from the domain is related to three numbers from the range, 16, 18, and 19. (Note, this is the graphical representation of the phone relationship.)

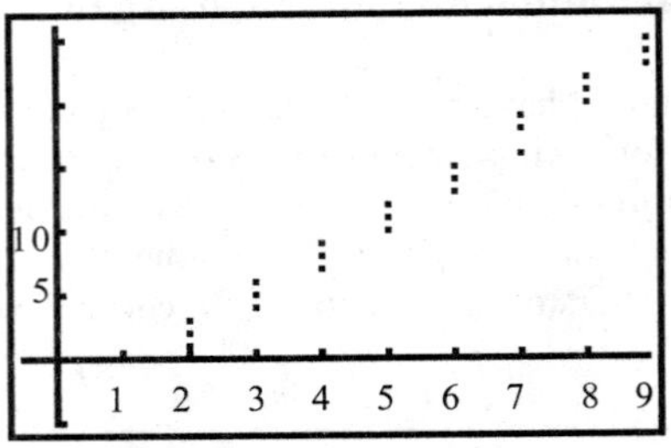

Figure 4.1.1

⁂

> **Definition of a Function**
> Consider the set of pairs of real numbers (x, y). If for each value of x there is one and only one corresponding value for y, then the set of all pairs of real numbers (x, y) is a **function**.

Function relationships have now been described primarily in three different representations. The numeric representation of a function relationship seems to be the way real-world relationships are commonly expressed because collected data usually comes in numeric form. The graphical representation of the function relationship can easily be developed from the numeric representation and is very useful in helping to describe relationships to people outside of mathematics. Increasing or decreasing trends are easily seen if the graphical representation is used. Maximum or minimum values as well as zeros and positive or negative values can be seen with a quick look at the graph, with no need to study the numeric information. The symbolic representation may be most useful to people who use and study mathematics. Most of the mathematics studied in high school and college centers on the symbolic forms of relationships. This section continues with more examples of differentiating between relations and functions; and finally, a new method for describing the symbolic representation is developed.

An example of the symbolic form of a function might be something like $-2\sqrt{x+3}$. This notation can be thought of as a short way of describing the set of all pairs of numbers $(x, -2\sqrt{x+3}\ )$. This idea connects especially well to the graphical representation of $-2\sqrt{x+3}$ because the coordinates of every point on the graph are represented by $(x, -2\sqrt{x+3}\ )$.

Example 4: Is the set of pairs of real numbers $(x, \pm x - 2)$, where $x \in (-\infty, \infty)$, a function?

Solution: The symbol $\pm$ means positive *and* negative, or $+x-2$ *and* $-x-2$; thus, for most values of x, there are two values for $\pm x - 2$. For example, if x has the value of 5, $\pm x - 2$ is $+5-2$ and $-5-2$; the relation has values of 3 and -7 when x is 5. This is contradictory to the definition of a function; therefore, the set of pairs of real numbers $(x, \pm x - 2)$ is not a function. The graph of $\pm x - 2$ is shown below; as you can see for nearly all x's in the domain, there are two corresponding values of the relation.

$\pm x - 2$

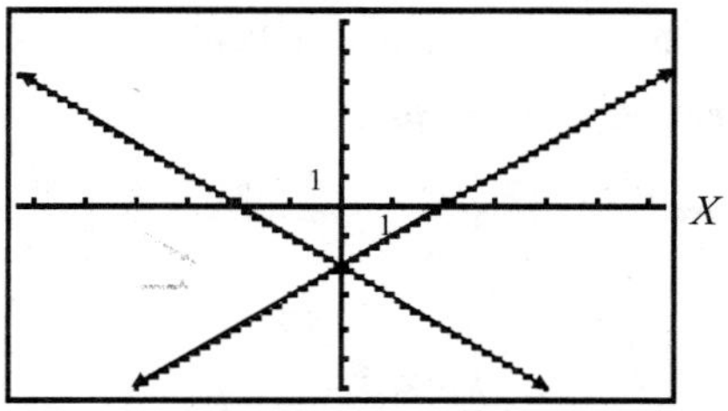

Figure 4.1.2

⁂

Example 5: Is the set of pairs of numbers { (1, 5), (2, 5), (3, 5), (4, 5) } a function? (It might be helpful to think of the second number as the expression $5x^0$.)

Solution: Yes. For every first number in each pair, there is only one corresponding second number; thus, it is a function. For example, for the first number 3, there is only one second number 5. For the first number 4, there is only one second number 5, etc.

⁂

Example 6: Is the set of pairs of real numbers $(x, 0.3x^2 + 2x - 5)$ a function? The domain is all real numbers; that is, x is an element of the interval $(-\infty, \infty)$.

Solution: Each element x taken from the domain causes $0.3x^2 + 2x - 5$ to be one and only one real number. Yes, $(x, 0.3x^2 + 2x - 5)$ is a function. This can be observed in the complete graph below because for each x in the window, you can only find one value for the expression for that x-value.

$0.3x^2 + 2x - 5$

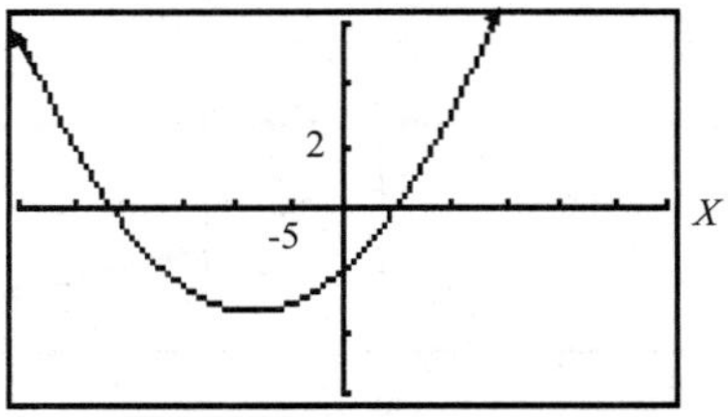

Figure 4.1.3

⁂

> **Definition of a Function, Again**
> If for every point on the graphical representation of a relationship, each x-coordinate from the domain has one and only one relation-coordinate, then the graph represents a **function**.

$0.3x^2 + 2x - 5$

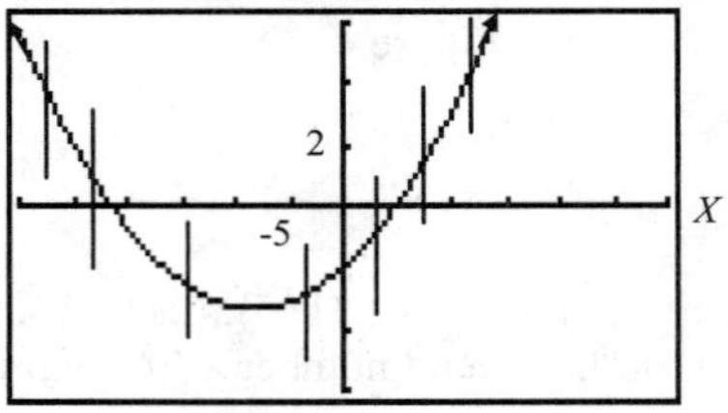

Figure 4.1.4

In the graphical representation of the function in Figure 4.1.4, every x-coordinate has exactly one relation-coordinate; therefore, it is a function. Imagine a vertical line moving from the left most point in the domain to the right most point; the vertical line only intersects the graph once for every x in the domain. Thus, for every x in the domain, there is only one relation-coordinate.

The graph of $\pm\sqrt{x}$ below is not the graph of a function because for every value of x in the domain except 0, there are two values for the relation. A vertical line can be drawn that crosses the graph twice. For example, if a vertical line is drawn through $x = 9$, the expression $\pm\sqrt{x}$ has two values, 3 and –3.

$\pm\sqrt{x}$

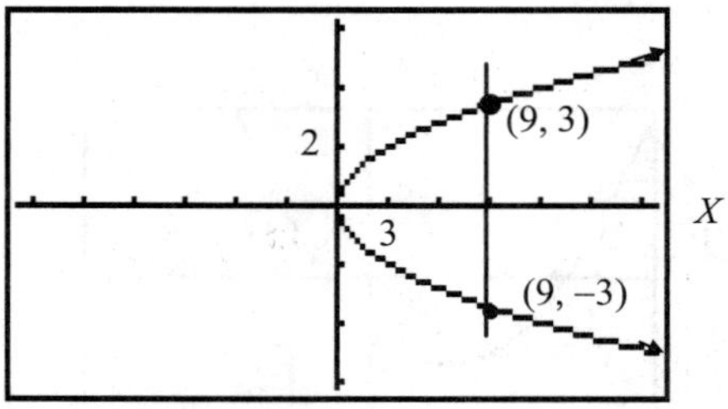

Figure 4.1.5

To identify whether or not a graph represents a function, this **vertical line test** can be used. If a vertical line can be drawn anywhere in the domain that intersects the graph more than once, the relation is not a function. That is, for the value of x where the vertical line intersects the graph more than once, there are two or more relation values. Thus, the relation does not represent a function.

Since a function is a set of pairs of numbers, these pairs can easily be put in numeric form with x in the first row and the function values in the second row. A function can be represented in numeric form as well as graphical, symbolic, and set form (a set of pairs of numbers). For example, the function $x + 2$ with a domain of $\{1, 2, 3, 4\}$ is:

Symbolic Representation

$$x + 2$$

Numeric Representation

x	1	2	3	4
$x+2$	3	4	5	6

x	$x+2$
1	3
2	4
3	5
4	6

Graphical Representation

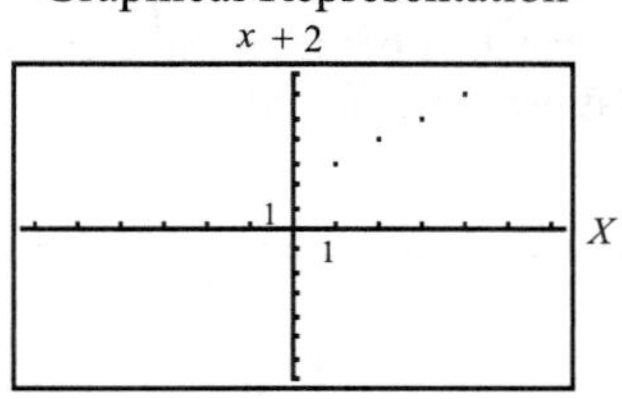

Set Representation
$\{(1, 3), (2, 4), (3, 5), (4, 6)\}$

Written Representation
"the sum of any number and 2"

Functional Notation

In mathematics, it is desirable to use symbols to represent numbers, functions, sets, etc. For example, the symbol 5 can represent a grouping of five things. The symbol x usually represents a set of real numbers. It is very common to symbolize functions using letters as well.

The set of ordered pairs of numbers $\{ (x, x + 2)$ where x is a real number $\}$ is a function. This set of ordered pairs of numbers can be symbolized by, for example, the letter f. However, **since the function is expressed in terms of the variable x, it is normal to represent this function f as $f(x)$.** This does not mean f times x. It is read f of x and represents the function. *This notation tells you the name of the function is f and it contains the variable x.* In this text, the letter x has been used frequently; however, it does not make any difference what letter is used. Graphing calculators many times *require* that the variable in the function be x. Further, graphing calculators often require the function names be Y1, Y2, etc. But this depends on what calculator you are using.

As another shortcut, the function $\{ (x, x + 2)$ where x is a real number $\}$ can be symbolized as $f(x)$. Further, it is useful to describe the function as $f(x) = x + 2$. For now, think of $f(x) = x + 2$ as a concise way to describe the function $x + 2$. **Read the above statement as "f of x is $x + 2$," or "the function f is $x + 2$."**

This new notation allows for further analysis of functions. One of the advantages to the new notation is the ability to describe one point on the graphical representation without making a complete graphical representation and to describe one pair of related numbers from the numeric representation without making a complete numeric representation. There is a very simple notation used to indicate a point on the graph or one pair of numbers from the numeric representation. If $g(x) = x^2 + 3x - 2$ and you want to look at the point when x is 5, replace the x with the number 5 like $g(5)$ (read "g of 5"). This means that the variable in function g is to be replaced with 5 and then evaluated. The value of the function when $x = 5$ is 38, and is identified as $g(5)$. This is the point (5, 38) and is on the graph of $g(x)$ and in the numeric representation as well. That is, the pair of related numbers (5, 38) is the same as (5, $g(5)$).

Example 7: If $h(x) = x - 4$, find $h(-6)$ and $h(13)$.

Solution: $h(-6)$ means that the variable in function h is to be replaced with −6, and the value of the function is to be calculated. Since $h(x) = x - 4$, then $h(-6) = (-6) - 4$ or −10, $h(-6) = -10$. A point on the graph of h is (−6, −10). Since $h(x) = x - 4$, then $h(13) = (13) - 4$ or 9, $h(13) = 9$. A point on the graph of h is (13, 9). The graph below shows the two points that were just found.

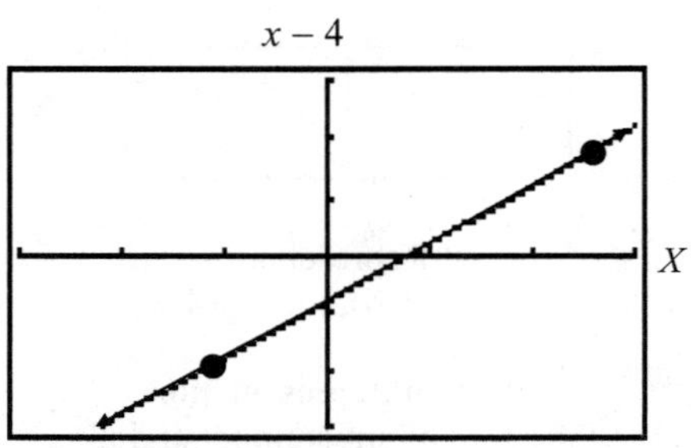

Figure 4.1.6

⁂

Example 8: If $f(z) = |z^2 - 3z + 2|$, find $f(3)$ and $f(0)$.

Solution: $f(3)$ means that the variable in the function f is to be replaced with a 3, and the value of the function is to be calculated. Likewise with 0.

Since $f(z) = |z^2 - 3z + 2|$
then $f(3) = |(3)^2 - 3(3) + 2|$
or $f(3) = 2$.

Since $f(z) = |z^2 - 3z + 2|$
then $f(0) = |(0)^2 - 3(0) + 2|$
or $f(0) = 2$.

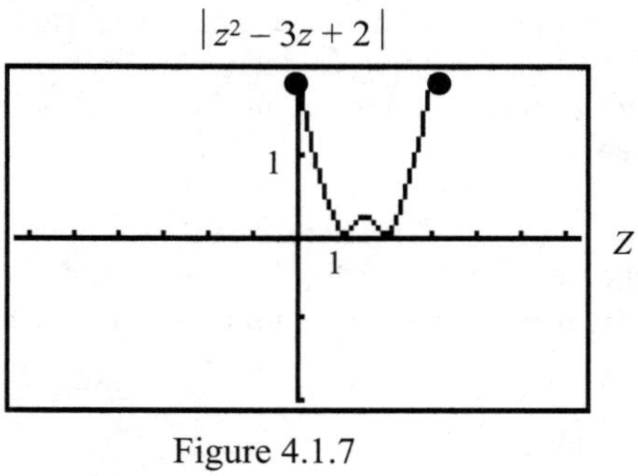

Figure 4.1.7

Most calculators use function notation and the screens below show the same work done above. The only difference on the calculator is that the name of the function must be Y_1 through Y_0.

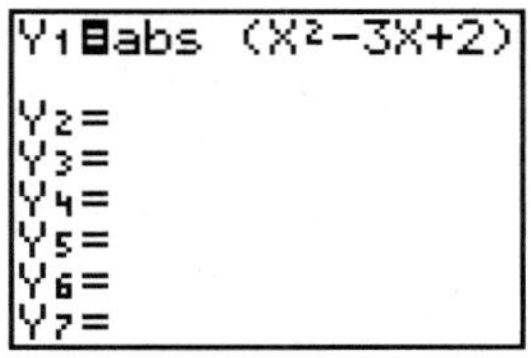

Figure 4.1.8a

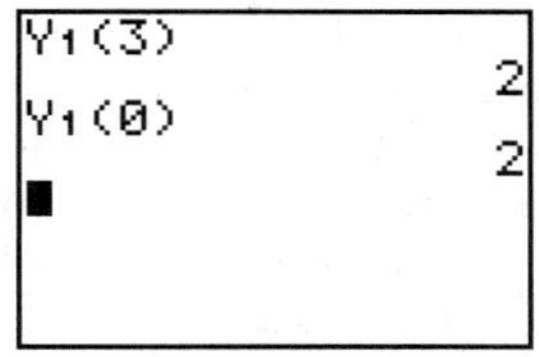

Figure 4.1.8b

The function-axis (vertical-axis) in the above graphs is labeled just as all graphs in Chapters Two and Three have been labeled. Since $h(x)$ is the same as $x - 4$, it is more common to label the function-axis with $h(x)$ instead of $x - 4$ as shown below. It is easier to write $h(x)$ and it does not take as much space as writing out the full expression or function. For the function $f(z)$, the vertical axis is normally labeled with $f(z)$ instead of $|z^2 - 3z + 2|$. For the graph of $f(z)$, the horizontal axis needs to be labeled with a z as shown below.

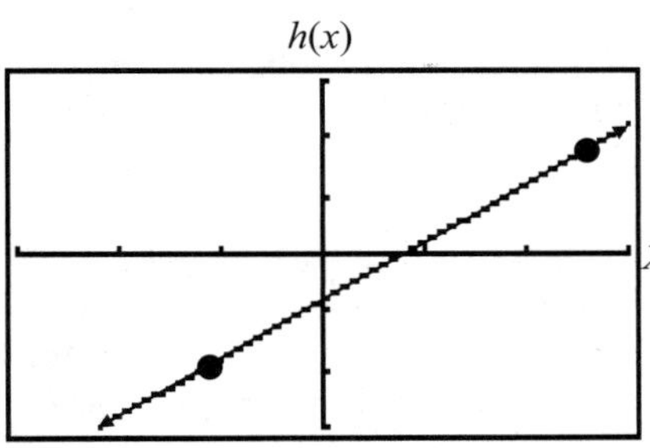

Figure 4.1.9a

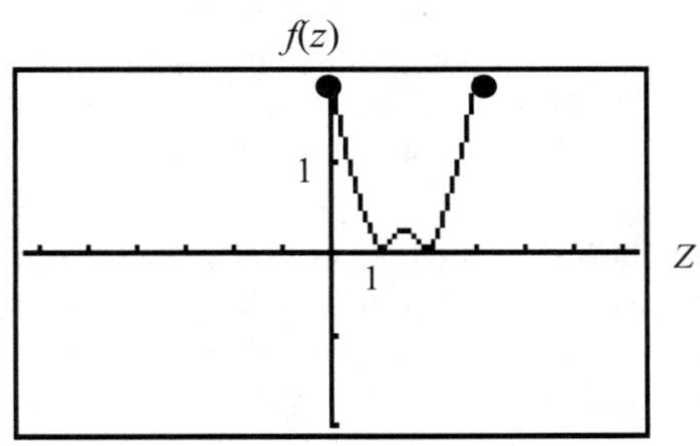

Figure 4.1.9b

A New Operation

The developer of the Polaris Amphitheater in Columbus, Ohio, sent a flyer to residents within a 10-mile radius of a proposed outdoor theater stating that comparably sized amphitheaters report the average car carries 3.3 guests and the average number of cars per event is 2500. The function $2500e$, or giving it a name, $N(e) = 2500e$ models the total number of cars at e events. For example, after 17 events, the total number of cars at the amphitheater is $N(17) = 2500 \cdot 17 = 42500$.

The function that models the total number of people attending an event is $3.3c$, or giving it a name, $P(c) = 3.3c$, where c is the number of cars at an event. This models the total number of people attending an event in c cars. For example, if 1000 cars are parked at an event, there are ($P(1000) = 3.3 \cdot 1000 = 3300$) 3,300 people in attendance. The developer of the outdoor amphitheater had to justify to the investors in the amphitheater when or whether they would make a profit from their investment. One of the things the developer projected was how many people would be expected to attend the first summer's events. The function $P(c) = 3.3c$ models the number of people from c cars and $N(e) = 2500e$ models the number of cars per e events. If the developer has scheduled 60 events for the first summer, how many people are expected for the first summer? While you may be able to figure out the answer to this question without the new notation, the new notation is used below to prepare you for more sophisticated examples.

> $P(c)$ represents the number of people attending one event in c cars, and $N(e)$ represents the number of cars at e events. But you want the number of people at e events. Since you want the total number of people, start with the people function $P(c) = 3.3c$. But c represents the number of cars at 1 event and you need the number of cars at e events. This is $N(e) = 2500e$. So, replace c with $2500e$ and simplify the new function that represents the number of people at e events.
>
> $$P(c) = 3.3c$$
>
> 
>
> $$\longrightarrow P(N(e)) = 3.3(2500e) = 8250e \longrightarrow$$
>
> 
>
> $$(P \circ N)(e) = 8250e$$
>
> By replacing the function variable c (which represents cars) with the function $N(e)$ (which represents cars) you have a new function $(P \circ N)(e) = 8250e$ that models the number of people in terms of the function variable e, the number of events. Finally, to answer the question "How many people are expected to attend 60 events?" find $(P \circ N)(60) = 8250 \cdot 60 = 495{,}000$.

The new operation is called the **composition of functions** and will be used mainly in Chapters 7 - 10 and in later courses. The symbol used to imply the composition of two functions is $\circ$. For example, to imply that you want to "plug" the function $f(x) = x + 4$ into the function $g(x) = 7x^2 - 3x + 5$, the notation is $(g \circ f)(x)$. Of course, you also know that $g(f(x))$ means to replace the variable in $g(x)$ with $f(x)$.

Example 9: The fry cook in the college cafeteria earns \$100 per week plus \$0.25 for each burger fried. Thus, the model for his weekly salary is $S(b) = 0.25b + 100$, where b represents the number of burgers fried for the week. He is interested in buying some CD's with his salary. The number of CD's (N) he can buy at \$9 each depends on his salary. That is, $N(S) = \frac{S}{9}$. Find a function that models the number of burgers that must be fried to buy N CD's.

Solution: You are looking for the number of CD's that can be purchased for b burgers. Since you want the number of CD's that can be purchased for flipping b burgers, start with the number of CD's function $N(S) = \frac{S}{9}$. But S represents the salary in dollars and you want salary in terms of burgers. So, replace S with $S(b) = 0.25b + 100$ (salary in burgers) and simplify the new function for the number of CD's that can be purchased for flipping b burgers.

$$N(S) = \frac{S}{9}$$

$$N(S(b)) = \frac{0.25b + 100}{9}$$

$$(N \circ S)(b) = \frac{0.25b + 100}{9}$$

The composition is $(N \circ S)(b) = \frac{0.25b+100}{9}$. This means, for example, If the cook fries 160 burgers, he can buy $(N \circ S)(160) = \frac{0.25 \times 160 + 100}{9}$ = about 15 CD's.

⁂

***Example 10*:** Find the composition $(f \circ g)(x)$, if $f(x) = x^2 - 5x + 3$ and $g(x) = x + 2$.

Solution: $(f \circ g)(x)$ means find $f(g(x))$; $f(g(x))$ means $f(x + 2)$, or $(f \circ g)(x) = (x + 2)^2 - 5(x + 2) + 3$.

⁂

4.1 EXERCISES

Many times you can work the exercises by algebraic, numeric, or graphical methods. You can learn about mathematics no matter which method you use. Part of the learning process is for you to decide which method is best for you to use. Please read the text carefully before you do the exercises. Read the **next assigned section** after you finish the assignment below.

A Quick Look To The Future

-4. Describe what you think $f(x) + g(x)$ means.

-3. Describe what you think $f(x) \times g(x)$ means.

-2. Give an example from arithmetic showing that a sum can be written as a product.

-1. Can $x + y + z$ be written as a product? If yes, what product?

Previous Topics Revisited

1. What is the zero of $2\sqrt{x-1} - 2$?
2. What is the minimum value of $\sqrt{x+2} - 19$?
3. What is the normal domain of $0.2\sqrt{x+4} - 2$?
4. Find the zeros of $3|x-1| - 2$.
5. What is the range of $2|x-1| + 3$?
6. Find the zeros of $2x^2 - 5x - 4$.

New Topic Exercises

Are the sets of ordered pairs of numbers in Exercises 7 - 9 functions? Why or why not?

7. $\{ (-1, 3), (1, 4), (2, 6), (1, 5) \}$
8. $\{ (-2, 5), (-1, 5), (0, 5), (1, 5), (2, 5), (3, 5) \}$
9. $\{ (0.1, 2), (0.2, 3), (0.3, 4) \}$

Do the following pairs of numbers represent functions? Why or why not? Use the normal domain for each expression.

10. $\{(x, 0x + 2)\}$
11. $\{(x, \frac{1}{2}x^2 - 4x + 5)\}$

12. $\{(x, |x-3|+4.1)\}$

13. $\{(x, 2x \pm 6)\}$

14. $\{(x, -3\sqrt{x+4} - 7)\}$

Do the following graphs represent functions? Why or why not?

15.

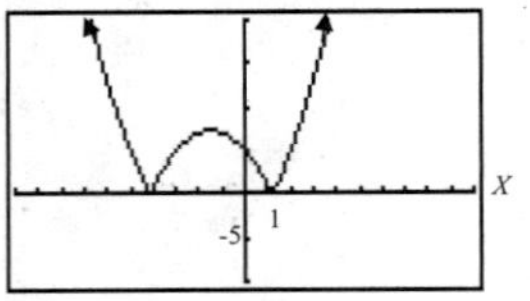

16.

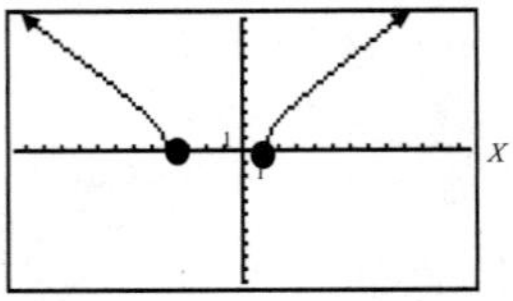

17.

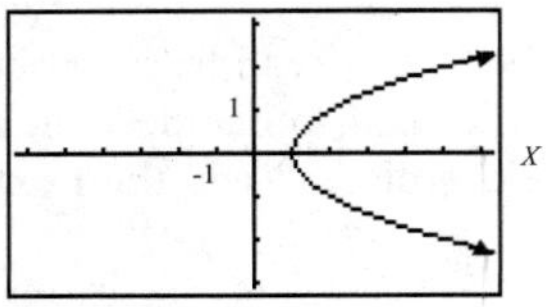

18.

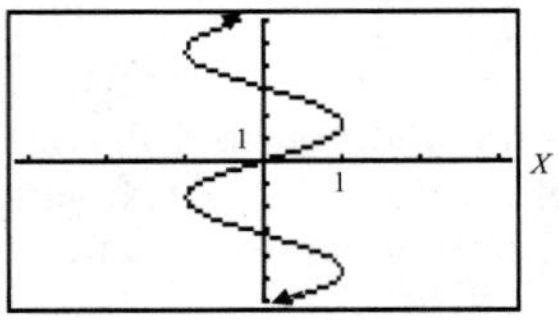

For the function $f(x) = 3x^2 + 8x - 2$, find the following:

19. $f(-1), f(-3), f(4)$

20. $f(a), f(b)$

21. $f(1) + f(-5)$

22. $f(-2), f(2)$, and the points $(-2, ?)$ and $(2, ?)$ on the graph of f.

For the function $g(x) = |x^2 - 1|$, find the following:

23. $g(1)$, $g(-1)$, and $g(2000)$

24. $g(a)$, $g(b)$

25. $g(-1) - [g(2)]^2$

26. $g(-3)$, $g(3)$, and the points $(-3, ?)$ and $(3, ?)$ on the graph of g.

Use the function $q(x) = |2x-7| - \sqrt{x+4}$ in Exercises 27 and 28.

27. $q(0)$, $q(4)$, $q(-4)$, $q(1)$, $q(\pi)$

28. Show all of the above points on the graph of $q(x)$.

29. One point on the graph of $v(x)$ is $(-3, 2)$, what is $v(-3)$?

30. The zero of $p(x)$ is 17, what is $p(17)$?

31. The numeric representation of $f(t)$ is:
What is $f(0)$?

t	–4	–1	0	2	5
$f(t)$	12	–3	9	4	97

What is $f(5)$?

32. Find $(h \circ j)(x)$ for the following functions.

a. $h(x) = \sqrt{x-5}$, $j(x) = x^2 + 5$

b. $h(x) = 2x^2 - 3x + 1$, $j(x) = x - 7$

c. $h(x) = 2x - 1$, $j(x) = \frac{1}{2}x + \frac{1}{2}$

33. How should the vertical-axis be labeled when the graph of $m(z) = z + 2$ is graphed?
How should the horizontal-axis be labeled?

34. Use set notation to describe any function of your choosing.

35. Draw the graphical representation of any function of your choice.

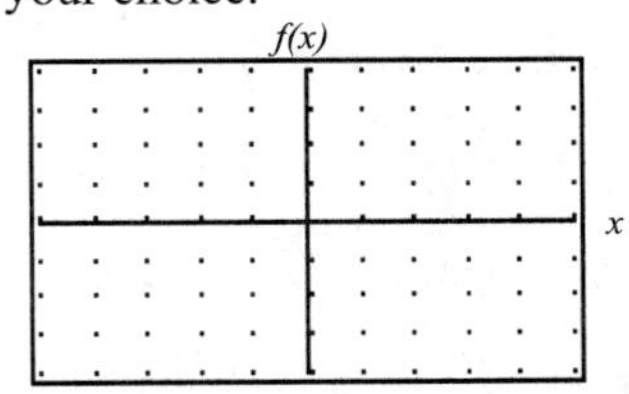

36. Draw the graphical representation of any relation that is NOT a function.

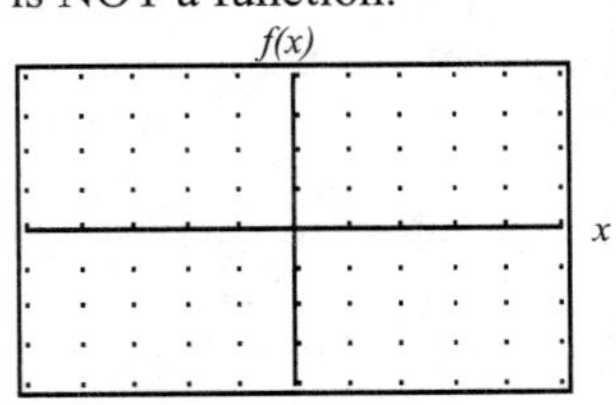

37. If $f(x) = x + 2$, find the value of $f(f(f(f(-3))))$

38. Make-up functions for $f(t)$, $g(z)$, and $h(w)$

Find $f(7)$ and one point on the graph of f.

Find $g(-2)$ and one point on the graph of g.

Find $h(0)$ and one point on the graph of h.

From Mathematics to English

39. After reading this section, make a list of questions that you want to ask your instructor.

40. Continue in your daily journal and make an entry. In addition to your normal entry on thoughts about the mathematics in this section, list at least two positive comments about what you have learned about this topic.

41. In paragraph format, summarize the material in this section of the text in your daily journal.

42. Describe how your classroom instructor made this topic more understandable and clear.

43. After reading the text and listening to your instructor, what do you not understand about this topic?

44. For the function $f(x) = -(x+3)^2 - 4$, explain how you might find $f(-3)$ without using pencil and paper

45. Explain why the set $\{(-1, 2), (-3, 2), (5, 2)\}$ is a function.

46. Explain why the set $\{(2, -1), (2, -3), (2, 5)\}$ is not a function.

47. Give an example of any relation in graphical form.

48. Is every function a relation? Explain.

49. If $f(x) = 2x + 4$, explain what $f(a)$ means.

50. What do the symbols $f(x)$ mean to you? What is f and what is x?

51. Describe what the notation $(g \circ f)(x)$ means to you.

52. If $f(x) = 2x + 6$ and $g(x) = -5x + 4$, make-up a meaning for the notation $(f + g)(x)$ and $(fg)(x)$.

From English to Mathematics

In the next six exercises, convert the English statement to an equivalent mathematical statement.

53. the function f is: a number squared minus fifteen

54. the function g is: the square root of a number plus four

55. the square root of a number plus four is the function p

56. the function R is the number x plus the number x squared

57. f of 5 is 9

58. f of seven is thirteen

Explorations

59. In a function, for every x there must be only one corresponding value of the function. Draw the graphs of two functions where this is true, <u>and</u> for every value of the function there is only one corresponding value for x.

60. Below are examples of relations that are not functions. Explore with your calculator and figure a way to graph each of the relations on your calculator. Transfer each graph to graph paper.

$\pm x + 3$ $\quad -2x \pm 1$ $\quad \pm\sqrt{x}$ $\quad \pm x^2 - 2$ $\quad \pm|x + 2| - 3$

61. Calculate $\dfrac{f(x+.01) - f(x)}{.01}$ when x is 1, 2, 3, and finally 4, for each function below:

a. $f(x) = 5x - 7$ b. $f(x) = 8x - 1$

62. Using a sentence structure, describe the placement of the graph of $g(x)$ with respect to $f(x)$.

Hint 1: put both graphs on the same window before answering the question.
Hint 2: if you store $f(x)$ in Y_1, then Y_2 can be written as $Y_1(x + 4)$

a. $f(x) = \sqrt{x} + 1$ $\quad g(x) = f(x+4)$

b. $f(x) = x^2 - 2$ $\quad g(x) = f(x+4)$

c. $f(x) = -|x| + 3$ $\quad g(x) = f(x+4)$

d. $f(x) = -x^2 + 3$ $\quad g(x) = f(x+4)$

63. Using a sentence structure, describe the placement of the graph of $g(x)$ with respect to $f(x)$.

Hint 1: put both graphs on the same window before answering the question.
Hint 2: if you store $f(x)$ in Y_1, then Y_2 can be written as $Y_1 + 4$

a. $f(x) = \sqrt{x} + 1$ $\quad g(x) = f(x) + 4$

b. $f(x) = x^2 - 2$ $\quad g(x) = f(x) + 4$

c. $f(x) = -|x| + 3$ $\quad g(x) = f(x) + 4$

d. $f(x) = -x^2 + 3$ $\quad g(x) = f(x) + 4$

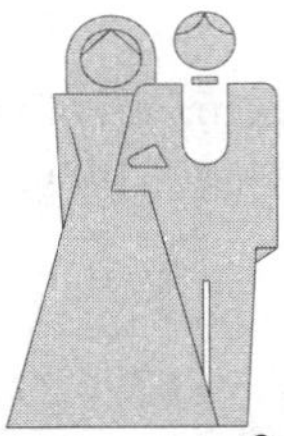

4.2 Addition and Subtraction of Polynomial Functions

The Wedding Plans

In making plans for his wedding to Jennifer, Bill found a reception hall for \$950. The management of the reception hall also **charges** \$1 per person for valet parking ($950 + 1x$, where x is the number of guests). The caterer estimated the **cost** of the meal per person at \$12.95 plus \$4 per person for wine ($12.95x + 4x$). The musicians **charge** \$350 and the caterer offered a rebate of \$1.50 per person if Bill paid at least a month before the wedding ($350 - 1.5x$). What is the cost of Bill's share of the wedding celebration if x people attend? The answer to this question can be symbolized as $(950 + 1x) + (12.95x + 4x) + (350 - 1.5x)$, where x is the number of people attending the reception. If Bill could just simplify this polynomial, he could simplify his wedding plans and make the bride's parents happy.

> Please note that the discussion that follows is symbolic in nature. You will likely understand the mathematics better and remember it longer if you use a visual development of the method for adding and subtraction. A visual method can be found in the StudyCard e-activity on adding and subtracting polynomials.

This section is a brief review of adding and subtracting polynomial functions. However, the notation used to imply addition or subtraction is new and you will be encouraged to check your addition or subtraction answers as demonstrated in Example 1.

Suppose $f(x)$ is $x^2 + 5x - 4$ and $g(x)$ is $3x^2 - 7x - 3$. The mathematical notation used to imply the addition of these two polynomial functions is $(f + g)(x)$. Likewise, if subtraction is desired, the notation is $(f - g)(x)$. Since polynomial functions represent real numbers it would make sense to add or subtract polynomial functions by using the properties of real numbers, and the rules for addition and subtraction developed earlier.

Example 1: If $f(x) = x^2 + 5x - 4$ and $g(x) = 3x^2 - 7x - 3$, find $(f + g)(x)$.

Solution: $(f + g)(x)$ means $(x^2 + 5x - 4) + (3x^2 - 7x - 3)$.

$[x^2 + 5x + (-4)] + [3x^2 + (-7x) + (-3)]$	rule for subtraction
$x^2 + 5x + (-4) + 3x^2 + (-7x) + (-3)$	associative property
$x^2 + 3x^2 + 5x + (-7x) + (-4) + (-3)$	commutative property
$(x^2 + 3x^2) + (5x + (-7x)) + ((-4) + (-3))$	associative property
$(1 + 3)x^2 + (5 + (-7))x + ((-4) + (-3))$	distributive property
$4x^2 + (-2x) + (-7)$	rules for addition
$4x^2 - 2x - 7$	rule for subtraction

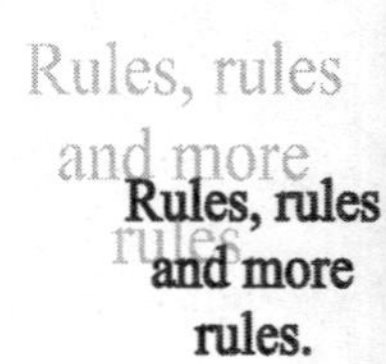

Below are two methods for checking to see if the sum $4x^2 - 2x - 7$ is correct; both methods are done with technology. The first method compares the numeric representations of the problem shown in Y_5 and of the answer obtained above shown in Y_6. If the numeric representations are the same, then the symbolic representations are the same and the answer is correct. The second method uses the calculator's logic system.

The problem is entered on the home screen followed by an equal sign followed by the answer obtained above. The number 1 means the statement is true; a 0 means the statement is false. A true statement means the answer is correct.

Method One

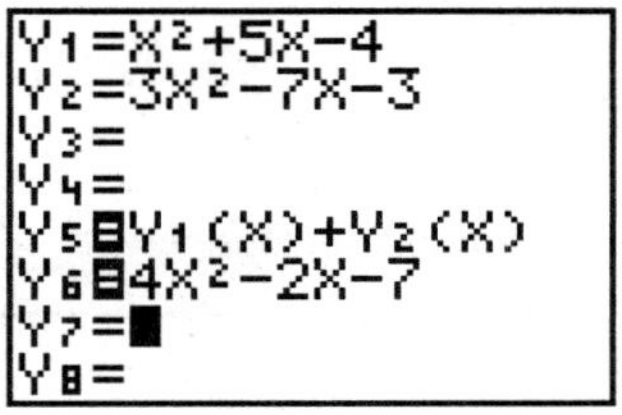

Figure 4.2.1a

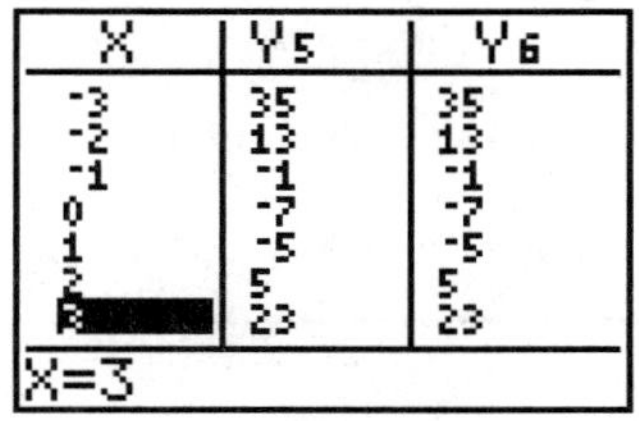

Figure 4.2.1b

The correct answer is stored in Y_5 and the answer being checked is stored in Y_6. Note that Y_1 and Y_2 are in the "off" state. Also, the correct function notation is not needed in Y_5, it can be written as $Y_1 + Y_2$.

Method Two

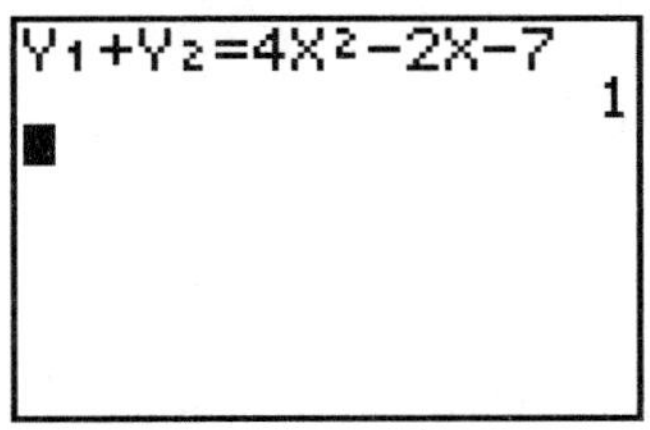

Figure 4.2.1c

Method Two causes the calculator to compare the numeric value of $Y_1 + Y_2$ to the numeric value of $4x^2 - 2x - 7$, where the value of x is what is currently stored in your calculator's memory. The number 1 indicates that the numeric value of the problem equals the numeric value of the answer. ***This method requires that the value of x be in the normal domain of the problem.***

⁂

If the analytical process in Example 1 is *studied carefully*, it can be seen that the $1x^2$ and the $3x^2$ were simplified by adding 1 + 3. The $5x$ and the $(-7x)$ were simplified by adding 5 + (–7). Finally, the (–4) and (–3) were added.

By using similar reasoning, $(-6x^2 + 13x - 5) + (2x^2 - 15x + 7)$ can be added by taking:

$[(-6) + 2]$ times x^2 plus

$[13 + (-15)]$ times x plus

$(-5) + 7$.

That is: $(-6 + 2)x^2 + (13 + (-15))x + (-5 + 7)$

$-4x^2 + (-2)x + 2$ or

$-4x^2 - 2x + 2$.

Terms that have identical variable parts can be simplified by adding their coefficients. These are called **similar terms** or **like terms**.

Addition of Polynomials

To add polynomials: add similar terms, then add these sums together.

Example 2: Add: $(-5x^3 + 5x^2 - 6x + 7) + (-2x^3 - 9x^2 - 3x - 9)$.

Solution: Add similar terms; $-5x^3$ and $-2x^3$; $5x^2$ and $-9x^2$; $-6x$ and $-3x$; 7 and -9.

$-5x^3 + (-2x^3) + 5x^2 + (-9x^2) + (-6x) + (-3x) + 7 + (-9)$

$-7x^3 + (-4x^2) + (-9x) + (-2)$

$-7x^3 - 4x^2 - 9x - 2$

⁂

Using a graphical approach, there is a third method you can use to check the answer to the above polynomial addition problem or any other operation.

Graph the problem $(-5x^3 + 5x^2 - 6x + 7) + (-2x^3 - 9x^2 - 3x - 9)$. Using the same coordinate system, also graph the answer $-7x^3 - 4x^2 - 9x - 2$.

If the two graphical representations are identical, the two functions must be equivalent in symbolic form. Likewise, if the numeric representations of two functions are identical, then the symbolic representations are equivalent.

Shown below are the graphs of the problem and the answer:

$(-5x^3 + 5x^2 - 6x + 7) + (-2x^3 - 9x^2 - 3x - 9)$ and $-7x^3 - 4x^2 - 9x - 2$.

Method Three:

$(-5x^3 + 5x^2 - 6x + 7) + (-2x^3 - 9x^2 - 3x - 9)$

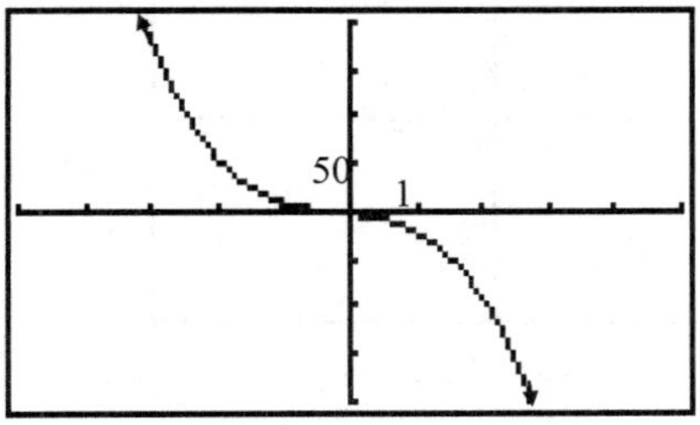

Figure 4.2.2a

$-7x^3 - 4x^2 - 9x - 2$

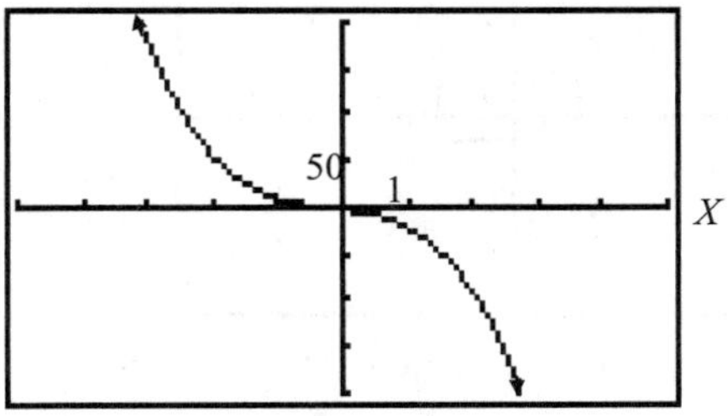

Figure 4.2.2b

Notice that the shape of the third degree polynomial is different than any other function studied thus far.

Below are the numeric representations for $(-5x^3 + 5x^2 - 6x + 7) + (-2x^3 - 9x^2 - 3x - 9)$ and $-7x^3 - 4x^2 - 9x - 2$.

Table 4.2.1

x	−3	−2	−1	0	1	2	3
$(-5x^3 + 5x^2 - 6x + 7) + (-2x^3 - 9x^2 - 3x - 9)$	178	56	10	−2	−22	−92	−254
$-7x^3 - 4x^2 - 9x - 2$	178	56	10	−2	−22	−92	−254

Hopefully, it is obvious that the numeric representations of the problem as well as the answer are identical. The conclusion is that the answer must be correct.

Since the two graphs/tables are identical, the two polynomial expressions are equivalent. It may not be obvious that the above graphs are identical; however, if a calculator is used to graph the two functions in sequence, the second graph will not show because it will be the same graph as the first function. When trace is used, if you jump from graph to graph the coordinates of the cursor will **NOT** change. This indicates the graphs are the same graph; thus, the problem and the answer are the same.

Example 3: Subtract: $(7x^2 - 5x - 8) - (-3x^2 + 6x + 3)$.

Solution: Use the rule of subtraction inside grouping symbols first:
$[7x^2 + (-5x) + (-8)] - (-3x^2 + 6x + 3)]$.

Since the second polynomial is being subtracted from the first, use the rule of subtraction again: $[7x^2 + (-5x) + (-8)] + [(+3x^2) + (-6x) + (-3)]$.

The problem has now been expressed in terms of addition; thus, it is now possible to add similar terms. The similar terms are $7x^2$ and $3x^2$; $-5x$ and $-6x$; -8 and -3.

Add similar terms:
$(7x^2 + 3x^2) + (-5x + [-6x]) + (-8 + [-3])$
$10x^2 + (-11x) + (-11)$
$10x^2 - 11x - 11.$

A check of the work is shown below by graphing both polynomial expressions.

$(7x^2 - 5x - 8) - (-3x^2 + 6x + 3)$

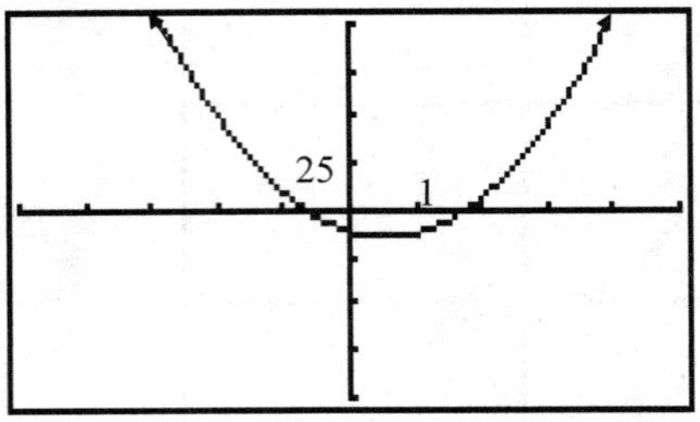

Figure 4.2.3a

$10x^2 - 11x - 11$

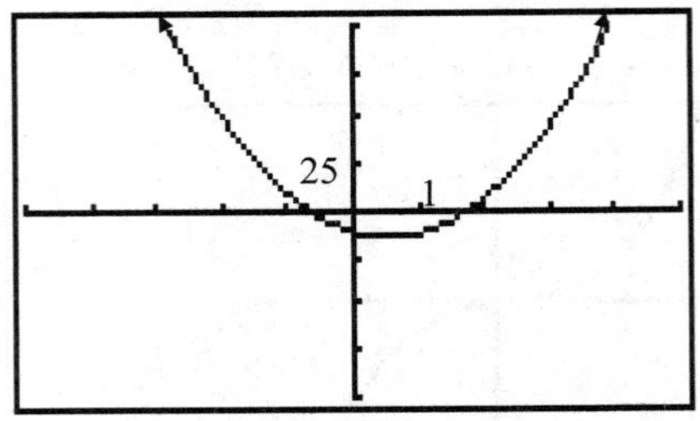

Figure 4.2.3b

A numerical check also confirms that the answer is correct.

Table 4.2.2

x	–3	–2	–1	0	1	2	3
$(7x^2 - 5x - 8) - (-3x^2 + 6x + 3)$	112	51	10	–11	–12	7	46
$10x^2 - 11x - 11$	112	51	10	–11	–12	7	46

⁂

Example 4: Simplify: $(2x^3 - 5x^2) - (-5x^3 - 3x^2) - (-6x^3 - x^2)$.

Solution: Use the rule of subtraction inside grouping symbols first:
$(2x^3 + [-5x^2]) - (-5x^3 + [-3x^2]) - (-6x^3 + [-x^2])$.

Use the rule of subtraction on the polynomials that are being subtracted.
$(2x^3 + [-5x^2]) + (+5x^3 + [+3x^2 1) + (+6x^3 + [+x^2])$.

Add the similar terms; $2x^3$, $5x^3$, and $6x^3$; $-5x^2$, $3x^2$, and x^2.
$2x^3 + 5x^3 + 6x^3 + (-5x^2) + 3x^2 + x^2$, or $13x^3 + (-x^2)$

The simplified form is $13x^3 - x^2$.
You are encouraged to check the work graphically or numerically. The logic check is:

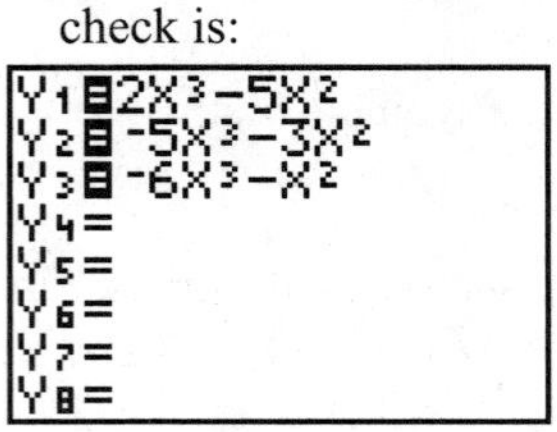

Figure 4.2.4a

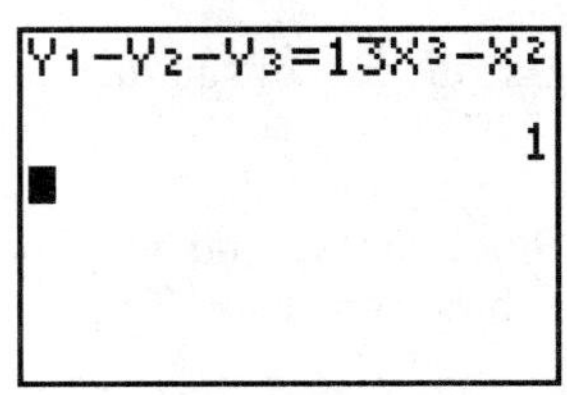

Figure 4.2.4b

⁂

Because of the use of technology in doing mathematics, many of the algebraic manipulation (like addition and subtraction of polynomials) skills learned to do mathematics with pencil and paper are not nearly as important. Other skills have taken their place. For example, instead of learning how to graph a function by hand, now you must learn how to find a window on your calculator that shows the complete graph of the function. You must learn how to enter functions on the calculator. Relative to the material in this section, the additions and subtractions similar to the above examples have diminished use. In their place you will find that you will need to do more additions and subtractions like those in the examples that follow.

Example 5: Find $(f-g)(x)$ if $f(x) = 6\sqrt{x-1}$ and $g(x) = 9\sqrt{x-1}$.

Solution: $(f-g)(x)$ is $6\sqrt{x-1} - 9\sqrt{x-1}$. The subtraction is an exact match with the pattern established in the distributive property $ab - ac = a(b - c)$; that is, $6\sqrt{x-1} - 9\sqrt{x-1} = \sqrt{x-1}(6-9) = \sqrt{x-1}(-3) = -3\sqrt{x-1}$. The difference of functions f and g is $(f-g)(x) = -3\sqrt{x-1}$. To confirm that the difference is correct, below is a numerical check. The correct answer is stored in Y_4 and the answer found above is stored in Y_5.

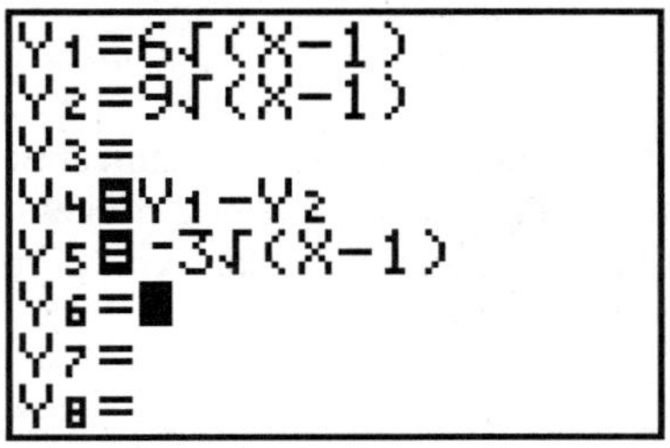

Figure 4.2.5a

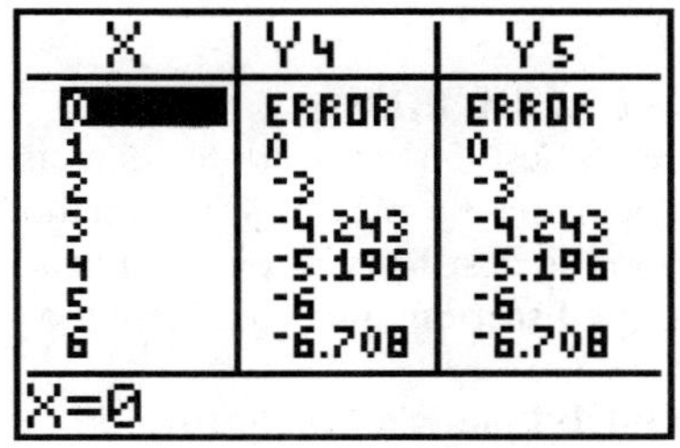

Figure 4.2.5b

The numeric representations of the problem and the answer are identical; therefore, the symbolic representations are the same.

Example 6: Find the sum of *f* and *g* if $f(x) = -7|x+2|$ and $g(x) = 3|x+2|$.

Solution: $(f+g)(x) = -7|x+2| + 3|x+2|$. The sum is an exact match to the pattern established in the distributive property $ab + ac = a(b+c)$. The sum is $-7|x+2| + 3|x+2| = |x+2|(-7+3) = -4|x+2|$. $(f+g)(x) = -4|x+2|$.

Example 7: Find $(f+g-h)(x)$, where $f(x) = 3|x-4|$, $g(x) = 2\sqrt{x+1}$, and $h(x) = 8|x-4|$.

Solution: $(f+g-h)(x)$ is $3|x-4| + 2\sqrt{x+1} - 8|x-4|$. The only terms that are similar are the first and last; thus, they are the only terms that can be simplified. $(f+g-h)(x)$ is $-5|x-4| + 2\sqrt{x+1}$.

Example 8: How do you transform the equation $7x + 4 = -3(x-4)$ to get zero on the right?

Solution: Add the term $3(x-4)$ to both sides of the equation and simplify the right side.

$7x + 4 + 3(x-4) = -3(x-4) + 3(x-4)$ This is like adding $-3b + 3b$.

$7x + 4 + 3(x-4) = 0$

Example 9: How do you transform the equation $-3x^2 + 2 = 9(x-1) - 5(x+3)$ to get zero on the right?

Solution: To get zero on the right side of the equation, use the subtraction and addition properties from Chapter One. Subtract $9(x-1)$ from both sides and add $5(x+3)$ to both sides. $-3x^2 + 2 - 9(x-1) + 5(x+3) = 9(x-1) - 5(x+3) - 9(x-1) + 5(x+3)$ now simplify $-3x^2 + 2 - 9(x-1) + 5(x+3) = 0$

⁂

4.2 EXERCISES

Many times you can work the exercises by algebraic, numeric, or graphical methods. You can learn about mathematics no matter which method you use. Part of the learning process is for you to decide which method is best for you to use. Please read the text carefully before you do the exercises. Read the **next assigned section** after you finish the assignment below.

A Quick Look To The Future

-4. If a second-degree polynomial function is multiplied be a third degree function, what do you suspect the degree of the product is?

-3. Can 2×3 + 5×3 be factored?

-2. $x^2 + 3x + 2$ is the same thing as $x(x+3) + 2$. Would you agree or disagree that $x(x+3) + 2$ is written in product form?

-1. While you can't simplify the sum $-2|x| + x^2$, can you graph the sum?

Previous Topics Revisited

1. $f(x) = x^8 - x^{21}$, find $f(-1)$.
2. $g(x) = (x+2)(x-1)$, find $g(-3)$.
3. $h(t) = t^2 - 2t - 5$, find $[h(-1)]^2$.
4. Is { (1,3), (2,3), (6,3), (−1,2) } a function? Why or why not?
5. Is $\pm\sqrt{x^2 - 15}$ a function? Why or why not?
6. What is the minimum value of $2\sqrt{x+3} - 5$?

New Topic Exercises

Do the indicated operations if $f(x) = -3x^2 - 6x + 3$, $g(x) = -x + 2$, and $h(x) = 4x^3 - 5x^2 - 7x + 13$.

7. $(f+g)(x)$
8. $(f-h)(x)$
9. $(f+g+h)(x)$
10. $(g-h)(x)$
11. $(h-g)(x)$
12. $(h-g+f)(x)$
13. Does $(g-h)(x) = (h-g)(x)$?

14. Is $\left(7x^3 - 13x^2 + 8x - 3\right) - \left(2x^3 + 4x^2 - x - 9\right)$ the same as $5x^3 - 9x^2 + 7x - 12$?

15. Is $12x^2 - 12x^2$ the same as $12x^0$?

16. Is $5x - x$ the same as 5?

17. For what values of x does $3x^2 + 7x^2$ equal $10x^2$?

18. If $(f + g)(x)$ is $3x^2 - 5x + 2$, find $f(x)$ and $g(x)$. (Note: there is more than one answer.)

19. Find two functions whose sum is $5x^3 - 4x^2 + 7x - 1$.

20. Find two functions whose difference is $5x^3 - 4x^2 + 7x - 1$.

21. Find four functions whose sum is $-3x^2 + x - 2$.

22. If $(f + g - h)(x)$ is $-4x^3 + 3x^2 - 5$, find $f(x)$, $g(x)$, and $h(x)$.

23. $f(x) = 2x^2 - 3x + 5$ and $g(x) = 7x - 1$

    a. find $(f + g)(2)$.
    b. find $(f - g)(-1)$
    c. find $[f(-2) + g(1)]^2$
    d. Does $f(3) + g(3)$ equal $(f + g)(3)$?
    e. Does $f(-1) - g(2)$ equal $(f - g)(-1 - 2)$?

24. Find $(f + g)(x)$ and $(f - g)(x)$.

    a. $f(x) = -2|x + 4|$ and $g(x) = 5|x + 4|$
    b. $f(x) = -7\sqrt{x - 5}$ and $g(x) = 3\sqrt{x - 5}$
    c. $f(x) = 6\left(3x^5\right)$ and $g(x) = 9\left(3x^5\right)$
    d. $f(x) = -3\left(\dfrac{2x}{x + 1}\right)$ and $g(x) = -7\left(\dfrac{2x}{x + 1}\right)$
    e. $f(x) = -7 \cdot 3^{x-4}$ and $g(x) = 5 \cdot 3^{x-4}$

From Mathematics to English

25. After reading this section, make a list of questions that you want to ask your instructor.

26. Continue in your daily journal and make an entry. In addition to your normal entry on thoughts about the mathematics in this section, list at least two positive comments about what you have learned about this topic.

27. In paragraph format, summarize the material in this section of the text in your daily journal.

28. Describe how your classroom instructor made this topic more understandable and clear.

29. After reading the text and listening to your instructor, what do you not understand about this topic?

30. Is the graph of $(j + k)(x)$ identical to $(k + j)(x)$? Why?

31. If the graphical representations of two functions are identical, why are the symbolic representations of the functions equivalent? Use paragraph format in answering this question.

32. Describe the relationship between the graphs of $f(x) = x$ and $g(x) = 2$ and the graph of $(f + g)(x)$.

33. For any two functions $f(x)$ and $g(x)$, does $(f - g)(x) = (g - f)(x)$? Why?

34. Describe how to find $(f + g)(x)$ when $f(x) = -12x^2 + 3x - 6$ and $g(x) = -4x + 3$. Do not actually find $(f + g)(x)$.

35. Describe how to find $(f - g)(x)$ when $f(x) = -12x^2 + 3x - 6$ and $g(x) = -4x + 3$. Do not actually find $(f - g)(x)$.

36. If $f(x) = -12x^2 + 3x - 6$ and $g(x) = -4x + 3$, describe how to find $[f(2) - g(5)]$. Use English -- not mathematics.

From English to Mathematics

In the next three exercises, convert the English statement to an equivalent mathematical statement.

37. the sum of function f and function g

38. three times the function f of x

39. the difference of functions f and g

Explorations

40. Using a calculator, store any linear function in Y_1, any quadratic function in Y_2, and the square root function $\sqrt{x+2}$ in Y_3.

    Describe the behavior of the functions $Y_1 + Y_3$.

    Describe the behavior of $Y_2 + Y_3$.

    Describe the behavior of the functions $Y_1 - Y_3$.

    Describe the behavior of $Y_2 - Y_3$.

    Include increasing/decreasing, max/min, positive/negative, zeros, and domain/range in your analysis of the behavior.

4.3 Multiplication of Polynomial Functions

The Weird Carpenter

Carla the carpenter loves perfect squares and prime numbers. She decides to build a deck on her house with dimensions of "a perfect square x plus the prime number 7 by the same perfect square x minus the prime number 3." What is the area of Carla's deck expressed symbolically? Since the area of a rectangle is length times width, the area is $(x + 7)(x - 3)$. If Carla knew more about mathematics, she would realize that the area can also be expressed as $x^2 + 4x - 21$ -- perhaps a simpler way of describing the area, or not. However, you should note that the area of a deck as a quadratic function and it comes from multiplying two linear functions.

Just as with addition and subtraction of polynomial functions, multiplication of polynomial functions requires the properties and rules of real numbers. The notation is similar to that used to denote addition or subtraction of polynomials. To indicate multiplication of two polynomial functions $f(x)$ and $g(x)$, use the notation $(fg)(x)$. For example, if $f(x) = 2x^2$ and $g(x) = x - 1$, $(fg)(x)$ means $f(x) \cdot g(x)$ or $2x^2(x - 1)$. Just like addition and subtraction of functions creates a new function called the **sum** or **difference**, multiplication of functions creates a new function called the **product**.

Please note that the discussion that follows is symbolic in nature. You will likely understand the mathematics better and remember it longer if you use a visual development of the method for multiplication. A visual method can be found in the StudyCard e-activity on multiplication of polynomials.

Since polynomial functions represent real numbers, it would make sense that to multiply polynomial functions you will use the properties and rules for multiplication developed in Chapter One.

Multiplication of Terms

Example 1: If $f(x) = 2x^4$ and $g(x) = -5x^3$, find $(fg)(x)$.

Solution: The problem is: $(2x^4)(-5x^3)$

change the grouping:	$2x^4(-5)x^3$	associative property
change the order:	$2(-5)x^4x^3$	commutative property
change the grouping:	$(2 \cdot (-5))(x^4x^3)$	associative property
multiply:	$(-10)(x^7)$	product property
change the grouping:	$-10x^7$	associative property

$(fg)(x)$ is $-10x^7$

⁂

Studying the above example reveals a quick way of multiplying terms. *The result of using the rules and properties* in this example illustrates the following idea.

> **Multiplication of Terms**
>
> Multiply the coefficients together using rules of multiplication of real numbers and multiply the bases together using the product property of exponents. Multiply these results together.

Example 2: Find $(fgh)(x)$ if $f(x) = -4x^5$, $g(x) = -3x^4$, and $h(x) = -x^6$.

Solution: Use the multiplication rule, multiply $(-4)(-3)(-1)$ and $(x^5)(x^4)(x^6)$ and then multiply these results together.

$(-4)(-3)(-1)$ is -12
$(x^5)(x^4)(x^6)$ is x^{15}
the product $(fgh)(x)$ is $-12x^{15}$

⁂

Multiplication of a Term and a Polynomial

The distributive property is used to multiply a term and a polynomial. Because a term represents a real number and a polynomial is a sum or difference of real numbers, the distributive property applies directly to this operation.

Example 3: If $f(x) = 2x^3$ and $g(x) = -4x^2 + 3x - 1$ find $(fg)(x)$.

Solution: The distributive property directs you to multiply the real number $2x^3$ times each real number $-4x^2$, $3x$, and 1.

$$\mathbf{2x^3}(-4x^2) + \mathbf{2x^3}(3x) - \mathbf{2x^3}(1)$$

The problem has now been restated as a product of terms. Multiply the terms together using the product rule for terms, $(fg)(x)$ is $-8x^5 + 6x^4 - 2x^3$

The numeric check below shows that the answer is correct.

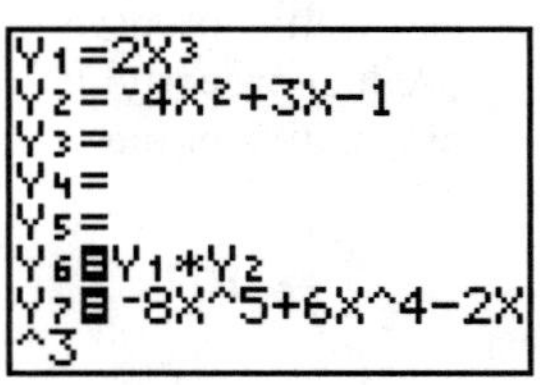

Figure 4.3.1a

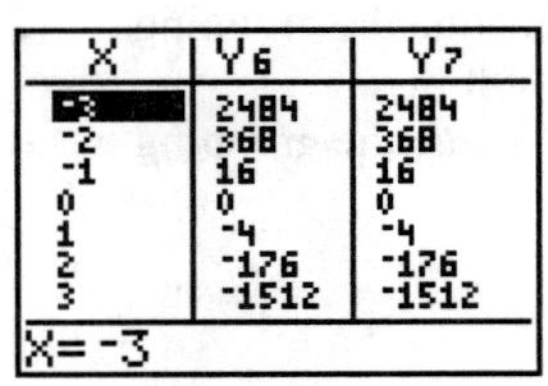

Figure 4.3.1b

Y_6 contains the correct answer and Y_7 the above answer. Since the numeric representations are the same, the symbolic representations are also the same. The answer is correct.

⁂

Multiplication of terms should be done without writing any intermediate steps. You should be able to look at the problem and write the product. Set this as a goal for working the exercises at the end of this section. This is also true for multiplying a term times a polynomial. When the exercises are assigned, try to do the product mentally and write only the product.

Example 4: If $f(x)$ is $-5x^4$ and $g(x)$ is $9 - 2x^2 + 4x^3 - 7x^4$, find $(fg)(x)$.

Solution: Multiply $-5x^4$ times 9, $2x^2$, $4x^3$, and $7x^4$

$$(-5x^4)9 - (-5x^4)2x^2 + (-5x^4)4x^3 - (-5x^4)7x^4$$
$$-45x^4 - (-10x^6) + (-20x^7) - (-35x^8)$$
$$-45x^4 + 10x^6 - 20x^7 + 35x^8$$
$$(fg)(x) \text{ is } -45x^4 + 10x^6 - 20x^7 + 35x^8$$

⁂

Multiplication of a Polynomial and a Polynomial

A polynomial represents a real number, for example $2x - 3$ is a real number. If the polynomial $(2x - 3)$ is multiplied by another polynomial like $(x^2 + 4x + 3)$, $(2x - 3)$ can be interpreted as a real number and $(x^2 + 4x + 3)$ as a sum of real numbers. This means the distributive property can be used.

Example 5: Multiply $f(x)$ times $g(x)$ when $f(x)$ is $(2x - 3)$ and $g(x)$ is $(x^2 + 4x + 3)$.

Solution: The distributive property allows for the product of a real number times a sum or difference of real numbers; thus, multiply the real number $(2x - 3)$ times the sum of real numbers $(x^2 + 4x + 3)$.

$$(fg)(x) = \mathbf{(2x - 3)}x^2 + \mathbf{(2x - 3)}4x + \mathbf{(2x - 3)}3$$

Now the problem has been restructured to be the product of a term times a polynomial, so multiply the term times the polynomial -- three times over.

$$(fg)(x) = 2x\cdot\mathbf{x^2} - 3\cdot\mathbf{x^2} + 2x\cdot\mathbf{4x} - 3\cdot\mathbf{4x} + 2x\cdot\mathbf{3} - 3\cdot\mathbf{3}$$

Now, it is the product of terms. Perform the multiplication.

$$(fg)(x) = 2x^3 - 3x^2 + 8x^2 - 12x + 6x - 9 \qquad \text{Now add similar terms.}$$
$$(fg)(x) = 2x^3 + 5x^2 - 6x - 9$$

A check using the calculator's logic shows the answer is correct.

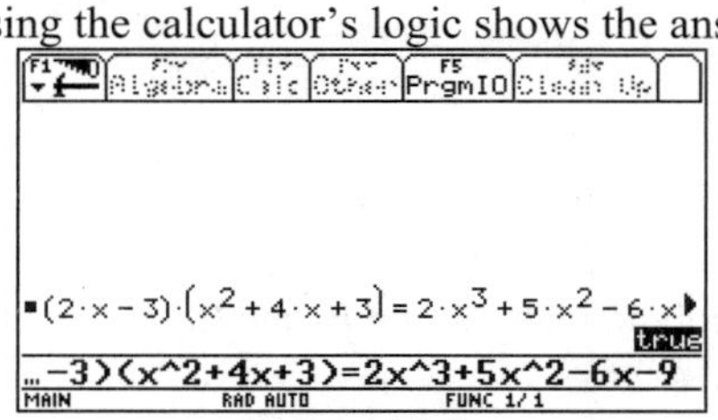

Figure 4.3.2

⁂

Because you must perform many multiplication's in the mathematics classes, it would be convenient to have a quicker method that can be used to multiply polynomials. An analysis of the above problem leads to the idea:

Multiplication of Polynomials

Multiply every term of the first polynomial times every term of the second polynomial, then add the products together.

Example 6: Multiply $(2x - 3)$ times $(x^2 + 4x + 3)$.

Solution: Think of $2x - 3$ as $2x + (-3)$ and multiply $2x$ and (-3) times x^2, $4x$, and 3; then add the results together.

$$\mathbf{2x}\cdot x^2 + \mathbf{2x}\cdot 4x + \mathbf{2x}\cdot 3 + \mathbf{(-3)}\cdot x^2 + \mathbf{(-3)}\cdot 4x + \mathbf{(-3)}\cdot 3$$
$$2x^3 + 8x^2 + 6x - 3x^2 - 12x - 9$$
$$2x^3 + 5x^2 - 6x - 9$$

⁂

Example 7: Find $(hk)(x)$ if $h(x) = (3x - 5)$, and $k(x) = (4x^2 - 2x - 1)$.

Solution: Think of the polynomials as $3x + (-5)$ and $4x^2 + (-2x) + (-1)$, then use the rule for multiplication.

$$\mathbf{3x}\cdot 4x^2 + \mathbf{3x}\cdot(-2x) + \mathbf{3x}\cdot(-1) + \mathbf{(-5)}\cdot 4x^2 + \mathbf{(-5)}\cdot(-2x) + \mathbf{(-5)}\cdot(-1)$$
$$12x^3 + (-6x^2) + (-3x) + (-20x^2) + 10x + 5$$
$$(hk)(x) \text{ is } 12x^3 - 26x^2 + 7x + 5$$

⁂

Special Products

Four polynomial products are used so often that it is highly desirable to learn a method for finding the product of these polynomials in your head! The first of four special products is the product of conjugates. **Conjugates** are binomials of the form $(a + b)(a - b)$. The binomials have **identical first terms** and **the second terms are opposites.** The binomial $(a + b)$ is called the **conjugate** of $(a - b)$ and the binomial $(a - b)$ is called the **conjugate** of $(a + b)$. Either binomial can be written first; that is, $(a + b)(a - b)$ is the same as $(a - b)(a + b)$ because of the commutative property of multiplication.

Multiplying Conjugates

Example 8: Multiply $(a + b)(a - b)$.

Solution: Think of the polynomials as $(a + b)$ and $(a + [-b])$ and use the rule for multiplication of polynomials.

$$\mathbf{(a + b)}\cdot a + \mathbf{(a + b)}\cdot[-b]$$
$$a\cdot\mathbf{a} + b\cdot\mathbf{a} + a\cdot\mathbf{[-b]} + b\cdot\mathbf{[-b]}$$
$$a^2 + ab - ab - b^2$$
$$a^2 - b^2$$

⁂

The previous example gives rise to a mental method for multiplying conjugates. Think of a as symbolizing the first term of any conjugate and b as symbolizing the second term of any conjugate.

> **Multiplication of Conjugates**
>
> To multiply the conjugates $(a + b)(a - b)$, square the first term and subtract the square of the second term; the product is $a^2 - b^2$.

Example 9: Multiply $(2x - 5)(2x + 5)$.

Solution: Square $2x$ and subtract the square of 5, or $4x^2 - 25$.

Just to remind you of how you can check your symbol manipulation work; graph the problem and graph the product on the same screen. If the graphs are the same, the answer is correct. The check is shown below.

$(2x - 5)(2x + 5)$

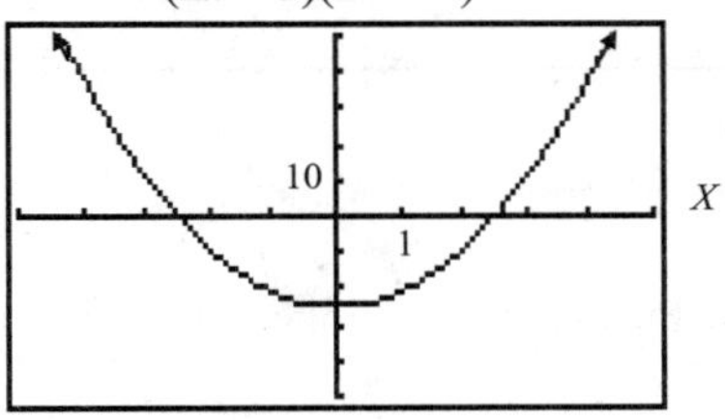

Figure 4.3.3a

$4x^2 - 25$

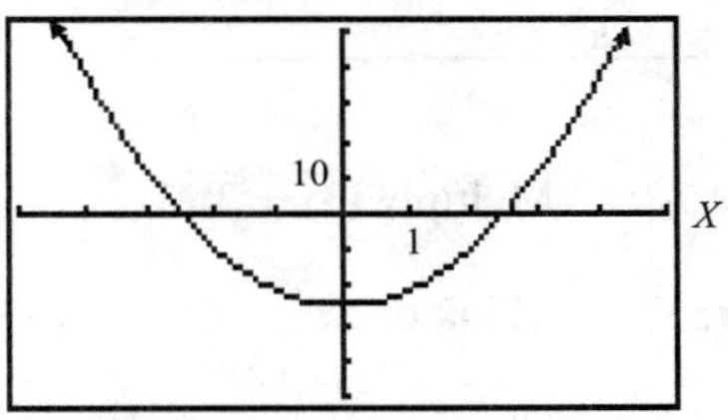

Figure 4.3.3b

⁂

Example 10: Multiply $(3x + 7)(3x - 7)$.

Solution: Square $3x$ and subtract the square of 7, or $9x^2 - 49$. ⁂

The FOIL Method

The second product that needs a mental method is the product of any two binomials where the first terms in each binomial are *similar* and the last terms in each binomial are *similar*. For example, in the binomials $(2x + 3)$ and $(5x + 7)$, $2x$ and $5x$ are similar, as are 3 and 7. While binomials can be multiplied using the method developed earlier in this section, it is helpful to learn the mental method that is developed in the next problem.

Example 11: Multiply $(2x + 3)(4x + 5)$.

Solution: Multiply every term of the first polynomial times every term of the second polynomial.

$(2x + 3)(4x + 5)$ is... $\underset{F}{2x \cdot 4x} + \underset{O}{2x \cdot 5} + \underset{I}{3 \cdot 4x} + \underset{L}{3 \cdot 5}$

(F) Observe that the first term of the first binomial is multiplied times the first term of the second binomial.
(O) Observe that the outer most terms are being multiplied.
(I) Observe that the inner most terms are being multiplied.
(L) Observe that the last term of the first polynomial is multiplied times the last term of the second binomial.

next... $8x^2 + 10x + 12x + 15$

finally... $8x^2 \; + \; 22x \; + \; 15$

F OI L

Multiplication of Binomials (FOIL METHOD) To multiply two binomials:

(F) Multiply the first term times the first term and write the product.

(O) Multiply the outer most terms together and store the product in your head.

(I) Multiply the inner most terms together and add to the product stored in your head; then write the sum.

(L) Multiply the last term of the first binomial times the last term of the second binomial and write the product.

Example 12: Multiply $(3x + 2)(5x - 3)$.

Solution: Think of $5x - 3$ as $5x + (-3)$ and use the rule for multiplication.

F $15x^2$

O + I $(-9x + 10x)$ or x

L -6

The product is $15x^2 + x - 6$. Then check is shown below:

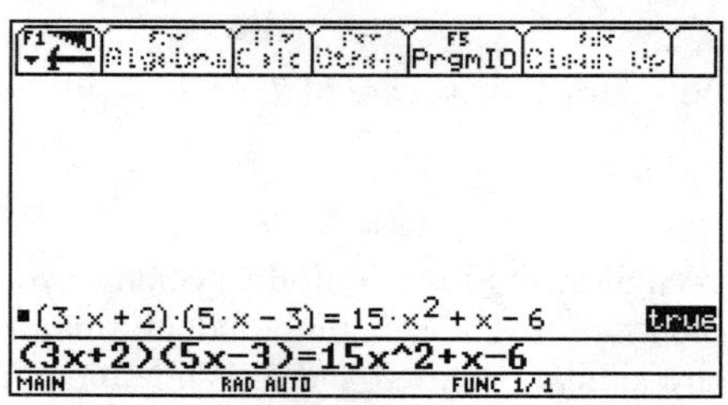

Figure 4.3.4b

Example 13: $f(x) = x - 4$ and $g(x) = x + 3$, find $(fg)(x)$.

Solution: Think of $x - 4$ as $x + (-4)$ and use the rule for multiplication.

F x^2

O + I $\{3x + (-4x)\}$ $-x$

L -12

The product $(fg)(x)$ is $x^2 - x - 12$.

The numerical representations are the same, the analytical work is correct.

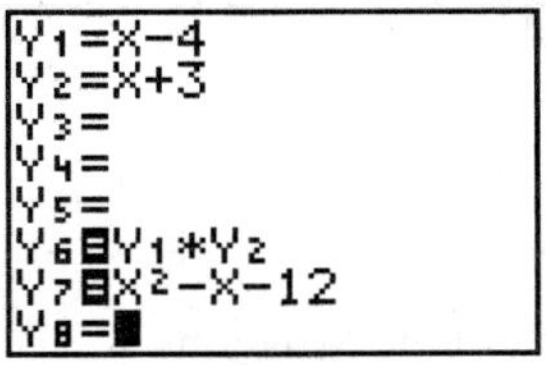

Figure 4.3.5a

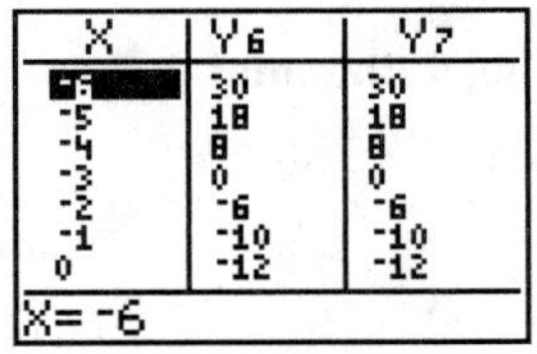

Figure 4.3.5b

⁂

Example 14: Multiply $(2x - 3)(x - 5)$.

Solution: Think of $2x - 3$ as $2x + (-3)$ and think of $x - 5$ as $x + (-5)$.

Use the FOIL method: F $2x^2$

$\{-10x + (-3x)\}$ O + I $-13x$

L 15

The product is $2x^2 - 13x + 15$.

Table 4.3.1

x	−2	−1	0	1	2	3	4	5	6
$(2x - 3)(x - 5)$	**49**	**30**	**15**	**4**	**−3**	**−6**	**−5**	**0**	**9**
$2x^2 - 13x + 15$	**49**	**30**	**15**	**4**	**−3**	**−6**	**−5**	**0**	**9**

Because the numeric representations are identical, the function $(2x - 3)(x - 5)$ is the same as the function $2x^2 - 13x + 15$. The symbol manipulation is correct.

⁂

Binomial Squared

The third special product results from squaring a binomial. Think of all binomials as being symbolized by $(a + b)$, where a represents the first term and b represents the second term. The objective is to simplify $(a + b)^2$ in your head.

Example 15: Simplify $(a + b)^2$ by squaring $a + b$.

Solution: Since $(a + b)^2$ means $(a + b)(a + b)$ use the FOIL method to develop a mental method
for squaring a binomial.

$(a + b)^2$ is $(a + b)(a + b)$ which is:

F		a^2
O + I	$ab + ab$	$2ab$
L		b^2

The product is $a^2 + 2ab + b^2$ and is called a **perfect square trinomial**, because it comes from squaring a binomial just like 25 is called a perfect square because it comes from squaring an integer.

⁂

Squaring a Binomial

$(a + b)^2$ is $a^2 + 2ab + b^2$

a^2	square the first term
$2ab$	double the product of the two terms
b^2	square the last term
$a^2 + 2ab + b^2$	add the results together

Example 16: If $f(x)$ is $x + 4$, find $[f(x)]^2$.

Solution: Use the squaring rule.

x^2	square the first term
$8x$	double the product of the two terms $(2{\cdot}x{\cdot}4)$
16	square the last term
$x^2 + 8x + 16$	

The analytical work is correct as shown in the check.

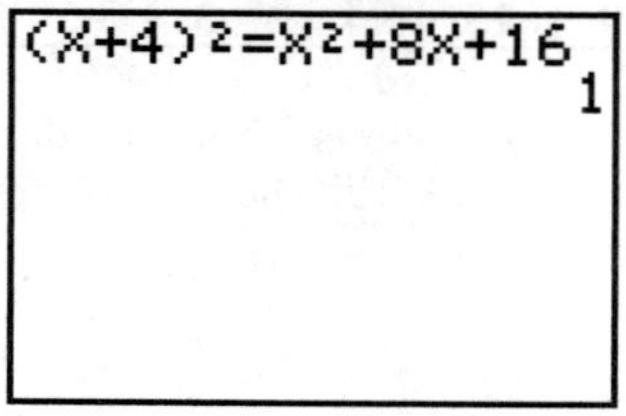

Figure 4.3.6

Example 17: Simplify $(x - 3)^2$.

Solution: Think of $x - 3$ as $x + (-3)$ and use the squaring rule.

x^2	square x
$-6x$	multiply 2, x, and (-3)
9	square (-3)

The perfect square trinomial is $x^2 - 6x + 9$.

The numeric check shows the answer is correct because the numeric representations of the problem function and the answer function are the same.

Table 4.3.2

x	−2	−1	0	1	2	3	4	5
$(x - 3)^2$	**25**	**16**	**9**	**4**	**1**	**0**	**1**	**4**
$x^2 - 6x + 9$	**25**	**16**	**9**	**4**	**1**	**0**	**1**	**4**

Example 18: Square the linear function $3x - 5$.

Solution: Think of $3x - 5$ as $3x + (-5)$ and use the rule for squaring.

$9x^2$	square $3x$
$-30x$	double $3x$ times (-5) $[2 \cdot 3x \cdot (-5)]$
25	square (-5)

The square of the linear function $3x - 5$ is the perfect square trinomial is $9x^2 - 30x + 25$.

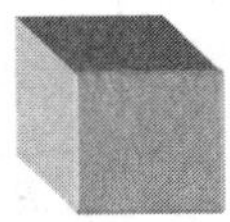

Plus or Minus

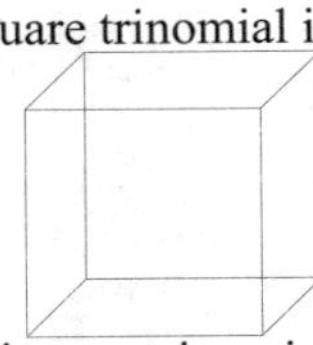

Sum or Difference of Cubes

The last special product that should be done mentally contains a binomial times a trinomial. The trinomial must be related to the binomial in the following way: if the binomial is symbolized as $(a + b)$, where a *represents* the first term of the binomial and b the second term, then:

The first term of the trinomial must be	a^2
The second term must be the opposite of the product ab	$-ab$
The last term must be	b^2

So, the last special product is $(a + b)$ times $(a^2 - ab + b^2)$. Below is the calculation of the product.

Example 19: Multiply $(a + b)(a^2 - ab + b^2)$.

Solution: The rule used to multiply polynomials tells you to multiply every term of the first polynomial times every term of the second polynomial.

$\mathbf{a}a^2 + \mathbf{a}(-ab) + \mathbf{a}b^2 + \mathbf{b}a^2 + \mathbf{b}(-ab) + \mathbf{b}b^2$	multiply
$a^3 + (-a^2b) + ab^2 + a^2b + (-ab^2) + b^3$	simplify
$a^3 + b^3$	simplify

The results can be generalized to conclude that for every binomial multiplied times the related trinomial as above, the product is the first term of the binomial cubed plus the second term of the binomial cubed.

Multiplication of a Binomial Times the Related Trinomial

$(a + b)$ times $(a^2 - ab + b^2)$ is the product $a^3 + b^3$.

Example 20: Is $(x^2 + 3x + 9)$ the related trinomial of $(x + 3)$?

Solution: No! The middle term must be the opposite of 3 times x $(-3x)$.

Example 21: Multiply the linear function $x - 5$ and the quadratic function $x^2 + 5x + 25$.

Solution: Since x^2 is the first term of the binomial squared, $5x$ is the opposite of the first term times the last term of the binomial, and 25 is the last term of the binomial squared, it fits the pattern. The product is x cubed plus (-5) cubed or $x^3 + (-125)$ or $x^3 - 125$. The product of the linear function $x - 5$ and the quadratic function $x^2 + 5x + 25$ is the cubic function $x^3 - 125$. The numeric check confirms that the product is correct.

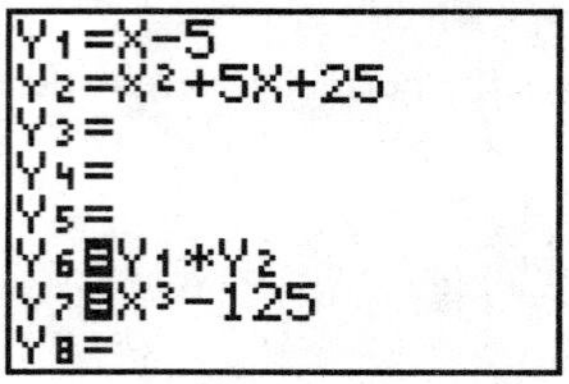

Figure 4.3.7a

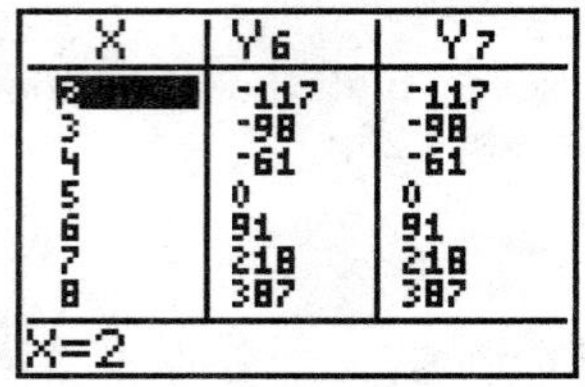

Figure 4.3.7b

⁂

Example 22: Multiply $(2x + 3)(4x^2 - 6x + 9)$.

Solution: The polynomials match the required conditions; thus, the product is

$(2x)^3 + 3^3$ or $8x^3 + 27$.

⁂

You should be able to do most multiplication of functions mentally. You should be able to check your work using technology, and you ought to recognize that multiplication of polynomial functions gives a product of higher degree than the functions being multiplied.

4.3 EXERCISES

Many times you can work the exercises by algebraic, numeric, or graphical methods. You can learn about mathematics no matter which method you use. Part of the learning process is for you to decide which method is best for you to use. Please read the text carefully before you do the exercises. Read the **next assigned section** after you finish the assignment below.

A Quick Look To The Future

-4. Write $3^2 - 5^2$ as a product.

-3. $x^2 - 3x - 6$ and $x^2 - 3(x - 2)$ are equivalent expressions. Is $x^2 - 3(x - 2)$ written as a product?

-2. What does adding $0\sqrt{x}$ to the linear function $3x + 4$ do to the graph of $3x + 4$?

-1. What is the initial size of a 500 K file being downloaded at a rate of 2.5 K per second?

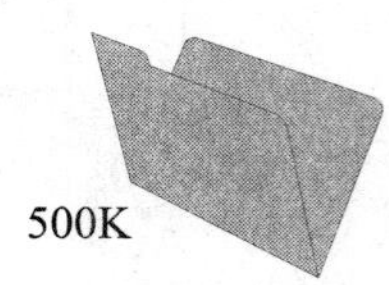

Previous Topics Revisited

Given: $f(x) = 3x^2 - 7x - 4$ and $g(x) = -2 - 5x + x^2$

1. Find $(g - f)(x)$
2. Find $(f + g)(x)$
3. Find $(f + g - f)(x)$
4. Find $f(-3.5)$
5. Find $g(-100)$
6. If $h(x) = |x^2 - 6x + 2|$, is $h(x)$ a function?

New Topic Exercises

In Exercises 7 - 44, multiply the functions and confirm your answers with technology.

7. $(-5x^3)(-2x^5)(-2x^9)$
8. $(-x)(-x)(-x)(-x)(-x)(-x)$
9. $(-x)^6$
10. $5(-3x^6)(-2x^3)(x)(-x)^2$
11. $-3x^2(2x^3 - 5x^2 + 4x - 8)$
12. $5x^4(2 - 3x + 4x^2 - x^3 + 7x^4)$
13. $(3x^4 - 2x^{17} + 5x^{32})(-2x)$
14. $-3x^2[6x^4 + (-6x^4)]$
15. $(x - 3)(x^3 - x^2 + x - 1)$
16. $(x + 2)(x^6 - x^4 + x^2 - 1)$
17. $(2x - 1)(x^2 - 3x + 4)$
18. $(2x - 5)(2x + 5)$
19. $(3x + 4)(3x - 4)$
20. $(11 - 2x)(11 + 2x)$
21. $(5x + 9)(5x - 9)$
22. $[(x + 1) + 3][(x + 1) - 3]$
23. $(x^3 - 5)(x^3 + 5)$
24. $(2x + 7)(x + 4)$
25. $(2 + 3x)(5 + x)$
26. $(5x + 2)(3x - 7)$
27. $(2a - 3)(4a + 1)$
28. $(x - \frac{1}{2})(x + \frac{1}{4})$
29. $(-2x + 3)(5x - 2)$
30. $(5r - 3)(2r - 1)$
31. $(3x - 8)(2x - 5)$
32. $(5x - 9)(5x - 9)$
33. $(3x^2 - 2)(x^2 + 5)$
34. $(4x^3 - 5)(3x^3 - 7)$
35. $(x - 9)^2$
36. $(2x - 1)^2$
37. $(1 + 2x)^2$
38. $(5t - 3)^2$
39. $(3x + 12)^2$
40. $(x - 1)(x^2 + x + 1)$
41. $(2x + 3)(4x^2 - 6x + 9)$
42. $(t - 5)(t^2 + 5t + 25)$
43. $(x^2 + 6x + 36)(x - 6)$
44. $(x - 1)(x^2 - 2x + 1)$

In Exercises 45 - 48, find any two functions whose product is given.

45. $7x^4$

46. $x^2 - 9$

47. $x^2 + 7x + 12$

48. $x^3 + 8$

49. Confirm that $(x + 2)(x + 2)$ is not the same thing as $x^2 + 4$.

50. Confirm that $(x - 3)(x + 5)$ is not the same thing as $x^2 - 15$.

51. Confirm that $(x + 4)^2$ is not the same thing as $x^2 + 16$.

52. Confirm that $(5x^2)(7x^2)$ is not the same thing as $35x^2$.

From Mathematics to English

53. After reading this section, make a list of questions that you want to ask your instructor.

54. Continue in your daily journal and make an entry. In addition to your normal entry on thoughts about the mathematics in this section, list at least two positive comments about what you have learned about this topic.

55. In paragraph format, summarize the material in this section of the text in your daily journal.

56. Describe how your classroom instructor made this topic more understandable and clear.

57. After reading the text and listening to your instructor, what do you not understand about this topic?

58. What does the statement $(a + b)^2 = a^2 + 2ab + b^2$ mean to you?

59. Explain why $(a + b)^2$ does *NOT* simplify to $a^2 + b^2$. (Hint: Use a counter-example).

60. List the steps, in English, used to multiply $(2x - 3)(4x^3 - 2x^2 + 7x - 1)$.

61. Develop a rule for cubing a binomial.

62. Explain why $(x + a)(x + b)$ does not equal $x^2 + ab$.

63. Describe what $(x + a)(x - a) = x^2 - a^2$ means to you.

64. Describe the relationship between the numeric representations of $(x + a)(x - a)$ and $x^2 - a^2$.

From English to Mathematics

In the next three exercises, convert the English statement to an equivalent mathematical statement.

65. the sum of x and thirty-two, times the difference of twice x and four

66. the difference of x and 7 times the sum of the product of x and 2, and 14

67. the sum of four thirds of a number squared and seven

Explorations

68. Graph each pair of the following functions and identify the x-intercepts (the zeros) of each.

a.	$(x-1)(x+2)$	a′.	x^2+x-2
b.	$(x-2)(x+3)$	b′.	x^2+x-6
c.	$(x-3)(x+4)$	c′.	x^2+x-12
d.	$(x-1.5)(x+3)$	d′.	$x^2+1.5x-4.5$
e.	$(x-4)(x+1)$	e′.	x^2-3x-4
f.	$(x-2)(x+1/2)$	f′.	$x^2-(3/2)x-1$
g.	$(x-6)(x+1/2)$	g′.	$x^2-(11/2)x-3$
h.	$(x-5)(x+1/4)$	h′.	$x^2-(19/4)x-5/4$

In general, where does the graph of $(x-\text{a})(x+\text{b})$ cross the x-axis?

69. Explore for the zeros of the following functions: List the zeros and support your answers with the graphs showing the zeros.

a.	$(2x+1)(x-3)$	b.	$(2x+3)(x-3)$
c.	$(2x+5)(x-3)$	d.	$(2x+7)(x-3)$
e.	$(5x-3)(3x+4)$	f.	$x(x+3)$
g.	$x^2(x+3)$	h.	$(x-3)^2$
i.	$(x+1)(x-2)(x+4)$	j.	$x(x+3)(3x-1)(2x-11)$

70. Explore for the zeros of the following product functions. List the zeros and support your answers with the graph.

a.	$(x^2-2)(x^2-3)$	b.	$(2x+3)(x-3)^2$
c.	$\lvert x-5\rvert\sqrt{x-1}$	d.	$\sqrt{2x+7}(x-5)$
e.	$\lvert x-1\rvert(x+1)^2$		

A zero is a zero is a zero.

A zero is a zero is a zero.

4.4 Factoring: Common Factors, Grouping, and Difference of Squares

Factoring means that you are to write a polynomial function as a product of functions. Factoring is used to manipulate fractions analytically. Knowing how to factor means that with a little more work, you will be able to reduce a fraction, add, subtract, multiply, and divide fractions. You will also have the basic skills necessary to simplify a complex fraction and solve selected equations. Fractions come from rational functions and rational functions are analyzed in Chapter Eight.

To factor or not to factor, that is the question!

Polynomial functions can be factored analytically -- similar to what you may have done in earlier courses -- or they can be factored with technology. If you are planning to use technology, you must first develop a method for finding the zeros of the function. A new method for finding zeros will be developed in this chapter.

Common Factor Polynomials

The distributive property in the form $ab + ac = a(b + c)$ is used to factor polynomials with a common factor. The terms ab and ac each share the factor a. The distributive property allows for moving the common factor outside the group of remaining factors and it provides a pattern for factoring polynomials like $2x + 4$.

Example 1: Factor the function $2x + 4$.

Solution:

$2x + 4$ given

$2x + 2\cdot 2$ *think* of the given polynomial as this polynomial

$2(x + 2)$ take out the 2 factor by applying the distributive property

⁂

Example 2: Factor $6x^2 - 8x + 12$.

Solution:

$6x^2 - 8x + 12$ given

$2\cdot 3x^2 - 2\cdot 4x + 2\cdot 6$ *think* of the polynomial like this and apply the distributive property

$2(3x^2 - 4x + 6)$ is the factored form

⁂

Example 3: Factor $(x + 1)17x^2 - (x + 1)3x + (x + 1)$.

Solution:

$\mathbf{(x + 1)}17x^2 - \mathbf{(x + 1)}3x + \mathbf{(x + 1)}$ given

$\mathbf{(x + 1)}17x^2 - \mathbf{(x + 1)}3x + \mathbf{(x + 1)}\cdot 1$ multiplicative identity

$\mathbf{(x + 1)}17x^2 - \mathbf{(x + 1)}3x + \mathbf{(x + 1)}1$ think of the polynomial as in the pattern $\boldsymbol{ab} + \boldsymbol{ac} + \boldsymbol{ad}$ and factor with the distributive property

$(x + 1)(17x^2 - 3x + 1)$ factored form

You can confirm that your answer is equivalent to the problem by making the numeric representation for both. If the numeric representations are the same, the answer and the problem represent the same function.

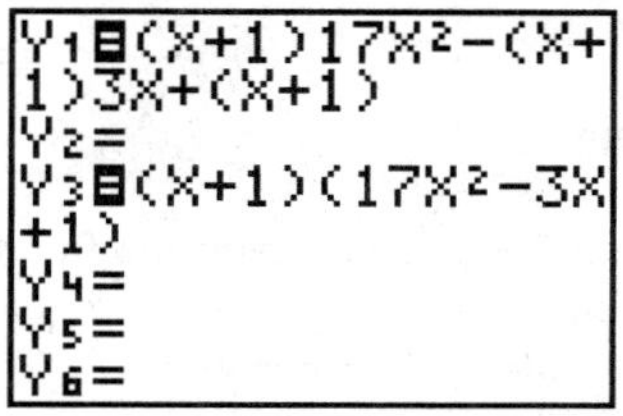

Figure 4.4.1a

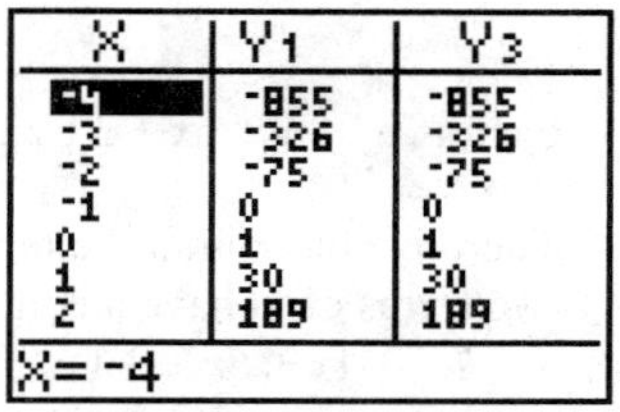

Figure 4.4.1b

Figure 4.4.1 confirms that the factored form of the function is correct.

⁂

Unlike arithmetic, where only integers are used as factors, *under certain conditions*, fractions can be used when factoring. If fractions were always used when factoring, every polynomial could be factored an infinite number of different ways. This would not necessarily produce anything of interest. However, in the problems that follow, the coefficient of the leading term of the polynomial will be factored by the distributive property requiring that fractions be used.

Example 4: Re-write $2x^2 - 5x + 4$ by factoring out 2.

Solution: Think of $2x^2 - 5x + 4$ as $\mathbf{2}\cdot x^2 - \mathbf{2}\cdot\frac{5}{2}x + \mathbf{2}\cdot 2$; then factor out **2**.

$2(x^2 - \frac{5}{2}x + 2)$ is the factored form, or $2(x^2 - 2.5x + 2)$ is also factored form.

This procedure is used in one analytical method for solving quadratic equations in Chapter Ten. This procedure is also used to find the maximum or minimum point on a quadratic function in Chapter Ten.

⁂

Example 5: Re-write $5x^2 + 4x + 1$ by factoring out the factor 5.

Solution: Think of $5x^2 + 4x + 1$ as $\mathbf{5}\cdot x^2 + \mathbf{5}\cdot\frac{4}{5}x + \mathbf{5}\cdot\frac{1}{5}$; then factor out **5**.

$5(x^2 + \frac{4}{5}x + \frac{1}{5})$ is the factored form, or $5(x^2 + 0.8x + 0.2)$ is factored form.

⁂

Example 6: Re-write $6x^2 + 23x + 20$ by factoring out the common factor of 6.

Solution: Think of $6x^2 + 23x + 20$ as $\mathbf{6}\cdot x^2 + \mathbf{6}\cdot\frac{23}{6}x + \mathbf{6}\cdot\frac{20}{6}$; then factor **6**.

$6(x^2 + \frac{23}{6}x + \frac{20}{6})$ is the factored form, or $6(x^2 + 3.833x + 3.333)$

is factored form using approximate values.

⁂

Factoring by Grouping

Another method of factoring is called **factoring by grouping**. Factoring by grouping is similar to factoring a common factor. It is normally used to factor a polynomial with four terms and no common factor to all four terms. Again, the distributive property will be used.

Example 7: Factor $2x^2 + x + 6x + 3$ by grouping.

Solution: There is no common factor in all four terms; however, the first two terms and the last two terms each have a common factor.

$= 2x^2 + 1x + 6x + 3\cdot 1$	identity property
$= x(2x + 1) + 3(2x + 1)$	distributive property
$= x(\mathbf{2x + 1}) + 3(\mathbf{2x + 1})$	***This form is NOT factored: remove the common factor*** $2x + 1$
$= (2x + 1)(x + 3)$	distributive property

The graphing calculator can be an aid to factoring some polynomials. To develop a method by which it can be used, the graph of the above polynomial is shown below with the zeros labeled. See if you can make a connection between the zeros of the polynomial and the factors of the polynomial as you look at the next few examples.

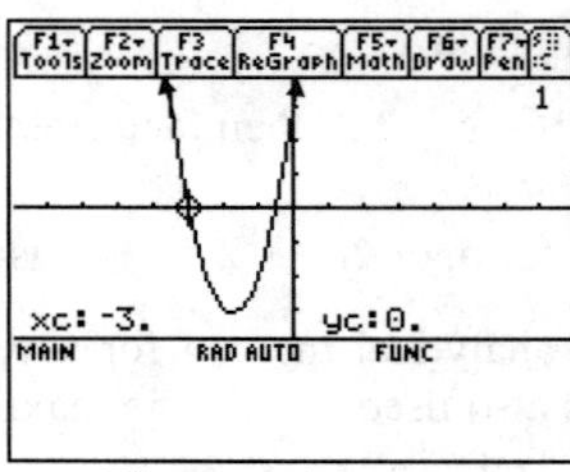

Figure 4.4.2a

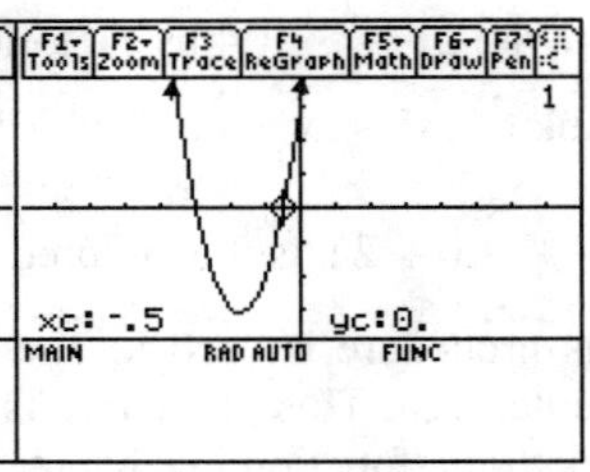

Figure 4.4.2b

Observe that the polynomial has a factor of $(x + 3)$ and a zero of -3.

Also note that the polynomial has a factor of $(2x + 1)$ and a zero of $-\frac{1}{2}$.

⁂

Example 8: Factor $12x^2 - 8x + 15x - 10$.

Solution: Factor the common factor from the first two terms and the last two terms.

$4x(3x - 2) + 5(3x - 2)$	factor the common factor again
$(3x - 2)(4x + 5)$	finished

To help you make the connections between the zeros of the polynomial and the factors of the polynomial, the graph of the polynomial is below with the zeros marked.

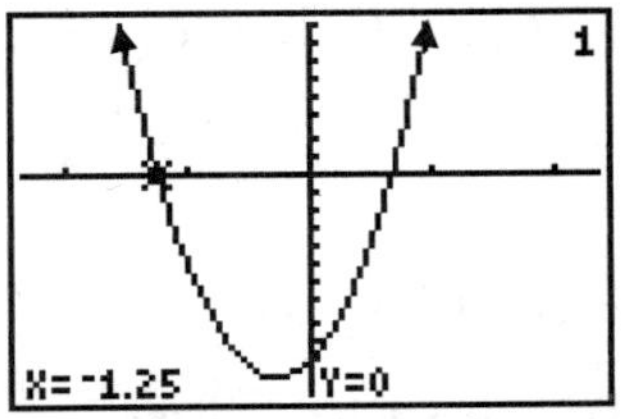

Figure 4.4.3a

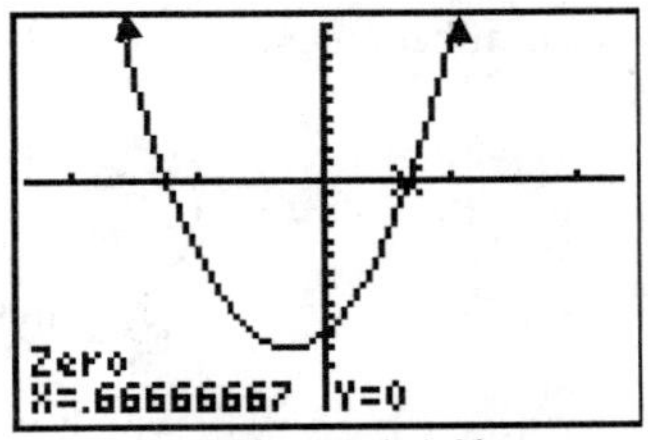

Figure 4.4.3b

As in the last example, observe that the polynomial has factors of $(3x - 2)$ and $(4x + 5)$ along with zeros of $\frac{2}{3}$ and $-\frac{5}{4}$.

⁂

Example 8 has a new feature of the calculator not mentioned before in this text; however, it will be used extensively from now on. Instead of using trace, which may or may not give exact zeros, the zero finder will give you the zeros. This feature, the zero finder (On some calculators it is called the root finder.), in conjunction with the feature that converts decimal numbers to their equivalent reduced fraction form, can be used to find the exact rational zeros. It will not find exact irrational zeros. Below is an example of the process of finding the exact zero $\frac{2}{3}$ from the graph in Figure 4.4.4a.

Step 1 Produce the graphical representation of the function with the zeros displayed.

Step 2 Use the zero finder to put the zero in address x. (On the TI-82 it is called the root finder.)

Step 3 Convert the decimal zero in address x to a fraction.

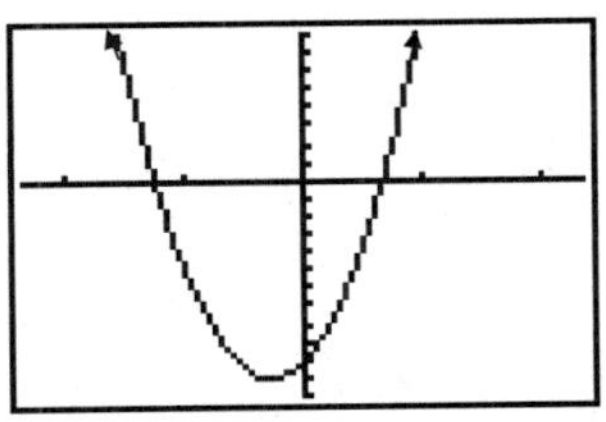

Figure 4.4.4a

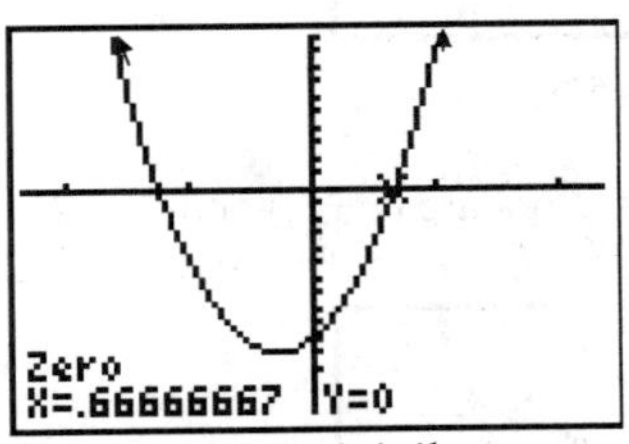

Figure 4.4.4b

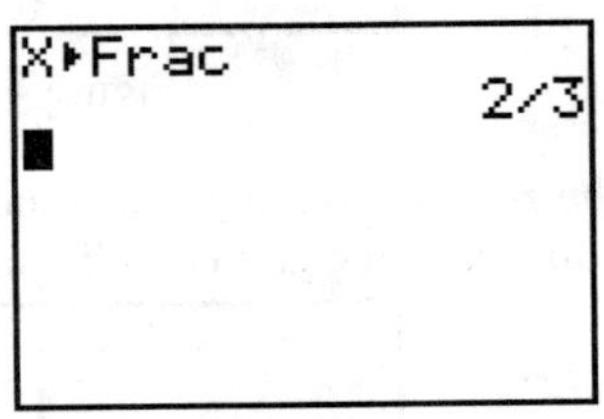

Figure 4.4.4c

Difference of Squares

The third type of polynomial in this section is called the **difference of two squares** and it has the form $x^2 - a^2$, where x represents any algebraic expression and a is a real number. The graph of the function $f(x) = x^2 - a^2$ can be used as an aid in the factoring process. This polynomial is relatively simple to factor analytically, but it is important that a visualization of the factoring process be developed as well as the analytical method of factoring.

One of the mental methods for multiplying polynomials is that of multiplying conjugates. Conjugates are polynomials of the form $(a + b)$ and $(a - b)$. The product of these conjugates is $a^2 - b^2$. The pattern for conjugates is: $(a + b)(a - b) = a^2 - b^2$.

If the pattern is reversed, it becomes: $a^2 - b^2 = (a + b)(a - b)$. It gives a method for factoring the difference of squares.

From $a^2 - b^2 = (a + b)(a - b)$,

$a^2 - b^2$	is a polynomial and it is equal to:
$(a + b)(a - b)$	which is a product of polynomials that is equivalent to
$a^2 - b^2$.	Thus, $(a + b)$ and $(a - b)$ are factors of $a^2 - b^2$.

Example 9: Factor $x^2 - 4$.

Solution: Think of $x^2 - 4$ as $x^2 - 2^2$ and use the pattern.

$a^2 - b^2 = (a + b)(a - b)$	is the pattern to follow
$x^2 - 2^2 = (x + 2)(x - 2)$	$(x + 2)(x - 2)$ is factored form

The graphical representation of the function $f(x) = x^2 - 4$ provides interesting information about the factors. The zeros of the function appear in the factors of the polynomial.

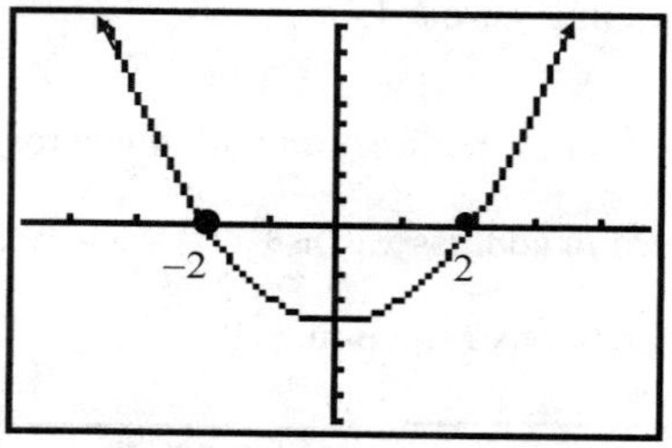

Figure 4.4.5

The zeros are –2 and 2, and the factors are $(x + 2)$ and $(x - 2)$.
Notice also the graphs of the factors $h(x) = x + 2$, $g(x) = x - 2$.

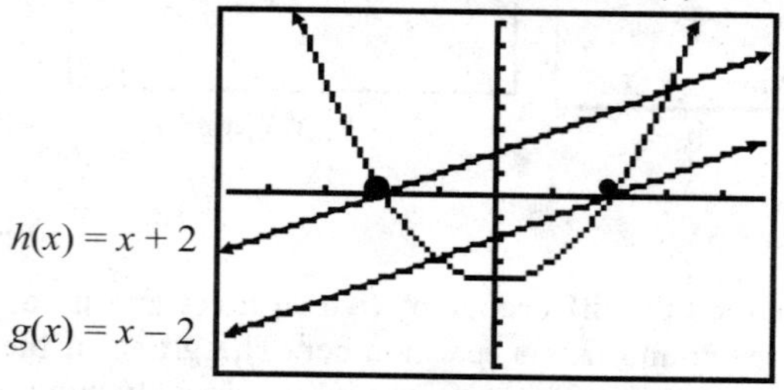

Figure 4.4.6

The zero of the function $h(x) = x + 2$ is –2 and the zero of $g(x) = x - 2$ is 2. The polynomial $x^2 - 4$ has zeros of –2 and 2, and factors of $(x + 2)$ and $(x - 2)$.

⁂

Observe that the opposite of the zero is a part of the factor. This leads to the conclusion that to factor $x^2 - 4$, or any other difference of squares polynomial, you should graph the related function and identify the zeros; and finally, the factors are (x – a zero)(x – a zero).

Example 10: Factor $x^2 - 25$.

Solution: The quickest way of factoring $x^2 - 25$ is to use the pattern.
The pattern is $a^2 - b^2 = (a + b)(a - b)$.
The factors are $x^2 - 5^2 = (x + 5)(x - 5)$. Again, review the graph of the related function $f(x) = x^2 - 25$.

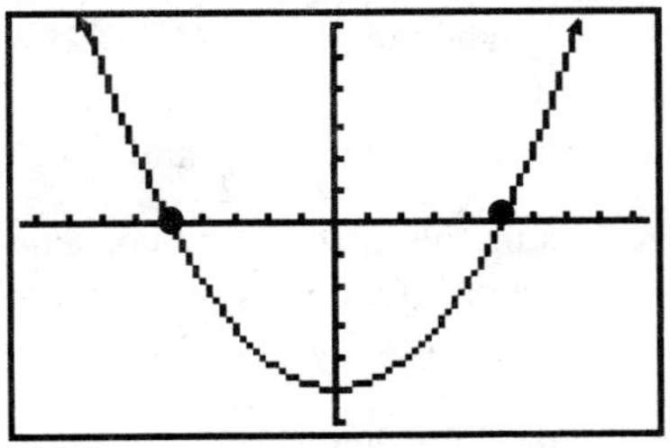

Figure 4.4.7

The zeros are –5 and 5 and the factors are $(x - (-5))$ and $(x - 5)$, or $(x + 5)(x - 5)$.

⁂

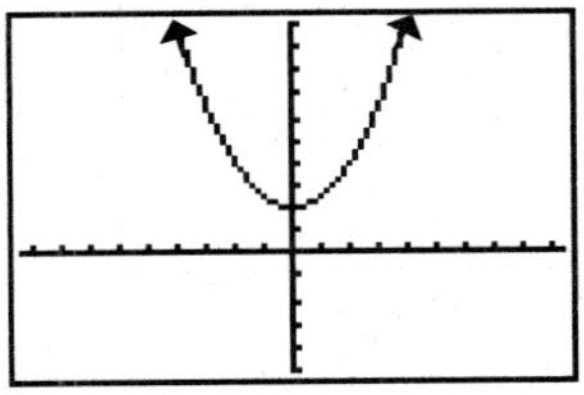

Figure 4.4.8

Since the difference of squares $a^2 - b^2$ can be factored, one of the questions always asked by beginning students is whether the sum of squares can be factored. *Based on the last two examples, if the graph of the related function has zeros, it appears to be factorable.* To confirm that the sum of squares *cannot* be factored, study the graph of the sum of squares $x^2 + 4$ in Figure 4.4.8. The graph has no real zeros! It is meaningless to say the factors are (x – a zero)(x – a zero); when there are no real zeros.

Example 11: Factor the polynomial function $f(x) = 4x^2 - 49$.

Solution: This polynomial is a little different than the form $x^2 - b^2$; however, if it is thought of as $(2x)^2 - 7^2$, it is the same form and can be factored by the pattern $x^2 - b^2 = (x + b)(x - b)$.
$(2x)^2 - 7^2 = (2x + 7)(2x - 7)$ is the factored form.

The graph of the related function $f(x) = 4x^2 - 49$ does not have zeros at 7 and –7! However, the graph of the function $f(x) = 4x^2 - 49$ as well as the graphs of the functions $g(x) = 2x + 7$ and $h(x) = 2x - 7$ show that the zeros are connected to the factors.

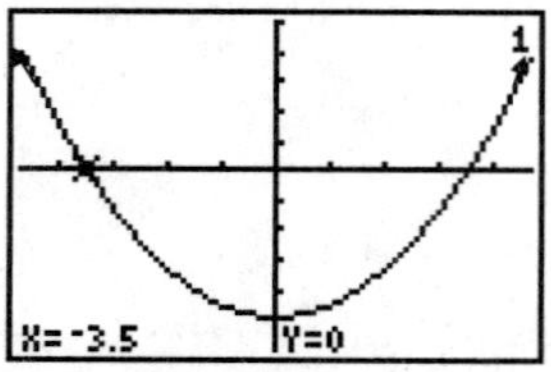

Figure 4.4.9a

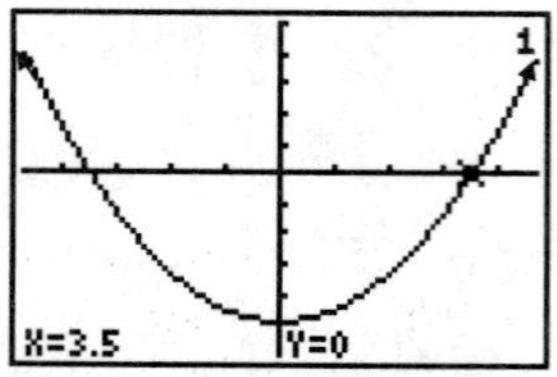

Figure 4.4.9b

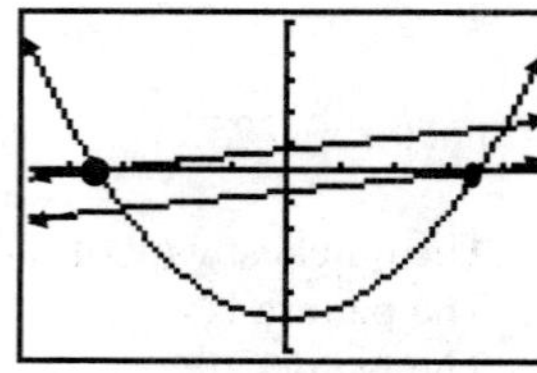
Figure 4.4.9c

The zero of $g(x) = 2x + 7$ is $-\frac{7}{2}$ and the zero of $h(x) = 2x - 7$ is $\frac{7}{2}$.

The zeros of the function $f(x) = 4x^2 - 49$ are $-\frac{7}{2}$ and $\frac{7}{2}$.

The numerator of each zero is the opposite of the constant term in the factor and the denominator of the zero is the coefficient of x of each factor.

⁂

<u>This example forms the basis for the graphical method for factoring.</u>

Zero Factor Connection
If the factors of a polynomial are $(ax + b)(ax - b)$, the zeros are $-\frac{b}{a}$ and $\frac{b}{a}$.
Conversely, if the zeros are $-\frac{b}{a}$ and $\frac{b}{a}$, the factors are $(ax + b)$ and $(ax - b)$.

To factor with technology, use the graphical representation to find the zeros, then use the zeros to find the probable factors, and finally, check the factors by multiplying them together to verify that their product is the given polynomial. The following example will confirm this idea. The opposites of the numerators in the zeros become the constants in the factors. The denominators of the zeros become the coefficients of x in each factor.

Example 12: Factor $9x^2 - 16$.

Solution: If the pattern is to be used to factor the polynomial, think of the polynomial as $(3x)^2 - 4^2$. This factors into $(3x + 4)(3x - 4)$. If the factoring is done graphically, are the zeros $-\frac{4}{3}$ and $\frac{4}{3}$? Below is the graph showing the zeros.

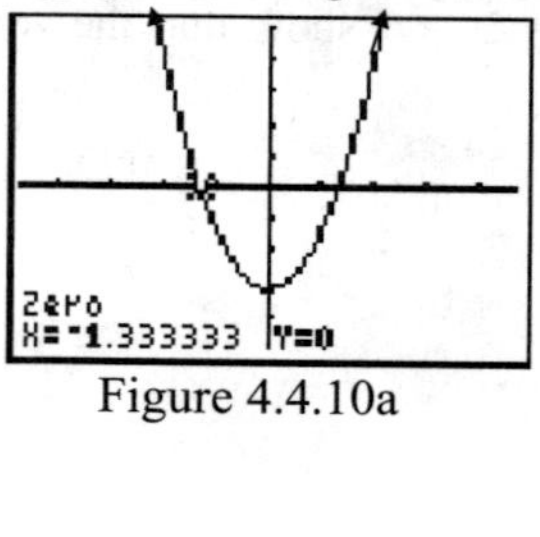

Figure 4.4.10a

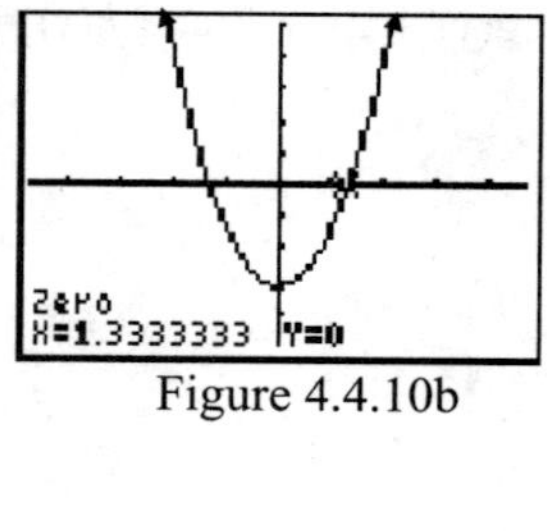

Figure 4.4.10b

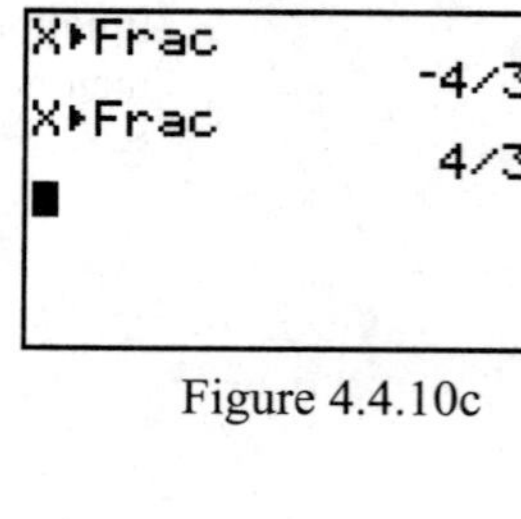

Figure 4.4.10c

As can be seen on the graph, the zeros are in fact, $-\frac{4}{3}$ and $\frac{4}{3}$.

This example verifies the notion that if the zeros are in the form $-\frac{b}{a}$ and $\frac{b}{a}$, the factors are $(ax + b)$ and $(ax - b)$ respectively.

⁂

Example 13: Factor $6x^2 + 25$.

Solution: There is no common factor and the graph of the function $f(x) = 6x^2 + 25$ shows no real zeros; therefore, it does not factor using real numbers as coefficients in the factors.

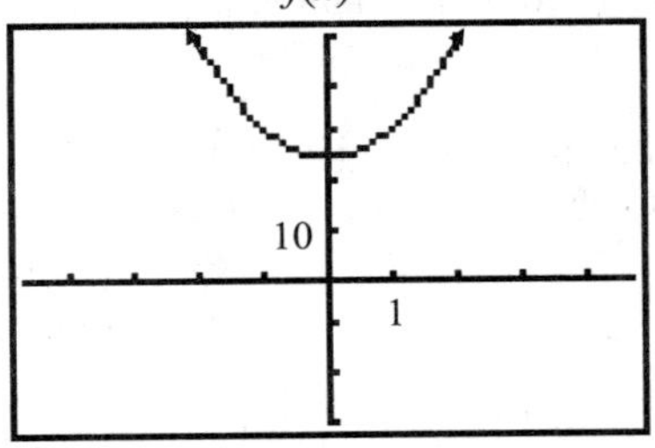

Figure 4.4.11

⁂

The intent of using technology to help you factor is not to replace the analytical and mental methods of factoring. If you will note the examples, the analytical or mental method is usually used and this is followed by confirmation with technology. The intent of using technology to factor is that it is an option you have, should you want to use it. It probably is faster to use the analytical method, but what do you do if you forget the analytical method? Or, what if the factoring problem is exceedingly difficult like $105x^2 + 16x - 336$? Might the analytical method take too long? Knowing the connection between the zeros and the factors is, in the long run, much more important than knowing how to factor. If you know this connection, then you also know how to factor.

4.4 EXERCISES

Many times you can work the exercises by algebraic, numeric, or graphical methods. You can learn about mathematics no matter which method you use. Part of the learning process is for you to decide which method is best for you to use. Please study the text carefully before you do the exercises. Read the **next assigned section** after you finish the assignment below.

A Quick Look To The Future

-4. How do you know that $x^2 + 3x - 4$ and $(x + 4)(x - 1)$ represent the same number?

-3. If you add two absolute value functions of the form $d|x+e|+f$, do you expect the sum to look like a V? Why?

-2. If you change your position by 3 feet in one second, how fast is your position changing?

-1. If you start at 3 on the y-axis and then changed the y-coordinate by 1 and the x-coordinate by 2, what new point would you be at?

Previous Topics Revisited

1. Multiply $(2x - 3)(2x + 3)(x + 1)(x - 1)$.
2. Find two functions whose product is the function $x^2 + 3x + 2$.
3. What is the degree of the product of a 2nd degree and a 3rd degree polynomial?
4. Find two functions whose difference is the function $x^2 + 3x + 2$.
5. What is the degree of the sum of a 2nd degree and a 3rd degree polynomial?
6. If $f(x) = x^2 - 3x$, is the graph of $f(x + 6)$ to the right or left of $f(x)$?

New Topic Exercises

Perform a complete factorization on the polynomials below. (Factoring completely means that each polynomial factor should not be factorable.) Do not use fractions.

7. $16x^2 - 32x + 48$
8. $ax^2 + ax - a$
9. $abx^3 - abx^2 + xab - 2ab$
10. $3x^2 + 12x - x - 4$
11. $10x^2 - 5x + 4x - 2$
12. $5x^2 - 10x + 9x - 15$
13. $12x^2 - 9x - 16x + 12$
14. $x^2 - 36$
15. $x^2 - 64$
16. $9 - x^2$
17. $x^4 - 1$
18. $4x^2 - 625$
19. $16x^2 - 81$

Factor the coefficient of x^2 from all three terms.

20. $2x^2 - 6x + 1$
21. $-3x^2 + x + 2$
22. $5x^2 - 7x + 15$
23. $-4x^2 + 6x - 5$

Find the zeros as integers or as reduced fractions in Exercises 24 - 28.

24. $h(x) = (x + 3)(x - 4)$
25. $q(x) = (x - 2)(x + 5)(x - 7)(x + 1)$
26. $f(x) = 3x + 4$
27. $g(x) = (2x - 5)(3x + 7)$
28. $j(x) = (5x - 3)(2x + 5)(6x - 1)$

29. Find ***any*** polynomial whose zeros are -5 and 5.
30. Find ***any*** polynomial with integer parameters whose zeros are $-\frac{4}{5}$ and $\frac{4}{5}$.

31. Find a zero and a factor of $600x - 1200$.

32. If 0.333333... is a zero of a polynomial function, what is one possible factor of the function?

33. If 0.25 is a zero of a polynomial, what is one possible factor of the polynomial?

34. If 0.5 is a zero of a polynomial, what is one possible factor of the polynomial?

35. If 0.666666... is a zero of a polynomial, what is one possible factor of the polynomial?

36. If 0.125 is a zero of a polynomial, what is one possible factor of the polynomial?

37. What do the following polynomials have in common?

$$2x + 4$$
$$13x + 26$$
$$-5x - 10$$

38. Find 3 different polynomial functions that have 3 as a zero.

39. Find 3 different polynomial functions that have 3 as the only zero.

40. Find 3 different polynomial functions that have zeros of –4 and 4.

From Mathematics to English

41. After reading this section, make a list of questions that you want to ask your instructor.

42. Continue in your daily journal and make an entry. In addition to your normal entry on thoughts about the mathematics in this section, list at least two positive comments about what you have learned about this topic.

43. In paragraph format, summarize the material in this section of the text in your daily journal.

44. Describe how your classroom instructor made this topic more understandable and clear.

45. After reading the text and listening to your instructor, what do you not understand about this topic?

46. Describe in paragraph format how to factor $6x^2 + 3x - 4x - 2$ graphically.

47. Explain the relationship between the factors $(ex + f)$ and $(ex - f)$, and the zeros of the polynomial function $g(x) = (ex + f)(ex - f)$.

48. List reasons why you know a particular polynomial cannot be factored. Examples may be used.

49. Can the graphing calculator find the exact zeros of the polynomial function $f(x) = \left(x - \sqrt{3}\right)\left(x + \pi\right)$? Explain.

50. Describe how you might factor the polynomial function $f(x) = 2x^5 + 5x^4 - 26x^3 - 65x^2 + 72x + 180$. Do not actually factor.

From English to Mathematics

In the next four Exercises, convert the English statement to an equivalent mathematical statement.

51. $(x + 2)$ and $(x - 2)$ are factors of $x^2 - 4$

52. the factors of $x^2 - 9$ are greater than 17

53. seven and nine are factors of sixty-three

54. factors of $x^2 - 6x - 7$ are less than 12

Explorations

55. For the functions listed below, find the zeros and propose what you think are possible factors of the polynomial function.

Polynomial Function	Zeros	Possible Factors
$f(x) = x^2 + x - 2$	______	______
$f(x) = x^2 - 2x - 3$	______	______
$g(x) = x^2 + 6x + 8$	______	______
$g(x) = x^2 - 4x - 21$	______	______
$f(x) = x^2 - 9x + 20$	______	______
$f(x) = x^2 - 3x - 18$	______	______
$q(x) = 6x^2 + 5x + 1$	______	______
$q(x) = x^2 + 5x$	______	______
$f(x) = 3x^2 + 2x$	______	______
$f(x) = 5x^2 - 9x$	______	______

56. Using a graphical method, factor the polynomials below.

a. $x^3 - 2x^2 - 5x + 6$

b. $2x^3 + 3x^2 - 2x - 3$

c. $x^3 + x^2 - 37x + 35$

d. $x^4 - 10x^2 + 9$

$x^2 + 3x - 4 = (x + 4)(x - 1)$

4.5 Factoring the Trinomial

The function under consideration is the general form of the quadratic function $f(x) = ax^2 + bx + c$, where the function parameters a, b, and c have no common integer factor; if a common integer factor exists, it is factored with the distributive property. The exponent on x in the first term must be twice the exponent on x in the second term. As with previous polynomials, the analytical method of factoring the quadratic trinomial may be the best choice; however, you are given the option of factoring with technology. As before, knowing how factors are related to the zeros of the function may be more important in the long run.

From reviewing multiplication of binomials, you know that when binomials of the form $(ax + b)$ and $(ax - b)$ are multiplied, the product is a binomial, for example, the conjugates $(2x + 3)(2x - 3) = 4x^2 - 9$. The product of binomials of the form $(ax + b)$ and $(cx + d)$ yields a trinomial, for example, $(2x - 1)(x + 4) = 2x^2 + 7x - 4$. Since conjugates have already been analyzed in the previous section, all that remains is to look at trinomials whose factors are binomials -- and not conjugates.

Consider the graphs of the polynomial functions:

$f(x) = (x + 2)(x - 3)$ or $f(x) = x^2 - x - 6$, and $g(x) = (x + 1)(x - 5)$ or $g(x) = x^2 - 4x - 5$.

$f(x)$

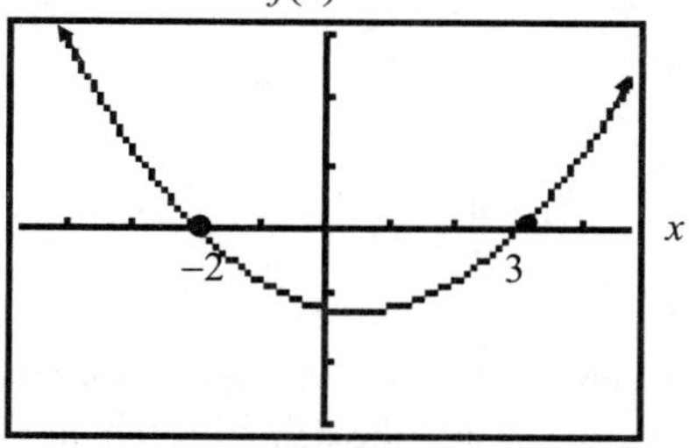

Figure 4.5.1a

$g(x)$

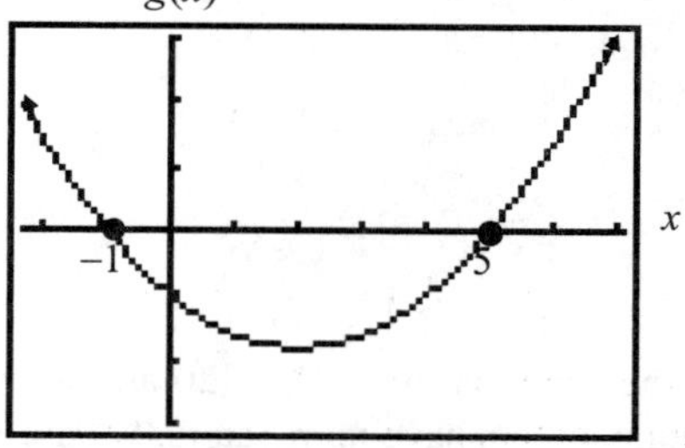

Figure 4.5.1b

As can be seen in Figure 4.5.1a, the polynomial $x^2 - x - 6$ has zeros of -2 and 3, and factors of $(x + 2)$ and $(x - 3)$ as shown above. In Figure 4.5.1b, the polynomial $x^2 - 4x - 5$ has zeros of -1 and 5, and factors of $(x + 1)$ and $(x - 5)$ as shown above. Just like the factors of the difference of squares polynomial were related to the zeros of the polynomial, the factors of the trinomial are related to the zeros of the polynomial function as well. That is, if the zeros are a and b, the factors are $(x - a)$ and $(x - b)$.

Example 1: Factor $f(x) = x^2 - 7x + 10$.

Solution: Graph the polynomial $x^2 - 7x + 10$ and find the zeros. If the zeros are integers, write the factors and check to confirm that their product is $x^2 - 7x + 10$. If the zeros are not integers, the polynomial cannot be factored because when the coefficient of x^2 is one, the zeros must be integers. Why? An integer window or decimal window may be more useful the using the zero finder.

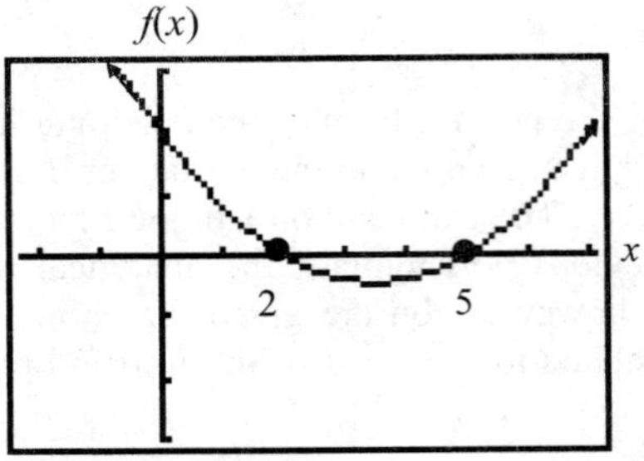

Figure 4.5.2

The zeros are 2 and 5. The factors are $(x - 2)$ and $(x - 5)$.

Check: $(x - 2)(x - 5) = x^2 - 7x + 10$.
F OI L

Yes, $x^2 - 7x + 10$ factors to $(x - 2)(x - 5)$.

⁂

Consider the trinomial whose factors are $(2x + 5)$ and $(3x - 4)$. Multiplying the factors, the trinomial is $6x^2 + 7x - 20$. Find the zero of each factor by graphing the functions $f(x) = 2x + 5$ and $g(x) = 3x - 4$, or find the zeros analytically. Using the analytical method developed in beginning algebra, the zeros are found to be:

$$2x + 5 = 0 \qquad 3x - 4 = 0$$

$$2x = -5 \qquad 3x = 4$$

$$x = -\frac{5}{2} \qquad x = \frac{4}{3}$$

The zeros of the trinomial $6x^2 + 7x - 20$ are rational numbers! Not only are the zeros rational numbers, but the rational numbers have numerators that are factors of -20 and the denominators are factors of 6. Is this always the case? That is, for the *factorable* trinomial $ax^2 + bx + c$, do the zeros have numerators that are factors of c and the denominators factors of a? To test this idea, two examples are analyzed below.

Example 2: Find the zeros of $15x^2 - 7x - 4$.

Solution: Graph the function $f(x) = 15x^2 - 7x - 4$ and find the zeros with the zero finder and convert them to reduced fractions

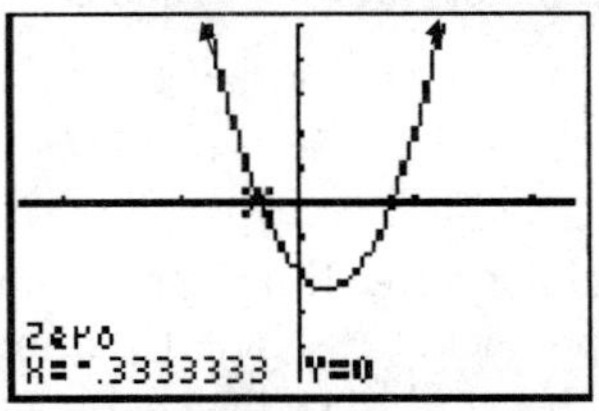

Figure 4.5.3a

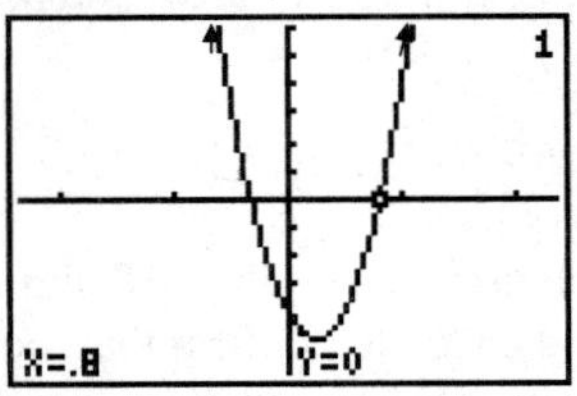

Figure 4.5.3b

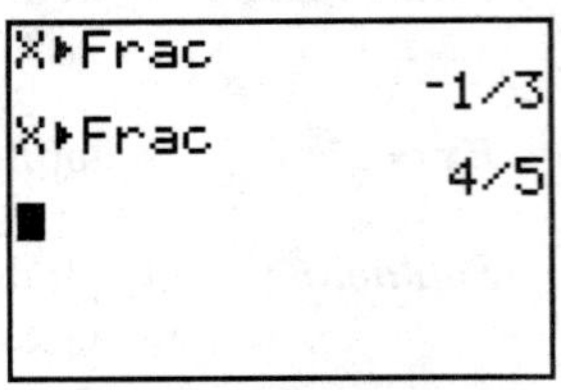

Figure 4.5.3c

In reduced fraction form, the zeros are $-\frac{1}{3}$ and $\frac{4}{5}$.

A quick check verifies that they are zeros.

$15\left(-\frac{1}{3}\right) - 7\left(-\frac{1}{3}\right) - 4 = 0$

Try it yourself on your calculator.

$15\left(\frac{4}{3}\right) - 7\left(\frac{4}{3}\right) - 4 = 0$

Yes, the zeros are rational numbers, the numerators are factors of –4, and the denominators are factors of 15.

⁂

Example 3: Find the zeros of $6x^2 + 23x + 20$.

Solution: Graph the function $f(x) = 6x^2 + 23x + 20$ and use the zero finder to find the zeros, and then convert them to reduced rational numbers.

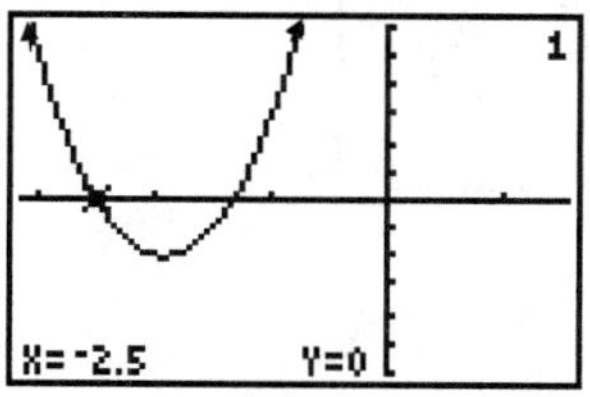

Figure 4.5.4a

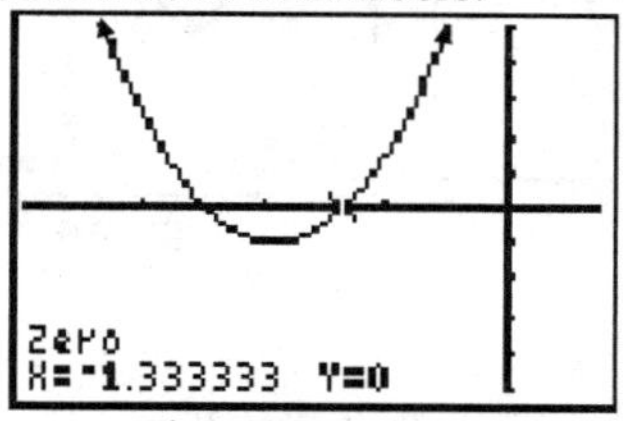

Figure 3.5.4b

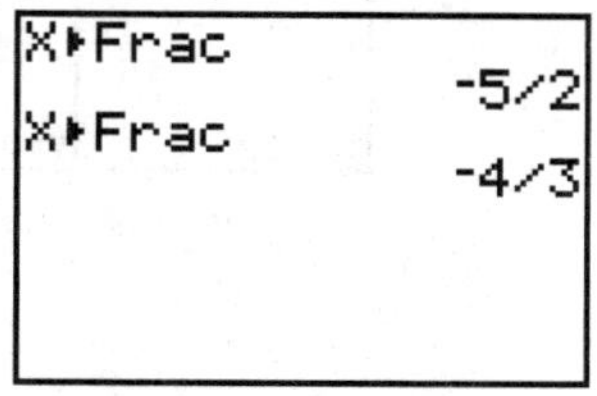

Figure 4.5.4c

In reduced fraction form, the zeros are $-\frac{5}{2}$ and $-\frac{4}{3}$. You are encouraged to check these with your calculator! Yes, the zeros are rational numbers, the numerators are factors of 20, and the denominators are factors of 6.

⁂

If the *zeros are integers*, factors can be found, as we did at the beginning of this section. If the *reduced fractional zeros* of a function are known, can the factors be found? To develop a technique for finding factors from zeros, the graph of $(x + 4)(3x – 2)$ is shown below.

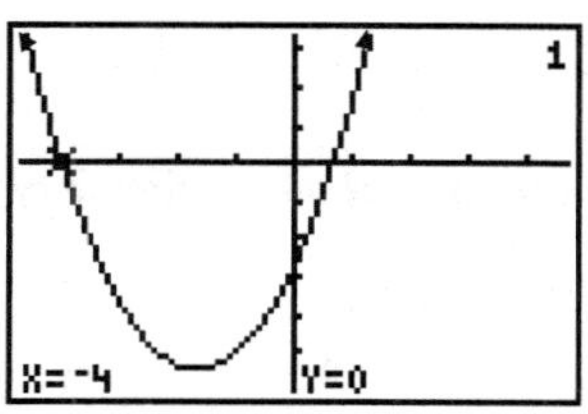

Figure 4.5.5a

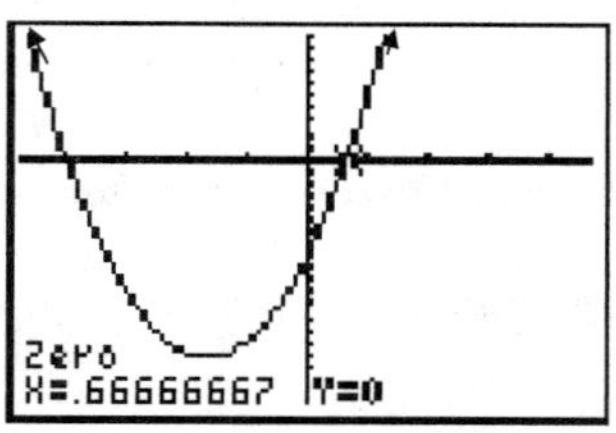

Figure 4.5.5b

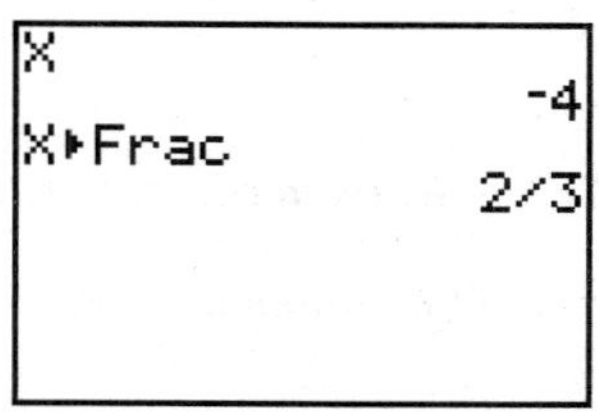

Figure 4.5.5c

Since $(x + 4)(3x – 2) = 3x^2 + 10x – 8$, the factors of $3x^2 + 10x – 8$ are known. The zeros are –4 and $\frac{2}{3}$. The integer zero –4 relates to the factor $(x + 4)$ as was discussed earlier. The reduced rational zero $\frac{2}{3}$ comes from the factor $(3x - 2)$. The denominator of the rational zero $\frac{2}{3}$ is the coefficient of x, and the ***opposite of the numerator*** is the constant. Just as in the last section, the factor $(ax + b)$ has a zero of $-\frac{b}{a}$. The opposite of the constant term is the numerator of the zero and the coefficient of x is the denominator.

Example 4: Factor $6x^2 + 23x + 20$.

Solution: Find the zeros in reduced rational form and proceed as outlined above.

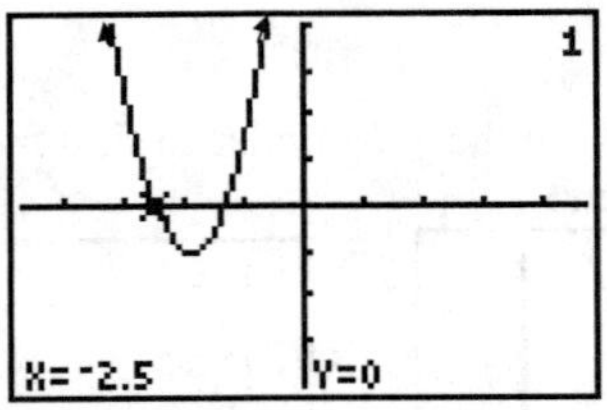

Figure 4.5.6a

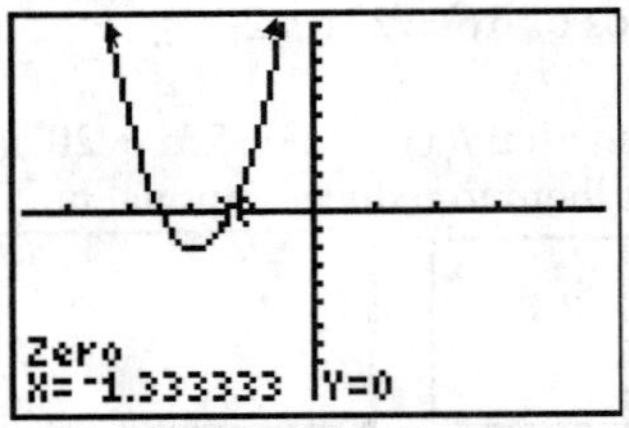

Figure 4.5.6b

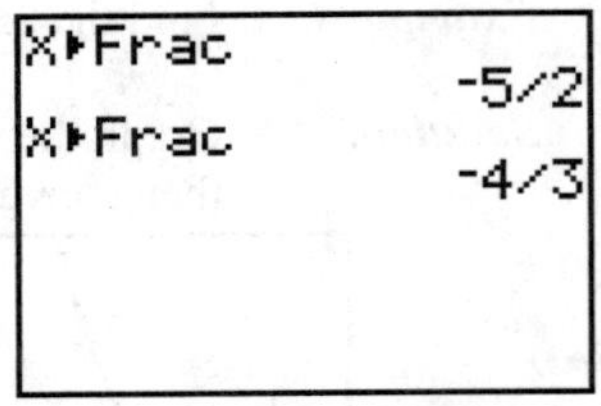

Figure 4.5.6c

The reduced fractional form of the zeros must have a numerator that is a factor of the constant term of the trinomial and the denominator must be a factor of the coefficient of x^2. In reduced fraction form, the zeros are $-\frac{5}{2}$ and $-\frac{4}{3}$. The denominators are the coefficients of x in each factor, and the opposite of the numerators make-up the constants in each factor. The factors are $(2x + 5)$ from the zero $-\frac{5}{2}$ and $(3x + 4)$ from the zero $-\frac{4}{3}$. A check can be performed to confirm that the factors are correct; multiply the factors. The product should be $6x^2 + 23x + 20$.

⁂

PLEASE NOTE! ***The reduced fractional form of the zeros must have a numerator that is a factor of the constant term of the trinomial and the denominator must be a factor of the coefficient of x^2.***

Example 6: Factor $f(x) = x^2 + x + 1$.

Solution: Find the reduced rational form of the zeros and convert them to factors.

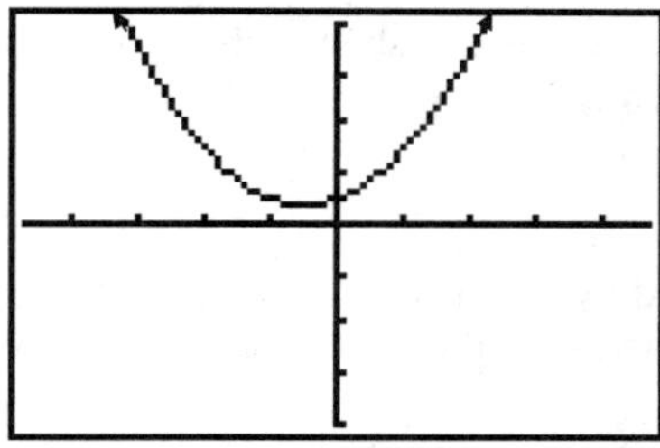

Figure 4.5.7

As can be seen from the graph, there are no real zeros. The polynomial does not factor with integer coefficients and constants.

⁂

Example 6: Factor $2x^2 - 5x - 4$.

Solution: Find the reduced rational form of the zeros and convert them to factors.

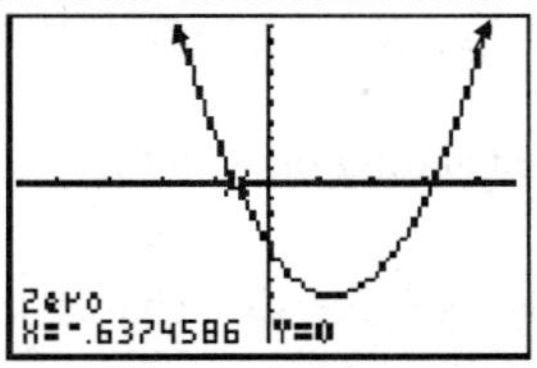

Figure 4.5.8a

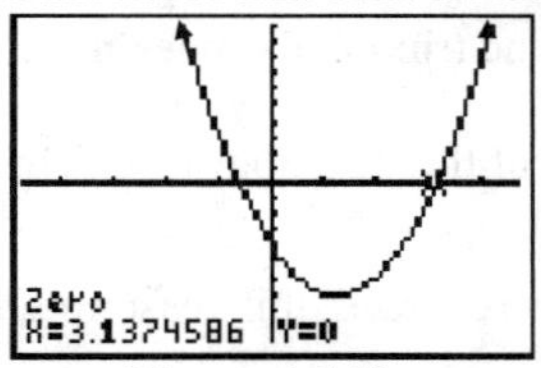

Figure 4.5.8b

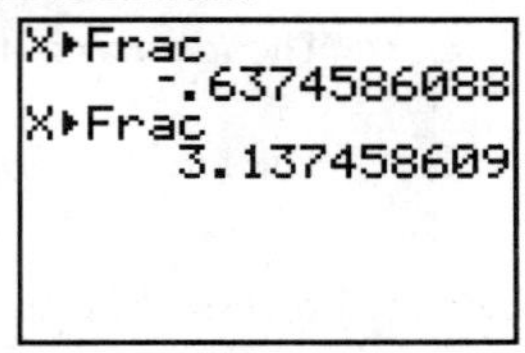

Figure 4.5.8c

The trinomial does not factor because the zeros cannot be converted to fractions -- see Figure 4.5.8c. As you will discover in Chapter Ten, the zeros are, in fact, irrational numbers. Irrational numbers, by definition, are numbers that cannot be written as fractions of two integers. The conclusion must be that the function $2x^2 - 5x - 4$ cannot be written if factored form.

⁂

Example 7: Factor $12x^2 - 13x - 35$.

Solution: Find the reduced rational zeros and convert them to factors. Remember that the numerators of the reduced rational zeros must be factors of 35 and the denominators of any possible reduced rational zeros must be factors of 12.

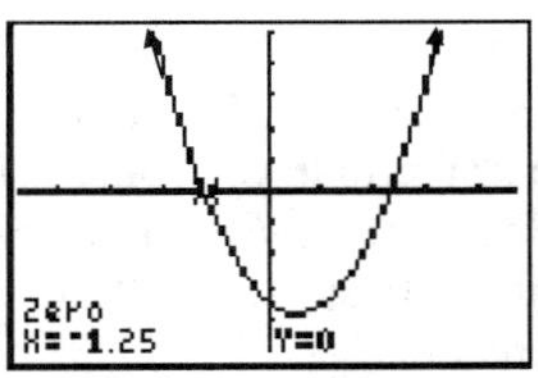

Figure 4.5.9a

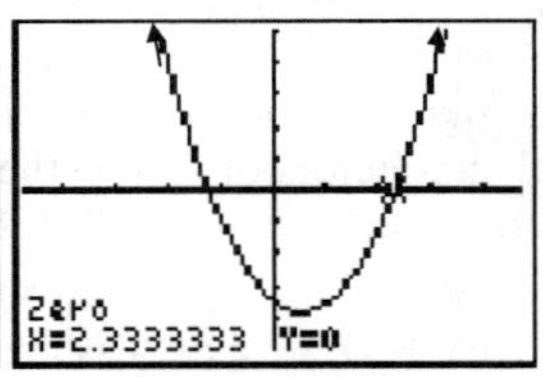

Figure 4.5.9b

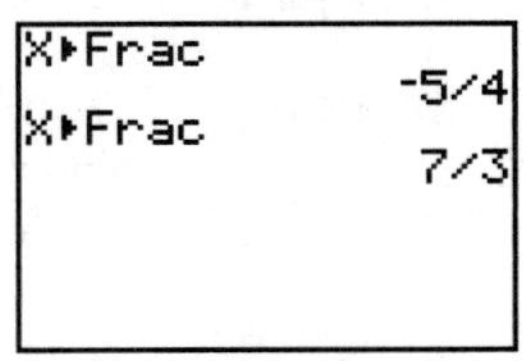

Figure 4.5.9c

The zeros are the rational numbers $-\frac{5}{4}$ and $\frac{7}{3}$. The factors become $(4x + 5)$ and $(3x - 7)$. A check shows that the factors are correct: $(3x - 7)(4x + 5) = 12x^2 - 13x - 35$.

⁂

Trinomials can also be factored by an analytical method. This method is derived from the FOIL method for multiplying binomials. Basically, the FOIL method is reversed in order to factor. Consider the following reasoning:

$$\begin{array}{l} (x + 5)(x - 3) = x^2 \ + 2x - 15 \\ \qquad\qquad\qquad\quad\ \text{F} \quad \text{OI} \quad \text{L} \end{array}$$

Some observations are important:

1. The *first term* of the trinomial comes from multiplying the *first terms* of the binomials.
2. The *last term* of the trinomial comes from multiplying the *last terms* of the binomials.
3. The *middle term* of the trinomial comes from adding the *outer terms plus the inner terms.*

When the process of FOIL is reversed, this reasoning is also reversed. That is, to factor a trinomial:

1. The *first term* of the trinomial factors into the *first terms* of each binomial.
2. The *last term* of the trinomial factors into the *last terms* of each binomial.
3. The factors chosen for the first and last terms of the binomials must satisfy the "outer plus inner" test. If the "outers plus inners" do not equal the middle term of the trinomial, try other factors of the first and/or last term of the trinomial.

The sample problems below will demonstrate the analytic method.

Example 8: Factor $x^2 - 5x - 24$ by the analytic method.

Solution: The trinomial factors into two binomials with factors of x^2 as the first term of each binomial and factors of -24 as the last term in each binomial.

$(x - 6)(x + 4)$	fails the OI test $[-6x + 4x \neq -5x]$
$(x - 12)(x + 2)$	fails the OI test $[-12x + 2x \neq -5x]$
$(x - 8)(x + 3)$	is correct.

Check: $(x - 8)(x + 3) = x^2 - 5x - 24$

⁂

Example 9: Factor $3x^2 - 25x - 18$ by the analytical method.

Solution: The trinomial factors into two binomials with factors of $3x^2$ as the first term of each binomial and factors of -18 as the last term in each binomial.

$(3x - 9)(x - 2)$ fails the OI test $[-9x + 6x \neq -25x]$

$(3x + 6)(x - 3)$ fails the OI test $[6x + (-9x) \neq -25x]$

$(3x + 2)(x - 9)$ is correct.

Check: $(3x + 2)(x - 9) = 3x^2 - 25x - 18$

⁂

Example 10: Factor $6x^2 - 23x + 20$ analytically.

Solution: The trinomial factors into two binomials with factors of $6x^2$ as the first term of each binomial and factors of 20 as the last term in each binomial.

$(6x - 5)(x - 4)$ fails the OI test $[-5x + (-24x) \neq -23x]$

$(2x - 4)(3x - 5)$ fails the OI test $[-12x + (-10x) \neq -23x]$

$(3x - 4)(2x - 5)$ is correct.

Check: $(3x - 4)(2x - 5) = 6x^2 - 23x + 20$

While the above analytical method for checking is probably the fastest, do not forget that you can check your work graphically and numerically as well. Below are the graphical representations of $(3x - 4)(2x - 5)$ and $6x^2 - 23x + 20$.

$(3x - 4)(2x - 5)$

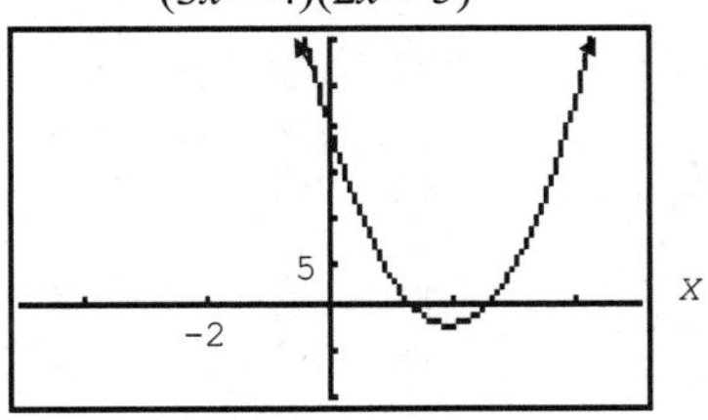

Figure 4.5.10a

$6x^2 - 23x + 20$

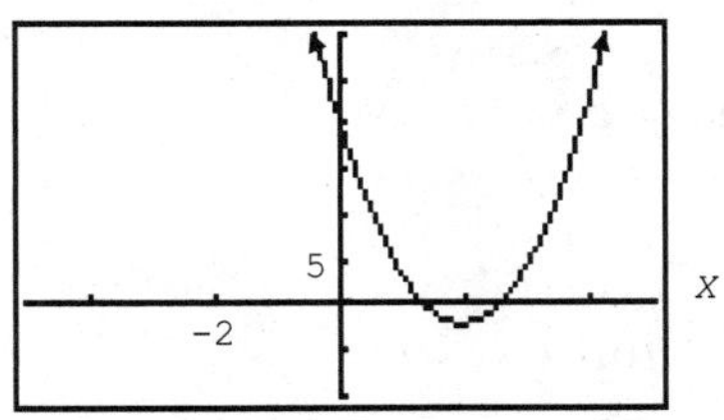

Figure 4.5.10b

The graphs are identical; therefore, $6x^2 - 23x + 20$ correctly factors into $(3x - 4)(2x - 5)$. Another method for checking your factoring is to compare numerical representations. If the numerical representations are the same, the factoring is correct. Below are the numeric representations of the problem and the answer. As you can see, they are identical also.

Table 4.5.1

x	–2	–1	0	1	2	3	4	5
$6x^2 - 23x + 20$	**90**	**49**	**20**	**3**	**–2**	**5**	**24**	**55**
$(3x - 4)(2x - 5)$	**90**	**49**	**20**	**3**	**–2**	**5**	**24**	**55**

⁂

4.5 EXERCISES

Many times you can work the exercises by algebraic, numeric, or graphical methods. You can learn about mathematics no matter which method you use. Part of the learning process is for you to decide which method is best for you to use. Please read the text before you do the exercises. Read the **next assigned section** after you finish the assignment below.

A Quick Look To The Future

-4. What kind of graph is the sum of a linear and quadratic function?

-3. What is the value of the function $3x - 4$ when x is 0? What point on the graph is this?

-2. How many lines can you draw through the point (0, 2), that rise at a rate of $\frac{1}{2}$?

-1. Is $2x + 3$ equivalent to $2(x - 1) + 5$?

Previous Topics Revisited

1. Factor $16x^4 - 32x^3 + 8x^2$
2. Factor $12x^2 - 28x + 15x - 35$
3. Factor $9x^4 - 121$
4. Multiply $(x - 3)(x^3 + 2x^2 - 5x + 1)$
5. Multiply $3x^2(2x^2 - 6x + 5)$
6. If $f(x) = 7x^2 - 13x + 4$ and $g(x) = 2x^2 - 4x - 1$, find $(g - f)(x)$.

New Topic Exercises

Factor the following trinomials in Exercises 7 - 18. If a polynomial cannot be factored exactly, it is a prime polynomial. For the non-factorable trinomials, indicate that they are prime polynomials. Use the method of your choice.

7. $x^2 - 2x - 3$
8. $x^2 + 12x + 32$
9. $x^2 - 15x + 36$
10. $2x^2 + 9x + 4$
11. $x^2 - 2x + 5$
12. $3x^2 - 11x + 10$
13. $7x^2 + 8x + 1$
14. $8x^2 + 14x - 15$
15. $-x^2 + 4x - 7$
16. $12x^2 - 17x + 6$
17. $10x^2 + 3x - 1$
18. $3x^2 + 2x - 2$

19. Find the zeros of $3x^2 + 2x - 1$, $12x^2 + 8x - 4$, $9x^2 + 6x - 3$, and $21x^2 + 14x - 7$.

20. Find the zeros of $10x^2 + 15x - 100$, $2x^2 + 3x - 20$, $6x^2 + 9x - 60$, and $8x^2 + 12x - 80$.

21. If the zeros of a trinomial $ax^2 + bx + c$ are $-\frac{1}{2}$ and $\frac{3}{5}$, what are possible factors of a?

22. If the zeros of a trinomial $ax^2 + bx + c$ are $\frac{4}{9}$ and 3, what are possible factors of c?

23. If the zeros of a trinomial $ax^2 + bx + c$ are -2 and 13, what are possible factors of a?

24. If the zeros of a trinomial $ax^2 + bx + c$ are -8 and 7, what are possible values for b?

25. If the zeros of a polynomial are –2, 4, –1, and $\frac{2}{3}$, is the polynomial a trinomial of the form $ax^2 + bx + c$? If not, what might the polynomial be?

From Mathematics to English

26. After reading this section, make a list of questions that you want to ask your instructor.

27. Continue in your daily journal and make an entry. In addition to your normal entry on thoughts about the mathematics in this section, list at least two positive comments about what you have learned about this topic.

28. In paragraph format, summarize the material in this section of the text in your daily journal.

29. Describe how your classroom instructor made this topic more understandable and clear.

30. After reading the text and listening to your instructor, what do you not understand about this topic?

31. If a trinomial is prime, make a list of possible reasons how you know it can not be factored.

32. Describe a plan of how you would factor $x^4 - 5x^2 + 4$.

33. Describe any relationship between d and the zeros of the trinomial $dax^2 + dbx + dc$. Give a rationale for your response.

34. Research college algebra textbooks for a theorem called the "Rational Zero Theorem." Write a brief paragraph describing the Rational Zero Theorem.

35. If the zeros of a quadratic function are irrational, why can't the function be factored?

36. What method would you use to factor $35x^2 - 22x - 24$? Why?

37. If there are no zeros of a quadratic function, why can't the function be factored?

From English To Mathematics

In the next three exercises, convert each English statement to an equivalent mathematical statement.

38. the product of seven plus a number and two minus the number

39. the area of a rectangle, "$x + 3$" by "$x - 7$"

40. a number x plus three, times the difference of x and 1

Explorations

41. Pick numbers at random for the parameters a, b, and c in the trinomial of the form $ax^2 + bx + c$ and try to find the factors. How many of the 10 polynomials you select are factorable?

42. What are the x-intercepts in the following functions? Verify them graphically.

a. $f(x) = (x-2)(x+4)(x-6)$

b. $f(x) = (x+7)(x-16)(x+2)$

c. $g(x) = (x^2-1)(x+7)(x-3)$

d. $g(x) = (x+2)^5$

e. $h(x) = (x+1)(x-1)(x+2)(x-2)(x+3)(x-3)$

f. $h(x) = (2x+1)(3x-5)(4x-3)$

g. $k(x) = (x-1)^2(x+4)^2$

h. $p(x) = (x+5)^2(2x-7)^3$

43. Find the zeros of the following functions.

a. $f(x) = (x-5)(x-1)(x+3)$

b. $f(x) = |(x-5)(x-1)(x+3)|$

c. $f(x) = |x-5|(x-1)(x+3)$

d. $f(x) = (x-5)|x-1|(x+3)$

e. $f(x) = (x-5)(x-1)|x+3|$

f. $f(x) = |(x-5)(x-1)|(x+3)$

g. $f(x) = (x-5)|(x-1)(x+3)|$

4.6 Function Operations from a Graphical Perspective

In business, many types of employees are paid a weekly wage according to a plan called a sliding pay scale. They may earn a weekly fixed amount in addition to, for example, 20% commission on sales from \$0 to \$1,000 and 30% commission on sales over \$1000. What function did you study in Chapters Two or Three that has this behavior -- a 30% rate of change on one branch and a 40% rate of change on another branch? None of them! As it turns out, the function that models the salary of this person is the sum of absolute value functions. This is a topic that is usually not included in textbooks because, traditionally, sums of absolute value functions are too complicated to study when you try to add them symbolically; however, you will not be adding them symbolically here either -- you will be studying sums graphically.

Analyzing the arithmetic operations of functions (adding, subtracting, and multiplying) from a purely symbolic perspective leaves much to be desired. Did you ever wonder why the arithmetic operation of addition of $|x + 3|$ and $|x - 7|$ is usually not studied? Is there a good reason why you do not study addition or subtraction of functions like $\sqrt{x-3}$ and $2|x+5|$? Why have you never analyzed the product of functions like $x^2 - 3$ and $\sqrt{x+1}$? The reason is that when you only study functions from a symbolic perspective, there is nothing to study in the above examples. When functions are represented symbolically, there are two choices for the analysis: 1) to simplify or 2) to solve equations/inequalities containing the functions. However, you cannot simplify $\sqrt{x-3}$ plus or minus $2|x+5|$. You cannot simplify $x^2 - 3$ times $\sqrt{x+1}$. If you cannot simplify, then the analytical methods for solving equations and inequalities containing arithmetic combinations of functions quite frequently cannot be used either.

The graphing calculator has allowed for a numeric and a graphical analysis of functions like $f(x) = |x-5| + 3\sqrt{x+4}$. With a graphing calculator, you can use it to find when the function is increasing or decreasing; find maximum or minimum values; find when it is positive, negative, or zero; calculate average rates of change; determine the domain and range; explore the general behavior of the function; and study the connections between the function parameters in the symbolic representation and behavior of the graphical representation. *You can solve equations and inequalities containing any combination of functions studied.* By analyzing functions of this nature, you will discover an amazing variety of behaviors. You will also find that knowing about this behavior and how to create it will be very useful when you use functions to model data that comes from real-world situations.

In the remainder of this section you will find several examples of the analysis of functions formed by sums, differences, and/or products of the elementary functions introduced in Chapter Two. Each of the examples analyzed demonstrates a function that has applications as a model for real-world data. Modeling real-world data is a significant use of mathematics by professional people in many different fields.

Analysis of $|x+3| + |x-2|$ - "The Sum of Two Absolute Value Functions"

To analyze any function, it is helpful to have the numeric representation and/or the graphical representation. Both a complete table and a complete graph are below. Remember the trace key on the calculator will give you the numeric representation the calculator uses to create the graphical representation.

Table 4.6.1

x	−6	−5	−4	−3	−2	−1	0	1	2	3	4	5
$\lvert x+3\rvert + \lvert x-2\rvert$	11	9	7	5	5	5	5	5	5	7	9	11

Right away, in the table, you see a new kind of behavior. The function remains constant when x is on the interval [−3, 2]. The graphical representation will help you visualize this behavior.

$|x+3| + |x-2|$

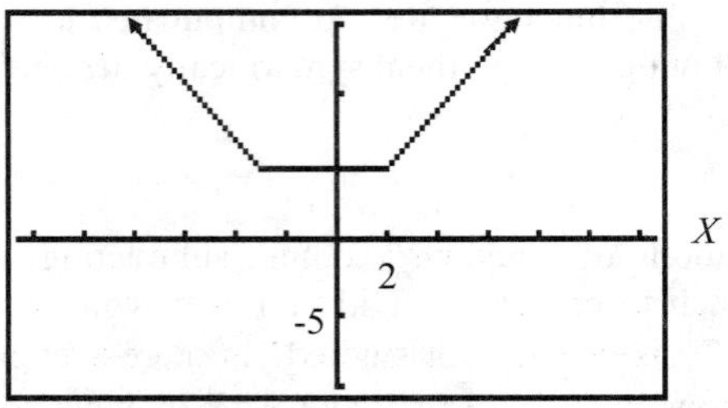

Figure 4.6.1

Increasing/decreasing   If you move the trace cursor to the left side of the viewing window, and then trace to the right (so that you can read it from left to right), you will notice that the function values drop until x is −3. The function is decreasing on the interval (−∞, −3). By either looking at the numeric representation above or using the trace key, you can see that the function remains at a constant value of 5 when x is on the interval [−3, 2]. When x is to the right of 2, the function increases; that is, the function is increasing on the interval (2, ∞). Please study the relationship between the numbers (parameters) in the symbolic representation and the behavior of increasing and decreasing.

Maximum/minimum   The complete graph in Figure 4.6.1 shows that the function goes toward infinity at both ends of the viewing domain. Since this trend is established, there is no maximum. The numeric representation or trace shows that the smallest value of the function, or minimum is 5.

Positive/negative/zeros   The numeric representation suggests and the graph shows that the function is never negative or zero! The function is positive when x is on the interval (−∞, ∞). By looking at the symbolic representation, can you give a reason why the function is always positive?

Domain/range   The symbolic representation of the function shows no division nor square roots; therefore, the normal domain is all real numbers. The numeric representation and the graphical representation both show that the range only includes numbers 5 and larger. Using an analytical approach, can you explain why the function never drops below 5?

Rate of change   Looking at the numeric representation helps most when trying to find the average rate of change. The x values are changing by 1; thus, the rate of change can be found by calculating the change in the function. When x is less than −3, the function changes at a rate of −2 units for the function while x changes by 1, making the rate of change a constant value of −2 while $x \in (-\infty, -3)$. When x is on the interval [−3, 2], the rate of change is zero. It changes from 5 to 5 to 5, etc., a zero change. Finally, when x is larger than 2, the function changes at a constant rate of 2 (from 7 to 9, from 9 to 11, etc.).

Further Exploration on Your Own

It is strongly suggested that you try exploring with your calculator to see if you can make "connections" between the numbers in the symbolic representation (parameters) and the behavior of a function that "looks like" the sum of two absolute value functions of the form $|x + a|$, where a is a real number. For example, after you do your investigation of functions like $|x + a| + |x + b|$, you should be able to look at the symbolic representation and find the minimum value of the function. From this you should know the range. You should know where to find the sharp corners of the graphical representation, and you should know when the function is constant by looking at the symbolic form. Finally, you *may* be able to explain why the rate of change is always –2 for the left piece of the graph and 2 for the right piece.

To make your investigation even more interesting, try adding one or two more absolute value functions of the form $|x + a|$ to the sum you are already investigating. Can you think of any relationships in the real world that behave like the above models?

Analysis of $|x + 3| - |x - 2|$ - "The Difference of Two Absolute Value Functions"

To analyze any function, it is worth while to see the numeric representation and/or the graphical representation. Both a complete table and a complete graph are below. Remember the trace key on the calculator will give you the numeric representation the calculator uses to create the graphical representation.

Table 4.6.2

x	–6	–5	–4	–3	–2	–1	0	1	2	3	4	5
$\lvert x+3\rvert - \lvert x-2\rvert$	–5	–5	–5	–5	–3	–1	1	3	5	5	5	5

Right away you see the same new kind of behavior from the last problem. The function remains constant when x is on the interval $(-\infty, -3]$ or on the interval $[2, \infty)$. This kind of behavior does not exist in the elementary functions studied in Chapter Two. The graphical representation will help visualize the behavior.

$|x + 3| - |x - 2|$

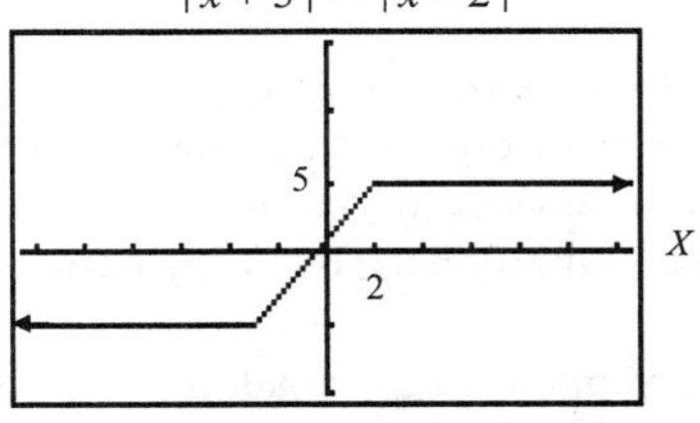

Figure 4.6.2

<u>Increasing/decreasing</u> The numeric representation and the graphical representation both show very clearly that the function is constant for a major part of the domain. It is increasing in value only when x is on the interval $(-3, 2)$. When x is left of –3, the function is the constant value of –5. To the right of 2, the function is a constant value of 5.

<u>Maximum/minimum</u> Either trace or the numeric representation of the function shows that the smallest the function gets is –5 and the biggest it gets is 5. These are the minimum and maximum values.

<u>Zeros</u> The numeric representation suggests a zero between –1 and 0. Trace and a decimal window show the zero to be –0.5. You can verify this analytically because the definition of a zero is a value for x that causes the function to be 0. Replacing x with –0.5 in $|x+3| - |x-2|$ confirms that it simplifies to 0. Don't forget that this evaluation process can be done on the calculator as well.

<u>Positive/negative</u> Since the function is zero when x is –0.5, it must be either positive or negative everywhere else. A quick glance at the graphical representation shows you that it is positive when x is to the right of –0.5 and negative to the left of –0.5. The function is negative when x is on the interval $(-\infty, -0.5)$ and it is positive when x is on the interval $(-0.5, \infty)$.

<u>Domain/range</u> The normal domain must be all real numbers because the symbolic representation shows that there are no operations that cause the function to be non-real. Since the minimum value of the function is –5 and the maximum is 5, the range is $[-5, 5]$.

<u>Rate of change</u> The rate of change of the function when x is on the interval $(-\infty, -3] \cup [2, \infty)$ is zero because the function does not change. If it is not changing, the rate of change is 0. The rate of change when x is on the interval $[-3, 2]$ is 2. If you use the numeric representation above, you can see that as x changes by 1, the function is changing by 2. For example, when x changes from –3 to –2, the function changes from –5 to –3. This is an increase of 2. The same thing happens when x changes from –1 to 0. The function changes from –1 to 1, an increase of 2.

Further Exploration on Your Own

It is strongly suggested that you try exploring with your calculator to see if you can make "connections" between the numbers in the symbolic representation and the behavior of functions that "look like" the difference of two absolute value functions of the form $|x + a|$, where a is a real number. For example, after you do your investigation of functions like $|x + a| - |x + b|$, you should be able to look at the symbolic representation and find the minimum and maximum values of the function; from this you should know the range. You should know where to find the sharp corners of the graphical representation, you should know when the function is constant by looking at the symbolic form. Finally, you *may* be able to explain why the rate of change is 2 for the middle piece of the graph.

As you continue your investigation, try changing from a difference to a sum and observe the change in the behavior. Try making the coefficient of one of the absolute value functions something other than 1. How would the graphical representation change if you added a number to the function? How would the graphical representation change if you multiplied the entire function by –1?

To make your investigation even more interesting, try adding and subtracting one or two more absolute value functions of the form $|x + a|$ to the difference you are already investigating. When you do this, can you tell where the sharp corners will be? Does the graph have an interval when it is constant? Why or why not? Can you think of any relationships in the real world that behave like the above models?

The final example of an analysis is for the sum of an absolute value function and a square root function. This example should give you some insight on how to create a function with a domain of your choosing. This will be helpful when you do the modeling projects.

Analysis of $|x-3|-4+0\sqrt{x}$ - "Any Function Plus the Square Root Function"

As in previous examples, below are the numeric and graphical representations of the function.

Table 4.6.3

x	−2	−1	0	1	2	3	4	5	6	7	8
$\|x-3\|-4+0\sqrt{x}$	NR	NR	−1	−2	−3	−4	−3	−2	−1	0	1

$|x-3|-4+0\sqrt{x}$

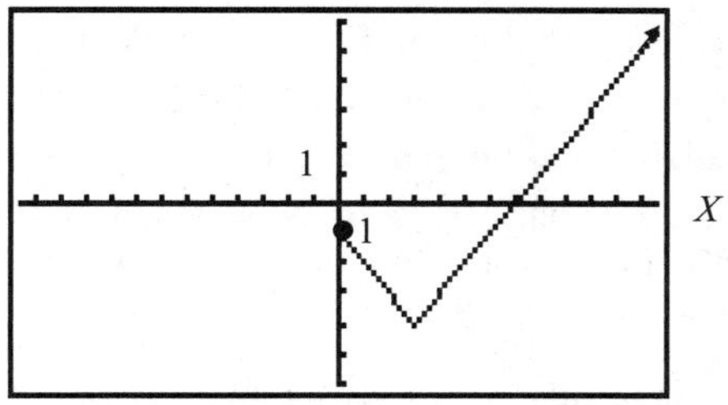

Figure 4.6.3

Increasing/decreasing The numeric representation shows the function decreasing until x has a value of 3. At this point the function values start to increase and the trend continues to the right. The conclusion is that the function is decreasing when x is on the interval (0, 3) and it is increasing when x is on the interval (3, ∞).

Maximum/minimum While you can see the minimum on the graphical representation, you cannot tell what it is. If you use trace and a decimal window you can find it, but you are using the calculator's numeric representation. You may as well look at the above numeric representation. The smallest value or minimum of the function is −4. There is no maximum to the right of the minimum because the function is increasing. An interesting event happens when x is 0. The function is −1, and this is higher than other function values around it -- yet it is not larger than function values to the right of 6. This value of −1 is called a ***local* maximum**. This is a new idea and usually left for discussion in a later course.

Rate of change The numeric representation or an integer window are always good sources of information for finding the rate of change. Looking at the table, when x is left of 3, the rate of change is −1. Why? When x is to the right of 3, the rate of change is 1. For example, as x changes from 3 to 4, the function changes from −4 to −3, a change of 1. As x changes from 4 to 5, the function changes from −3 to −2, a change of 1. As x changes from 5 to 6, the function changes from −2 to −1, again, a change of 1. Conclusion: the average rate of change of the function is −1 when x is on the interval [0, 3], and the rate of change of the function is 1 when x is on the interval [3, ∞).

Zeros From the numeric representation, you can see the zero is 7. This is the only zero because the function is increasing to the right of 3 and so it will never return to 0 again.

Positive/negative The numeric representation shows fairly clearly that the function is positive when x is on the interval (7, ∞) and it is negative when x is on the interval [0, 7).

Domain/range Since the minimum value of the function is −4, the range must be all real numbers −4 and larger, [−4, ∞). The normal domain of the function gives rise to an interesting situation. The function is the sum of an absolute value function and a square root function. A typical absolute value function has a domain of $(-\infty, \infty)$ and this square root function has a domain of $[0, \infty)$. The sum of the two functions has a domain of $[0, \infty)$, the same as the square root function.

Further Exploration on Your Own

This example is significant because the absolute value function has had its domain altered while the shape of the graph has remained the same. Hopefully, you have already come to expect that different symbolic representations of functions have different graphical representations. In the case of $|x-3|-4+0\sqrt{x}$, the graphical representation is still a **V** with a different domain. Explore a little with your calculator by entering any function of your choosing and then add the function $0\sqrt{x}$. Find the domain of your function. The beauty of adding $0\sqrt{x}$ is that you are adding 0 to the function. This is why the graph does not change, just the domain. Try graphing any function on your calculator. Now add $0\sqrt{x}$ to it and graph it again. Note that there is no change in shape, only a change in the domain. See the Explorations exercises for a further investigation of this idea.

A brief reminder: you know by now that each type of function has a graphical representation that is unique to that function. A linear function will never have a graphical representation that looks like a quadratic function, and a quadratic function will never have a graphical representation that looks like a square root function. etc. When you perform arithmetic operations ($+ - \div \times$) on these basic functions, you create a whole new group of functions that on occasion will have some behavior of each of the basic functions, yet still maintain their own identity. As you do the homework exercises, try to discover this basic-like behavior.

4.6 EXERCISES

Many times you can work the exercises by algebraic, numeric, or graphical methods. You can learn about mathematics no matter which method you use. Part of the learning process is for you to decide which method is best for you to use. Please read the text before you do the exercises. Read the **next assigned section** after you finish the assignment below.

A Quick Look To The Future

-4. What value of x causes $2x + 4$ to be zero?

-3. Describe how you might graph a linear function without a calculator.

-2. Rewrite $3x + 7$ in the form of $d(x + e) + f$, where d, e, and f are parameters.

-1. What does $F = \frac{9}{5}C + 32$ mean to you?

Previous Topics Revisited

1. Factor $18x^2 + 3x - 28$

2. What are the zeros of $(2x - 5)(x + 4)(3x + 7)(x - 13)(4x + 15)$?

3. List two polynomial functions with zeros of $\frac{2}{3}$ and -8.

4. Factor $x^8 - 1$

5. Without doing any paper or calculator work, find the zero of $2x + 4000$.

6. Multiply $(x - 5)(x + 2)$. When you multiply two linear functions, what kind of function is the product?

New Topic Exercises

Identify the behavior -- when increasing, when decreasing, the maximum or minimum, the average rate of change between 1 and 2, when positive, when negative, the zeros, the normal domain, and the range -- for each of the following functions.

7. $2|x+3|+|x-2|$
8. $2|x+3|+|x-2|-|x+6|$
9. $|x-1|\sqrt{x+2}$
10. $-x^2+3|x-1|$

Find the normal domain of the following functions.

11. $x+1+0\sqrt{x}$
12. $-x^2+4+0\sqrt{x}$
13. $-|x|+3+0\sqrt{x}$
14. $x+1+0\sqrt{x+2}$
15. $-x^2+4+0\sqrt{x+2}$
16. $-|x|+3+0\sqrt{x+2}$

17. Is the graph of $x-3+2\sqrt{x}$ a straight line? Justify your response.

18. Is the graph of $x^2-2x-3+2\sqrt{x}$ a parabola?

19. Is the graph of $|x-3|\cdot|x-3|$ a parabola?

20. The graphical representation of the function $f(x)=|x-5|+3\sqrt{x+4}$ looks like a check mark (√).
    a. Why doesn't the graph exist left of –4?
    b. Is the graph linear for large x? Explain.
    c. When does the graph have a sharp corner?
    d. Explain why you think there is a little hook at the left end of the graph?
    e. What piece of the function looks like a square root function?
    f. What piece of the function looks like an absolute value function?

From Mathematics to English

21. After reading this section, make a list of questions that you want to ask your instructor.

22. Continue in your daily journal and make an entry. In addition to your normal entry on thoughts about the mathematics in this section, list at least two positive comments about what you have learned about this topic.

23. In paragraph format, summarize the material in this section of the text in your daily journal.

24. Describe how your classroom instructor made this topic more understandable and clear.

25. After reading the text and listening to your instructor, what do you not understand about this topic?

26. List three reasons why you should study mathematics.

27. Describe how you can make the domain of a linear function $[3, \infty)$.

28. Describe how you can make a broken line with corners at –2, 3, and 5?

29. Describe the relationship between the domains of $f(x) = \sqrt{x-2}$ and $g(x) = \sqrt{-x+5}$ and the sum $(f + g)(x)$.

30. Describe the relationship between the domains of $f(x) = \sqrt{x-2}$ and $g(x) = \sqrt{-x+5}$ and the difference $(f - g)(x)$.

31. Describe the relationship between the domains of $f(x) = \sqrt{x-2}$ and $g(x) = \sqrt{-x+5}$ and the product $(fg)(x)$.

Explorations

32. Create five different functions that are sums, differences, products, and/or quotients of the basic parent functions studied in Chapter Three (linear, quadratic, absolute value, square root, and exponential). List the symbolic representation and include a complete graphical representation. Describe any parent-like behavior in the function.

33. When is $(x - 3)$ positive or zero?

    a. What is the domain of $-2x+3+0\sqrt{(x-3)}$?

    b. What is the domain of $|x+4|-5+0\sqrt{(x-3)}$?

    c. What is the domain of $x^2-3x+2+0\sqrt{(x-3)}$?

34. An airplane flying from Columbus to New York (≈ 550 miles) has the following flight plan: $h(d) = -|.03d - 3.3| - |.03d - 12| + 15.1$, where d is the ground distance from Columbus and $h(d)$ is the height of the airplane at distance d. Approximate answers are acceptable.

a. When is the plane increasing in height?

b. When is the plane at a constant height?

c. When is the plane decreasing in height?

d. When is the plane at a zero height?

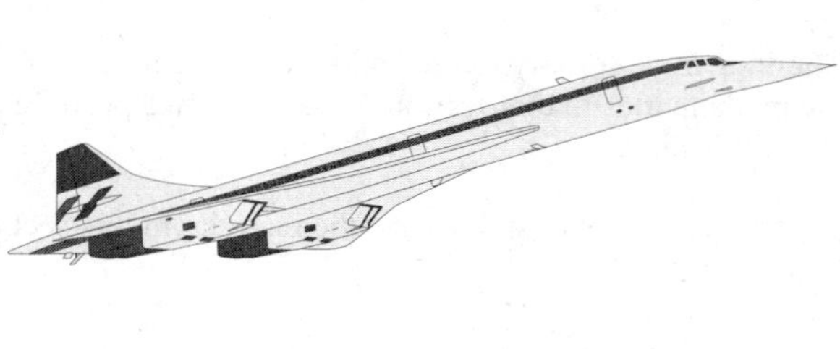

e. What is the maximum height of the plane?

f. How fast is the plane rising near take-off? (What is the average rate of change?)

g. What is the rate of change in the middle of the flight?

h. What is the average rate of descent of the plane?

i. What is the range of the flight plan model? (The mathematical concept of range.)

CHAPTER FOUR TEST

1. Does the set { (–1,4), (6,1), (2,3), (3,4) } represent a function?

2. Does the set { $(x, x^2 \pm 2)$ } represent a function?
(*x* is a real number)

3. Does the graph to the right represent a function?

$\pm\sqrt{2-x}$

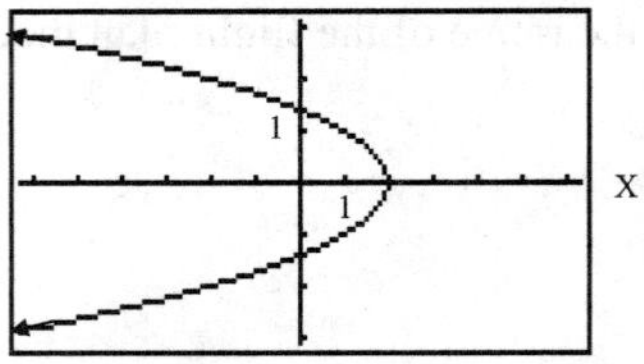

4. If $f(x) = 2|x - 67| + 43$, find $f(-8)$.

5. If $g(x) = -4\sqrt{x-16} + 3$, find $g(20)$.

6. If $h(x)$ is a function and $h(3)$ is 11, name one point on the graph of $h(x)$.

7. Is the graph of $f(x) = |x - 2|$ above or below the graph of $f(x) - 3$?

Given $f(x) = x^2 - 6x + 2$, $g(x) = -x^3 - 2x^2 + 4x - 1$, and $h(x) = x - 2$, do the indicated operations in Questions 8 - 11.

8. $(g - f)(x)$ 9. $(f + g - h)(x)$ 10. $(h \cdot f)(x)$ 11. $[h(x)]^2$

Multiply and simplify the product in Questions 12 - 16.

12. $(-3x^2)(2x)(-x^4)$ 13. $(x - 1)(x^2 + x + 1)$ 14. $(3x - 5)(3x + 5)$

15. $(x^2 - 3)(x^2 + 1)$ 16. $(3x - 2)^2$

17. Factor $96x^2 - 164x + 63$.

18. What is the maximum value of the function $f(x) = -|x-5| - |x+3|$?

19. What is the rate of change of the function $g(x) = |x-2| - |x+3|$ when *x* is on the interval $(-3, 2)$?

20. What is the domain of the function $f(x) = \dfrac{3\sqrt{x-1}}{\sqrt{x-1}}$?

CHAPTER FIVE
ADVANCED ANALYSIS OF THE LINEAR FUNCTION

5.0 Introduction and Review of the Function $dx + e$

Health Care and the GNP

Below are the numeric and graphic representations of money spent on health care (h) represented as a percent of the United States Gross National Product (GNP) as a function of time. If you are interested in a national health care program, one of the many arguments you can use to justify your position might be to project, based on the current data trends remaining constant, what is going to happen to health care spending if no national plan is implemented. To project into the future based on past history, you should have the symbolic representation of the data below in Table 5.0.1. To help find the symbolic representation, the graphical representation of the data will give you clues as to what model you should use.

Table 5.0.1

Time (t)	1960	1965	1970	1975	1980	1985	1990	1995	2002	2003
% of GNP(h)	5.3	5.9	7.3	8.3	9.2	10.5	12.0	13.5	14.9	15.3

Source: U.S. Dept. of Health and Human Services

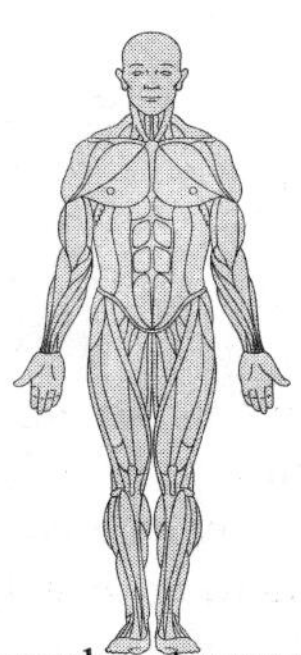

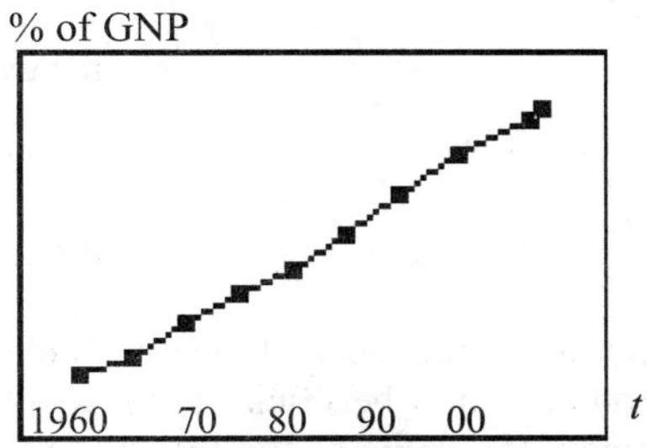

Figure 5.0.1

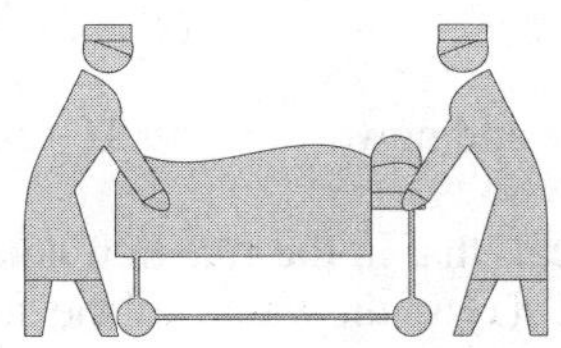

The graphical representation strongly suggests that a linear model of health care spending as a percent of the GNP is warranted. Knowing that the model is linear means you must find values for the parameters d and e in the linear function $dx + e$. You may already know how to find them based on Chapter Three material. In this chapter, you will learn more about linear functions and how to find the symbolic form.

The **linear function** is a polynomial function of the form $f(x) = dx + e$, with d and e being the known real-number parameters. If the value of d is zero, the function becomes $0x + e$ or just the function e. Since e is a real number and real numbers have a constant value, the function $f(x) = e$ is called a **constant function**. The constant function has a graph that is a straight line; thus, it is a special case of the linear function $f(x) = dx + e$. When the value of d is not zero, the function $f(x) = dx + e$ is called a **linear function** because the graph of the function is a straight line, as shown in Chapter Three.

One of the most important ideas developed in this chapter is how you can convert information about the graphic or numeric representation of the linear function to the symbolic representation of the function. You must be able to convert information from one form to another. The chapter ends with applications of the linear function.

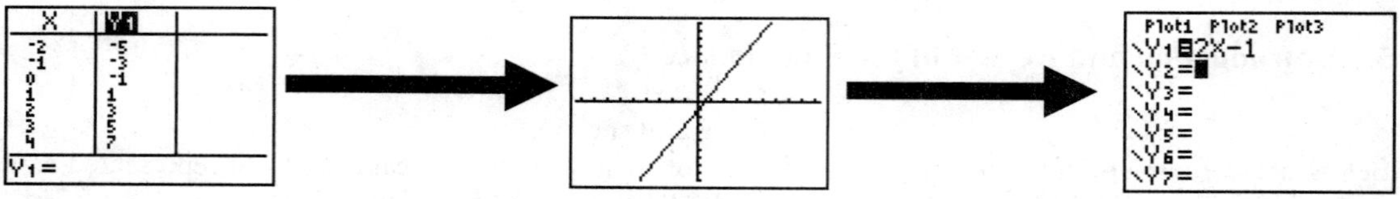

Quick Review and New Notation

Recall from Section 3.1 that the graph of a linear function either rises, falls, or remains constant. When the coefficient of x is positive the graph rises, when the coefficient of x is negative, the graph falls. When the graphical representation of a function rises, mathematicians say that the function is increasing. When the graphical representation of a function falls, the function is decreasing. The linear function has no minimum or maximum value, and the normal domain and range are all real numbers.

In this chapter, a slightly different notation is used to represent functions. In previous chapters, the symbolic representation of a function was denoted as a set of ordered pairs of numbers, for example, $\{ (x, 3x - 5) \}$ for all x in the domain of $3x - 5$. On occasion, the shortened notation $3x - 5$ was used. The notation $f(x) = 3x - 5$ has also been used. Finally, another notation that can be used to show the symbolic representation of a linear function is $y = 3x - 5$. The function has been symbolized by:

	$3x - 5$,	the expression notation;
	$f(x) = 3x - 5$,	the functional notation;
and now,	$y = 3x - 5$,	the y notation.

Recall that in the function notation $f(x)$, f is the name of the function and x is the variable in the function, and $f(x)$ is the value of the function at x. It is beneficial to think of the new notation $y = 3x - 5$, as a statement telling you that ***y* represents the value of the function at *x***. At this time, it may be counter-productive to think of the new notation as an equation. It is the symbolic representation of a linear function. The concept of equations is developed in Chapter Six.

The coordinates of the points on the graph can be represented by $(x, 3x - 5)$, $(x, f(x))$, and now by (x, y). The simplest representation may be (x, y). However, just because it is the quickest to write does not mean that it is always the notation that should be used. Sometimes a mathematical topic may be easier to understand if one of the other notations is used. Identifying the vertical axis as the ***y*-axis** or the ***f*(*x*)-axis** is easier to write than the **3*x* – 5-axis**. Selected topics in mathematics require the use of the functional notation, while other topics require the use of the y notation.

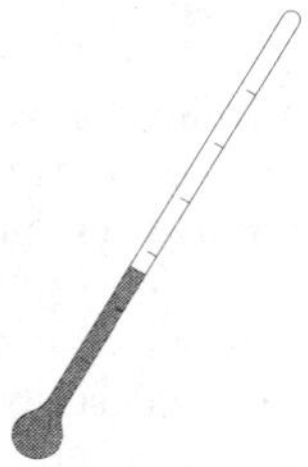

The Linear Function in Action - The Temperature Conversion

Scientists have a tendency to measure the temperature of an object using the Centigrade scale, and some non-science fields use the Fahrenheit scale. This provides a constant need to be able to convert from one system to the other. How is it done?

The linear function $F = \frac{9}{5}C + 32$ models the relationship between the two scales. The most common use of the function is to convert from Centigrade to Fahrenheit. If your concern is with using this model to convert, the numeric representation or trace on an integer window is particularly useful because both show instant access to pairs of equivalent temperature readings. On the other hand, if you are the programmer of the time and temperature displays often found at banks, you will need the symbolic representation of the conversion model.

Table 5.0.2

C	0	5	10	15	20	25	30	35	40
$F = \frac{9}{5}C + 32$	32	41	50	59	68	77	86	95	104

If the temperature is 25°C, the equivalent Fahrenheit temperature is 77°. If the temperature is 68°F, the equivalent Centigrade reading is 20°. For a degree by degree conversion, trace on an integer window works really well. Table works well also (with $\Delta x = 1$). Try it.

When you finish this chapter, you will know how to take linear data and write the symbolic representation of that data.

5.1 Rate of Change, Initial Conditions, and the Zero -- Slope and Intercepts of the Linear Function

Depending on the user-friendliness of a World Wide Web site (or your web browser) when you download files, you may find interesting and useful information on the download screen. For example, in 2002 when downloading a 5.2 meg file, the screen information showed how fast the data was being transferred to a computer, the size of the software, and the time to download. You may recall from Chapter Three that these numbers describe the very behavior of the linear function. How fast the data is being transferred is the rate of change. The size of the file is a new behavior called the initial condition. The **initial condition** is the value of the function when the function variable is zero. Thus, since the function variable is time, the full size of the software is yet to be transferred at time zero. The time it takes to download is the zero of the function modeling the downloading process; that is, at time t when there is nothing left to download, the model must be zero.

While many situations in the world outside the mathematics classroom naturally tend to have the rate of change and the initial condition known, many other times data points are known. With this data, the rate of change can be found as follows. Consider two data points from a linear function; point 1 or P_1 is (2, –3) and point 2 or P_2 is (4, 5). Imagine moving from the first point to the second point. How much does the y-coordinate change? It changes from –3 to 5; the function has *increased* by 8 units from P_1 to P_2. How much does the x-coordinate change? It changes from 2 to 4; it has increased in value also, by 2 units. Refer to Figure 5.1.1 to visualize this change in both y and x.

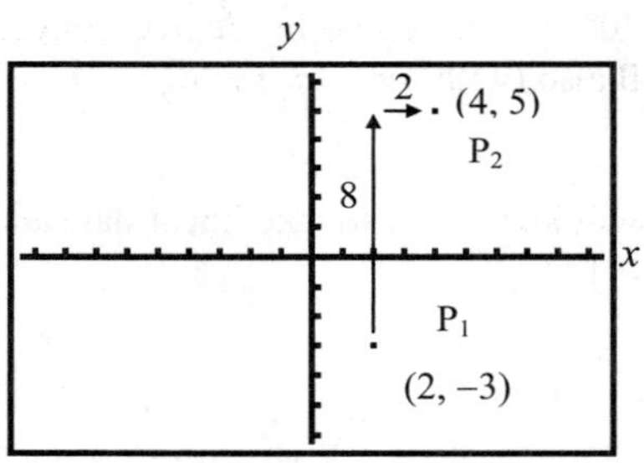

Figure 5.1.1

Consider two points on the graph of another linear function, $P_1(-5, 3)$ and $P_2(4, -2)$. Again, imagine moving from P_1 to P_2. How much does the y-coordinate change? The y-coordinate changes from 3 to –2. It has decreased by 5 units, or it has changed by –5 units. The x-coordinate increased by 9 units (from –5 to 4). Figure 5.1.2 shows these changes.

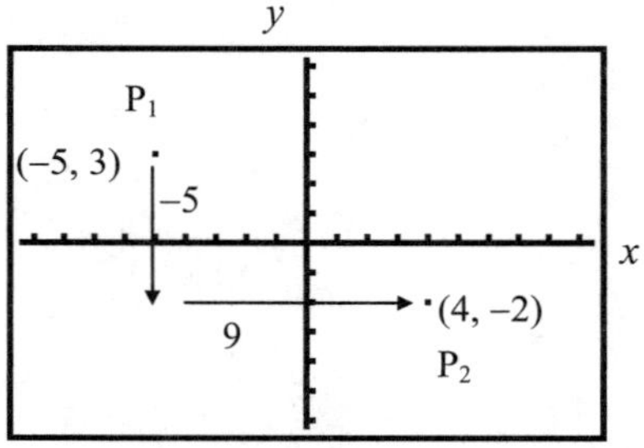

Figure 5.1.2

The information about the change in the y-coordinate and the change in the x-coordinate gives a method for measuring the rate of change of the line. The rate of change of the linear function is sometimes called the **slope** of the graph. The slope can be thought of as showing the amount of vertical change for a corresponding horizontal change. As movement is made from one point on the line to another, like if the vertical change is 2 and the horizontal change is 2 then the slope (rate of change) is known. Like the slope of the roof on a house, if the builder makes the roof rise 6 feet for 12 feet of horizontal change, the direction (or slope) of the roof is known.

Slope of a Line

$$\text{slope} = \frac{\text{change in the } y\text{-coordinate}}{\text{change in the } x\text{-coordinate}}$$

from *any* point on the line to *any* other point on the line.

The symbol commonly used to designate the slope is a lower case m. The notation for change in the y-coordinate is Δy, and the notation for change in the x-coordinate is Δx. Thus, the notation for the slope of a line is: $m = \frac{\Delta y}{\Delta x}$. You should recognize this number called slope as the average rate of change of a function. For a linear function, the **rate of change** is called the **slope**. From work done in Sections 2.3 and 3.1, you recall that in a linear function the rate of change is constant. This is why you can calculate the slope using *any* two points of your choosing. In many of the calculations of average rate of change, the change in x (Δx) is 1 (because of the integer window). As you can see from the above definition, Δx does not have to be 1. However, it may be easier to calculate when Δx is 1 because an integer window automatically sets Δx at 1.

Example 1: The electric company, has the following charges for the *kWh* hours used (200, \$23) and (1800, \$138). What is the electric rate they charge?

Solution: From the first point to the second point, the y-coordinate changes from 23 to 138, or it increases by 115 ($\Delta y = 115$). The x-coordinate changes from 200 to 1800, or it increases by 1600 ($\Delta x = 1600$). The slope (rate) is $\frac{115}{1600}$. If the calculation starts at point two, the change in y from the second point to the first is –115 ($\Delta y = -115$); it changes from 138 to 23 or a decrease of 115. The x changes from 1800 to 200, or a decrease of –1600 ($\Delta x = -1600$). The slope (rate) is $\frac{-115}{-1600}$, or also $\frac{115}{1600}$. ***It does not make any difference whether the calculation is done from point one to two, or from point two to one!***

The number $\frac{-115}{-1600}$ may be recognizable to you in decimal form. It is 0.071875, or 0.072, or \$0.072. It is the cost of one *kWh* hour of electricity!!

⁂

Example 2: What is the slope of the graph of $y = \frac{1}{4}x - 3$?

Solution: Find any two points on the graph of $y = \frac{1}{4}x - 3$ and use the slope formula.

When $x = 4$, y is: $\frac{1}{4}x - 3 = \frac{1}{4}(4) - 3 = -2$

or $P_1 = (4, -2)$.

When $x = -8$, y is: $\frac{1}{4}x - 3 = \frac{1}{4}(-8) - 3 = -5$

or $P_2 = (-8, -5)$.

The change in $y = -3$; $\Delta y = -3$ (from -2 to -5 is a decrease of 3).

The change in $x = -12$; $\Delta x = -12$ (from 4 to -8 is a decrease of 12).

The slope $= \frac{\Delta y}{\Delta x} = \frac{-3}{-12} = \frac{3}{12} = \frac{1}{4}$. Of course, if you remember your work from Sections 2.2, 2.3, and 3.1, you know the slope of the graph of a linear function is the coefficient of x, ¼.

⁂

Interpreting the slope of a line as the rate of change of y with respect to x, shows you how fast y is changing. For example, if the slope is $\frac{3}{1}$, the y-coordinate (the function) is changing *three* times as fast as the x-coordinate. The y-coordinate changes 3 units, while the x-coordinate changes by 1 unit. If the slope is $\frac{4}{3}$, the y-coordinate changes 4 units as the x-coordinate changes by 3.

Example 3: Line one has a slope of $\frac{7}{8}$ and line two has a slope of $\frac{3}{4}$; which line is rising faster?

Solution: If the slope of line two is rewritten as $\frac{6}{8}$, the slopes are more easily compared. Line one rises 7 units while x increases by 8 units, and at the same time line two only rises 6 units as x increases by 8 units. The conclusion has to be that the line with a slope of $\frac{7}{8}$ is rising faster.

⁂

Example 4: If the slope of a line is $-\frac{1}{2}$, how fast is it rising?

Solution: The fraction $-\frac{1}{2}$ can be thought of as $\frac{-1}{2}$; that is, as x increases by 2 units the *line drops by 1 unit.* It is not rising at all; it is falling at a rate of 1 unit for every increase of 2 units in x.

⁂

Example 5: Is the slope of every line positive or negative?

Solution: NO! The line joining (–1, –3) and (4, –3) has a slope that is neither positive nor negative. From the first point to the second point, the change in y is 0 (from –3 to –3 = 0). The change in x from the first point to the second is 5 (from –1 to 4). The slope is $\frac{0}{5}$, which is **ZERO.** Zero is neither positive nor negative. Secondly, consider the line joining (3, 5) and (3, –2). The change in y is –7 (from 5 to –2). The change in x is 0 (from 3 to 3 = 0). The slope is $\frac{-7}{0}$. A real number divided by zero is undefined. The slope of the line joining (3, 5) and (3, –2) is undefined: not positive, not negative, and not zero.

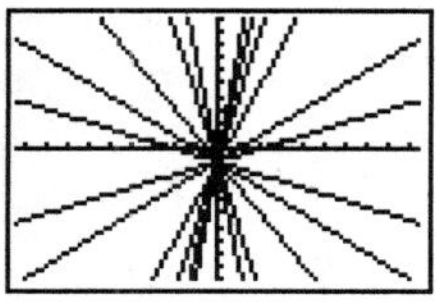

⁂

Example 6: Find the slopes of the graphs of $y=\frac{1}{2}x+5$, $y=1x+0$, $y=2x-4$, $y=-\frac{1}{2}x+4$, $y=-1x+1$, and $y=-2x+5$.

Solution: Find points on the graph of each function to calculate the slope. Δy and Δx are calculated *from* point one *to* point two.

Function	Point one	Point two	Δy	Δx	m
$y=\frac{1}{2}x+5$	(4, 7)	(–2, 4)	–3	–6	$\frac{1}{2}$
$y=1x+0$	(–2, –2)	(3, 3)	5	5	1
$y=2x-4$	(–1, –6)	(–2, –8)	–2	–1	2
$y=-\frac{1}{2}x+4$	(–4, 6)	(10, –1)	–7	14	$-\frac{1}{2}$
$y=-1x+1$	(3, –2)	(–5, 6)	8	–8	–1
$y=-2x-5$	(1, –7)	(0, –5)	2	–1	–2

Hopefully you know all this work is not necessary, provided you remember the connection between the behavior of *rate of change* and the numbers (parameters) in the symbolic representation of the linear function. In the above table, compare m with the coefficient of x for each function. An obvious conclusion is quickly reached if you have not already made it.

⁂

In general: *The slope of the graph of a linear function in simplified form is the coefficient of the variable in the function.*

Example 7: What is the slope of the graph of $y = 2x - 1$?

Solution: The slope is the coefficient of x. It is 2. ($m = 2$)

⁂

Two points on the graph of a linear function that are of special interest are the points where the graph crosses each axis.

> An **intercept** is a point common to the graph of a function and the x-axis or y-axis.

In Section 3.1, the y-intercept was found by graphical methods; the function was graphed and trace was used to find the intercept. An analytic method also works: when the graph intersects the y-axis ($f(x)$-axis), the value of x is 0; replace x with 0 and evaluate the function for the y-intercept.

> The ***x*-intercept** is a zero of the function.

The x-intercept is a zero because the x-intercept is a value of x that makes the function ZERO.

Example 8: Find the zero of $y = \frac{2}{3}x + 14$.

Solution: Find the graphical representation of $y = \frac{2}{3}x + 14$ and use the zero finder to locate the x-coordinate of the point where the graph (the function) is zero.
Figure 5.1.3 shows the zero.

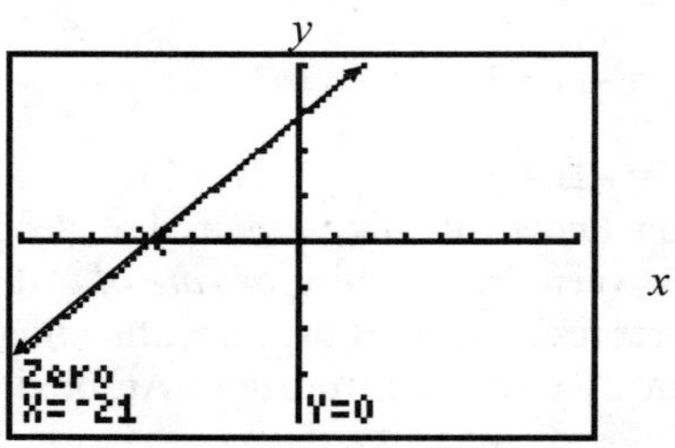

Figure 5.1.3

⁂

Example 9: Find the y-intercept of $y = 2x - 3$.

Solution: Replace x with 0 and calculate y: $2(0) - 3 = -3$.

The y-intercept of $y = 2x - 3$ is -3. Trace and a decimal window will always show you the y-intercept because a zero-value for x is in the numeric representation used by the calculator.

⁂

Example 10: Find the y-intercept of $y = \frac{2}{3}x + 4$.

Just plug it in. Plug it in.

Solution: Replace x with 0 and calculate y: $\frac{2}{3}(0) + 4 = 4$.

The y-intercept of $y = \frac{2}{3}x + 4$ is 4.

⁂

Just as slope has a special symbol m that represents the value of the slope, the symbol b represents the y-intercept.

From Examples 9 and 10, where the y-intercept is calculated, the y-intercept appears to always be the constant term e of the symbolic representation of linear function $dx + e$. Verify that this is true by finding the y-intercept of several linear functions.

Example 11: Prove that the y-intercept of $y = dx + e$ is e.

Solution: When $y = dx + e$ crosses the y-axis, the x-coordinate is 0 because every x-coordinate on the y-axis is 0. By substitution, when $x = 0$, then $y = 0x + e$ or $y = e$. Thus, the point $(0, e)$ is on the graph of $y = dx + e$ **and** on the y-axis. Therefore, e is the y-intercept.

⁂

The *y*-intercept

The number e in the linear function $dx + e$ is the y-intercept and is commonly symbolized with the letter $\boldsymbol{b}$.

Since the coefficient of x is the slope of the graph of a linear function and the constant term is the y-intercept, and since the symbol commonly used for the slope is m and the y-intercept is b, the linear function can be written as $y = mx + b$ or $f(x) = mx + b$.

Slope-Intercept Form of the Linear Function

$y = mx + b$ is the **slope-intercept form** of the symbolic representation of every linear function.

From the symbolic representation of the linear function $y = 2x + 5$, the slope and y-intercept of the graphical representation are immediately known; the slope is 2 and the y-intercept is 5. For $f(x) = \frac{1}{2}x - 3$, $m = \frac{1}{2}$ and $b = -3$.

Example 12: Find the x-intercept (the zero), the y-intercept, and the slope of $y = 2x - 3$.

Solution: The slope is the coefficient of x, which is 2. The y-intercept is the constant –3. To find the zero, graph $y = 2x - 3$ and use the zero finder to find the x-intercept. See the graph below for the x-intercept (zero).

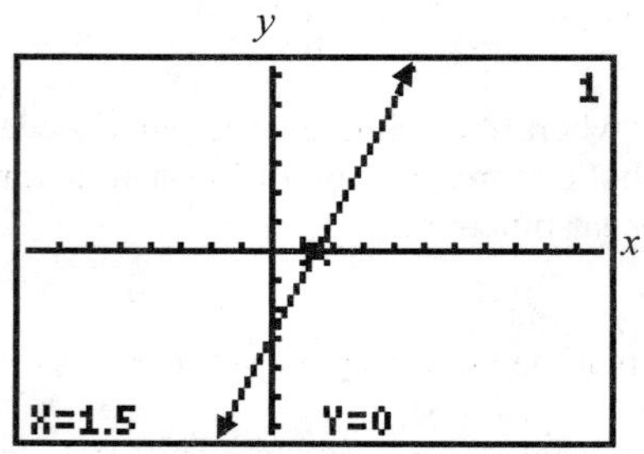

Figure 5.1.4

⁂

As you read materials from other courses, or as you read the newspaper or magazines, or as you watch TV information programs, you should notice when linear relationships are described you are almost always given the rate of change (slope) and many times the initial condition (the y-intercept). Typical linear relationships of this type are developed in the next few examples.

Example 13: "A typical 14 ounce bottle of ketchup at Max and Erma's restaurant is used at the average rate of 0.6 ounces per day." Write this English statement of a linear function in symbolic mathematical form.

Solution: The ketchup is used as a function of time (t). At time 0, the amount of ketchup is 14 ounces; thus, the initial condition is known. This is the value of b in the linear function $y = mx + b$, because it happens when time is zero. Ketchup being used at a rate of 0.6 ounces per day is an exact description of rate of change. Since the amount of ketchup is decreasing, the slope (rate of change) is –0.6. It is the value of m in $y = mx + b$. Use substitution in $y = mx + b$; replace m with –0.6, b with 14, the function variable with t, and the name of the function with K. The symbolic representation of the ketchup problem is $K = -0.6t + 14$, where K is the amount of ketchup left in the bottle after t days.

⁂

Example 14. "Farmer Tom's father-in-law willed him 640 acres of land and Tom then bought more land at a rate of 60 acres per year." Write this English statement of a linear function in symbolic mathematical form.

Solution: The farmland is being bought as time (t) passes. At time 0, the amount of land is 640 acres (the initial condition). This is the value of b in the linear function $y = mx + b$.

Land being bought at a rate of 60 acres per year is an exact description of rate of change. Since the amount of land is increasing, the slope (rate of change) is positive 60. This is the value of m in $y = mx + b$. Use substitution in $y = mx + b$, replace m with 60, b with 640, the function variable with t, and the name of the function with L. The symbolic representation of the farm land problem is $L = 60t + 640$, where L is the amount of land owned by Tom after t years.

⁂

5.1 EXERCISES

Many times you can work the exercises by algebraic, numeric, or graphical methods. You can learn about mathematics no matter which method you use. Part of the learning process is for you to decide which method is best for you to use. Please read the text carefully before you do the exercises. Read the **next assigned section** after you finish the assignment below.

A Quick Look To The Future

-4. Do you think it is possible to graph a linear function by hand as fast as your can with a graphing calculator?

-3. What does $2(x - 3) + 4$ simplify to?

-2. Make-up an application that would use $2x + 1.5$ in some way.

-1. If $f(x) = 2x + 4$ is a linear function, what is $7 = 2x + 4$?

Previous Topics Revisited

For exercises 1 – 3, use the function $f(x) = |x+4| - 2\sqrt{x+1} - 3$ and find the following:

1. The zeros.
2. The domain.
3. The average rate of change of the function as x changes from 1 to 2.
4. Factor $72x^2 - 41x - 45$
5. Factor $144x^2 - 225$
6. Multiply $(2x^3 - 3x^2 - 2x + 1)(x - 2)$

New Topic Exercises

7. Find the change in y (Δy) and change in x (Δx) from the point $(-1, -3)$ to the point $(-4, 2)$.

8. Find the change in y and change in x from $(-4, 2)$ to $(-1, -3)$.

9. Find the slope of the line that contains the points $(0, 5)$ and $(-1, 6)$.

10. Find the slope of the line that contains the points $(-3, -2)$ and $(-1, -4)$.

11. Find the slope of the line through $(0, 2)$ and $(2, 0)$.

12. Find the slope of the line through $(1, 1)$, $(2, 3)$, and $(-2, -5)$.

13. What is the slope of the line joining (4, –2) and (–6, –2)?

14. What is the slope of the line joining (–3, 2) and (–3, –8)?

15. What is the slope of the graphical representation of $y = \frac{2}{3}x + 42$?

16. What is the slope of the graphical representation of $f(x) = -5.36x - 13$?

17. What is the slope of the line $y = 0x - 23.89$.

18. What is the slope of the line $f(x) = 5$?

19. Find the slope and the y-intercept of $y = 2x - 79$. What is Δy? What is Δx? (Hint: There are many possible answers for Δy and Δx.)

20. Find the slope and y-intercept of $y = 0.6x - 19.2$. What is Δy? What is Δx?

21. Where does the graphical representation of the function $\frac{1}{2}x - 50$ intercept the function-axis?

22. Does the graph of $y = 19$ intersect the x-axis?

23. What is the x-intercept of $y = \frac{2}{3}x - 19$?

24. What is the zero of $y = -\frac{3}{5}x + 7$?

25. Is $y = 2x + 19$ an increasing or decreasing function?

26. Is $y = -\frac{3}{4}x - 5$ an increasing or decreasing function?

27. Is $y = 4$ increasing or decreasing?

28. On Route 64 in West Virginia near Sandstone Mountain, a road sign indicates that the incline down the mountain is a 7% grade. What is the slope of the road? On the Okinawa Expressway leaving Naha City to the north, there is a road sign indicating a 3½% incline. What is the slope of the road? How fast is the road rising?

29. The graph of $y = \sqrt{2x-3}\sqrt{2x-3}$ is a straight line. What is the slope of the line?

30. The roof truss on a house has a horizontal brace of 12 feet and a vertical support of 12 feet. What is the slope of the roof? How fast is the roof rising?

31. A popular ski resort advertises a mountain with an 8-mile ski run. If the mountain is 1.5 miles high and the distance from the bottom center of the mountain to the foot of the mountain is 7.86 miles, what is the slope of the mountain?

32. Find five different linear functions with y-intercepts of 7.

33. Find five different linear functions with x-intercepts of 0.

34. Find any linear function with an x-intercept of 7 and a y-intercept of 7.

35. What is the rate of change in a line whose slope is $\frac{3}{4}$? (Express the rate of change in percent form.)

36. If the slope of a line is negative and the line crosses the x-axis at –4, is the line above or below the x-axis to the right of this x-intercept?

37. If the zero of a linear function is 5.2 and the graphical representation of the function has a positive slope, is the function above or below the x-axis to the left of 5.2?

38. Give an example of a linear function whose graphical representation is not ever in quadrant III. (Below the x-axis and left of the y-axis.)

39. Give an example of a linear function whose graphical representation is only found in two quadrants.

40. If a linear function is negative only when $x \in (2, \infty)$, is it increasing or decreasing?

41. Find a linear function that is positive only to the right of 1,000,000 (when $x > 1{,}000{,}000$).

42. Rewrite each English description of a linear function in symbolic mathematical form. Identify the variable in the function and the name of the function by telling what they represent. Be specific.

    a. "The Norris Clothing store has 500 dresses at the beginning of an inventory cycle and sells an average of 8 per day."

    b. "A land developer has 32 properties and increases her holding by 6 properties per year."

    c. "Wilber, the rural mail carrier, has 4,620 pieces of mail to deliver and can deliver them at an average of 12 pieces per minute."

    d. "Nancy is reading a 250,000 word book and she can read approximately 350 words per minute."

    e. "Elizabeth the nurse set the drip rate on a 1,000 ml IV bottle at 2 ml per minute."

    f. "The Yellow Taxi Company of Tiro, Ohio, charges $2.25 plus $2.40 per mile."

    g. "A file is being downloaded at a rate of 1.2K per second and the file is 401K bytes in size."

From Mathematics to English

43. After reading this section, make a list of questions that you want to ask your instructor.

44. Continue in your daily journal and make an entry. In addition to your normal entry on thoughts about the mathematics in this section, list at least two positive comments about what you have learned about this topic.

45. In paragraph format, summarize the material in this section of the text in your daily journal.

46. Describe how your classroom instructor made this topic more understandable and clear.

47. After reading the text and listening to your instructor, what do you not understand about this topic?

48. Slope is rate of change of y with respect to x, as was discussed earlier in this section. Explain a slope of 3 in terms of a rate of change.

49. The graph of $y = 3\sqrt{x+2}$ from $x = 1$ to $x = 2$ may look like a line. Explain why it is not a line.

50. Is $0y = x + 2$ a linear function? Explain.

51. The function $y = 0.5x + 1$ has a slope of 0.5. Explain what a slope of 0.5 means.

52. Is there a linear relationship between the distance traveled s in time t when traveling at a constant rate of change? Explain your answer in paragraph form.

53. Normally, when you think of the direction of a line, you think in terms of an angle measured in degrees. In this section, the direction is measured by slope. List advantages and disadvantages of using slope. Describe real-world situations where slope is used.

54. Give your own definition of slope.

55. Describe a linear relationship (function).

56. What is an intercept?

57. What is the average rate of change of a function?

From English to Mathematics
In the next three exercises, convert the English statement to an equivalent mathematical statement.

58. two thirds of the number x plus one hundred is the linear function y

59. the linear function $f(x)$ is 4 of any number x minus thirty seven

60. y is a function of three fourths of x plus fifteen

Explorations

61. Graph the linear functions $y = mx$; where $m = 0, 1, 2, 3, 4, 5,$ and 6 on a large piece of paper. As the slope increases by equal increments, does the angle the line makes with the x-axis increase by equal increments? (You will need a protractor.)

62. In Section 4.6, you learned that when you add an absolute value function to a square root function, a new function is developed that has different behavior than the absolute value or square root functions. Is this true when linear functions are added? Answer this question yourself by finding the relationship between Y_1, Y_2, and Y_3. Also, in the exercises below, find the slope and *y*-intercept of Y_1, Y_2, and Y_3. Explain the relationships you find in paragraph format.

a. $Y_1 = 2x - 1$ $\quad Y_2 = x + 3$ $\quad Y_3 = Y_1 + Y_2$

b. $Y_1 = (\frac{1}{2})x + 2$ $\quad Y_2 = -x + 5$ $\quad Y_3 = Y_1 + Y_2$

c. $Y_1 = x - 3$ $\quad Y_2 = -x + 2$ $\quad Y_3 = Y_1 + Y_2$

d. $Y_1 = (\frac{1}{4})x + 2$ $\quad Y_2 = (\frac{1}{4})x + 3$ $\quad Y_3 = Y_1 + Y_2$

e. $Y_1 = -3x + 4$ $\quad Y_2 = 3x - 4$ $\quad Y_3 = Y_1 + Y_2$

f. $Y_1 = 2x - 5$ $\quad Y_2 = -x + 5$ $\quad Y_3 = Y_1 + Y_2$

g. $Y_1 = 0x + 2$ $\quad Y_2 = 2x - 1$ $\quad Y_3 = Y_1 + Y_2$

5.2 Slope-Intercept Method of Graphing

While it is relatively simple to graph the linear function using the graphing calculator, many times it is just as simple to graph the linear function by hand. Using the methods presented in this section, you will be able to graph the linear function by hand nearly as fast as you can with a calculator. The slope-intercept method of graphing is many times used in other mathematics classes, and the concepts presented here are assumed to be understood in these courses.

The function $y = mx + b$ has a graph that crosses the y-axis at b; thus, the y-intercept is known. The first step in graphing the linear function by hand is to plot the y-intercept. Second, the coefficient of the variable x is the slope of the graph; thus, the slope of the graphical representation of the function is known. Since slope is the change in y (Δy) divided by the change in x (Δx) from any point on the graph to any other point on the graph, and since one point on the graph has already been plotted, the slope can be used to plot another point on the graph of the linear function. In particular, *from the y-intercept,* change the y-coordinate by an amount equal to the numerator of the slope and change the x-coordinate by an amount equal to the denominator of the slope; that is, *from the y-intercept point* move up or down by the amount Δy and then left or right the amount Δx. Plot the second point.

The graph of $y = \frac{1}{2}x - 3$ has a y-intercept of –3; thus, it is plotted first. From the y-intercept, move up 1 then to the right by 2, and plot the second point.

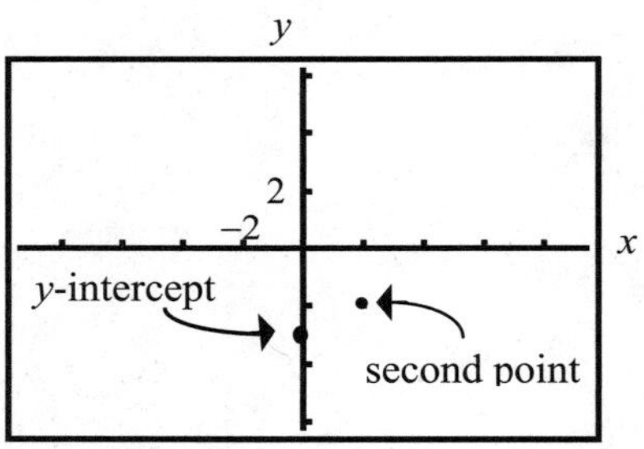

Figure 5.2.1

The graph of the linear function is a straight line; thus, using a straight edge, sketch all of the other points in the viewing window as shown in Figure 5.2.2.

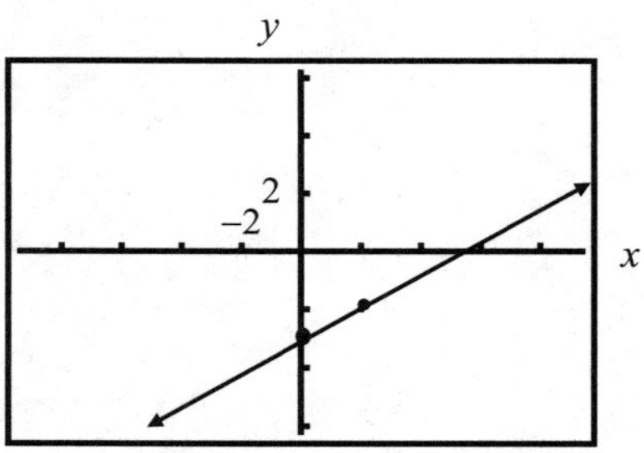

Figure 5.2.2

Graphing a linear function by this method is called **graphing by the slope-intercept method**.

Example 1: Graph $y = -\frac{3}{4}x + 2$ without the aid of technology.

Solution: The function is in the form $y = mx + b$. The y-intercept is 2 and the slope is $-\frac{3}{4}$; or, $\Delta y = -3$ and $\Delta x = 4$. As shown below, plot the y-intercept at (0, 2) on an appropriate viewing window. Then, *from the y-intercept,* move down 3 and right 4 and plot a second point on the graph of $-\frac{3}{4}x + 2$. Using a straight edge, plot all of the remaining points in the viewing window. The graph is shown below in Figure 5.2.3.

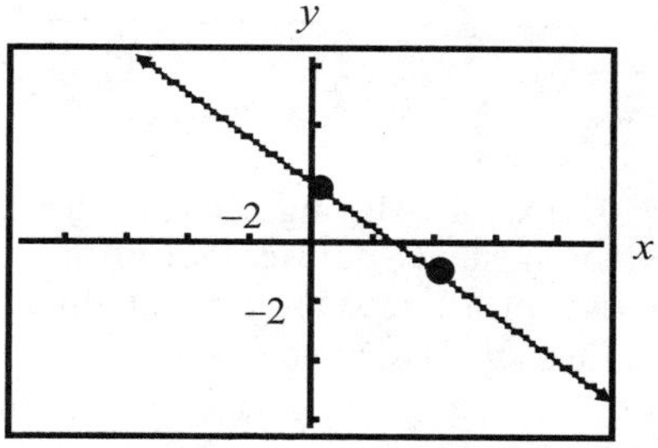

Figure 5.2.3

⁂

Example 2: Find the graphical representation of $y = 2x - 5$ without the aid of technology.

Solution: The function is in the form $y = mx + b$; thus, the y-intercept is at (0, –5) and the slope is 2. Since 2 is the same number as $\frac{2}{1}$, $\Delta y = 2$ and $\Delta x = 1$. The slope of 2 can also be thought of as any quotient whose value is 2; for example, $\frac{4}{2}$ or $\frac{-6}{(-3)}$. The values of Δy and Δx are different each time, but the points plotted are always on the graph of the function. See the solution below in Figure 5.2.4.

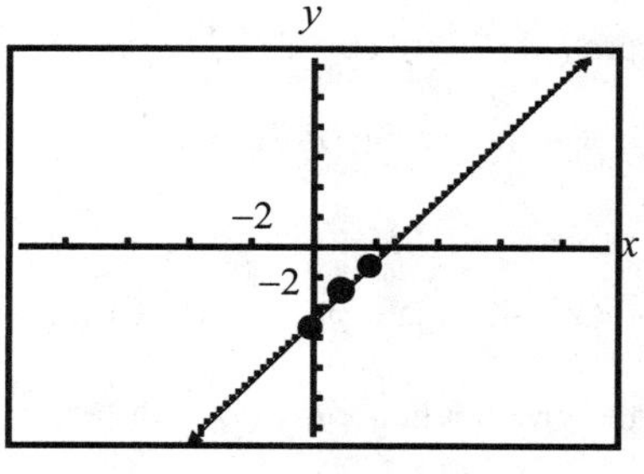

Figure 5.2.4

⁂

Example 3: Graph $y = 6$. (Hint: $y = 0x + 6$.)

Solution: The slope is 0; thus, $\Delta y = 0$ and Δx can be any non-zero number. Plot the y-intercept (0, 6). From the intercept, move up or down zero and left or right by any amount; then, plot the second point. Use a straight edge to sketch in the remaining points in the viewing window. The solution is shown below in Figure 5.2.5, with a Δx of 5.

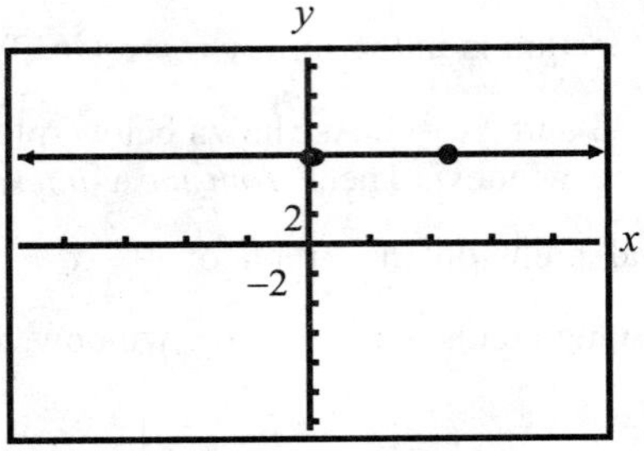

Figure 5.2.5

⁂

5.2 EXERCISES

Many times you can work the exercises by algebraic, numeric, or graphical methods. You can learn about mathematics no matter which method you use. Part of the learning process is for you to decide which method is best for you to use. Please read the text carefully before you do the exercises. Read the **next assigned section** after you finish the assignment below.

A Quick Look To The Future

-4. Is (2, –3) a point on the graph of $y = 4(x - 2) - 3$?

-3. If your weekly wages are based on an hourly rate, is there a linear relationship between time on the job and your wages?

-2. At the point where $f(x) = 2x - 1$ and $g(x) = -x + 3$ intersect, does $f(x) = g(x)$?

-1. If you agree that $7 < 12$, would you also agree that $7 \cdot (-2) < 12 \cdot (-2)$?

Previous Topics Revisited

1. What is the slope of the line joining the points (–1, 2) and (9, –12)?

2. What is Δy and Δx from (–1, –4) to (6, –3)?

3. Is the graph of $f(x) = x + 4$ increasing or decreasing?

Given the function $f(x) = |x - 4| - 2|x + 3|$, answer Exercises 4 and 5.

4. What is the slope of the graph when $x \in (-\infty, -3]$?

5. What is the rate of change of the graph when $x \in [4, \infty)$?

6. Factor $6x^2 + 5x - 8$

New Topic Exercises

In the next twelve exercises, do not use a calculator. Graph each function on an appropriate viewing window. You should check your work with a calculator.

7. $y = 2x - 5$
8. $f(x) = -\frac{3}{4}x + 2$
9. $y = -4x - 400$
10. $f(x) = x + 25$
11. $y = -x + 3$
12. $f(x) = x$
13. $y = 700$
14. $f(x) = 1.2x - 13$
15. $y = -3.6x + 2000$
16. $f(x) = \left(\frac{1}{3}\right)x - 250$

How should the horizontal and vertical axes be labeled in Exercises 17 and 18?

17. $z = 2k - 3$.
18. $f(x) = 0.01x + 500$

From Mathematics to English

19. After reading this section, make a list of questions that you want to ask your instructor.

20. Continue in your daily journal and make an entry. In addition to your normal entry on thoughts about the mathematics in this section, list at least two positive comments about what you have learned about this topic.

21. In paragraph format, summarize the material in this section of the text in your daily journal.

22. Describe how your classroom instructor made this topic more understandable and clear.

23. After reading the text and listening to your instructor, what do you not understand about this topic?

24. What is a good viewing window in which to graph $y = x + 3000$? Why?

25. What is a good viewing window in which to graph $y = 0.0001x$? Why?

26. If $y = 30x + 250$ is graphed on a viewing domain of [–10, 10], what is a good viewing range? Explain your response.

27. If $y = 0.01x + 0.25$ is graphed on a viewing domain of [–10, 10], what is a good viewing range? Explain your response.

28. Describe how you will graph the linear function $y = \frac{2x+5}{3}$ without technology.

29. Explain why the graph of $y = 0.002x + 3$ looks like a horizontal line when graphed on a standard window of [–10, 10] by [–10, 10]?

30. How can you graph the straight line $x = 4$? Describe the graph of $x = 4$.

31. Describe the graph of $y = 2x - 3 + 0\sqrt{x}$.

32. Describe the graph of $f(x) = -\frac{1}{2}x + 3 + 0\sqrt{x+2}\sqrt{-x+3}$.

From English to Mathematics
In the next three exercises, convert the English statement to an equivalent mathematical statement.

33. y is a function of seven times the number x, plus forty-seven

34. the number of deaths D is a function of the death rate 0.34 times the number of living L

35. the severity of a sunburn S is a function of the amount of time t lying in the sun

Explorations

36. What is a good viewing window for the graph of $y = 4500x + 70000$? Make certain the graph does not look horizontal or vertical. Find a good viewing window for the graph of $y = \frac{1}{100}x - 0.003$. Make certain the graph is distinct from the x-axis.

37. What geometric properties do all of the graphical representations of the following linear functions have in common?

$y = 2x - 5$ $\quad$ $y = 2x + 4$ $\quad$ $y = 2x$ $\quad$ $y = 2x - 0.1$

38. Graph $y = 0.000004x + 2587600$; show the x and y intercepts.

5.3 Point-Slope Form: $y = m(x - x_1) + y_1$

As you discovered in Section 5.1, many times linear relationships are described by the slope (rate of change) and the *y*-intercept (initial condition). Many times they are not. Often, linear relationships are described by a numeric representation.

The Electric Company

Below in Table 5.3.1 is the numeric representation of the 1992 proposed non-summer residential electric charges (E_p post-1992) put before the North Carolina Utilities Commission by the North Carolina Power Company. The function variable is in kilowatt-hours used (k) and the current electric charges (E_c pre-1992), as well as the then proposed electric charges (E_p) are in dollars.

Table 5.3.1

k	200	500	700	1000	1500	2000	3000
E_c	21.37	42.69	56.90	78.21	113.74	149.26	220.31
E_p	24.36	48.16	64.02	87.82	127.48	167.14	246.46

The North Carolina Power customers can see that if they use 1000 *kWh* of electricity in a month, their charges changed from \$78.21 to \$87.82. But what if they use 854 *kWh* or 4236 *kWh* per month, how did the charges change? What is the rate at which customers were charged before the rate request? What was the rate after it was granted? Is the rate increase constant across all usage levels? Well, it seems there are many questions. A bigger question may be "Can mathematics help answer these questions?" Actually, mathematics can answer these and other questions about the rate increase request. The first step is to find the symbolic representation of the data. If the data is linear, you know from Section 5.1 that the slope (rate of change) and the *y*-intercept (initial condition) are needed to create the symbolic representation. You can see that neither of these numbers is shown in Table 5.3.1. You may not even know if the data is linear from just looking at the table. The point of the discussion is that unlike in Section 4.1 where the rate of change and the initial conditions (the *y*-intercept) were given, we seem to need another way of getting the symbolic representation of the function that models the electric charges? This section starts with the development of another form of the linear function called the **point-slope form** of a linear function. Before it is developed, a look at the graphical representation of both the current (pre-1992) electric charges E_c and the proposed charges (post-1992) E_p in Figure 5.3.1 show you that the data appears to be linear and this means that the rate of change of the proposed post-1992 rate change is constant throughout the problem domain. Figure 5.3.1c shows the graph of the pre-1992 electric charges.

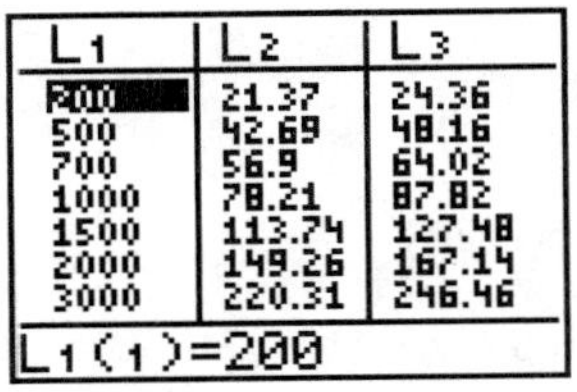

Figure 5.3.1a

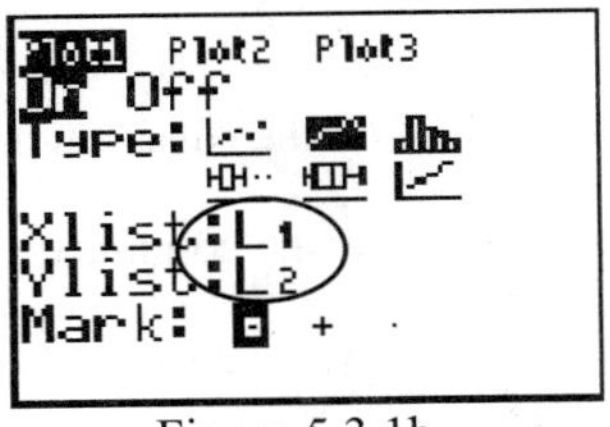

Figure 5.3.1b

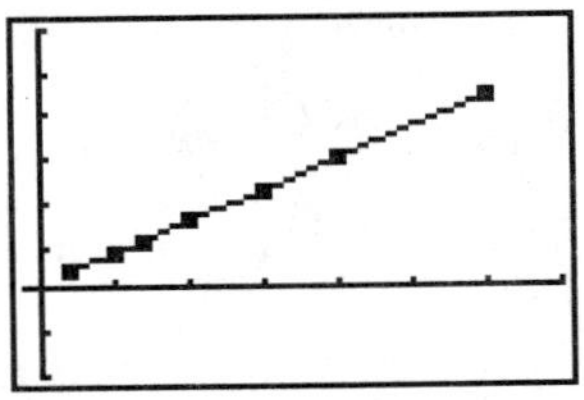

Figure 5.3.1c

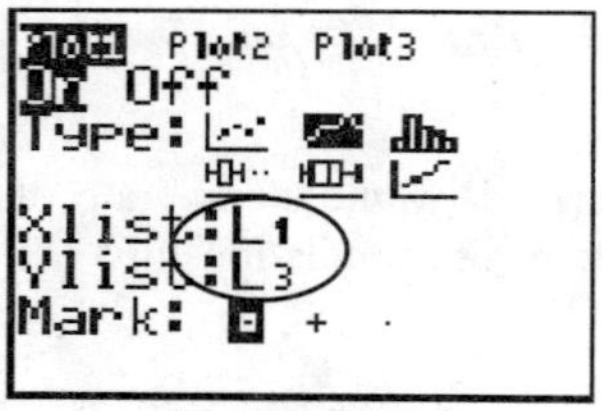

Figure 5.3.1d

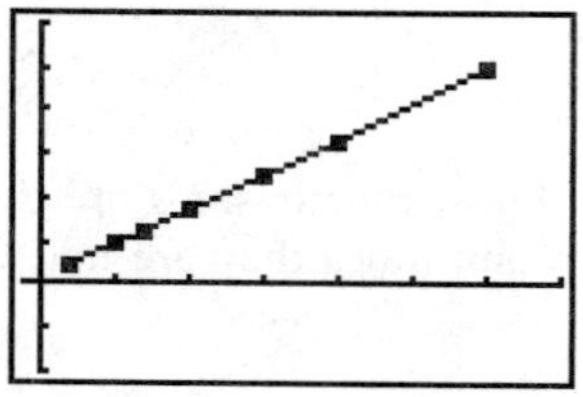

Figure 5.3.1e

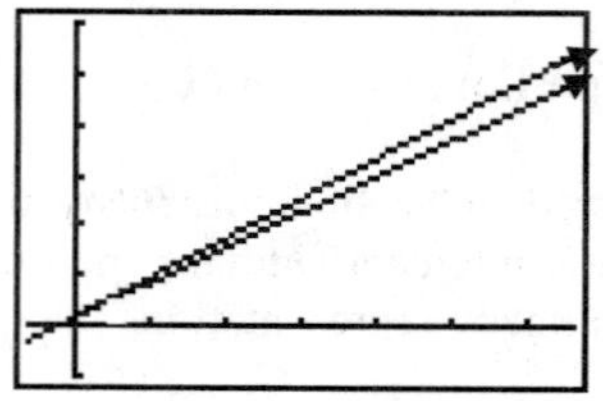

Figure 5.3.1f

Figure 5.3.1e shows the post-1992 charges, and Figure 5.3.1f, the graphs of both. How do you create a linear model for the data when all you have given are several data points? This question is answered below.

In Figure 5.3.2, suppose you know what the y-intercept and the slope are $(0, b)$ and m, respectively. Further, you know that the point (x_1, y_1) is *any* point on the graphical representation of the function other than the y-intercept.

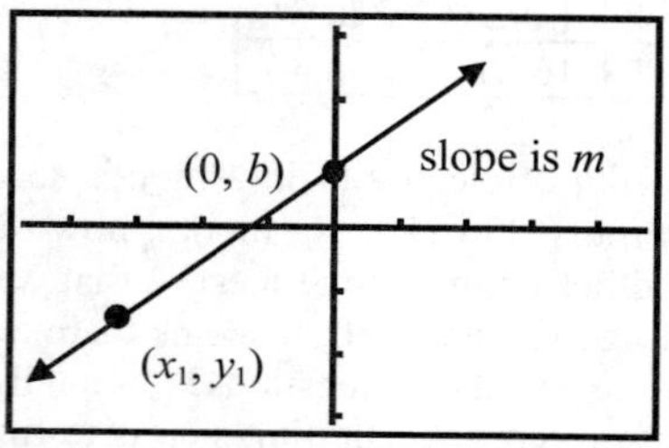

Figure 5.3.2

Since the slope and y-intercept are known, they can be substituted into the slope-intercept form; thus, the symbolic representation of the function is $y = mx + b$.

If two points on the graph are given: $(0, b)$ and (x_1, y_1), it is possible to find Δy and Δx. $\Delta y = b - y_1$ and $\Delta x = 0 - x_1$ thus, m is $\dfrac{b - y_1}{0 - x_1}$ or $\dfrac{b - y_1}{-x_1}$.

If $m = \dfrac{b - y_1}{-x_1}$ then $-mx_1 = b - y_1$ by multiplying both sides of the equation by $(-x_1)$ using the multiplication property from Chapter One.

And $y_1 - mx_1 = b$, by adding y_1 to both sides of the equation using the addition property from Chapter One. So,

$$b = y_1 - mx_1$$

and finally, from $y = mx + b$ replace b with $y_1 - mx_1$

$$y = mx + b$$

$$y = mx + y_1 - mx_1 \qquad \text{by the substitution property}$$

or $$y = mx - mx_1 + y_1 \qquad \text{by the commutative property}$$

and $$y = m(x - x_1) + y_1 \qquad \text{by the distributive property}$$

The point-slope form of a linear function is:

$$y = m(x - x_1) + y_1$$

m is the slope of the graph and (x_1, y_1) is <u>any</u> point on the graph.

In order to use the point-slope form $y = m(x - x_1) + y_1$ to obtain the symbolic representation of the linear function in a problem, the slope (m) and any point (x_1, y_1) must be known. That is, from the data in a problem, the slope (m) and any point (x_1, y_1) must be determined in order to get the symbolic form. Once they are known, replace the symbols m, x_1, and y_1 with the values found in the data from the application.

Example 1: What is the symbolic representation of the pre-1992 electric bill?

Solution: Using Table 5.3.1 choose any point (x_1, y_1) from the numeric representation of the pre-1992 function (E_c), for example (1000, 78.21). Now calculate the slope by choosing any two points and then calculate ΔE_c and Δk. From the point (1000, 78.21) to the point (2000, 149.26), the electric charges change by 71.05 and the kWh used change by 1000. The slope is $\dfrac{\Delta E_c}{\Delta k} = \dfrac{71.05}{1000} = 0.07105$.
Using substitution into $y = m(x - x_1) + y_1$, replace the function name with E_c, replace m with 0.07105, replace the function variable with k, and replace x_1 and y_1 with the values 1000 and 78.21 respectively. The symbolic representation of the current electric charges is $E_c = 0.07105(k - 1000) + 78.21$. If you are interested in simplifying, the function simplifies to $E_c = 0.07105k + 7.16$.

⁂

Example 2: What is the symbolic representation of the post-1992 electric bill (E_p) ?

Solution: Using Table 5.3.1 choose any point (x_1, y_1) from the numeric representation of the function (E_p), for example (1000, 87.82). Now calculate the slope by choosing any two points and then calculate ΔE_p and Δk. From the point (1000, 87.82) to the point (2000, 167.14), the electric charges change by 79.32 and the kWh used change by 1000. The slope is $\dfrac{\Delta E_p}{\Delta k} = \dfrac{79.32}{1000} = 0.07932$. Using substitution into $y = m(x - x_1) + y_1$, replace the function name with E_p, replace m with 0.07932, replace the function variable with k, and replace x_1 and y_1 with the values 1000 and 87.82 respectively. The symbolic representation of the post-1992 electric charges is $E_p = 0.07932(k - 1000) + 87.82$. If you are interested in simplifying, the function simplifies to $E_p = 0.07932k + 8.50$.

⁂

Example 3: If the slope of a linear function is $\dfrac{2}{3}$ and the graph crosses the y-axis at –8, what is the symbolic representation?

Solution: The point-slope form need not be used because m and b are given, not m and (x_1, y_1). Since m and b are given, $m = \frac{2}{3}$ and $b = -8$, replace m and b with these values in the slope-intercept form $y = mx + b$. The symbolic representation of the function is $y = \frac{2}{3}x - 8$.

⁂

Example 4: What is the symbolic representation of the graph in Figure 5.3.3?

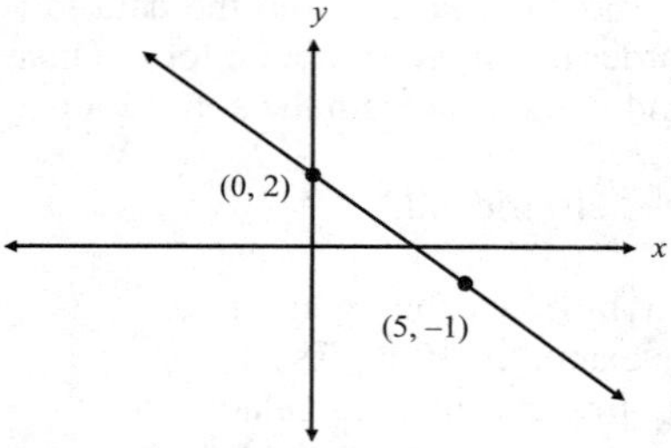

Figure 5.3.3

Solution: The point (0, 2) is on the graph; thus, the y-intercept is known; it is 2 ($b = 2$). A second point on the graph is also known (5, –1). With two points known, the slope can be found. The y-coordinate decreases from 2 to –1; therefore, $\Delta y = -3$. The x-coordinate increases from 0 to 5; therefore, $\Delta x = 5$.

The slope $= -\frac{3}{5}$, $m = -\frac{3}{5}$.

From $y = mx + b$, substitute the known numbers to get the symbolic form $y = -\frac{3}{5}x + 2$. Another method for solving this problem is to use the point-slope form; since m has been evaluated and a point (5, –1) is known, replace m with $-\frac{3}{5}$, x_1 with 5, and y_1 with –1. The symbolic representation is $y = -\frac{3}{5}(x - 5) - 1$. You are encouraged to verify that these two answers are the same symbolic representation.

⁂

Example 5: The numeric representation of a linear function is below in Table 5.3.2. What is the symbolic representation of this linear function?

Table 5.3.2

x	–6	–3	6	8
y	–2	0	6	$7\frac{1}{3}$

Solution: Several data pairs are given. To find the value of the slope, only two pairs are required. From the pairs of numbers (–6, –2) and (–3, 0), $\Delta y = 2$ (y increases from –2 to 0) and $\Delta x = 3$ (x increases from –6 to –3). Thus, $m = \frac{2}{3}$. Using m and a point such as point (6, 6), the symbolic representation comes from the point-slope form

$y = m(x - x_1) + y_1$. Replace m, x_1, and y_1 with the values $\frac{2}{3}$, 6, and 6 to get $y = \frac{2}{3}(x - 6) + 6$. This simplifies to $y = \frac{2}{3}x + 2$.

⁂

Parallel and Perpendicular Lines.

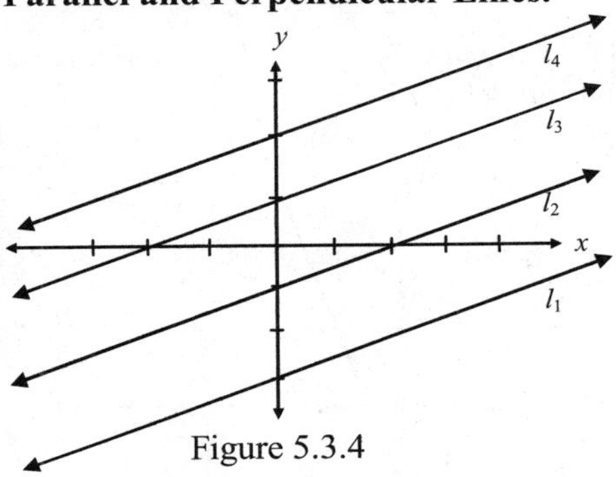

Figure 5.3.4

If two linear functions are graphed at the same time, they may meet to form an angle or they may not intersect. If they do not intersect, they are **parallel** (||). If they intersect, the angle formed may be a right angle (90°). This means the lines are **perpendicular** (⊥). The case where the angle formed by the lines is not a right angle will be studied later. Now, parallel and perpendicular lines will be analyzed. As will be shown, by using slopes to measure the direction of the graph of the linear function, it can be decided when graphs are parallel, perpendicular, or neither.

It Figure 5.3.4, all of the graphs have slopes of $\frac{1}{2}$. If the slopes are all the same, then the directions are all the same. If the directions are the same, the graphs are parallel.

> **In general:** If $m_1 = m_2$, then $l_1 \parallel l_2$ and
>
> If $1_1 \parallel l_2$, then $m_1 = m_2$
>
> where m_1 and m_2 are the slopes of non-vertical lines l_1 and l_2.

In Figure 5.3.5, lines l_1 and l_3 are perpendicular. The slopes of the lines are opposite in sign and reciprocals, for example –2 and ½. If multiplied together, their product is –1. Lines l_1 and l_2 intersect but do not form a right angle. They are not perpendicular, nor is the product of their slopes –1.

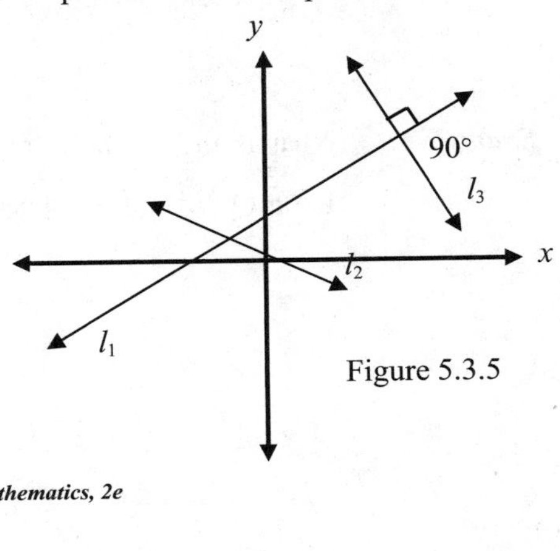

Figure 5.3.5

In general: If l_1 is perpendicular to l_2, then $m_1 m_2 = -1$ and,

If $m_1 m_2 = -1$, then l_1 is perpendicular to l_2

where m_1 and m_2 are the slopes of non-vertical lines l_1 and l_2.

No proof is given in this text.

Example 6: The graph of $y = 2x - 5$ is parallel to a line through (–2, 5). What is the symbolic representation of the line through (–2, 5) and parallel to $y = 2x - 5$?

Solution: When two lines are parallel, they have equal slopes. Since $y = 2x - 5$ is in slope-intercept form, the coefficient of x is the slope; thus, $m = 2$. The point (–2, 5) on the graphical representation of the desired line is given. From the point-slope form of the linear function, $y = m(x - x_1) + y_1$ replace m, x_1, and y_1 with the values 2, –2, and 5 respectively. The symbolic representation is $y = 2(x + 2) + 5$ or $y = 2x + 9$. That is, $y = 2x + 9$ is parallel to $y = 2x - 5$ and passes through (–2, 5). Use trace and a decimal window to confirm this.

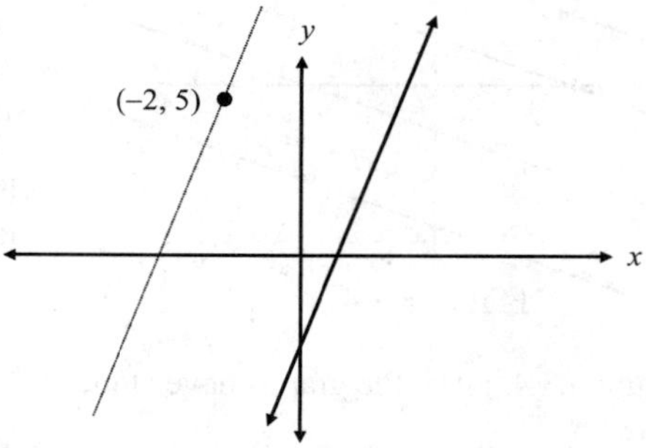

Figure 5.3.6

⁂

Example 7: Are the graphs of $y = \frac{1}{2}(x - 3) + 2$ and $y = -2x + 7$ perpendicular?

Solution: The first linear function is in point-slope form; thus, you know it has a slope of $\frac{1}{2}$. The second function is in slope-intercept form; thus, you know it has a slope –2. The product of the slopes is $\frac{1}{2}(-2) = -1$. Yes, the lines are perpendicular.

⁂

Example 8: What is the symbolic representation of a linear function that is perpendicular to $y = \frac{3}{4}(x + 3) - 17$ and has a y-intercept of 259?

Solution: The given line is in point-slope form; thus, you know it has a slope of $\frac{3}{4}$. The slope of the desired line must be $-\frac{4}{3}$; that is, it is opposite in sign and the reciprocal of $\frac{3}{4}$. The value of b is given to be 259. Now, both m and b have been determined. Using the slope-intercept form, the symbolic representation is $y = -\frac{4}{3}x + 259$.

⁂

When data in a linear relationship is collected, it must include either the rate of change (slope) and initial conditions (the y-intercept), or the slope and any data point, if the symbolic representation of the relationship is to be developed. Many times you can find this information from the numeric or graphic representation of the data. This section has developed methods by which you can create a symbolic representation of linear relationships.

5.3 EXERCISES

Many times you can work the exercises by algebraic, numeric, or graphical methods. You can learn about mathematics no matter which method you use. Part of the learning process is for you to decide which method is best for you to use. Please read the text carefully before you do the exercises. Read the **next assigned section** after you finish the assignment below.

A Quick Look To The Future

-4. If you are given a data relationship and you suspect it is linear, how can you decide for certain?

-3. Is $x = 3$ an equation or a function?

-2. Describe when your height was larger than 2 feet.

-1. What would you guess the solution to $|x| = 2$ to be?

Previous Topics Revisited

Without the aid of a calculator or computer, graph the next three functions.

1. $y = -\frac{1}{2}x + 83$
2. $f(x) = 0.1x - 47$
3. $d = 30t + 50$
4. What is the slope of the line joining (–200, 300) and (–201, 305)?
5. Is $y = -60x + 7$ increasing or decreasing?
6. What is the minimum value of the function $f(x) = \sqrt{x+1} \cdot |x-2|$?

New Topic Exercises

7. What is the symbolic representation of the graph of the linear function below?

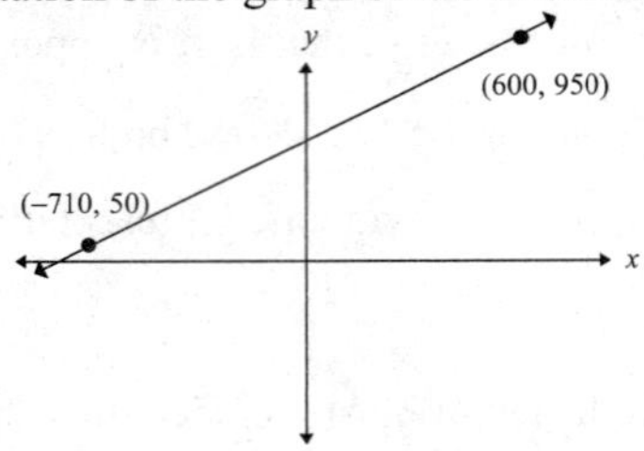

8. What is the symbolic representation of the linear function expressed below in numeric form? The domain is the set of all real numbers; just a few of the elements of the domain are displayed.

x	−2	0	1	20
$f(x)$	−688	−608	−568	192

9. What is the symbolic representation of the graph of the linear function below? Make the domain [2, 6].

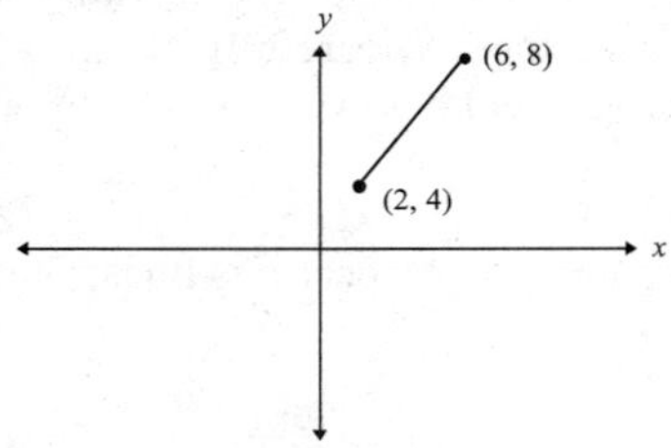

10. Find the symbolic representation of the linear function when the graph passes through the points (−50, 50) and (200, −200).

11. What is the symbolic representation of the linear function whose graph has a slope of −0.3 and a $f(x)$-intercept of 17?

12. Is the graphical representation of the function below a linear function? Why or why not?

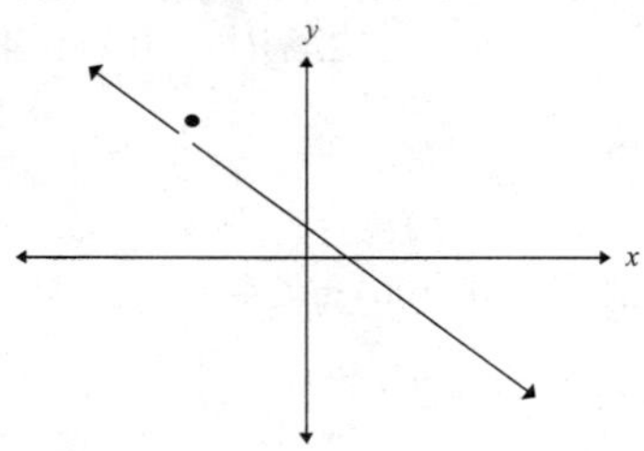

13. What is the symbolic representation of the linear function whose graph passes through (−0.1, 0.1) and (0.6, 0.6)?

14. Are the points (2, 8) and (−6, −16) on a line parallel to $y = 3x - 5$?

15. Find the symbolic representation of the linear function whose graph crosses the x-axis at 14 and the y-axis at 14.

16. What is the symbolic representation of the linear function perpendicular to $y = -\frac{1}{2}x + 3$ and through (420, 560)?

17. What is the symbolic representation of the linear function whose graph has a slope of –65 and passes through (–4, 2)?

18. Are the points (–6, 20) and (4, –10) on a line perpendicular to $y = \frac{1}{3}x + 50$?

19. What is the slope of a line perpendicular to $y = -\frac{1}{4}x + 4$?

20. Can the following data be modeled with a linear function? If yes, what linear function? The function variable is time (t) in hours and the function (S) is the number of pages of the textbook read and studied.

t	0.25	0.5	1	2	4	6	10
S	5.75	11.5	23	46	92	138	230

21. Can the following data be modeled with a linear function? If yes, what linear function? The function variable is US dollars (d) and the function (P) is 1993 Mexican Pesos.

d	17	25	80	100	200	500
P	49 130	72 250	231 200	289 000	578 000	1 445 000

22. Taxi company A advertises the following fare plan: "$1.40 plus $1.75 per mile." Taxi company B publishes the following mileage chart:

miles	1	2	3	4	5	6	10	12	14	16	20	25
fare	4.7	5.9	7.1	8.3	9.5	10.7	15.5	17.9	20.3	22.7	27.5	33.5

a. What is the symbolic representation of the cab fare function for the "A" company?

b. What is the symbolic representation of the cab fare function for the "B" company?

c. What is taxi fare to travel 2.5 miles with company A? With company B?

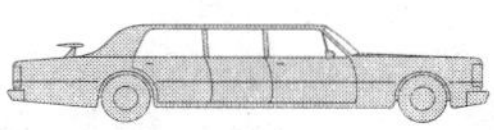

d. What is taxi fare to travel 15 miles with company A? With company B?

e. How far can you travel in an "A" taxi for $30? In a "B" taxi?

f. Make numeric representations of both taxi fare functions with a Δx of 0.1 mile. At about what distance is the cost of taxi A the same as taxi B?

g. If the taxi companies have 900 trips per week at an average distance of 5 miles each, which fare schedule will generate the most income for the company?

23. The post office hired an efficiency engineer to study the mail sorting process done by a postal employee. She watched the postal worker sorting a pile of mail and gathered the following data, where t is time and S is the amount of mail left to be sorted.

t	0 sec	5 sec	20 sec	1 min	5 min	10 min	20 min	30 min	40 min
S	6000	5990	5960	5880	5400	4800	3600	2400	1200

a. How fast is the mail being sorted?
b. What mathematical term is used to describe how fast the mail is being sorted?
c. How much mail is being sorted?
d. What mathematical term is used to describe the amount of mail to be sorted at 0 seconds?
e. What is the function that models the sorting process?
f. How much mail is left to be sorted after 120 seconds? After 8 minutes?
g. How much mail has been sorted after 120 seconds? After 8 minutes?
h. What is the zero of the function that models the sorting process?
i. How long does it take for all of the mail to be sorted?

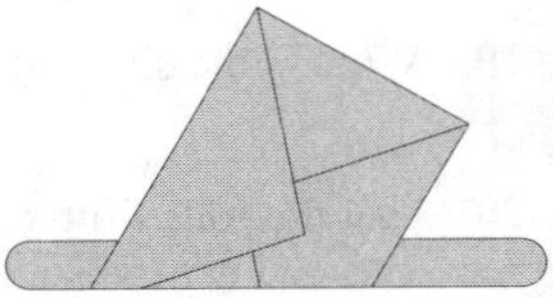

24. Find the symbolic representation of any linear function with a slope of $\frac{5}{2}$.

25. Find the symbolic representation of any linear function whose numeric representation is:

x	...	−3	...
y	...	5	...

26. Find the symbolic representation of any linear function that crosses the y-axis at −8.

27. The base of an isosceles triangle is on the y-axis from (0, 0) to (0, 6). Find the symbolic representation of any two linear functions that contain the equal sides of this triangle.

28. Find the symbolic representation of any linear function parallel to $f(x) = -3.58x + 9$.

29. Find the symbolic representation of any linear function that is perpendicular to $y = 2.5x - 3.11$.

From Mathematics to English

30. After reading this section, make a list of questions that you want to ask your instructor.

31. Continue in your daily journal and make an entry. In addition to your normal entry on thoughts about the mathematics in this section, list at least two positive comments about what you have learned about this topic.

32. In paragraph format, summarize the material in this section of the text in your daily journal.

33. Describe how your classroom instructor made this topic more understandable and clear.

34. After reading the text and listening to your instructor, what do you not understand about this topic?

35. The model for the current electricity charges for "The Electric Company Problem" is $E_c = 0.07105k + 7.16$. Explain the meaning of the slope and E_c-intercept in the context of this problem.

36. What information about the graphical representation of a linear function is needed to find the symbolic representation?

37. If the only thing known about the graph of a linear function is its slope, can one distinct symbolic representation be found? Explain your answer.

38. Can more than one symbolic representation be found for a line through (2, 1)? Explain your answer.

39. Can a single symbolic representation be found for a line whose y-intercept is 6? Explain.

40. If the graph of $y = ex + f$ is perpendicular to $y = gx + h$, what is the relationship between e and g?

41. If the graph of $y = ex + f$ is parallel to $y = gx + h$, what is the relationship between e and g? Are f and h related?

42. Make a list of applications of the linear function. Include background material and uses of the linear function.

43. What is the meaning of the words "point-slope form of a linear function?"

44. What conditions must be met before two lines can be perpendicular?

45. Describe how you put the function $y = -2(x+4) - 5$ in slope-intercept form. Do not actually put it in slope-intercept form.

46. Describe how you can put $y = 2x + 5$ in point-slope form. Do not actually do it.

47. If you memorize the point-slope form of a linear function, do you understand it? Explain.

From English to Mathematics

In the next three exercises, convert the English statement to an equivalent mathematical statement.

48. A is a function of two-thirds of S minus four

49. A depends on two-thirds of S minus four

50. the consumption of resources C of an American person is thirty times the consumption c of a person living in India

Explorations

51. The y-intercept of a linear function is symbolized by b and the x-intercept by a. Find the symbolic representation of a linear function with x-intercept a and y-intercept b. Put it in a simple form.

5.4 The Linear Function as a Mathematical Model

You have learned two ways of developing the symbolic representation of a linear function from information about the graphic or numeric representation of the function. The first is to use the slope-intercept form of the linear function, and second is to use the point-slope form of the linear function. Both are symbolic representations of the linear function. The goal of this section is to apply what you know about linear functions to linear relationships. That is, when you want to solve a problem that involves a linear relationship, you must know the symbolic representation of the relationship. You can develop the symbolic form from information in the problem and then solve a variety of problems. Many times data from outside the classroom does not exactly match the linear model; in this case you may use the linear function as a mathematical model to *approximate* the data.

The few examples that follow demonstrate the type of analysis that is expected of you. All of the tools needed to analyze these real-world problems have been presented in this text. One of the most important skills needed is the ability to convert English statements into a mathematical model -- a linear function. The ability to take numeric or graphical data and express it in symbolic form is also an important tool. Finally, a thorough understanding of the behavior of the linear function is necessary.

Give Me the Money

Below are the estimated wages for a server working at the Blue Point Café in Duck, NC. He is paid a salary plus he earns an average of $3.50 in tips per person served. The data is based on estimations made by the Chamber of Commerce. The variable in the wage function is the number of people served p and the function name is wages w.

Table 5.4.1

p	10	15	20	25	30	35	40	45	50	55
w	75.00	92.50	110.00	127.50	145.00	162.50	180.00	197.50	215.00	232.50

Example 1: Specify a reasonable problem domain for the Café Problem.

Solution: You know the server can't serve less than 0 people, and it may be reasonable to assume that he earns his base salary with out serving any people. On the high end, if it is the middle of tourist season, he may be asked to work overtime and may serve as many as 30 people per shift for a 7-day week. So, a reasonable problem domain may be $p \in [0, 210]$.

⁂

Example 2: What kind of function does the data in the Café Problem represent?

Solution: You might first think that the best way to decide what function could best represent the data is to look at the graph of the data. So Figure 5.4.1 shows the graphical representation of the data.

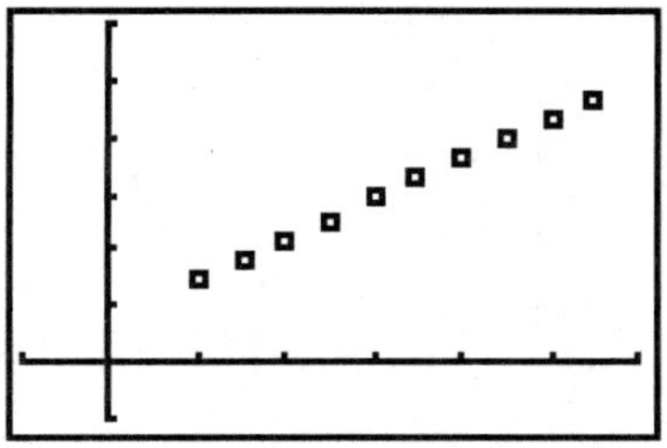

Figure 5.4.1

Certainly the data "looks" linear; thus, the linear function may be a good model of the data. Another approach is to look at the rate of change from data-point to data-point. The defining feature of the linear function is a constant rate of change. So, Figure 5.4.2 shows the rate of change from data-point to data-point.

L1	L2	L3 3
10	75	3.5
15	92.5	3.5
20	110	3.5
25	127.5	3.5
30	145	3.5
35	162.5	3.5
40	180	3.5

L3 ="ΔList(L2)/ΔL

Figure 5.4.2

The rate of change is a constant 3.5. This is conclusive evidence that the data is linear. Further proof that the data is linear is given in the original statement of the problem when you were told "he earns an average of $3.50 in tips per person served." You were told the data is linear!

⁂

Example 3: What model can be used to represent the Café Problem data symbolically?

Solution: Since it is linear and the slope (rate of change) was given, you need one more piece of information – either the initial condition or a point on the graph. You were not given the wages at 0 people served (initial conditions). But many points were given, pick one and use the point-slope form of a linear function $w = m(p - p_1) + w_1$. The point (10, 75) is a good choice and the slope was given as 3.5. The model of the data is

$w = 3.5(p - 10) + 75$, or in simplified form, it is $w = 3.5p + 40$.

⁂

Example 4: What is the initial condition in the Café Problem?

Solution: By representing the model as $w = 3.5p + 40$, you immediately know that at 0 people served, the wages are $40.

⁂

Example 5: What are the approximate wages after 62 people are served?

Solution: Replace p with 62 and evaluate the function. The wages would be \$257.00

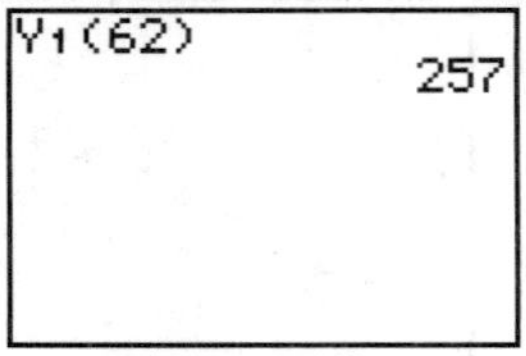

Figure 5.4.3

Example 6: What is the model of the Café Problem where the domain is [0, 70]?

Solution: Add the function $0\sqrt{-t(t-70)}$ to $w = 3.5p + 40$.

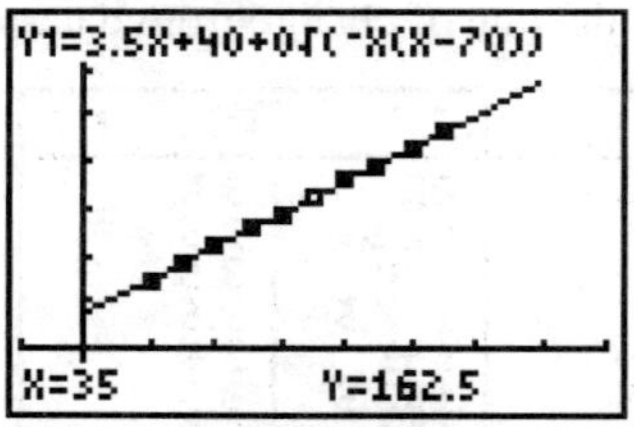

Figure 5.4.4

College Cost

Below in Table 5.4.2 is a data relationship from The College Board on the yearly cost of tuition and fees to attend a public four-year college. It is difficult to tell if the relationship is linear

Table 5.4.2

t	1980	1981	1982	1983	1984	1985	1986	1987	1988	1989	1990	1991	1992
C	804	909	1031	1148	1228	1318	1414	1537	1646	1781	1908	2137	2315

Source: The College Board

Table 5.4.2 by only looking at the numeric representation. One of the behaviors of the linear function is that the rate of change of the cost with respect to time must be constant. Thus, one approach to analyzing this data is to calculate the slope (rate of change) and see if it is constant. The slope is $\frac{\Delta C}{\Delta t}$ and the data shows a Δt of 1 throughout. A quick calculation shows the values for ΔC in Table 5.4.3 below. With the rate of change (slope) being $\frac{\Delta C}{\Delta t} = \frac{\Delta C}{1} = \Delta C$, you can see that the relationship is not an exact linear relationship.

Table 5.4.3

t	1980	1981	1982	1983	1984	1985	1986	1987	1988	1989	1990	1991	1992
C	804	909	1031	1148	1228	1318	1414	1537	1646	1781	1908	2137	2315

$\Delta C = 105$ $\Delta C = 117$ $\Delta C = 123$ $\Delta C = 135$ $\Delta C = 178$

The slopes from point to point vary from 105 to 178, certainly not a constant rate of change. However, look at the graphical representation of the data relationship in Figure 5.4.5. When you look at the bigger picture, the data looks much like a straight line. It may be a fair assumption to model the data with a linear

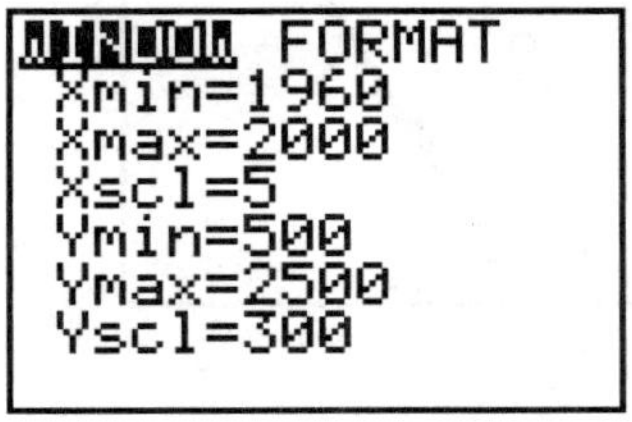

Figure 5.4.5a

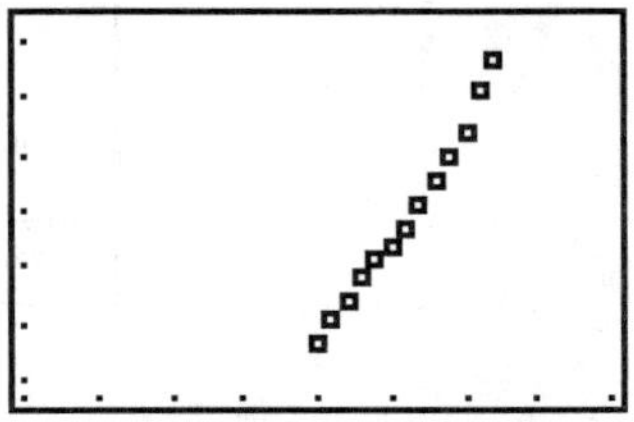

Figure 5.4.5b

function. But to get the symbolic representation of the linear function that models the cost of an education what slope should you use? The slope is different from point to point. Instead of guessing what the slope should be, try taking the mean of the slopes for all points.

Example 6: What is the average of the slopes in the college cost problem?

Solution: As discussed above, the slope from point to point is ΔC, find all ΔC's and find the mean.

C	804	909	1031	1148	1228	1318	1414	1537	1646	1781	1908	2137	2315

105 122 117 80 90 96 123 109 135 127 229 178

The mean is $\dfrac{105+122+117+80+90+96+123+109+135+127+229+178}{12} = 126$.

$\dfrac{\Delta C}{\Delta t} = \dfrac{126}{1} = 126$. A reasonable slope to use in the model of a linear function is 126.

⁂

Example 7: What linear function can be used as a mathematical model of the college cost problem?

Solution: The point-slope form of the linear function can be used. The slope is know as 126 and a data pair near the middle of the data may be a good choice for (x_1, y_1). You may also want to consider using the mean of the t values and the mean of the C values for a reasonable data point. The mean of the domain values is 1986 and the mean of the range values is 1475; thus, a good data point may be (1986, 1475). From the point-slope form $y = m(x - x_1) + y_1$, replace the function name with C, the slope with 126, the function variable with t, x_1 with 1986, and y_1 with 1475. The model for the college cost problem is $C = 126(t - 1986) + 1475$. You may simplify it to $C = 126t - 248761$. The data and the model are shown in Figure 5.4.6.

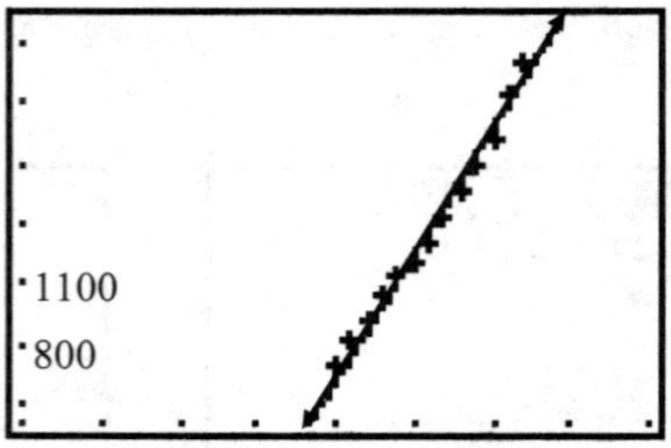

Figure 5.4.6

⁂

Example 8: Using the model $C = 126t - 248761$, what was the cost of tuition and fees for one year at a public four-year college in the year 1998?

Solution: Replace t with 1998 and evaluate: $C = 126(1998) - 248761 = \$2987$.

⁂

Example 9: What is the problem domain of $C = 126t - 248761$?

Solution: The zero of the model is approximately 1974. Verify this. Thus, the model is not good in 1974 because you know that tuition was not 0 in 1974. This means that you may want to start the problem domain in 1978, for example, and perhaps end it in the year 2000. A reasonable problem domain is [1978, 2000].

⁂

Real world data may not match the exactness of a linear function. When it doesn't match the linear function exactly -- use approximate values for the slope, y-intercept, and specific data points. One reasonable approximation might be the mean of the appropriate data.

5.4 EXERCISES

Many times you can work the exercises by algebraic, numeric, or graphical methods. You can learn about mathematics no matter which method you use. Part of the learning process is for you to decide which method is best for you to use. Please read the text carefully before you do the exercises. Read the **next assigned section** after you finish the assignment below. All numeric answers must have an error ≤ 0.01.

A Quick Look To The Future

-4. If $f(x) = g(x)$, and $f(x) = 2x + 3$ and $g(x) = -x + 1$, does $2x + 3 = -x + 1$?

-3. If $x + 1 > 2x - 1$, when would you expect the graph of $x + 1$ to be above the graph of $2x - 1$?

-2. Does the graph of $g(x) = |x - 2| + 3$ ever equal 3?

-1. What do you call the statement $A = lw$?

Previous Topics Revisited

1. Find the symbolic representation of the linear function below. The domain is all real numbers; a subset is shown below.

x	–5	0	8	100
y	–210	–60	180	2940

2. Find the symbolic representation linear function shown below.

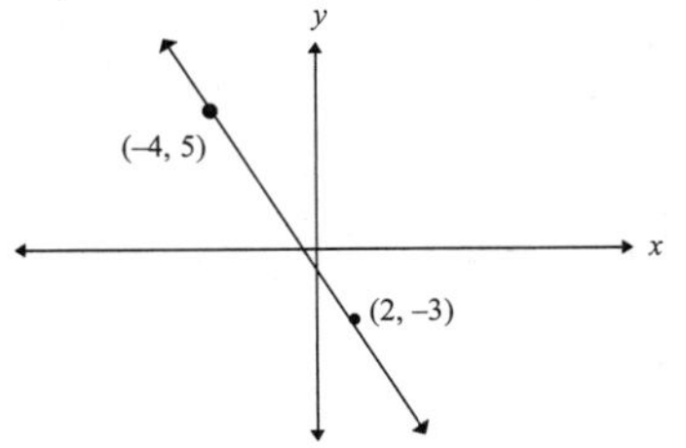

3. Is $y = \frac{3}{5}x + 16$ perpendicular to $y = \frac{5}{3}x - 16$?

4. Graph $y = 7x - 53$ without the aid of technology.

5. Graph $f(x) = -\frac{1}{4}(x-3) + 5$ without the aid of technology.

6. Find the slope of the line joining (–3, 5) and (258, –3).

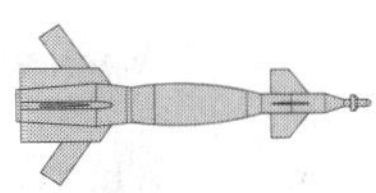

New Topic Exercises

On a world-wide level, "Six times as much public research money goes for research on weapons as for research on health protection." World Military and Social Expenditures - 1989, Ruth Leger Sivard, p. 5.

7. Let H represent the amount of money spent on health research and W the amount of money spent on weapons research.

    a. Does the quotation tell you if the function that models weapon research is increasing or decreasing? If yes, is the weapons research function increasing or decreasing?

    b. Find the y-intercept (W-intercept), the slope of the graph of the function, and the symbolic representation of the function that models the relationship between weapon and health research spending.

    c. If \$2.5 billion is spent on health research, how much is spent on weapons research?

    d. If \$36 billion is spent on weapons research, how much is spent on health protection research?

    e. Is there a maximum value of the weapon research function? (assuming a normal domain)

Below is a graph of atmospheric concentrations of chlorofluorocarbons. Chlorofluorocarbons and other trace gases are called the **greenhouse gases**. The increased concentration of these gases in the atmosphere is the reason for world-wide concern for the unnatural rise in global temperatures. These trace gases also contribute to the reduction of ozone in the stratosphere, which shields the earth's surface against ultraviolet radiation. This reduction in ozone causes an increase in skin cancer. The linear model representing the growth of these gases only applies to recent times. Before the industrial age, there was little or no increase in the level of these gases in the atmosphere.

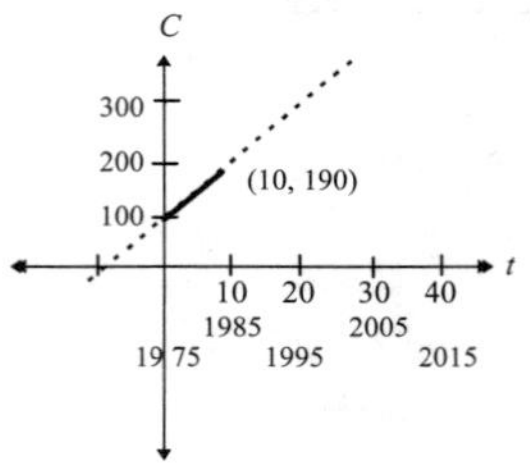

Source: "Population, Resources, Environment: An Uncertain Future", Robert Repetto, p. 28, figure 6; Population Reference Bureau, Inc.

8. The 1975 level of 100 is an arbitrary setting; that is, for example, the level in 1985 of 190 simply means that the concentration is 1.9 times the concentration in 1975 ($190 = 1.9 \times 100$).

   a. What is the slope, C-intercept, and symbolic representation of the linear function that models the chlorofluorocarbons in the atmosphere? Do not use calendar years in the model. Use time as it starts at 0 in 1975.

   b. When will the concentration be 4 times what it was in 1975?

   c. What will the greenhouse gas concentration be in the year 2025?

   d. Using a problem domain of $[-5, 60]$, what are the minimum and maximum values of the function (from part a) that models the chlorofluorocarbons in the atmosphere? (In calendar years, this is 1970 to 2035.)

   e. What is the zero of your model of the chlorofluorocarbons in the atmosphere?

   f. What problem domain will cause your model to never be negative?

   g. What is the symbolic representation of a model of the chlorofluorocarbons that has a domain of $[-5, 60]$?

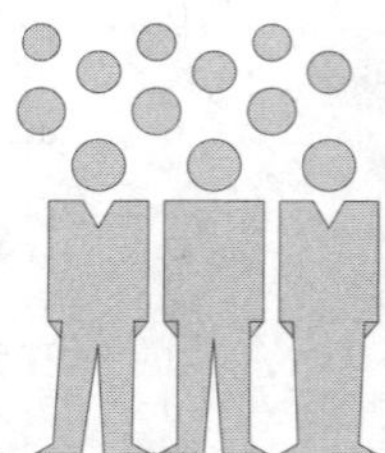

9. The human population of earth from 1750 to 1850 is shown in numeric form as:

t	1750	1800	1850
P	800,000,000	1,000,000,000	1,200,000,000

Instead of using such large numbers, let time start in 1750 as 0 and express the population in billions. The numeric representation becomes: **(continued)**

t	0	50	100
P	0.8	1.0	1.2

a. How fast were humans populating the earth during this 100-year period? (What is the rate of change?) Write the answer in percent form as well as decimal form.

b. Find the symbolic representation of the linear function that models the growth of the human population during this time period. Restrict the domain to [0, 100].

c. If, in fact, human population is modeled with the linear function from part b with an unrestricted domain, what was the population in 1950? What was the actual population in 1950? What is your explanation for the difference between the projected and the actual?

d. What is the range of the population model in part b?

e. Using the model in part b with an unrestricted domain, estimate the population in the calendar year 1000?

f. Using the model in part b with an unrestricted domain, in what calendar was the population 0.5 billion?

10. The numeric representation of the world population below shows the human population with time in calendar years.

t	1950	1994	2000
P	2,800,000,000	5,400,000,000	6,100,000,000

As in the previous exercise, it may be better if time is measured at 0 in 1950 and population in billions. This is shown below.

t	0	44	50
P	2.8	5.4	6.1

a. Find the average rate of change and express it in decimal and percent form.

b. What is the P-intercept?

c. What is the symbolic representation of the linear function that models human population. Restrict the domain to [–20, 70]?

d. Using the model from part c, find the population in 1940 and in 2001.

e. In what year was the population 6.2 billion?

f. Find the maximum and minimum values of the function in part c.

11. "With the intensive application of pesticides since World War II, the number of insect species resistant to chemicals began a meteoric rise. Before 1945, about a dozen species {(1945, 12)} were known to have developed resistance to pre-DDT insecticides, whereas by 1960 as many as 137 species {(1960, 137)} were known to be resistant to the new pesticides. The government and

industry response to resistance was to spray more intensively and more frequently, creating more resistant species and thus aggravating the problem they were alleged to solve. There are now more than 440 insect species {(1990, 440)} resistant to insecticides."

Source: EarthRight: Every Citizen's Guide by H. Patricia Hynes, p. 7.

Using the information in the above paragraph; that is, the three points (t, I) {(1945, 12), (1960, 137), (1990, 440)}, answer the questions below.

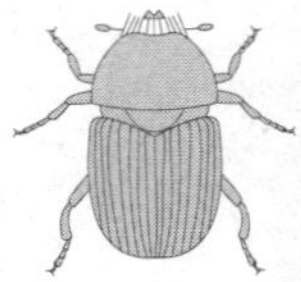

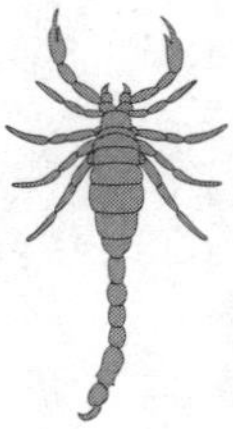

a. What is the slope of the line joining the points (1945, 12) and (1960, 137)?

b. What is the slope of the line joining the points (1960, 137) and (1990, 440)?

c. What was the rate of growth from 1945 to 1960? (Express in percent form.)

d. What was the rate of growth from 1960 to 1990? (Express in percent form.)

e. Has the growth of insects resistant to chemical pesticides followed an exact linear growth pattern?

f. Using the average (mean) t-coordinate, the average I-coordinate, and the average slope, find a linear function that models the data. Use calendar years.

g. Using the model, how many insects will be resistant to pesticides in 2010?

h. In what year will 700 species of insects be resistant to pesticides?

12. The gas tank on a typical mid-sized import car is 12.8 gallons. If a reasonable driver averages 32 miles per gallon, answer the following questions about the gasoline left in the tank (G) and the miles driven (d).

a. If a function is developed to model the amount of gas left in the tank, what is the slope of this function?

b. What mathematical term is used to describe the number 12.8 from the above problem?

c. What is the problem domain?

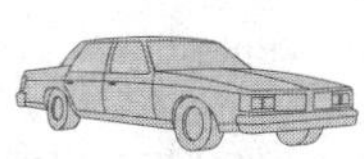

d. What function can be used to model the gas (G) left in the tank at m miles driven? Restrict the domain to the problem domain.

e. Using your model, what is the zero? What is the physical significance of the zero?

f. How much gasoline is left after driving 250 miles?

g. If 8.2 gallons of gasoline are left in the tank, how far has the car been driven?

h. What is the model for gasoline left in the tank if the driver gets 38 miles per gallon (mpg)?

i. What is the problem domain at 38 mpg?

j. At 38 mpg, what is the zero of the model?

13. A window washer at the Dallas-Ft. Worth airport was observed washing one window in 12 seconds. The concourse under her responsibility has 1,873 windows.

    a. Is the number 12 the slope of a linear function that models the number of windows left to wash? If no, what number represents the rate in which she is washing windows?

    b. If a function is used to model the number of windows left to wash, what mathematical term is used to describe the number 1,873?

    c. What linear function can be used to model the number of windows (W) left to wash at time t?

    d. What is the problem domain of the model?

    e. What is a linear function that models the number of windows left to wash, where the domain is the problem domain?

    f. Will the window washer finish the entire concourse on her 8-hour shift if she has two twenty minute breaks and an hour for lunch?

    g. How many windows are left to wash after she has been working for 3 hours?

    h. What model can be used for the window washer if there are 935 windows in the concourse?

    i. What model can be used for the original concourse, but the washer washes one window every 15 seconds?

    j. What is the problem domain for the model in part h?

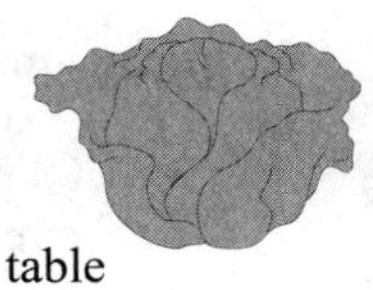

14. At many restaurants, servers are paid a set weekly salary plus tips. At the Duck Deli, servers are paid $80 per week plus tips. A serious job seeker calls the local Restaurant Association and discovers that servers average $3.50 in tips per table served. Below is a table he made in an attempt to project what he might gross per week should he take the job as server at the Duck Deli.

Tables Served (T)	20	30	35	40	50	65	90
Salary (S)	150	185	202.50	220	255	307.50	395

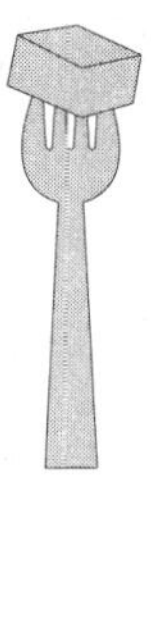

    a. What mathematical term is used to describe the number $3.50 per table?

    b. What mathematical term is used to describe the $80?

    c. List two points that are on the graphical representation of the salary model.

    d. What is the symbolic representation of his salary model?
    e. If he serves 48 tables during the week, what will be his gross salary?

    f. If his gross salary is $300.50, how many tables did he serve during the week?

    g. How many tables does he serve if his gross salary is $80?

    h. Is the salary model an increasing or decreasing function?

From Mathematics to English

15. After reading this section, make a list of questions that you want to ask your instructor.

16. Continue in your daily journal and make an entry. In addition to your normal entry on thoughts about the mathematics in this section, list at least two positive comments about what you have learned about this topic.

17. In paragraph format, summarize the material in this section of the text in your daily journal.

18. Describe how your classroom instructor made this topic more understandable and clear.

19. After reading the text and listening to your instructor, what do you not understand about this topic?

20. List several examples of real-world data where the linear function cannot be used as a mathematical model.

21. Suppose you drop a stone from the top of a building. Would the linear function be a good mathematical model of the speed of the stone? That is, is the relationship between the elapsed time and the speed of the stone related by a linear function? Explain.

22. Is the following data related in a linear function? Why or why not?

Year	1940	1950	1960	1970	1980	1990	1994	2006
National Debt in Billions	51	257	291	381	909	3113	3500	8400

23. List the advantages and disadvantages of using the linear function to model real-world situations.

24. Describe a "linear model".

Explorations

25. From Exercise 14 - the waiter problem - graph the salary model followed by a model that shows a wage of \$90 plus tips. Of \$100 plus tips. Of \$110 plus tips. How are all of the graphs related?

    From Exercise 14 - the waiter problem - graph the salary model followed by a model that shows a wage of \$80 plus tips at a rate of \$3.75 per hour. At \$4.00 per hour. At \$4.25 per hour. At \$4.50 per hour. How are the graphs related?

26. From Exercise 12 - the gas tank problem - graph the gas model followed by a model that shows a tank size of 13 gallon. Of 13.5 gallons. Of 16 gallons. How are all of the graphs related?

    From Exercise 12 - the gas tank problem - graph the gas model followed by a model that shows a rate of 35 mpg. Of 37 mpg. Of 40 mpg. Of 43 mpg. How are the graphs related? What is the problem domain for each of these models?

CHAPTER FIVE TEST

1. Find the slope of the line containing the points (–4, 2) and (–5, –3).
2. What is the slope of the graphical representation $f(x) = 4x - \frac{2}{3}$?
3. Is $y = -\frac{1}{2}x + 4$ increasing or decreasing over its domain?
4. How fast is the function $y = 2x - 5$ changing with respect to x?
5. Can you see the graph of $y = 6x - 500$ in the viewing window [–10, 10] by [–50, 50]?
6. Does the graph of $y = 2x - 1000000$ cross the x-axis?
7. From the partial numeric representation of a linear function, find the symbolic representation.

x	–20	43	30
$f(x)$	160	–155	–90

8. From the graphical representation of a linear function, find the symbolic representation.

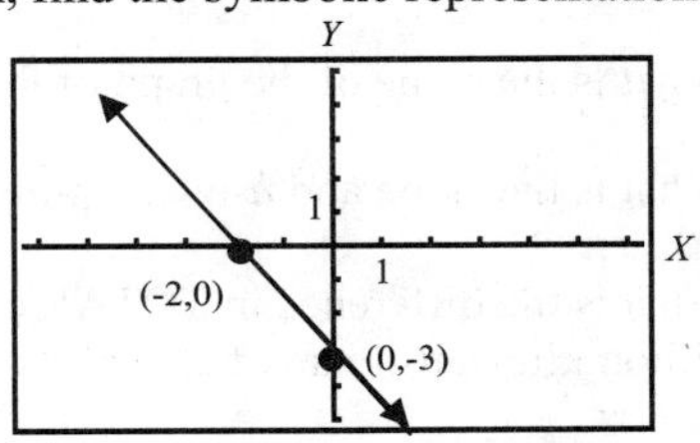

9. Are the points (5, 8) and (1, 12) on a line parallel to $y = -x + 4$?
10. Find the symbolic representation of the line perpendicular to $y = 4x - 1$ and through (0, 9).
11. Are the graphs of $y = 3x - 5$ and $y = 4x + 2 - x$ parallel, perpendicular, or neither?
12. Are the graphs of $y = -x + 2$ and $y = 3x + 2 - 2x$ parallel, perpendicular, or neither?
13. What is the slope of every line that is parallel to the graph of $y = \frac{1}{4}x + 10$?
14. What is the slope of every line that is perpendicular to the graph of $y = 0x - 4$?

15. Find the symbolic representation of the line that is perpendicular to the graph of $y = \frac{1}{2}x - 4$ and passes through $(-8, 3)$.

"An 80-megawatt solar thermal electric plant built in the desert east of Los Angeles in 1989 converts an extraordinary 22% of the incoming sunlight into electricity. It does so at a cost of 8 cents per kilowatt-hour -- a third less than the 12 cents per kilowatt-hour cost of power from a new nuclear plant."
Source: *State of the World*, Lester R. Brown, 1990 by Worldwatch Institute.

With x representing the number of kilowatt-hours of electricity generated, s the cost of producing x kilowatt-hours of electricity from the solar generator, and n the cost of producing x kilowatt-hours of electricity from the nuclear plant, answer the following questions.

16. What linear model can be used to represent the cost of x kilowatt-hours of electricity from the solar plant?

17. What linear model can be used to represent the cost of x kilowatt-hours of electricity from the nuclear plant?

18. What is the slope of the graph of the solar model?

19. What is the slope and n-intercept of the nuclear model?

20. What is the difference in TOTAL cost between the solar and nuclear plants for producing 1 billion kilowatt-hours of electricity at each plant?

CHAPTER SIX
EQUATIONS AND INEQUALITIES

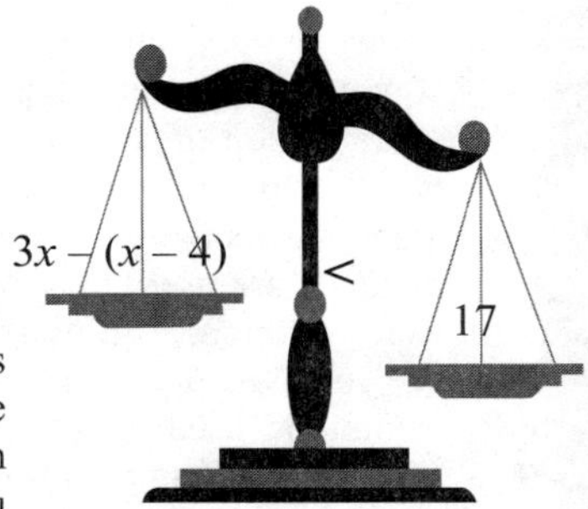

6.0 Introduction

In Chapter One, numeric equations like $2(4 - 7) = 9 - 15$ were studied to introduce the properties of equality. In this chapter, equations and inequalities containing functions are examined. In the last four chapters, the concept of the function has been emphasized. You learned that functions can be written in symbolic form, graphical form, numeric form, and in written form. Further, you learned that functions may be described by using function notation and the y-notation. When this notation is used, the function takes on the appearance of what you may think of as an equation. However, it is more useful to think of an **equation** as a statement of equality between two functions. For example, the linear function $3x - 15$ and the constant function 9, when equal, form the equation $3x - 15 = 9$. This equation is a description of the point on the graph of the function $3x - 15$ when the function is 9. Or looking at the equation another way, it describes a related pair of numbers from the numeric representation of the function $3x - 15$ when the function is 9. If you look at the meaning of a more complicated equation like $2x - 5 = \sqrt{x+3} + 4$, it is the statement of equality between the two functions $2x - 5$ and $\sqrt{x+3} + 4$. This equation is describing when the function $2x - 5$ and the function $\sqrt{x+3} + 4$ are equal; that is, this equation is just a simple way of describing the point where the graphical representation of $2x - 5$ and the graphical representation of $\sqrt{x+3} + 4$ intersect.

It may be useful to think of an **inequality** as a statement relating two functions of different value. For example, the statement $-2x + 4 < 3(x + 2) - 5$ describes when the graphical representation of the function $-2x + 4$ is less than (below) the graphical representation $3(x + 2) - 5$. Given any two functions and a particular value for x, one function must be smaller than, equal to, or greater than the other function. Remember, the symbol $<$ means *is less than.* Three other symbols of inequality will be studied in this chapter as well:

$>$ is greater than

$\leq$ is less than *or* equal to

$\geq$ is greater than *or* equal to.

The goal of this chapter is for you to learn to solve equations and inequalities. To **solve** an equation or inequality means to use a process (algorithm) to actually find the x-coordinate of the point of intersection (when solving an equation) or the x-coordinates of the points where one function is below or above the other function (when solving an inequality). There is a fundamental difference in the thought process between how you solve an equation or inequality graphically or numerically and how you solve it using algebra. If you solve algebraically, you ask the questions: What algebraic properties (see Chapter One) can I use to get the variable by itself on one side of the equation or inequality? Or, what value for the variable makes the equation a true statement? To solve numerically or graphically, you ask the question: What should I do to get a table or graph that shows the solution? This idea may be new to you; the process is developed in this chapter.

A numerical method for solving equations and inequalities will be introduced first. Hopefully this method will provide a connection between the arithmetic you already know and the algebra you are now learning. The graphical approach to solving equations is presented second. This will allow you to visualize the process. After the visualization of the solution of an equation or inequality is completed, the next step will be to develop some algebraic techniques that will also solve equations and inequalities. If the algebraic method *can* be used, it will normally give the exact solution. If the graphical or numerical methods are used, they will many times give an approximate solution, or perhaps, quite often lead to an exact solution. You will need to know how all methods can be used to solve equations and inequalities. As you will discover, sometimes the analytical method (using algebra and other mathematical tools) cannot be used to solve equations and inequalities. Sometimes (although much less frequently) the graphical or numerical methods cannot be used. <u>You must decide which method will work and which method is best for you.</u>

In this chapter, approximate solutions must have an error that is less than or equal to 0.01. Error is a part of your everyday life; however, you may have normally ignored it until now. Error will be discussed in more detail in Section 6.1.

Equation Solving in Action

Did you ever find yourself saying, "Let's see now. . . What grade do I need on the final exam to get a "B" in this course?" Well, equation solving can answer your question. Consider the following situation: The total points assigned for the course is 580. That is, 100 points come from explorations, 200 points come from lab projects, 80 points from writing assignments, 50 points from group quizzes, 100 points from three midterms, and 50 points from the final exam. You have accumulated 416 points before the exam is given. How many points do you need on the exam to get a "B"? The function that models your grade is

$G = \dfrac{\text{earned points}}{\text{assigned points}} \cdot 100$. Since you will earn $416 + x$ points and the assigned points are 580, the model becomes $G = \dfrac{416 + x}{580} \cdot 100$. If the lowest possible "B" grade is 80%, then the equation required is: $80 = \dfrac{416 + x}{580} \cdot 100$, where x is the score you earn on the final exam. The solution to the equation is the minimum grade you can get on the final in order to earn a "B" in the course.

The problem above can also be solved as an inequality. You must earn *at least* 80%, *at least* means greater than or equal to; thus, your grade: $\dfrac{416 + x}{580} \cdot 100$, must be ≥ 80. Thus, you can solve the inequality $\dfrac{416 + x}{580} \cdot 100 \geq 80$. As Chapter Six unfolds, you will learn how to solve equations and inequalities by numerical, graphical, and analytical methods.

6.1 Solving Equations Containing the Linear Function

The Beetles Are Coming

Suppose that you raise red raspberry plants and you notice that on June 28 there are 17 Japanese Beetles on your small set of plants. The next day your see 4 more and the day after 4 more, etc. The function that models the number of beetles on your raspberry plants is $17 + 4x$, where x is time in days starting at 0 on June 28. This model may also be written as $4x + 17$.

You should recognize the function $y = dx + e$ as a linear function with a rate of change of d and initial condition of e. Solving the equation $dx + e = f$ is just another way of asking the question: When does the graph of $dx + e$ have (equal) a value of f? Another way of looking at solving equations is to find when (the value of x) the numeric representation of the function is f. That is, consider the point (–3, 5) (This means 3 days before June 28 there were 5 beetles on your raspberry plants.) on the graphical representation of the function $4x + 17$ or as a data pair in the numeric representation of $4x + 17$, as shown below in Figures 6.1.1 and 6.1.2.

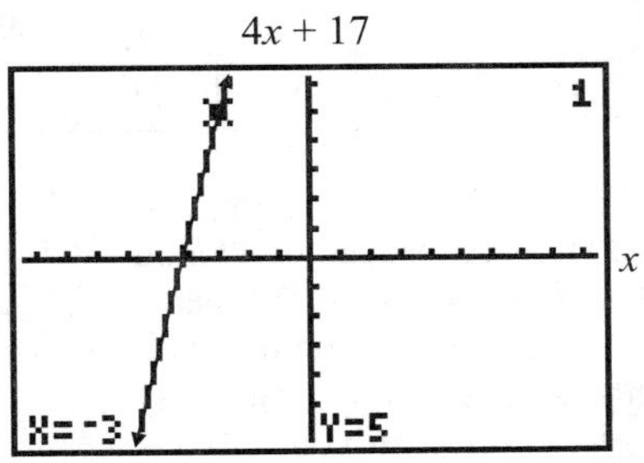

Figure 6.1.1

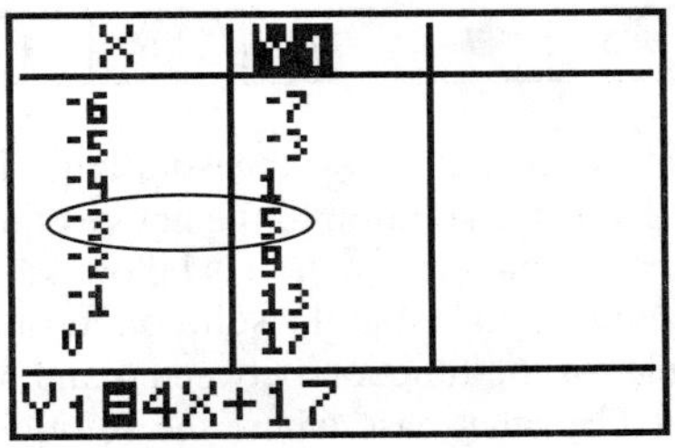

Figure 6.1.2

This pair of numbers describes the solution to the equation $4x + 17 = 5$, where –3 is the solution. That is, the value of –3 causes the function to be 5.

Of course an equation containing the linear function may be considerably more complicated, such as an equation that has a linear function on both sides of the equal sign, like $-3(x + 5) - 9 = 4x + 4$. The idea behind solving the equation remains the same; find a value for x that makes the linear function on the left have the same value as the linear function on the right? If both functions are graphed and the intersection of the graphs is on the calculator screen, you have found the solution. Why? Because at the point of intersection, the x value causes both function to be the same (equal).

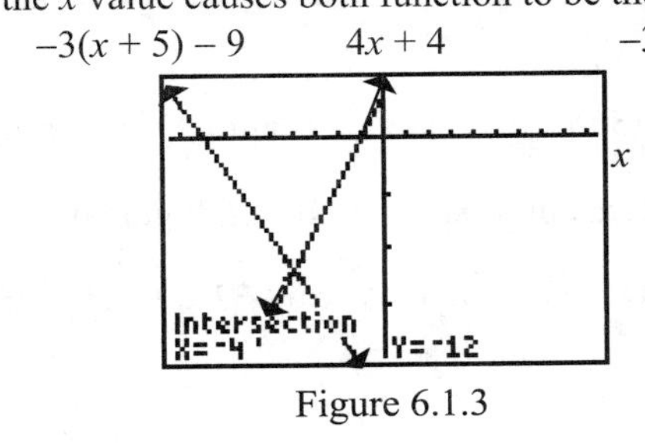

Figure 6.1.3

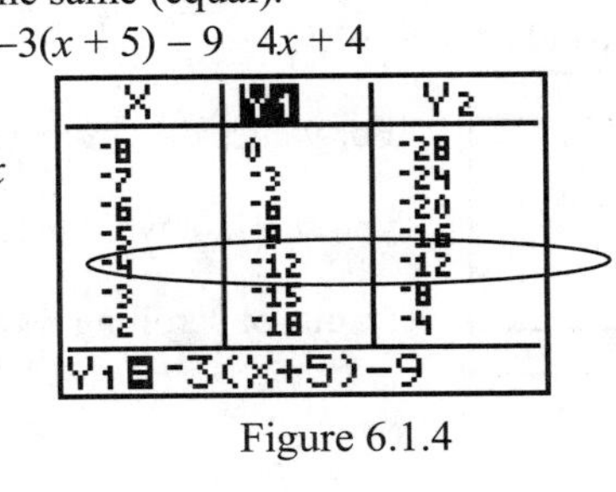

Figure 6.1.4

Thus, you may now see a method for solving equations. If the equation is simple, you may want to try to trace to the solution. (Caution: This may not give exact answers.) If the equation is a little more complicated, you may want to try to find the intersection with the intersection finder on your calculator. Also demonstrated above is the numerical method -- finding the numeric representation that shows a value for x that makes both functions the same. Still another technology-based method will be developed later and a method using algebra will end the possible methods for solving an equation. It will not suffice for you to learn one method. You will find that any one method will either not solve all types of equations or may be very difficult to use on some types of equations.

The Numerical Method

The objective in solving equations is to find all values for the variable that make the equation a true statement. That is, in solving the equation $2x + 5 = 13$, you know $2x + 5$ can have an infinite number of values because the linear function $y = 2x + 5$ has a range of all real numbers. Since the goal is to make a true statement, look for a value for x that yields the element 13 from the range. The numerical approach to solving equations is to make a table and look for a value of x that causes the function $2x + 5$ to be 13. Below the table shows the exact solution to be 4 - a value for x that causes the function to be 13.

Table 6.1.1

x	1	2	3	**4**	5	6	7	8	9
$2x + 5$	7	9	11	**13**	15	17	19	21	23

π $\quad -6$

$\sqrt{3}$ $\quad 0$

$\frac{1}{2}$

What if the table does not show the solution? This is possible for a couple of reasons. First, what if the solution to the above equation were 4.5 instead of 4? You would have observed that the function is below 13 when x is 4 and above 13 when x is 5; therefore, you conclude that the solution must be between 4 and 5 and then you must make a table for the function between 4 and 5 with steps in x of 0.1. This may give the answer. The other case where the solution does not show in your table is when the solution is smaller than your first number for x or larger than your biggest number for x. You will be able to see this because the function will either be larger than 13 or smaller than 13 on your table. Make a new table where the function values are closer to 13 and hope to find the solution in the new table. This whole process may seem long; however, a graphing calculator can make the table for you as fast as you can enter the function. The beauty of the table is that once it is set, you can scroll forward or backward until you find the value of x that causes the function to be 13. If you can only come close to 13, you can set a new table with a Δx of 0.1 or smaller and try again for the value of x that causes the function to be 13.

Trying a second equation may help you understand the numerical method. Below is the solution to the equation $-2(x + 5) + 7 = -32$. Start with a table where *xmin* is a reasonable guess. For example, start the table with -5 (*xmin*) and let Δx be 1.1

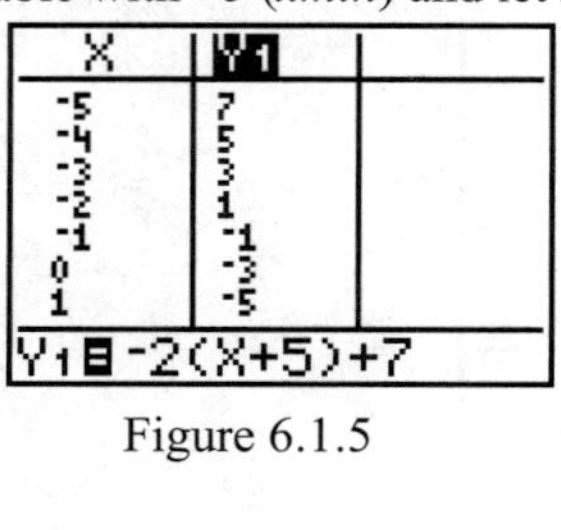

Figure 6.1.5

For values of x from -5 to 1, the function is not -32. You can see that if you scroll down, the function gets closer to -32. Figure 6.1.6 shows the results of scrolling down until numbers near -32 are on the table.

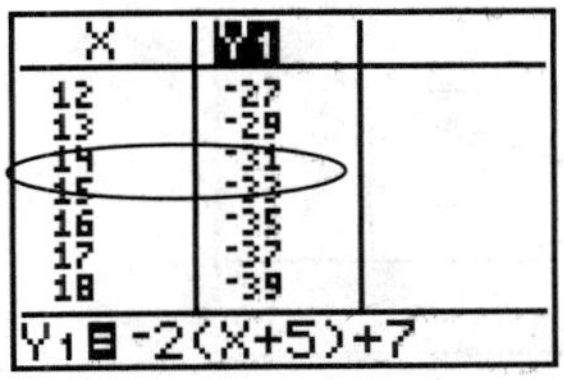

Figure 6.1.6

You can see that this table has x values that cause the function to jump over the function value of –32. Refine this table by using an *xmin* of 14 and a Δx of 0.1 This may show a better solution than the approximate value of 14 or 15.

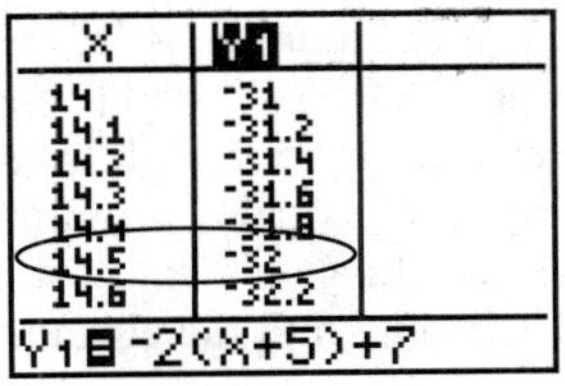

Figure 6.1.7

Figure 6.1.7 shows that if x is 14.5, the function is exactly –32. The equation has been solved by the numerical method.

The Trace Method

A small search party is looking for a missing camper in a 150 acre wooded area. They can cover 2 acres in 6 minutes. The function that models the amount of area left to search at time t is $A = -\frac{2}{6}t + 150$. Why? How much time will it take to search until 80 acres are left? To find out, solve the equation $80 = -\frac{2}{6}t + 150$. Using the trace method requires that you graph the function $A = -\frac{2}{6}t + 150$. Use trace until the value of A is 80. If you can't get an exact value of 80 for A from the trace cursor, either use the approximate value or change to a decimal window. Figure 6.1.8 shows an approximate solution of about 208.5 minutes and Figure 6.1.9 shows the trace cursor on the exact solution of 210 minutes.

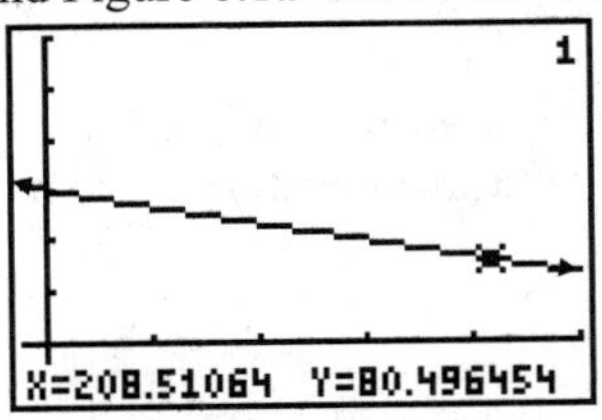

Figure 6.1.8

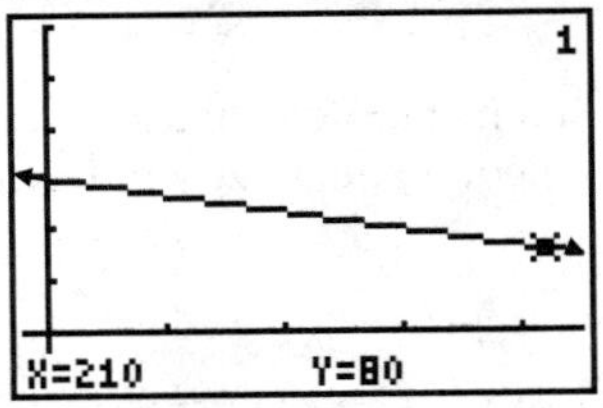

Figure 6.1.9

The Intersection Method

The Taxi Fare

Suppose that taxi company A has a fare schedule of \$1.20 per mile plus a \$3 usage fee. Taxi company B charges \$0.90 per mile plus a \$5 usage fee. The model of the taxi fare for company A is $F_A = 1.2m + 3$ and for company B it is $F_B = 0.9m + 5$. How far can you travel so that the fares are equal? The answer to this question can be found by solving the equation $1.2m + 3 = 0.9m + 5$. What you really want to know is when F_A equals F_B. Remember, this means you must find a value for m that causes F_A to equal F_B.

F_A and F_B

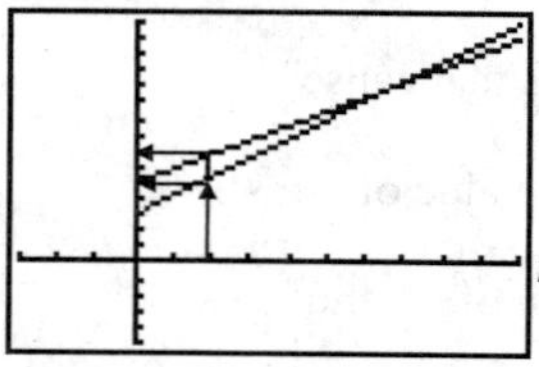

Figure 6.1.10a

F_A and F_B

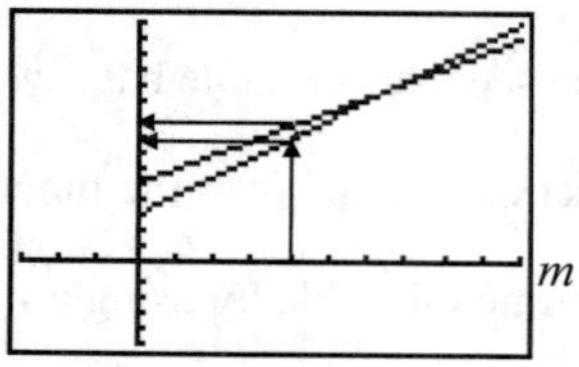

Figure 6.1.10b

F_A and F_B

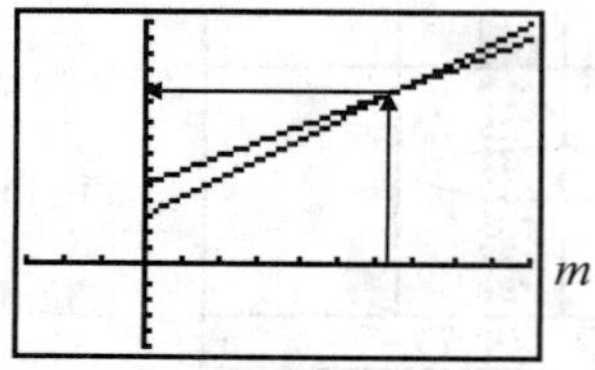

Figure 6.1.10c

Please notice in Figure 6.1.10a that at 2 miles, F_A (2) = 5.4 and F_B (2) = 6.8. Then in Figure 6.1.10b, at 4 miles driven, F_A (4) = 7.8 and F_B (4) = 8.6, but at 6.67 miles driven, F_A (6.67) = 11 and F_B (6.67) = 11. That is, the only distance traveled (6.67 miles) that gives the same taxi fare ($11.00) is at the intersection of the two fare models. Thus, at the intersection of the two graphs, F_A equals F_B and the m-coordinate at that point is the value of m that causes F_A to equal F_B. Figure 6.1.11a shows you that if you drive approximately 6.67 miles the taxi fare is $11 for both companies. The solution to the equation is

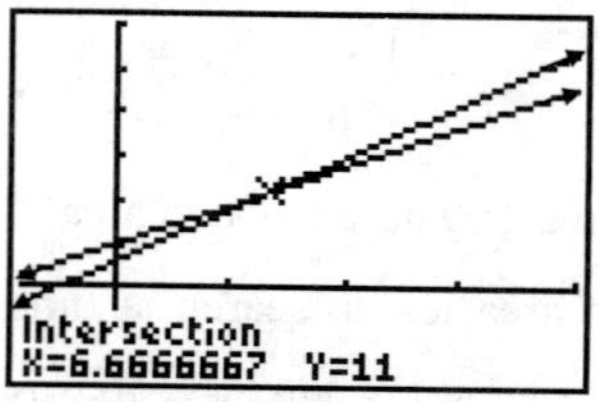

Figure 6.1.11a

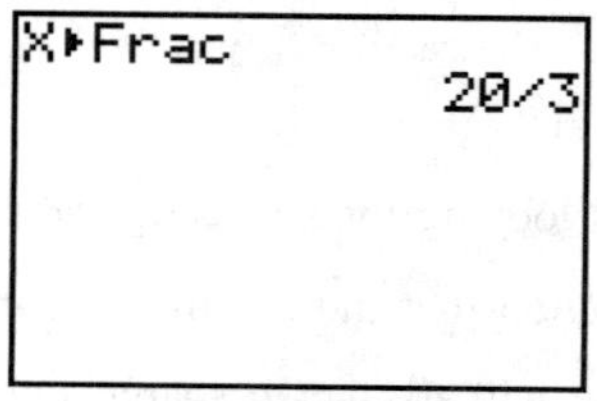

Figure 6.1.11b

approximately 6.67 miles, or as shown in Figure 6.1.11b, the solution is $\frac{20}{3}$ or $6\frac{2}{3}$ miles.

The Zeros Method

The last technology-based method combines what you learned in Chapters Two, Three, and Four about finding the zeros of a function and what you learned about manipulating the symbols in an equation in Chapter One.

As a sample of how to solve with the zeros method, solve the equation used above; that is, solve the equation $1.2m + 3 = 0.9m + 5$. To use the zeros method, use the properties of equality from Chapter One to transform the above equation so zero is on one side of the equation and all symbols on the other.

$1.2m + 3 = 0.9m + 5$ subtract $0.9m$ and 5 from both sides

$1.2m + 3 - 0.9m - 5 = 0$ no need to simplify unless you want to

Recall that using the subtraction property of equality does not change the solution to the equation, thus solving the equation $1.2m + 3 - 0.9m - 5 = 0$ is the same as solving equation $1.2m + 3 = 0.9m + 5$. The left side of the transformed equation is the function $1.2m + 3 - 0.9m - 5$, and the question you are being asked to answer is when is the function $1.2m + 3 - 0.9m - 5$ equal to 0? This is not new! You learned how to find zeros with the "zero finder" on the calculator. Your calculator may use the word "root" instead of "zero". This is a name mathematicians use for the solution to any equation. Since the **root** of the equation is the same as the "zero" of the function on the left, you are using the zeros method for solving the equation by finding the zero of the related function.

The roots of an equation are the zeros of the related function.

Figure 6.1.12 shows the approximate zero of the function $1.2m + 3 - 0.9m - 5$ and Figure 6.1.13 shows the exact zero.

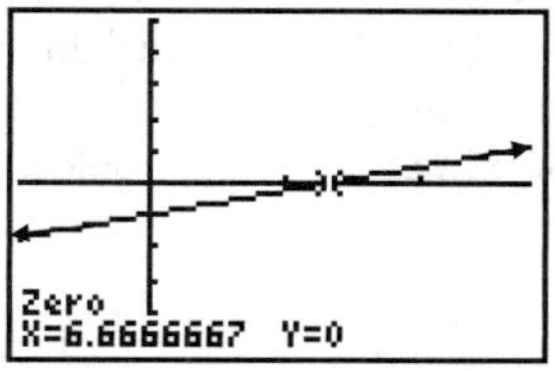

Figure 6.1.12

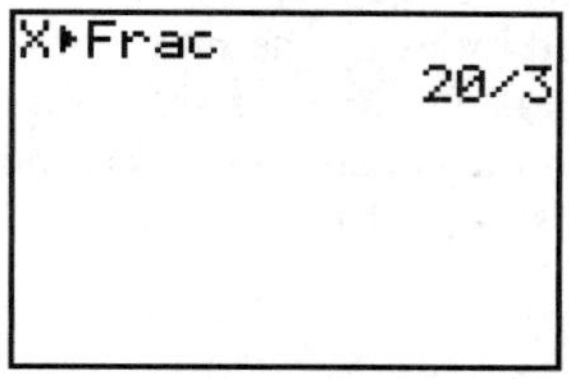

Figure 6.1.13

Perhaps the best feature of this method is that you always know where to find the real solution to the equation - it is always on the x-axis.

The Analytical Method

The Electric Company Rate

If the Carolina Power Company charges \$0.07 per *kWh* (kilowatt hour) used plus a monthly service fee of \$9.50 and the Ohio Power Company charges \$0.08 per *kWh* used plus a \$3.75 monthly service charge, what number of *kWh* used will cause the monthly electric bill from Carolina to be the same as from Ohio? The mathematical model of the Carolina Power electric bill is $E_C = 0.07k + 9.5$, and the Ohio Power model is $E_O = 0.08k + 3.75$. Since you want to know what value of k makes the electric bill from Carolina Power ($0.07k + 9.5$) equal the electric bill from Ohio Power ($0.08k + 3.75$), solve the equation $0.07k + 9.5 = 0.08k + 3.75$. The equation can be solved by using the properties of equalities studied in Chapter 1. As a quick reminder they are listed below.

Addition Property:	If $a = b$ then $a + c = b + c$
Subtraction Property:	If $a = b$ then $a - c = b - c$
Multiplication Property:	If $a = b$ then $ac = bc$ where $c \neq 0$
Division Property:	If $a = b$ then $\frac{a}{c} = \frac{b}{c}$ where $c \neq 0$

The equations

$$a = b$$

$$a + c = b + c$$

$$a - c = b - c$$

$$ac = bc$$

$$\frac{a}{c} = \frac{b}{c} \quad \text{are all equivalent equations } (c \neq 0).$$

If a number is added to, subtracted from, or a non-zero number is multiplied by, or divided into both sides of an equation, the original equation and the **transformed** equation have the same solutions. These basic properties of equality allow for an algebraic method for solving equations.

The algorithm used in solving equations with algebra is to isolate the variable on one side of the equation. This can be accomplished by using the properties of equality. With the variable by itself on one side of the equation, the solution is on the other side of the equal sign. When the algebraic method works, it can give the *exact* solution to the equation, or it can be approximate as shown below. (An **algorithm** is a set of steps used to accomplish a goal.)

To solve the equation:

$0.07k + 9.5 = 0.08k + 3.75$	subtract $0.08k$ from both sides
$0.07k + 9.5 - 0.08k = 0.08k + 3.75 - 0.08k$	simplify
$-0.01k + 9.5 = 3.75$	subtract 9.5 from both sides
$-0.01k + 9.5 - 9.5 = 3.75 - 9.5$	simplify
$-0.01k = -5.75$	divide both sides by -0.01
$\frac{-0.01k}{-0.01} = \frac{-5.75}{-0.01}$	simplify
$k = 575$	

If a customer in North Carolina and in Ohio use 575 *kWh* of electricity, their electric bills will be identical.

You should try each of these methods and select two or three methods to use on a regular basis. You must not assume you can always just use one method because any one of these methods may not solve all equations you encounter.

Example 1 Solve the equation $3(x - 6) - 5(x + 7) = -23$ numerically.

Solution: Make a table and look for a value of x that causes the function $3(x - 6) - 5(x + 7)$ to have a value of -23. The written page cannot contain a dynamic table as you may make on your calculator; instead, you will only see a static table as shown below.

Table 6.1.2

x	–10	–9	–8	–7	–6	–5	–4	–3	–2
$3(x-6)-5(x+7)$	–33	–35	–37	–39	–41	–43	–45	–47	–49

After looking at the behavior of the function, it appears that the solution is less than –10. Make another table.

Table 6.1.3

x	–20	–19	–18	–17	–16	**–15**	–14	–13	–12
$3(x-6)-5(x+7)$	–13	–15	–17	–19	–21	**–23**	–25	–27	–29

The solution is –15, it is a value for x that causes the function to be –23.

⁂

Example 2: Solve the same equation with the trace method.

Solution: Graph the function $y = 3(x-6) - 5(x+7)$ and trace to a point where it has a value of –23.

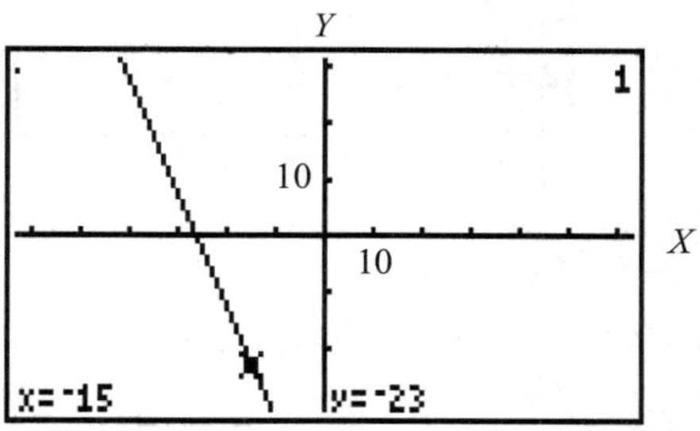

Figure 6.1.13

The trace cursor is on the point where the function is –23. The solution to the equation $3(x-6) - 5(x+7) = -23$ is –15. The window is zoom-int (the integer window).

⁂

Error

You may want to think of error as the difference between your answer and the exact answer. Further, since you may want to interpret error as a zero or positive number, find the absolute value of the difference between your solution and the exact solution.

> **Error:** The absolute value of the difference between the exact root (R) and your approximation of the root (r).
>
> $$error = |R - r|$$
>
> where R is the exact root of an equation and r is your approximation of R.

Since the calculator does not always give exact values for the solution (root) to an equation, as a general rule, you must express all roots with an error less that or equal to 0.01. The exceptions are 1) if the solution (root) is less than 0.01, or 2) if the solution is a very large number. In these two cases, you may approximate the root to whatever makes sense in the equation being solved.

Caution: When the free moving cursor (not trace) is used, the numbers displayed on the calculator's screen are called **screen coordinates**; they are not exact numbers. In most cases these numbers should not be used.

Example 3: Solve the equation $\frac{1}{4}x - 12 = -20$ using the zeros method.

Solution: Add 20 to both sides of the equation to get the transformed equation $\frac{1}{4}x - 12 + 20 = 0$. Find the zero of the function $f(x) = \frac{1}{4}x - 12 + 20$ with the zero finder on the calculator. The exact solution is –32.

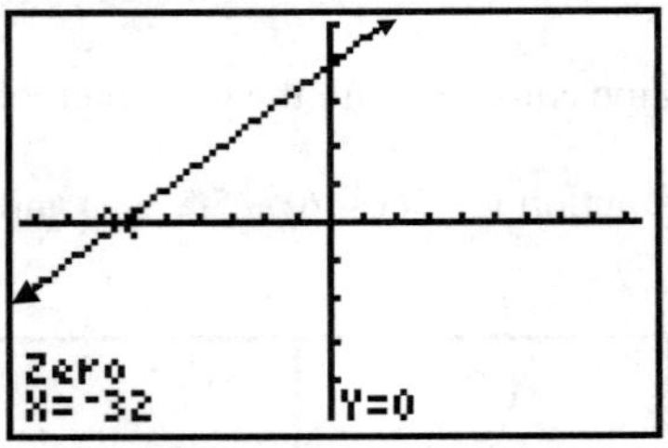

Figure 6.1.14

⁂

Example 4: Solve $3.5x + 2.1 = -5.9$ using the analytical method.

Solution:

$3.5x + 2.1 = -5.9$ subtract 2.1 from both sides

$3.5x + 2.1 - 2.1 = -5.9 - 2.1$ simplify

$3.5x = -8$ divide both sides by 3.5

$\frac{3.5x}{3.5} = \frac{-8}{3.5}$ simplify

$x = -2.29$

–2.29 is an approximate solution and $\frac{-16}{7}$ is the exact solution. The solution is also called the root of the equation.

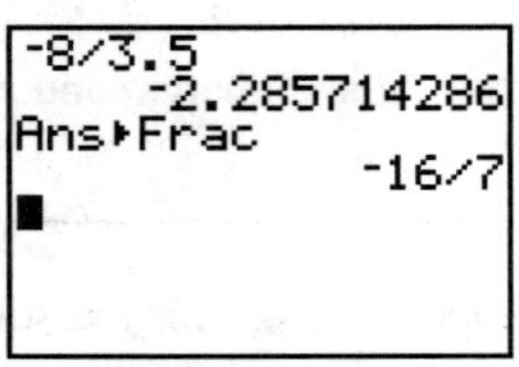

Figure 6.1.15

⁂

Example 5: Solve $5.3 - \sqrt{7}x = 12.6$ with an error ≤ 0.01. Use the zeros method.

Solution: Subtract 12.6 from both sides of the equation and then graph the related linear function $y = 5.3 - \sqrt{7}x - 12.6$. In entering the number $\sqrt{7}$ times x on the calculator, you must put a times sign between them or parentheses around one of the factors. Find the zero with the zero finder.

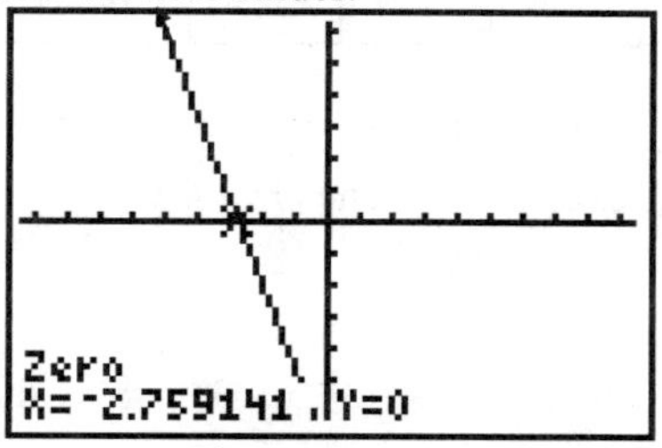

Figure 6.1.16

The solution to the equation $5.3 - \sqrt{7}x = 12.6$ is –2.76, where the error ≤ 0.01

⁂

Example 6: Find the exact solution to $5.3 - \sqrt{7}x = 12.6$.

Solution: From $5.3 - \sqrt{7}x = 12.6$ subtract 5.3 from both sides

$5.3 - 5.3 - \sqrt{7}x = 12.6 - 5.3$ simplify

$-\sqrt{7}x = 7.3$ divide both sides by $-\sqrt{7}$

$\dfrac{-\sqrt{7}x}{-\sqrt{7}} = \dfrac{7.3}{-\sqrt{7}}$ simplify

$x = \dfrac{-7.3}{\sqrt{7}}$ the exact solution is $\dfrac{-7.3}{\sqrt{7}}$.

⁂

Example 7: Solve $3x - 5 = x + 2$ with an error ≤ 0.01 using the numerical method.

Solution: Subtract x and 2 from both sides of the equation to yield the equivalent equation $2x - 7 = 0$. Find the zero of the related function $y = 2x - 7$. The zero is shown below in the second table.

Table 6.1.4

x	–1	0	1	2	**3**	**4**	5	6
$y = 2x - 7$	–9	–7	–5	–3	–1	1	3	5

The data shows that there is a zero between 3 and 4. At 3 the function is negative and at 4 it is positive; some place between 3 and 4 it must be 0. Make a new table between 3 and 4 using a Δx of 0.1.

Table 6.1.5

x	3	3.1	3.2	3.3	3.4	**3.5**	3.6	3.7
$y = 2x - 7$	–1	–0.8	–0.6	–0.4	–0.2	**0**	0.2	0.4

The value of x that causes the function to be 0 is 3.5. The solution to the equation $3x - 5 = x + 2$ is exactly 3.5. (The error is 0.)

⁂

Example 8: Solve the equation $5(x - 2) = -3(2x + 3) - \pi$ by the intersection method.

Solution: The intersection method works because the only value of x that causes the function $y = 5(x - 2)$ and the function $y = 3(2x + 3) - \pi$ to be equal is the x-coordinate of the intersection of the graphs. The intersection finder is used to locate the intersection.

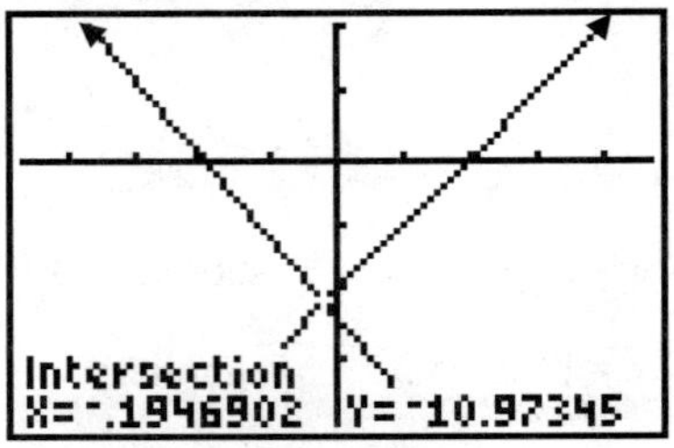

Figure 6.1.17

If the calculator values are rounded to the hundredths place, the error will be less than 0.01. The solution to the equation $5(x - 2) = -3(2x + 3) - \pi$ is –0.19.

⁂

6.1 EXERCISES

Many times you can work the exercises by algebraic, numeric, or graphical methods. You can learn about mathematics no matter which method you use. Part of the learning process is for you to decide which method is best for you to use. Please read the text carefully before you do the exercises. Read the **next assigned section** after you finish the assignment below. All numeric answers must have an error ≤ 0.01.

A Quick Look To The Future

-4. Every point on the graph of $y = x + 1$ can be symbolized as (x, y) or $(x, x + 1)$. Describe all points below the graph of $y = x + 1$.

-3. If $|x - 1| = 2$, does this imply that $x - 1 = -2$?

-2. Describe a relationship where as one variable increases, the related variable get proportionately larger.

-1. What happens to the thickness of a sheet of paper every time you fold the paper in half?

Previous Topics Revisited

For every fast-food hamburger from Central American cattle, about 55 square feet of rain forest is destroyed. *Earth Right - Every Citizen's Guide*, H. Patricia Haynes, p. 217, 1990.

If n is the number of hamburgers sold and A is the area of rain forest destroyed to make grazing land for the cattle used to make the hamburgers, answer the first three questions:

1. Express the rain forest area destroyed as a function of the number of hamburgers sold.
2. How many acres of rain forest must be used to produce the beef for 10,000 hamburgers?
3. How many hamburgers are sold for every 10,000 square feet of rain forest destroyed?
4. What linear function contains the points $(-2, 5)$ and $(-3, -7)$?
5. What is the linear function whose graph has a slope of $-\frac{2}{3}$ and contains the point $(0, -2)$?
6. What is the slope and y-intercept of the graph of $y = \frac{1}{4}x - 7$?

New Topic Exercises

Solve the following equations by any method of your choice.

7. If it costs \$2,400 to pave the driveway to your home with asphalt and each year it costs \$325 to re-seal the driveway with blacktop sealer, the function that models the total cost of the driveway is $C = 325t + 2400$. Find when the driveway will cost \$3,000. Find when it will cost \$3,500. Find when it will cost \$4,200.

8. If the cost of using concrete on the same driveway in Exercise 7 is \$3,650 and it costs \$0 to maintain the drive year after year, the function that models the total cost of the concrete driveway is $C = 0t + 3650$. In what year does the cost of the asphalt driveway equal the cost of the concrete driveway?

9. The Cleanit Car Company (CCC) pays each member of its towel drying crew a weekly salary of \$50 plus \$.75 per car. The crew, being nearly graduated from college, realizes that with special care of each car they can average \$.45 per car in tips. How many cars must be washed at the Cleanit Car Company for a towel dryer to gross \$200 per week? The function that models gross salary is $S(c) = 0.75c + 0.45c + 50$.

10. The Better Cleanit Car Company (BCCC) pays each member of its towel drying crew a weekly salary of \$80 plus \$.55 per car. This crew also has nearly graduated from college and realizes that with special care of each car they can average \$.71 per car in tips. How many cars must be washed at the Better Cleanit Car Company for a towel dryer to gross \$200 per week? The function that models gross salary is $S(c) = 0.55c + 0.71c + 80$.

11. What number of cars must be washed for the gross salary of a CCC employee to equal the salary of a BCCC employee?

12. Archeologists hired college students to simulate the building of a Hopewell Indian burial mound. Each of 150 students was given a 2 cubic foot basket in which to transport soil. One student could increase the volume of the mound by 2 cubic feet per 20 minutes. How long did it take for all 150 students to build a mound with a volume of 50,000 cubic feet? How long would it take if only 35 students worked? How long would it take for all 150 students to build a 140,000 cubic foot mound?

13. At a New Years Eve party, the caterer supplied a bowl with approximately 1000 chips. During the party, guest #1 ate the chips at a rate of 22 chips per minute (cpm), guest #2 at 13 cpm, guest #3 at 26 cpm, and the caterer added chips to the bowl at an average rate of 47 cpm. Are the number of chips increasing or decreasing? When are there 200 chips left in the bowl? When is the bowl empty? If guest #2 doubles her eating rate, when will the bowl be empty?

14. To prepare for a mathematics conference, hotel workers must take 5000 chairs from the storage room and set them up for the opening session. David sets up 10 chairs every 3 minutes. Diana sets up 12 chairs every 4 minutes, and Debby sets up 7 chairs every 2 minutes. How long does it take to be half finished? How long does it take to have no chairs left to set up?

15. The ABC Painting Company is doing a job cost estimation for a house with 4200 square feet surface area. Painter *A* paints and prepares at a rate of 125 square feet per hour. Painter *B* paints and prepares at a rate of 65 square feet per hour. Painter *C* works at 23 square feet per hour. (Painter *C* is a trainee and is painter *B*'s son.) How long will it take to paint 75% of the house? How long will it take so that there is 0 square feet left to paint? If the ABC Painting Company charges $10.50 per person per painting hour, how much should they estimate as the labor cost for the job?

16. Downloading software from a Web site transfers to your computer at a rate of 1.2K per second. The software being downloaded is 391K is size. How long will it take to download the software? If you log in at a faster rate of 2.4K per second, how long will it take? What download rate would cause the software to download in 1 minute?

17. $5.7x + 2.9x - 1.4 = \sqrt{6}$

18. $\sqrt{3}x + 2 = 1.5$

19. $\pi x = 2$

20. $3.2(1.5x + 7.2) - 1.5(x - 5.1) = 7(x + 3) - 5$

21. $\sqrt{2}(x - \sqrt{5}) = x + \frac{1}{2}$

22. $-(-[x + 4] - 13) = 2x - 5$

23. $0.002x + 15 = 300$

24. $500x = 5000$

25. $\sqrt{7}(x - \pi) = 2.5(x + \sqrt{2.9})$

26. $2(x + 1) = 2x + 1$

27. $65000(x - 5) = 42000x$

28. $(\frac{1}{4})x + \frac{1}{2} = (\frac{3}{4})x$

29. $0.80 = \dfrac{416 + x}{580}$

30 List any linear equation (other than $x = 5$) that has a solution of 5.

31. List any linear equation (other than $x = -5$) that has a solution of -5.

32. List any linear equation (other than $x = \pi$) that has a solution of π.

From Mathematics to English

33. After reading this section, make a list of questions that you want to ask your instructor.

34. Continue in your daily journal and make an entry. In addition to your normal entry on thoughts about the mathematics in this section, list at least two positive comments about what you have learned about this topic.

35. In paragraph format, summarize the material in this section of the text in your daily journal.

36. Describe how your classroom instructor made this topic more understandable and clear.

37. After reading the text and listening to your instructor, what do you not understand about this topic?

38. What is the most important idea presented in this section? Why?

39. Which method would you use to solve the equation $0.002x - 1.05 = 3.001$? Why?

40. Describe how to solve the equation $2x - 5 = 4.1$ with an error less than 0.01, using the graphical method.

41. Describe how to solve the equation $2x - 5 = 4.1$ with an error less than 0.01, using the numeric method.

42. Which method would you use to solve the equation $3x = -15$? Why?

43. Describe the relationship between solving an equation and finding the zeros of a function.

44 Discuss why you know some linear equations have no solution and other linear equations have an infinite number of solutions. Give examples of each type.

45. Using the analytic (algebraic) method for solving equations, describe the steps you would use to solve the equation $-3(x+5) = 17.6$. Do not actually solve.

46. Using the graphical method for solving equations, describe the steps you would use to solve the equation $-3(x+5) = 17.6$. Do not actually solve.

47. What is an equation?

48. What is the "solution" to an equation?

49. Why should you check a solution to an equation?

50. If you memorize a method for solving an equation, do you understand what a solution to an equation is? Explain.

51. Can every equation be solved using technology? Explain.

52. Can every equation be solved using algebra or other mathematics? Explain.

From English To Mathematics
In the next three exercises, convert the English statement to an equivalent mathematical statement.

53. three times a number x is fifty-two

54. the length is three times the width

55. the product of seven and a number x is four minus the number x

Explorations
56. Using interval notation (except for zeros), describe when each of the linear functions has a value less than zero, zero, and greater than zero. Approximate numbers are acceptable as long as the error is ≤ 0.01

		$f(x) < 0$	$f(x) = 0$	$f(x) > 0$
a.	$f(x) = 4x + 7$			
b.	$f(x) = -3x + 2$			
c.	$f(x) = 0.06x - 9$			
d.	$f(x) = -53x + 80$			
e.	$f(x) = (\frac{1}{4})x + 2$			
f.	$f(x) = -(\frac{3}{4})x + 5$			
g.	$f(x) = \pi x + 1$			

57. Solve the following linear equations by a method of your choice. If you use the algebraic method, show your work; if you use the graphical method, show the graph used to yield the solution.

a. $-2[x - 5(3x + 4) - 6] + 6[-4 + 7(9x - 1) - 2(x - 6)] = 0$

b. $0.003x - 4.932(x + 79) = -23\{5x - 2[-7.4x + (-5)] - 32.69\}$

c. $-420x - 250(32x - 57) - [-49x - 21(17x + 89) - 430x] = 425(84x - 7)$

d. $0 = 0.24x - 0.11(-88 + 14x) - 7\{2 - 5[3 - 5(2x - 1) - 9.3] + 12.5\}$

6.2 Solving Inequalities Containing the Linear Function

What Is My Net Salary?

If you are planning to work at a restaurant where you expect an average of \$3.50 in tips per person served plus \$100 in weekly salary, you should know that the model of your weekly gross salary is $S_G = 3.5p + 100$. If payroll deductions amount to about 25% of your gross salary, then your net salary can be modeled by the function $S_N = 0.75(3.5p + 100)$. A typical question you may want to ask yourself is "How many people must I serve if I need to net \$250 per week?" In Section 6.1, you learned to use the model to make the equation $250 = 0.75(3.5p + 100)$. The solution to this equation is shown in Figure 6.2.1 to be 67 people.

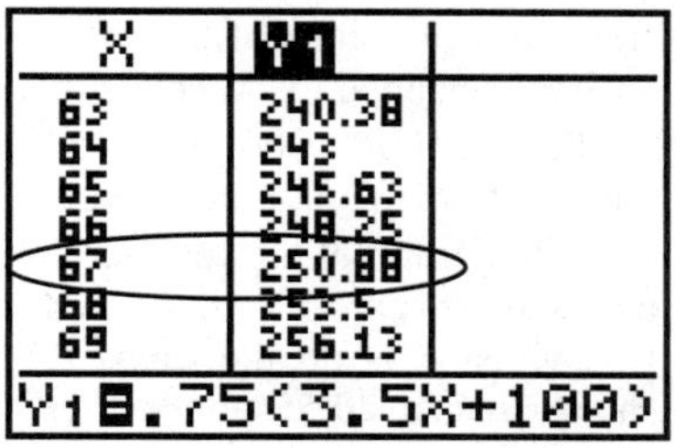

X	Y1
63	240.38
64	243
65	245.63
66	248.25
67	250.88
68	253.5
69	256.13

Y1=.75(3.5X+100)

Figure 6.2.1

While the numeric representation of the function that models your net salary shows the answer to your question, it also shows the answer to two other questions. How many people must you wait on to earn less than \$250 and how many people must you wait on to earn more than \$250? Each of these statements can be expressed in symbolic mathematical form as $0.75(3.5p + 100) < 250$ and $0.75(3.5p + 100) > 250$. These statements are called **inequalities**. Since they contain linear functions, they are linear inequalities. Figure 6.2.1 shows the solution to these inequalities. Hopefully, it is clear that if you serve over 67 people your net pay will be larger than 250; thus, the solution to $0.75(3.5p + 100) > 250$ is $(67, \infty)$. Figure 6.2.1 also shows you that if you serve under 67 people your net pay is less than 250; therefore, the solution to $0.75(3.5p + 100) < 250$ is $(-\infty, 67)$. Please be aware that the normal domain of the function $S_N = 0.75(3.5p + 100)$ is being used when the above answers are written.

A quick review: To solve inequalities, the solution to the related equation is found first and then numbers either above or below the solution to the equation are the solution to one or the other inequality. If your original question is how many people must I serve to net \$250 or larger than \$250, the solution is $[67, \infty)$. The inequality now becomes $0.75(3.5p + 100) \geq 250$. Likewise, the solution to $0.75(3.5p + 100) \leq 250$ is $(-\infty, 67]$.

The inequality symbols are less than (<), greater than (>), less than or equal to (≤), and greater than or equal to (≥). If you think algebraically, solving an inequality means using the properties of inequalities to isolate the variable on the left side of the inequality so that you can find values for the variable in the inequality that make it a true statement. If you think graphically, your thinking is totally different. This is explained in the next problem.

The Evacuation of the Outer Banks

In early August, there may be as many as 300,000 people on the Outer Banks of North Carolina. During a hurricane, people are encouraged to leave the island and head back to the mainland. There are two major exits. Suppose the Route 158 exit will allow for 4200 people per hour and the Route 64 exit 3100 people per hour. This means that during an evacuation of the Outer Banks, the function that models the number of people left on the island is $P_L = -4200t - 3100t + 300000$. A reasonable problem domain may be around [0, 40] hours, and the normal domain is all real numbers.

If the evacuation officials need to know when less than half of the people are left on the island, the model becomes the inequality $-4200t - 3100t + 300000 < 150000$. Using the same idea behind the zeros method for solving equations, subtract 150000 from both sides of the inequality. This means the answer to the question posed by the officials has now been changed to "When is the transformed model $-4200t - 3100t + 300000 - 150000$ less than zero?" If you visualize the solution in your mind you should see that you want to know when the graph of the function $-4200t - 3100t + 300000 - 150000$ is below the t -axis. To find when it is below the t-axis you must first find when it is on the t-axis as shown in Figure 6.2.2. The figure shows that at 20.55 hours after the call for evacuation, half of the people remain on the island. Inspection of the graph also shows that the transformed function is less than zero, or negative, (When less

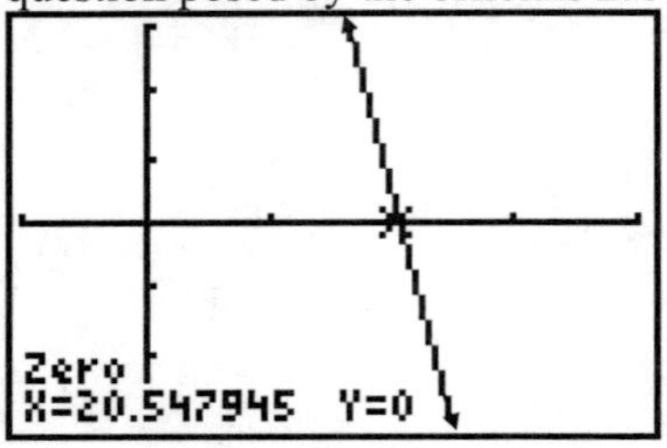

Figure 6.2.2

than half of the people remain.) to the right of the zero of the function. Thus, on the interval $(20.55, \infty)$, the function $-4200t - 3100t + 300000 - 150000$ is negative.

The algorithm for finding when the graphical representation is above or below the horizontal axis may be a simpler process than finding when the graph of the original model $(-4200t - 3100t + 300000)$ is below the number 150,000 as shown in Figure 6.2.3 below. As you can see, the solution to the problem is the same. The graphical representation of the original model is below 150,000 when t is greater than 20.55 hours.

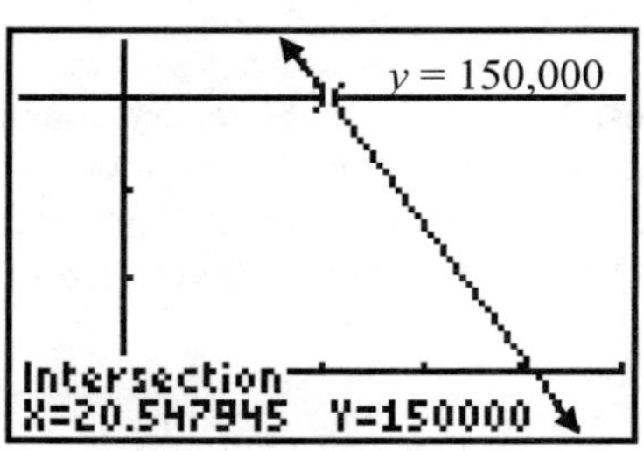

Figure 6.2.3

The zeros method for solving inequalities requires that you find the zero of the transformed function followed by identifying when the graph is either above or below the horizontal axis. Of course, the other methods of solving equations will also help solve the linear inequalities.

Example 1: Solve the inequality $\frac{1}{2}x - 3 < 0$ by the numerical method.

Solution: Look at the numeric representation of $y = \frac{1}{2}x - 3$ to see when the function values are smaller than zero. The numeric representation shows that if x is less than 6, the

X	Y2	
3	-1.5	
4	-1	
5	-.5	
6	0	
7	.5	
8	1	
9	1.5	

Y2=.5X-3

Figure 6.2.4

function $\frac{1}{2}x - 3$ is less than 0. The inequality has been solved. The solution to $\frac{1}{2}x - 3 < 0$ is $(-\infty, 6)$.

⁂

Example 2: Solve the inequality $-5x + 7 \le 3x - 5$ using the intersection method.

Solution: The intersection method requires that you graph the function $-5x + 7$ and the function $3x - 5$ and answer the question "When is the graphical representation of $-5x + 7$ below or on the graphical representation of $3x - 5$?"

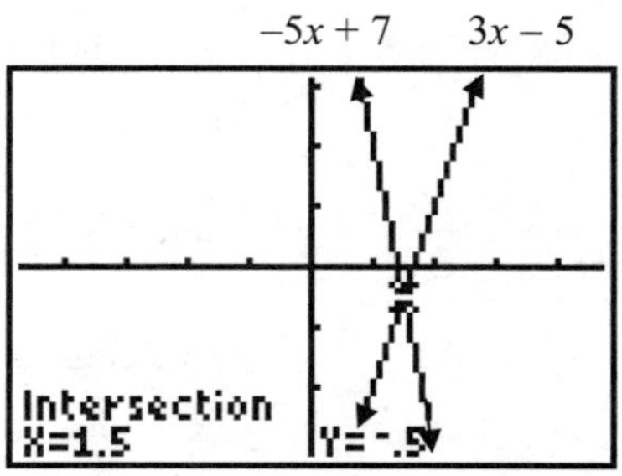

Figure 6.2.5

The solution to $-5x + 7 \le 3x - 5$ is $[1.5, \infty)$ because the graph of $-5x + 7$ is on or below the graph of $3x - 5$ when x is 1.5 and larger.

⁂

To solve an inequality algebraically, you need the properties of inequality from Section 1.1. A listing of the properties is below:

If $a < b$ then $a + c < b + c$

and $\quad a - c < b - c$

and $\quad ac < bc \quad$ where $c > 0$

and $\quad \frac{a}{c} < \frac{b}{c} \quad$ where $c > 0$

and $\quad ac > bc \quad$ where $c < 0$

and $\quad \frac{a}{c} > \frac{b}{c} \quad$ where $c < 0$ Each property also applies for $>$, $\geq$, and $\leq$.

Inequalities can be solved algebraically as well as numerically and graphically. The algebraic method (often called the **analytical method**) requires the use of the inequality properties. Linear inequalities are solved by a process similar to that used for solving linear equations algebraically. However, it is recommended to always isolate the variable on the left side of the inequality. You naturally read English and graphs from left to right; thus, by isolating the variable on the left side of the inequality, it is easier to read and understand. Do you know what values for x make the inequality $-4 < x$ true by just reading it? Possibly not. If the same question is asked of the inequality $x > -4$, now do you know what values of x cause the inequality to be a true statement? Hopefully you know the solution to be $(-4, \infty)$!

Example 3: Solve $2x + 5 > 0$ algebraically.

Solution: From $2x + 5 > 0$ — subtract 5 from both sides

$2x + 5 - 5 > 0 - 5$ — simplify

$2x > -5$ — divide both sides by 2 or multiply both sides by $\frac{1}{2}$

$\frac{2x}{2} > \frac{-5}{2}$ — simplify

$x > -2.5$ — x belongs to the interval $(-2.5, \infty)$.

⁂

Example 4: Solve $-3x + 4 \geq 0$ analytically.

Solution: From $-3x + 4 \geq 0$ — subtract 4 from both sides

$-3x + 4 - 4 \geq 0 - 4$ — simplify

$-3x \geq -4$ — divide both sides by -3 and change the $\geq$ to $\leq$

$\frac{-3x}{-3} \leq \frac{-4}{-3}$ — simplify

$x \le \frac{4}{3}$ note that the $\ge$ becomes $\le$ because of division by a negative

x belongs to the interval $\left(-\infty, \frac{4}{3}\right]$.

⁂

Example 5: Solve $-5x + 7 \le 3x - 5$ algebraically.

Solution: From $-5x + 7 \le 3x - 5$ subtract 7 and $3x$ from both sides

$-5x + 7 - 7 - 3x \le 3x - 5 - 7 - 3x$ simplify

$-8x \le -12$ divide both sides by -8 and change $\le$ to $\ge$

$\frac{-8x}{-8} \ge \frac{-12}{-8}$ simplify

$x \ge \frac{3}{2}$ x belongs to the interval $\left[\frac{3}{2}, \infty\right)$.

⁂

Example 6: Solve the inequality $2.4(x - 1) \le 3.7x$ analytically and verify numerically.

Solution: From $2.4(x - 1) \le 3.7x$ simplify the left side

$2.4x - 2.4 \le 3.7x$ add 2.4 to both sides

$2.4x - 2.4 + 2.4 \le 3.7x + 2.4$ simplify the left side

$2.4x \le 3.7x + 2.4$ subtract $3.7x$ from both sides

$2.4x - 3.7x \le 3.7x + 2.4 - 3.7x$ simplify

$-1.3x \le 2.4$ divide both sides by -1.3

$\frac{-1.3x}{-1.3} \ge \frac{2.4}{-1.3}$ simplify

$x \ge -1.85$

The solution to $2.4(x - 1) \le 3.7x$ is $[-1.85, \infty)$. The confirmation of the solution is shown in Table 5.2.1 where the numerical method has been used.

Table 6.2.1

x	−1.88	−1.87	−1.86	−1.85	−1.84	−1.83
$2.4(x-1) - 3.7x$	.044	.031	.018	.005	−.008	−.021

⁂

As with solving equations, you have several choices of methods. You will find that the technology-based methods will not solve all inequalities, nor will the algebraic method solve all inequalities. Further, one method is not always easier or always harder than any other method. You must learn how to select an appropriate method as you do the assigned exercises below.

6.2 EXERCISES

Many times you can work the exercises by algebraic, numeric, or graphical methods. You can learn about mathematics no matter which method you use. Part of the learning process is for you to decide which method is best for you to use. Please read the text carefully before you do the exercises. Read the **next assigned section** after you finish the assignment below. All numeric answers must have an error ≤ 0.01.

A Quick Look To The Future

-4. Without using a calculator or pencil and paper, what is the solution to $|x+2| = -4$?

-3. What is another name for the area of a rectangle formula $A = lw$?

-2. The human population of earth is doubling about every 40 years. When will there be four times as many people on earth as right now?

-1. If x^2 is $x \cdot x$, and x^3 is $x \cdot x \cdot x$, what is $x^2 \cdot x^3$?

Previous Topics Revisited

1. Solve $3x - 5 = 2(4x + 7)$
2. Solve $-1.3x + 7.99 = 4.2(x - 2)$
3. Solve $0.002x - 5.999 = .003(x + 4.7)$

Individuals in the United States create 4.3 pounds of waste each day. By comparison, the average Japanese, 2.4 pounds.

Let G represent the amount of garbage created by an individual in time t, where t is measured in days. Write a function with respect to time for garbage generated by the American, and write a function for the Japanese.

4. Graph both garbage functions on the same screen with a viewing domain of [0, 7300] (7300 days = 20 years using a 365-day year) for t, or create a table for both functions. Using the table or graph, what is the *difference* between garbage generated by the American and the Japanese after 10 years? After twenty years?

5. How much total garbage is generated by the American and by the Japanese in an estimated life of 72 years?

6. Find the x and y intercepts of $2x + 3y = 6$.

New Topic Exercises

Solve the following inequalities by a method of your choice. Use interval notation to describe the answers.

7. A hotel manager estimates that a 350 sheet roll of toilet paper is used at a rate of 32 sheets per day. How many days from putting a new roll on the dispenser will it have more than 80 sheets left? What if the manger bought paper with 600 sheets, how many days from putting a new roll on the dispenser will it have more than 80 sheets left? How long will this roll contain paper?

8. In planning her wedding reception, Jennifer discovered that reception hall A charges \$1000 for the use of the facility, \$35 per person for food, and \$12 per person for limited drinks. Reception hall B charges \$1,300 for the use of the facility, \$29 per person for food, and \$15 per person for limited drinks. For how many wedding guests is the charge for hall A less than the charges for hall B? For how many guests is the charge for hall B greater than for hall A? If hall A raises the food rate to \$38 per person and hall B to \$35, for how many quests is the charge for hall A less than the charge for hall B?

9. Taxi company AA charges a fare of \$1.25 per mile plus \$2. Taxi company AAA charges a fare of \$4 plus \$0.95 per mile. How many mile can you ride in taxi AA and have a fare that is less than what taxi AAA charges? If you were the owner of taxi company AAAA and most all of you fares come from trips under 7 miles, would you use the fare schedule of company AA or AAA? Why?

10. College bookstore A buys a textbook from the publisher for \$27.50 and marks it up by 25%. The gross profit made by the bookstore from selling n books can be modeled by the function $G_A = (0.25 \cdot 27.5)n$. How many books must be sold to gain more than \$5,000 in gross profit?

11. Bookstore B buys the same book from a black market dealer for \$18.25 and marks up the book by 38%. The function that models the gross profit from n books sold by bookstore B is $G_B = (0.38 \cdot 18.25)n$. How many books must be sold so that the gross profits are larger than \$5,000?

12. $\frac{1}{4}x + 2.6 > 0$

13. $1.5(x+7) - 2(x-3) \geq x$

14. $-1(2x-5) < -(3x-5)$

15. $0 > 2x + 4$

16. $5x + 2.5(x-3) < 5x$

17. $x > x$

18. $-x < x$

19. $0.002(x - 3.99) \leq 0$

20. $6000(2x-15) \geq 1200$

21. $4.7x - 3.1(x - 14.9) \leq 0.2(5.1x + 17)$

22. $43x - \pi \geq 95x$

23. $\sqrt{5} - 13x \geq 9$

24. $\sqrt{6} \cdot \sqrt{8} - 3x \leq 15.6$

25. Find any "greater than" linear inequality (except $x > -4$) that has a solution set of $(-4, \infty)$.

26. Find any "greater than" linear inequality that has a solution set of $(-\infty, 4)$.

27. Find any "less than or equal to" linear inequality (except $x \leq 5$) with a solution set of $(-\infty, 5]$.

28. If any linear function $f(x)$ is a decreasing function and has a zero of –5, is $f(x) < 0$ to the left of –5 or to the right of –5?

29. If any linear function $f(x)$ is an increasing function and has a zero of –17, is $f(x)$ negative to the left of –17 or to the right of –17?

From Mathematics to English

30. What is a linear inequality?

31. What is the solution to a linear inequality? Describe numerically. Describe graphically.

32. If you memorize how to solve a linear inequality, do you understand what the solution to a linear inequality is? Explain.

33. Can you solve every linear inequality with technology? Why?

34. Explain why the inequality symbol must be reversed when both sides of the inequality are multiplied or divided by a negative number.

35. Which method of solving inequalities is your favorite? Why?

36. Give an example of a linear inequality with no solution. (Think graphically.)

37. Describe how to solve the inequality $-3(2x+5) \le 4x+2$ graphically and then describe how to solve it algebraically. (Do not actually solve.)

38. After reading this section, make a list of questions that you want to ask your instructor.

39. Continue in your daily journal and make an entry. In addition to your normal entry on thoughts about the mathematics in this section, list at least two positive comments about what you have learned about this topic.

40. In paragraph format, summarize the material in this section of the text in your daily journal.

41. Describe how your classroom instructor made this topic more understandable and clear.

42. After reading the text and listening to your instructor, what do you not understand about this topic?

43. What is the most important idea presented in this section? Why?

From English to Mathematics

In the next three exercises, convert the English statement to an equivalent mathematical statement.

44. four plus the number x is more than zero

45. the rectangle with width w and length l has a perimeter less than seventy four

46. a number x is at least twenty

47. a number x is at most twenty

48. the area of a circle with radius r is under seventeen

Explorations

49. Solve the following linear inequalities. If you use the algebraic method, show your work; if you use the graphical method, show the graph used to solve the inequality.

a. $-6.2x + 4.78[-23.5x + 41.3(15x - 7.22) - 12.58x] + 15.25x \geq 0$

b. $536(-17x + 429) \leq -\{2x - 13[-43x + 39(0.002x - 51) - 14x] + 64\}$

c. $-2\{-2[-2(-2x + 3)]\} + 3\{-6[-5(-7x + 4)]\} > 34$

d. $-3\{-5x + 7(-8x + 2) - 13(2x - 9)\} + 5x(-13) < 196(x - 2)$

50. In this section, you learned to solve linear inequalities graphically and numerically as well as algebraically. While solving more complicated inequalities involving quadratic, absolute value, or square root functions by algebraic methods requires new algebraic properties, solving them graphically or numerically requires no more than what you have already learned in this section. Try solving the following inequalities by graphical or numerical methods.

a. $\frac{1}{2}x^2 - 3x - 5 \leq 0$

b. $-3|x - 1| \leq -4$

c. $2\sqrt{x+5} - 7 > 0$

d. $(x - 3)(x + 2) \geq 0$

e. $-\frac{1}{4}|x-3| \cdot |x+4| + 2 \geq 0$

51. Try solving the non-linear inequalities by the graphical method. The algebraic method just learned may not apply. Show the graph of the function used to solve the inequality.

a. $2x + 5 > 0$

b. $|2x+5| > 0$

c. $(2x + 5)^2 > 0$

d. $\sqrt{2x+5} > 0$

e. $(2x + 5)^3 > 0$

6.3 Solving Inequalities and Equations Containing the Absolute Value Function

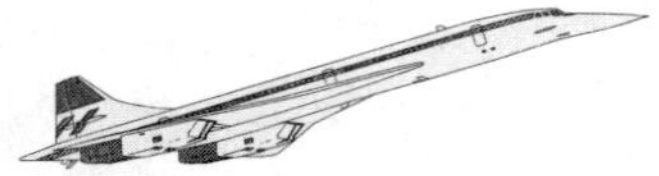

How High is the Airplane?

The on-board computer controls the height of an airplane on a short 40-minute flight from Cincinnati to Columbus by using the model $h = -1150|t - 20| + 23000$. The navigator decides it is necessary to control the domain and make it [0, 40], by adding the function $0\sqrt{-(t(t-40))}$ to the height function (see Section 4.6). Suppose a passenger has a pressure sensitive package that should not be above 20,000 feet for more than 10 minutes; how long is the plane above 20,000 feet? Looking at this question analytically, the answer is the solution to the inequality $-1150|t - 20| + 23000 > 20000$. Below is the graph of the height function and the function $h = 20000$.

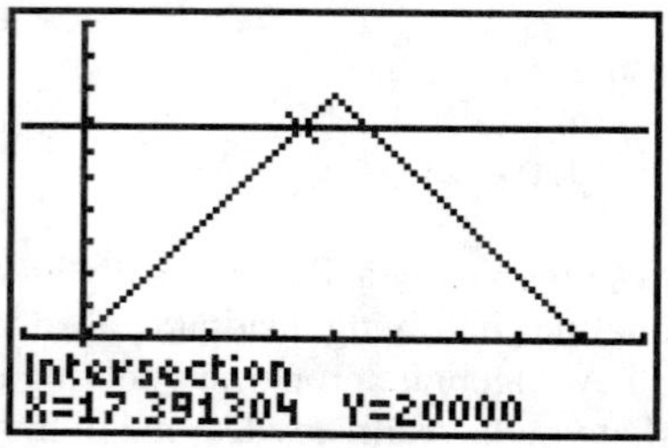

Figure 6.3.1a

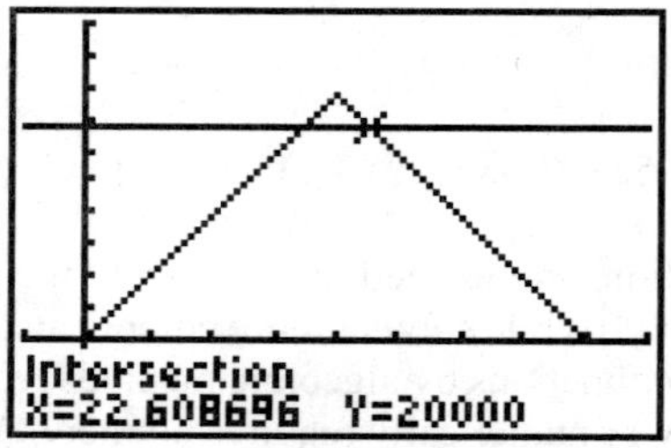

Figure 6.3.1b

The intersection method of solving inequalities shows that the plane is greater than 20,000 feet from 17.39 minutes to 22.61 minutes of the flight; this is less than 10 minutes. The passenger can relax.

Solving an inequality containing an absolute value function like $-1150|t - 20| + 23000 > 20000$ requires no new technology based method! The intersection method is demonstrated in the above sample. The zeros method and the numeric method can also be used as demonstrated in the next two illustrations.

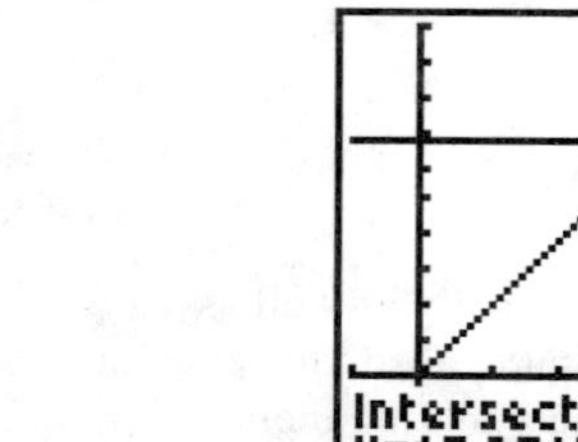

Figure 6.3.2

Take Your Medicine

The level of many medications, like Imipramine, in the blood steam can be modeled by absolute value functions because the doctor wants the medication level to rise at a constant rate, level off, and then the blood level must decrease at a constant rate. The function $L(t) = -25(|t - 3| + |t - 17|) + 500$ models the level of the medication when a patient is on the medication for a 20 week period. $L(t)$ is measured in nanograms per milliliter of blood and t is in weeks. One piece of data the doctor must know is when the level is, for example, 125 nanograms per *ml.* This is simply asking when the function $L(t) = -25(|t - 3| + |t - 17|) + 500$ has a value of 125. In other words, the answer to the question is the solution to the equation $-25(|t - 3| + |t - 17|) + 500 = 125$. A relatively nice method for solving any equation is the zeros method. To solve by the zeros method, you must transform the equation so that 0 is on one side of the equation and a function on the other side. Subtract 125 from both sides of the equation to get the transformed equation $-25(|t - 3| + |t - 17|) + 500 - 125 = 0$. Graph the function $y = -25(|t - 3| + |t - 17|) + 500 - 125$ and find the zeros. The zeros are shown in Figures 6.3.3a and b.

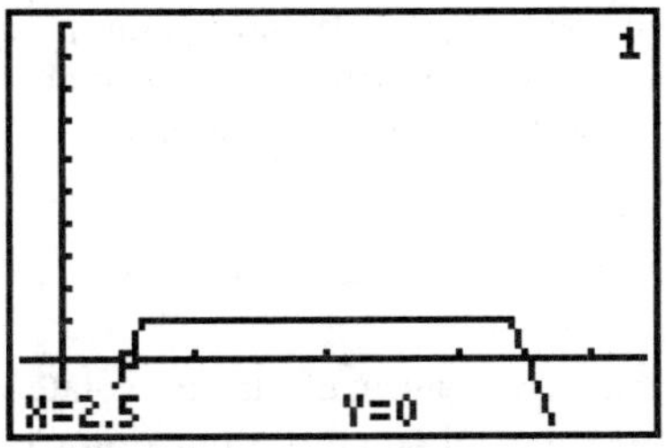

Figure 6.3.3a

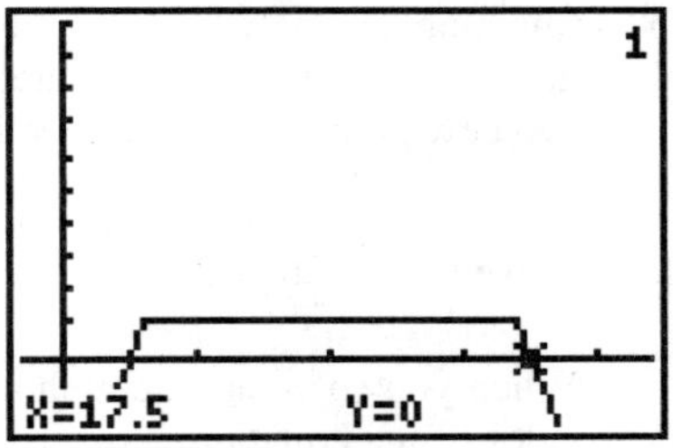

Figure 6.3.3b

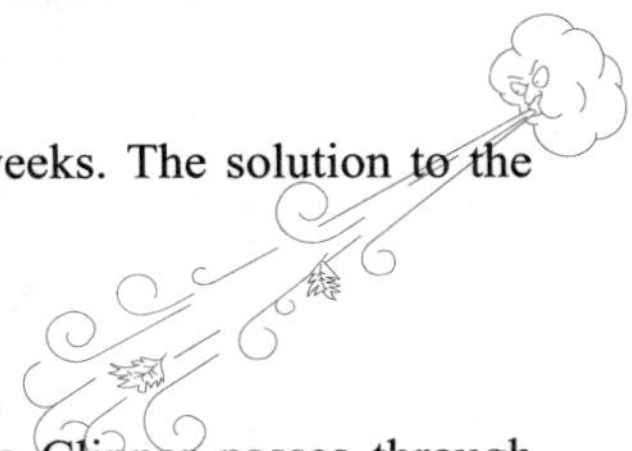

The blood level is 125 nanograms per *ml* after 2½ weeks and again at 17½ weeks. The solution to the equation $-25(|t-3|+|t-17|)+500=125$ is the set of two numbers {2½, 17½}.

The Alberta Clipper

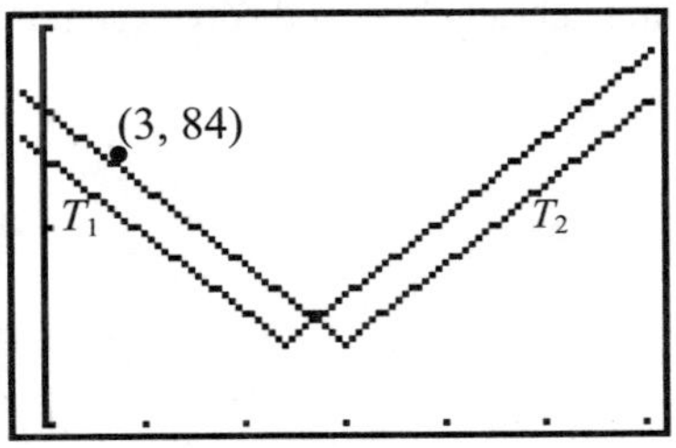

Figure 6.3.4

If a weather system called an Alberta Clipper passes through your area, the temperature will quickly drop, reach a minimum, and then rise again. For example, if the temperature before the clipper passes through is 84°, it may drop to 60° in a 12-hour period and then rise again. The model of this type of temperature change is the function $T_1 = 2|t-12|+60$, where the starting time is not in clock hours but just hours. For this problem, assume a noon arrival. If the temperature starts its decline at 3:00 PM (3 hours later) on a summer day, the model can be adjusted to $T_2 = 2|t-15|+60$. If you also want to change the function so that it has a reasonable problem domain of 24 hours past 3:00 PM, add the function $0\sqrt{-(t-3)(t-27)}$ to $T_2 = 2|t-15|+60$ and you now have a model that starts at 3:00 PM and stops 24 hours later. It is $T_3 = 2|t-15|+60+0\sqrt{-(t-3)(t-27)}$ and it is displayed in Figure 6.3.5.

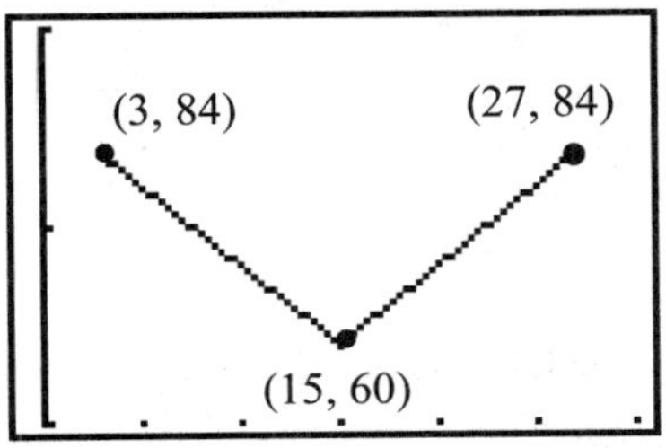

Figure 6.3.5

If it is your responsibility to predict when the outside temperature will be 70° or below, you are being asked to solve the inequality $2|t-15|+60 \le 70$. The numerical method will very nicely show when the model is 70° or below. If you are following along on your own calculator, you may use the function $T_2 = 2|t-15|+60$, or you may use the function where the domain has been restricted to the problem domain $T_3 = 2|t-15|+60+0\sqrt{-(t-3)(t-27)}$.

Table 6.3.1

t	9 PM	10 PM	11 PM	12 PM	1 am	2 am	3 am	4 am	5 am	6 am	7 am	8 am	9 am
t	9	10	11	12	13	14	15	16	17	18	19	20	21
T	72	70	68	66	64	62	60	62	64	66	68	70	72

The table shows that the temperature should be 70° or below when $t \in [10, 20]$; in clock hours, this is 10 PM to 8 am the next morning. This inequality containing an absolute value function has just been solved by the numerical method.

The only method remaining to analyze is the algebraic method of solving equations and inequalities containing the absolute value function. Using algebra, you must review the concept of absolute value and you must first learn two new properties about inequalities and the absolute value function.

Example 1: Solve the equation $2|x-3|=7$ analytically (algebraically).

Solution: When you solve an equation analytically, you normally try to isolate x on one of the sides of the equation. None of the properties of equality studied in Chapter One or in beginning algebra will help. The concept of absolute value can be used to develop an argument for how to solve the equation.

Divide both sides by 2 $\qquad |x-3|=\frac{7}{2}.$

Under what conditions can $|x-3|$ equal $\frac{7}{2}$? You should know of two ways, if $x-3$ is $\frac{7}{2}$ or if $x-3$ is $-\frac{7}{2}$! Rewrite this idea in algebraic form and solve each of the equations.

$$x-3=\frac{7}{2} \quad \text{and} \quad x-3=-\frac{7}{2}$$

$$x=3+\frac{7}{2} \quad \text{and} \quad x=3-\frac{7}{2}$$

$$x=\frac{13}{2} \quad \text{and} \quad x=-\frac{1}{2}$$

The solution set is $\left\{\frac{13}{2},-\frac{1}{2}\right\}$.

⁂

Example 2: Solve the equation $3|x-5|-4=0$ analytically.

Solution: To solve the equation, isolate the absolute value on the left side of the equation.

$3|x-5|-4+\mathbf{4}=0+\mathbf{4}$ add 4 to both sides

$3|x-5|=4$ simplify

$\frac{3|x-5|}{3}=\frac{4}{3}$ divide both sides by 3

$|x-5|=\frac{4}{3}$ simplify

Argue as follows: there are two ways that the absolute value of $(x-5)$ can equal $\frac{4}{3}$, if $(x-5)$ is a positive $\frac{4}{3}$ or if $(x-5)$ is a negative $\frac{4}{3}$.

$$x-5=\frac{4}{3} \qquad\qquad x-5=-\frac{4}{3}$$

solve each equation:

$$x=\frac{4}{3}+5 \qquad\qquad x=-\frac{4}{3}+5$$

$$x=\frac{4}{3}+\frac{15}{3} \qquad\qquad x=-\frac{4}{3}+\frac{15}{3}$$

$$x=\frac{19}{3} \quad \text{and} \quad x=\frac{11}{3}$$

The solution set is $\left\{\frac{19}{3}, \frac{11}{3}\right\}$.

⁂

Example 3: Solve the equation $|2x-5|+8=14$.

Solution: To solve the equation graphically, subtract 14 from both sides to produce an equation with 0 on one side. The transformed equation is $|2x-5|-6=0$. Graph the related function $y=|2x-5|-6$ and find the zeros.

$y=|2x-5|-6$

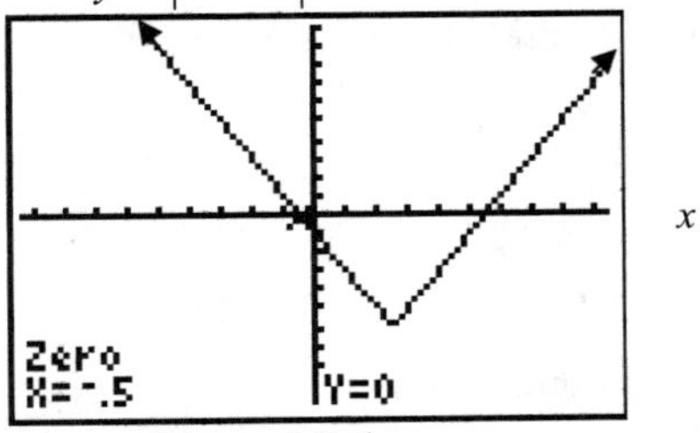

Figure 6.3.6

The zeros are –0.5 and 5.5; thus, the solution to the equation is the set {–0.5, 5.5}.

⁂

> If you choose to use the algebraic method for solving an equation containing an absolute value function, then use the following idea.
>
> If $|f(x)|=a$ where $a \geq 0$, then $f(x)=a$ and $f(x)=-a$.

The algebraic method for solving inequalities with absolute values requires the use of two properties that have not been studied in this text. Further, one of the properties uses a three-sided inequality. You can solve a three-sided inequality by using the properties from Chapter One; the same properties are listed again in this chapter. The three-sided inequality is shortened notation for the inequalities $a<x$ **and** $x<b$.

The x represents all real numbers that make $a < x$ **and** $x < b$ true statements. For example, consider the inequalities $2 < x$ **and** $x < 7$. The values for x that make *both* of the statements true are numbers between 2 and 7. For example, 3 makes both statements true; $2 < 3$ **and** $3 < 7$. The abbreviated notation is the three sided inequality $2 < x < 7$; it means $2 < x$ **and** $x < 7$. Since the numbers that make the statements true are between 2 and 7, the statement $2 < x < 7$ is commonly read as "x is between 2 and 7." To solve the inequality, the variable needs to be isolated in the middle.

Example 4: Solve the inequality $-4 < 2x - 1 < 7$.

Solution: The x must be isolated in the middle. From the original statement, add 1 to all three sides.

$-4 + 1 < 2x - 1 + 1 < 7 + 1$ now simplify all sides

$3 < 2x < 8$ finally, divide by 2

$\frac{-3}{2} < \frac{2x}{2} < \frac{8}{2}$ simplify

$\frac{-3}{2} < x < 4$

The solution is all real numbers between $\frac{-3}{2}$ and 4.

In interval notation, the solution set is $\left(\frac{-3}{2}, 4\right)$.

⁂

Example 5: Solve the inequality $-12 \le x + 7 \le -1$.

Solution: Subtract 7 from all three sides.

$-12 - 7 \le x + 7 - 7 \le -1 - 7$ then simplify

$-19 \le x \le -8$ the solution is all real numbers between and including -19 and -8.

The solution set is all numbers in the interval $[-19, -8]$.

⁂

To help you understand the less than and the greater than properties - the properties that allow you to use algebra to solve inequalities containing the absolute value function - study the graph of $y = |x|$ and the graph of $y = a$; where $a > 0$ very carefully.

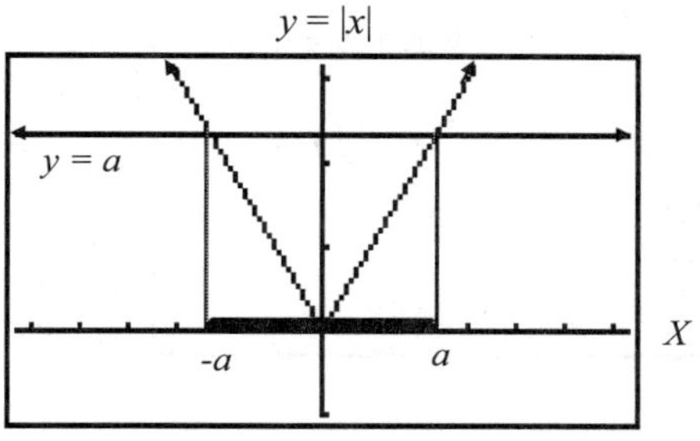

Figure 6.3.7a

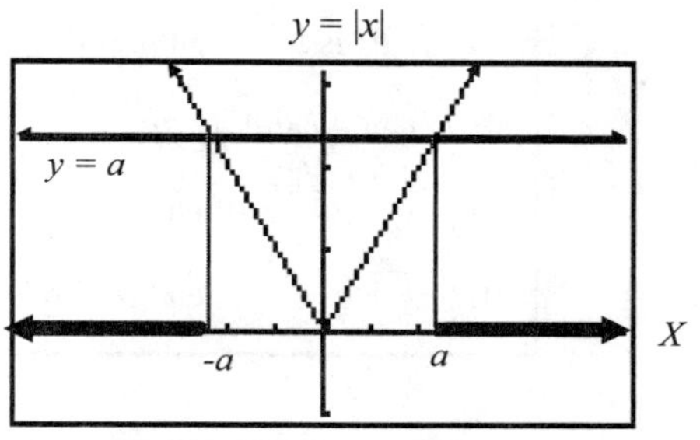

Figure 6.3.7b

What is significant about the above visualizations is that they show values of x that make the inequality $|x| < a$ a true statement are *between* a and $-a$. The values of x that make the inequality $|x| > a$ a true statement are *smaller* than $-a$ or larger than a. Another way of looking at the situation is that when the graph of $|x|$ is below (less than) the graph of a, x is on the interval $(-a, a)$ and when the graph of $|x|$ is above (greater than) the graph of a, x is on the interval $(-\infty, -a) \cup (a, \infty)$.

The Less Than Property

$|x|$ is less than the positive constant a only when x is *between* $-a$ and a.

If $|x| < a$, then x belongs to the interval $(-a, a)$.

If $|x| \le a$, then x belongs to the interval $[-a, a]$.

The Larger Than Property

$|x|$ is larger than the positive constant a when x is less than $-a$ or when x is larger than a.

If $|x| > a$ then x belongs to the interval $(-\infty, -a) \cup (a, \infty)$.

If $|x| \ge a$ then x belongs to the interval $(-\infty, -a] \cup [a, \infty)$.

This idea can be further generalized to replace the function x in the properties with any linear function $dx + e$.

The Less Than Property Again

$|\, dx + e \,|$ is less than the positive constant a only when $dx + e$ is *between* $-a$ and a.

If $|\, dx + e \,| < a$, then $-a < dx + e < a$.

If $|\, dx + e \,| \le a$, then $-a \le dx + e \le a$

The Larger Than Property Again

$|\, dx + e \,|$ is larger than the positive constant a when $dx + e$ is less than $-a$ or when $dx + e$ is larger than a.

If $|\, dx + e \,| > a$ then $dx + e < -a$ or $dx + e > a$.

If $|\, dx + e \,| \ge a$ then $dx + e \le -a$ or $dx + e \ge a$.

Each of these properties converts an inequality with absolute values to an *equivalent* inequality or inequalities without absolute values. When the inequality no longer contains an absolute value, it can be solved with the properties developed earlier.

Example 6: Solve $-2|x-5| > -6$ using algebra.

Solution: Before any property can be applied, divide both sides by -2 so that an absolute value is isolated on the left side. Division by -2 allows for the inequality being solved to match the pattern of the properties.

$\frac{-2\|x-5\|}{-2} > \frac{-6}{-2}$	simplify and change the inequality to <
$\|x-5\| < 3$	the > inequality has now become a < inequality; now use the less than property
$-3 < x-5 < 3$	solve by adding 5 to all three sides
$-3+\mathbf{5} < x-5+\mathbf{5} < 3+\mathbf{5}$	simplify
$2 < x < 8$	x is all real numbers between 2 and 8.

The solution in interval notation is $(2, 8)$.

⁂

Example 7: Solve $|2x - 3| \le 7.2$ analytically.

Solution: The inequality is in the proper form for the less than property. Use it.

$-7.2 \le 2x - 3 \le 7.2$	add 3 to all three sides
$-7.2 + 3 \le 2x - 3 + 3 \le 7.2 + 3$	simplify
$-4.2 \le 2x \le 10.2$	divide by 2
$\frac{-4.2}{2} \le \frac{2x}{2} \le \frac{10.2}{2}$	simplify
$-2.1 \le x \le 5.1$	x is all real numbers between and including -2.1 and 5.1 or $x \in [-2.1, 5.1]$.

⁂

Example 8: Solve the inequality $-3|x + 2.6| \le -9.6$ analytically.

Solution: To use a property, a positive number must be on the right side of the inequality.

$\frac{-3\lvert x+2.6\rvert}{-3} \le \frac{-9.6}{-3}$	divide both sides by -3
$\lvert x + 2.6\rvert \ge 3.2$	simplify ($\le$ becomes $\ge$) & use the greater than property
$x + 2.6 \le -3.2$ *or* $x + 2.6 \ge 3.2$	subtract 2.6 and simplify
$x \le -5.8$ *or* $x \ge 0.6$	rewrite with intervals
$(-\infty, -5.8] \cup [0.6, \infty)$	

When x belongs to the interval $(-\infty, -5.8] \cup [0.6, \infty)$, the inequality $-3|x + 2.6| \le -9.6$ is a true statement.

⁂

Example 9: Solve the inequality $|x - 4| \ge -6$.

Solution: The inequality does not meet the conditions of the greater than property; thus, it cannot be used. However, if 6 is added to both sides, the left side becomes an absolute value function. The inequality becomes $|x - 4| + 6 \ge 0$. Graph the related absolute value function $y = |x - 4| + 6$ to visualize when it is greater than or equal to 0.

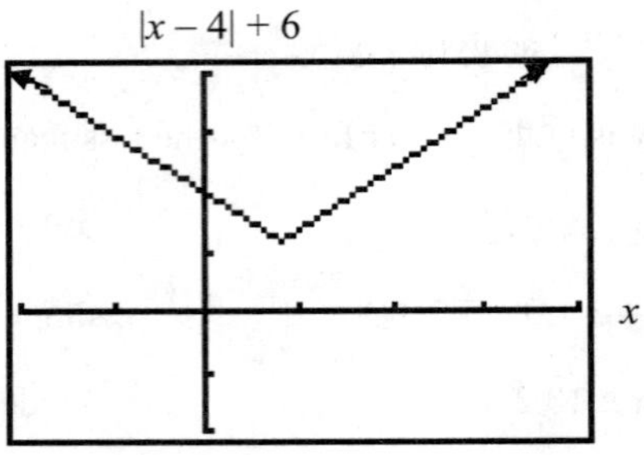

Figure 6.3.8

From the graph, it is clear that no matter what x is, the function is always greater than 0. The solution is all real numbers or $(-\infty, \infty)$.

⁂

Example 10: Solve the inequality $|2x + 5| - 3 > 0$ analytically.

Solution: To make the inequality match the condition of the greater than property, add 3 to both sides. The inequality becomes:

$\lvert 2x + 5\rvert > 3$	apply the property to yield inequalities without absolute values
$2x + 5 < -3$ *or* $2x + 5 > 3$	solve each inequality
$2x < -8$ *or* $2x > -2$	
$x < -4$ *or* $x > -1$	the solution is all numbers in the interval $(-\infty, -4) \cup (-1, \infty)$.

⁂

Example 11: Solve the equation $|x-5| - |x+3| = 4$ using the numerical method.

Solution: Make a complete table for the function $|x-5| - |x+3| - 4$ and find the zeros.

Table 6.3.2

x	–5	–4	–3	–2	**–1**	0	1	2	3	4	5	6
$\lvert x-5\rvert - \lvert x+3\rvert - 4$	4	4	4	2	**0**	–2	–4	–6	–8	–10	–12	–12

The table is complete because it demonstrates the established trends on the left and right, and all interesting behavior is displayed. The zero is –1; thus, the solution to the equation $|x-5| - |x+3| = 4$ is –1.

⁂

Example 12: Solve the equation $-4|x+3|-9=0$.

Solution: To solve the equation algebraically (analytically), isolate the absolute value on the left side of the equation.

$$-4|x+3| = 9 \qquad \text{add 9 to both sides}$$

$$|x+3| = -\frac{9}{4} \qquad \text{divide both sides by } -4$$

The only time the equation can be a true statement is if the absolute value of $(x+3)$ has a value of $-\frac{9}{4}$; however, the absolute value of any number can never equal $-\frac{9}{4}$ because the absolute value is zero or larger. There is no solution. The solution set is empty. The graph below verifies that $-4|x+3|-9$ NEVER equals zero!

$y=-4|x+3|-9$

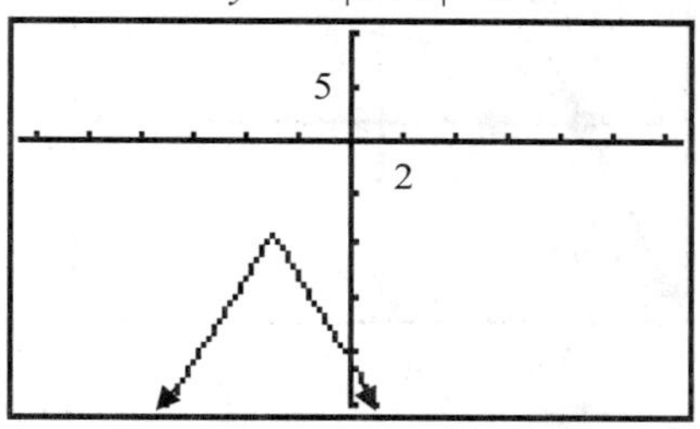

Figure 6.3.9

***

Example 13: Solve the inequality $|x-3|-5>0$.

Solution: Since there is a zero on the right side and an absolute value function on the left side, graph the absolute value function and use the zeros method for solving the inequality. Because the problem is to find when the function is greater than zero, use the graph to find what values of x cause $|x-3|-5$ to be bigger than 0. See Figure 6.3.10 below.

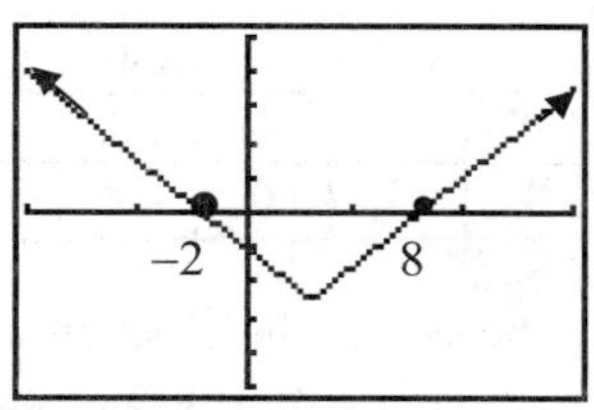

Figure 6.3.10

The zeros of the absolute value function are -2 and 8. You can see that the function is greater than 0 when x is less than -2 or when x is larger than 8. The solution is the interval $(-\infty, -2) \cup (8, \infty)$. Reminder: $\cup$ is the union of the sets. Numbers in the union belong to the first set *or* the second set *or* to both sets. The solution set contains numbers in $(-\infty, -2)$ *or* $(8, \infty)$. That is, $|x - 3| - 5$ is greater than zero when x belongs to the set $(-\infty, -2) \cup (8, \infty)$.

⁂

Example 14: Solve the inequality $-2|x + 5| + 3 > 0$.

Solution: Because an absolute value function is on the left and 0 is on the right, graph the related function $y = -2|x + 5| + 3$ and use the zeros method for solving. From this information you can see that the function is greater than zero when x is between the zeros.

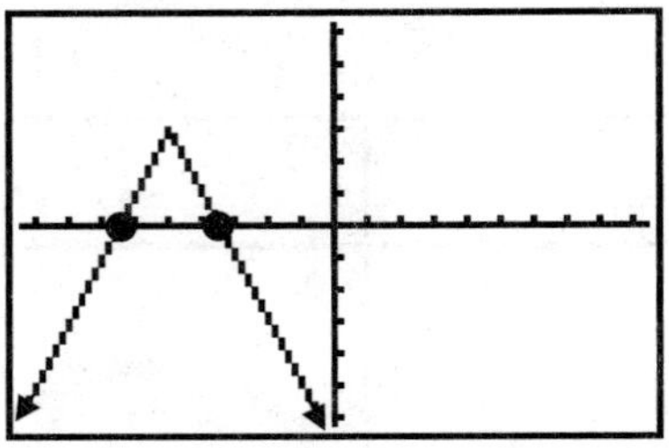

Figure 6.3.11

The function has zeros at –6.5 and at –3.5. The solution is the interval $(-6.5, -3.5)$. That is, $x \in (-6.5, -3.5)$.

⁂

Example 15: Solve the inequality $|x+2|+|x-2| \le 6$.

Solution: While the numerical method may not be as fast as the graphical method, it is still a good idea to consider it once in awhile. Subtract 6 from both sides to get 0 on the right. Now find the zeros and when the function $|x+2|+|x-2|-6$ is ≤ 0.

Table 6.3.3

x	–5	–4	**–3**	**–2**	**–1**	**0**	**1**	**2**	**3**	4	5
$\lvert x+2\rvert+\lvert x-2\rvert-6$	4	2	**0**	**–2**	**–2**	**–2**	**–2**	**–2**	**0**	2	4

The table shows that the function is either 0 or negative when x is on the interval $[-3, 3]$. The solution to the inequality is $x \in [-3, 3]$. You may want to graph the function to confirm this is correct.

⁂

Example 16: Solve the inequality $-2.1|x - 4.3| \geq 6.8$. As usual, use an error of less than 0.01 if appropriate.

Solution: The number zero needs to be on the right side of the inequality. Subtract 6.8 from both sides. The new inequality is $-2.1|x - 4.3| - 6.8 \geq 0$. Graph the related absolute value function $f(x) = -2.1|x - 4.3| - 6.8$ and find the zeros. Next, find when the related function is positive.

$-2.1|x - 4.3| - 6.8$

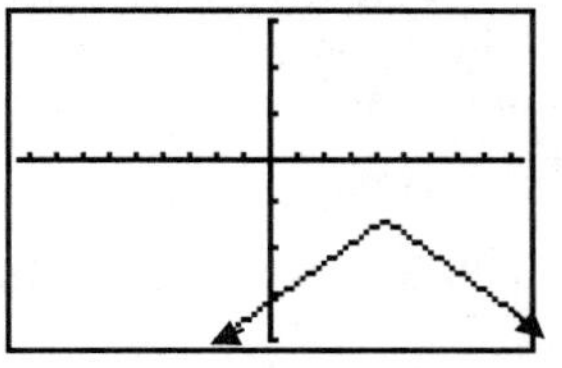

Figure 6.3.12

It can be seen from the graph that the related function is *never* larger than zero. It is always less than zero. That is, the inequality has no values that make it a true statement. The solution set is empty. $x \in \varnothing$

⁂

Example 17: Solve the inequality $-2.1|x - 4.3| \leq 6.8$. As usual, use an error of less than 0.01 if appropriate.

Solution: A zero should be on the right side of the inequality. Subtract 6.8 from both sides. The new inequality is $-2.1|x - 4.3| - 6.8 \leq 0$. Graph the related absolute value function $y = -2.1|x - 4.3| - 6.8$ and find the zeros. Next, find values for x that make the related function negative. The graph of the related function is displayed in Figure 6.3.12 above. As can be seen from the graph, the function is ALWAYS negative. Every real number makes the statement a true statement. The solution set is $(-\infty, \infty)$.

⁂

6.3 EXERCISES

Many times you can work the exercises by algebraic, numeric, or graphical methods. You can learn about mathematics no matter which method you use. Part of the learning process is for you to decide which method is best for you to use. Please read the text carefully before you do the exercises. Read the **next assigned section** after you finish the assignment below. All numeric answers must have an error ≤ 0.01.

A Quick Look To The Future

-4. Describe the phrase "is proportional to."

-3. Is 2^x ever negative?

-2. Describe as best you can the number $3^{\frac{1}{2}}$.

-1. If you cut a sheet of paper in half and then lay one sheet on the other and then keep on doing this, will the paper ever equal a thickness of two inches?

Previous Topics Revisited

1. Solve $3(x-1) \le 5(2x+4)$
2. Solve $-5.6(x+4.2) > 1$
3. Solve $0 \ge 3.5x - 2.88$
4. Solve $3(x-1) = 5(2x+4)$
5. Solve $0 = 3.5x - 2.88$

Over 20,000 high school students were asked to identify a human population growth function in graphical form. Below is information about the response to the question. The twelfth grade students were non-science students.

GRADE	PERCENT WITH CORRECT RESPONSE
9	16
10	18
12	24

Source: Population Education Interchange. February 1988. vol. 17 #1. Population Reference Bureau, Inc., Washington, DC.

6. Can this data be modeled by a linear function? If yes, what is the linear function; if no, why not?

New Topic Exercises
Solve the inequalities or equations.

7. A commercial airliner leaves Columbus, Ohio and heads straight for Chicago, Illinois. The flight takes 60 minutes; the first 30 minutes it ascends at a rate of 900 feet per minute reaching an altitude of 27,000 feet and the second 30 minutes it descends at a rate of 900 feet per minute reaching Chicago after 60 minutes. When is the plane above 20,000 feet? When is the plane below 10,000 feet?

8. In reference to the "Medication Problem", the doctor may want the patient to be on the medication longer than 20 weeks and the level of the drug more than 125 nanograms per *ml* of blood. The model for the level of Imipramine in a patient with a four week phase-in and phase-out period, a level of 180 nanograms per *ml*, and for a duration of 52 weeks is $I(t) = -22.5(|t-4| + |t-48|) + 1170$. The model with a domain restricted to [0, 52] is $I(t) = -22.5(|t-4| + |t-48|) + 1170 + 0\sqrt{-t(t-52)}$. When is the level of the drug above 100 nanograms? When is it below 180 nanograms? When is the level 180 nanograms?

9. Like the fast moving Alberta Clipper weather front, a slower moving clipper may lower the temperature at a lower rate. The model of the temperature during a slow moving front with an arrival temperature of 80° is $T = \frac{2}{3}|t-24| + 64$. Since the effects of this front last for 48 hours, the model with a restricted domain of [0, 48] is $T = \frac{2}{3}|t-24| + 64 + 0\sqrt{-t(t-48)}$. How long is the temperature (T) below 80°? How many hours after the arrival of the clipper does the temperature reach a minimum of 64°? When is the temperature above 70°?

10. $|x - 3| - 1 < 0$, $2|x - 3| - 1 < 0$, $-2|x - 3| - 1 > 0$

11. $|x + 4| \geq 5$, $|x + 3| \geq 5$, $|x + 2| \geq 5$, $|x + 1| \geq 5$

12. $|x + 3| > -4$, $|x + 3| < -4$, $-|x + 3| > -4$, $-|x + 3| < -4$

13. $-1.5|2.2x + 4.3| \geq -1.9$
14. $0.003|x - 0.03| \leq 0.5$
15. $-5 \leq 2x + 7 \leq 9$
16. $3 \leq 5x - 3 \leq 20$
17. $-5|x - 3| + 4 = 1$
18. $0.23|x - 1.5| = 1.6$
19. $2|x + 4| = -12$
20. $24|x + 17| - 26 = 47$
21. $|x-5| + |x-2| = 0$
22. $3|x-1| = 2|x+4|$
23. $-4|x+5| + 2|x+1| = -3|x-4|$
24. $|x+3| > |x-5|$
25. $3|x-1| \geq x+2$
26. $|x-4| > |x-3| + \sqrt{x-1}$
27. $-x^2 + 6|x+2| \leq |x-5|$
28. $|x-4| < |x+2| - 5$
29. $-x^2 + 4 \leq -5|x-1| + 3$

30. In the inequality $|x + 3| + ? < 0$, what value can the "?" be replaced with so the inequality has no solution? (Hint: There are many answers.)

31. In the inequality $-2|x + 4| + ? > 0$, describe all possible numbers that "?" can be replaced with so there is no solution.

32. In the inequality $-2|x + 4|$? 0, what inequality symbol can "?" be replaced with so the inequality has no solution?

33. Create an equation containing an absolute function that has a solution of -3.

34. Find an equation containing an absolute function that has a solution of -4 and -2.

35. Develop an inequality containing an absolute function that has a solution of $[-4, -2]$.

36. Make an inequality containing an absolute function that has a solution of $(-\infty, -4] \cup [-2, \infty)$.

37. Write any inequality containing an absolute function that has no solution.

38. Create any inequality containing an absolute function that has $(-\infty, \infty)$ as a solution.

From Mathematics to English

39. In paragraph format, explain the less than property.

40. List each step needed to solve the inequality $-3|x + 4| - 2 > -5$.

41. Which method for solving inequalities (technological or algebraic) is the best for you? Why?

42. Explain how it is possible for an inequality containing the absolute value function to have a single number as a solution.

43. Explain the greater than property.

44. Describe the solution to an inequality containing the absolute value function, from a graphical perspective.

45. Describe the solution to an inequality containing the absolute value function, from a numerical perspective.

46. After reading this section, make a list of questions that you want to ask your instructor.

47. Continue in your daily journal and make an entry. In addition to your normal entry on thoughts about the mathematics in this section, list at least two positive comments about what you have learned about this topic.

48. In paragraph format, summarize the material in this section of the text in your daily journal.

49. Describe how your classroom instructor made this topic more understandable and clear.

50. After reading the text and listening to your instructor, what do you not understand about this topic?

From English to Mathematics

In the next three Exercises, convert the English statement to an equivalent mathematical statement.

51. the absolute value of any number minus three is at least four

52. three is smaller than the absolute value of any number plus five

53. the absolute value of a number x is positive

Explorations

54. Given an inequality in the form: $d|x + e| + f > 0$, discuss how the values of d, e, and f determine the solution set.

55. In this section, you learned about the behavior of functions containing absolute values of the form $d|x + e| + f$. Explore the behavior of more complicated absolute value functions by solving the following equations and inequalities. (A graphical or numerical approach is suggested!)

a. $|x - 1| + |x + 3| < 16 - 4|x - 8|$

b. $|x + 4| + |x - 5| > |x - 3| + 2|x + 7| - 28$

c. $-|x + 3| \cdot |x - 2| = 0$

d. $|2x + 9| - |x - 7| \geq -|3x - 2| + 9$

e. $|3x - 8| + 2|4x + 1| \leq |x + 6| \cdot |3x - 1|$

f. $|2x^2 - 6x - 3| = 0$

g. $|x^3 - x| < -1$

6.4 Formulas and Direct Variation

Formulas

One form of an equation is a **formula**. A formula can contain several variables (sometimes called parameters), each symbolizing a real-world entity. The purpose of this section is to use formulas as you may have done before. For example, to find the area of a circle, you must use the formula $A = \pi r^2$ and replace the parameter r with the radius to find the area. In this section you will also use the properties of equality to transform a given formula to one that can be used on a calculator or computer. For example, if you want to find the area of several circles with radii of 3, 4, 6, and 8 centimeters, it can be done very easily on the calculator as shown in Figure 6.4.1. The reason the calculator can calculate the area is that

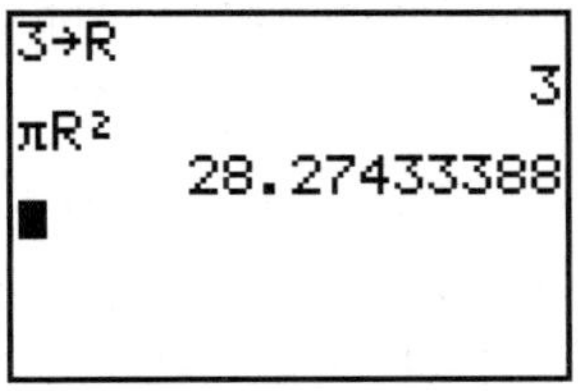

Figure 6.4.1a

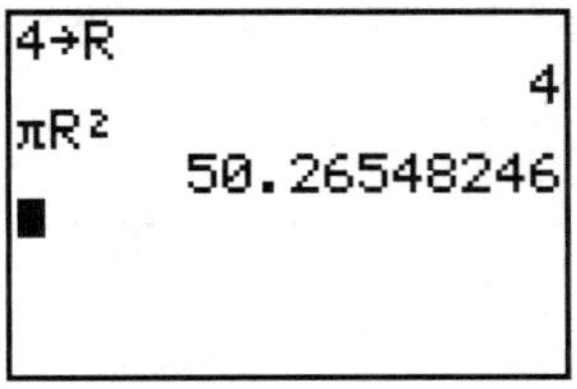

Figure 6.4.1b

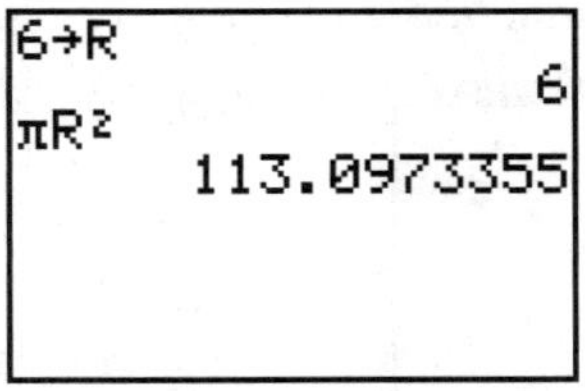

Figure 6.4.1c

the formula πr^2 is the area A. That is, the formula already has the parameter you are trying to calculate isolated on one side of the equation. Most calculators or computer software (there are exceptions) cannot be used to calculate the radius when you enter the area and the formula $A = \pi r^2$ as demonstrated in Figure 6.4.2. The first line tells the calculator what the area A is and the third line tells the calculator the formula for the area which includes the radius. It does not know what to do. It responds with a 0, not the radius. On the other hand, if you isolate the variable r in the formula and enter the function $\sqrt{\frac{A}{\pi}}$, it will display the number 1.954410048 - the approximate value of the radius in a circle of area 12.

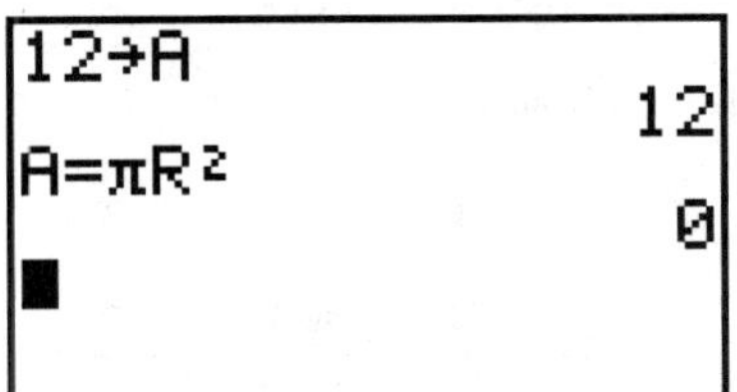

Figure 6.4.2

There is another way of answering the original question of finding the area of circles of radii 3, 4, 6, and 8. Since $A = \pi r^2$ is a function, enter πx^2 in address Y_1 and look at the numeric representation of the function as shown in Figure 6.4.3.

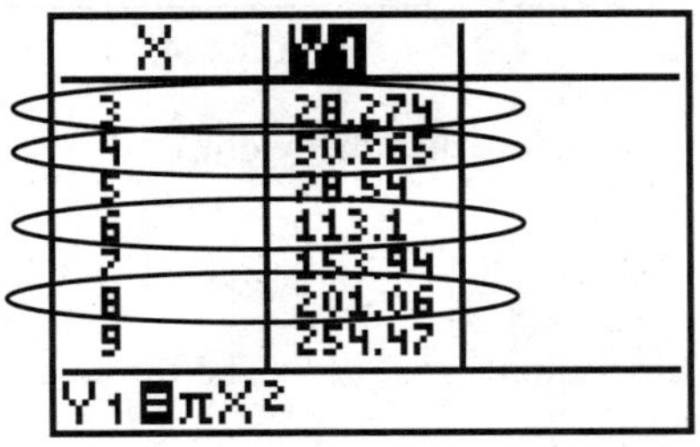

Figure 6.4.3

In any given formula, you will need to isolate one of the variables on one side of the equation using the properties from Chapter One. Frequently, the given formula cannot be used in the available form because it is in the wrong format. For example, you may need to have the calculator calculate a value for l in the formula $A = l \cdot w$. With most software used today, the computer must have l isolated on one side of the equal sign before it can use the formula; thus you must isolate the variable l algebraically. Some mathematicians and scientists call this process of isolating a variable on one side of the equality **solving the formula** for the variable.

Example 1: From the equation (formula) $P = 2(l + w)$, isolate w. The formula is used to find the perimeter of a rectangle whose length is l and width is w. Find the width of a rectangle whose perimeter is 50 inches and the length is 10, 9, 8, 7, and 6.

Solution: The w can be isolated by using the properties of equality. From the original formula, multiply both sides by $\frac{1}{2}$.

$\frac{1}{2} \cdot P = \frac{1}{2} \cdot 2(l + w)$ simplify both sides

$\frac{1}{2}P = l + w$ subtract l from both sides

$\frac{1}{2}P - l = l + w - l$ simplify

$\frac{1}{2}P - l = w$ The w has now been isolated.

The equation can also be written as $w = \frac{1}{2}P - l$. The quickest way of finding the width of the rectangles with perimeter 50 and length of 10, 9, 8, 7, and 6 is to treat $w = \frac{1}{2}P - l$ as a function where P is 50, and the function variable is l. The symbolic representation of the width function is $w = \frac{1}{2}(50) - l$, or $w = 25 - l$.

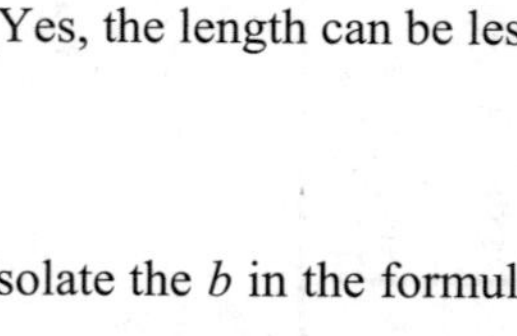

Figure 6.4.4a

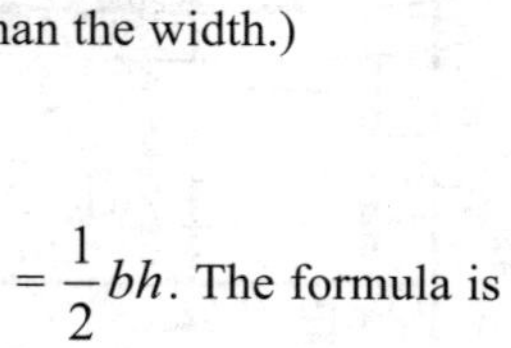

Figure 6.4.4b

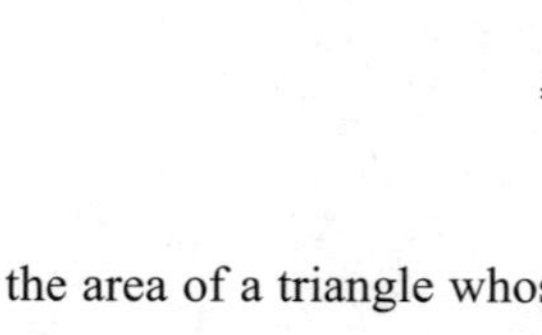

Figure 6.4.4c

(Yes, the length can be less than the width.)

⁂

Example 2: Isolate the b in the formula $A = \frac{1}{2}bh$. The formula is for the area of a triangle whose base is b and height is h.

Solution: Start the process by multiplying both sides by 2.

$2 \cdot A = 2 \cdot \frac{1}{2}bh$ simplify

$2A = bh$ divide by h

$\frac{2A}{h} = \frac{bh}{h}$ simplify

$\frac{2A}{h} = b$ the problem is finished

or

$$b = \frac{2A}{h}$$

⁂

Example 3: The formula used to describe the distance traveled by an object in free fall is $S = \frac{1}{2}gt^2$. S represents the distance traveled, g is the symbol for the constant acceleration due to the earth's gravity (32 ft/sec^2), and t is time in seconds. Isolate g.

Solution: From the original equation, multiply by 2.

$\mathbf{2} \cdot S = \mathbf{2} \cdot \frac{1}{2}gt^2$ simplify

$2S = gt^2$ divide both sides by t^2

$\frac{2S}{t^2} = \frac{gt^2}{t^2}$ simplify

$\frac{2S}{t^2} = g$ g is now isolated

A problem like this may be encountered in a beginning physics course. It can be used to experimentally find the local value of g.

⁂

Example 4: The formula for finding the maturity value of a simple interest investment is $M = P(1 + RT)$, where M is the maturity value, P is the principal invested, R is the annual interest rate, and T is time in years. Isolate R and then find values for R when P is \$10,000, M is \$12,000, and T is ½ year, ¾ year, and 1 year.

Solution: From $M = P(1 + RT)$, divide both sides by P.

$$\frac{M}{P} = \frac{P(1+RT)}{P}$$

simplify and subtract 1 from both sides $\frac{M}{P} - 1 = RT$

multiply both sides by $\frac{1}{T}$ $\qquad \frac{1}{T} \cdot \left(\frac{M}{P} - 1\right) = RT \cdot \frac{1}{T}$

simplify the right side $\qquad R = \frac{1}{T} \cdot \left(\frac{M}{P} - 1\right)$

Now replace the parameters P and M with \$10,000 and \$12,000 respectively and then calculate R when T is ½ year, ¾ year, and 1 year.

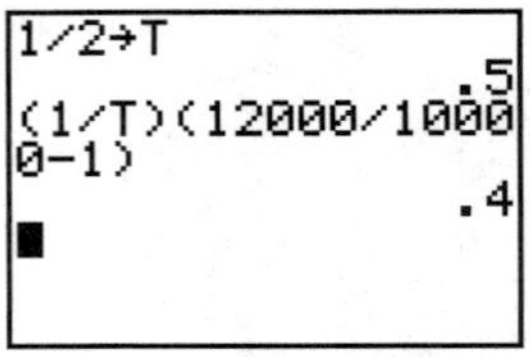

Figure 6.4.5a

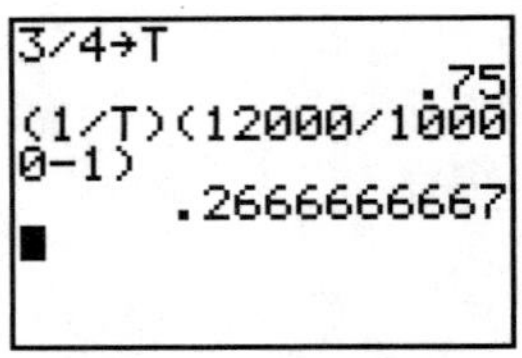

Figure 6.4.5b

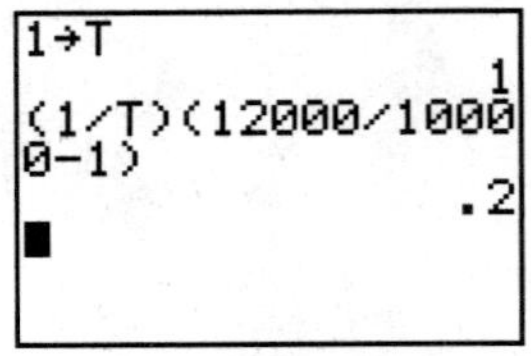

Figure 6.4.5c

The calculations show that the interest rate for half a year is 40% (.4), for three quarters of a year it is 26.67%, and for one year it is 20%.

⁂

Direct Variation

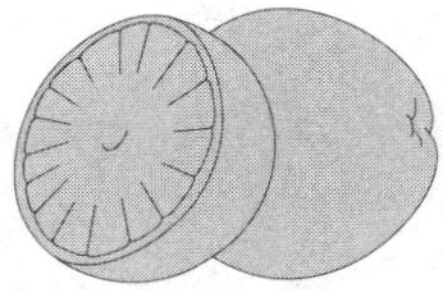

Fresh Kiwi

If you are working the cash register at the local grocery store and a customer buys 8 kiwi, what do you charge if the kiwi are sold "6 for a dollar"? One approach to solving this problem is to argue that the rate at which the management is charging is \$1 per 6. That is, the rate of change of the total charge function for the kiwi is $\frac{1}{6}$. This rate does not change (the rate is constant) regardless of the number of kiwi sold. This means that a linear function can be used to model the total charge in terms of the number of kiwi sold. The slope (rate of change) is given and if the customer buys no kiwi (the initial condition) the charge is \$0; that is, the linear function that describes the total charges is $C = \frac{1}{6}k + 0$. This comes from the slope-intercept form of the linear function ($y = mx + b$). The charge can now be calculated to be \$1.33.

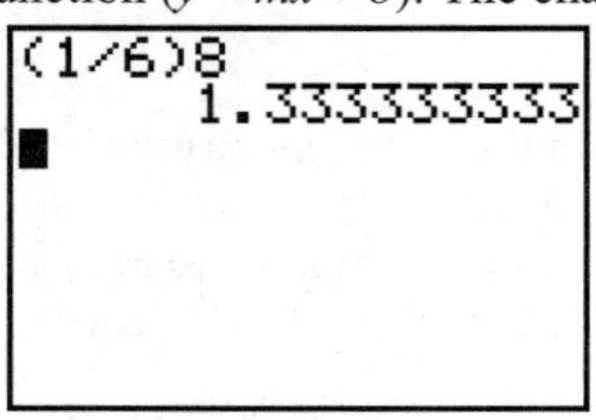

Figure 6.4.6

Another way of looking at this problem is that the stated rate of 6 kiwi for \$1 is nothing more than one point on the graph of the total charge function (k, C); it is the point (6, 1). The point (0, 0) is also on the graph because if you buy 0 kiwi, you pay \$0. Using the point-slope form of a linear function

($y = m(x - x_1) + y_1$) the model for the total charges of kiwi is $C = \frac{1}{6}(k - 6) + 1$. Probably not a good choice for the model since $C = \frac{1}{6}k + 0$, or $C = \frac{1}{6}k$ looks so much simpler. Of course, if you simplify $C = \frac{1}{6}(k - 6) + 1$, you get $C = \frac{1}{6}k$.

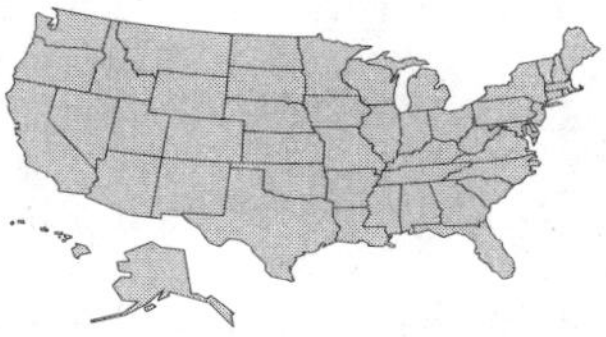

The Map

On a relatively small map of the country, the scale is marked 1 inch is 250 miles. If you want to travel from point *A* to point *B* on the map, how many miles must you drive if the measured distance on the map is 3.2 inches? The marked scale can be thought of as the fraction $\frac{250 \text{ real miles}}{1 \text{ scaled mile}}$; you may recognize this as the rate of change of a function that models the distance traveled on the road in terms of the distance on the map. Since moving 0 inches on the map means 0 miles on the road, the initial conditions indicate that the *y*-intercept in the slope-intercept form of the linear function is 0. The rate of change is the slope and the linear function that models the real-world distance traveled is $D_r = \frac{250}{1}D_m$ or just $D_r = 250D_m$, where D_m is the distance on the *m*ap and D_r is the distance traveled in the *r*eal world. To find the actual distance traveled when the map distance is 3.2 inches, use the function $D_r = 250D_m$ and replace D_m with 3.2. The distance traveled in the real world is 250(3.2), or 800 miles.

Please notice that in each of these examples, the function is of the form $y = mx$, where m is a constant. Functions of this form describe a relationship called a **direct variation**. The requirements of the direct variation relationship demand that initial conditions always be (0, 0); that is, when x is 0, y must be 0. The constant (m) is sometimes called the **constant of variation** instead of the slope or rate of change. The statement $y = mx$ is many times read as y varies directly as x; meaning that y equals a constant times x. Because the constant is used in several types of variation - not just linear - the common notation is k.

> The Direct Variation Relationship is: $y = kx$ where k is the constant of variation.

The direct variation relationship $y = kx$ comes from a linear function; thus, it is a **linear direct variation**.

Another way of expressing a linear direct variation is as a proportion. If the related pair of numbers (x_1, y_1) come from the linear variation $y = kx$, then $y_1 = kx_1$, or $k = \frac{y_1}{x_1}$. If the related pair of numbers

(x_2, y_2) come from the linear variation $y = kx$, then $y_2 = kx_2$, or $k = \frac{y_2}{x_2}$. Since k is constant for the entire domain of the direct variation, then $\frac{y_1}{x_1}$ must equal $\frac{y_2}{x_2}$, or $\frac{y_1}{x_1} = \frac{y_2}{x_2}$. This relationship is called a **proportion** and can be used to solve linear direct variation problems without finding the constant of variation.

Example 5: The recipe for 60 chocolate chip cookies calls for 2¼ cups of flour. How many cups of flour are needed for 100 cookies of the same size?

Solution: This is a linear relationship (direct variation). Use the proportion $\frac{y_1}{x_1} = \frac{y_2}{x_2}$ where x_1 and y_1 is the known related pair (60, 2.25) and x_2 is 100 cookies. The proportion becomes $\frac{2.25 \text{ cups}}{60 \text{ cookies}} = \frac{y \text{ cups}}{100 \text{ cookies}}$, solve for y by multiplying both sides by 100.

$$y = 100 \cdot \frac{2.25}{60} = 3.75 \text{ cups of flour.}$$

⁂

When y varies directly as x^2, the model of the function changes to $y = kx^2$, and is called a **quadratic direct variation**. When y varies directly as $\sqrt{x}$, the model of the function becomes $y = k\sqrt{x}$, and is called a **square root direct variation**.

Example 6: The velocity of an object (such as a ball) in free fall (ignoring wind resistance) varies directly as the square root of the distance traveled. If measurements indicate that after falling 4 feet the velocity is 16 feet per second, what is the velocity after 7 feet, 10 feet, and 50 feet?

Solution: Since the relationship is a direct variation, it must have the form $y = k\sqrt{x}$. Using variables that have appropriate meaning, it might be written as $v = k\sqrt{d}$. To find the constant of variation, use the given data of v is 16 when d is 4. The model becomes the equation $16 = k\sqrt{4}$. Solve it for k. The solution is 8. The model of velocity during free fall is $v = 8\sqrt{d}$. The calculator can now be used to find the velocity when the ball has fallen 7, 10, and 50 feet. A sample calculation is shown in Figure 6.4.7. The calculation shows that at 7 feet into the free fall, the ball has a velocity of 21.17 feet per second. The remaining calculations show that at 10 feet from the start of the fall, the ball has a velocity of 25.3 feet per second, and at 50 feet, 56.57 feet per second.

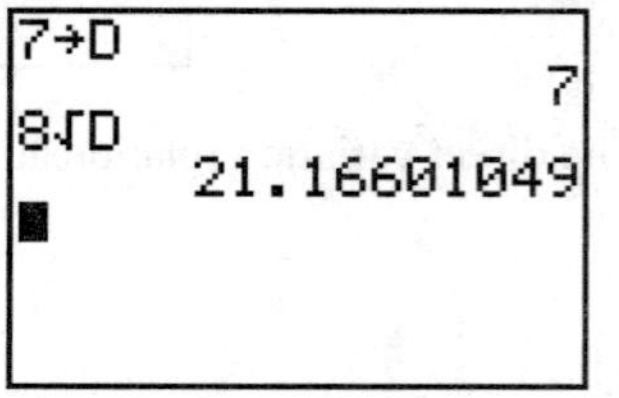

Figure 6.4.7

⁂

Example 7: The surface area of a sphere is directly related to the square of the radius of the sphere. If the surface area of the Earth is approximately 197,359,000 square miles and the radius is around 3963 miles, find the surface area of a spherical balloon when its radius is 2 inches, 4 inches, 6, inches, and 8 inches.

Solution: The direct variation model is $y = kx^2$; however, using appropriate symbols, it is $S = kr^2$. Using the known data, the model becomes the equation $197359000 = k3963^2$. To find *k*, solve the equation. $197359000 = k(15705369)$ The solution is $12.5663 = k$. The model of surface area of a sphere is $S = 12.5663r^2$. Since this is a quadratic function (See Section 3.2.), the numeric representation will show the surface area of the balloon at 2, 4, 6, and 8 inches of radius.

The surface area at a radius of 2 inches is 50.265 square inches. At a 4 inch radius the surface is 201.06 square inches. etc. The difference in the units of miles in the original data and of inches in the problem is not important because when the equation is solved for *k*, the units cancel just like terms cancel.

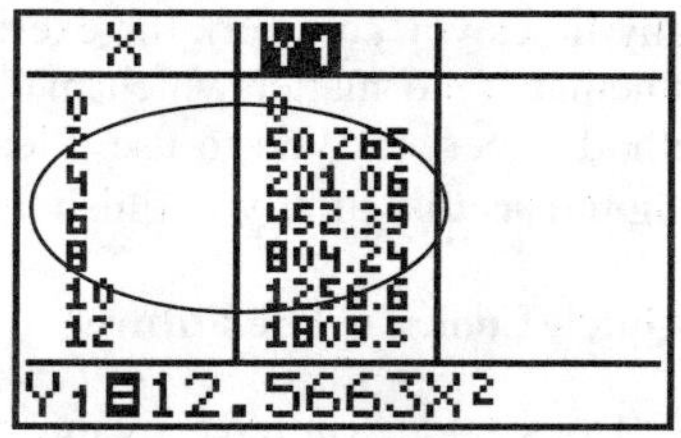

Figure 6.4.8

***

Example 8: If David can mow 5000 square feet of lawn in 35 minutes, how many square feet can he mow in 4 hours? (4 hours = 240 minutes)

Solution: This is a linear direct variation problem. It can be solved using the proportion method.

From the proportion $\frac{y_1}{x_1} = \frac{y_2}{x_2}$, let (x_1, y_1) be the point (35, 5000) and the point (x_2, y_2) is then $(240, y_2)$. The proportion becomes $\frac{5000}{35} = \frac{y_2}{240}$, solve for y_2 by multiplying both sides of the equation be 240. The solution is 34,285.71 square feet.

Alternate Solution: The problem can also be solved by the linear direct variation formula. From $y = kx$, find the constant of variation. It may make more sense to use variables like $A = kt$ so that you can keep track of time and area.

$5000 = 35k$, divide both sides by 35

$k = 142.857$

The model now becomes $A = 142.857t$. Replace *t* with 240 and find *A*.

$A = 142.857(240) = 34{,}285.68$. (Slightly different due to a rounding error.)

***

Example 9: If simple interest varies directly with the amount of money invested, and an investment of \$10,000 earns \$1,200, how much will a \$50,000 investment earn in the same time?

Solution: Like the last example, this problem can be worked with either method. To use the direct variation formula, use reasonable variables like $I = kP$, where P is the principal invested and I is the interest earned. To find k, substitute the known values to get the equation $1200 = 10000k$. Divide both sides by 10000 to find k; $k = 0.12$. The formula now becomes $I = 0.12P$. Replace P with 50,000 and find I. $I = 0.12(50000) = \$6000$.

⁂

6.4 EXERCISES

Many times you can work the exercises by algebraic, numeric, or graphical methods. You can learn about mathematics no matter which method you use. Part of the learning process is for you to decide which method is best for you to use. Please read the text carefully before you do the exercises. Read the **next assigned section** after you finish the assignment below. All numeric answers must have an error ≤ 0.01.

A Quick Look To The Future

-4. Does $\left(\frac{1}{2}\right)^x$ ever have a value of zero?

-3. Why does $\frac{3^5}{3^5}$ equal 1?

-2. If 3^{x+1} has the same value as 3^2, what is x? (Pencil and paper or a calculator is not needed.)

-1. Does money invested at 8% compounded annually grow exponentially?

Previous Topics Revisited

Solve the review exercises using interval notation when describing solutions to the inequalities.

1. $2|x - 2.3| - 4.66 \ge -3.1$
2. $|3x + 4| \le 2$
3. $|x - 5.99| < 5.99$
4. $3x - 4.5 > -2.1(x - 7)$
5. $-2x + 4.2 \le 7(2x - 3.1)$
6. $-2x + 4.2 = 7(2x - 3.1)$

New Topic Exercises

7. The outside Fahrenheit temperature (T_F) can be determined by the number of times a cricket chirps (c) in one minute. The model is $T_F = \frac{7}{30}c + 40$. What is the outside temperature if a cricket chirps 80 times in one minute? How many times does a cricket chirp in one minute if the outside temperature is 70°? 80°? 75°? 65°?

8. The height of a commercial airplane on a 50-minute flight can be modeled by $h = -1120|t - 25| + 28000$, where h is in feet and t in minutes. How high is the plane 10 minutes into the flight? How far into the flight is the plane when it is at 10,000 feet? 15,000 feet? 20,000 feet, 26,000 feet?

9. The "formula" that models the conversion between the Fahrenheit and Celsius temperature is $F = \frac{9}{5}C + 32$. Find F when C is 100°. Find C when F is 32°, 60°, 72°, and 100°.

10. Human population in recent times can be **approximated** by the linear model $P = 100{,}000{,}000t - 194{,}000{,}000{,}000$, where t is the calendar year and P is human population.

    a. What will the human population be in the year 2010?
    (Hint: enter 194,000,000,000 on the calculator as 1.94 EE 11 and enter 100,000,000 as 1 EE 8. EE is the scientific notation key on the calculator.)

    b. When will the human population be **at least** 7,000,000,000? 8,000,000,000? 10,000,000,000?

11. The model for the number of insects resistant to pesticides is
$I = 9t - 17490$, where t is the calendar year and I is the number of insects resistant to pesticides.

    a. How many insects will be resistant in the calendar year 2100?

    b. When will there be over 600 insects resistant to pesticides? 800? 900? 1200?

12. $S = vt$ is the formula for distance (S) traveled at an average velocity v in time t.

    a. How many hours does it take to travel 700 miles at an average velocity of 65 mph?

    b. What average velocity would be required to reach the moon (240,000 miles) in 6 hours?

    c. What average velocity would be required to reach the sun (93,000,000 miles) in 2 days? (Express the velocity in miles per hour.)

For the following formulas, isolate the variable noted and then find the value of that variable under the conditions given.

13. $C = 2\pi r,\ r$. The circumference of a circle. Find r when C is 2, 4, 6, 12, 18, and 24.

14. $P = l + l + w + w,\ l$. The perimeter of a rectangle. Find l when P is 40 and w is 3, 5, 7, 9, and 15.

15. $A = \frac{1}{2}h(b + B),\ b$. The area of a trapezoid. Find b when h is 4, B is 9, and A is 30, 40, 50, 60, and 75.

16. $V = \frac{4}{3}\pi r^3,\ \pi$. The volume of a sphere. Find π when V is 100 and r is 2.879. Find π when V is 200 and r is 3.6278. Find π when V is 300 and r is 4.15283. Find π when V is 900 and r is 5.989418.

17. $A = \frac{1}{2}bh,\ h$. The area of a triangle. Find h when A is 100 and b is 12, 19, 30, 65, and 90.

18. $S = v_0t + \frac{1}{2}gt^2,\ v_0$. The distance traveled during projectile motion. Find v_0 when S is 100 feet, g is 32 feet per sec^2, and t is 1, 2, 3, 4, 5, 6, 7, 8, 9, and 10 seconds.

19. $C = \frac{5}{9}(F - 32)$, F. Conversion from Fahrenheit to Centigrade. Find F when C is $-20°$, $-10°$, $0°$, $30°$, $40°$, and $100°$.

Solve the following linear direct variation problems.

20. If it takes 45 minutes to stain one window shutter, how many shutters can be stained in 360 minutes?

21. If sales tax is \$0.11 on a \$2 purchase, what is the tax on a \$37 purchase?

22. If the scale on a map is 1.5 inches = 20 miles, how many miles is a car trip of 17 inches?

23. If a photographic copier can copy 15 pages in 10 seconds, how long will it take to copy 850 pages?

24. If you can drive 300 miles in 6 hours, how many miles can you drive in 2 hours? 5 hours? 8 hours?

25. If the grocery store sells 6 bars of soap for \$0.90, how much do 10 bars of soap cost?

26. If 1 gallon of gasoline costs \$1.39, how many gallons can you buy for \$15?

Solve the following variation problems.

27. The area of a circle varies directly as the square of the radius. A circle of radius 3 *cm* has an area of 28.2743 cm^2. What is the radius of a circle with area of 100 in^2? What is the radius if the area is 200 in^2? 500 in^2? 1000 ft^2? 1500 mi^2? What is the area if the radius is 2 feet? 6 meters? 5 miles?

28. The velocity at which a car is traveling when the brakes are applied strong enough to cause skid marks on the pavement varies directly as the square root of the length of the skid mark. A skid mark of 36 feet is caused by a car traveling at 33 miles per hour (mph). What is the length of the skid mark if the car's velocity is 50 mph? 60 mph? 75 mph? 90mph? 120 mph? How fast must a car be traveling for the skid mark to be 40 feet? 60 feet? 80 feet? 100 feet? 200 feet? 300 feet?

29. The distance traveled during free fall (caused by being dropped) varies directly with the square of the length of time falling. A stone dropped 20 feet has been falling for 1.118 seconds. How far will the stone fall in 2 seconds? 4 seconds? 8 seconds? 12 seconds? 30 seconds? How long will it take for a water balloon to fall 10 feet? 16 feet? 24 feet? 50 feet? 200 feet?

30. If a coiled spring with a weight attached is hanging from a support is pulled downward and released it will bounce up and down with a period (t) that varies directly as the square root of the size of the weight (w). For a particular spring, a weight of 10 pounds will cause the spring to bounce with a period of 2 seconds. For the same spring, what is the period if a 3 pound weight is set in motion? A 5 pound weight? A 12 pound weight? What weight will cause a period of 1 second? 3 seconds? 6 seconds?

From Mathematics to English

31. List any and all of the formulas you are familiar with. Do not use any reference books. Describe what each formula does.

32. In sentence format, describe how to isolate the variable C in the formula $F = \frac{9}{5}C + 32$.

33. Explain the formula $y = m(x - x_1) + y_1$.

34. What is the formula $y = mx + b$ used for?

35. Is $m = \frac{\Delta y}{\Delta x}$ a formula? Why?

36. Write out a problem that can be solved using the idea of variation or proportion. Solve the problem.

37. Describe why the relationship between light intensity and distance to the light source is not a direct variation if the intensity of light decreases as the distance from the light source increases.

38. What is a direct variation relationship?

39. After reading this section, make a list of questions that you want to ask your instructor.

40. Continue in your daily journal and make an entry. In addition to your normal entry on thoughts about the mathematics in this section, list at least two positive comments about what you have learned about this topic.

41. In paragraph format, summarize the material in this section of the text in your daily journal.

42. Describe how your classroom instructor made this topic more understandable and clear.

43. After reading the text and listening to your instructor, what do you not understand about this topic?

From English to Mathematics

In the next three exercises, convert the English statement to an equivalent mathematical statement.

44. distance equals rate times time

45. circumference is pi times the diameter

46. area is pi times the radius squared

Explorations

47. Fill in the table below, where: $c^2 = a^2 + b^2$

a	b^2	C
5		12
3		8
20		48
4		5

a^2	b	C
	1	3
	2	5
	6	20
	15	60

a	b	c^2
3	4	
5	12	
6	8	
22	9	

What is the purpose of the mathematical model $c^2 = a^2 + b^2$?

Does it model more than just what you listed in the previous questions? If so, what?

48. Use the graphing calculator to graph:

a. $3x - 5y = 6$

b. $x^2 - 6y + 2x = 0$

c. $-2y + 3\sqrt{x-4} = 3y + 2$

Draw the graph on paper and label each graph.

CHAPTER SIX TEST

All numeric answers must have an error ≤ 0.01.

Solve the equations in Questions 1 - 5.

1. $4.3x - 2.9x + 5.2 = 0$
2. $5(x - 2) - 3(2x - 4) = 7(x - 1) + 12$
3. $(\sqrt{2})x - 5 = 0$
4. $3(x - 1) = 3x - 5$
5. $0 = \pi x - 9$

Solve the inequalities/equations in Questions 6 - 14, use interval notation to describe the answers to inequalities.

6. $-5x + 40 < 0$
7. $3.6x - 12.9 \geq 0$
8. $-2(x - 3) + 7 \leq 3(x + 2) - 5$
9. $-0.003(x + 4.001) > 0$
10. $-6 \leq 7x + 2 \leq 0$
11. $-3|x + 2| - 5 < 0$
12. $-3|x - 2| + 5 \geq 0$
13. $5.6|x - 7.1| = 4.5$
14. $3 > 2|x - 4| + 1$
15. If d varied directly with v, and v is 60 when d is 180, find d when v is 30.
16. Isolate x_1 in: $A = \dfrac{x_1 + x_2}{2}$
17. Isolate x in: $y = mx + b$

The cost of producing electric power from windmills is about 6.4 cents or \$0.064 per kilowatt-hour. If the meter fee to the power company is \$20 per month, y is your total monthly electric bill in dollars, and x is the monthly kilowatt-hours you used, answer the following questions.

18. What linear mathematical model can be used to calculate your monthly electric bill?
19. If you use 1200 kilowatt-hours during the month, what is the electric bill?
20. How much electricity will you get for a total monthly bill of \$100?

CHAPTER SEVEN
ADVANCED ANALYSIS OF THE EXPONENTIAL FUNCTION

7.0 Introduction

People Population

Below is the numeric representation of the relationship between time t and the number of people P on Earth. If you know the Earth will, at most, support life for 10 to 12 billion people, and it is your job to let national and world government officials know of the population growth, how should you proceed? If you must make a projection of the number of people on Earth in the next few years, or when the population will reach 12 billion, or what will happen if the rate of growth were lowered to 0.1%, you really need to have the symbolic representation of a function that models the number of humans on Earth. To help find this model, the numerical and graphical representations of the data should help.

Table 7.0.1

t (year)	1000	1750	1800	1850	1900	1950	1994	1999	2001	2006
P (Human Population)	0.2	0.8	1	1.2	1.7	2.8	5.4	6	6.2	6.6

(Population is in billions)

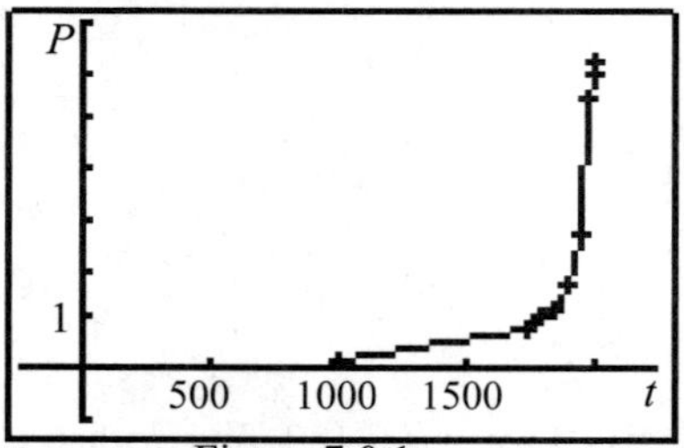

Figure 7.0.1

The graphical representation suggests a data relationship that you first encountered in Chapters Two and Three. The function that models this kind of data is the exponential function.

You have now finished your investigation of the linear function (in Chapter Five) and are ready to move on to investigate the exponential function. You will learn about the behaviors of the basic exponential function as well as formalize your understanding of the new topic of geometric transformations. This is a new concept that will be revisited in the next three chapters. Transformations should be of immense help when you develop mathematical models in lab projects. The study of the exponential function in Section 7.1 provides motivation for simplification of exponential expressions in Section 7.2 and solving equations in Section 7.3. Chapter Seven ends with applications of the exponential function. It includes the exponential function as a model population growth of E. Coli bacteria and humans. The growth of money invested in a compound interest account can also be modeled with an exponential function of the type analyzed in this chapter.

The Exponential Function in Action - Cancer Is a Problem

Some types of breast cancer cells can increase their numbers exponentially. Typical breast cancer cells may double in number every three months. The reason for the somewhat vague words like "some types", "may double", etc., is that there are several types of breast cancers and the doubling rates are not always exact. On average, you might conclude that a doubling time is three months. Another problem is that the growth in numbers of cells is sometimes restricted by the amount of nutrition they can get from your blood and limitations in the physical space in which they grow. When restrictions are put on the growth of the size of the cancer cell population, the doubling time will change. The approximate size of a single breast cancer cell is 9×10^{-9} cubic centimeters (cc), or 0.000000009 cc. To give you an idea of sizes measured in cubic centimeters, consider the table below.

object	approximate size in cubic centimeters	
average cell size	0.000000009	(9×10^{-9})
the eye of a needle	0.0007	(7×10^{-4})
a flea	0.0004	(4×10^{-4})
a pin head	0.008	(8×10^{-3})
the stem on your wristwatch	0.125	(1.25×10^{-1})
a small pea	0.5	(5×10^{-1})
a peanut m&m	1.5	(1.5×10^{0})
your nose	8	(8×10^{0})

Below is the graphical representation of the data generated under the condition of starting with a cell of size 9×10^{-9}cc and doubling in size every three months. The window is [-20, 120] by [-5, 25]. The scatter plot connects the data with a line.

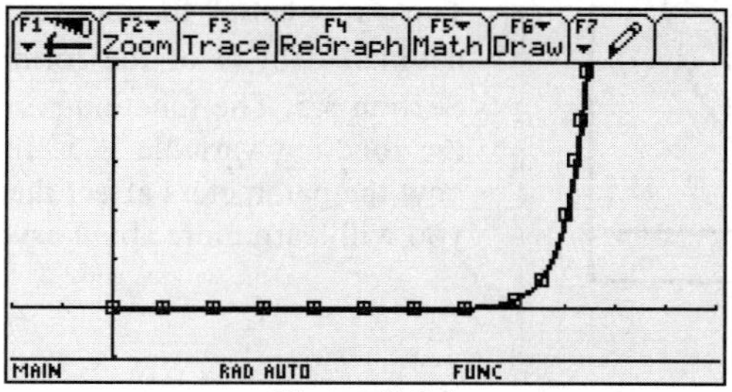

Figure 7.0.2

The graphical representation of the data for growth of a cancerous tumor very clearly shows why early detection of breast cancer is crucial. Please note that the size of the tumor is quite small until about the 70th month. The relationship looks like the functions studied in Section 3.5. Remember, the above model is approximate; the actual growth rate of a cancer will vary with time and conditions, so you should consider the model as representing *average* cell growth.

7.1 The Exponential Function

The National Debt

Table 7.1.1 shows the growth of the national debt for the last 50 years. Is it reasonable to model the data using a function already studied? Does the data look linear or have a constant rate of change? Does the data look like a V and have a constant rate of change for each branch of the V? Does the data look like a parabola or a square root function? If you find the graphical representation of the data, you can answer these questions. The graph is shown below in Figure 7.1.1. As you can tell from the graph of the data, the shape is not linear, quadratic, absolute value, nor square root. But it does look like another shape from Chapters Two and Three. If the graph of the data relationship is not something you recognize, it must be some other kind of function. Is there any function in symbolic form that can be written that looks like the data in graphical form? You may recall from Chapter Two data relationships called the exponential relationship. Figure 7.1.2 shows the graph of the data and the graphical representation of the exponential function $D = 1.101^{t-1910} + 225$. While it is unlikely that any function can model the data exactly, the function shows a remarkable likeness in behavior to the actual data.

Table 7.1.1

t (year)	1940	1950	1960	1970	1980	1990	2002	2006
D (National Debt)	51	257	291	381	909	3113	6000	8400

(Debt is in billions.)

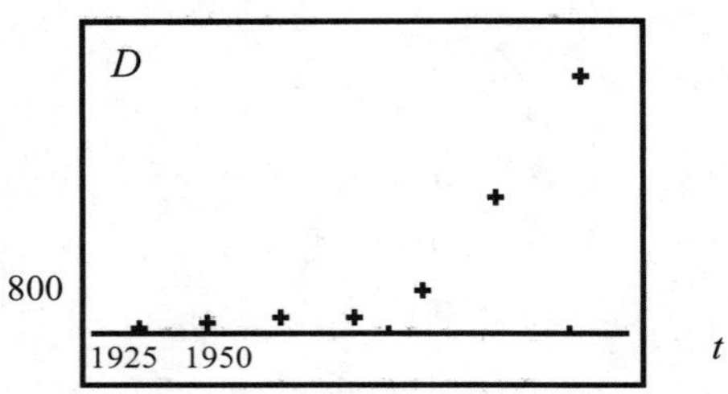

Figure 7.1.1

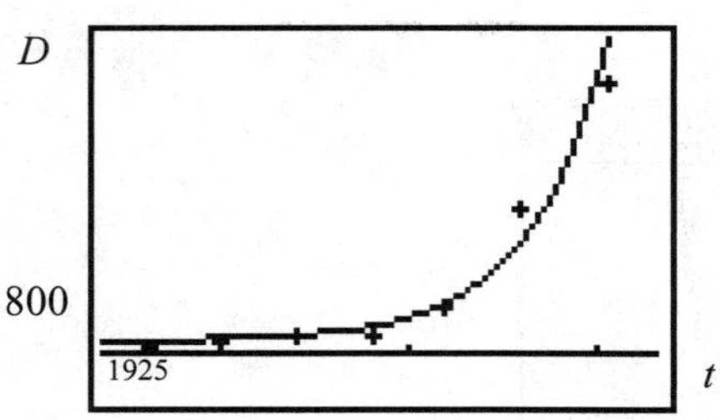

Figure 7.1.2

The exponential function that models the growth of the national debt is of the form $y = d \cdot b^{x+e} + f$ studied in Section 3.5. The function parameters are d, b, e, and f, and the function variable is x. In this section you will study how the parameters affect the behavior of the function and you will learn more about asymptotic behavior.

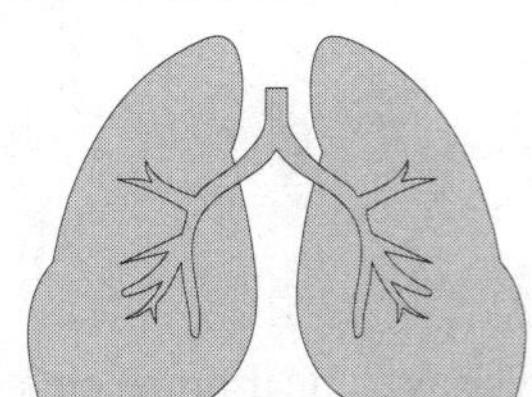

Mycobacterium Tuberculosis

Below is the data relationship between time and the population of tuberculosis bacteria growing under unrestricted conditions. **Unrestricted growth** conditions means that the bacteria have all the space and food needed, and no predators to kill them. This is something like what might happen in a petri dish in a medical laboratory. Time is in hours and the bacteria population is in thousands. Is the number of bacteria increasing or decreasing? Are they increasing in population at a constant rate of change, or not? Is there a maximum or

Table 7.1.2

t	0	6	12	18	24	30	36	42	48
B	5	10	20	40	80	160	320	640	1280

minimum number of bacteria? Can the number of bacteria ever be less than zero (negative). As you think about the answers to these questions also think about what type of function matches these behaviors. For example, can you think of a linear function that is increasing, yet never negative? Well, maybe, if you restrict the domain. So if you think the data relationship is linear, how can you find a linear function where the rate of change is not constant? You can't. Have you ever seen an absolute value or quadratic function where they only increase? Again, maybe, if you restrict the domain. Assuming this, can you create a quadratic that increases as fast at the data relationship? No, you can't. If not any of these functions, then what? Hopefully you have guessed that it is the exponential function.

Why do you want to model the data relationship? Well, if you are in a medical field, knowing how long it takes for an unknown bacteria to double may tell you the kind of bacteria you are analyzing. Suppose you know that it takes 100 billion of a bacteria to cause a health problem to a human, wouldn't you want to know how long it takes for the population to reach 100 billion? What if it is your job to slow the growth rate of a bacteria; what function parameter can you change to model the new relationship?

To see how the symbolic representation of the model can be created, consider the following argument.

Table 7.1.3

t	0	6	12	18	24	30	36	42	48
B	5	10	20	40	80	160	320	640	1280

In order to get 5 (000) bacteria at time zero (0), you could argue the model is the function is $B = 5$, but this function will generate the correct data only for the first data pair. You can see that at 6 hours the population has doubled. Doubling is like multiplying by 2; but be careful, if you multiply 5 times 2 you will get the correct number of bacteria, but time has changed by 6 hours -- not 1. You want to get 10 bacteria as time changes by 6 hours, so the model is <u>not</u> $B = 5\times 2t$.When t is 6, the exponent should be 1 and 2^1 will double the number of bacteria to 10. So, Δt is 6, based on the data, but the exponent must change by 1. The arithmetic operation that does this is division by 6 or multiplication by $\frac{1}{6}$. Thus, you have a proposed model of $B = 5 \cdot 2^{\frac{1}{6}t}$. Is this correct? To find out, check the value of the function at 12 hours, 18 hours, etc. Yes, it is correct.

If you make a slight modification of the parameters in the exponential function $y = d \cdot b^{x+e} + f$, you can model the population growth of the mycobacterium tuberculosis. The model is $B = 5 \cdot 2^{\frac{1}{6}t}$, where the function parameters e and f are 0, and a new parameter is shown as the coefficient of the function variable x or t. The graphical representation of the data relationship shown in Figure 7.1.3 below has the same shape and behaviors as the national debt function in Figure 7.1.2. The window is [–10, 50] by [-50, 500] and the function is entered on the calculator as 5 ∗ 2 ^ ((1 / 6) x).

$$B = 5 \cdot 2^{\frac{1}{6}t}$$

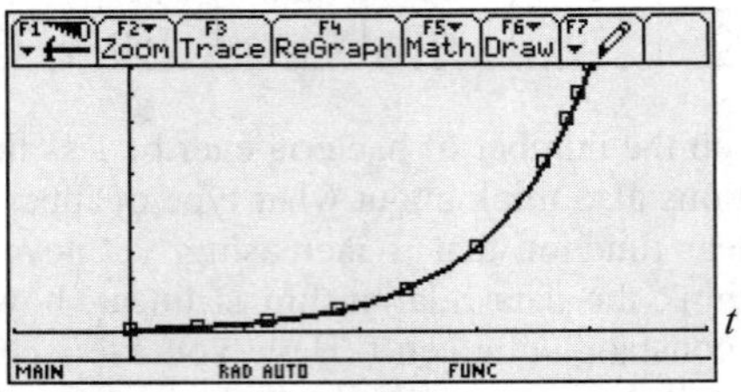

Figure 7.1.3

The reason for studying data relationships found outside the mathematics classroom is that scientists, engineers, business persons, teachers, etc. actually must be familiar with the functions that model these relationships. The models include the exponential function as well as the linear, quadratic, absolute value, and square root functions already introduced. A second, and probably more important reason for starting each section with problems is that you can learn mathematics by relating a situation you understand to the mathematics being presented in this text. For example, you know that the human population is increasing; therefore, you also know that the model of human population is increasing. You know that the human population has always been zero or larger; thus, you know the model is never negative (0 or larger). Keep these ideas in mind as you study the behaviors of the exponential function below.

The Standard Form of an Exponential Function is:

In $y =$ notation it is written: $y = d \cdot b^{x+e} + f$

or in functional notation it is written: $g(x) = d \cdot b^{x+e} + f$

b must be positive and not 1, and x represents a real number. The function parameters are b, d, e, and f, $d \neq 0$, $b > 0$, $b \neq 1$.

The basic definition of any function requires that both the function variable, as well as the function, be real numbers. The base b cannot be negative because under certain circumstances when a negative number is raised to a real exponent, the value of the function is not a real number; for example, $(-2)^{\frac{1}{2}}$ is not a real number. Even though you may not know what $(-2)^{\frac{1}{2}}$ means yet, if -2 is raised to the exponent $\frac{1}{2}$ with a calculator, an error message will be displayed. A more complete meaning of $(-2)^{\frac{1}{2}}$ is developed in Chapter Ten and a more thorough meaning of $2^{\frac{1}{2}}$ is explained in Chapter Nine.

Before the analysis of the first exponential function, it is necessary to look at what kind of numbers can be in the domain of the exponential function. Just as a reminder, 3^4 means $3 \cdot 3 \cdot 3 \cdot 3$, which, if simplified, becomes 81. In Chapter One, examples were given for a number raised to an integer exponent, not for a number raised to a rational or irrational exponent. Chapter Nine contains a definition of a number raised to a rational (fractional) exponent. The definition of raising a number to an irrational exponent is beyond the scope of this text. However, to analyze the exponential function, it is necessary to have an idea of what it means to raise a number to a *real* (rational or irrational) exponent.

To get an idea of what it means to raise a number to an exponent like $\frac{1}{2}$ or π, consider the following reasoning:

You know that 2^0 equals 1.

You know that 2^1 equals 2.

It would follow that $2^{\frac{1}{2}}$ must be a number between 1 and 2 because the exponent $\frac{1}{2}$ is between the exponents of 0 and 1.

$$2^0 = 1$$

$$2^{\frac{1}{2}} = ?$$

$$2^1 = 2$$

The calculator will find the value of $2^{\frac{1}{2}}$ by entering the keystrokes 2 ^ .5 ENTER. The value of $2^{\frac{1}{2}}$ is approximately 1.414. The same argument can be made for any other desired exponent. For example, look at $2^{\sqrt{3}}$.

$$2^1 = 2$$

$$2^{\sqrt{3}} = ?$$

$$2^2 = 4$$

Since $\sqrt{3}$ is between 1 and 2, you would expect $2^{\sqrt{3}}$ to be between 2 and 4. A calculator will show that $2^{\sqrt{3}}$ is approximately 3.322.

Analysis of $B = 5 \cdot 2^{\frac{1}{6}t}$ - The Tuberculosis Problem

As can be seen in the graph below of the tuberculosis bacteria function, there does not appear to be much in the way of any interesting behavior. One of the characteristics of the exponential function is the asymptotic behavior briefly introduced in Section 3.5. It is found in other types of functions studied later.

$B = 5 \cdot 2^{\frac{1}{6}t}$

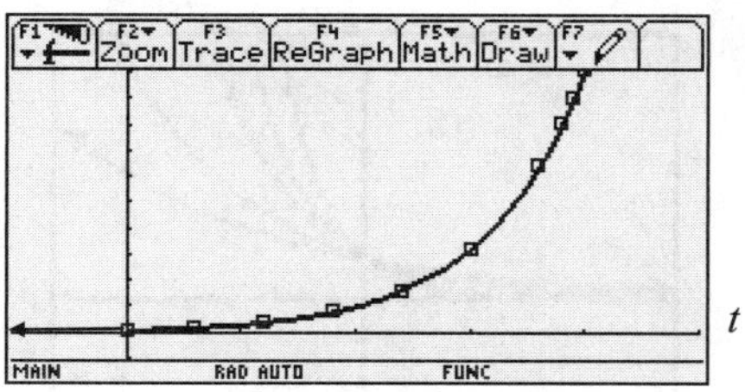

Figure 7.1.4

The window is [−10, 50] by [−50, 500] and the function is entered on the calculator as 5 ∗ 2 ^ ((1 / 6) x).

Increasing/decreasing A quick look at the numerical or graphical representation shows that the function is increasing. It may "look" constant for small x values, but if you trace on the graph (remember to trace from left to right), you will see the values of the function are increasing.

Maximum/minimum Since the function is increasing for all values of x, there can be no maximum. If the graph is read backward from *right to left*, the function values seem to get lower and lower. As discussed before, the function never gets to zero; therefore, there is no minimum. Remember if a reasonable problem domain is used rather than the normal domain, there will be a minimum at the beginning, and a maximum at the end of the domain.

Positive/negative Because the graph is always above the time-axis and never reaches zero, the function is always positive and never negative.

Zeros If the function is always positive, there can be no zeros.

Never zero. Never negative.

Rate of change The average rate of change is not a constant number. For example, as x changes from 6 to 7 hours, the average rate of change is 1.225. As x changes from 50 to 51, the average rate of change is 197.494. You must conclude that there is no constant rate of change.

Asymptote The new type of behavior mentioned above is that the function gets closer and closer to a horizontal line. It **approaches** the t-axis as t becomes smaller and smaller. You should confirm this statement by tracing to the left on the graph of the function. In mathematical terminology, the function approaches the t-axis as t approaches $-\infty$. The line the graph is approaching is called the **asymptote** and the function's new behavior is called **asymptotic behavior**.

Domain/range The symbolic representation shows no division by zero or square roots of a negative. You can conclude that the normal domain is all real numbers because there is no arithmetic operation that causes the function values to be non-real. Since the model approaches 0 for small t, and it goes to infinity for large t, the range is $(0, \infty)$.

Analysis of the Function Parameters

Do all exponential functions have behavior similar to the models above? If you graph several exponential functions of the form $y = d \cdot b^x$, where b is a number larger than 1 and d is positive, you may be convinced that they all behave the same. Behaving the same implies: the function is increasing, no maximum or minimum, a variable rate of change, no zero, always positive, and the same domain and range. Below are several graphs with $b > 1$. (Tic marks = 1.) The functions are entered like $Y_1 = 2 \wedge x$.

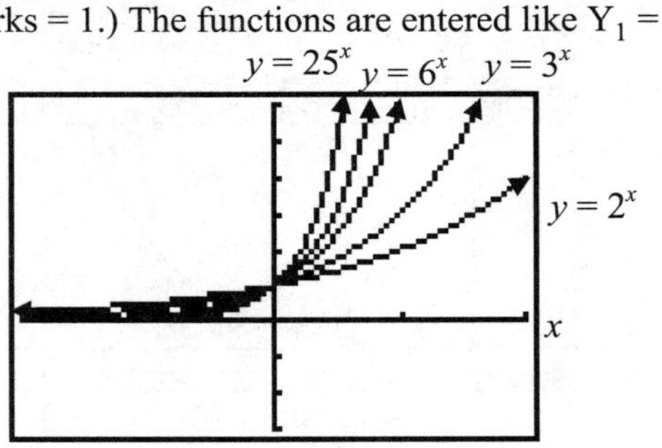

Figure 7.1.5

What if the base number b is between 0 and 1? What does the graph look like? Below are the graphs of $y = 0.9^x$, $y = 0.7^x$, $y = 0.5^x$, $y = 0.3^x$, and $y = 0.1^x$ on the same window as Figure 7.1.5

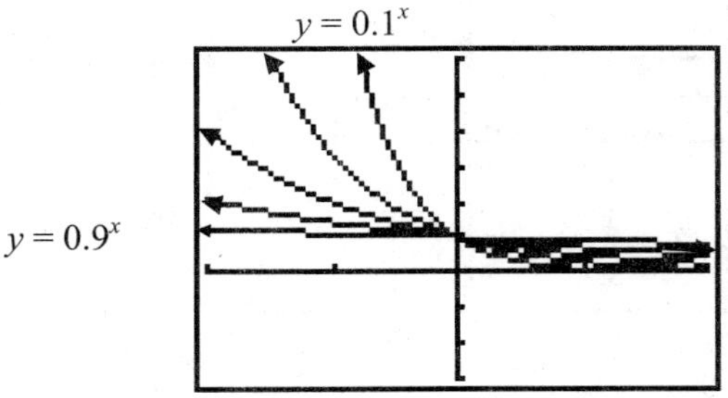

Figure 7.1.6

The general shape is the same, but the end behavior is reversed. For large x, the function approaches zero, and for small x, the function approaches ∞. The y-intercept has not changed. You may also note that the closer the base (b) is to 1, the flatter the graph. Currently, the function parameter b tells you if the function is increasing or decreasing. In the absence of parameters e and f, it also tells you the horizontal asymptote is the x-axis; thus, these functions have no zeros and the range is $(0, \infty)$. All this information from one parameter!

Of course, there are four function parameters for the exponential function. You have just looked at the function parameter b. You are strongly encouraged to experiment with many values for b so that you can make a connection between this parameter and the related behavior of the function. The remaining parameters are left for you to learn about when you do the exercise set at the end of the chapter. Each parameter controls a behavior. You must understand the relationship between a parameter and the resulting behavior.

Transformations

In this chapter, for the first time, you will investigate what happens to the graph of a function when you perform arithmetic operations on the function. The function you perform the operation on, called the **parent function**, can be any function; however, since this section in on the exponential function most examples will contain the exponential function as the parent. The arithmetic operations include adding a number to the parent function, multiplying a parent times a number, taking the opposite of the parent, and replacing the variable in the parent with a linear function like $(x + e)$, where e is a real number. The idea is to alter the symbolic representation of the parent function using these operations and then predict what happens to the graphical representation of the parent. These operations are called **geometric transformations** and they are very useful in creating a mathematical model that matches real-world data like you find in the lab projects or from Section 2.1. To find out how a transformation alters the graphical representation, use the graphing calculator to examine the results of changes in the parent function. Below are the graphs of $y = 3^x$, the parent function, and $y = 3^x + 5$ and $y = 3^x - 7$, the **transformed functions**.

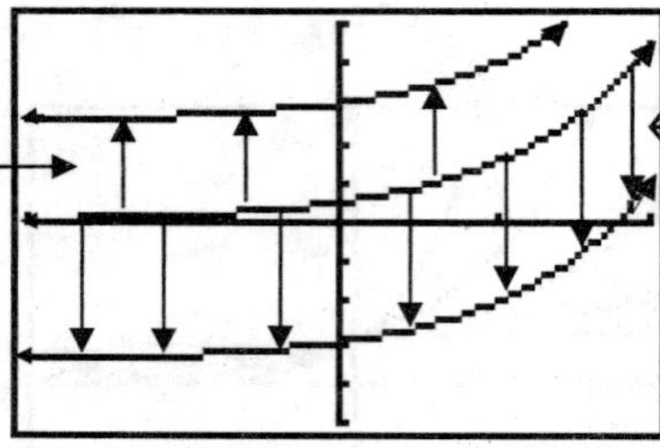

Figure 7.1.7

You can see that the number added to the parent function appears to move the entire graph of the parent function up when the number being added is positive and down when the number being added is negative. Another way of looking at this is that the vertical distance between the graphs of the parent and the transformed functions is always the same. Please verify this with your calculator. Graph the functions and put the trace cursor anywhere on the parent function, jump to the transformed function and note the changes in the value of the function (the value of y). If you jump from $y = 3^x$ to $y = 3^x + 5$, you will note a change in the function value (y) of 5 for each x value you try. Try this for many values of x until you have convinced yourself that adding a number to the symbolic representation of a function will cause the graph of the transformed function to "appear" to move up or down.

Now is an excellent time to remind you of the function notation developed for the first time in Chapter Four. If you identify the parent function as $f(x)$, then the transformed function can be written as $f(x) + 5$ or $f(x) - 7$. On calculators, this means you can call the parent function Y_1 and $f(x) + 5$ can then be entered in Y_2 as $Y_1(x) + 5$ (or just $Y_1 + 5$), and in Y_3 you can store the function $f(x) - 7$ as $Y_1(x) - 7$ (or just $Y_1 - 7$).

Another approach to verifying that the graphical representation moves up or down is to look at the numeric representation of the parent and transformed functions as shown in the table below.

Table 7.1.3

x	−3	−2	−1	0	1	2	3	4
$y = 3^x$	0.04	0.11	0.33	1	3	9	27	81
$y = 3^x + 5$	5.04	5.11	5.33	6	8	12	32	86

As shown above, adding 5 to $y = 3^x$ causes all transformed values to be exactly 5 units larger than the parent!

It is common to say that the graph of $y = 3^x + 5$ has been shifted (moved) up 5 units from the graph of $y = 3^x$. The graph of $y = 3^x - 7$ has been shifted down 7 units from the graph of $y = 3^x$. This movement is referred to as a **vertical shift** because the entire graph has been moved in a vertical direction. *Every* individual point on the original graph appears to have been shifted up or down to its new position. On your next laboratory project, take advantage of transformations to help you develop the symbolic representation of the function.

To find out what happens to the graph of the parent function when the variable x is replaced with the linear function $x + e$, consider the graphs below. To make the idea a little more clear, the parent function chosen is the absolute value function $y = |x|$. The transformed functions are $y = |x + 3|$ and $y = |x + 8|$. If you use function notation, the parent function becomes $f(x) = |x|$, and the transformed functions become $f(x+3)$ and $f(x+8)$. On calculators the notation is $Y_1 = |x|$, $Y_2 = Y_1(x+3)$, and $Y_3 = Y_1(x+8)$. Each tic mark is 2 on the graph in Figure 7.1.8 below. When you compare the graphs, you discover that replacing the function variable with $x + e$ does not shift the graph of the parent up! What is happening is that the transformed functions appear to have been shifted to the LEFT. This transformation is called a **horizontal shift**. Replacing the function variable with the linear function $x + e$ will shift any function to the left by an amount e if e is positive. Please confirm with your calculator that the horizontal shift is to the right if e in negative.

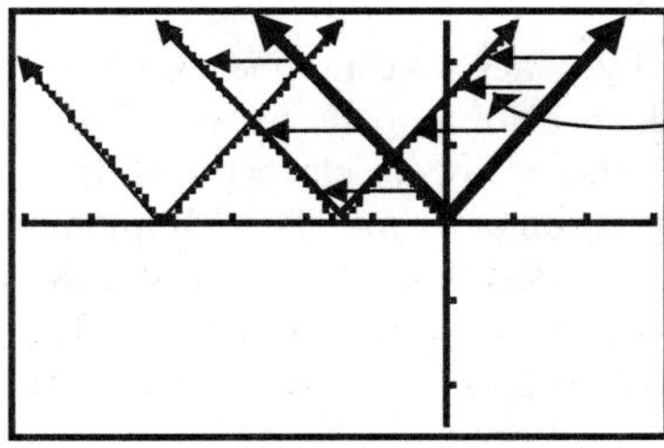

Figure 7.1.8

The graphs are identical in shape; however each of the transformed functions is in a new position to the left of the parent function.

In addition to vertical and horizontal shifts of any function, you should look at what happens when a function is multiplied by a number. Consider the functions $Y_1 = 10^x$ (the parent) and $Y_2 = 2\cdot 10^x$ (the transformed function), the y-coordinate of $2\cdot 10^x$ is 2 times the y-coordinate 10^x. Since the y-coordinate measures the height (distance above/below the x-axis) of a graph, $2\cdot 10^x$ is twice as high as 10^x. This notation is confirmed in Figure 7.1.9 below. When $x = 1$, the parent graph is at 10 and the graph of $2\cdot 10^x$ is at 20. When $x = 1.5$, the parent graph is at 31.6 and the graph of $2\cdot 10^x$ is at 63.2. Because each y-coordinate of the parent graph is multiplied by 2 to get the corresponding y-coordinate of the transformed graph, the new graph acts as if it has been stretched. Actually, this kind of transformation is called a **vertical stretch**. The window used in Figure 7.1.9 is [–4, 2] by [–6, 25].

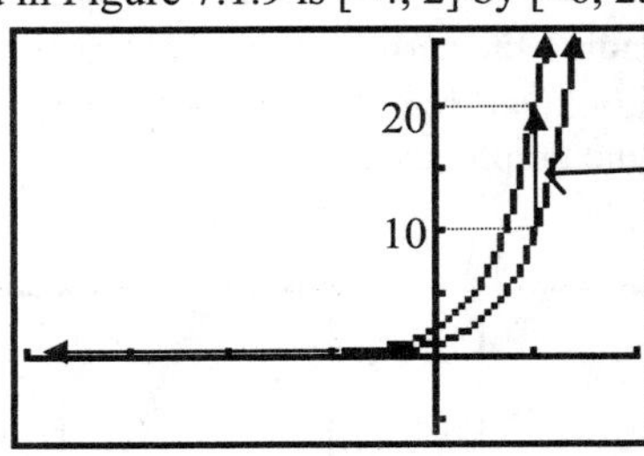

Figure 7.1.9

The graphs are obviously not 10 units apart when x is any other number. Try your trace key to find the <u>vertical distance</u> between the graphs when x is –0.5, 0, or 2. Or for example, if you use function notation $Y_1(2) = 100$ and $Y_2(2) = 200$. The vertical distance between the parent and transformed function is now 100 units -- not 10 units. However, the transformed function is still 2 times as far from the x-axis as is the parent function. A numeric argument will also show that the transformed function is twice as far from the x-axis as the parent function. For each value of x you see that the transformed function is twice the parent.

Table 7.1.4

x	–2	–1	0	1	2	3	4	5
10^x	0.01	0.1	1	10	100	1000	10000	100000
$2\cdot 10^x$	0.02	0.2	2	20	200	2000	20000	200000

As you can guess, if the number multiplied by the parent function is $\frac{1}{2}$, for every x in the domain, the value of the transformed function will be one-half the corresponding value of the parent function. This transformation is called a **shrink**, because the transformed function is smaller in value or closer to the x-axis.

If the parent function is multiplied by a negative number, each y-coordinate on the new function, in addition to being stretched or shrunk, is also opposite in sign. The ***visual effect*** is that the transformed graph has been flipped over and is farther away (or closer) from the x-axis than the parent function. It is possible to do more than one transformation at a time! The "flipping over" effect is caused by multiplying by –1. This transformation is called a **reflection about the x-axis**. Figure 7.1.10 below demonstrates behavior changes in the graph of $y = 2^x$ to the graph of $y = -2^x$. The parent function has been reflected about the x-axis. When the parent graph of 2^x is 2, the graph of the transformed function -2^x is at –2.

$y = 2^x$

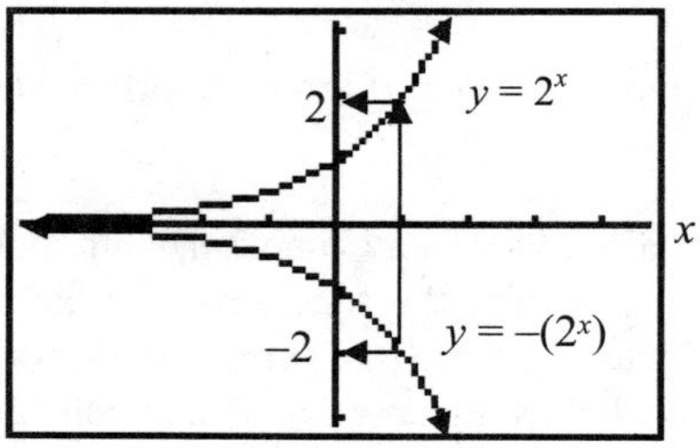

Figure 7.1.10

If the parent exponential function does not have a zero, the transformed function may have a zero. As was shown in Chapter Six, zeros of functions become important when solving equations. The graphical or numerical process of finding zeros is the same as it was in Chapter Three except you must have an error of ≤ 0.01. Below is an example of finding the zero of $y = -(2)^{x-3} + 5$. The analytical method for finding the zero cannot be done until you study logarithms in Chapter Fourteen. Using the root (or zero) finder on the calculator, the zero is marked on the graph below.

$y = -(2)^{x-3} + 5$

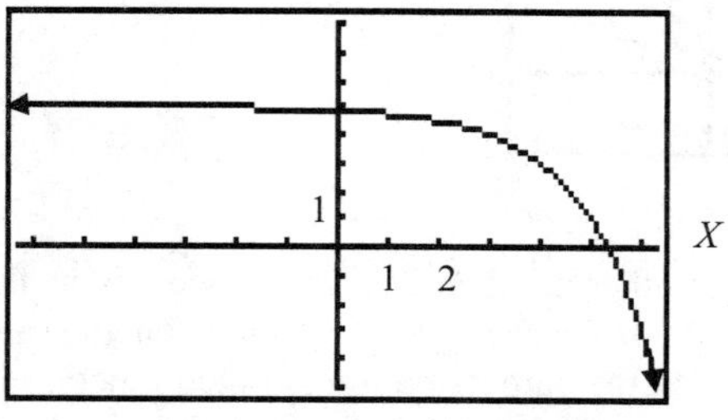

Figure 7.1.11a

$y = -(2)^{x-3} + 5$

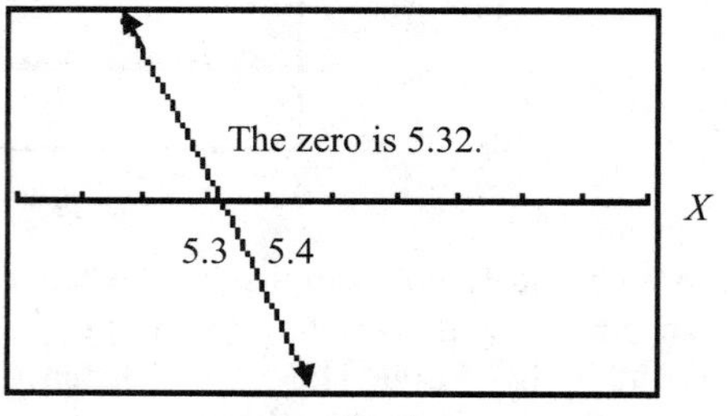

Figure 7.1.11b

Finding the zero by the numerical method is shown below in the numerical representation.

Table 7.1.5

x	2	3	4	5	6	7
$y = -(2)^{x-3} + 5$	4.5	4	3	1	–3	–11

Between 5 and 6, the function changes from positive to negative. To find the zero with an error ≤ 0.01, make another table starting around 5.3 and with Δx of 0.01.

Table 7.1.6

x	5.30	5.31	5.32	5.33	5.34
$y=-(2)^{x-3}+5$	0.075	0.041	0.007	–0.03	–0.06

The zero is between 5.32 and 5.33; pick any number between these for an error less than 0.01. The zero is 5.325.

Example 1: Is $y=(-4)2^{x+1}$ increasing or decreasing?

Solution: When the base is larger than 1, the function increases; however, the –4 times the function reflects the parent graph about the x-axis. The graph gets smaller and smaller as x gets larger and larger, as confirmed in the graph below. The function is decreasing over its normal domain.

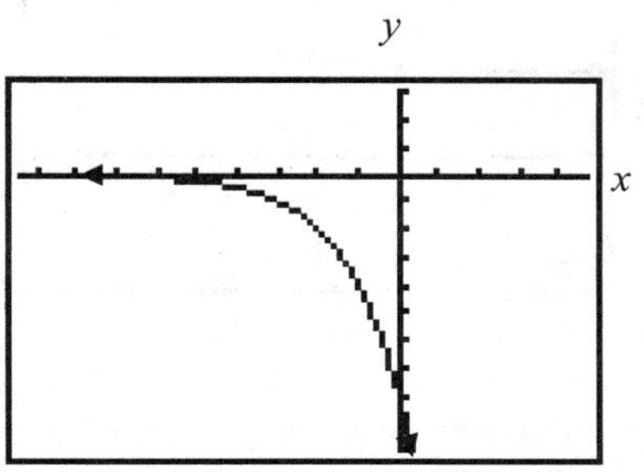

Figure 7.1.12

⁂

Example 2: Find the zeros of the function $f(x) = 2^{x+3} - 5$.

Solution: Graph the function and use the zero finder to find the zero. The zero is –0.678 with an error of less than 0.01.

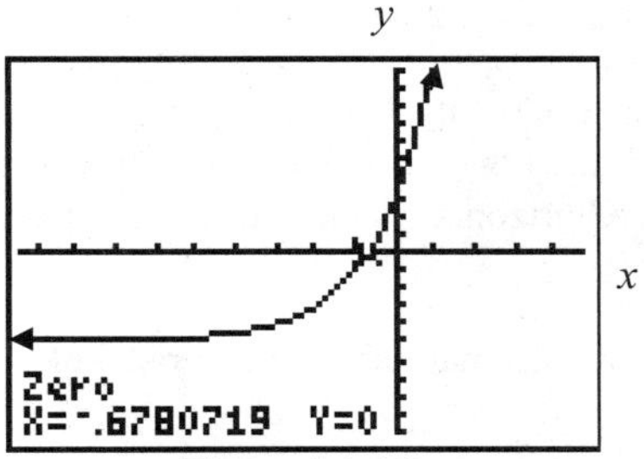

Figure 7.1.13

⁂

Example 3: What is the range of the function $y = 2^{x+3} - 5$?

Solution: Referring to Figure 7.1.13 above, the horizontal asymptote is –5; thus, the range is all numbers above –5, that is, $(-5,\infty)$. This is confirmed by reasoning that the range of the parent graph $y = 2^x$ is $(0, \infty)$ and the transformed graph of $y = 2^{x+3} - 5$ has been translated down 5 units; therefore, the range is $(-5,\infty)$. Remember, $y = -5$ is the horizontal asymptote. The function approaches but never reaches -5; therefore, the range is not $[-5, \infty)$.

Example 4: Find the y-intercept of $y = 2.63^x + 4.5$.

Solution: Every exponential graph in the form b^x studied has had a y-intercept at 1. Thus, $y = 2.63^x$ must also cross at 1. The translation of this function is up 4.5; therefore, the y-intercept must be 4.5 units above 1. The y-intercept is 5.5. A graph of the function confirms this reasoning.

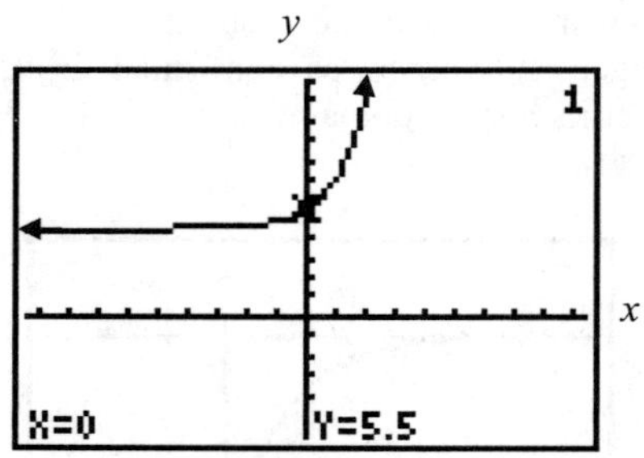

Figure 7.1.14

To find the y-intercept numerically, replace x with 0 and find the value of the function. $y = 2.63^0 + 4.5 = 1 + 4.5 = 5.5$ Of course if you have the function stored in Y_1, for example, you can find the y-intercept using function notation $Y_1(0)$ on the home screen.

Example 5: What is the range of $y = 2.63^x + 4.5$?

Solution: Since the function is the transformed function $y = 2.63^x$; where the horizontal asymptote is 0, the transformation is a vertical shift of up 4.5 making the transformed function have a horizontal asymptote at 4.5; thus, the range is $(4.5, \infty)$.

Example 6: Describe the transformations of the graph of $y = 3^x$ that will yield the graph of $y = 3^{x+1} + 4$.

Solution: To get the graph of $y = 3^{x+1} + 4$ from the graph of $y = 3^x$, shift (move) the graph of $y = 3^x$ left 1 unit and then up 4 units. The transformations are a horizontal shift of -1 and a vertical shift of 4.

Example 7: Describe the transformations of the graph of $y = 3^x$ that will yield the graph of $y = -2 \cdot 3^x + 4$.

Solution: Reflect the graph about the x-axis, then stretch $y = -3^x$ by a factor of 2, and finally, move this graph up by 4 units. The transformations are a reflection about the x-axis, a stretch of 2 and a vertical shift of 4.

⁂

Example 8: Develop an exponential function that is decreasing and has a horizontal asymptote at 6.

Solution: If the base of the exponential function is on the interval (0, 1), the function will decrease. Or, if the base is on the interval $(1, \infty)$ and a transformation of a reflection is given to the function, it will be decreasing. To put the horizontal asymptote at 6, move the graph up 6 by giving it a vertical shift of 6. One possible function that is decreasing and has a horizontal asymptote of 6 is $y = -(2)^x + 6$.

⁂

Example 9: Create an exponential function that has a y-intercept of –4.

Solution: Since all exponential functions of the form $y = b^x$ cross the y-axis at 1, give the function a vertical shift of –5. From 1 to –4 is –5 units. One possible function is $y = 3^x - 5$.

⁂

7.1 EXERCISES

Many times you can work the exercises by algebraic, numeric, or graphical methods. You can learn about mathematics no matter which method you use. Part of the learning process is for you to decide which method is best for you to use. Please read the text carefully before you do the exercises. Read the **next assigned section** after you finish the assignment below.

A Quick Look To The Future

-4. Simplify $(x \cdot x \cdot x \cdot x)\,(x \cdot x \cdot x \cdot x)\,(x \cdot x \cdot x \cdot x)\,(x \cdot x \cdot x \cdot x)\,(x \cdot x \cdot x \cdot x)$.

-3. Is $6 = 2^{x+1} - 4$ an equation or a function?

-2. Give an example of something that grows exponentially.

-1. What is the strength of a solution that is 12% alcohol?

Previous Topics Revisited

1. Isolate the m in the formula $y = m(x - x_1) + y_1$.

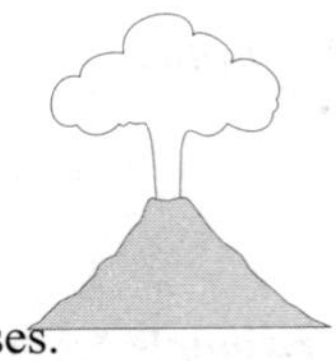

2. The concentration of greenhouse gases is approximated by the linear function $c = 9t + 100$, where t is time in years since 1975 and c is the concentration of gases as compared to a concentration of 100 in 1975. Isolate the variable t.

3. From the greenhouse gases problem, find the current yearly rate of change of gases. Express it in percent form.

4. Solve $3|x + 4| - 3 \geq 2$

5. Solve $-|x - 3.5| + 1.1 \leq -0.5$

6. Solve $3(x + 4.2) - 13.8 > 5x + 7.9$

New Topic Exercises

In this section you learned that the function parameter b in the function $y = d \cdot b^{x+e} + f$ controls the behavior of increasing and decreasing. Describe the behavior the function parameter f controls after graphing all of the functions in Exercise 7.

7 $y = 2^x + 1,\ y = 2^x + 3,\ y = 2^x + 5,\ y = 2^x + 7,\ f(x) = 2^x - 2,\ f(x) = 2^x - 4,\ f(x) = 2^x - 6$

Describe the behavior the function parameter e controls after graphing all of the functions in Exercise 8.

8. $y = 2^x,\ y = 2^{x+2},\ y = 2^{x+4},\ y = 2^{x+6},\ f(x) = 2^{x-1},\ f(x) = 2^{x-3},\ f(x) = 2^{x-5}$

Describe what behavior the function parameter d controls after graphing all of the functions in Exercises 9 and 10.

9. $f(x) = 2 \cdot 3^x,\ f(x) = 4 \cdot 3^x,\ f(x) = 7 \cdot 3^x,\ y = \frac{9}{10} \cdot 3^x,\ y = \frac{3}{5} \cdot 3^x,\ y = \frac{1}{7} \cdot 3^x$

10. $f(x) = -2 \cdot 3^x,\ f(x) = -4 \cdot 3^x,\ f(x) = -7 \cdot 3^x,\ y = -\frac{9}{10} \cdot 3^x,\ y = -\frac{3}{5} \cdot 3^x,\ y = -\frac{1}{7} \cdot 3^x$

11. Using the model for the size of the breast cancer tumor $S = (9 \times 10^{-9}) \cdot 2^{\frac{1}{3}t}$, find the size of the tumor after 5 years of growth.

12. How long does it take for the tumor to grow to the size of a peanut m&m (1.5 cc)?

13. If drugs reduce the doubling rate to every 5 months and the size of the tumor when the rate change takes place is 1×10^{-3} cc, the model becomes $S = (1 \times 10^{-3}) \cdot 2^{\frac{1}{5}t}$. How long does it take for the tumor to become the size of a small pea (0.5 cc)?

14. One model for the national debt problem is $D = 7.4172 \times 10^{-61} \times 1.0763^t$, where t is time in calendar years and D is the national debt in billions. Using this model, when will the debt be 9000 billion dollars (9 trillion)?

15. Using the model for the national debt problem of $D = 1.101^{t-1910} + 225$, where t is time in calendar years and D is the national debt in billions; when will the debt be 9000 billion dollars (9 trillion)?

16. The model for the growth of the tuberculosis bacteria problem is $T_B = 5000 \cdot 2^{\frac{1}{6}t}$ where the initial number of bacteria is 5000 and time is in hours. How long does it take for the 5000 bacteria to become 50,000 bacteria?

17. How many bacteria are there at 100 days into the growth period?

Identify the following functions as increasing or decreasing over the domain.

18. $y = 6^x$

19. $y = \left(\frac{1}{6}\right)^x$

20. $f(x) = -(6)^x$

21. $f(x) = -\left(\frac{1}{6}\right)^x$

22. $f(x) = 2000^{x+3}$

23. $y = \left(\frac{1}{4}\right)^{x-2} + 5$

For each function, find the zero, when the function is positive, when negative and the y-intercept.

24. $y = 2.1136^x$

25. $y = -2\left(\frac{1}{3}\right)^{x+2} + 5$

26. $y = 7^x - 7$

27. $y = \left(\frac{1}{2}\right)^{2x}$

28. $y = 3 - 2^x$

29. $y = 2.5(5.6)^{x+1.1} - 3.9$

Find the range and the horizontal asymptote of the following functions.

30. $y = -2 \cdot 5^{x-3} + 7$

31. $y = \left(\frac{1}{2}\right)^{x+4} - 3$

32. $f(x) = -\left(\frac{1}{4}\right)^{x-2} + 5$

33. $f(x) = 3 \cdot 5^{x+6} - 11$

Develop an exponential function that satisfies the criteria listed:

34. Is increasing and has a horizontal asymptote at 4.

35. Is decreasing and has a horizontal asymptote at –3.

36. Crosses the y-axis at 4 and is decreasing.

37. Has a zero of 2.

38. Is never positive.

39. Passes through the point (2, 3).

Describe the transformations of the first graph that will yield the second graph.

40. $y = 3^x$ $\qquad$ $y = 3^{x-1} + 2$

41. $y = \left(\frac{1}{2}\right)^x$ $\qquad$ $y = \left(\frac{1}{2}\right)^{x+3} - 1$

42. $y = 3^x$ $\qquad$ $y = 4 \cdot 3^x - 5$

43. $f(x) = \left(\frac{1}{4}\right)^x$ $\qquad$ $f(x) = -2 \cdot \left(\frac{1}{4}\right)^{x+3}$

44. $f(x) = 5^x$ $\qquad$ $f(x) = \left(-\frac{1}{2}\right)5^{x-2} + 6$

Find the symbolic representation of an exponential function that contains the sequence of data pairs.

45.

x	1	2	3	4	5	6	7	...
y	1	2	4	8	16	32	64	

46.

x	1	2	3	4	5	6	...
$f(x)$	1	3	9	27	81	243	

47.

x	1	2	3	4	5	...
y	1	0.5	0.25	0.125	0.0625	

48. With a parent function of $f(x) = 2x^2 - 3x$, find the symbolic representation of the transformed function that has been:

a. stretched by a factor of 2

b. shrunk by a factor of $\frac{1}{3}$

c. moved to the right by 4 units

d. moved up 5 units

e. reflected about the x-axis

f. reflected about the x-axis, moved up 2 units, and moved to the right by 3 units

From Mathematics to English

49. The function $y = 2^x$ increases as does the function $y = 2x$. Discuss any differences in "how" they increase.

50. Under what conditions on d and b does the function $f(x) = db^x$ increase? Decrease?

51. Explain why $y = 2^{x+3}$ and $y = 2^x \cdot 2^3$ have identical graphical representations.

52. Discuss why an exponential function of the type studied in this chapter cannot have more than one zero.

53. For the parent exponential function $y = 5^x$, describe the transformation of the parent that give the graphs of the functions below:

a. $y = 5^x + 3$ b. $y = 5^{x-3}$

c. $y = 5^{x+2} - 7$ d. $y = -(5)^{x-5} + 2$

54. Explain why the graph of $f(x) = 2^{|x|}$ is symmetric with respect to the $f(x)$-axis.

55. What does exponential growth mean?

56. If you memorize "f is the vertical shift," does this mean that you understand what a vertical shift is? Why?

57. How do you know that the relationship between time and the height of a person is not an exponential relationship?

58. What is a "horizontal shift?"

59. What is a "reflection about the x-axis?"

60. After reading this section, make a list of questions that you want to ask your instructor.

61. Continue in your daily journal and make an entry. In addition to your normal entry on thoughts about the mathematics in this section, list at least two positive comments about what you have learned about this topic.

62. In paragraph format, summarize the material in this section of the text in your daily journal.

63. Describe how your classroom instructor made this topic more understandable and clear.

64. After reading the text and listening to your instructor, what do you not understand about this topic?

Explorations

65. You are familiar with using the transformations of $y = 2^x$ to obtain the graph of $y = d \cdot 2^{x+e} + f$. Investigate the transformations below where the base is not 2.

a. What transformation of $y = 3^x$ yields the graph of $y = 3^{2x-4}$?

b. What transformation of $y = \left(\frac{1}{2}\right)^x$ yields the graph of $y = \left(\frac{1}{2}\right)^{3x-6}$?

c. What transformation of $y = 4^x$ yields the graph of $y = 4^{7x-14}$?

d. What transformation of $y = (0.6)^x$ yields the graph of $y = (0.6)^{4x-8}$?

Describe the transformation of $y = b^x$ that yields the graph of $y = b^{gx+h}$.

66. What happens to the graph of a function when it is transformed by the absolute value function? You saw earlier that the arithmetic process of adding a number to a function "moves" the graph of the function up or down. Thus, the graph of the function has been transformed into a different graph. What happens to the graph of a function when the absolute value of the function is taken? How is the original graph transformed?

a. Compare the graphs of $y = \left(\frac{1}{2}\right)^x - 5$ and $y = \left|\left(\frac{1}{2}\right)^x - 5\right|$.

b. Compare the graphs of $y = -\left(\frac{1}{2}\right)^x$ and $y = \left|-\left(\frac{1}{2}\right)^x\right|$.

c. Compare the graphs of $y = 3^x - 4$ and $y = |3^x - 4|$.

d. Compare the graphs of $y = -3^{x+2} + 4$ and $y = |-3^{x+2} + 4|$.

67. What happens to the graph of a function when it is transformed by the squaring function? How is the original graph transformed?

a. Compare the graphs of $f(x) = \left(\frac{1}{2}\right)^x - 5$ and $[f(x)]^2$.

b. Compare the graphs of $f(x) = -\left(\frac{1}{2}\right)^x$ and $[f(x)]^2$.

c. Compare the graphs of $y = 3^x - 4$ and $y = [3^x - 4]^2$.

d. Compare the graphs of $y = -3^{x+2} + 4$ and $y = [-3^{x+2} + 4]^2$.

7.2 Simplifying Exponential Functions

What Do You Mean, Equality?

Please consider the graphical and numerical representations of the two exponential functions $f(x) = 2^x \cdot 2^3$ and $g(x) = 2^{x+3}$ shown in Figures 7.2.1 and 7.2.2. Both graphs are on the same window, [–9, 5] by [–3, 10].

$f(x) = 2^x \cdot 2^3$

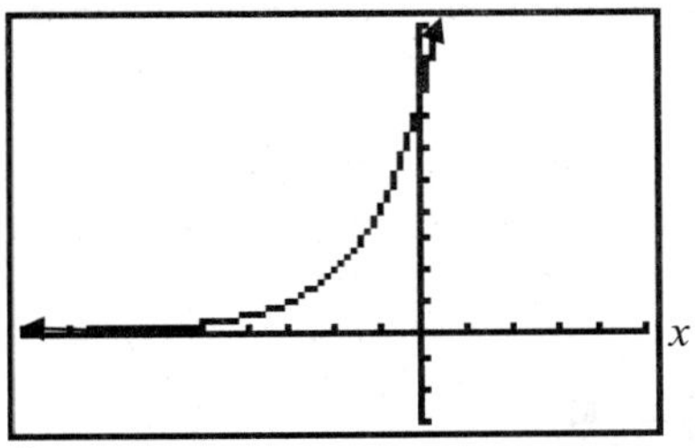

Figure 7.2.1

$g(x) = 2^{x+3}$

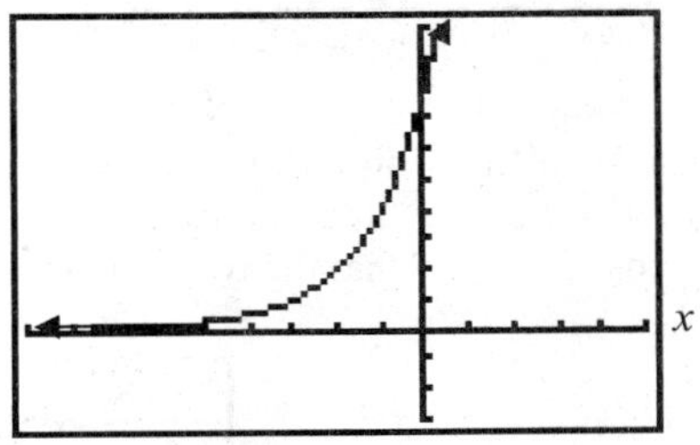

Figure 7.2.2

While it may be difficult to tell that the graphical representations are identical on the printed page, you can confirm that they are the same by graphing them yourself and then using trace. Notice the numerical representations also show that the two functions are the same as shown in Figure 7.2.3.

X	Y1	Y2
-3	1	1
-2	2	2
-1	4	4
0	8	8
1	16	16
2	32	32
3	64	64

Y2=2^(X+3)

Figure 7.2.3

Using calculator notation, $Y_1 = (2 \wedge x)(2 \wedge 3)$ and $Y_2 = 2 \wedge (x + 3)$, it is clear (see above) that the numerical representations are the same. If the graphical representations are the same and the numerical representations are the same, can you conclude that the symbolic representations are the same? The evidence strongly suggests that $2^x \cdot 2^3 = 2^{x+3}$.

What Makes them Equivalent?

Now look at the numerical and graphical representations of the two functions $f(x) = \dfrac{3^x}{3^2}$ and $g(x) = 3^{x-2}$ in Figures 7.2.4 and 7.2.5. The window used is [–5, 9] by [–3, 10].

$f(x) = \frac{3^x}{3^2}$

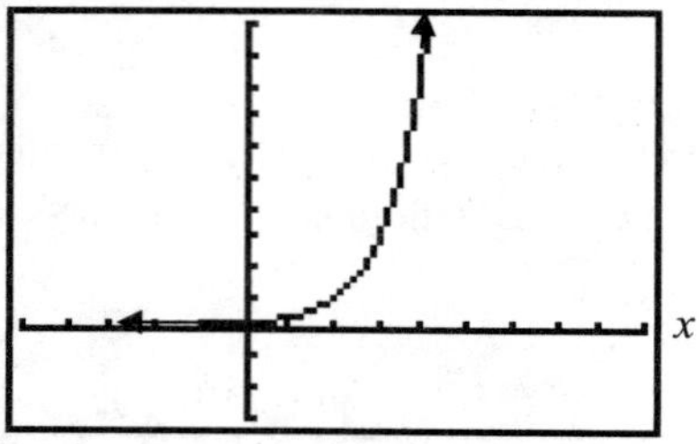

Figure 7.2.4

$g(x) = 3^{x-2}$

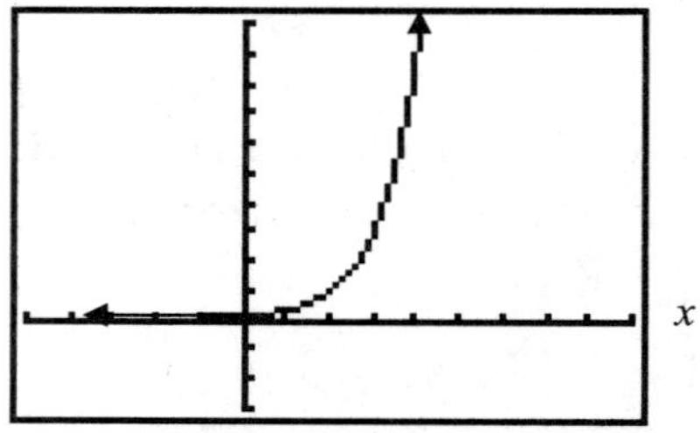

Figure 7.2.5

While it may be difficult to tell that the graphical representations are identical on the printed page, you can confirm that they are the same by graphing them yourself and then using trace. Notice the numerical representations also show that the two functions are the same as shown in Figure 7.2.6.

X	Y1	Y2
0	.11111	.11111
1	.33333	.33333
2	1	1
3	3	3
4	9	9
5	27	27
6	81	81

Y2■3^(X-2)

Figure 7.2.6

Using calculator notation, $Y_1 = (3 \wedge x) / (3^2)$ and $Y_2 = 3 \wedge (x - 2)$. It should be clear that the numerical representations are the same. If the graphical representations are the same and the numerical representations are the same, can you conclude that the symbolic representations are the same? The evidence strongly suggests that $\frac{3^x}{3^2} = 3^{x-2}$.

These two problems may make you wonder if, for example, $7^{x-3} \cdot 7^{2x+5} = 7^{x-3+2x+5} = 7^{3x+2}$? Likewise, you may wonder if $\frac{5^{4x+3}}{5^{x-2}} = 5^{(4x+3)-(x-2)} = 5^{3x+5}$? Is it always true that when exponential functions with the same base are multiplied that the product can be found by adding the exponents? Likewise, is it always true when exponential functions are divided that the quotient can be found by subtracting the exponents as shown in the problem above?

Does the same idea apply in arithmetic? That is, does $5^6 \cdot 5^7 = 5^{6+7} = 5^{13}$? Or does $\frac{19^8}{19^5} = 19^{8-5} = 19^3$? The calculator can confirm the notion as shown in Figure 7.2.7.

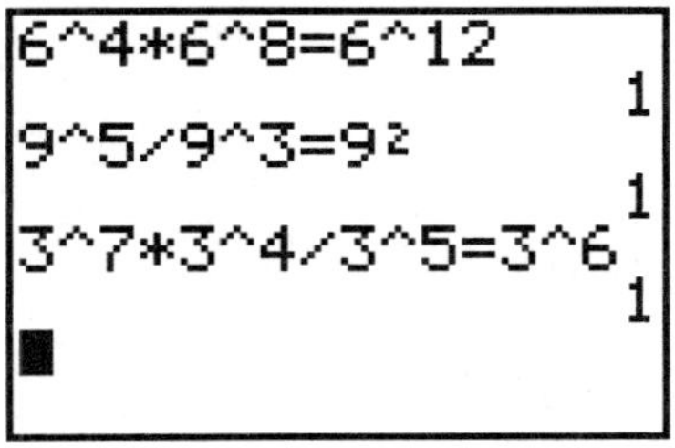

Figure 7.2.7

Recall that a calculator response of 1 means that the statement is true. The way these three statements are written in mathematics is $6^4 \cdot 6^8 = 6^{12}$, $\dfrac{9^5}{9^3} = 9^2$, and $\dfrac{3^7 \cdot 3^4}{3^5} = 3^6$. It certainly does appear that the ideas apply to arithmetic expressions as well as exponential functions. Based on this experimentation, the following generalizations are made.

Reminder: Positive Integer Exponent b^n means $\underbrace{b \cdot b \cdot b \cdot b \cdot b \ldots}_{n \text{ times}}$

The number n is a positive integer and is called the **exponent**. The number b is called the **base** and can be any real number for arithmetic expressions and $b > 0$ and $b \neq 1$ for exponential functions.

Reminder: A Negative Exponent b^{-p} means $\dfrac{1}{b^p}$ or $b^{-p} = \dfrac{1}{b^p}$

Where p is a positive integer and $b \neq 0$ for arithmetic expressions and $b > 0$ and $b \neq 1$ for exponential functions.

Throughout this text, it is understood that if the exponent is negative or zero, the base cannot equal zero. Further, even though a number raised to a real exponent is not defined in this text, assume that the exponent can be a real number as justified by the discussion in Section 7.1.

Quickie Example 1: $5^{-x} = \dfrac{1}{5^x}$ Note: the base does not change sign.

Quickie Example 2: $1^{-x} = \dfrac{1}{1^x}$ Note: the base does not change sign.

Quickie Example 3: $\dfrac{1}{2^x} = 2^{-x}$ Note: the base does not change sign, just the exponent changes sign.

Product Property of Exponents

$$b^n \cdot b^m = b^{n+m}$$

Where $b \neq 0$ for arithmetic expressions, and $b > 0$ and $b \neq 1$ for exponential functions.

Quickie Example 4: $x^2 \cdot x^5 = x^{2+5} = x^7$

Quickie Example 5: $x^{-6} \cdot x^9 = x^{-6+9} = x^3$

Quickie Example 6: $x^5 \cdot x^{-4} = x^{5+(-4)} = x^1 = x$

Quickie Example 7: $(-4)^{-7} \cdot (-4)^{-2} = (-4)^{-7+(-2)} = (-4)^{-9}$

> **Quotient Property of Exponents** $\dfrac{b^n}{b^m} = b^{n-m}$
>
> Where $b \neq 0$ for arithmetic expressions and $b > 0$ and $b \neq 1$ for exponential functions.

Quickie Example 8: $\dfrac{x^{-3}}{x^4} = x^{-3-4} = x^{-7}$

Quickie Example 9: $\dfrac{(-1)^6}{(-1)^{-4}} = (-1)^{6-(-4)} = (-1)^{10}$

Quickie Example 10: $\dfrac{x^{-3}}{x^{-4}} = x^{-3-(-4)} = x^1$

Can you simplify an exponential function raised to an exponent? For example, does $\left(3 \cdot 2^x\right)^4$ simplify? Of course you can use the idea that when a number (the function is a number) is raised to the exponent 4, it means to multiply that number by itself 4 times. Thus, the function simplifies to $3 \cdot 2^x \cdot 3 \cdot 2^x \cdot 3 \cdot 2^x \cdot 3 \cdot 2^x$, which can be written as $3 \cdot 3 \cdot 3 \cdot 3 \cdot 2^x \cdot 2^x \cdot 2^x \cdot 2^x$ by the commutative property for multiplication. The product property of exponents can now be used to rewrite this as $3^4 \cdot 2^{4x}$. Yes, $\left(3 \cdot 2^x\right)^4$ can be simplified. You should confirm this on your calculator by looking at the numerical representations of $\left(3 \cdot 2^x\right)^4$ and $3^4 \cdot 2^{4x}$. While you certainly can question whether $3^4 \cdot 2^{4x}$ looks simpler than $\left(3 \cdot 2^x\right)^4$, what you can't question is that both represent the same function. There is one more property of exponents that will help rewrite functions like $\left(3 \cdot 2^x\right)^4$ as functions like $3^4 \cdot 2^{4x}$. It is describe below in three parts.

> **I. Power Property of Exponents** $\left(b^n\right)^m = b^{n \cdot m}$
>
> Where $b \neq 0$ if n or m is 0 for arithmetic expressions and $b > 0$ and $b \neq 1$ for exponential functions.

Quickie Example 11: $\left(x^{3}\right)^{5} = x^{3\cdot 5} = x^{15}$

Quickie Example 12: $\left((-2)^{3}\right)^{-4} = (-2)^{3\cdot(-4)} = (-2)^{-12}$

Quickie Example 13: $\left(x^{-2}\right)^{-4} = x^{(-2)(-4)} = x^{8}$

> **II. Power Property of Exponents** $\left(a^{n} \cdot b^{m}\right)^{p} = a^{pn} \cdot b^{pm}$
>
> Where a and $b \neq 0$ if n, m, or p is 0 for arithmetic expressions, and $a > 0$, $b > 0$ and $a \neq 1$, $b \neq 1$ for exponential functions.

Quickie Example 14: $\left(6^{3} \cdot x^{-5}\right)^{4} = 6^{4\cdot 3} \cdot x^{4\cdot(-5)} \cdot = 6^{12} x^{-20}$

Quickie Example 15: $\left((-2)^{4} x^{-1}\right)^{2} = (-2)^{2\cdot 4} x^{2(-1)} =$

$= (-2)^{8} x^{-2}$

> **III. Power Property of Exponents** $\left(\dfrac{a^{n}}{b^{m}}\right)^{p} = \dfrac{a^{pn}}{b^{pm}}$
>
> Where a and $b \neq 0$ if n, m, or p is 0 for arithmetic expressions, and $a > 0$, $b > 0$ and $a \neq 1$, $b \neq 1$ for exponential functions.

Quickie Example 16: $\left(\dfrac{x^{5}}{x^{6}}\right)^{3} = \dfrac{x^{3\cdot 5}}{x^{3\cdot 6}} = \dfrac{x^{15}}{x^{18}} = x^{-3}$ or $\dfrac{1}{x^{3}}$

Quickie Example 17: $\left(\dfrac{4^{-2}}{5^{2}}\right)^{-3} = \dfrac{4^{(-3)(-2)}}{5^{(-3)(2)}} = \dfrac{4^{6}}{5^{-6}}$ or $4^{6} \cdot 5^{6}$

Example 1: If $f(x) = x^{4} \cdot x^{-3}$, simplify $\left(f(x)\right)^{2}$.

Solution: Simplify inside grouping symbols first by multiplying. Then, perform the power operation.

$\left(f(x)\right)^{2} = \left[x^{4}x^{-3}\right]^{2}$	given
$\left(f(x)\right)^{2} = \left[x^{1}\right]^{2}$	product property
$\left(f(x)\right)^{2} = x^{2}$	power property I

⁂

Example 2: $f(x) = \dfrac{x^4 x^9}{x^{13}}$, simplify it.

Solution: Simplify the numerator first, $f(x) = \dfrac{x^{13}}{x^{13}}$ product property

Does 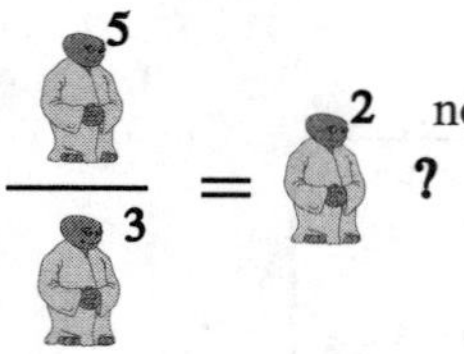

now divide $f(x) = x^0$ quotient property

$f(x) = 1$ definition of 0 exponent ⁂

Example 3: Simplify the function $f(x) = \dfrac{[x^{-4}x^9]^{-2}}{x^{15}x^{-6}}$.

Solution: Simplify the numerator and denominator separately, then divide.

$f(x) = \dfrac{[x^{-4}x^9]^{-2}}{x^{15}x^{-6}}$ given

$f(x) = \dfrac{[x^5]^{-2}}{x^9}$ product property

$f(x) = \dfrac{x^{-10}}{x^9}$ power property I

$f(x) = x^{-19}$ or $\dfrac{1}{x^{19}}$ quotient property

You may want you to use the definition of a negative exponent and replace the x^{-19} with $\dfrac{1}{x^{19}}$. It is optional in this text. Technology based confirmation that the simplified function is equivalent to the original function is shown in Figure 7.2.8.

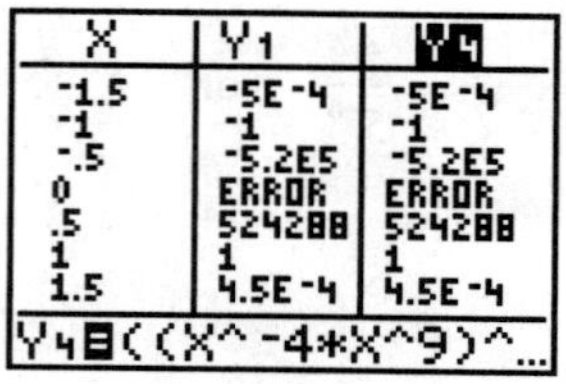

X	Y1	Y4
-1.5	-5E-4	-5E-4
-1	-1	-1
-.5	-5.2E5	-5.2E5
0	ERROR	ERROR
.5	524288	524288
1	1	1
1.5	4.5E-4	4.5E-4

Y4=((X^-4*X^9)^...

Figure 7.2.8a

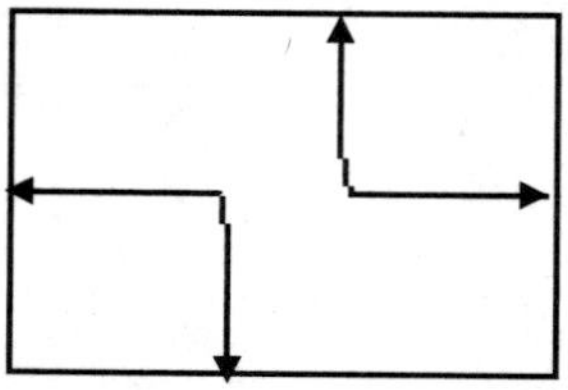

Figure 7.2.8b

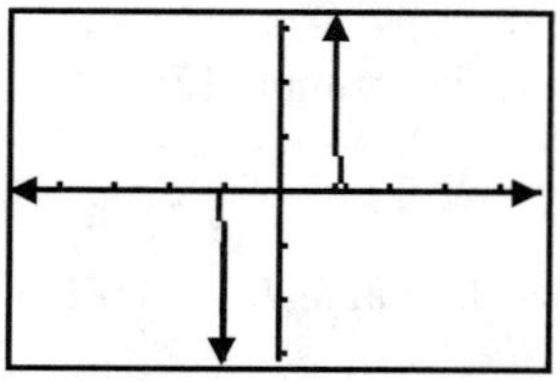

Figure 7.2.8c

Figure 7.2.8a shows that the numerical representations are the same and Figures 7.2.8b and c are the graphs of both functions -- one without the axes drawn -- one with the axes activated. The window is [–4.7, 4.7] by [–3.1, 3.1]. Since the numeric and graphic representations are the same, the answer is correct. Comment: these functions are not exponential functions. This is why the domain contains negative numbers. Why does the domain not contain 0?

⁂

Example 4: Simplify the exponential function $g(x) = \left(5^x\right)^2$.

Solution: Use power property I and multiply the exponents, x and 2, $g(x) = \left(5^x\right)^2 = 5^{2x}$.

⁂

Example 5: Simplify the exponential function $h(x) = 2^{x-1} \cdot 2^{x+4}$.

Solution: Use the product property and add the exponents $x - 1$ and $x + 4$.

$h(x) = 2^{x-1} \cdot 2^{x+4} = 2^{(x-1)+(x+4)} = 2^{2x+3}$

⁂

Example 6: Simplify the exponential function $q(x) = 3^{x^2} \cdot 3^{2x} - 1$.

Solution: Use the product property on the like bases. The −1 does not simplify.

$q(x) = 3^{x^2} \cdot 3^{2x} - 1 = 3^{x^2+2x} - 1$

Algebraic symbol manipulation work can be checked graphically or numerically by graphing the problem and your answer or by making a table for the problem and your answer. Below is a numerical check of the above work.

Table 7.2.1

x	...	−3	−2	−1	0	1	2	...
$3^{x^2} \cdot 3^{2x} - 1$	...	26	0	−0.667	0	26	6560	...
$3^{x^2+2x} - 1$	...	26	0	−0.667	0	26	6560	...

The numeric representations are the same; therefore, the functions are equivalent. The algebraic work is correct.

⁂

7.2 EXERCISES

Many times you can work the exercises by algebraic, numeric, or graphical methods. You can learn about mathematics no matter which method you use. Part of the learning process is for you to decide which method is best for you to use. Please read the text carefully before you do the exercises. Read the **next assigned section** after you finish the assignment below.

A Quick Look To The Future

-4. Can the graph of an exponential function like those studied in Section 7.1 cross a horizontal line more than once?

-3. If a pack of mice double their population every three months, can you model this behavior with a linear function?

-2. What is the value of the function $f(x)=\dfrac{1}{x-2}$ when x is 2.000000001?

-1. Graph $y=\dfrac{x+2}{x^2-4}$ on a decimal window. What do you see when $x = 2$?

Previous Topics Revisited

1. Is the function $f(x)=-2\left(\dfrac{1}{4}\right)^{x-3}+2$ increasing or decreasing? Why?

2. Find the zero of $y = 5^{x+4} - 2$.

3. Find the $f(x)$-intercept of $f(x)=\left(\dfrac{1}{2}\right)^{2-x}$.

4. From the intercept form of the linear function $\dfrac{x}{a}+\dfrac{y}{b}=1$, isolate b.

5. The linear function for approximating infant death rates (d) in developed countries is $d=\left(-\dfrac{7}{8}\right)t+45$, with $t = 0$ in 1950. Find t in calendar years when the death rate is 0.

6. Solve $-2|x + 4.22| + 13.8 \le 12.99$

New Topic Exercises

Simplify each function. Negative exponents are acceptable in the simplified form. Try confirming that your answers are correct by comparing numeric representations of the problem and your answer.

7. $f(x) = (x-2)(x-2)(x-2)(x-2)(x-2)(x-2)^2$

8. $g(x) = (x+3)^{-1}(x+3)^{-1}(x+3)^{-1}(x+3)^{-1}(x+3)^{-2}$

9. $h(x) = (2^{x+2})(2^{x+2})(2^{x+2})$

10. $i(x)=\dfrac{3^{x-4}}{5\cdot 3^{x-4}}$

11. $y=\dfrac{4\cdot 3^{2x-5}}{2\cdot 3^{x+4}}+7$

12. $f(x) = [(2x^4)(-3x^6)]^0$ (Note: Do not use [ ] or { } on your calculator as grouping symbols.)

13. $y = [(2x^4)(-3x^6)]^{-2}$

14. $y = [(2x^4)(-3x^6)]^3$

15. $g(x)=\dfrac{5x^{-9}}{20x^{-3}}$

16. $f(x)=\dfrac{(-3x^5)(-2x^{-3})}{6x^2}$

17. $y=\dfrac{x^{-2000}}{x^{1000}}$

18. $g(x)=\left(\dfrac{2x^{-8}}{x^2}\right)^{-3}$

19. $Y_1(x) = \dfrac{(2x^4)(-4x^{-3})^2}{(-8x^5)}$

20. $y(x) = \dfrac{3x^5 \cdot x^{-6} \cdot (-2x^2)^{-3}}{x^4 \cdot 3x^{-7} \cdot x^3}$

21. $f(x) = 2^x \cdot 2^{x^2-1}$

22. $y = [-5 \cdot 3^{x^2}]\{7 \cdot 3^{x+1}\}$

23. $f(x) = [-2 \cdot 5^{x^2+2}]^{-3}$

24. $g(x) = \dfrac{3^{x+1}}{3^{x-1}}$

25. $h(x) = \{2^x \cdot 2^x \cdot 2^x\}^2$

26. $i(x) = \left[\left(2^{2x-3}\right)\left(2^{-3x+2}\right)\right]^2$

27. $j(x) = \left(\dfrac{3^{-x+4}}{3^{-2x+5}}\right)3^{x-2} - 5$

28. $k(x) = 5^{x^2-2x+4} \cdot 5^{-x^2+3x-1} \cdot 5^{x^2+x+1} + 13$

29. Find any function that simplifies through the product property to $y = x^5$.

30. Find any function that simplifies through the quotient property to $y = x^5$.

31. Find any function that simplifies through the product and quotient property to $y = x^5$.

32. Find any function that simplifies through the power property I to $y = x^6$.

33. Find any function that simplifies through the power property II to $y = x^6$.

34. Find any function that simplifies through the power property III to $y = x^6$.

35. Find any exponential function that simplifies through the product property to $y = 7^{x-1}$.

36. Find any exponential function that simplifies through the quotient property to $y = 7^{x-1}$.

37. Find any exponential function that simplifies through the product property to $y = 3^{3x^2+5x-2}$.

38. Find any exponential function that simplifies through the quotient property to $y = 3^{3x^2+5x-2}$.

From Mathematics to English

39. Explain the quotient property of exponents.

40. Why can't the base be 0 in the definition of a zero power ($b^0 = 1$).

41. Give a counter-example to show that: $\dfrac{x^8}{x^{-6}} \neq x^2$.

42. Does $(2x^4)^3 = 2x^{12}$? Why?

43. If you memorize the product property, does this mean you understand it? Explain.

44. If you understand the quotient property, do you think you could use it in new and unusual situations? Why?

45. Why must you understand how to use the properties of exponents?

46. After reading this section, make a list of questions that you want to ask your instructor.

47. Continue in your daily journal and make an entry. In addition to your normal entry on thoughts about the mathematics in this section, list at least two positive comments about what you have learned about this topic.

48. In paragraph format, summarize the material in this section of the text in your daily journal.

49. Describe how your classroom instructor made this topic more understandable and clear.

50. After reading the text and listening to your instructor, what do you not understand about this topic?

From English to Mathematics

Write each of the following English statements as an equivalent mathematical statement.

51. the cube of x

52. the square of the square root of x

53. four-thirds of π times the radius cubed

54. the cube of the sum of x and 10

55. the fourth power of the difference of x and 3

Explorations

56. Graph $y = x^5 / x^3$ and $y = x^2$ on the same window. Are the graphs the same? Why?

Graph $y = x^2 \cdot x$ and $y = x^3$ on the same window. Are the graphs the same? Why?

Graph $y = 3x^{-2}$ and $y = 3 / x^2$ on the same window. Are the graphs the same? Why?

57. Graph the following functions:

$y = x^1$ $\quad$ $y = x^2$ $\quad$ $y = x^3$ $\quad$ $y = x^4$ $\quad$ $y = x^5 \ldots$

List the functions that have similar graphs.

Explain the relationship that exists between the exponent in the function and the general shape of the graph of the function.

7.3 Equations and Inequalities Containing the Exponential Function

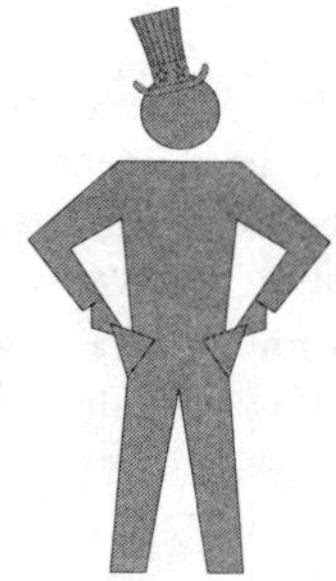

We All Owe the National Debt

Below in Table 7.3.1 is the amount of the national debt "theoretically" owed by each person in the United States for the years from 1950 to 2006. If each person would have paid the amount shown, the federal debt would have been \$0. The graphical representation of this data relationship shows the shape is exponential.

Table 7.3.1

t (time in calendar years)	1950	1960	1970	1980	1985	1990	2000	2006
N (National per-capita debt)	1688	1572	1807	3981	7614	12848	20163	28404

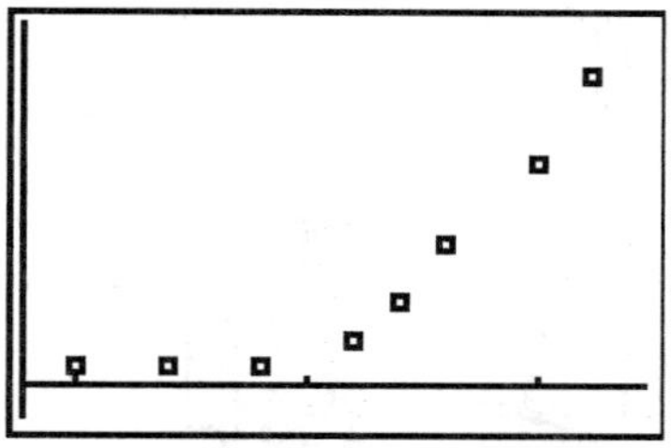

Figure 7.3.1a

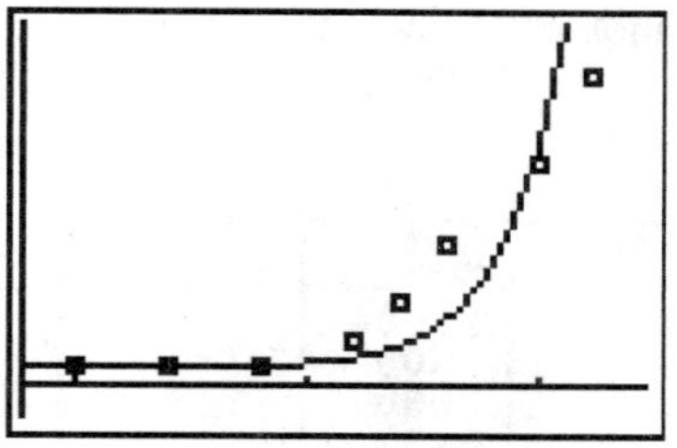

Figure 7.3.1b

Figure 7.3.1b shows the graph of the data relationship and the graph of the model of the per-capita national debt as $N = 1.165^{t-1935} + 1500$. A reasonable problem domain may be [1945, 2012]. Using this model, when was the national debt \$15,000 per person in the United States? To answer this question is to solve the equation $15000 = 1.165^{t-1935} + 1500$. Pick one of the technology-based methods -- say the intersection method. To solve by the intersection method, graph the functions Y_1 = 15000 and $Y_2 = 1.165^{t-1935} + 1500$ and find the intersection as shown in Figure 7.3.2 below.

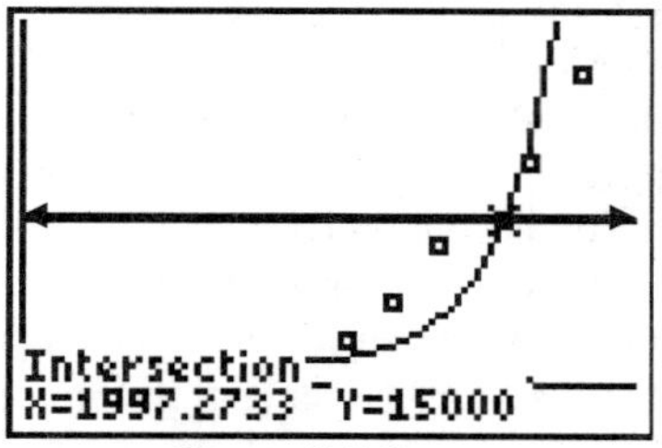

Figure 7.3.2

Using the same window as in Figure 7.3.1, the intersection method shows the solution to the equation to be around March of 1997.

The Book Value of My Car Keeps Dropping

Suppose that you bought a car for $12,000. One thing you know about the value of your car is it keeps dropping year after year. This decreasing in value is called **depreciation**. That is, your car depreciates in value as it ages. One method for calculating the value of your car is to assume that it depreciates in value at a certain rate -- like 20% per year. The model for the value of a car using this method is $V = V_0(1-r)^t$, where V is the value of the car at the end of year t, V_0 is the initial value of the car, and r is the annual rate of depreciation. Thus, the model for the value of your car becomes $V = 12000(1-0.2)^t$, where t is time in years. What is your car worth if you keep it for 5 years? This can be answered by solving the equation $V = 12000(1-0.2)^5$. There is not much to do -- just evaluate $12000(1-0.2)^5$. It is worth $3,932.16.

If you want to know when the car will have a value of $6,000, you must solve the equation $6000 = 12000(1-0.2)^t$. With the function $V = 12000(1-0.2)^t$ stored in Y_1, the numeric method shows a solution between 3 and 4 years in Figure 7.3.4a. With a new table starting at 3 and a Δx of 0.01, Figure 7.3.4b shows a solution between 3.10 years and 3.11 years. A solution with an error less that 0.01 is 3.105 years. That is, your car is worth $6,000 when it is 3.105 years old.

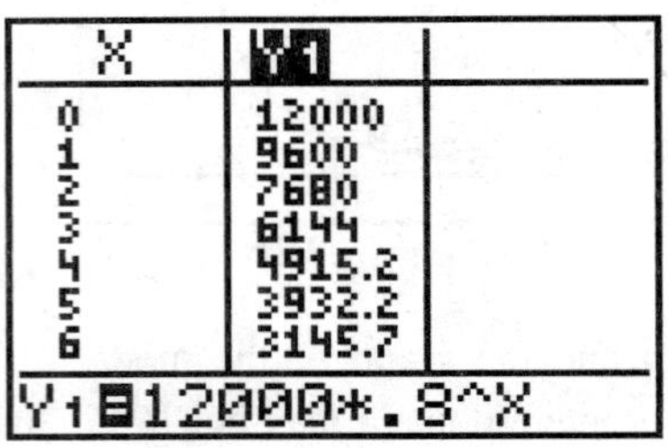

Figure 7.3.3a

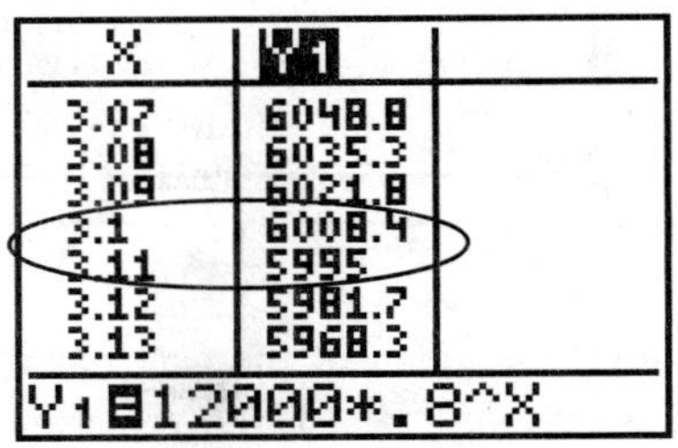

Figure 7.3.3b

How Hot Is the Coffee?

Sir Isaac Newton was very observant and could take his observations, in the form of data relationships, and develop mathematical models of the data. This is much like what you are asked to do in laboratory projects; however, Newton did not have the aid of the graphing calculator. One mathematical model he developed comes from a relationship you encounter daily -- the cooling of any hot object to the room temperature. The coffee you drink, the pizza you eat, the burner on the kitchen stove, the TV you watch, all cool to room temperature in a very predictable way. Newton's model of the cooling process is $y = (y_0 - y_1)0.3679^{rt} + y_1$. The function parameters are the initial temperature y_0, the room temperature y_1, and the rate of cooling r. The function variable is time t, and the temperature at time t is the function y. A cup of coffee in an un-insulated cup cools at a rate of around 15% per minute. If the initial temperature of the coffee is 160° and room temperature is 75°, how long does it take for the coffee to reach 80°? To answer this question, replace the function parameters with the given values and the function with 80°. This gives the equation $80 = (160-75)0.3679^{0.15t} + 75$; solve it.

$80 = (160 - 75)0.3679^{0.15t} + 75$ subtract 80 from both sides

$0 = (160 - 75)0.3679^{0.15t} + 75 - 80$ use the zeros method

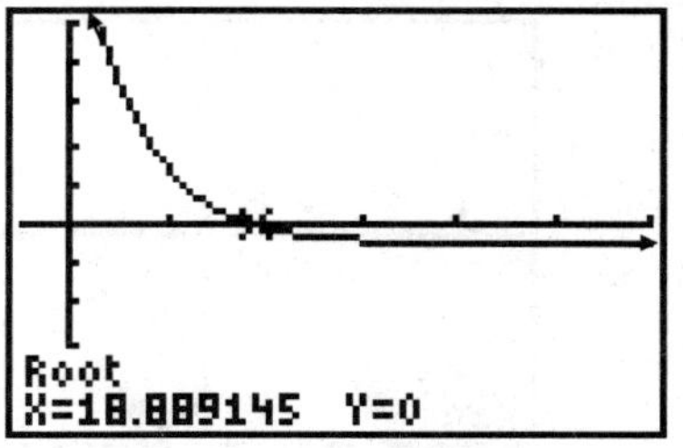

Figure 7.3.4

The answer to the question is that it takes 18.89 minutes for the coffee to reach 80°.

Because you are using technology-based methods, you can solve equations containing exponential functions that are a little more complicated than the previous examples. A few examples follow.

Example 1: Solve the equation $-3 \cdot 2^{x+5} + 3 = 7^{x-2} - 4$.

Solution: The difficult part may be to find the window that contains the solution. Probably the easiest method is the zeros method because you know that if there is a zero, it is on the *x*-axis. Add 4 and subtract 7^{x-2} from both sides of the equation to give the transformed equation $-3 \cdot 2^{x+5} + 3 - 7^{x-2} + 4 = 0$. Graph the function $f(x) = -3 \cdot 2^{x+5} + 3 - 7^{x-2} + 4$ and find the zero. It is –3.778. The window is

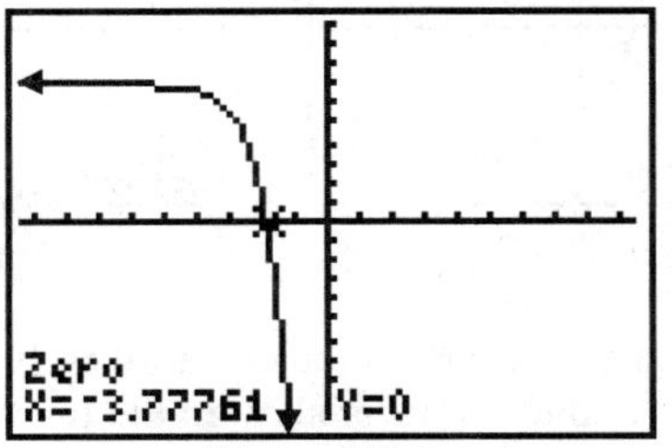

Figure 7.3.5

[–18, 18] by [–10, 10]. The reason you know that the window shows the complete graph is that the two strongest pieces of the function $f(x) = -3 \cdot 2^{x+5} + 3 - 7^{x-2} + 4$ are $-3 \cdot 2^{x+5}$ and -7^{x-2} , both of which are decreasing. Thus you should expect that the graph approach –∞ very quickly. Also, the 3 + 4 piece of the function suggests that the horizontal asymptote is at 7.

⁂

Example 2: Solve the equation $\left(\frac{6}{5}\right)^{-|x|} = 1$.

Solution: The trace method works well if the solution(s) is a simple rational number and you are using a decimal window. The problem is that you don't know what kind of solution to expect before you solve the equation. Figure 7.3.6 shows a complete graph of the

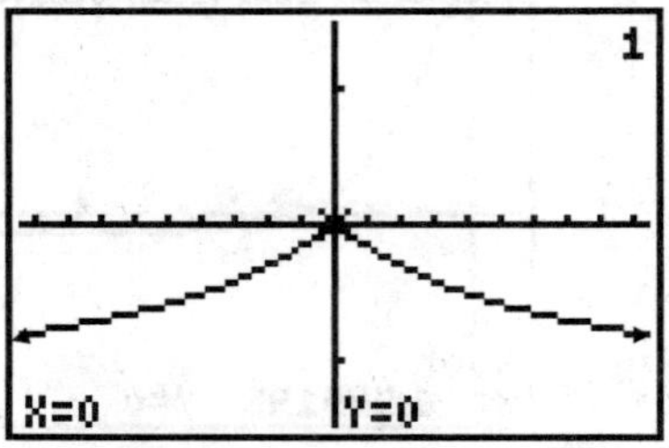

Figure 7.3.6

function $y = \left(\frac{6}{5}\right)^{-|x|} - 1$ on the window [–9.4, 9.4] by [–1.5, 1.5]. The trace cursor shows the solution to be 0.

Example 3: Solve the equation $2^{x-5} = 9\sqrt{x+3}$.

Solution: After experimenting with several windows in order to find a complete graph of the function $y = 2^{x-5} - 9\sqrt{x+3}$, the graph shows two possible solutions as suggested in Figure 6.3.7a on the window [–14.1, 14.1] by [–30, 30]. Figure 7.3.7b clearly

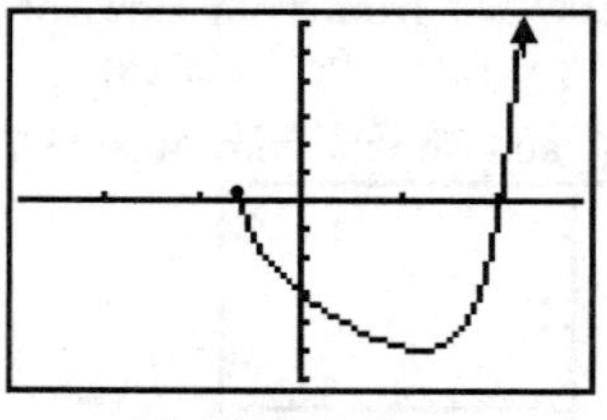

Figure 7.3.7a

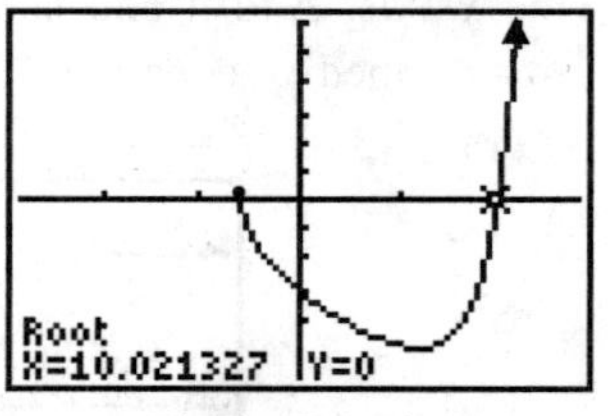

Figure 7.3.7b

displays one of the solutions to the equation as 10.02. Six or seven zoom-in's around the other possible solution never showed the function above the *x*-axis; however, you can see in Figure 7.3.8 that the numerical method does show a solution between –3.00 and –2.99. A reasonable approximation is –2.995. The equation does have two solutions 10.02 and –2.995.

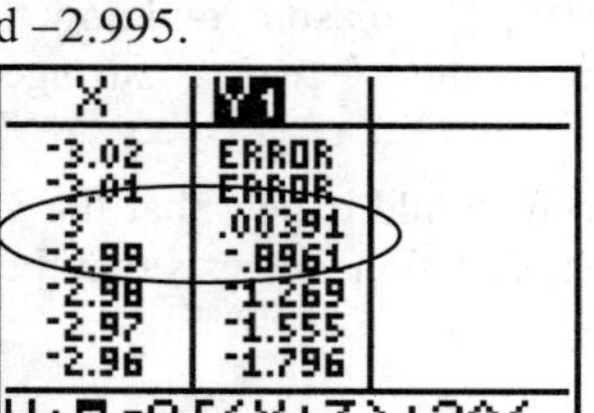

Figure 7.3.8

Unlike the equation in Example 3, sometime exponential equations can easily be solved using an analytical approach.

Example 4 Solve the equation $3^{x+1} = 81$.

Solution: Both sides of the equation must be simplified to the same base. That is, in this equation both sides can be simplified to 3 raised to an exponent, $3^{x+1} = 3^4$. Thinking algebraically, how is it possible for 3^{x+1} to equal 3^4? The only way is if the exponents are equal. Thus $x + 1 = 4$, or $x = 3$. The solution is 3. Please check this to confirm it is correct.

⁂

Example 5: Solve the equation $3 \cdot 2^{3x+5} = 6 \cdot 2^{x^2}$.

Solution: Using the same reasoning as above, start by dividing both sides of the equation by 3.

Think Algebraically!

$$\frac{3 \cdot 2^{3x+5}}{3} = \frac{6 \cdot 2^{x^2}}{3} \qquad \text{now simplify}$$

$$2^{3x+5} = 2 \cdot 2^{x^2} \qquad \text{use the product property of exponents}$$

$$2^{3x+5} = 2^{x^2+1}$$

The only way for this to be a true statement is if the exponents are equal.

$$3x + 5 = x^2 + 1$$

How do you solve this equation with algebra? If you transform the equation to get zero on one side, you can then find the zeros using what you learned in Chapter Three about the relationship between factors and zeros.

$$0 = x^2 - 3x - 4$$

$$0 = (x - 4)(x + 1)$$ The solution set is 4 and –1.

⁂

Example 6: Solve the inequality $2^{x-5} \le 9\sqrt{x+3}$ by the zeros method.

Solution: The first step in solving any inequality by the zeros method is to find the zeros. The zeros of the function $f(x) = 2^{x-5} - 9\sqrt{x+3}$ were found in Example 3 to be –2.995 and 10.012. The graphical representation of this function shows that the function is

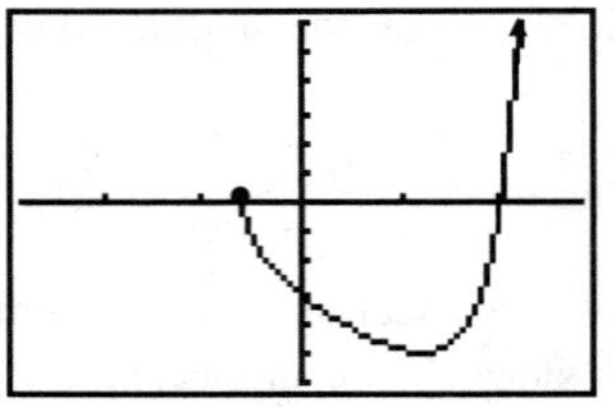

Figure 7.3.9

negative between the zeros; thus, the solution to the inequality is [–2.995, 10.012].

⁂

Example 7: Solve the inequality $-2 \cdot 3^{\sqrt{x+1}} > 3x^2 + 6x - 5$ by the intersection method.

Solution: A window showing the intersection of the functions on each side of the inequality must be found. The actual intersection must then be found and finally answer the question "When is the function $y = -2 \cdot 3^{\sqrt{x+1}}$ larger than (above) the function $y = 3x^2 + 6x - 5$"?

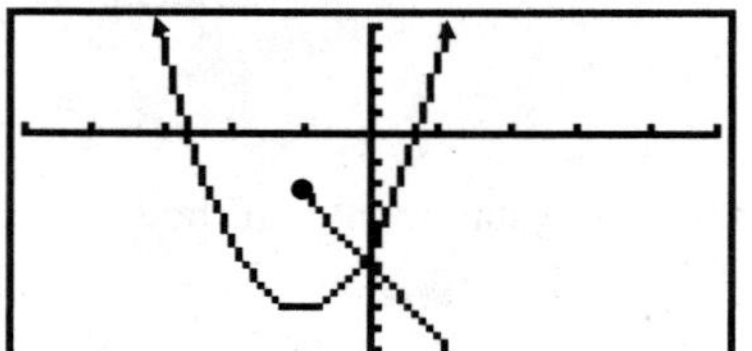

Figure 7.3.10a

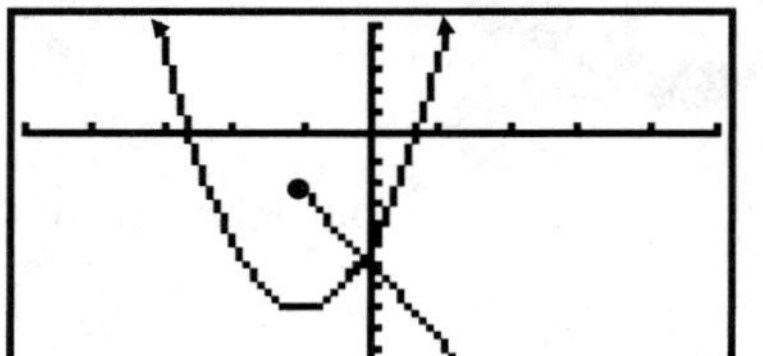

Figure 7.3.10b

The window is [–5, 5] by [–15, 5]. The graphs show that the exponential function is above the quadratic function when x is left of –0.112. You also know that the exponential function $y = -2 \cdot 3^{\sqrt{x+1}}$ has a normal domain of [–1, ∞); therefore, the exponential function is above the quadratic when x is on the interval [–1, –0.112). The solution to the inequality is [–1, –0.112).

⁂

7.3 EXERCISES

Many times you can work the exercises by algebraic, numeric, or graphical methods. You can learn about mathematics no matter which method you use. Part of the learning process is for you to decide which method is best for you to use. Please read the text carefully before you do the exercises. Read the **next assigned section** after you finish the assignment below.

A Quick Look To The Future

-4. Would you make more money if the relationship between hours worked and your salary was exponential rather than linear?

-3. Is $y = x^{-1}$ a linear function? A quadratic function? An exponential function? An absolute value Function? A square root function?

-2. Does $\frac{x^2}{x} = x$ for all real numbers?

-1. Does $\frac{2}{3} \cdot \frac{5}{7} = \frac{2 \cdot 5}{3 \cdot 7}$?

Previous Topics Revisited

1. Does $\left(5^x\right)^3$ equal 5^{x+3} ? Why or why not?
2. Are the numeric representations of $\frac{x^5}{x^{-2}}$ and x^3 the same? Why?
3. Simplify $\left(\frac{2^{x+3}}{2^{2x-5}}\right)^3$
4. What transformation of $y = 3^x$ will put the horizontal asymptote at $y = 4$?
5. Create an exponential function that is decreasing and has a horizontal asymptote at -5.
6. If the area of an oil spill varies directly as the cube of time in minutes and after 5 minutes the area is 16,000 square feet, what is the model of this relationship? What is the area after 33 minutes? How long will it take for the spill to reach 25 million square feet?

New Topic Exercises

Solve the following equations or inequalities.

7. Find the initial cost (value) of a car that has been depreciated by the model $V = V_0(1-r)^t$ when the value after 4 years is \$8,000 and the depreciation rate is 15%.
8. How many years will it take for a car that costs \$32,000 to have a value of less than \$6,000 if the rate of depreciation is 25%?
9. What rate of depreciation will cause a \$18,500 car have a value of \$4,000 after 8 years?
10. Using the model of the national debt developed in the "We All Owe the National Debt Problem," in what year did we each owe \$20,000?
11. In what year did each person in the U.S. "owe" about \$2,000 (on average) of the national debt?
12. If the rate of cooling of a solid burner on an electric stove is 10% per minute, with an initial temperature of the burner at 500° and room temperature at 70°, how long will it take for the temperature of the burner to reach 75°?

13. A solar heated storage area (a rock bin) cools at a rate of 5% per hour. If the temperature of the rock storage bin is 150° when the sun sets, how long will it take for the storage area to reach 80° when the surrounding temperature is 45°? How long will it take if the surrounding temperature is 30°?

14. What initial temperature must be obtained in the rock storage area if you want the storage temperature to be above 90° after 12 hours? Assume a temperature of 55° surrounding the storage area.

15. $4^{3x-5} = 16$

16. $3^{2x+7} = 9^{x-3}$

17. $2 \cdot 2^{x-3} = 4 \cdot 2^{3x+1}$

18. $2 \cdot 5^{2x-7} = 10 \cdot 5^{x^2}$

19. $3^{x+4} < 7$

20. $4^{2x-5} = 7^{x-1} - 3$

21. $\left(\frac{1}{2}\right)^{x+3} = \left(\frac{1}{5}\right)^{x+1} - 1$

22. $\left(\frac{3}{4}\right)^{2x-3} \geq 2^{x+4} + 3$

23. $2^{|x-1|} = (x-1)^2$

24. $\left(\frac{1}{5}\right)^{x^2} = x^2$

25. $3x + 4 = 2^{x+1} - 4$

From Mathematics to English

26. When $x \in [0, \infty)$, does the function x^2 rise faster than 2^x? Explain your answer.

27. Without solving the equation $2^{x-1} + 3 = -\left(\frac{1}{2}\right)^x + 1$ and without looking at the graphs of the functions on each side of the equation, how do you know there is no solution to the equation?

28. How many solutions to the equation $d \cdot b^{x+e} + f = 0$ are possible? Why?

29. Describe the solution to an exponential equation in graphical terms.

30. If you memorize how to solve an exponential equation, do you understand how to solve other exponential equations?

31. What is an exponential equation?

32. Describe how to solve the equation $2^x = 8$. Do not actually solve the equation.

33. After reading this section, make a list of questions that you want to ask your instructor.

34. Continue in your daily journal and make an entry. In addition to your normal entry on thoughts about the mathematics in this section, list at least two positive comments about what you have learned about this topic.

35. In paragraph format, summarize the material in this section of the text in your daily journal.

36. Describe how your classroom instructor made this topic more understandable and clear.

37. After reading the text and listening to your instructor, what do you not understand about this topic?

From English to Mathematics
Write each of the following English statements as an equivalent mathematical statement.

38. The function $3 \cdot 2^{x+1} - 3$ is 7.

39. The function 5^{x-1} is the same as the function $\left(\frac{2}{3}\right)^{x+1} - 2$.

40. The function 5^{x-1} is less than 6.

Explorations
Solve the following inequalities.

41. $2^{x+3} - 1 \geq 2x + 7$

42. $\left(\frac{1}{5}\right)^{x^2} < x^2$

43. $-|x+2| + 3 < \left(\frac{3}{2}\right)^{x-1} - 1$

7.4 The Exponential Function as a Mathematical Model

In the last section you briefly investigated uses of the exponential function in deprecation, the national debt, and in cooling. The applications of the exponential function in this section involve growth. Growth of bacterial population, human population, and growth of money will be investigated.

Bacterial Population Growth

A bacteria commonly found in the human body is called E. coli. It can double in number every 20 minutes. Suppose at time 0, two bacteria are observed. If growth is unrestricted, the numeric representation of the growth relationship is shown below. Unrestricted growth implies all the space needed is available, there are no predators, and all the food needed is available.

	time (x)	bacteria population (y)
	0	2
	20	4
	40	8
	60	16
	80	32
	100	64
	120	128
	140	256
	160	512
(3 hours)	180	1,024
(4 hours)	240	8,192
(5 hours)	300	65,536
(6 hours)	360	524,288
(7 hours)	420	4,194,304
(8 hours)	480	33,554,432
(9 hours)	540	268,435,456
(10 hours)	600	2,147,483,648

The number of bacteria is growing exponentially. You may recognize the sequence of numbers 2, 4, 8, 16..., as 2 raised to an integer exponent; thus, a first guess at a model might be 2^x. You will notice however, that the E. coli doesn't double every 1 minute. It doubles every 20 minutes. How can you alter the exponent so that it increases by 1 every time x increases by 20? Divide x by 20 (or multiply by 1/20).

Now the model is $2^{\frac{x}{20}}$. This will almost predict the correct number of bacteria. But what if you start with 17,000 bacteria instead of 2? How can you change the model to correctly predict the number of bacteria after x minutes? The answer is simple, multiply by 17,000. Finally, the symbolic representation for this type of growth is the exponential function $y = y_0 \cdot 2^{\frac{x}{20}}$, where y_0 is the initial number of bacteria at time zero, x is time in minutes, and y is the number of bacteria at time x. Knowing the symbolic representation allows you to solve a variety of problems related to bacterial populations. A few samples follow.

Using the symbolic representation as a model, if y_0 is 2 and time is 480 minutes (8 hours), the number of E. coli bacteria can be found.

$y = 2 \cdot 2^{\frac{480}{20}}$	The calculator keystrokes are:
$y = 2 \cdot 2^{24}$	2 × 2 ^ 24 ENTER
$y = 33{,}554{,}432$	The 2 bacteria at the beginning have increased their numbers to 33,554,432 in 8 hours.

If the bacteria colony at time zero is 1,000,000 and the number of bacteria after 3 hours (180 minutes) needs to be found, the symbolic representation of the mathematical model will allow you to calculate this number.

$y = 1000000 \cdot 2^{\frac{180}{20}}$	The keystrokes are:
$y = 1000000 \cdot 2^{9}$	1000000 × 2 ^ 9 ENTER

$y = 512{,}000{,}000$ bacteria after 3 hours.

Example 1: Find the number of E. coli bacteria after two days of unrestricted growth, if 3,000,000 are observed at time zero.

Solution: Two days is 48 hours, which is 2880 minutes. The number of E. coli bacteria is:

$y = 3000000 \cdot 2^{\frac{2880}{20}}$	The keystrokes are:
$y = 3000000 \cdot 2^{144}$	3000000 × 2 ^ 144 ENTER

$y = 6.69022356\,x10^{49}$

or $y =$ 69,022,356,000,000,000,000,000,000,000,000,000,000,000,000,000,000.

This number is approximate because the limits of the calculator were exceeded in the calculation process.

⁂

The graphical representation of the E. coli population growth model is shown below in Figure 7.4.1. The population at time zero is chosen to be 1,000,000. Since the y-coordinates are rather large, it is suggested that if you graph the exponential function for large x, care must be taken when specifying the viewing range. Relate the problem you are solving to the window you should use. The window used in Figure 7.4.1 is [−100, 300] by [−1E8, 1E9].

$$1000000 \cdot 2^{\frac{x}{20}}$$

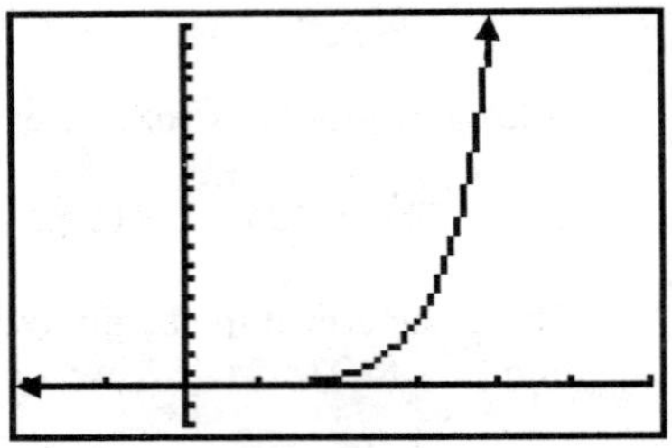

Figure 7.4.1

As you might expect, even though the symbolic form is not quite the same as you found in Section 7.1, the behavior of the function is typical for an exponential function. It is increasing on the domain; it has no maximum or minimum; and it is always positive -- thus, there are no zeros. The most interesting behavior is the average rate of change of the function. If you compare the average rate of change of the function for small values of time and then large values, you will find an immense difference. For example, the average rate of change between 20 and 21 minutes is 70,530. The E. coli are growing at a rate of 70,530 bacteria per minute. If you calculate the average rate of change between 120 and 121 minutes, it is 2,256,955. That is, the bacteria are growing at a rate of over 2 million per minute. From the graphical representation of the exponential function, the trace key on the graphing calculator can be used to estimate the number of bacteria at any given time. Likewise, the time can be estimated for any given number of bacteria.

Example 2: How long does it take for 500,000 E. Coli bacteria to become 20,000,000 bacteria under unrestricted growth conditions?

Solution: The exponential function is $y = 500000 \cdot 2^{\frac{x}{20}}$. One method for solving this problem is to graph the function and use the trace key to display the coordinates of points on the graph to find x when y = 20,000,000. The point is shown on the graph below.

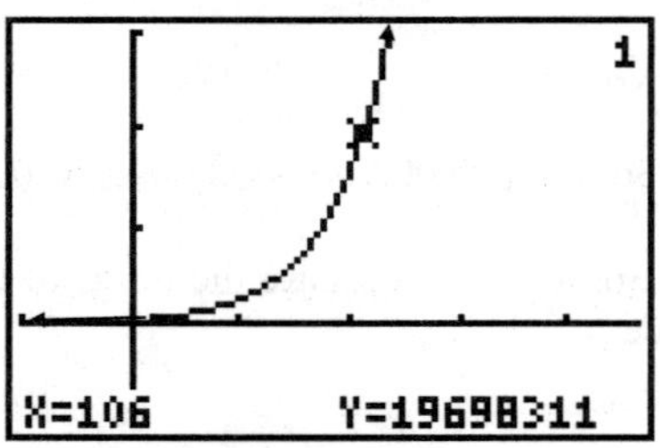

Figure 7.4.2

It takes approximately 106 minutes for 500,000 E. coli to grow to 20,000,000 E. coli bacteria. This problem can also be solved by using the mathematical model of E. coli growth as an equation. Substitute the known values and solve the equation using the technique developed in Chapter Six.

$20000000 = 500000 \cdot 2^{\frac{x}{20}}$ divide both sides by 500,000

$40 = 2^{\frac{x}{20}}$ subtract 40 from both sides

$0 = 2^{\frac{x}{20}} - 40$ graph the function $y = 2^{\frac{x}{20}} - 40$ and find the zero.

$y = 2^{\frac{x}{20}} - 40$

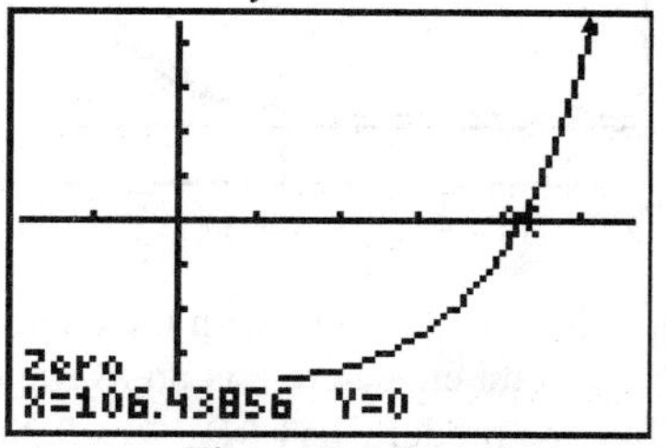

Figure 7.4.3

The solution with an error < 0.01 is 106.44. Bacteria cannot always grow unrestricted. Conditions in the habitat (the human body) such as lack of space, absence of food, and presence of natural enemies retards or controls growth.

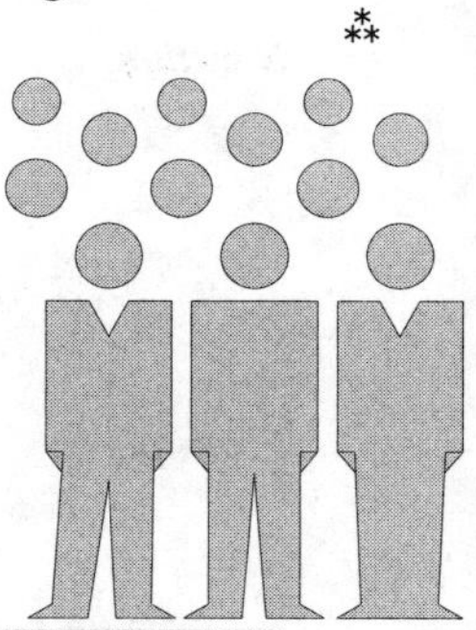

Human Population Growth

Displayed below is data on the human population of earth measured since the year 1000 AD. The human population data is in billions (1×10^9). Time is measured in calendar years.

Table 7.4.1

time (calendar)	1000	1750	1800	1850	1900	1950	1994	2001	2006
human population	.2	0.8	1.0	1.2	1.7	2.8	5.4	6.2	6.6

The human population numbers in the above numeric representation do not "look" as drastic in growth as the previously studied bacteria. Of course, time is measured in years instead of minutes and the word *billion* does not look as impressive as the long numbers generated in the bacteria table. None-the-less, human population is growing exponentially. You will be more convinced by looking at the graphical representation of the data.

The exponential function below that models human population growth uses an average annual growth rate of 1% and will model the current data from Table 7.4.1. It is $y = (1.01)^{x-1850} + 0.2$. The graph of the data relationship and the model are shown in Figure 7.4.4 below.

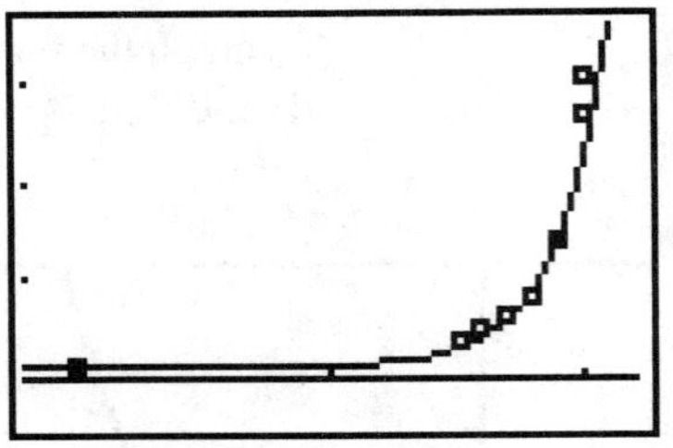

Figure 7.4.4

Usually, the form of the growth model is $y = y_0(1 + r)^x$, with the coefficient y_0 as the beginning population for the time period, $(1 + r)$ as the base number, where r is an average yearly rate of growth, x is time in years (NOT CALENDAR YEARS like the model $y = (1.01)^{x-1850} + 0.2$), and y is the population at time x.

Example 3: Find the human population in 2012 if an estimation of the annual rate of growth is 1.5%, and the population in 1950 was 2.8 billion people.

Solution: With an annual rate (r) of growth at 1.5% and an initial population (y_0) of 2.8 billion in 1950, the model becomes $y = 2.8(1+0.015)^{62}$ where time is 62 because from 1950 to 2012 is 62 years. To find the projected population, just calculate. 2.8 * 1.015 ^ 62 ENTER. It is 7.05 billion.

⁂

Example 4: The human population in 1991 was around 5.2 billion and the growth rate was about 1.8%. Under these conditions, how long would it take for the human population to reach 10 billion?

Solution: The exponential function is $y = 5200000000(1.018)^x$, where r is 0.018 and $(1 + r)$ is 1.018, x is 0 in 1991, and 5.2 billion can be entered on the calculator as 5.2 EE 9. One method for answering the question is the trace method. Graph the function and use the trace feature of the calculator to find a point on the graph when y = 10 billion (1 EE 10). A good viewing window is [0,50] by [0,12 billion].

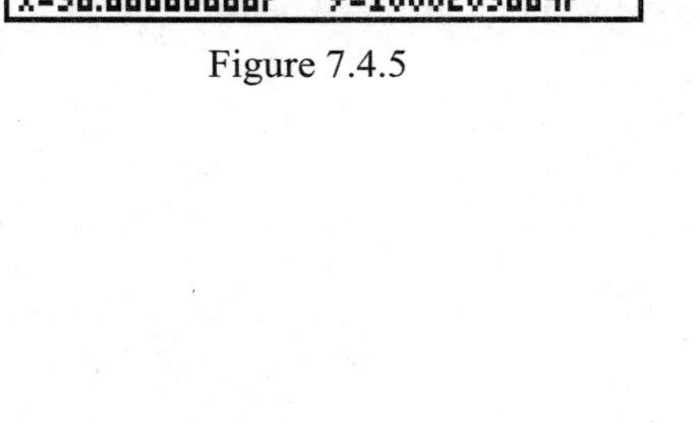

Figure 7.4.5

As can be seen from the graph, in approximately 36.7 years from 1991 (the year 2028), the population of earth will be about 10 billion people, assuming a continued growth rate of 1.8% and no change in the mortality rate. The problem can also be solved by the method developed in Chapter Six; when y is replaced with 10E9 (10 billion), the mathematical model becomes an equation. Solve it.

$10\text{E}9 = 5.2\text{E}9(1.018)^x$ subtract 10E9 from both sides

$0 = 5.2\text{E}9(1.018)^x - 10\text{E}9$ graph $y = 5.2\text{E}9(1.018)^x - 10\text{E}9$ and find the zero.

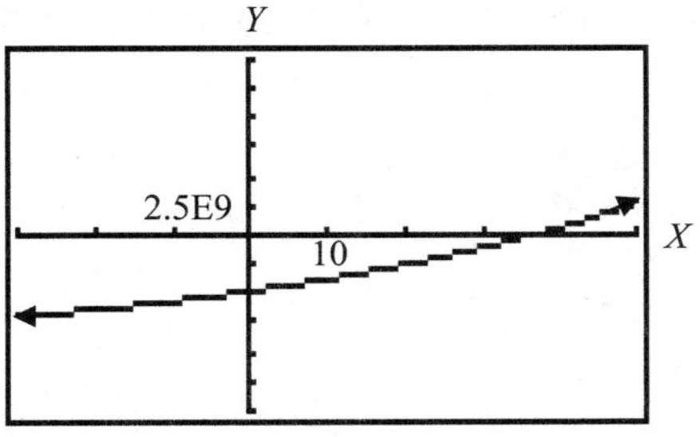

Figure 7.4.6a

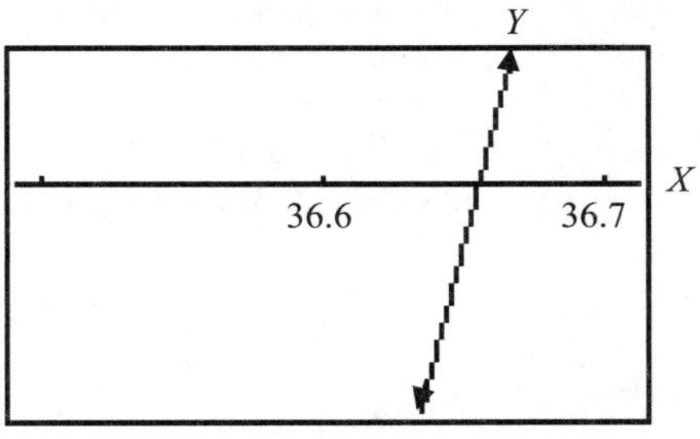

Figure 7.4.6b

Figure 7.4.6b shows a zoom-in with tic marks set at 0.1. As you can see from the graph on the right, the zero is between 36.6 and 36.7, a good estimation of the zero is 36.66 (the year 2028).

⁂

Literature on human population suggests that the earth cannot hold more than 10 to 11 billion people. It may be of further interest to note that it took approximately 1,000,000 years for humans to populate to 4 billion people; this happened in 1976 AD. Just 25 years later, the human population is over 6 billion and, as calculated in the above problem, it will take another 37 years to add another 4 billion people to this planet. The first 4 billion people in a million years -- the next 4 billion in 37 years.

Growth of Money

The last application of the exponential function studied here is the growth of money. Money grows exponentially if it is invested where the interest is compounded. Besides compound interest bank accounts, investing in land or property is another example of money growing exponentially. As with human population growth, the circumstances under which money grows may vary. Changing circumstances require a change in the exponential function used to model the growth. For the sake of simplicity, no changes in conditions in a particular investment are assumed. One other condition that will be used for the sake of simplicity is that the compounding period will be a year -- the time period used to add interest to the account or other investment.

The exponential function used to model the growth of money is the same function as for population growth because of the common use of the annual growth rate.

$y = y_0(1 + r)^x$, where x represents time in years, r is annual interest rate in decimal form, y_0 is the amount invested at the beginning of the investment, and y is the value of the investment after x years have passed.

Example 5: Make a numeric representation for an investment of \$10,000 with an annual rate of 12% for 10 years.

Solution: Given $y_0 = 10000$ and $r = 0.12$. The symbolic representation is $y = 10000(1.12)^x$. Use TblSet and start the table at 1 and make $\Delta x = 1$.

(in years) x	y (in dollars)
1	11,200.00
2	12,544.00
3	14,049.28
4	15,735.19
5	17,623.42
6	19,738.23
7	22,106,81
8	24,759.63
9	27,730.79
10	31,058.48

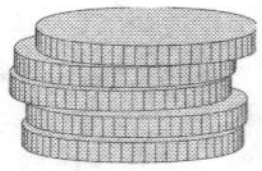

The numeric representation allows you to scan through the information and get a general idea of what the value of your investment will be as time passes. For example, after ten years your investment will more than triple.

⁂

Example 6: How long does it take \$1,000 to grow to \$2,000 at an annual interest rate of 12.5%?

Solution: Graph the function $y = 1000(1.125)^x$ and find a point on the graph where the y-coordinate is \$2,000. The corresponding x-coordinate will be the time for \$1,000 to grow to \$2,000. Figure 7.4.7a shows an approximate solution of 5.9 years and the

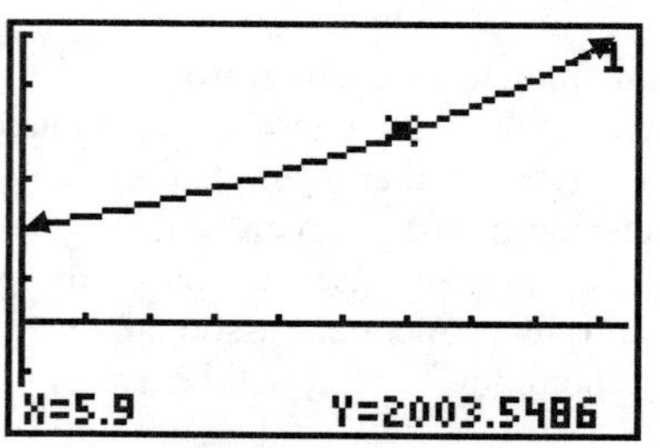

Figure 7.4.7a

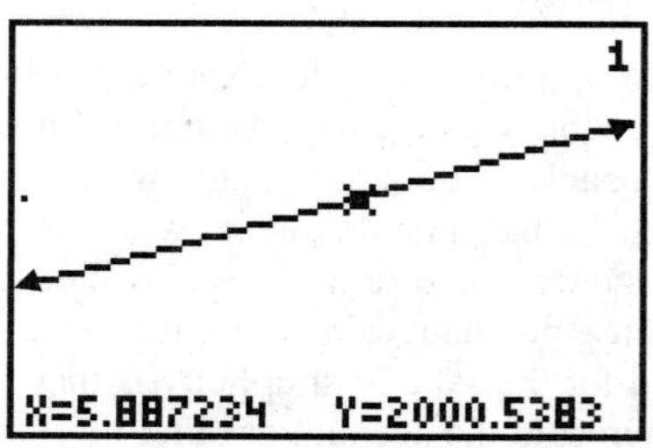

Figure 7.4.7b

zoom-in shows a better solution of 5.89 years. The zeros method also works quite well and is shown below. Replace y with 2000 and solve the equation.

$2000 = 1000(1.125)^x$	divide both sides by 1000
$2 = 1.125^x$	subtract 2 from both sides
$0 = 1.125^x - 2$	Graph the function $y = 1.125^x - 2$ and find the zero.

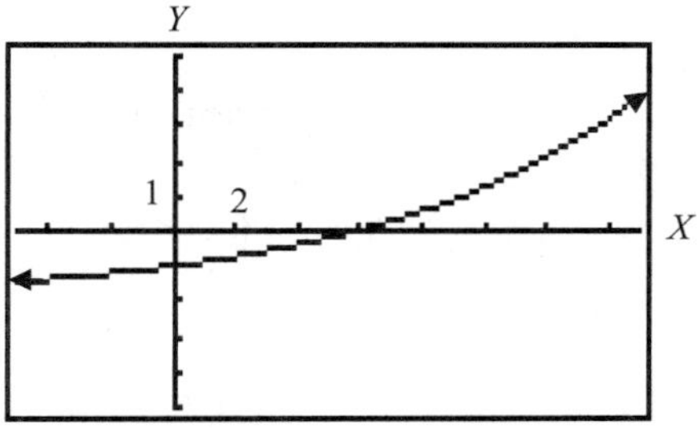

Figure 7.4.8a

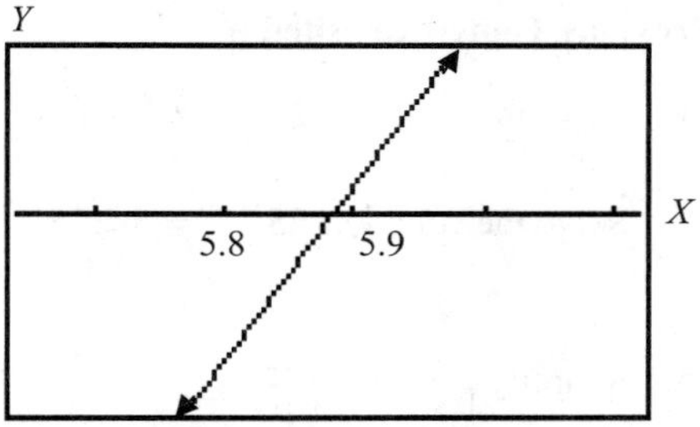

Figure 7.4.8b

As shown in the graphical representation in Figure 7.4.8b, the time it takes for \$1,000 to grow to \$2,000 is about 5.89 years. Finally, the problem can also be solved by the numerical method. To solve the equation $0 = 1.125^x - 2$, make the numerical representation of the function $y = 1.125^x - 2$ and find the zero.

Table 7.4.2

x	1	2	3	4	5	6	7
$y = 1.125^x - 2$	–0.875	–0.734	–0.576	–0.398	–0.198	0.27	0.281

Because the function changes sign between 5 and 6, there is a zero between 5 and 6. Now make a numeric representation with a Δx of 0.01.

Table 7.4.3

x	5.85	5.86	5.87	5.88	5.89	5.90	5.91
$y = 1.125^x - 2$	–.0082	–.0059	–.0035	–.0012	.0012	.0036	.0059

The function changes sign between 5.88 and 5.89; choose any number between the two. With an error of < 0.01, the solution is 5.885.

⁂

7.4 EXERCISES

Many times you can work the exercises by algebraic, numeric, or graphical methods. You can learn about mathematics no matter which method you use. Part of the learning process is for you to decide which method is best for you to use. Please read the text carefully before you do the exercises. Read the **next assigned section** after you finish the assignment below. All numeric answers must have an error ≤ 0.01.

A Quick Look To The Future

-4. What function, if any, have you studied that has a graph that is in two pieces?

-3. Does $\frac{4}{5} = \frac{4 \cdot 3}{5 \cdot 3}$?

-2. Do the 3's cancel in $\frac{2}{3} \div \frac{3}{5}$?

-1. Does $\frac{1}{2} + \frac{3}{4} = \frac{1+3}{2+4}$?

Previous Topics Revisited

1. Solve the equation $2^{x-3} - 4 = \sqrt{x+1} - 4$
2. Solve the equation $3^{2x-5} - 4 = \left(\frac{3}{5}\right)^x + 1$
3. Solve the equation $25^{3x-5} = 5 \cdot 125^{x+2}$
4. Simplify $-2\left(2x^{-3} \cdot x^4\right)^2$
5. Simplify $\left(\frac{2^{x^2-3x+5}}{2^{-x^2+5x-1}}\right)^3$
6. Create an exponential function that passes through the point (1, 2).

New Topic Exercises

7. Using the unrestricted population growth model for E. coli bacteria, find how long it takes for 1,000,000 bacteria to grow to 10,000,000. Remember: 1,000,000 on the calculator is 1 EE 6 and 10,000,000 is 1 EE 7, or you may symbolize millions with numbers like 1 and 10 .

8. If 1,000,000 E. coli bacteria in a petri dish are counted at 10:00 AM, how many are in the petri dish at noon?

9. How long does it take 4 billion E. coli bacteria to grow to 8 billion bacteria?

10. Bacillus stearothermophilus bacteria under unrestricted growth condition will double in population every 8 minutes. This means that the model for this growth is $y = y_0 \cdot 2^{\frac{x}{8}}$. How long will it take for 10,000 bacillus stearothermophilus bacteria to grow to 1,000,000 bacteria?

11. If 1E9 bacillus stearothermophilus bacteria are observed after 3 hours of growth, how many bacteria were there at the beginning of the growth period?

12. What annual growth rate would allow approximately 6.5 billion humans today to grow into 8 billion in 100 years?

13. Under the current human population growth rate of 1.8%, how long will it take for the current 6.5 billion people to become 50 billion people?

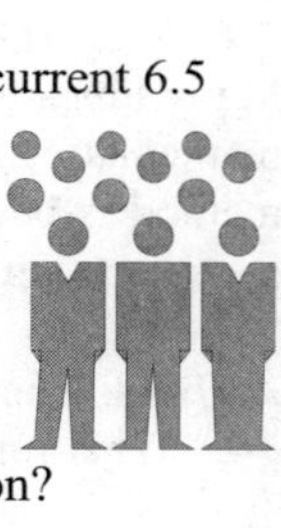

14. With a current population of 6.5 billion and growth rates of 1.8%, 1.4%, 1.0%, 0.6% and 0.2%, find the population of humans on earth in 10 years under each of the rates.

15. How many people will there be on earth in 5, 10, 15, 20, 25, and 30 years if the current growth rate is maintained at 1.8% and you assume initial conditions of 6.5 billion?

16. How long does it take $100,000 to grow to $200,000 at 8% compounded annually? How long does it take $1 to grow to $2 under the same conditions?

17. How much money is in an account after 50 years if the original investment is $10,000 at 25% compounded annually?

18. Find the account balance at the end of 5, 10, 15, 20, and 25 years if $5,000 is invested at 12%.

19. How much money should you invest in an account that earns 8% interest compounded annually if you want to have $250,000 in your account after 40 years?

20. Which account will earn more money a) $5,000 at 9½% for 7 years, or b) $5,000 at 10% for 6 years?

21. How much interest is earned on an investment of $10,000 at 14% for 30 years?

22. How long does it take for P dollars to become $2P$ dollars at 12% compounded annually?

23. Is a $1,000 investment at 12% compounded annually for 10 years, better or worse than a $1,100 investment at 12% compounded annually for 9 years?

24. How long will it take for a $1,000 investment at 8% compounded annually to be worth more that a $1,100 investment at 7% compounded annually?

25. What interest rate at annual compounding will cause an investment of P dollars to become $2P$ dollars in less than 5 years?

From Mathematics to English

26. Why won't there be 50 billion people on earth in 115 years? Refer to Exercise 13.

27. Research and then write a paragraph about E. coli bacteria in the human body.

28. The mathematical model for human population growth is $y = y_0 \cdot (1 + r)^x$. Discuss hat happens if $r = 0$.

29. Make a list of socially acceptable ways of lowering human population.

30. List possible conditions that you think will keep human population below 10 to 11 billion.

31. If you memorize the model for growth of the E. coli, do you know how to find the model for growth of Mycobacterium tuberculosis? Explain.

32. If you memorize the growth model for money, do you know how money grows if it is compounded quarterly? Explain.

33. What is a mathematical model?

34. Make a list of things that grow exponentially.

35. After reading this section, make a list of questions that you want to ask your instructor.

36. Continue in your daily journal and make an entry. In addition to your normal entry on thoughts about the mathematics in this section, list at least two positive comments about what you have learned about this topic.

37. In paragraph format, summarize the material in this section of the text in your daily journal.

38. Describe how your classroom instructor made this topic more understandable and clear.

39. After reading the text and listening to your instructor, what do you not understand about this topic?

Explorations

40. While the Escherichia coli (E. coli) bacteria can double in population every 20 minutes under optimum conditions, other living things double in times other than 20 minutes. For each creature listed below, find a symbolic representation of its growth function.

	Creature	Approximate time to double the population
a.	mouse	17 months
b.	human (current conditions)	39 years
c.	Mycobacterium tuberculosis	6 hours

CHAPTER SEVEN TEST

1. Is $y = \left(\frac{1}{4}\right)^{x+2} - 1$ increasing or decreasing over its normal domain?

2. What conditions on d cause the function $y = d \cdot 2^{x+4} + 3$ to decrease over its normal domain?

3. Find the zero and y-intercept of $y = -3^{x-2} + 5$.

4. Create an exponential function of the form $f(x) = d \cdot b^{x+e} + f$ that is increasing and has a horizontal asymptote at -3.

5. What is the vertical distance between the graphs of $y = 7^x$ and $y = 7^x - 5$?

6. Name the transformations of $y = 5^x$ that will yield $y = -3 \cdot 5^{x+4} - 2$?

7. How much farther from the x-axis is $y = 2^{x-3} + 4$ than is $y = 2^{x-3}$?

8. Simplify $\dfrac{(3x^4)(-2x^{-5})}{18x^{-3}}$

9. Simplify $\left[(-2)^4 x^{-3} \cdot (-2)^{-2} x^{-5}\right]^{-3}$

10. Simplify $\dfrac{3^{x^2-5x+1}}{3^{-2x^2+4x-5}}$

11. Simplify $3^{5x-3} \cdot 3^{2x+1}$

12. Simplify $\left(5^{2x+1}\right)^3$

13. Solve the equation $8^{x+2} = 2^{3x-5}$

14. Solve the equation $3 \cdot 5^{2x+1} = 15 \cdot 5^{x-7}$

15. Solve the equation $5^{x+1} - 3 = \left(\frac{7}{9}\right)^{x-2} + 1$

16. Solve the equation $|x-1| - 3 - \sqrt{x+2} = -3 \cdot 2^{3x}$

17. Using the mathematical model $y = y_0 \cdot 2^{\frac{x}{20}}$ for the unrestricted growth model of the E. Coli bacteria, how many minutes does it take for 50 bacteria to grow to 50,000,000 bacteria?

18. Using the mathematical model $y = y_0(1+r)^x$ for the unrestricted growth model for humans, how many years does it take for 6 billion humans to grow to 12 billion at an annual rate of 0.5% (1/2 percent)?

19. In 1950, the human growth rate was approximately 2% and there were 2.5 billion people on Earth. How many people were there is 1951? (Assuming the growth rate was constant.)

20. What annual interest rate would cause an investment of \$800 to become \$80,000 in four years? (Use the same mathematical model as for human growth).

CHAPTER EIGHT
ANALYSIS OF THE RATIONAL FUNCTION

8.0 Introduction

The Bright Light Bulb

Below is the numeric representation of the relationship between the intensity of light I and the distance d (in feet) from a light bulb. The data was collected with a light meter and a tape measure. The problem is to create a mathematical model of the data so that someone like a photographer can use the model to predict the light intensity at any point from the light source. To discover what the model is, the first step is to find the graphical representation of the data. This should suggest a general mathematical model. From this point, you will need to the find function parameters so that the model matches the data.

Table 8.0.1

d	.5	1	2	4	5	6	7	8	9
I	64	16	4	1	0.64	0.44	0.33	0.25	0.2

As you might suspect, there is a new behavior in the data. Much of what you learn in this chapter will help you develop the model for this data.

The Rational Function in Action **The Speed Check**

Did you ever notice a short white marker bar line drawn perpendicular to the pavement on long straight stretches of the freeway? Usually there are three or four of them with a constant distance between them. If you approximate the distance between the markers, you will find them to be spaced at quarter ($\frac{1}{4}$) mile intervals. If you are wondering what all this has to do with the rational function, read on. You may remember the "distance/velocity formula" used in earlier classes. It states that the distance traveled can be calculated by multiplying the average velocity times time. That is, $D = vt$. In the case of the white markers on the freeway, the distance is $\frac{1}{4}$ mile $\left(D = \frac{1}{4}\right)$. Substituting this known value, the "formula" becomes:

$$D = vt \qquad \text{or} \qquad \frac{1}{4} = vt \qquad \text{which is} \qquad v = \frac{\frac{1}{4}}{t}$$

The function $v = \dfrac{\frac{1}{4}}{t}$ is a rational function that models the average velocity in miles per hour. The problem with this model is that the police officer measures time in seconds and the model requires hours. If you want t to be entered in seconds yet v to be calculated in miles per hour, you can convert the seconds to hours by dividing t seconds by 3600 -- the number of seconds in an hour. The model now becomes $v = \dfrac{\frac{1}{4}}{t} = \dfrac{\frac{1}{4}}{\frac{t}{3600}} = \dfrac{1}{4} \cdot \dfrac{3600}{t} = \dfrac{900}{t}$; that is, $v = \dfrac{900}{t}$ where t is in seconds and v is in miles per hour. A highway patrol officer in a plane with a stopwatch can measure the time it takes for a motorist to travel a distance of $\frac{1}{4}$ mile and then use the velocity function to calculate the velocity. The officer can use the graph of the function and the trace key to find a value for v from the known value of t. On the graph below, the trace key shows that if the motorist travels the distance in 12 seconds, the velocity is 75 mph, and if she travels it in 9 seconds, she is traveling at an average speed of 100 mph.

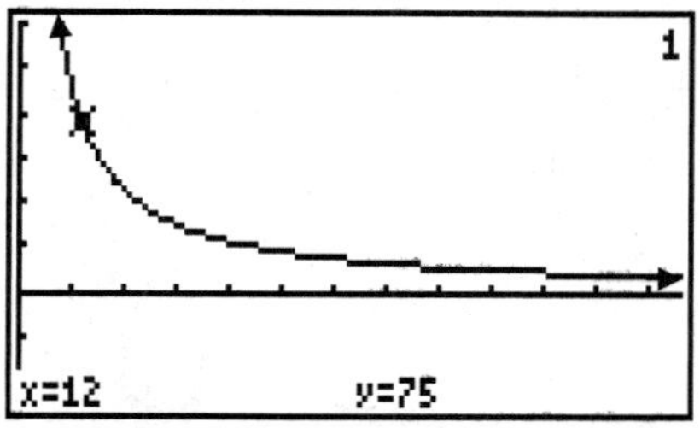

Figure 8.0.1a

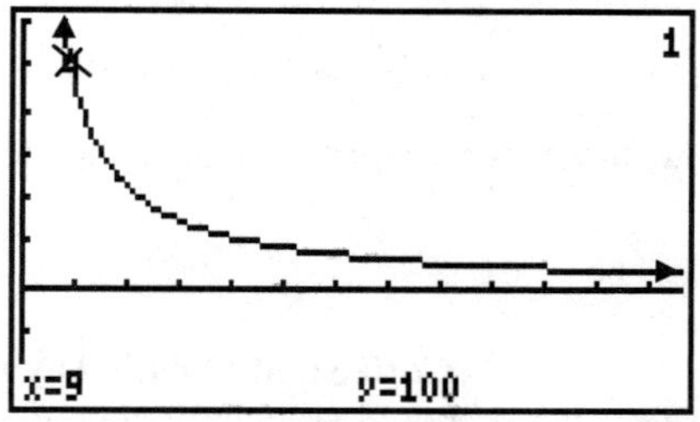

Figure 8.0.1b

If the officer has no graphing calculator, she can use the numeric representation of the velocity function as shown below.

Table 8.0.2

t	5	6	7	8	9	10	11	12	13	14	15
$\frac{900}{t}$	180	150	129	113	100	90	82	75	69	64	60

It could be that she may want to use a table where Δx is 0.1 instead of 1, provided she has a stopwatch that can be read to the nearest tenth of a second.

The rational function was not introduced in Chapter Three with the linear, quadratic, absolute value, exponential, and square root functions because it has a characteristic that is different than all of the other elementary functions studied -- its graph is split into pieces. All other functions studied have a behavior mathematicians call **continuity.** This means the entire graph can be drawn without lifting your pencil. As you will see in the next section, rational functions cannot be graphed in one continuous movement of the pencil. They are split into at least 2 pieces. The primary reason for this is that the domain is not one continuous set of real numbers. For example, the function $f(x) = \dfrac{1}{(x-2)(x+3)}$ has a domain of $(-\infty, -3) \cup (-3, 2) \cup (2, \infty)$. If you graph this function correctly, you will see three different pieces of the function, one per continuous piece interval that makes up the domain.

The rational function studied in Section 8.1 is the simplest rational function. To prepare you for more complex rational functions and the analytical method of solving rational equations, the middle three sections of Chapter Eight develops ideas on algebraic simplifications of rational functions. You will learn how to reduce, add, subtract, multiply, and divide rational expressions that appear in more complicated rational functions. The algebraic manipulations end with simplification of complex rational functions. In the last section of the chapter you will learn how to solve equations numerically, graphically, and analytically.

8.1 Rational Functions

Figure 8.0.1 shows the graph of the rational function $v = \frac{900}{t}$; however, the domain of the function has been restricted to $(0, \infty)$. Because of the restricted domain, you only see one piece of the graph. As you will see below, there are two pieces to the complete graph of functions like $v = \frac{900}{t}$.

Coffee and Cream Anyone?

A typical coffee mug will hold around 300 cubic centimeter (cc) of coffee. This will leave space for cream. The coffee situation is this: what percent of a coffee and cream mixture is coffee? This is sometimes called the **strength** of the mixture. Many restaurants provide cream in small containers. Since the containers are usually not full, an estimation of the amount of cream in one container is 6 cc. Suppose you put one container of cream in your coffee, the strength has changed from 100% coffee to something lower. The strength can be found by dividing the amount of coffee by the amount of mixture; thus, the mixture now has a strength of $\frac{300}{300+6}$. This number is the decimal number 0.980. Most people prefer to express the strength of a mixture as a percent. You can do this by multiplying by 100; therefore, the strength is 98.0% coffee. The calculation can be written as $100 \cdot \frac{300}{300+6}$. If you add two containers of cream, the strength has changed to $100 \cdot \frac{300}{300+6 \cdot 2}$, or 96.2%. Perhaps you want to cool your coffee fast so you may add three creams to you coffee. It now has a strength of $100 \cdot \frac{300}{300+6 \cdot 3}$, or 94.3%. If you don't like coffee but feel you must drink it at breakfast, you may want to put 4 creams in your coffee. It now has a strength of $100 \cdot \frac{300}{300+6 \cdot 4}$, or 92.6%. You can see a pattern developing for the strength of the coffee mixture. If you symbolize the number of creams with x, the function that models the strength of your coffee is $S_c = 100 \cdot \frac{300}{300+6x}$. If you are interested, you can simplify the model by factoring 300 into 6×50 and factoring $300 + 6x$ into $6(50 + x)$. The 6 factor cancels leaving the model $S_c = 100 \cdot \frac{50}{50+x}$ or $S_c = 100 \cdot \frac{50}{x+50}$.

The numeric representation of this function will confirm that the strengths given above are correct. Figure 8.1.1 is the numeric representation of $S_c = 100 \cdot \frac{50}{x+50}$ with a domain displayed similar to the above work.

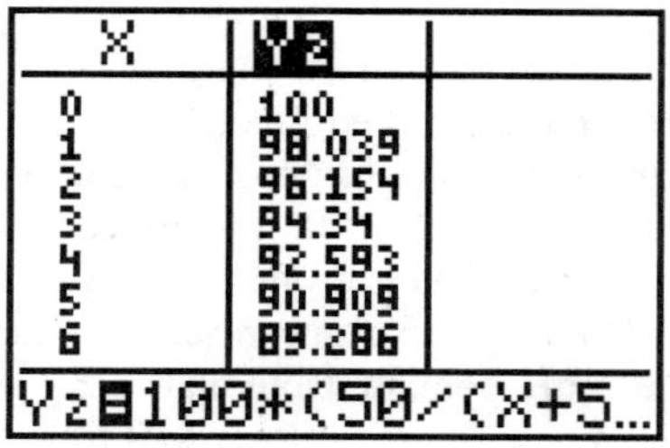

Figure 8.1.1

The two figures below show a complete graph on the normal domain in Figure 8.1.2a and a graph with the problem domain shown in Figure 8.1.2b, with windows of [–80, 14] by [–1000, 1000] and

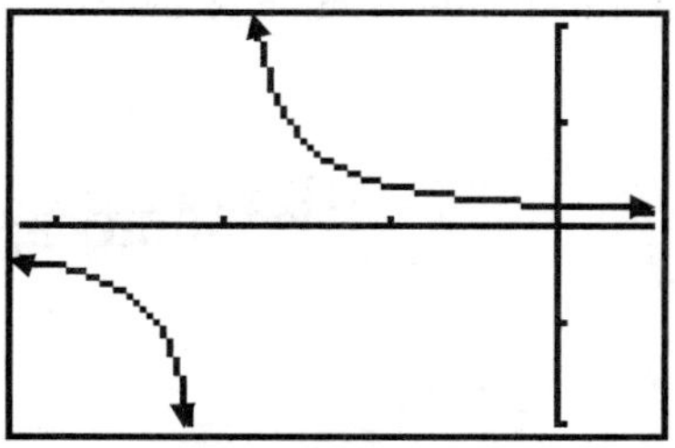

Figure 8.1.2a

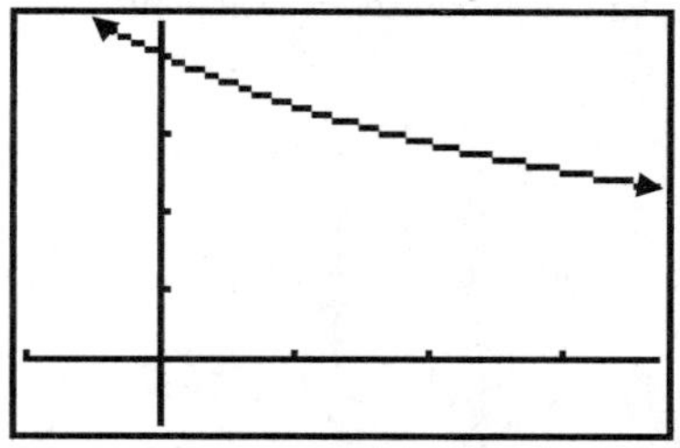

Figure 8.1.2b

[–10, 37] by [–20, 110] respectively. The two pieces of the function show clearly in Figure 8.1.2a.

The coffee and the speed check problems demonstrate that there is a use for an entirely new kind of function -- the rational function. Below is a discussion of the properties of the function.

A rational function is a quotient of polynomials. The parent or simplest rational function is $y=\frac{1}{x}$ or $f(x)=\frac{1}{x}$. In arithmetic you learned division by zero is undefined. This means in the function $\frac{1}{x}$, the value of x cannot be 0. If the domain of the function contains 0, the value of the function is undefined at that value. The definition of a function requires that the elements of the domain must be real numbers AND the corresponding values of the range must also be real numbers; thus, $x \neq 0$. In the rational function $\frac{1}{x}$, the behavior of the function as values of x get closer and closer to 0 proves to be one of the interesting characteristics of the function. To investigate the behavior of the parent rational function $\frac{1}{x}$, a numeric representation is developed below.

Table 8.1.1

x	–500	–10	–5	–1	–0.1	–0.01	–0.001	0.001	0.01	0.1	1	2	5	10	500
$\frac{1}{x}$	–.002	–0.1	–0.2	–1	–10	–100	–1000	1000	100	10	1	0.5	0.2	0.1	0.002

Observable Behavior
If you look for behavior similar to previously studied functions, you will note no maximum or minimum values. While you should notice that the function has positive and negative values, it is never 0! Graph the function and use trace to show you that it is never 0. As you read the graph from left to right, you very quickly see that it is decreasing; that is, the graph is dropping in value until x becomes 0. At this point the function does not exist. Then to the right of 0, you also can see the function dropping in value. That is, the function is decreasing throughout the entire domain. Note that as x gets closer to 0 from the left of 0, the function gets smaller and smaller. However, as x gets closer to 0 from the right of 0 (reading it from right to left), the function gets very large. Another characteristic is that as x gets larger, the function gets closer to 0. Likewise, as x gets smaller, the function gets closer to 0. Mathematicians use the word *approaches* to mean *gets closer*. The symbol used to denote *approaches* is →. From your work with interval notation, you learned that the symbol ∞ means increasing without bound, or increasing without limit. If these two concepts are put to use when the rational function is described, a description of the rational function $\frac{1}{x}$, becomes:

Is infinity approachable?

As x approaches 0 from the left, $\frac{1}{x}$ decreases without bound.

$$\left(\text{as } x \to 0 \text{ from the left, } \frac{1}{x} \to -\infty\right)$$

As x approaches 0 from the right, $\frac{1}{x}$ increases without bound.

$$\left(\text{as } x \to 0 \text{ from the right, } \frac{1}{x} \to \infty\right)$$

As x increases without bound, $\frac{1}{x}$ approaches 0.

$$\left(\text{as } x \to \infty, \frac{1}{x} \to 0\right)$$

As x decreases without bound, $\frac{1}{x}$ approaches 0.

$$\left(\text{as } x \to -\infty, \frac{1}{x} \to 0.\right)$$

From the numeric representation above, a graphical representation can be developed. It becomes impractical to try to plot all of the points from the above numeric representation because of the great differences in the y-coordinates. If a scale is chosen to display the point (0.01, 100), the point (–500, –0.002) would be so close to the x-axis that it would not be visible. If a scale is chosen to display (–500, –0.002), the point (0.01, 100) would be so close to the y-axis that it would not be displayed. The graphical representation below gives a complete graph of the function; that is, all interesting behavior is displayed. The window is zoom-decimal.

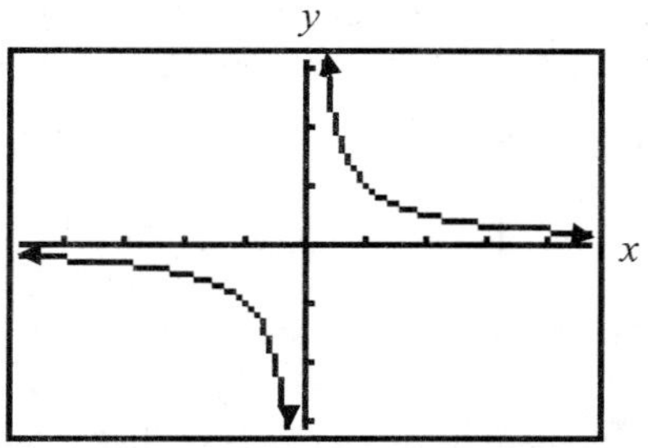

Figure 8.1.3

The parent rational function $y = \frac{1}{x}$ does not have a zero. It does not cross the x-axis -- the place where zeros occur. The function does not have a y-intercept. When a function crosses the y-axis, the value of x is 0. In the parent function $y = \frac{1}{x}$, 0 is not in the domain of the function; thus, it cannot cross the y-axis. The graph of the function gets close to the y-axis both on the left and on the right. Or, using the new terminology, the graph approaches the y-axis as x approaches 0. The y-axis is called a **vertical asymptote** for the function $\frac{1}{x}$. (An asymptote is a line the graph approaches.) The x-axis has the same characteristic. The graph of $y = \frac{1}{x}$ approaches the x-axis as $x \to -\infty$ and as $x \to \infty$. The x-axis is a **horizontal asymptote** for the rational function $\frac{1}{x}$. Most of the rational functions studied at this level will all have horizontal and vertical asymptotes. The concentration in this section will be to study rational functions in the form $f(x) = d\frac{1}{x+e} + f$.

What happens to the graph of $f(x) = \frac{1}{x}$ when a number is added to the function? What happens to the graph of the function when x is replaced with the linear function $(x + e)$? What happens to the graphical representation if the symbolic representation of the function $\frac{1}{x}$ is multiplied by a number? Can a rational function have zeros? Can a rational function cross the y-axis? The answer to these and other questions are in this section.

You are cautioned about using the graphing calculator to graph functions with vertical asymptotes. The graphing calculator plots a sample of data points from the numeric representation in succession from left to right, while at the same time, it also connects these points so that the graph takes on a "continuous" appearance. This means at the vertical asymptote, it will try to connect the point on the left of the real asymptote to the next point on the right of the asymptote. This line is not part of the graph. It shows on the calculator's screen as a nearly vertical line near the vertical asymptote. This vertical line does not represent either points on the graph of the function or the vertical asymptote; it is an error in the graphing procedure. If the vertical asymptote is at a simple decimal value like –3, 2, 1.6, etc., a decimal window is a wise choice in which to graph the

function because at the point where the vertical asymptote exists, the function does not exist; thus the calculator cannot connect a point on the graph to a point that does not exist, and it will not draw an erroneous vertical line.

Transformations Revisited

Notice the relationship between the graphs of the parent function $y = \frac{1}{x}$ and the transformed function $y = \frac{1}{x} + 3$ below in Figure 8.1.4. If you are using a calculator that accepts function notation, you can enter the transformed function as $Y_2 = Y_1 + 3$ provided you have stored the parent in Y_1.

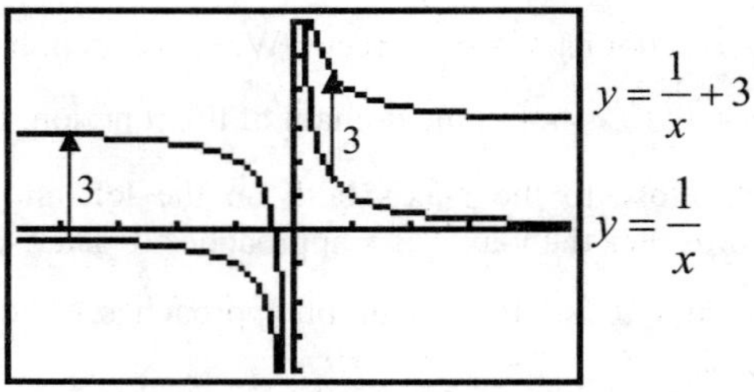

Figure 8.1.4

The graph of $y = \frac{1}{x} + 3$ "appears" to be the graph of $y = \frac{1}{x}$, but moved up by 3 units. Remember that the graph of $y = \frac{1}{x} + 3$ is the plot of sample points from the numeric representation, that is, the calculator did not move the parent graph up 3 units. <u>For descriptive purposes</u>, the graph of $y = \frac{1}{x} + 3$ is said to be the graph of $y = \frac{1}{x}$ moved up by 3 units. This is referred to as a **vertical shift or translation**. If the graph of $y = \frac{1}{x}$ is known, the graph of $y = \frac{1}{x} + 3$ can be obtained without technology by shifting ***every point*** in the parent function up 3 units. By similar reasoning, can the graph of $y = \frac{1}{x} - 5$ be found by moving the graph of $y = \frac{1}{x}$ down by 5 units? To find out, let the graphing calculator plot the function $y = \frac{1}{x} - 5$ and observe the graph to see if it appears to be the same graph as $y = \frac{1}{x}$ shifted down by 5 units. If you can use function notation, you only need change Y_2 to $Y_1 - 5$. It may appear to you that the graphs are not 5 units apart, but the trace cursor will convince you otherwise. Using the trace cursor, jump from one function to the other and for every x in the domain, you will find the y-values are always 5 units different.

The parent

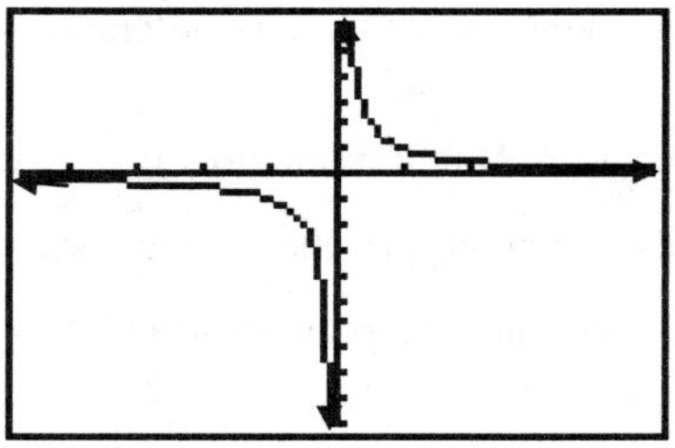

Figure 8.1.5a

The parent and transformed

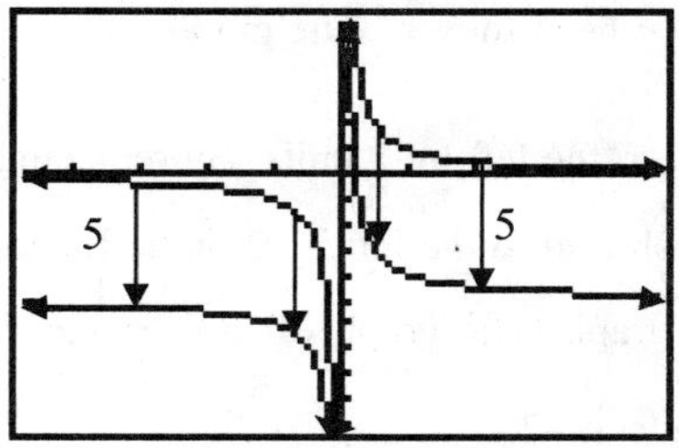

Figure 8.1.5b

In general, the plot of a rational function of the form $y = \frac{1}{x} + f$ is the same graph of $y = \frac{1}{x}$, but vertically shifted by an amount *f*. You can verify this on your calculator by graphing the parent function $y = \frac{1}{x}$ and the transformed function $y = \frac{1}{x} + f$, where *f* is any number. Trace on the parent and then jump to the transformed function. The difference between the *y*-coordinates will be *f*. As the graph of $y = \frac{1}{x}$ is shifted, the horizontal asymptote is also shifted. In $y = \frac{1}{x} - 5$ the horizontal asymptote is the line $y = -5$; that is, the graph of $y = \frac{1}{x} - 5$ approaches –5 as *x* approaches $-\infty$ or ∞.

In the parent rational function $y = \frac{1}{x}$, replace the variable *x* with $(x - 4)$ and also $(x + 2)$ to find what happens to the resulting graphical representations. Below are the graphs of $y = \frac{1}{(x-4)}$ and as well as the parent. If you are using function notation, let $Y_2 = Y_1(x - 4)$ and $Y_3 = Y_1(x + 2)$ because this function notation means "replace the variable in function Y_1 with $x - 4$ or $x + 2$."

$y = \frac{1}{x}$

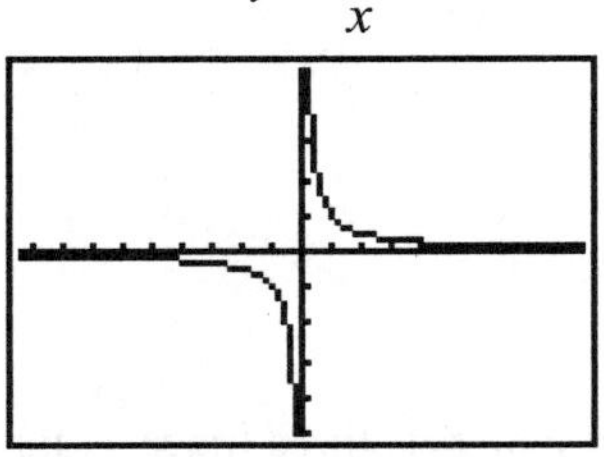

Figure 8.1.6a

$y = \frac{1}{(x-4)}$

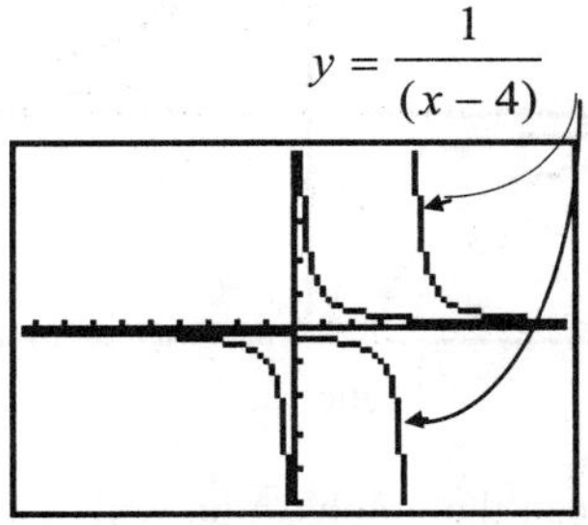

Figure 8.1.6b

$y = \frac{1}{x+2}$

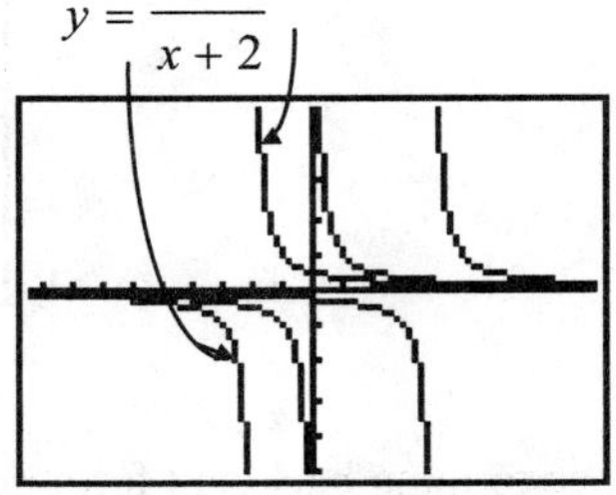

Figure 8.1.6c

As can be seen in the above graph, it appears that the entire graph of $y = \frac{1}{x}$ has been moved to the right by 4 units. The graph can be thought of as the graph of $y = \frac{1}{x}$ shifted right 4 units. The calculator did not move the graph of $y=\frac{1}{x}$; it plotted sample points from the numeric representation of $y=\frac{1}{(x-4)}$. It "appears" to have been moved. The graph of $y=\frac{1}{(x+2)}$ appears to be the same graph as $y=\frac{1}{x}$, but it has been shifted to the left by 2 units. By replacing x with $(x + 2)$ in the function $y=\frac{1}{x}$, the transformed graph has been shifted to the left by 2 units. It should also be noted that the vertical asymptote is shifted along with the graph. The graph of $y=\frac{1}{(x-4)}$ has a vertical asymptote at 4 and $y=\frac{1}{(x+2)}$ has a vertical asymptote at –2.

Example 1: Describe how to graph the function $y=\frac{1}{(x-2)}+3$ using transformations.

Solution: The variable in $y=\frac{1}{x}$ has been replaced with $(x - 2)$; thus, the graph of $y=\frac{1}{x}$ has been shifted right by 2 units. At the same time, 3 has been added to the function. This moves the graph up 3 units. To graph $y=\frac{1}{(x-2)}+3$, graph the function $y=\frac{1}{x}$, except shift the graph right 2 units and up 3 units. The new horizontal asymptote is at 3 and the new vertical asymptote is at 2.

⁂

Below are the graphs of $y=\frac{1}{x}$ and $y=3\cdot\frac{1}{x}$ (also written as $y=\frac{3}{x}$). Like the exponential function in Section 7.1, is the graph of $y=3\cdot\frac{1}{x}$ three times as far from the x-axis as is $y=\frac{1}{x}$? While you can't tell from looking at the figure, use your calculator and graph $Y_1=\frac{1}{x}$ and $Y_2=3Y_1$.

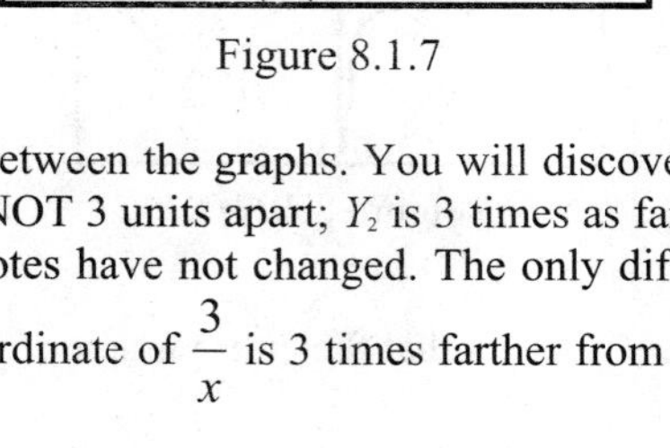

Figure 8.1.7

Use trace to jump back and forth between the graphs. You will discover that Y_2 is three times as far from the x-axis as is Y_1. The graphs are NOT 3 units apart; Y_2 is 3 times as far from the x-axis as is Y_1. Note that the horizontal and vertical asymptotes have not changed. The only difference between the graphs is that for each x in the domain, the y-coordinate of $\frac{3}{x}$ is 3 times farther from the x-axis than the y-coordinate of

$\frac{1}{x}$. This is called a **stretch** by a factor of 3 because the transformed function seems to have been pulled away from the x-axis.

Example 3: Is the y-coordinate of $\frac{5}{x}$ 5 times as far from the x-axis as the corresponding y-coordinate of $\frac{1}{x}$?

Solution: The problem can be solved two ways. Analytically, you know that every point on the graph of $y=\frac{1}{x}$ can be thought of as having coordinates $(x, \frac{1}{x})$. Every point on the graph of $y=\frac{5}{x}$ has coordinates of $(x, \frac{5}{x})$. Since the y-coordinate of $\frac{5}{x}$ is $5\cdot\frac{1}{x}$ and $\frac{1}{x}$ is the y-coordinate of every point on the graph of $y=\frac{1}{x}$, yes, the points on the graph of $y=\frac{5}{x}$ are 5 times as high as the points on the graph of $y=\frac{1}{x}$. Note: on the bottom branch of the graph, each point is 5 times lower. You can verify graphically, the graphs of each function can be put on the same coordinate system, and trace can be used to verify that $\frac{5}{x}$ is 5 times as far from the x-axis as is $\frac{1}{x}$.

⁂

The last transformation is the reflection of a graph about the x-axis. Does multiplying the parent function $y=\frac{1}{x}$ times –1 give a reflection about the x-axis? Instead of looking at the graphical representation to answer the question, use the numeric representations below.

Table 8.1.2

x	–3	–2	–1	1	2	3	4
$\frac{1}{x}$	**–0.33**	**–0.5**	**–1**	**1**	**0.5**	**0.33**	**0.25**
$-\frac{1}{x}$	**0.33**	**0.5**	**1**	**–1**	**–0.5**	**–0.33**	**–0.25**

You see that when $\frac{1}{x}$ is positive, $-\frac{1}{x}$ is negative, and vice versa. From a graphical perspective, this means that the graphs are mirror images with respect to the x-axis. This is called a **reflection about the x-axis**. You should try graphing these two functions on your calculator to verify that this statement is true.

Example 4: Find the zero of the function $f(x)=\frac{1}{x}-3$.

Solution: Graph the function making certain the x-intercept is in the window. Be careful that you do not interpret a possible erroneous vertical line as part of the graph. The graph is below with the zero marked. The zero is exactly $\frac{1}{3}$.

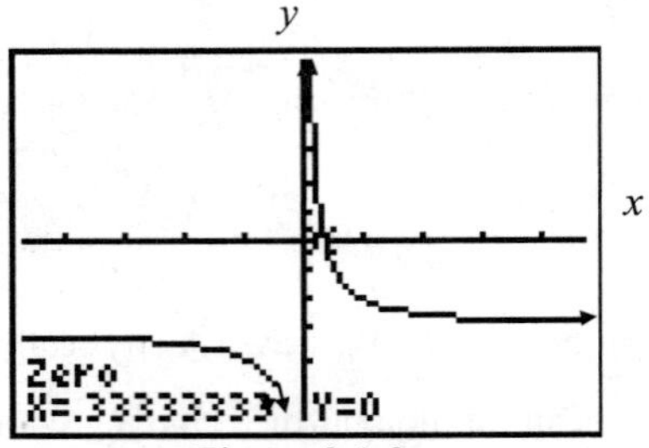

Figure 8.1.8

⁂

Example 5: When is $y = \dfrac{4}{(x+1)} - 2$ increasing and when is it decreasing?

Solution: As x gets larger, the graph is falling (decreasing) to the left of the vertical asymptote. As x gets larger, the graph is falling (decreasing) to the right of the vertical asymptote. The vertical asymptote is at –1. The graph is decreasing on the interval $(-\infty, -1)$ and decreasing on the interval $(-1, \infty)$. See the graph below. The function is never increasing. The function does not exist when x is –1.

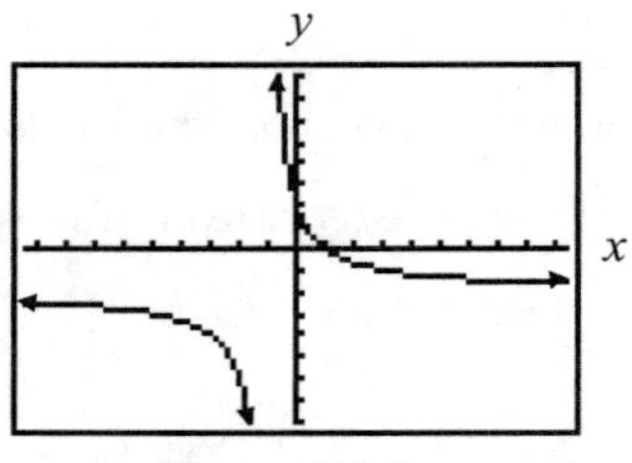

Figure 8.1.9

⁂

Example 6: What are the normal domain and range of $y = \dfrac{1}{x-6} + 4$?

Solution: The symbolic representation gives a clue as to the normal domain. If x is 6, the function is not real because of division by zero. Since there is no square root in the function there is no other value of x that will cause a non-real number. The normal domain is $(-\infty, 6) \cup (6, \infty)$. The range can be determined analytically by using what you know about transformations. The function $y = \dfrac{1}{x-6} + 4$ is the transformed function of $y = \dfrac{1}{x}$. Since $y = \dfrac{1}{x}$ has a horizontal asymptote at zero, the range does not include 0. The function $y = \dfrac{1}{x-6} + 4$ has been shifted up 4 units; thus, it has a horizontal asymptote at 4. With a horizontal asymptote at 4, the range is all real numbers except 4. The range is $(-\infty, 4) \cup (4, \infty)$.

⁂

8.1 EXERCISES

Many times you can work the exercises by algebraic, numeric, or graphical methods. You can learn about mathematics no matter which method you use. Part of the learning process is for you to decide which method is best for you to use. Please read the text carefully before you do the exercises. Read the **next assigned section** after you finish the assignment below. All numeric answers must have an error ≤ 0.01.

A Quick Look To The Future

-4. Does $\dfrac{2x}{x^2+x} = \dfrac{2}{x+1}$ for all x in the domain of $\dfrac{2x}{x^2+x}$?

-3. If $f(x) = \dfrac{1}{x}$ and $g(x) = x^2$, what do you think $fg(x)$ means?

-2. A carpenter puts a $\frac{3}{4}$ inch board on top of another $\frac{3}{4}$ inch board and the total thickness is $\frac{6}{8}$ inches. True or false?

-1. Can a horizontal line cross the graph of a rational function of the form $y = \dfrac{d}{x+e} + f$ more than once?

Previous Topics Revisited

1. If 20,000,000 E. coli bacteria are observed after 6 hours of unrestricted growth, how many bacteria were there at the beginning of the growth period?

2. Using the exponential growth function $y = y_0(1 + r)^x$ on a colony of 10 mice at time 0 and a growth rate of 4.2% per month, find the number of mice after 20 months under unrestricted growth conditions.

3. With approximately 6.5 billion people on earth today, find a growth rate that will allow 8.3 billion people in 1000 years. (Hint: Use the function $y = 6500000000(1 + x)^{1000}$.)

4. Solve the equation $\left(\dfrac{1}{4}\right)^{x-1} = \sqrt{x+3} - 2$

5. Solve the equation $3 \cdot 4^{2x+5} = 12 \cdot 2^{x-1}$

6. Simplify the function $\dfrac{3^{2x^2+4x-7}}{3^{x^2-5x-1}}$

New Topic Exercises

7. If the server at your favorite restaurant fills your very large coffee cup with only 250 cc of coffee, how can you change the coffee problem model $S_c = 100 \cdot \dfrac{300}{300+6x}$ to apply to this problem?

8. Using the coffee problem model, how can you change it if each coffee cream container has 8 cc of cream?

9. Using the original coffee problem model, how many creams must be added to make the strength 40%?

10. What is the strength of the coffee if 15 creams are added?

11. If you start with a very large container and 12 fluid ounces of lemon juice, what function can be used to model the strength of the lemon-water mix if water is added in a full 2-ounce container, one at a time?

12. Using the lemon strength model from Exercise 11, how many 2-ounce containers should be added to make the strength 10% lemon juice?

13. Using the lemon strength model from Exercise 11, what is the strength if 6 2-ounce containers are added?

14. Is the original coffee model increasing or decreasing? Where is the vertical asymptote?

15. What is the average rate of change of the coffee model between 1 and 2 creams added? What is the average rate of change of the coffee model between 10 and 12 creams added?

16. Does the coffee model have a zero? If yes, where?

17. Assuming a normal domain, when is the coffee model negative?

18. What is a reasonable problem domain for the coffee model?

Find the horizontal and vertical asymptotes in the following rational functions. Also specify the normal domain and range.

19. $y = \frac{3}{x} - 7$

20. $y = \frac{7}{(x+4)}$

21. $y = \frac{4}{(x-4)} - 4$

22. $y = \frac{-2}{(x+1.5)} + 3$

23. $f(x) = \frac{3}{x+14} - 18$

24. $f(x) = \frac{-6}{x-17} + 15$

25. $f(x) = \frac{1}{2-x} + 3$

Find the average rate of change of the following rational functions when x changes from 2 to 3.

26. $y=\dfrac{0.6}{(x+1)}-2$

27. $y=\dfrac{2}{(x-19)}+47$

28. $y=\dfrac{4}{(1-x)}-3.2$

29. $y=\dfrac{4}{(x-1)}+2$

Specify when each function is increasing and when it is decreasing. Use interval notation.

30. $f(x)=\dfrac{2.1}{(x+3)}-17$

31. $f(x)=\dfrac{-20}{(x+3)}+47$

32. $f(x)=\dfrac{5}{(x+3)}$

33. $f(x)=\dfrac{-1}{(x+3)}-2$

34. $f(x)=\dfrac{5}{(x-7.5)}+3$

Find the zeros and the y-intercepts.

35. $y=\dfrac{2.5}{(x-3.7)}-4.2$

36. $y=\dfrac{4.4}{(x+1.9)}+3.6$

37. $y=\dfrac{0.3}{(x-2.1)}-1.7$

38. $y=\dfrac{-4}{(x+2.1)}+4.8$

Describe the transformations of the parent function $y=\dfrac{1}{x}$ that will give the graph of the following transformed functions.

39. $y=\dfrac{1}{x}+4$

40. $y=\dfrac{1}{x+4}$

41. $y=\dfrac{4}{x}$

42. $y=\dfrac{-1}{x+3}-5$

43. $y=\dfrac{-2}{x+3}-6$

Develop a rational function that has the following horizontal and vertical asymptotes.

44. Horizontal at –3 and vertical at 4.

45. Horizontal at 5 and vertical at –1.

46. Vertical at –2, 3, and 5.

Develop the rational functions in Exercises 47 - 51.

47. Any rational function that is increasing on its domain.

48. Any rational function that is decreasing on its domain.

49. Any rational function that passes through the point (2, 1).

50. Any rational function that has a normal domain of $(-\infty, 7) \cup (7, \infty)$.

51. Any rational function that has a range of $(-\infty, 7) \cup (7, \infty)$.

52. For the function $y = \dfrac{2}{(x-2)} + 4$, as $x \to \infty$, $y \to$ what number?

53. For the function $y = \dfrac{2}{(x-2)} + 4$, as $x \to 2$, $y \to$ what number?

54. For the function $y = \dfrac{2}{(x-2)} - 5$, as $x \to \infty$, $y \to$ what number?

55. For the function $y = \dfrac{2}{(x-5)} + 4$, as $x \to 5$, $y \to$ what number?

56. In the function $y = \dfrac{d}{(x+e)} + f$, what values for d cause the function to be increasing on the normal domain.

57. In the function $y = \dfrac{d}{(x+e)} + f$, what values for d cause the function to be decreasing on the normal domain?

58. Where is the vertical asymptote and what is the normal domain for the function $y = \dfrac{d}{(x+e)} + f$?

59. Where is the horizontal asymptote and what is the range for the function $y = \dfrac{d}{(x+e)} + f$?

60. What values of f will cause the function $y = \dfrac{3}{(x+2)} + f$ to have a zero? What value of f will cause the function not to have a zero?

From Mathematics to English

61. Can a rational function of the form $y = \dfrac{d}{(x+e)} + f$ have more than one zero? Why?

62. What does the x represent in the coffee problem model?

63. What is a mathematical model?

64. What function types have horizontal asymptotes?

65. Describe a vertical asymptote.

66. Describe data from outside the mathematics classroom that you think is related by a rational function.

67. Give your definition of the transformation called a “stretch”.

68. Give your definition of the transformation called a “vertical translation”.

69. After reading this section, make a list of questions that you want to ask your instructor.

70. Continue in your daily journal and make an entry. In addition to your normal entry on thoughts about the mathematics in this section, list at least two positive comments about what you have learned about this topic.

71. In paragraph format, summarize the material in this section of the text in your daily journal.

72. Describe how your classroom instructor made this topic more understandable and clear.

73. After reading the text and listening to your instructor, what do you not understand about this topic?

Explorations

74. Make a list of everything you know about the function $y=\dfrac{-2}{(x+4.1)}-3.6$, that is, list the behaviors.

75. Each rational function in this section either increases or decreases both to the left and right of the vertical asymptote. Explore other rational functions on your calculator and find one that increases to the left of the vertical asymptote and decreases to the right of the vertical asymptote. What is your function?

76. Find a rational function that decreases to the left of the vertical asymptote and increases to the right of the vertical asymptote.

77. Describe the transformation of the graph of $y=\dfrac{1}{x}$ to get the graph of each function below.

a. $y=\dfrac{1}{2x+6}$ b. $y=\dfrac{1}{2x-6}$ c. $y=\dfrac{1}{5x-10}$

d. $y=\dfrac{1}{5x+10}$ e. $y=\dfrac{1}{2x-5}$

Where are the vertical asymptotes in each of the above graphs?

78. Below are several rational functions as well as rational functions that have been transformed by the absolute value function. Describe what the absolute value transformation does to the graph of a rational function.

a. $f(x)=\frac{1}{x}$ $g(x)=\left|\frac{1}{x}\right|$

b. $f(x)=\frac{1}{x}-3$ $g(x)=\left|\frac{1}{x}-3\right|$

c. $f(x)=\frac{1}{x}+3$ $g(x)=\left|\frac{1}{x}+3\right|$

d. $f(x)=\frac{1}{(x+3)}$ $g(x)=\left|\frac{1}{(x+3)}\right|$

e. $f(x)=\frac{-2}{(x+3)}$ $g(x)=\left|\frac{-2}{(x+3)}\right|$

f. $f(x)=\frac{-2}{(x+3)}-5$ $g(x)=\left|\frac{-2}{(x+3)}-5\right|$

In one or two sentences, summarize the effect the absolute value transformation has on the graph of a rational function.

79. Below are several rational functions as well as rational functions that have been transformed by squaring. Describe what the squaring transformation does to the graph of a rational function.

a. $f(x)=\frac{1}{x}$ $g(x)=[f(x)]^2$

b. $f(x)=\frac{1}{x}-3$ $g(x)=[f(x)]^2$

c. $f(x)=\frac{1}{x}+3$ $g(x)=[f(x)]^2$

d. $f(x)=\frac{1}{(x+3)}$ $g(x)=[f(x)]^2$

e. $f(x)=\frac{-2}{(x+3)}$ $g(x)=[f(x)]^2$

f. $f(x)=\frac{-2}{(x+3)}-5$ $g(x)=[f(x)]^2$

In one or two sentences, summarize the effect the squaring transformation has on the rational function.

80. Below are four fairly complicated rational functions. What is different about these functions is that they may have more than one vertical asymptote. Find the zeros of these functions.

a. $y=-2\frac{x}{(x+2)(x-6)}+3$

b. $f(x)=5-\frac{x-3}{x^2-2x-3}$

c. $y=\frac{(x+3)(x-2)(x+7)}{(x-5)(x+3)(x-2)}$

d. $f(x)=\frac{x^3+8x^2+x-42}{x^3-4x^2-11x+30}$

8.2 The Fundamental Property of Rational Functions

The rational functions studied in the last section were restricted to functions that can be put in the form $f(x)=\dfrac{d}{x+e}+f$. Rational functions are studied in greater detail in college algebra. To prepare you for further study of the rational function and related topics, the next three sections of this chapter concentrate on reducing, multiplying, dividing, adding, and subtracting rational functions many of which are more complicated than those studied in Section 8.1.

Are They Really the Same?

How can you tell if the functions $y=\dfrac{x+2}{x^2-4}$ and $y=\dfrac{1}{x-2}$ are the same or different? Hopefully you know to compare the numeric or graphic representations. Figure 8.2.1 shows that the two

X	Y1	Y2
-3	-.2	-.2
-2	ERROR	-.25
-1	-.3333	-.3333
0	-.5	-.5
1	-1	-1
2	ERROR	ERROR
3	1	1

Y1=(X+2)/(X²-4)

Figure 8.2.1

functions are the same on the interval $(-\infty, -2) \cup (-2, \infty)$. Neither function is real when x is 2.

It may be helpful to examine the two functions again with the numerators and denominators written in factored form; they are $y=\dfrac{(x+2)}{(x+2)(x-2)}$ and $y=\dfrac{1}{(x-2)}$. It appears that the $(x+2)$ factors in the first function canceled leaving the second function. In Figure 8.2.2 are the numeric representations of $y=\dfrac{x^2-1}{x^2-3x+2}$ and $y=\dfrac{x+1}{x-2}$.

X	Y1	Y2
-2	.25	.25
-1	0	0
0	-.5	-.5
1	ERROR	-2
2	ERROR	ERROR
3	4	4
4	2.5	2.5

Y1=(X²-1)/(X²-3...

Figure 8.2.2

Again, it may be helpful to factor all numerators and denominators. The functions become $f(x)=\dfrac{(x+1)(x-1)}{(x-2)(x-1)}$ and $g(x)=\dfrac{(x+1)}{(x-2)}$. It appears that the $(x-1)$ factors in the first function canceled leaving the second function. Figure 8.2.2 shows that the two functions are the same on the domain $(-\infty, 1) \cup (1, \infty)$. At $x = 2$, both functions are not real. There is an important event happening in these two problems. It appears that when like factors cancel the simplified function has a very slightly different normal domain than the more complex function. The canceling of factors gives rise to a generalization -- the cancellation property.

Cancellation Property

Given the function $\dfrac{f(x)\cdot h(x)}{g(x)\cdot h(x)}$ then $\dfrac{f(x)\cdot h(x)}{g(x)\cdot h(x)} = \dfrac{f(x)}{g(x)}$

on the normal domain of $g(x)$

If you rewrite the statement $\dfrac{f(x)\cdot h(x)}{g(x)\cdot h(x)} = \dfrac{f(x)}{g(x)}$ in reverse order, you also have a very useful property that will be used primarily when adding or subtracting rational functions. It is called the fundamental property of rational functions.

Fundamental Property of Rational Functions

Given the fraction $\dfrac{f(x)}{g(x)}$ then $\dfrac{f(x)}{g(x)} = \dfrac{f(x)\cdot h(x)}{g(x)\cdot h(x)}$

on the normal domain of $g(x)$

This property allows for the numerator *and* denominator of a rational function to be multiplied by the same function. The fundamental and cancellation properties are used in the remaining sections of this chapter.

If the numerator and denominator are in *factored form*, the cancellation property allows for a *factor* in the numerator to cancel the same *factor* in the denominator. Factors are polynomials that are being multiplied. It is not always clear that the **cancellation property does not apply to the terms that make up the polynomials** in the numerator and denominator; however, the property only allows for the cancellation of polynomials that are factors. One use of the cancellation property is to reduce fractions. In order to use this property, you must satisfy the conditions of the property, and have a function with the numerator and denominator in product form. If the numerator and denominator share identical factors, they can be canceled.

To Reduce a Rational Function

Completely factor the numerator and denominator and cancel all like factors.

Simplify the numerator and denominator if necessary and note the normal domain.

Example 1: Reduce the function $f(x) = \dfrac{x^2 - 4}{x + 2}$.

Solution: $f(x) = \dfrac{(x+2)(x-2)}{(x+2)}$ factor the numerator and denominator

$f(x) = \dfrac{\cancel{(x+2)}(x-2)}{\cancel{(x+2)}}$ cancel like factors

$f(x) = x - 2$ $\dfrac{x^2 - 4}{x + 2} = x - 2$ on the domain of $(-\infty, -2) \cup (-2, \infty)$.

⁂

Example 2: Reduce the function $f(x) = \dfrac{x^2 - x - 12}{2x + 6}$.

Solution: $f(x) = \dfrac{(x-4)(x+3)}{2(x+3)}$ factor the numerator and denominator

$f(x) = \dfrac{(x-4)\cancel{(x+3)}}{2\cancel{(x+3)}}$ cancel like factors

$f(x) = \dfrac{x-4}{2}$ $\dfrac{x^2 - x - 12}{2x + 6} = \dfrac{x-4}{2}$ on the domain of $(-\infty, -3) \cup (-3, \infty)$.

⁂

Example 3: Reduce the function $g(x) = \dfrac{12x^2 + 7x - 10}{16x^2 - 25}$.

Solution: $g(x) = \dfrac{(3x-2)(4x+5)}{(4x-5)(4x+5)}$ factor the numerator and denominator

$$g(x) = \frac{(3x-2)\cancel{(4x+5)}}{(4x-5)\cancel{(4x+5)}}$$ cancel like factors

$$g(x) = \frac{3x-2}{4x-5}$$ $\frac{12x^2+7x-10}{16x^2-25} = \frac{3x-2}{4x-5}$ on the domain of $(-\infty, -\frac{5}{4}) \cup (-\frac{5}{4}, \infty)$ and the domain of $\frac{3x-2}{4x-5}$.

⁂

The cancellation property allows for the cancellation of like, or identical factors. Sometimes, rational functions contain opposite factors -- one in the numerator and the other in the denominator. One way of identifying opposite polynomials is to replace the variable with a value from the domain and evaluate the polynomials. If the polynomials are opposites, their values will be opposite real numbers. For example, $x - 2$ and $2 - x$ are opposites. When x is replaced with any number, like 5, the polynomials have values of 3 and –3 respectively. This makes $x - 2$ and $2 - x$ opposites. Perhaps a stronger argument for accepting that $x - 2$ and $2 - x$ are opposites is to look at the numeric representations of both.

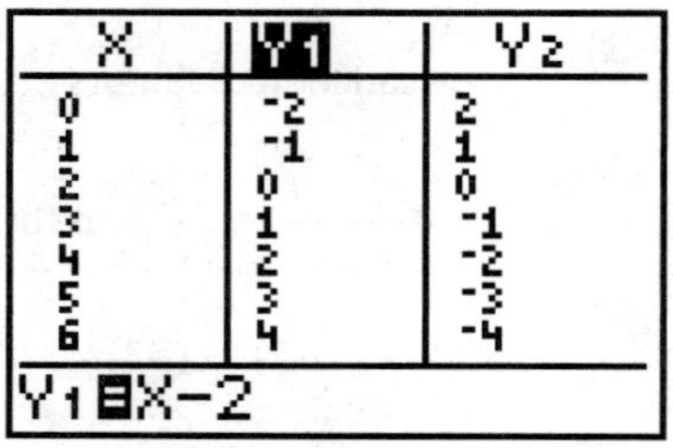

Figure 8.2.3

Of course, the proof is in the algebra, $x - 2 = -(2 - x)$ by algebraic simplification.

When opposites are divided, the quotient is –1. For example,

$$\frac{-3}{3} = -1, \quad \frac{(x-5)}{(5-x)} = -1, \quad \frac{(2-7)}{(7-2)} = -1, \quad \text{etc.}$$

Example 4: Reduce the function $g(x) = \dfrac{x-7}{14-2x}$.

Solution: $$g(x) = \frac{(x-7)}{2(7-x)}$$ factor the numerator and denominator

$$g(x) = \frac{\overbrace{\cancel{(x-7)}}^{-1}}{2\cancel{(7-x)}}$$ cancel opposites to get –1

$$g(x) = \frac{-1}{2}$$ $\frac{x-7}{14-2x} = \frac{-1}{2}$ on the domain $(-\infty, 7) \cup (7, \infty)$

⁂

Example 5: Reduce the function $h(x) = \dfrac{x^2 - 4}{2 + x - x^2}$.

Solution:

$$h(x) = \frac{(x+2)(x-2)}{(2-x)(1+x)}$$ factor the numerator and denominator

$$h(x) = \frac{(x+2)\overbrace{\cancel{(x-2)}}^{-1}}{\cancel{(2-x)}(1+x)}$$ divide opposite factors to get -1

$$h(x) = \frac{-1(x+2)}{(1+x)}$$ simplify the numerator and denominator

$$h(x) = \frac{-x-2}{x+1}$$

OR $h(x) = -\dfrac{x+2}{x+1}$

$\dfrac{x^2-4}{2+x-x^2} = \dfrac{-x-2}{x+1}$ on the domain $(-\infty, 2) \cup (2, \infty)$ and the domain of $\dfrac{-x-2}{x+1}$.

⁂

It may be worth your time to check your work graphically or numerically. You may have noted in the last five problems that the domain of the reduced function is only different by one or two numbers. When you graph the problem and your answer, the graphs should be the same except for the few numbers in the domains that are different. Below are the graphs of the Example 5 functions $y = \dfrac{x^2-4}{2+x-x^2}$ and $y = \dfrac{-x-2}{x+1}$ on a decimal window.

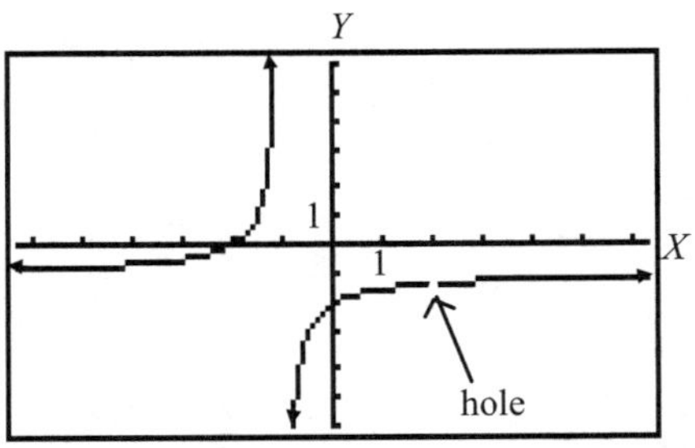

Figure 8.2.4a

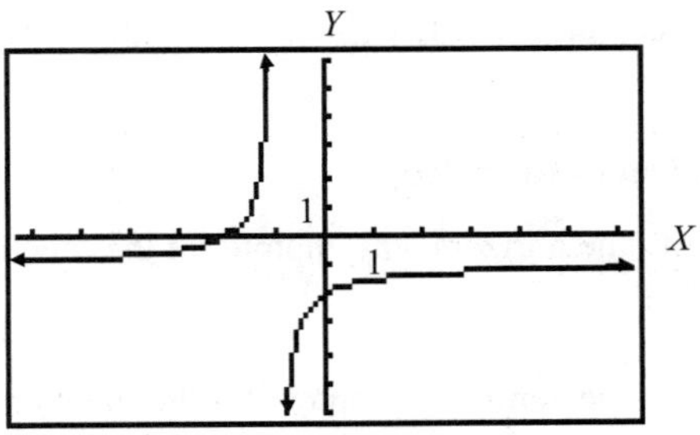

Figure 8.2.4b

The analytical work is correct! The graphs are the same except that the point $(2, -\frac{4}{3})$ is missing in the problem and it is part of the answer. A numeric check will show the same thing. The tables will be different only when x is 2.

An important observation here is that there is a hole in the graph at the value of x that makes the canceled factor zero.

Example 6: Find a rational function that has a hole in the graph at $x = 3$.

Solution: From looking at the above example, it appears that when like factors cancel, there is a hole in the graph at the value of x that makes the factor zero. One example of a function with a hole in the graph at 3 is $f(x) = \frac{1}{2}x + \frac{x-3}{x-3}$.

You are encouraged to verify this on your calculator. A decimal window must be used to see the holes.

⁂

8.2 EXERCISES

Many times you can work the exercises by algebraic, numeric, or graphical methods. You can learn about mathematics no matter which method you use. Part of the learning process is for you to decide which method is best for you to use. Please read the text carefully before you do the exercises. Read the **next assigned section** after you finish the assignment below. All numeric answers must have an error ≤ 0.01.

A Quick Look To The Future

-4. Are $\frac{x-1}{x+2}$ and $\frac{x-1}{x+2} \cdot \frac{x+3}{x+3} \cdot \frac{x+1}{x+1} \cdot \frac{x-5}{x-5}$ equal?

-3. Are $\frac{x-1}{x+2}$ and $\frac{x-1}{x+2} + \frac{x+3}{x+3} - \frac{x+1}{x+1} \cdot \frac{x-5}{x-5}$ equal?

-2. What value of x will cause $\frac{1}{x-2} = 0$ to be a true statement?

-1. What values of x make $\sqrt{x}$ a real number?

Previous Topics Revisited

1. Identify the horizontal asymptote for the function $y = \frac{2}{x+3} - 5$.

2. Identify the vertical asymptote for the function $y = \frac{-2}{x-7} + 25$.

3. Find a rational function that has a horizontal asymptote at 4 and a vertical asymptote at –3.

4. How long does it take for \$1,000 to grow to \$2,000 at 24% compounded yearly? (Hint: Use $y = y_0(1+r)^x$, *where* $y = 2000$, $y_0 = 1000$, $r = .24$.)

5. Using the growth model $y = y_0 \cdot 2^{\frac{x}{r}}$, find how long it takes for a population of 10 rats to grow to 100 rats if they can double there numbers at the growth rate of every 15 months.

6. Solve the equation $2^{x-5} - 7\sqrt{x+3} = x^2 + 3x$

New Topic Exercises

Reduce the following functions and identify the normal domain that makes the function and your answer equivalent.

7. $y = \dfrac{x^2 - 4}{2x + 4}$

8. $y = \dfrac{x^2 - 3x - 88}{x^2 - 11x}$

9. $f(x) = \dfrac{6}{3x^2 - 21x + 39}$

10. $f(x) = \dfrac{2x^2 - x - 15}{4x^2 - 25}$

11. $f(x) = \dfrac{6x^2 + 13x - 5}{12x^2 - 13x + 3}$

12. $g(x) = \dfrac{15x^2 - 11x - 14}{3x^2 - x - 2}$

13. $g(x) = \dfrac{15x^2 - 11x - 14}{14 - 10x}$

14. $g(x) = \dfrac{25x^2 - 4}{4 - 10x}$

15. $g(x) = \dfrac{x^2 - 3x - 2}{2 + 3x - x^2}$

16. $g(x) = \dfrac{(x+5)^6}{(x+5)^5}$

17. Where is the hole in the complex rational function $y = 5x^2 - 3x + 2 + \dfrac{x+3}{x+3}$?

Find <u>any</u> function that has a hole in the graphical representation when x is:

18. 4

19. 2

20. –3 and 5

21. 2, –4, and 6

22. Develop a rational function whose domain is all real numbers except 3 and –1.

From Mathematics to English

23. In the fraction $\dfrac{x+3}{x-5}$, explain why the x's do not cancel.

24. Do the 6's cancel in the fraction $\dfrac{2x+6}{3x+6}$? Why?

25. Explain why the following statement is incorrect: $\dfrac{x^2 - 5x + 3}{x^2 + 2x - 1} = \dfrac{-2}{1}$

26. Explain why the following statement is correct: $\dfrac{(x^2)(-5x)(3)}{(x^2)(2x)(-1)} = \dfrac{15}{2}\ (x \neq 0)$

27. Are $x^2 - 3x - 5$ and $5 + 3x - x^2$ opposites? Why? What is their quotient?

28. Explain the difference between the graphs of $y = \dfrac{x^2 - 25}{x + 5}$ and $y = x - 5$.

29. What does it mean to reduce a rational function?

30. Explain the cancellation property.

31. Would you call the statement $y = \dfrac{x - 5}{x + 2}$ an equation? Why?

32. After reading this section, make a list of questions that you want to ask your instructor.

33. Continue in your daily journal and make an entry. In addition to your normal entry on thoughts about the mathematics in this section, list at least two positive comments about what you have learned about this topic.

34. In paragraph format, summarize the material in this section of the text in your daily journal.

35. Describe how your classroom instructor made this topic more understandable and clear.

36. After reading the text and listening to your instructor, what do you not understand about this topic?

Explorations

37. Make up a list of 5 rational functions that are reducible. One of the 5 functions should have a numerator of degree four. Show the reduced form of the functions as well as the original form.

38. Rather than checking your answers to Exercises 7 - 16 in the answer key, graph the problem and your answer on the same screen. How do you know if your answer is correct from looking at the graphs? If you think you have the correct answer, are the graphs identical?

8.3 Multiplication and Division of Rational Functions

What's That Rule Again?

How do you multiply rational functions? Do you multiply numerators and denominators? Do you cross multiply? If you don't know, try a problem and let technology check to see if you are correct. For example, consider multiplying the rational functions $f(x) = \dfrac{2}{x-1}$ and $g(x) = \dfrac{-1}{x}$ when you use the rule "multiply numerators and denominators." Is the product $(fg)(x) = \dfrac{2(-1)}{(x-1)(x)}$? If you are interested in simplification of this function, please do so. It simplifies to $(fg)(x) = \dfrac{-2}{x^2 - x}$. To find if this is the product, store function f in Y_1, g in Y_2, symbolize the technology-based product in Y_3, and store the function fg in Y_4. Make either the graphical or numerical representations of Y_3 and Y_4. If they are the same, you have very strong evidence that the rule you used is correct. This process is displayed below in

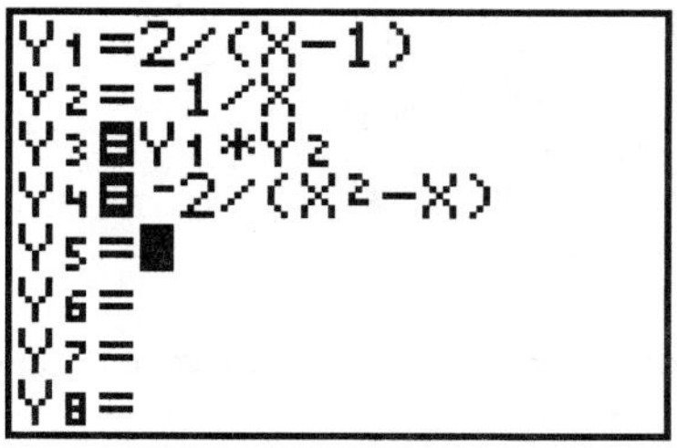

Figure 8.3.1a

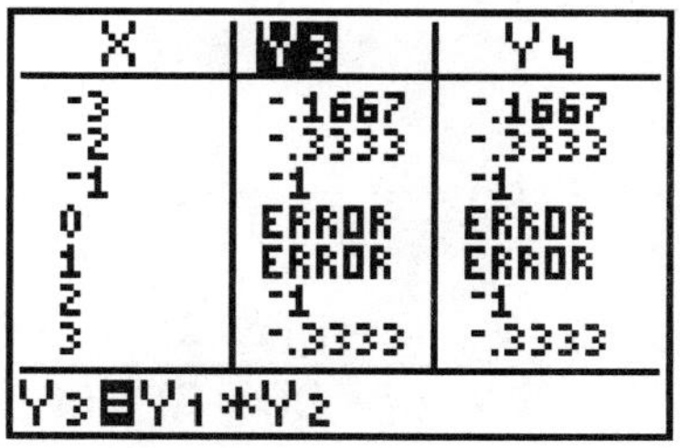

Figure 8.3.1b

Figure 8.3.1. The calculator holds the correct product in Y_3 and the proposed product is stored in Y_4. (Note that Y_1 and Y_2 are deselected because they are not needed.) Because the numeric representations are the same, you can conclude that the product is correct. That is, the correct method for multiplication of rational functions is:

> **Rule for Multiplication of Rational Functions**
>
> To multiply rational functions, multiply numerators and multiply denominators. Symbolically, the rule is:
>
> $$\frac{f(x)}{g(x)} \cdot \frac{h(x)}{k(x)} = \frac{f(x) \cdot h(x)}{g(x) \cdot k(x)}$$

The actual process of multiplying or dividing rational functions is very similar to the process of reducing a rational function. The process is developed from the rules for multiplication and division. These rules are the same rules you used in arithmetic to multiply or divide numeric fractions. The rule for division is:

Rule for Division of Rational Functions

To divide rational functions, replace the divisor (second function) with its reciprocal and replace the division sign with a multiplication sign. Proceed as in multiplication. Symbolically, the rule is:

$$\frac{f(x)}{g(x)} \div \frac{h(x)}{k(x)} = \frac{f(x)}{g(x)} \cdot \frac{k(x)}{h(x)}$$

Although these rules give the correct product or quotient, if the answer needs to be reduced, the first step in multiplication is NOT to multiply numerators and denominators. In cases where the product must be reduced, each numerator and denominator should be factored first. After doing this, the cancellation property can be used because the numerators and denominators are in factored form. When like factors have been canceled, the last step is to multiply numerators and denominators. By following this procedure, the product will always be in reduced form. Recall from Section 8.2, the process of reducing alters the normal domain.

Example 1: If $f(x) = \dfrac{x^2 + x - 2}{x^2 - 9}$ and $g(x) = \dfrac{x + 3}{2x - 2}$, find $(fg)(x)$.

Solution:

$$(fg)(x) = \frac{(x-1)(x+2)}{(x+3)(x-3)} \cdot \frac{(x+3)}{2(x-1)}$$ factor numerators and denominators

$$(fg)(x) = \frac{\cancel{(x-1)}(x+2)}{\cancel{(x+3)}(x-3)} \cdot \frac{\cancel{(x+3)}}{2\cancel{(x-1)}}$$ cancel like factors

$$(fg)(x) = \frac{x+2}{2(x-3)} \text{ OR } \frac{x+2}{2x-6}$$ multiply remaining factors

Remember, the process of canceling causes the domain to change. The original product and the final simplified form are only equivalent when $x \in (-\infty, -3) \cup (-3, 1) \cup (1, \infty)$.

⁂

Example 2: If $f(x) = \dfrac{2x^2 + 7x - 15}{3x^2 + 11x - 4}$ and $g(x) = \dfrac{9x^2 - 1}{6x^2 - 7x - 3}$, find $(fg)(x)$.

Solution:

$$(fg)(x) = \frac{(x+5)(2x-3)}{(x+4)(3x-1)} \cdot \frac{(3x+1)(3x-1)}{(2x-3)(3x+1)}$$ factor

$$(fg)(x) = \frac{(x+5)\cancel{(2x-3)}}{(x+4)\cancel{(3x-1)}} \cdot \frac{\cancel{(3x+1)}\cancel{(3x-1)}}{\cancel{(2x-3)}\cancel{(3x+1)}}$$ cancel like factors

$(fg)(x) = \dfrac{x+5}{x+4}$ multiply remaining factors

Is the answer correct? Probably, but a check will verify it. Below is a numeric check with the problem on line two and the answer on line three. As you can see, the analytical work is correct; however, the domains are different and they are only equivalent functions when $x \notin \left\{-\frac{1}{3}, \frac{1}{3}, \frac{3}{2}\right\}$

Table 8.3.1

x	–3	–2	–1	0	1	2	3
$\frac{2x^2+7x-15}{3x^2+11x-4} \cdot \frac{9x^2-1}{6x^2-7x-3}$	2	1.5	1.33	1.25	1.2	1.17	1.14
$\frac{x+5}{x+4}$	2	1.5	1.33	1.25	1.2	1.17	1.14

⁂

Example 3: If $h(x) = \dfrac{x^2+x-2}{x^2-16}$ and $k(x) = \dfrac{8-2x}{x+2}$, find $(hk)(x)$.

Solution:

$(hk)(x) = \dfrac{(x-1)(x+2)}{(x+4)(x-4)} \cdot \dfrac{2(4-x)}{(x+2)}$ factor

$(hk)(x) = \dfrac{(x-1)\cancel{(x+2)}}{(x+4)(x-4)} \cdot \dfrac{2(4-x)}{\cancel{(x+2)}}$ cancel like factors

$(hk)(x) = \dfrac{(x-1)(x+2)}{(x+4)\cancel{(x-4)}} \cdot \dfrac{2\overbrace{\cancel{(4-x)}}^{-1}}{(x+2)}$ replace opposites with –1

$(hk)(x) = \dfrac{(x-1)(-1)(2)}{(x+4)}$ OR $\dfrac{-2x+2}{x+4}$ multiply remaining factors

The original product and the final simplified product are equivalent when

$x \in (-\infty, -2) \cup (-2, 4) \cup (4, \infty)$.

⁂

Example 4: If $f(x) = \dfrac{x^2-x-6}{x^2+3x-4}$ and $g(x) = \dfrac{x^2-5x+6}{x^2+2x-8}$, find $(f/g)(x)$.

Solution: Replace the divisor with its reciprocal and division with multiplication.

$$(f/g)(x) = \frac{x^2 - x - 6}{x^2 + 3x - 4} \cdot \frac{x^2 + 2x - 8}{x^2 - 5x + 6}$$

$$(f/g)(x) = \frac{(x-3)(x+2)}{(x+4)(x-1)} \cdot \frac{(x+4)(x-2)}{(x-2)(x-3)} \quad \text{factor}$$

$$(f/g)(x) = \frac{\cancel{(x-3)}(x+2)}{\cancel{(x+4)}(x-1)} \cdot \frac{\cancel{(x+4)}\cancel{(x-2)}}{\cancel{(x-2)}\cancel{(x-3)}} \quad \text{cancel like factors}$$

$$(f/g)(x) = \frac{x+2}{x-1} \quad \text{multiply the remaining factors}$$

The above work is probably correct; however, a graphical check is shown below. A decimal window is used and the graphs are the same except at the values for x not allowed in the original quotient but allowed in the simplified quotient.

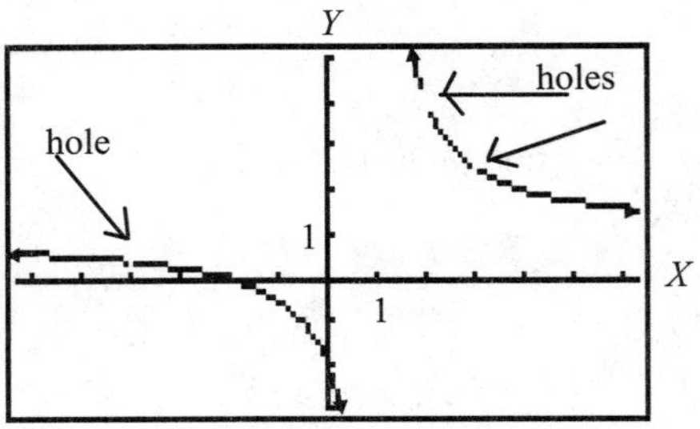

Figure 8.3.2a

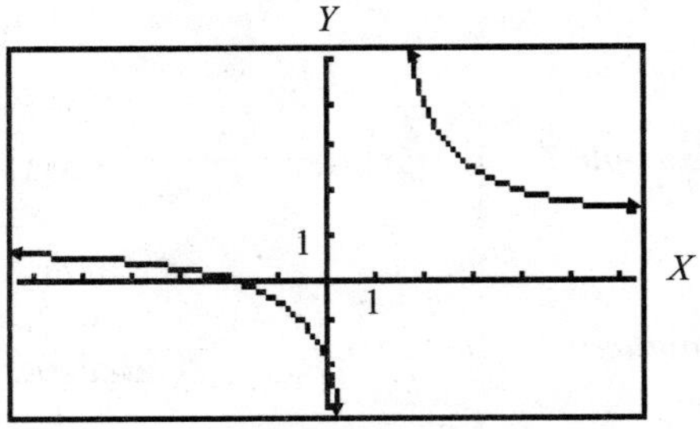

Figure 8.3.2b

The graphical representations are the same except at –4, 2, and 3, where the problem function has holes in the graph and the answer function does not.

⁂

Example 5: If $h(x) = \dfrac{1-x^2}{6x^2 + 13x - 5}$ and $k(x) = \dfrac{x^2 + 2x + 1}{2x^2 + 3x - 5}$, find $(h/k)(x)$.

Solution: Replace the divisor with its reciprocal and then / with ·

$$(h/k)(x) = \frac{1-x^2}{6x^2 + 13x - 5} \cdot \frac{2x^2 + 3x - 5}{x^2 + 2x + 1}$$

$$(h/k)(x) = \frac{(1-x)(1+x)}{(2x+5)(3x-1)} \cdot \frac{(2x+5)(x-1)}{(x+1)(x+1)} \quad \text{factor}$$

$$(h/k)(x) = \frac{(1-x)\cancel{(1+x)}}{\cancel{(2x+5)}(3x-1)} \cdot \frac{\cancel{(2x+5)}(x-1)}{\cancel{(x+1)}(x+1)} \quad \text{cancel like factors}$$

$$(h/k)(x) = \frac{(1-x)(x-1)}{(3x-1)(x+1)} \text{ OR } \frac{-x^2+2x-1}{3x^2+2x-1} \qquad \text{multiply}$$

The un-simplified quotient and the simplified quotient are equivalent when

$$x \in \left(-\infty, -\frac{5}{2}\right) \cup \left(-\frac{5}{2}, 1\right) \cup \left(1, \infty\right)$$

⁂

8.3 EXERCISES

Many times you can work the exercises by algebraic, numeric, or graphical methods. You can learn about mathematics no matter which method you use. Part of the learning process is for you to decide which method is best for you to use. Please read the text carefully before you do the exercises. Read the **next assigned section** after you finish the assignment below.

A Quick Look To The Future

-4. Is $\frac{2}{x} - \frac{x-3}{x}$ the same thing as $\frac{2}{x} + \frac{3-x}{x}$?

-3. Is $\frac{1}{x} + 4 = \frac{2}{x-1}$ a function or an equation?

-2. Can $\sqrt{-x}$ be a real number?

-1. Are $\sqrt{4 \cdot 3}$ and $\sqrt{4} \cdot \sqrt{3}$ the same number?

Previous Topics Revisited

1. Reduce $\dfrac{6x^2-7x-5}{3x^2+7x-20}$

2. Reduce $\dfrac{81-4x^2}{2x^2-7x-9}$

3. Reduce $\dfrac{(x-1)(2-x)(3x+2)(4x-5)}{(x-2)(3x+2)^2(5-4x)(1-x)}$

4. Where is the vertical asymptote of $f(x) = \dfrac{1}{x-3} + 4$?

5. Where is the horizontal asymptote of $f(x) = \dfrac{3}{x+2} + 5$?

6. How long does it take 1,000,000 E. coli bacteria to double under unrestricted growth? How long does it take 50 E. coli to double?

New Topic Exercises

Find $(fg)(x)$.

7. $f(x)=\dfrac{x^2-x-2}{2x^2-3x-2}$ and $g(x)=\dfrac{2x^2-5x-3}{x^2+x}$

8. $f(x)=\dfrac{x^2-1}{x^2+5x+6}$ and $g(x)=\dfrac{x^2-x-6}{x^2-4x+3}$

9. $f(x)=\dfrac{4-x^2}{x^2-x-6}$ and $g(x)=\dfrac{3x-x^2}{x^2-x-2}$

10. $f(x)=\dfrac{6x^2+13x-28}{4x^2-7x+3}$ and $g(x)=\dfrac{8x^2+14x-15}{2x^2+11x+14}$

11. $f(x)=\dfrac{4x^2-9}{2x^2-x-10}$ and $g(x)=\dfrac{2x^2+3x-20}{3-2x}$

Find $(f/g)(x)$ in simplified form.

12. $(f/g)(x)=\dfrac{x+4}{2x+10}\div\dfrac{x^2-16}{x^2-25}$

13. $(f/g)(x)=\dfrac{6x^2+11x-10}{x^2-6x+5}\div\dfrac{3x^2+10x-8}{x^2-4x-5}$

14. $(f/g)(x)=\dfrac{2x^2-3x-2}{3x^2+x-4}\div\dfrac{4x^2-1}{6x^2+5x-4}$

15. $(f/g)(x)=\dfrac{6-5x+x^2}{3x^2+2x}\div\dfrac{x^2-5x+6}{2x^2+4x}$

16. $(f/g)(x)=\dfrac{6x^2-x-12}{15x^2+x-2}\div\dfrac{2x^2-x-3}{10x^2-11x-6}$

17. What is the normal domain of the un-simplified product and of the simplified product in Exercise 11?

18. What is the normal domain of the un-simplified quotient and of the simplified quotient in Exercise 14?

19. What is the first operation that should be performed in $\dfrac{a}{b}\div\dfrac{c}{d}\cdot\dfrac{e}{f}$?

20. What value for a will cause the product $\dfrac{1}{a}\cdot\dfrac{3}{5}$ to be an integer?

From Mathematics to English

21. Does $\dfrac{x}{x+2}\cdot\dfrac{x^2-4}{x^2}=\dfrac{x-2}{x}$? Why?

22. Describe any rule you might use to multiply two rational functions.

23. Describe any rule you might use to divide two rational functions.

24. Does the domain of the product of two rational functions always change when you multiply? Explain.

25. Describe how to multiply $\frac{x}{x+2}$ and $\frac{x+2}{x^2+3x}$. Do not actually multiply.

26. Is the mathematical statement in Exercise 21 an equation?

27. After reading this section, make a list of questions that you want to ask your instructor.

28. Continue in your daily journal and make an entry. In addition to your normal entry on thoughts about the mathematics in this section, list at least two positive comments about what you have learned about this topic.

29. In paragraph format, summarize the material in this section of the text in your daily journal.

30. Describe how your classroom instructor made this topic more understandable and clear.

31. After reading the text and listening to your instructor, what do you not understand about this topic?

Explorations

32. Find two functions whose product is:

a. $\frac{x-4}{x+3}$ b. $\frac{x^2-4}{x}$ c. $\frac{-1}{x^2-4x-5}$

33. Find two functions whose quotient is:

a. $\frac{x+1}{x-1}$ b. $\frac{x^2-3x+2}{-4}$ c. $\frac{x^2+5x-24}{x^2-2x-35}$

34. How are the graphical representations of $\frac{1}{x-1}\cdot\frac{x^2-1}{2}$ and $\frac{x+1}{2}$ different? How are they the same?

8.4 Addition and Subtraction of Rational Functions and Simplification of Complex Rational Functions

Do You Add Rational Functions By Adding Numerators and Denominators?
You have a marvelous tool you can use to help answer this question, the calculator. Even at an arithmetic level, it will show you whether this idea for adding numeric fractions is correct. For example, study the example in Figure 8.4.1 of adding $\frac{1}{3}+\frac{5}{6}$. If you use the rule "add numerators and denominators" the sum is $\frac{1+5}{3+6}$ or $\frac{6}{9}$. The calculator shows the sum to be $\frac{7}{6}$, these are not the same number. One of the answers is incorrect.

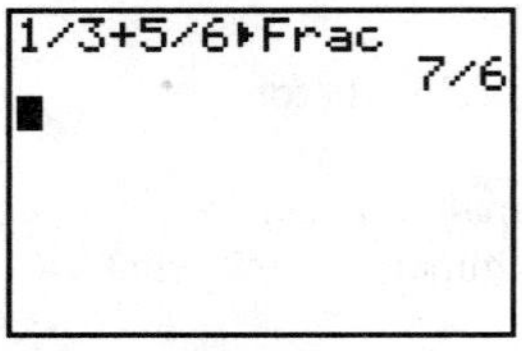

Figure 8.4.1

The calculator will also allow you to investigate a strategy for adding rational functions. Using the same idea -- to add rational functions "add numerators and denominators" -- the calculator will show you whether your reasoning is correct. For the figures below, Y_3 is the calculator sum of the functions $Y_1=\frac{2}{x-3}$ and $Y_2=\frac{5}{x^2-9}$, and Y_4 is the "human" sum found by adding "numerators and denominators".

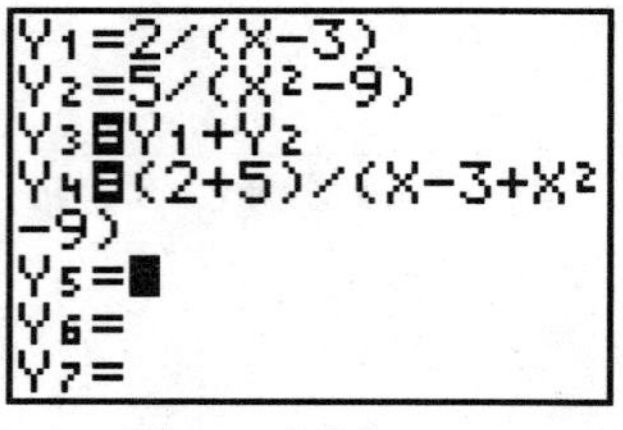

Figure 8.4.2a

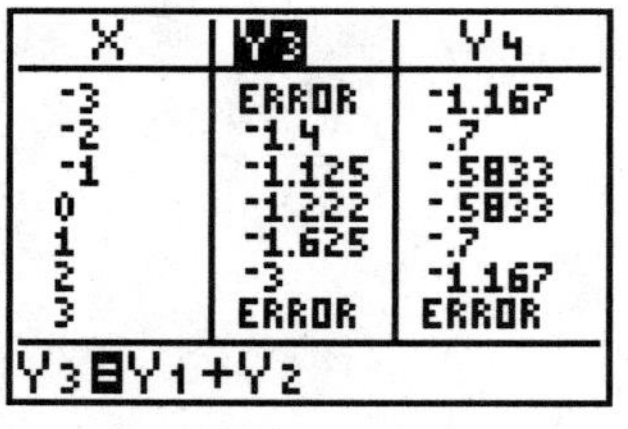

Figure 8.4.2b

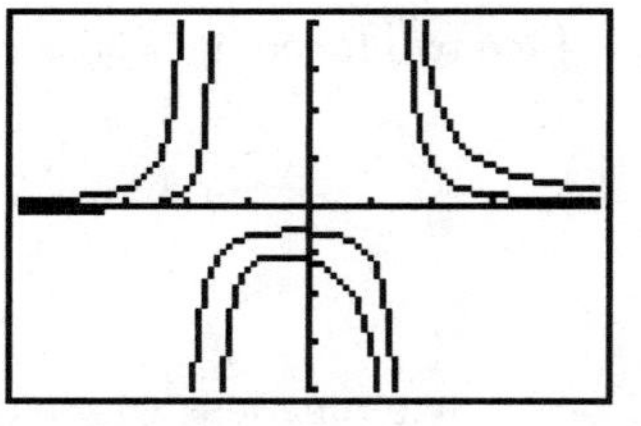
Figure 8.4.2c

Since the numeric and graphic representations are not the same, you can conclude that the symbolic representations are not the same. That is, $\frac{2}{x-3}+\frac{5}{x^2-9}$ is not equivalent to $\frac{2+5}{x-3+x^2-9}$. If the sum cannot be found by "adding numerators and denominators", how can you find the sum? Perhaps the addition rule below will work.

The operation of addition or subtraction of rational functions is similar to addition or subtraction of arithmetic fractions. The rules from arithmetic still apply.

Rule for Addition of Rational Functions

To add rational functions with identical denominators, add the numerators and leave the denominator the same.

Symbolically: $\dfrac{f(x)}{h(x)} + \dfrac{g(x)}{h(x)} = \dfrac{f(x)+g(x)}{h(x)}$

Rule for Subtraction of Rational Functions

To subtract rational functions with identical denominators, subtract the numerators and leave the denominator the same.

Symbolically: $\dfrac{f(x)}{h(x)} - \dfrac{g(x)}{h(x)} = \dfrac{f(x)-g(x)}{h(x)}$

If the denominators are not the same, the fundamental property of rational functions must be used to make the denominators identical. It is listed again below.

Fundamental Property of Rational Functions

Given the function $\dfrac{f(x)}{g(x)}$ then $\dfrac{f(x)}{g(x)} = \dfrac{f(x)\cdot h(x)}{g(x)\cdot h(x)}$

where $h(x) \neq 0$

To develop a method for adding and subtracting rational functions, a series of examples follows. The examples lead to a four-step method for finding a sum or difference.

Example 1: If $f(x) = \dfrac{2x+5}{7}$ and $g(x) = \dfrac{3x-2}{7}$, find $(f+g)(x)$.

Solution: Since the denominators are identical, add the numerators and keep the denominator as 7.

$(f+g)(x) = \dfrac{2x+5}{7} + \dfrac{3x-2}{7}$ Add

$(f+g)(x) = \dfrac{2x+5+3x-2}{7}$ simplify the numerator

$(f+g)(x) = \dfrac{5x+3}{7}$ check the work with your calculator

This is easy.

Example 2: If $h(x) = \dfrac{2x+5}{x}$ and $k(x) = \dfrac{3x-2}{x}$, find $(h + k)(x)$.

Solution: The denominators are identical, so add the numerators, keeping the denominator as x.

$(h+k)(x) = \dfrac{2x+5}{x} + \dfrac{3x-2}{x}$ add

$(h+k)(x) = \dfrac{2x+5+3x-2}{x}$ simplify the numerator

$(h+k)(x) = \dfrac{5x+3}{x}$ check the work with your calculator

Example 3: If $s(x) = \dfrac{2x+5}{x+1}$ and $t(x) = \dfrac{3x-2}{x+1}$, find $(s + t)(x)$.

Solution: The denominators are the same, so add the numerators.

$(s+t)(x) = \dfrac{2x+5}{x+1} + \dfrac{3x-2}{x+1}$ add

$(s+t)(x) = \dfrac{2x+5+3x-2}{x+1}$ simplify the numerator

$(s+t)(x) = \dfrac{5x+3}{x+1}$ check the work with your calculator

To check, store the calculator's sum in Y_3 and the above sum in Y_4 as shown in Figure 8.4.3 below.

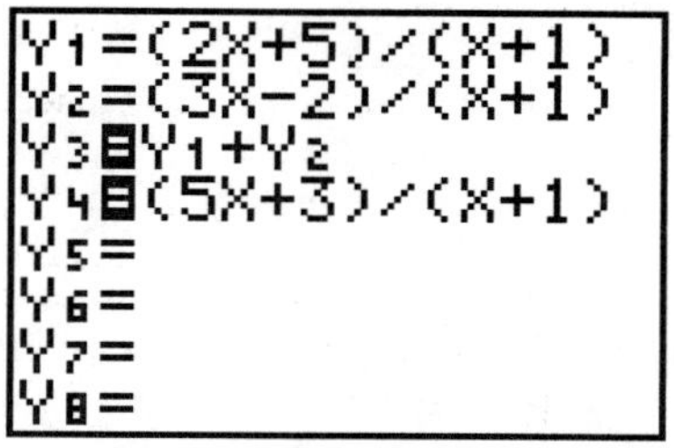

Figure 8.4.3a

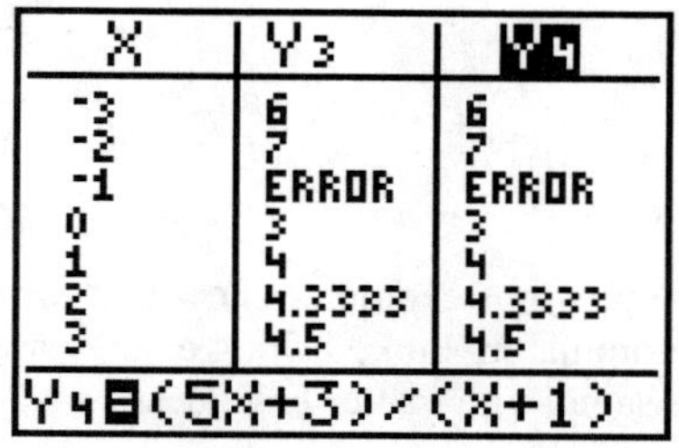

Figure 8.4.3b

The numeric representation of the correct sum is in Y_3 and the numeric representation above sum is in Y_4. Since the sample numeric representations are the same, you can safely conclude that the answer is correct.

⁂

Example 4: If $j(x) = \dfrac{2x+5}{(x-3)^2(x+4)}$ and $k(x) = \dfrac{3x-2}{(x-3)^2(x+4)}$, find $(j+k)(x)$.

Solution: Since the denominators are identical, add the numerators while keeping the denominator $(x-3)^2(x+4)$.

$$(j+k)(x) = \frac{2x+5}{(x-3)^2(x+4)} + \frac{3x-2}{(x-3)^2(x+4)} \qquad \text{add}$$

$$(j+k)(x) = \frac{2x+5+3x-2}{(x-3)^2(x+4)} \qquad \text{simplify the numerator}$$

$$(j+k)(x) = \frac{5x+3}{(x-3)^2(x+4)} \qquad \text{check the work with your calculator}$$

⁂

Example 5: If $f(x) = \dfrac{2x+5}{(x-3)^2(x+4)}$ and $g(x) = \dfrac{3x-2}{(x-3)^2(x+4)}$, find $(f-g)(x)$.

Solution: Since the denominators are identical, subtract the numerators while keeping the denominator $(x-3)^2(x+4)$.

$$(f-g)(x) = \frac{2x+5}{(x-3)^2(x+4)} - \frac{3x-2}{(x-3)^2(x+4)} \qquad \text{subtract}$$

$$(f-g)(x) = \frac{2x+5-(3x-2)}{(x-3)^2(x+4)} \qquad \text{simplify the numerator}$$

$$(f-g)(x)=\frac{-x+7}{(x-3)^2(x+4)}$$ check the work on the calculator

⁂

The previous examples show that rational functions can be added or subtracted no matter how messy the denominators look, as long as they are identical. The major concern for you is to make the denominators the same -- no matter how messy -- so that the functions can be added.

Example 6: If $f(x)=\dfrac{1}{2x}$ and $g(x)=\dfrac{3}{x}$, find $(f+g)(x)$.

Solution: The denominators are not identical. The ***only*** method available to make them the same is the fundamental property of rational functions. That is, the denominators can be made the same ***only by multiplying*** the top and bottom of a rational function by any non-zero function. If the second function is multiplied by 2 (top and bottom), the second denominator is changed to $2x$. Use the fundamental property on the second function.

$$(f+g)(x)=\frac{1}{2x}+\frac{3}{x}$$ make the denominators the same

$$(f+g)(x)=\frac{2}{2x}+\frac{3\cdot\mathbf{2}}{x\cdot\mathbf{2}}$$ the denominators are now the same; simplify the numerators.

$$(f+g)(x)=\frac{1}{2x}+\frac{6}{2x}$$ add

$$(f+g)(x)=\frac{7}{2x}$$ check the work with your calculator

⁂

Example 7: If $h(x)=\dfrac{4}{x}$ and $j(x)=\dfrac{2}{x^2}$, find $(h-j)(x)$.

Solution: The denominators are not the same; use the fundamental property on the first function to make it x^2.

$$(h-j)(x)=\frac{4}{x}-\frac{2}{x^2}$$ make the denominators the same

$$(h-j)(x)=\frac{4\cdot\boldsymbol{x}}{x\cdot\boldsymbol{x}}-\frac{2}{x^2}$$ simplify

$$(h-j)(x)=\frac{4x}{x^2}-\frac{2}{x^2}$$ subtract

$$(h-j)(x)=\frac{4x-2}{x^2}$$ check the work with your calculator

Example 8: $(f-g)(x)=\dfrac{x+2}{x(x-1)}-\dfrac{x}{x-1}$, simplify it.

Solution: The denominators are not the same, but if the fundamental property is used on the second fraction, the denominators can be the same.

$$(f-g)(x)=\frac{x+2}{x(x-1)}-\frac{x\cdot \boldsymbol{x}}{(x-1)\cdot \boldsymbol{x}}$$ simplify the numerator

$$(f-g)(x)=\frac{x+2}{x(x-1)}-\frac{x^2}{(x-1)\cdot x}$$ subtract fractions

$$(f-g)(x)=\frac{x+2-x^2}{x(x-1)}$$ simplify the numerator

$$(f-g)(x)=\frac{2+x-x^2}{x(x-1)}$$ check the work with your calculator

In every example above, the denominators were in factored form. *When the denominators are not factored, the first step in adding or subtracting is to factor them. This allows for the use of the fundamental property to make the denominators identical.* As will be shown in the next few examples, there are four (4) basic steps used to add or subtract rational functions.

They are:

1. Factor the denominators
2. Make the denominators identical
3. Simplify the numerators
4. Add or subtract.

In a few cases, it will be necessary to add a fifth step -- reduce the sum or difference of the functions.

Example 9: $(f+g)(x)=\dfrac{3}{x^2-1}+\dfrac{2x}{x+1}$, simplify it.

Solution: $(f+g)(x)=\dfrac{3}{(x-1)(x+1)}+\dfrac{2x}{(x+1)}$ 1. Factor the denominators

$$(f+g)(x)=\frac{3}{(x-1)(x+1)}+\frac{2x\cdot\mathbf{(x-1)}}{(x+1)\cdot\mathbf{(x-1)}}$$ 2. Make the denominators identical

$$(f+g)(x)=\frac{3}{(x-1)(x+1)}+\frac{2x^2-2x}{(x+1)(x-1)}$$ 3. Simplify the numerator

$$(f+g)(x)=\frac{2x^2-2x+3}{(x+1)(x-1)}$$ 4. Add the fractions

⁂

Example 10: $(f-g)(x)=\dfrac{x+1}{x^2+3x-4}-\dfrac{x-3}{x^2-1}$, simplify it.

Solution: $$(f-g)(x)=\frac{x+1}{(x+4)(x-1)}-\frac{x-3}{(x+1)(x-1)}$$ 1. Factor the denominators

$$(f-g)(x)=\frac{(x+1)\cdot\mathbf{(x+1)}}{(x+4)(x-1)\cdot\mathbf{(x+1)}}-\frac{(x-3)\cdot\mathbf{(x+4)}}{(x+1)(x-1)\cdot\mathbf{(x+4)}}$$ 2. Make the denominators equal

$$(f-g)(x)=\frac{x^2+2x+1}{(x+4)(x-1)(x+1)}-\frac{x^2+x-12}{(x+1)(x-1)(x+4)}$$ 3. Simplify the numerators

$$(f-g)(x)=\frac{x^2+2x+1-(x^2+x-12)}{(x+4)(x-1)(x+1)}$$ 4. Subtract

$$(f-g)(x)=\frac{x+13}{(x+4)(x-1)(x+1)}$$

⁂

Example 11: $(f+g)(x)=\dfrac{3}{x-2}+5$, simplify it.

Solution: $$(f+g)(x)=\frac{3}{1\cdot(x-2)}+\frac{5}{1}$$ 1. Factor the denominators

$$(f+g)(x)=\frac{3}{1\cdot(x-2)}+\frac{5\cdot\mathbf{(x-2)}}{1\cdot\mathbf{(x-2)}}$$ 2. Make the denominators identical

$$(f+g)(x)=\frac{3}{1\cdot(x-2)}+\frac{5x-10}{1\cdot(x-2)}$$ 3. Simplify the numerators

$$(f+g)(x)=\frac{5x-7}{x-2}$$ 4. Add and check the work

⁂

Example 12: Simplify the difference $(f-g)(x)=\frac{-2}{x-5}-3$.

Solution:

$$(f-g)(x)=\frac{-2}{1\cdot(x-5)}-\frac{3}{1}$$ 1. Factor the denominators

$$(f-g)(x)=\frac{-2}{1\cdot(x-5)}-\frac{3\cdot\mathbf{(x-5)}}{1\cdot\mathbf{(x-5)}}$$ 2. Make the denominators identical

$$(f-g)(x)=\frac{-2}{1\cdot(x-5)}-\frac{3x-15}{1\cdot(x-5)}$$ 3. Simplify the numerators

$$(f-g)(x)=\frac{-2-(3x-15)}{x-5}$$ 4. Subtract

$$(f-g)(x)=\frac{-3x+13}{x-5}$$ 5. You are encouraged to do a graphical check of the above analytical work. The graphic or numeric representations of the problem and the answer should be identical.

⁂

Simplifying Complex Rational Functions

A **complex rational function** contains one or more rational functions in the numerator and/or denominator of a rational function. You learned about the properties used to simplify rational functions in Sections 8.2 through 8.4, and you will also use them to simplify complex rational functions. Below are several examples of simplification of complex rational functions.

Example 13: Simplify the complex fraction $\dfrac{\frac{1}{4}+6}{3-\frac{1}{2}}$.

Solution: If the denominators within the complex fraction are factored, you can see what number to multiply times the numerator $\frac{1}{4}+6$ and the denominator $3-\frac{1}{2}$ in order to <u>eliminate those denominators</u>. The numerator ***<u>and</u>*** denominator of the complex fraction must be multiplied by 2·2, or 4. When multiplied by 4, the fraction $\frac{1}{4}$ becomes 1 and the $\frac{1}{2}$ becomes 2.

$$\frac{\left(\frac{1}{4}+6\right)\cdot \mathbf{4}}{\left(3-\frac{1}{2}\right)\cdot \mathbf{4}}$$

The distributive property must be used both on the top (numerator) and bottom (denominator) because a monomial is being multiplied by a binomial.

$$\frac{\frac{1}{4}\cdot 4+6\cdot 4}{3\cdot 4-\frac{1}{2}\cdot 4}$$

Simplify

$$\frac{1+24}{12-2}$$

The complex fraction simplifies to $\frac{25}{10}$, and this reduces to $\frac{5}{2}$.

In the previous problem, the fundamental property was used to simplify the complex fraction. This is the method that will be used on rational functions. If you ask yourself, "What can be multiplied by the numerator AND the denominator, so that the denominators of the rational functions within the complex rational function cancel?," you will have simplified the complex rational function when you multiply.

Example 14: Simplify the complex rational function $\dfrac{\frac{1}{x^2}+6}{3-\frac{1}{x}}$.

Solution: If the denominators of the rational functions within the complex rational function are factored, it can be seen that to ***eliminate these denominators***, the numerator ***and*** denominator of the complex rational function must be multiplied by $x\cdot x$, or x^2. When multiplied by x^2, the function $\frac{1}{x^2}$ becomes 1 and $\frac{1}{x}$ becomes x.

$$\frac{\boldsymbol{x^2}\cdot(\frac{1}{x^2}+6)}{\boldsymbol{x^2}\cdot(3-\frac{1}{x})}$$

The distributive property must be used both on the top and bottom because a monomial is being multiplied by a binomial.

$\dfrac{x^2 \cdot \dfrac{1}{x^2} + x^2 \cdot 6}{x^2 \cdot 3 - x^2 \dfrac{1}{x}}$, simplify the numerator and denominator to get $\dfrac{1+6x^2}{3x^2 - x}$.

This fraction cannot be reduced. The normal domains of the un-simplified and simplified functions are both all real numbers except 0 and $\dfrac{1}{3}$.

⁂

Example 15: Simplify the complex rational function $\dfrac{\dfrac{x}{x-1}}{\dfrac{x^2}{x^2-1}}$.

Solution: If the denominators within the complex rational function are factored, it can be seen that to *eliminate these denominators*, the numerator *and* denominator of the complex function must be multiplied by $(x + 1)(x - 1)$. When multiplied by $(x - 1)(x + 1)$, the function $\dfrac{x}{x-1}$ becomes $x(x + 1)$ and $\dfrac{x^2}{x^2-1}$ becomes x^2.

$$\frac{\dfrac{x}{(x-1)} \cdot \mathbf{(x+1)(x-1)}}{\dfrac{x^2}{(x^2-1)} \cdot \mathbf{(x+1)(x-1)}}$$

Simplify $\dfrac{\dfrac{x}{\cancel{(x-1)}} \cdot \mathbf{(x+1)\cancel{(x-1)}}}{\dfrac{x^2}{\cancel{(x-1)(x+1)}} \cdot \mathbf{\cancel{(x+1)(x-1)}}}$ to $\dfrac{x(x+1)}{x^2}$

The complex rational function is now a simple rational function, but it should be reduced to $\dfrac{x(x+1)}{x^2} = \dfrac{x+1}{x}$.

Reminder: The original function has a normal domain of all real numbers except –1, 0, and 1, and the simplified rational function has a normal domain of all real numbers except 0. This means that the numeric, graphic, or symbolic representations of the un-simplified and the simplified functions are equal only when $x \in (-\infty, -1) \cup (-1, 1) \cup (1, \infty)$, where neither is real at 0.

⁂

Example 16: Simplify the complex rational function $\dfrac{\dfrac{2x}{x^2 - 3x - 4}}{\dfrac{x}{x+1} + 1}$.

Solution: If the denominators in the complex function are factored, it can be seen that to *eliminate these denominators*, the numerator *and* the denominator of the complex function must be multiplied by $(x + 1)(x - 4)$. When multiplied by $(x + 1)(x - 4)$, the function $\dfrac{2x}{(x+1)(x-4)}$ becomes $2x$, and $\dfrac{x}{x+1}$ becomes $x(x - 4)$. The distributive property must be used on the bottom because a monomial is being multiplied by a binomial.

$$\frac{\dfrac{2x}{x^2-3x-4}\cdot(x+1)(x-4)}{\left(\dfrac{x}{x+1}+1\right)\cdot(x+1)(x-4)}$$

$$\frac{\dfrac{2x}{\cancel{(x-4)(x+1)}}\cdot\cancel{(x+1)(x-4)}}{\dfrac{x}{\cancel{x+1}}\cancel{(x+1)}(x-4)+1\cdot(x+1)(x-4)}$$

$$\frac{2x}{x(x-4)+(x+1)(x-4)}$$

$$\frac{2x}{x^2-4x+x^2-3x-4}$$

$$\frac{2x}{2x^2-7x-4}$$

Are the normal domains of the un-simplified and simplified functions the same?

⁂

8.4 EXERCISES

Many times you can work the exercises by algebraic, numeric, or graphical methods. You can learn about mathematics no matter which method you use. Part of the learning process is for you to decide which method is best for you to use. Please read the text carefully before you do the exercises. Read the **next assigned section** after you finish the assignment below.

A Quick Look To The Future

-4. List three methods for solving any equation.

-3. Can $\sqrt{x}+1$ ever have a value of 0?

-2. Is $\sqrt{\dfrac{3}{4}}$ the same thing as $\dfrac{\sqrt{3}}{\sqrt{4}}$?

-1. Is $2\sqrt{5}+4\sqrt{5}$ the same thing as $6\sqrt{5}$?

Previous Topics Revisited

1. Simplify the product $(fg)(x) = \dfrac{3x^2+x-2}{x^2-1} \cdot \dfrac{x^2-2x+1}{9x^2-4}$

2. Simplify the quotient $(f/g)(x) = \dfrac{4x+10}{x^2-4x-5} \div \dfrac{4x^2-25}{x^2-5x}$

3. Simplify the quotient $(f/g)(x) = \dfrac{2x^2-x-6}{x^2-3x-4} \div \dfrac{4x^2-9}{x^2-2x-8}$

4. Reduce $f(x) = \dfrac{9x^2-16}{9x^2+24x+16}$

5. Reduce $f(x) = \dfrac{4-x^2}{x-2}$

6. Identify the horizontal and vertical asymptotes in the graph of $y = \dfrac{-2}{x+4} - 3$.

New Topic Exercises

Find the sum of the rational functions and identify a normal domain that makes the un-simplified and simplified forms of the sum equivalent.

7. $(f+g)(x) = \dfrac{3}{x-1} + \dfrac{5}{x+1}$

8. $(f+g)(x) = \dfrac{x+1}{x^2-4} + \dfrac{3}{x+2}$

9. $(f+g)(x) = \dfrac{2x+3}{x-5} + \dfrac{x^2-2}{x^2-2x-15}$

10. $(f+g)(x) = \dfrac{3}{x+1} + \dfrac{5}{1}$

Find the difference of the rational functions and identify a normal domain that makes the un-simplified and simplified forms of the difference equivalent.

11. $(f-g)(x) = \dfrac{3x+1}{2x+3} - \dfrac{x^2-3x+2}{4x^2-9}$

12. $(f-g)(x) = \dfrac{2}{x-1} - \dfrac{1}{1-x}$

13. $(f-g)(x) = \dfrac{5x-3}{3x+1} - \dfrac{2x-1}{3x-1}$

14. $(f-g)(x) = \dfrac{-2}{x-5} - \dfrac{7}{1}$

Simplify the complex rational functions and identify a normal domain that makes the un-simplified and simplified forms of the functions equivalent.

15. $f(x) = \dfrac{\frac{2}{3}+5}{4-\frac{3}{5}}$

16. $f(x) = \dfrac{\frac{6}{x^2}-3x}{2+\frac{4}{x}}$

17. $f(x) = \dfrac{\dfrac{x+2}{x^2-4}}{\dfrac{x}{x-2}+3}$

18. $f(x) = \dfrac{\dfrac{x+1}{x^2-5x+6}-x}{2x+\dfrac{3}{x-2}}$

19. Give a counter-example to show that the statement $\dfrac{a}{b}+\dfrac{c}{c}=\dfrac{a+c}{b+d}$ is incorrect.

20. Give a counter-example to show that the statement $\dfrac{a}{b}-\dfrac{c}{c}=\dfrac{a-c}{b-d}$ is incorrect.

21. Do the functions $y=\dfrac{3}{x+1}+\dfrac{2}{x+1}$ and $y=\dfrac{5}{x+1}$ have identical graphs? Why?

22. What operation should be done first, addition or multiplication? $\dfrac{a}{b}+\dfrac{c}{c}\cdot\dfrac{e}{f}$

23. At what values of x are the functions $y=\dfrac{x+5}{x^2-25}$ and $y=\dfrac{1}{x-5}$ different?

24. At what values of x are the functions $f(x)=\dfrac{x-1}{x^2-1}+\dfrac{x+6}{x+6}$ and $g(x)=\dfrac{1}{x+1}+1$ different?

25. Find just one example where the incorrect rule for adding rational functions "add the numerators and the denominators" to get the "sum", gives the correct sum.

26. What is the normal domain of the function in Exercise 9?

27. Find two functions whose sum is $(f+g)(x)=\dfrac{x+1}{x-5}$.

28. Find two functions whose difference is $(f-g)(x)=\dfrac{x+1}{x-5}$.

From Mathematics to English

29. Describe how to add $(f+g)(x)=\dfrac{3}{x-6}+\dfrac{5}{x+6}$. Do not actually add.

30. Describe the fundamental property of rational functions.

31. Does the domain of a sum of two rational functions change when you add? Explain.

32. Is the statement $\dfrac{3}{x}+\dfrac{5}{x}=\dfrac{8}{x}$ an equation? Why?

33. After reading this section, make a list of questions that you want to ask your instructor.

34. Continue in your daily journal and make an entry. In addition to your normal entry on thoughts about the mathematics in this section, list at least two positive comments about what you have learned about this topic.

35. In paragraph format, summarize the material in this section of the text in your daily journal.

36. Describe how your classroom instructor made this topic more understandable and clear.

37. After reading the text and listening to your instructor, what do you not understand about this topic?

Explorations

38. The difference of $\frac{2x}{x^2-1}-\frac{1}{x+1}$ is $\frac{1}{x-1}$. Are the graphs of $\frac{2x}{x^2-1}-\frac{1}{x+1}$ and $\frac{1}{x-1}$ identical graphs?

39. Explore the graphs of:

a. $\frac{1}{x+1}$

b. $\frac{1}{x+1}+\frac{1}{x+3}$

c. $\frac{1}{x+1}+\frac{1}{x+3}+\frac{1}{x+5}$

d. $\frac{1}{x+1}+\frac{1}{x+3}+\frac{1}{x+5}+\frac{1}{x+7}$

Summarize your analysis of these graphs. What common element do they share? How are they different?

8.5 Solving Equations and Inequalities Containing the Rational Function and Inverse Variation

Laundry Detergent and Peanuts

Suppose you own the XYZ Chemical Company and one of your products is powered laundry detergent. Instead of selling the detergent as a 100% detergent mixture, you decide to mix pulverized bleached peanut shells in with your 100% detergent and sell it as a mixture.

In making the mixture, you start with 20 cups of pure detergent and you put in 1 cup of peanut shells. The strength of the mixture is now $S_d = \frac{20}{20+1}$, or 95% detergent. You may recall in Section 7.1 that to get the strength in % form you must multiply by 100; that is, the model for the strength of the detergent directly in % form is $S_d = 100 \cdot \frac{20}{20+1}$. You test your detergent and discover that it still cleans clothes nearly as well as the 100% detergent. You add another cup of peanut shells to the original mixture, the strength is now $S_d = 100 \cdot \frac{20}{20+1\cdot 2}$, or 91%. You add another cup and the strength is now $S_d = 100 \cdot \frac{20}{20+1\cdot 3}$, or 87%. As you see the pattern for the strength develop, you see that if x cups of peanut shells are added the model for the strength of the detergent mixture is $S_d = 100 \cdot \frac{20}{20+1\cdot x}$, or just $S_d = 100 \cdot \frac{20}{x+20}$.

If you know that your detergent will clean sufficiently well at a 70% strength, how many cups of peanut shells should be added to the 20 cups of detergent to maintain this level? This question can be answered by solving the equation $70 = 100 \cdot \frac{20}{x+20}$. This equation can be solved by any of the four technology based methods. The analytic (algebraic) method will be developed later in this section. The method below is the zeros method. Start by subtracting 70 from both sides of the equation to produce 0 on one side and a function on the other side.

$$0 = 100 \cdot \frac{20}{x+20} - 70$$

Graph the function $y = 100 \cdot \frac{20}{x+20} - 70$ and find the zero. To help find the graph, recall that it has a vertical asymptote at –20 and a horizontal at –70. Below in Figure 8.5.1 is the graph on the window [–60, 34] by [–250, 150] with the zero displayed. If 8.57 cups of peanut shells are mixed with 20 cups of

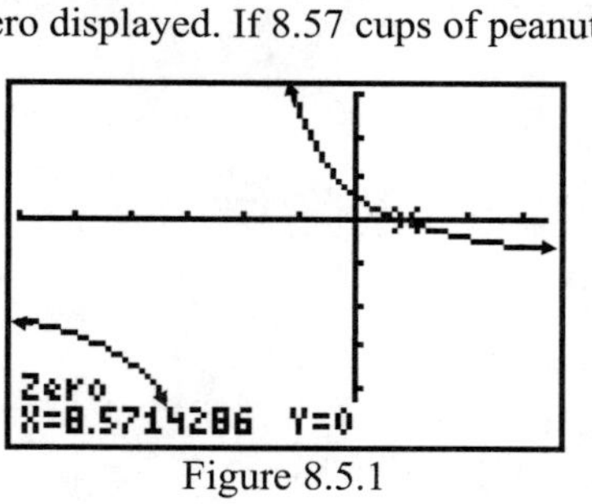

Figure 8.5.1

detergent, the strength of the mixture is 70% detergent.

For advertising reasons, you decide that you want more than an 80% strength. How many cups of peanut shells can you add and be truthful in your advertisement? The question can be answered by solving the inequality "strength > 80". Since the strength is modeled by $S_d = 100 \cdot \frac{20}{x+20}$, solve the inequality $100 \cdot \frac{20}{x+20} > 80$. Using the same window as above, the graphs of $S_d = 100 \cdot \frac{20}{x+20}$ and $S_d = 80$ are below. The intersection method for solving inequalities requires the intersection of the two graphs and you must see when the strength function is above (>) 80. The graph shows that the strength is

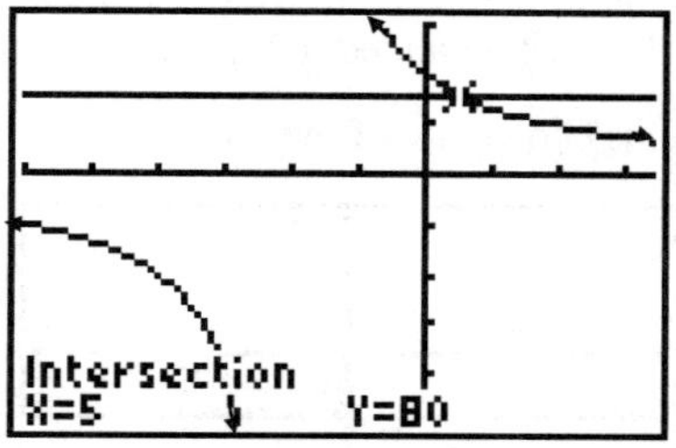

Figure 8.5.2

greater than 80 when x is less than 5. Since the problem domain is $[0, \infty)$, x must be on the interval $[0, 5)$ to make the strength greater than 80%.

The 2-Cycle Engine

Owners of a leaf blower, weed trimmer, lawn mower, or chain saw know that you must mix oil with the gasoline for these tools and all tools driven by a 2-cycle engine. Toys like gas powered and radio controlled model race cars, airplanes, and boats also use 2-cycle engines. The direction manual that comes with the tool or toy gives the "strength" of the gasoline solution using ratios. For example, it may indicate that the engine requires 16 parts gasoline to 1 part oil, or 40 parts gasoline to 1 part oil. A ratio is nothing but a fraction; that is, the ratio 16:1 is the fraction $\frac{16}{1}$. To mathematically analyze the gas-oil solution problem, it will be helpful to rewrite the ratios as oil to gasoline instead of gasoline to oil, and it will also be helpful to write these ratios (fractions) as percents. So, the tool or toy instruction manual will read something like 16:1 and you will invert this to 1:16, which is the fraction $\frac{1}{16}$, which is the percent number 6.25%. Or 32:1 becomes 1:32 or the fraction $\frac{1}{32}$ and the percent number 3.125%.

If you start with 1 gallon of gasoline, how much oil should you use to make a 5% (or $\frac{1}{20}$ or 1:20) oil solution? To solve this problem it is helpful to use the unit of measure used on the oil container -- ounces. One gallon of gasoline is 128 fluid ounces, and the "strength" of the oil-gas solution is the amount of oil divided by the total amount of solution. If you add 1 ounce of oil to 128 ounces of gasoline the strength is

$\frac{1}{128+1}$, or in percent form this is $100 \cdot \frac{1}{128+1}$, or 0.78% oil. If you add 2 ounces of oil to 128 ounces of gasoline, the strength is $100 \cdot \frac{2}{128+2}$, or 1.54% oil. Adding 3 ounces of oil makes the strength $100 \cdot \frac{3}{128+3}$, or 2.29% oil. Hopefully, you can see a pattern developing in the calculation of the strength of the 2-cycle gasoline solution. It is $S_g = 100 \cdot \frac{x}{128+x}$, where x ounces of oil are added to 128 ounces (1 gallon) of gasoline.

To make a 5% oil solution, replace the function name with 5 and solve the equation. The solution to the equation shows that 6.74 ounces of oil must be added to 1 gallon of gasoline. The solution is found by the intersection method. That is, the graphs shown are of the functions $S_g = 100 \cdot \frac{x}{128+x}$ and $S_g = 5$.

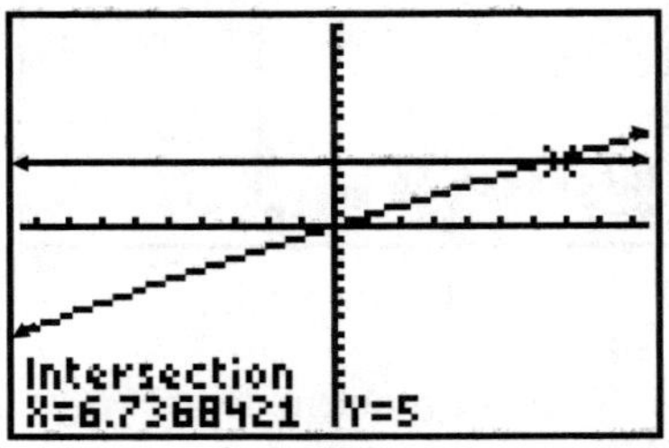

Figure 8.5.3

How much oil can you add to the 128 ounces of gasoline if the instruction manual allows an oil-gasoline solution of between 5 and 10% oil? This can be solved quite nicely with the intersection method. That is, if you think graphically, perhaps you can visualize the solution as all values for x that cause the graph of $S_g = 100 \cdot \frac{x}{128+x}$ to be between $S_g = 5$ and $S_g = 10$ as shown in Figure 8.5.4.

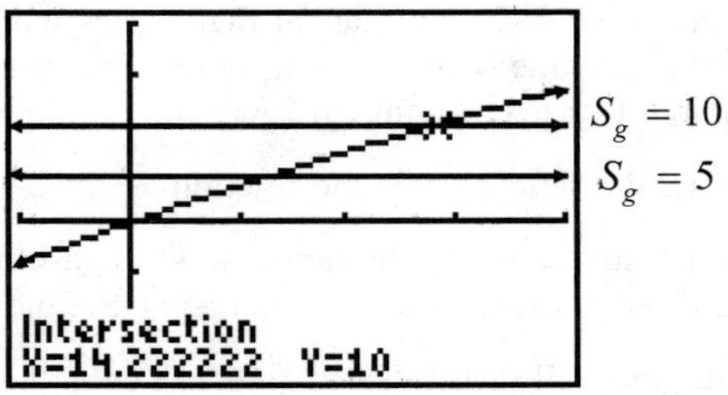

Figure 8.5.4

Figure 8.5.3 shows that the function $S_g = 100 \cdot \frac{x}{128+x}$ is above 5 when x is to the right of 6.74 and Figure 8.5.4 shows that the graph of $S_g = 100 \cdot \frac{x}{128+x}$ is below 10 when x is left of 14.22. The conclusion must be that if $x \in (6.74, 14.22)$ the strength function is between 5 and 10%.

The technology-based methods for solving rational equations are basically the same as for solving any other equation. There is however, a caution when using the zeros method. The erroneous vertical line that is sometimes drawn by the calculator near the vertical asymptote may look like a zero. Be careful to ignore this misleading information. Also, when you use the zeros approach in conjunction with the numerical method of solving equations, you look for a change in sign in the numerical representation to help you find the zero of the function. Rational functions can have a change of sign at vertical asymptotes. This can also be misleading information. There is no zero at a vertical asymptote -- usually just a change in sign. Finally, if you use a decimal window to try to eliminate the erroneous vertical line drawn by the calculator near the vertical asymptote, you may also find that the graph stops in the middle of the screen. This is an incorrect graphical representation. Any rational function of the form $y = d \cdot \frac{1}{x+e} + f$ tends toward ∞ or $-\infty$ on either side of the vertical asymptote. It does not stop "in the middle of the screen."

Below are a few examples of solving equations and inequalities using technology-based methods. The analytical method follows.

Example 1: Solve the equation $\frac{3}{x-2} + 5 = -17.3$.

Solution: Transform the equation so that 0 is on the right. That is, add 17.3 to both sides so that the zeros-method can be used.

$$\frac{3}{x-2} + 5 + 17.3 = 0$$

Graph the function $y = \frac{3}{x-2} + 22.3$, and find the zeros. From what you learned in Section 8.1, recall that the horizontal asymptote is at 22.3, and the vertical asymptote is at 2; thus, choose the window with this in mind. That is, a small decimal window for x values and a $ymax$ above 22.3. The window below is [–4.7, 4.7] by [–10, 40].

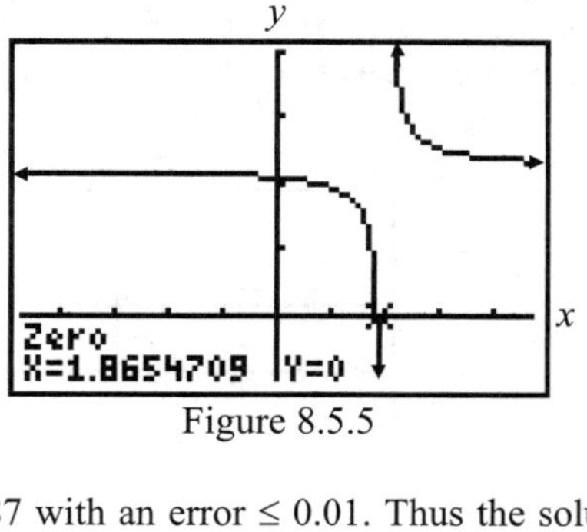

Figure 8.5.5

The zero is 1.87 with an error ≤ 0.01. Thus the solution to the equation is 1.87 with an error ≤ 0.01.

⁂

Example 2: Solve the inequality $\frac{-2}{x+3} + \frac{5}{x-3} < 7.1$.

Solution: Subtract 7.1 from both sides to get 0 on the right side. The left side of the inequality becomes the rational function $y = \frac{-2}{x+3} + \frac{5}{x-3} - 7.1$.

Graph the function and find the values for x that make the function less than 0. The window is [–9.4, 9.4] by [–15, 3].

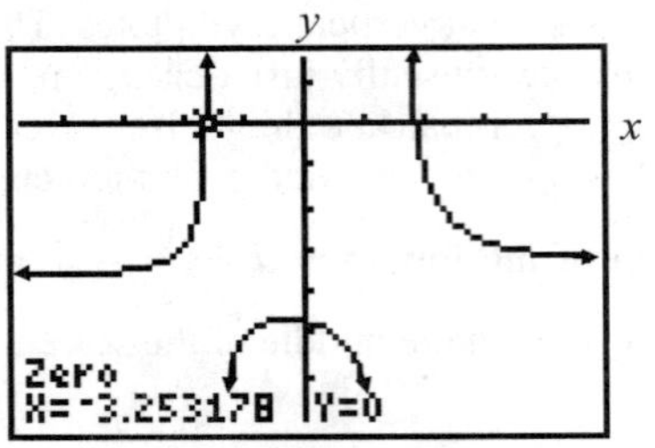

Figure 8.5.6a

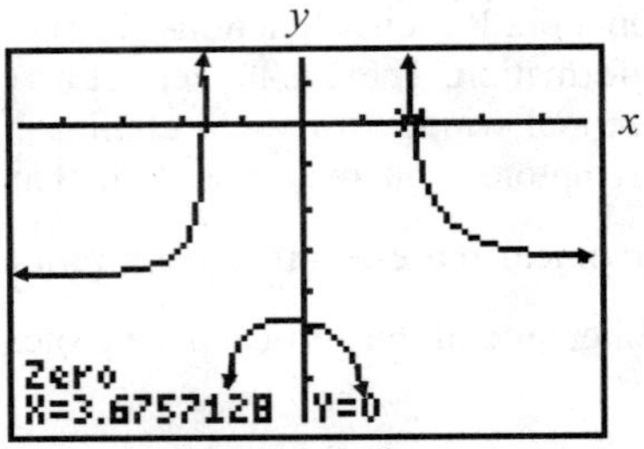

Figure 8.5.6b

The zeros are –3.25 and 3.68 with error ≤ 0.01. The function is less than zero to the left of –3.25 and it is less than zero to the right of 3.68. At –3 and 3 the function is not real and the graph shows that the function is less than zero between the vertical asymptotes at –3 and 3. Further inspection of the graph shows that the only time it is positive is between each zero and the vertical asymptote near that zero. The solution to the inequality is (–∞, –3.25) ∪ (–3, 3) ∪ (3.68, ∞).

⁂

Example 3: Solve the equation $\frac{x+3}{5x-10} = \frac{1}{x-2}$.

Solution: Transform the equation by subtracting $\frac{1}{x-2}$ from the right side and graph the function on the left side of the equation. The equation becomes $\frac{x+3}{5x-10} - \frac{1}{x-2} = 0$.

Now graph the function $y = \frac{x+3}{5x-10} - \frac{1}{x-2}$. Values for x that cause the function to be 0 are solutions to the equation. The window is [–4.7, 4.7] by [–0.5, 0.5].

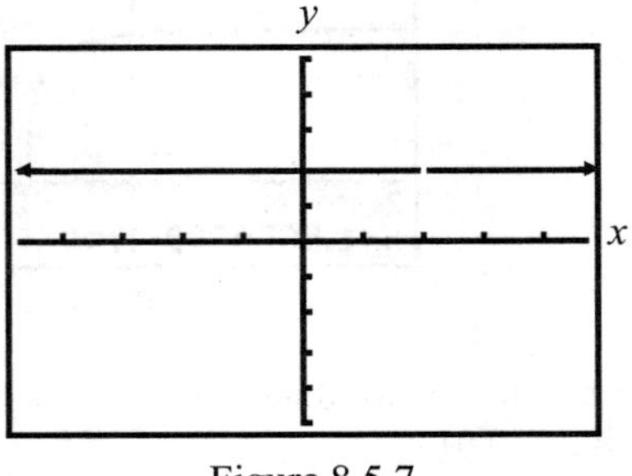

Figure 8.5.7

As can be seen above, the graph never has a value of zero. The solution set to the equation is empty; that is, there are no real solutions. This can be written as $x \in \varnothing$

⁂

Example 4: Solve the inequality $\dfrac{3x-7}{x+1} > 9$ by the intersection method.

Solution: Graph the functions $f(x) = \dfrac{3x-7}{x+1}$ and $g(x) = 9$; find the intersection and when the graph of $f(x)$ is greater than (above) $g(x)$. Below are the graphs on the window [–4.7, 4.7] by [–15, 15].

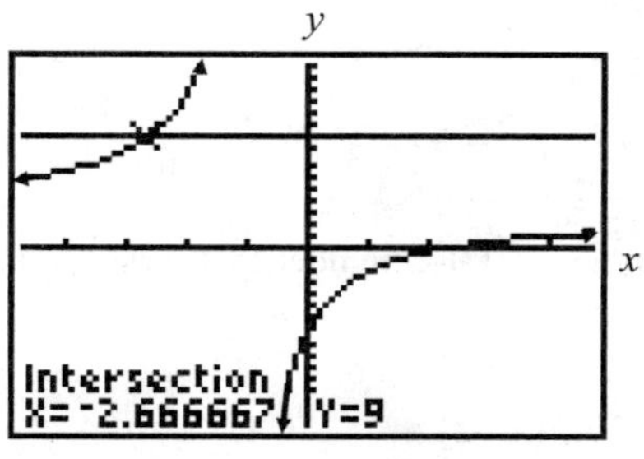

Figure 8.5.8a

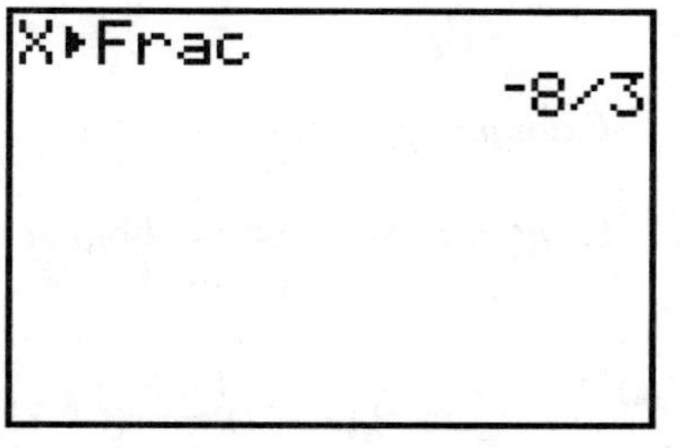

Figure 8.5.8a

When x is $-\frac{8}{3}$, $f(x)$ is 9, and between $-\frac{8}{3}$ and -1 $f(x)$ is greater than 9. Remember, at -1 there is a vertical asymptote; thus, the function approaches ∞. The solution to the inequality is $\left(-\frac{8}{3}, -1\right)$

⁂

Analytical Method

Equations containing rational functions can be solved analytically by clearing both sides of the equation of any and all rational functions in the equation. For example, if an equation contains a function with a denominator of $(x + 3)$, it can be cleared by multiplying both sides by $(x + 3)$. In the equation $\dfrac{2x+5}{x+3} = 7$, if both sides are multiplied by $(x + 3)$, the $(x + 3)$ cancels.

$$\mathbf{(x+3)}\frac{2x+5}{(x+3)} = 7\ \mathbf{(x+3)}$$

$$2x + 5 = 7(x + 3)$$

The resulting equation is similar to those solved in Chapter Six. If an equation contains functions with denominators of $(x + 2)$ and $(x - 1)$, both sides can be multiplied by $(x + 2)$ **AND** $(x - 1)$. This would cause both denominators to cancel.

Do not forget that the domain of function $y = \dfrac{2x+5}{x+3}$ does not contain –3. However, when you eliminate the denominator, the transformed function no longer contains the rational function. This does not change the fact that the original equation contains the function $y = \dfrac{2x+5}{x+3}$ where x may NOT be –3 and the transformed equation contains functions that have domains that include all real numbers. The analytical method may cause the transformed equation to have solutions that are not solutions of the original equation.

Caution!

Example 5: Solve the equation $\dfrac{3}{x-2} + 5 = -17.3$ analytically.

Solution: Multiply both sides by $(x - 2)$. Remember that 2 is not in the domain of the rational function.

$$\mathbf{(x-2)}\left(\frac{3}{x-2} + 5\right) = -17.3\mathbf{(x-2)}$$

$$\cancel{(x-2)}\frac{3}{\cancel{(x-2)}} + (x-2)\cdot 5 = -17.3(x-2) \qquad \text{simplify}$$

$$3 + 5x - 10 = -17.3x + 34.6$$

$$5x - 7 = -17.3x + 34.6 \qquad \text{add } 17.3x \text{ to both sides}$$

$$22.3x - 7 = 34.6 \qquad \text{add 7 to both sides}$$

$$22.3x = 41.6 \qquad \text{divide both sides by 22.3}$$

$$x = 1.87 \qquad \text{with error less than 0.01}$$

Example 6: Solve the equation $\dfrac{x+3}{5x-10} = \dfrac{1}{x-2}$ analytically.

Solution: Since $5x - 10$ is $5(x - 2)$, multiply both sides by 5 and $(x - 2)$, where 2 is not in the domain of the function in the original equation.

$$\mathbf{5(x-2)}\frac{x+3}{5(x-2)} = \frac{1}{(x-2)}\mathbf{5(x-2)}$$

$$\cancel{5(x-2)}\frac{x+3}{\cancel{5(x-2)}}=\frac{1}{\cancel{(x-2)}}\cancel{5(x-2)}$$

$$x+3=5$$

$$x=2$$

Recall that $x \neq 2$. While 2 is a solution to $x + 3 = 5$, it is not a solution to the original equation because the domains of the functions in the original equation do not contain 2. The conclusion you must draw is that the original equation has no solution. This same equation was solved graphically in the example 3 above. The graph had no zeros; thus, it has no solution.

⁂

Example 7: Solve the equation $\dfrac{5x-26}{x-7}=\dfrac{9}{x-7}+3$ analytically.

Solution: If both sides are multiplied by $(x - 7)$ the denominators will cancel, leaving an equation without fractions. Remember, $x \neq 7$ in the functions in the equation.

$$(x-7)\frac{5x-26}{(x-7)}=\left(\frac{9}{(x-7)}+3\right)(x-7)$$

The right side is a binomial; thus, $(x - 7)$ must be multiplied by both terms.

$$(x-7)\cdot\frac{5x-26}{(x-7)}=\frac{9}{(x-7)}\cdot(x-7)+3\cdot(x-7)$$

$$\cancel{(x-7)}\cdot\frac{5x-26}{\cancel{(x-7)}}=\frac{9}{\cancel{(x-7)}}\cdot\cancel{(x-7)}+3\cdot(x-7)$$

$$5x-26=9+3(x-7)$$

$$5x-26=9+3x-21$$

$$5x-26=3x-12$$

$$5x-26=-12$$

$$2x=14$$

$$x=7$$

While 7 is a solution to $5x - 26 = 9 + 3(x - 7)$, 7 is not in the domain of the functions in the original equation. Again, you must conclude that the original equation has no solution. This can be verified by solving the equation graphically. The zeros method uses the function $y = \frac{5x-26}{x-7} - \frac{9}{x-7} - 3$ which has no zeros, only a hole at $x = 7$.

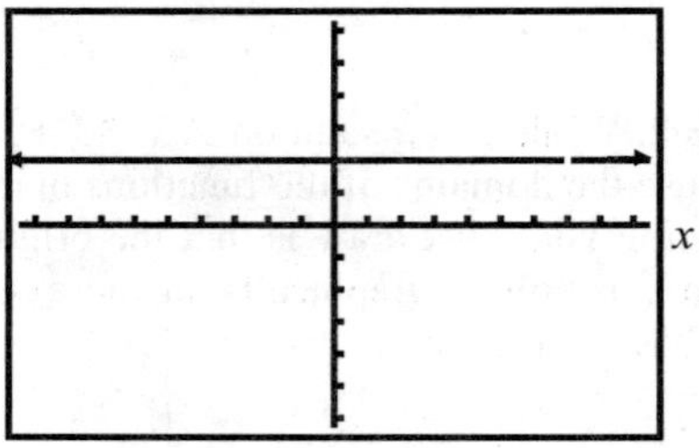

Figure 8.5.9

The window is a decimal window. As can be seen, there are no zeros, which means the equation has no real solution.

⁂

Inverse Variation

In Section 6.5, you learned about a special kind of function called direct variation functions. You used three types; the linear direct variation $y = kx$, quadratic direct variation $y = kx^2$, and square root direct variation $y = k\sqrt{x}$. Many data relationships do not have the characteristics of the direct variation functions, that is, in direct variation as x increases so does y. For example, as you move farther away from a light (the distance is increasing), the intensity of the light decreases. This does not match the pattern of the direct variation. Or for example, gravitational attraction between two objects becomes less as the distance between the two objects increases. This does not match any of the direct variation relationships.

The second kind of variation function is the inverse variation relationship. The basic form of this function is $y = \frac{k}{x}$ and is read "y varies inversely as x. Like before, k is the constant of variation.

Example 8 If the average velocity of a car as it travels between two fixed points varies inversely with time, and the velocity is 100 miles per hour (mph) when it takes 9 seconds to travel between the fixed points, what is the inverse variation model? What is the velocity when the car takes 12 seconds to travel the fixed distance? How long does it take to travel the fixed distance at a velocity of 120 mph?

Solution: Since the relationship between velocity and time is inverse variation, the model is $v = \frac{k}{t}$. The constant of variation can be found by replacing v with 100 and t with 9.

Solve the equation $100 = \frac{k}{9}$ for k. Multiply both sides by 9 and the solution is $k = 900$. The model for this relationship is $v = \frac{900}{t}$. To find the velocity of the car if it takes 12 seconds, solve the equation $v = \frac{900}{12}$. It is 75 mph. To find the time it takes for the car to travel the distance at 120 mph, solve the equation $120 = \frac{900}{t}$. Multiply both sides by t and divide both sides by 120. The solution is 7.5 seconds.

⁂

***Example* 9:** The number of oscillations that a pendulum makes is inversely related to the square root of the length of the pendulum. If a ½ foot long pendulum goes through 1.273 oscillations in 1 second, what is the inverse variation model. If the pendulum has 6 oscillations in one second, how long is the pendulum? If the pendulum is 50 feet long, how many oscillations per second will it have?

Solution: The model is $N = \frac{k}{\sqrt{l}}$ and the constant of variation can be found by substituting ½ for l, and 1.273 for N. The model becomes $1.273 = \frac{k}{\sqrt{0.5}}$, multiply both sides by $\sqrt{0.5}$ to solve for k. The value for k is $1.273\sqrt{0.5}$, or approximately 0.9. The model for l measured in feet is $N = \frac{0.9}{\sqrt{l}}$. To find how long a pendulum is needed to create 6 oscillations per second, replace N with 6 and solve the equation for l. The equation is $6 = \frac{0.9}{\sqrt{l}}$, solve for l. Since you have not solved equations like this before using algebra, use a technology-based method. If you experiment with a pendulum you would learn that the length is quite small when there are 6 oscillations per second. The zeros method shows the solution to be 0.0225 foot. The function graphed is $y = \frac{0.9}{\sqrt{x}} - 6$.

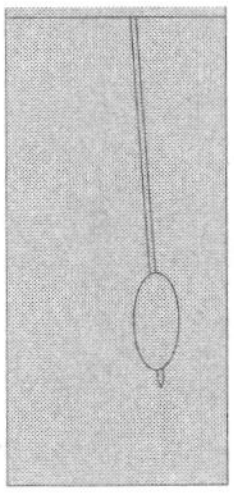

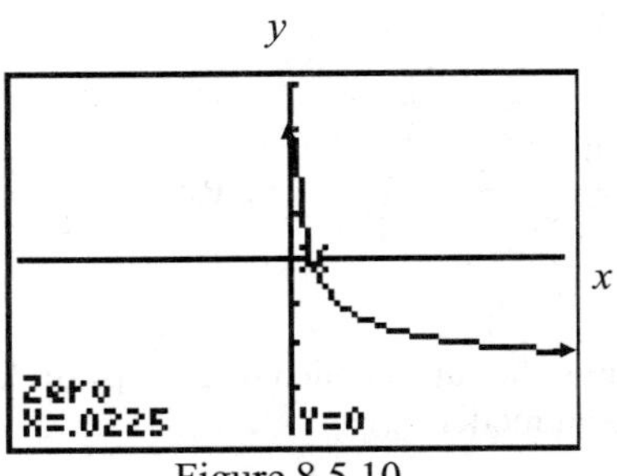

Figure 8.5.10

The window is [–0.1, 0.1] by [–20, 20]. Finally, a 50-foot long pendulum will oscillate $N = \frac{0.9}{\sqrt{50}} = 0.127$ times in 1 second (or 1 time in 7.87 seconds).

⁂

***Example* 10:** The intensity of light varies inversely with the distance from the light squared. Find the model of this relationship if at 10 feet from the light source, the intensity is 60 lumens.

Solution: The model is $I = \dfrac{k}{d^2}$ and k can be found from $60 = \dfrac{k}{10^2}$; k is 6000. The model now becomes $I = \dfrac{6000}{d^2}$.

8.5 EXERCISES

Many times you can work the exercises by algebraic, numeric, or graphical methods. You can learn about mathematics no matter which method you use. Part of the learning process is for you to decide which method is best for you to use. Please read the text carefully before you do the exercises. Read the **next assigned section** after you finish the assignment below. All numeric answers must have an error ≤ 0.01.

A Quick Look To The Future

-4. What number can you square to get 13?

-3. Does $\sqrt{4}\cdot\sqrt{3}$ equal $4\sqrt{3}$?

-2. Do the following calculations: $\left(\sqrt{5}\right)^2$, $\left(\sqrt{7}\right)^2$, $\left(\sqrt{13}\right)^2$, and $\left(\sqrt[3]{5}\right)^3$.

-1. Use your calculator to evaluate $3^{\frac{1}{2}}$ and $\sqrt{3}$. Is there anything unusual in the answers?

Previous Topics Revisited

1. Add $\dfrac{4}{x+2}+\dfrac{3}{x-2}$
2. Subtract $\dfrac{5}{x+4}-7$
3. Simplify $\dfrac{3x-\dfrac{5}{x-3}}{\dfrac{2}{x-3}+5x}$
4. Multiply $\dfrac{x^2-9}{x^2-x-6}\cdot\dfrac{x^2-2x-8}{x^2-2x-3}$
5. Divide $\dfrac{x^2-3x-10}{x^2+x-2}\div\dfrac{x^2-9x+20}{x^2-2x-3}$
6. Reduce $\dfrac{(x-3)^2(x+4)^2}{(x+2)(x^2+x-12)}$

New Topic Exercises

7. How many cups of peanut shells should be added to 20 cups of detergent to make the strength of the mixture 65%? (See the detergent model S_d.)

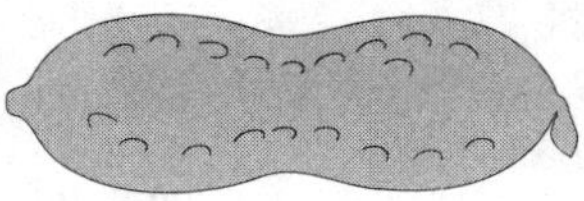

8. What is the strength of the detergent mixture when 60 cups of peanut shells are used with the 20 cups of detergent?
9. If the strength is to be at most 62% detergent, how many cups of peanut shells can be added?
10. What function parameter must be changed so that the detergent model applies when 50 cups of detergent are used to make the mixture?
11. With a base of 50 cups of detergent, how many cups of peanut shells must be added to make a 90% mixture?

12. With a base of 200 cups of detergent, how many cups of peanut shells must be added to make a 90% mixture?

13. Using the 2-cycle model, predict what percent oil is a mixture that is 64 parts gasoline and 1 part oil?

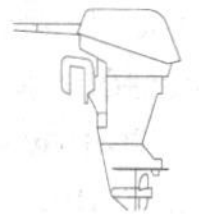

14. How many ounces of oil are needed to make one gallon of gasoline a 2% oil-gas solution?

15. How many ounces of oil are needed to make one gallon of gasoline an 80 part gas to 1 part oil solution?

16. If the oil-gas solution must be between 1 and 2% for a model airplane engine to run correctly, how much oil can be added to the gasoline?

17. If you mix an oil-gas solution starting with 264 ounces (2 gallons) of gasoline, what function parameter in the 2-cycle model must be changed?

18. How much oil should be used in the 2 gallon gasoline solution if the strength must be 16 parts gasoline to 1 part oil?

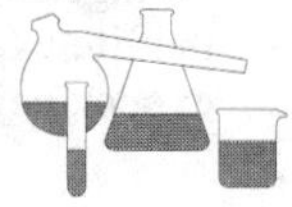

Solve the following equations or inequalities.

19. $\frac{2.5}{x+5} = \frac{7.9}{x}$

20. $\frac{x}{x+2} = \frac{2}{3}$

21. $\frac{x}{x+2} \geq \frac{2}{3}$

22. $\frac{9}{x^2-9} - \frac{3}{x-3} = \frac{-4}{x+3}$

23. $\frac{3.11}{x+7.6} = \frac{2.05}{x}$

24. $\frac{4}{x-2} + 3 = \frac{3x-6}{x-2}$

25. $\frac{4}{x-2} + 3 < \frac{-3}{x+2} - 5$

26. $\frac{x}{2x+4} - \frac{1}{x+2} = 1$

27. $\frac{x}{x+3} = \frac{2x}{x+3} + 3$

28. $\frac{2}{x-1} + 4 \leq 3$

29. $\frac{3}{x+2} + \frac{5}{x-1} = \frac{-4x}{(x+2)(x-1)}$

30. $\frac{1}{x} + 0.26 = 3.9$

31. $\frac{3x-5}{x-4} = \frac{7}{x-4}$

32. Boyle discovered if you hold the temperature of a gas constant, the pressure of the gas varies inversely with the volume of the gas. (Imagine a balloon full of air that can be squeezed to decrease the volume -- what happens to the pressure in the balloon?) If the pressure of a gas is 200 pounds per square inch when the volume is 50 cubic inches, what is the pressure if the volume is changed to 100 cubic inches? What is the volume at 400 pounds per square inch of pressure?

33. Resistance to the flow of electricity through a wire varies inversely as the square of the diameter of the wire. If a wire of diameter 0.08 inch has a resistance of 16.9 Ω (ohms), what is the model for resistance? What is the resistance in the wire of diameter 0.25 inch? If the resistance through the wire measures 200 Ω, what is the diameter of the wire?

34. The life expectancy of an incandescent light bulb varies inversely with the 12th power of the applied voltage. If the life expectancy of a light bulb is 750 hours when run at 120 volts, what is the model of the life expectancy? What is the life expectancy when run at 110 volts? If the bulb has a life expectancy of 2000 hours, what is the voltage applied to the bulb?

35. In the following equation, replace the "?" with a value so that the equation has no solution.
$\frac{2}{x-6}+?=1$

36. In the previous equation, what value of "?" will cause the equation to have a solution?

37. Find a rational function in standard form $\left(y=\frac{d}{x+e}+f\right)$ that is identical to $y=\frac{x+3}{x-1}$.

From Mathematics to English

38. Recalling the material on rational functions in Section 7.1 explain why the equation $\frac{1}{x-3}+4=4$ has no solution. There is no need to solve the equation by graphical or algebraic methods.

39. Solve the equation: $\frac{x+3}{x-2}=\frac{5}{x-2}$. What is the domain of the function $y=\frac{x+3}{x-2}$?

Can this equation have a solution of 2? Why?

40. Can a rational equation of the form $\frac{a}{x-d}+b=c$ have more than one solution? Why?

41. What is the *most possible* number of solutions to the equation: $\frac{a}{x-c}+b=\frac{d}{x-f}+e$? Explain.

42. What is the difference between a function and an equation?

43. Describe the zeros method of solving a greater than inequality.

44. After reading this section, make a list of questions that you want to ask your instructor.

45. Continue in your daily journal and make an entry. In addition to your normal entry on thoughts about the mathematics in this section, list at least two positive comments about what you have learned about this topic.

46. In paragraph format, summarize the material in this section of the text in your daily journal.

47. Describe how your classroom instructor made this topic more understandable and clear.

48. After reading the text and listening to your instructor, what do you not understand about this topic?

Explorations

49. Find values for *a, b, c, d, e,* and *f* such that the equation $\frac{a}{x-c}+b=\frac{d}{x-f}+e$ has: no solution, 1 solution, and 2 solutions. There is no need to find any of the solutions.

50. Solve the equation $\frac{2}{x-1}=\frac{5}{x+2}$ by a graphical method; however, DO NOT USE the related function $y=\frac{2}{x-1}-\frac{5}{x+2}$ or the function $y=\frac{5}{x+2}-\frac{2}{x-1}$ to find the solution!

51. While solving the next few equations analytically is difficult or impossible, try solving them graphically or numerically.

a. $\frac{-|x-5|}{x+1}\div\frac{\sqrt{x+2}}{x-1}-\frac{x^2-6x-8}{\sqrt{x^2-4}}=0$

b. $\left|\frac{2}{x}-3\right|=0$

c. $\frac{-|x-5|}{x+1}\cdot\frac{\sqrt{x+2}}{x-1}-\frac{x^2-6x+4}{\sqrt{x^2-4}}=0$

d. $0=\left(\frac{2}{x-4}-3\right)^2$

e. $\frac{-2}{x-1}+3=2^{x+1}+7$

f. $2^{\frac{1}{x-3}}=0$

g. $\frac{2^x-3}{x}=-2$

h. $\frac{-2}{x-1}+3=2^{x+1}+7$

CHAPTER EIGHT TEST

1. Where is the vertical asymptote of the function $y=\frac{2}{x-1}+3$?
2. Where is the horizontal asymptote of the function $y=\frac{-4}{x+3}-6$?
3. Is $y=\frac{2}{x}+3$ increasing or decreasing on its domain?
4. Find the zero of $y=\frac{-2.6}{x+3.1}-4.7$.
5. Where are the vertical and horizontal asymptotes of the function $y=\frac{a}{x+b}+c$, where *a*, *b*, and *c* are constant real numbers?
6. Reduce $\frac{2x^2-x-15}{9-x^2}$.
7. What is the normal domain of $\frac{2x^2-x-15}{9-x^2}$?
8. Is the function $y=\frac{x-4}{x+2}$ in reduced form? Why?
9. Is the normal domain of $\frac{x+2}{x^2-4}$ the same as $\frac{1}{x-2}$?
10. Multiply $\frac{3x^2+11x-4}{x^2+3x-4}\cdot\frac{x^2-4x+3}{3x^2-11x+3}$.
11. Divide $\frac{x^2+x-2}{3x+3}\div\frac{x-1}{5x+5}$.
12. Add $\frac{2}{x-3}+5$.
13. Subtract $\frac{1}{x-1}-\frac{1}{x-1}$.
14. Simplify $\dfrac{\frac{1}{x}-2}{3-\frac{1}{x^2}}$.
15. Does $\frac{3}{x-2}=0$ have a solution? If yes, what?
16. Solve $\frac{2.1}{x}+3.6=-4.2$.
17. Solve $\frac{-3}{x+2}-5=1$.
18. Solve $\frac{x}{x+5}+2=\frac{2x+5}{x+5}$.
19. What value of *b* will cause the equation $\frac{2}{x+4}+b=3$ to have no solution?
20. What is the relationship between the solution to the equation $\frac{-2}{x+1}-4=0$ and the zero of the function $y=\frac{-2}{x+1}-4$?

CHAPTER NINE
ADVANCED ANALYSIS OF THE SQUARE ROOT FUNCTION

8.0 Introduction

The Long Skid Mark

The highway patrol and other traffic enforcement agencies assume that when the driver of a car applies the brakes hard and fast enough, the length of the skid mark (in feet) made by the car determines the velocity (in miles per hour) of the car when the brakes were applied. To confirm that this is a true statement, the results of a skid experiment are shown below. The data shows the relationship between the length of a skid mark and the velocity of the car used to make the skid mark.

Table 9.0.1

l	0	10	20	30	40	50	60	70	80	90	100
V	0	17	25	30	35	39	43	46	49	52	55

To find a model for the data, it would be very helpful to look at the graphical representation of the data relationship. If the shape is recognizable then you have a clue as to the symbolic representation of the model. If the shape is not recognizable, then you will need to try combinations of known functions or research for a new kind of function that behaves like the data relationship. Below are the graphs of the data and the data connected on the window [–20, 120] by [–10, 75]. Recalling the mathematics studied in

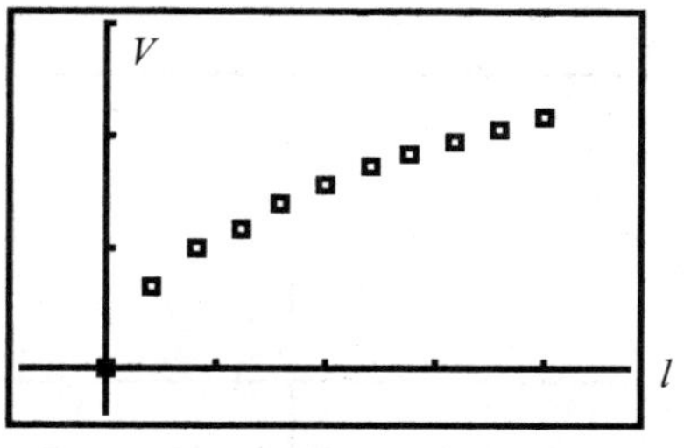

Figure 9.0.1a

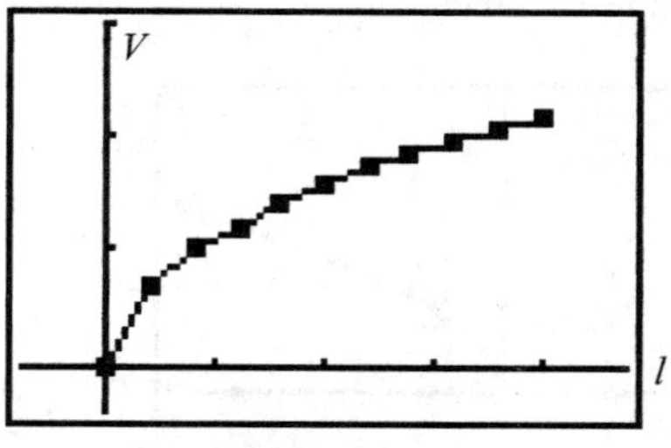

Figure 9.0.1b

Section 3.4, you should recognize the shape as a square root function like those studied in Chapter Three. Not only does it look like a square root function of the form $V = d\sqrt{l+e}+f$, you can also see that the function parameters e and f are 0 and parameter d is positive. Below is the graph of the data and the function $V = 5.5\sqrt{l}$ on the same window as above.

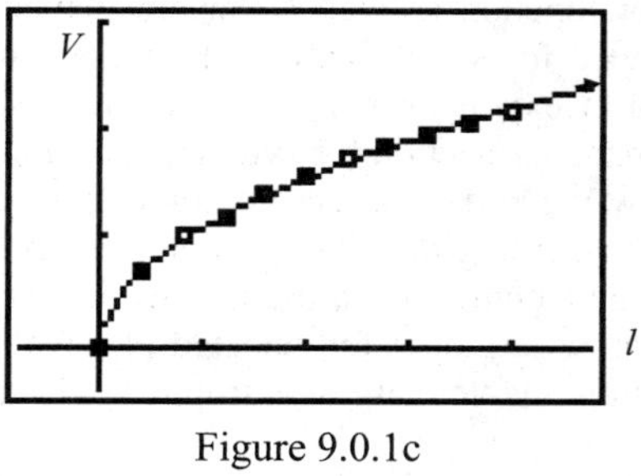

Figure 9.0.1c

As a result of this experimentation, traffic accident investigators now have a mathematical model they can use to determine if the driver of the car in the accident was speeding.

(This is the clown, not the teacher.)

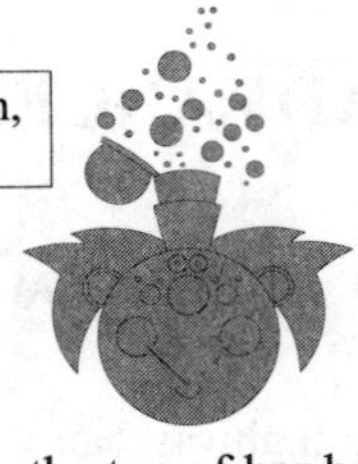

The Square Root Function in Action

Your mathematics teacher, being the intelligent person that she is, agrees to let the class clown drop a water balloon on her head from the second story balcony of the mathematics building, provided he earns an "A" on the "Memory Model Laboratory Project." The teacher, a former class clown in her college days at Otterbein College, knows that the balloon will not break until its velocity reaches 30 feet per second (f/s). She has made a quick mental estimation on the distance between the top of her head (she stands 5' 7") and chest level of a person standing on the balcony, and has guessed it to be 13 feet. As it turns out, the class clown finally realizes how much he can learn about mathematics by using the graphing calculator, and he aces the lab project. Does the teacher get wet? Does the class clown have a trick up his sleeve? Perhaps a little physics research and a graph of the function that relates height to velocity is in order.

A beginning physics text will show you that the function that models the velocity of an object dropped near the surface of the earth is $v = \sqrt{2gh}$, where g is the acceleration due to gravity and it has been measured to be 32 feet per second per second. Replacing g with 32 and simplifying the square root, the function becomes $v = 8\sqrt{h}$. Figure 9.0.2 shows the graphical representation of the function with the cursor at 13 feet. It looks like the teacher made a good choice. She got her student to realize the value of mathematics and did not get wet in the deal. You might wonder what would have happened if the student had studied physics? He would have learned that had he raised his arms above his head (about a two-foot difference), the teacher would have had a bad hair day. The graph below confirms that the velocity is over 30 f/s at a height of 15 feet.

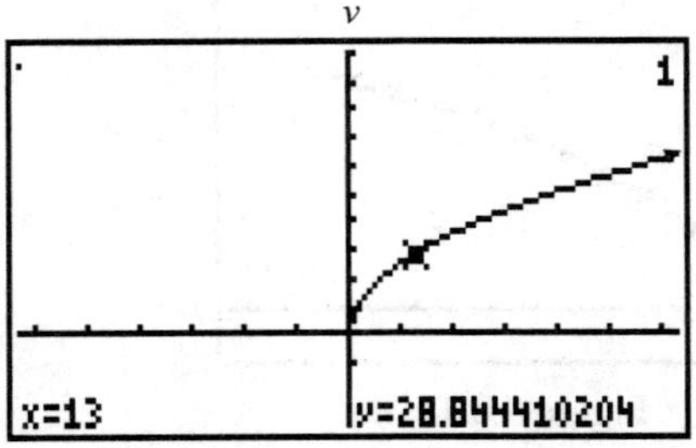

Figure 9.0.2

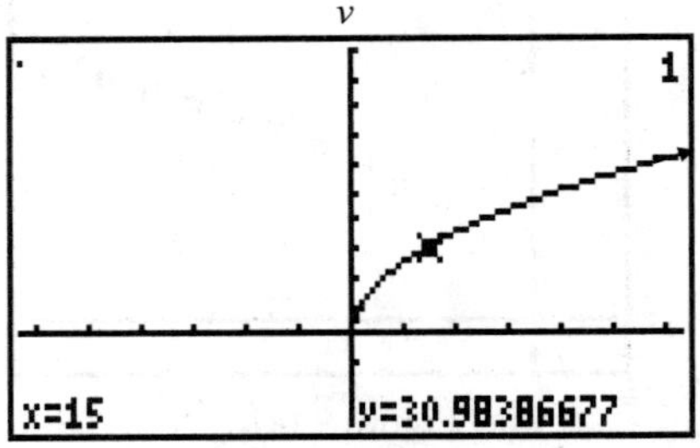

Figure 9.0.3

Square roots were studied briefly in Chapters Two and Three; however, you simply evaluated the square root of a number and you saw shapes from data relationships that look like the square root function. In Chapter Three, you were introduced to the square root function, and this chapter is devoted to a more thorough study of the square root function as well as several topics related to the function. From the square root function comes the need for simplification of square root expressions, which is studied in Sections 9.2 and 9.3. A new kind of exponent that is related to the square root function is described in Section 9.4 -- the fractional exponent. In Section 9.5, you are asked to solve equations containing square root functions. As always, no new method is needed to solve these equations by the technology-based methods. The analytical method for solving the equations is also developed; this method requires a new property and some special algebraic techniques. In the last section of Chapter Nine you will investigate the use of the square root function as a model for selected physical situations, including the area of a region under the square root function, the velocity of a dropped object, the period of a pendulum, and the distance to the horizon.

9.1 The Square Root Function

> **The Square Root Function is:**
>
> $$y = d\sqrt{x+e}+f$$
>
> Where the function parameters are *d*, *e*, and *f* and $d \neq 0$.

Changing from the Normal Domain to the Problem Domain

The growth of money problem studied in Section 7.4 has the function $y = y_0(1+r)^t$ as a model. The model has a normal domain of $(-\infty, \infty)$. However, if you invest money for a 20 year period, the problem domain is [0, 20], not $(-\infty, \infty)$. How can you change the model so that the domain is [0, 20]? You have already learned a little about changing the domain in Section 4.6. In Section 4.6, you learned that if you perform the operations of addition, subtraction, or multiplication of functions to form a new function, the domain of the sum, difference, or product is the intersection of the domains of the functions being operated on. Thus, if $f(x) = 2x - 5$ with a domain of $(-\infty, \infty)$, and $g(x) = \sqrt{x-3}$ with a domain of $[3, \infty)$ are added to get the function $(f+g)(x) = 2x - 5 + \sqrt{x-3}$, the domain of the sum function is $[3, \infty)$. (Check this on your calculator.) Function $(f+g)(x)$ is no longer a linear function, nor is it a square root function. By multiplying the square root part of the function by 0, it now behaves like the linear function $2x + 5$ (It is $2x+5+0\sqrt{x-3}$.) with a domain of $[3, \infty)$.

To force the domain of $y = y_0(1+r)^t$ to be $[0, \infty)$, add the function $0\sqrt{t}$. Thus, the function $y = y_0(1+r)^t + 0\sqrt{t}$ is the money model with a domain of $[0, \infty)$. The next step is to stop the domain at 20. What function can you think of that has a domain that stops at 20? Certainly no function studied in this text behaves like this. You were introduced to the function $f(x) = d\sqrt{x+e}+f$ in Chapter Three; however, you know that the domain is $[-e, \infty)$. That is, the smallest number in the domain is $-e$. There is a simple "manipulation" you can do to the function to make the domain all real numbers less than *e*. Change the function to $f(x) = d\sqrt{-x+e}+f$. This function has a domain of $(-\infty, e]$. Confirm this by looking at the numerical (or graphical) representations of several functions in this form where you enter the function parameters *d*, *e*, and *f*. Finally, the function $\sqrt{-t+20}$ has a domain of $(-\infty, 20]$; thus, if you add (or subtract or multiply) this function to $0\sqrt{t}$, the sum (difference or product) of the functions has a domain of [0, 20]. That is, $y = y_0(1+r)^t + 0\left(\sqrt{t}\sqrt{-t+20}\right)$ has a domain of [0, 20]. This idea can be extended to any function.

Example 1: One reasonable model for the amount of garbage generated per person per day in the United States is $g = 0.0457t - 86.88$, where *t* is time in calendar years and *g* is the amount of garbage generated by each person per day in pounds. The model has a domain of (–∞, ∞); however, a reasonable problem domain is [1940, 2012]. How can you make the model have a domain of [1940, 2012]?

Solution: The function $\sqrt{t-1940}$ has a domain of [1940, ∞) and the function $\sqrt{-t+2012}$ has a domain of (–∞, 2012]; thus, if you add the function $0\sqrt{t-1940}\cdot\sqrt{-t+2012}$ to the model $g = 0.0457t - 86.88$, the sum is the linear model with a domain of [1940, 2012]. The final model is $g = 0.0457t - 86.88 + 0\sqrt{t-1940}\cdot\sqrt{-t+2012}$. Below in Figure 9.1.1 is the graph of the function that models the garbage problem on a window of [1930, 2020] by [–5, 8].

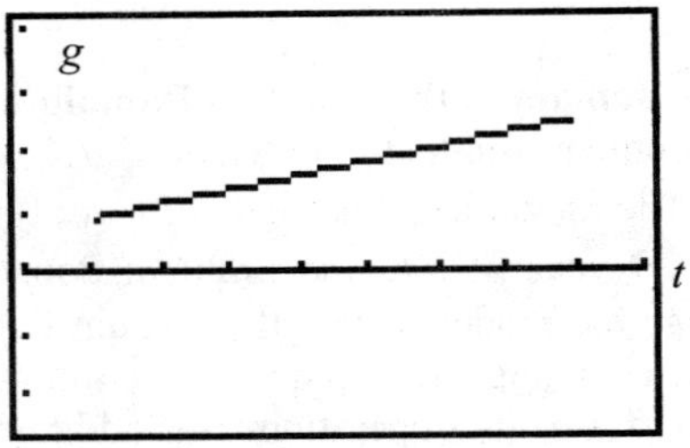

Figure 9.1.1

⁂

The Parent Square Root Function

Consider the parent square root function $f(x) = \sqrt{x}$ in its three representations:

Table 9.1.1

x	0	1	2	3	4	6	9	16	50
$f(x)$	0	1	1.414	1.732	2	2.450	3	4	7.071

and

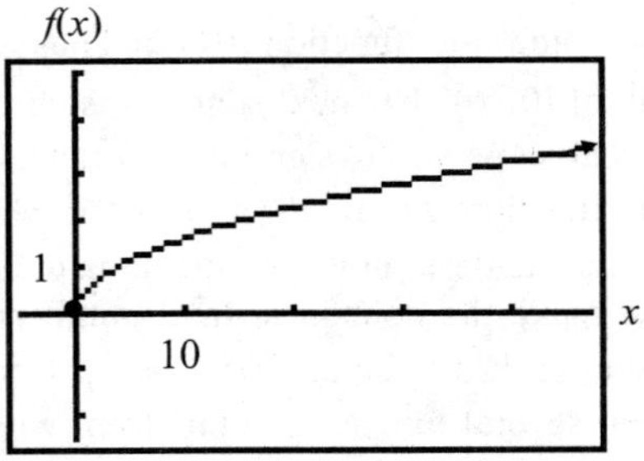

Figure 9.1.2

From the numeric or graphic representations, the following observations are made:

- As x gets larger in value, the parent square root function increases in value (it is increasing).
- The minimum value of the function is 0; therefore, the range is [0, ∞) and the function is never negative.

- The domain is $[0, \infty)$.
- The zero of the function is at $x = 0$.
- The rate of change is not constant.
- Finally, near the beginning of the domain, the square root function is relatively curved compared to larger elements of the domain where it is relatively linear.

Transformations Again

The square root function $y = d\sqrt{x+e}+f$ begs to be investigated to see if the transformation concepts of reflection, stretching, shrinking, and horizontal and vertical shifts apply, just as they have applied to all of the earlier functions studied. Keeping in mind the graph of the parent square root function in Y_1, observe the graphs of the functions $Y_2 = \sqrt{x} - 3$ and $Y_3 = \sqrt{x} + 4$ as shown in Figure 9.1.3b.

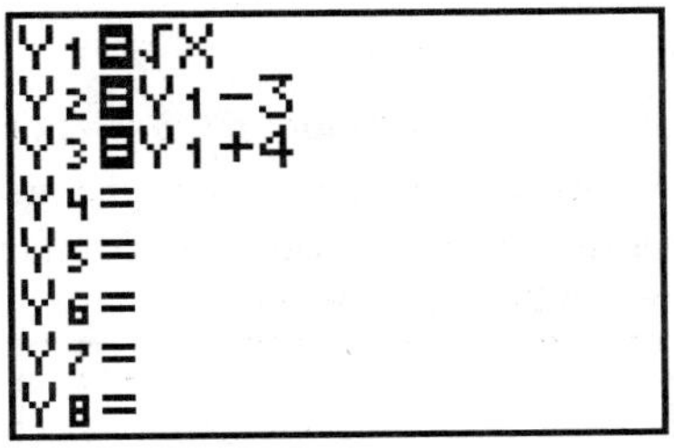

Figure 9.1.3a

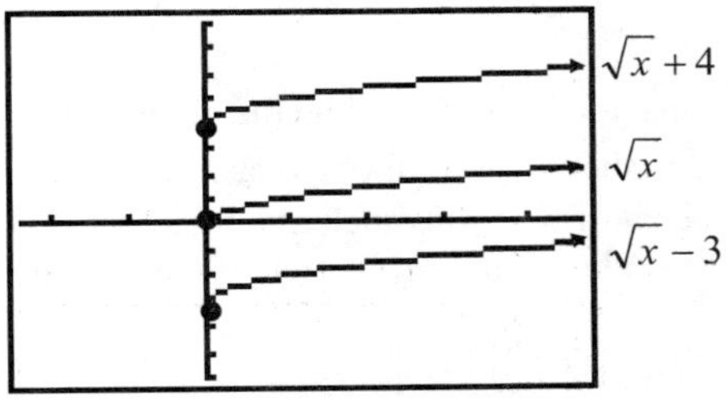

Figure 9.1.3b

You should enter these three functions on your calculator and with the trace cursor jump back and forth between the parent and one of the transformed functions. You will discover that the function $Y_3 = \sqrt{x} + 4$ is exactly 4 units farther from the x-axis than the parent function for every x in the domain. Likewise, the transformed function $Y_2 = \sqrt{x} - 3$ $(Y_1 - 3)$ is exactly 3 units below the parent function. If these graphs are an indication of all square root graphs, you can conclude that vertical shifts do apply to the parent square root function. You should enter the parent function on your calculator as well as several functions of the form $y = \sqrt{x} + f$ to convince yourself of this conjecture. Of course, the parent can be any square root function, not just $y = \sqrt{x}$. The transformation of a vertical shift will apply.

Two examples of horizontal shifts are shown below in Figures 9.1.4a and b with the functions $Y_2(x) = \sqrt{x+4}$ and $Y_3(x) = \sqrt{x-5}$. On calculators accepting function notation, replace the function variable with $x + 4$ or $x - 5$ as shown below.

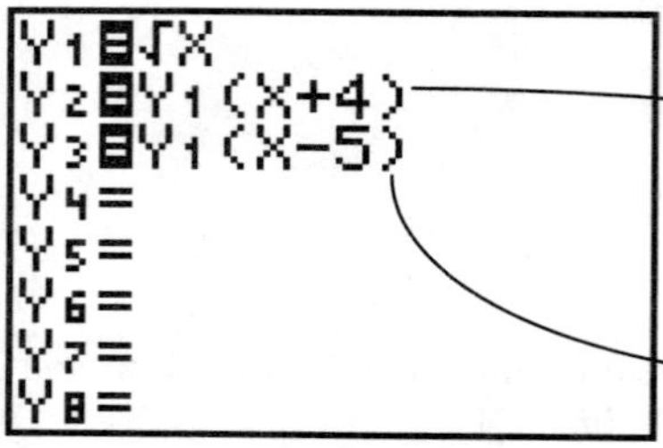

Figure 9.1.4a

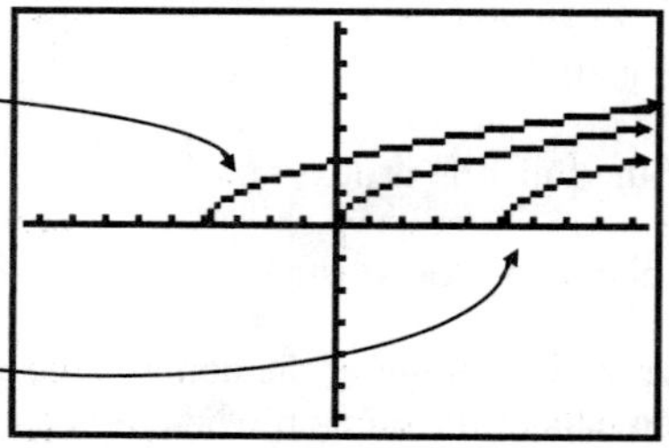

Figure 9.1.4b

From these two graphs it appears that the concept of horizontal shifts also applies to the parent square root function. The graph of $Y_2 = \sqrt{x+4}$ is the identical graph to $Y_1 = \sqrt{x}$ except that it has been shifted to the left by 4 units. Every point on the graph of $Y_2 = \sqrt{x+4}$ is 4 units to the left of the corresponding point on the graph of $Y_1 = \sqrt{x}$. The same is true for the graph of $Y_3 = \sqrt{x-5}$; every point on $Y_3 = \sqrt{x-5}$ is 5 units to the right of every corresponding point on the graph of $Y_1 = \sqrt{x}$. The graph of $Y_3 = \sqrt{x-5}$ has been shifted to the right by 5 units.

The function $y = 3\sqrt{x}$ should be the graph of $y = \sqrt{x}$ stretched by a factor of 3. Each point on the graph of $y = 3\sqrt{x}$ should be 3 times as far from the x-axis as the corresponding point on the graph of $y = \sqrt{x}$. When you confirm an idea such as a transformation like a stretch on your calculator, you may want to use a decimal window. The verification of the stretch is shown on the graph below.

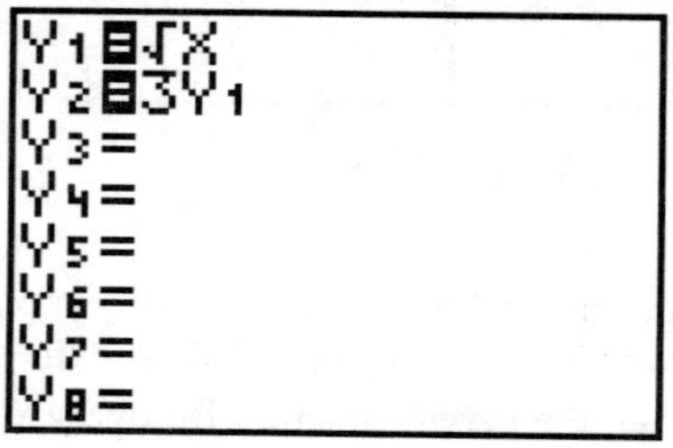

Figure 9.1.5a

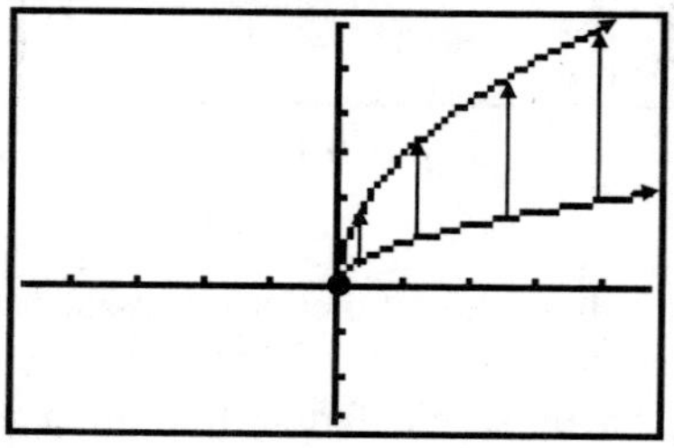

Figure 9.1.5b

The distance between the graphs is not constant like a vertical shift. The distance between the transformed function and the x-axis is 3 times the distance from the parent to the x-axis. Try confirming this notion with your calculator.

Another commonly used transformation is the reflection about the x-axis. Does it work on the square root function? A simple way of answering the question is to try it! In the graphs below, you see the parent function $f(x) = \sqrt{x}$ and the transformed function $f(x) = -\sqrt{x}$. How are they different? It looks like the transformed function is a mirror image of the parent. To verify this, put the trace cursor on a point on the parent function and then jump to the transformed function. You will see that the function values are opposites.

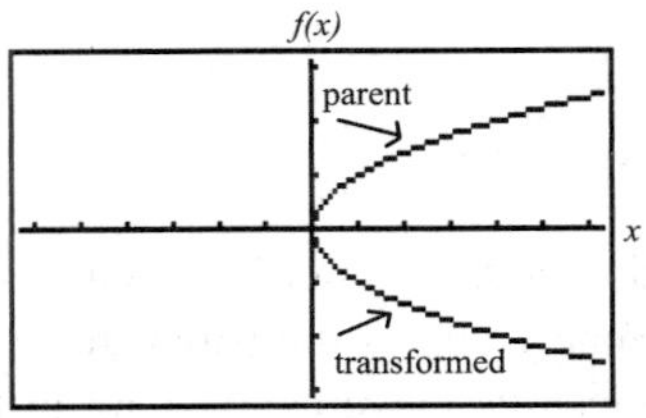

Figure 9.1.6a

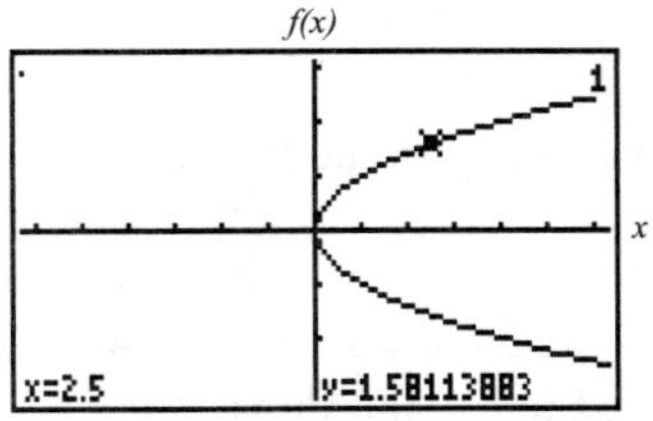

Figure 9.1.6b

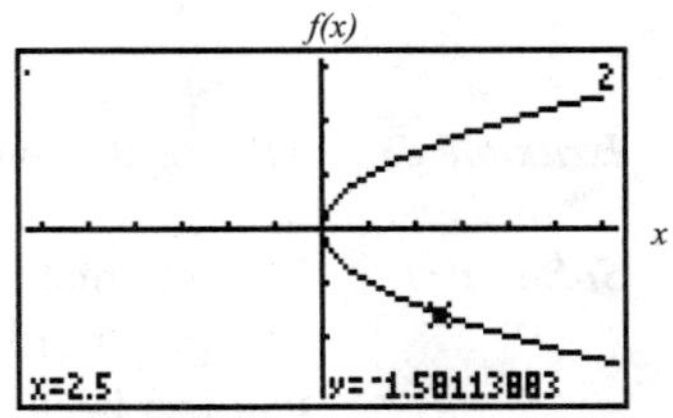

Figure 9.1.6c

You are encouraged to verify that for every point on the parent there is a corresponding point on the transformed function directly below and the same distance from the x-axis. Using the table feature may be a good choice.

Example 2: How is the graph of $y=\sqrt{x+3}-2$ different from the graph of $y=\sqrt{x}$?

Solution: The graph of $y=\sqrt{x+3}-2$ is identical to $y=\sqrt{x}$ but it has been shifted horizontally 3 units to the left and shifted vertically down 2 units. Confirmation is shown below in Figure 9.1.7.

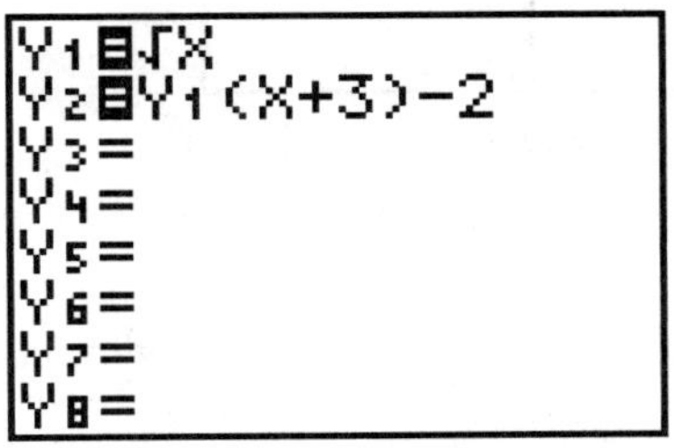

Figure 9.1.7a

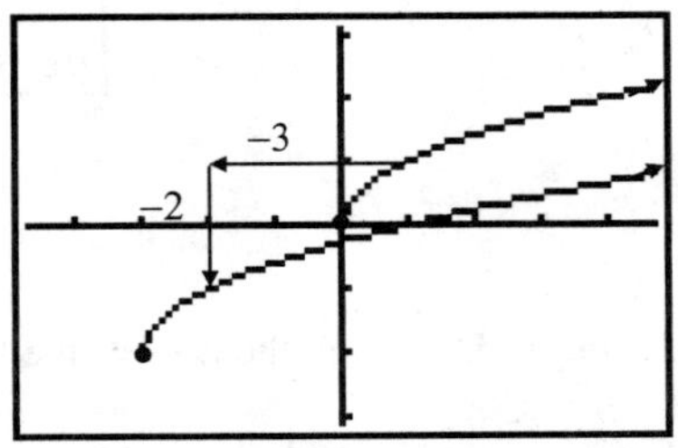

Figure 9.1.7b

***

Example 3: Does the function $y=-\sqrt{x-1}+4$ increase or decrease on its domain?

Solution: Multiplying the parent square root function by a -1 reflects it about the x-axis. Since the parent square root function increases throughout the domain, the function $y=-\sqrt{x-1}+4$ decreases throughout the domain. The other two parameters are not related to increasing or decreasing. The numeric representation of the function confirms this reasoning.

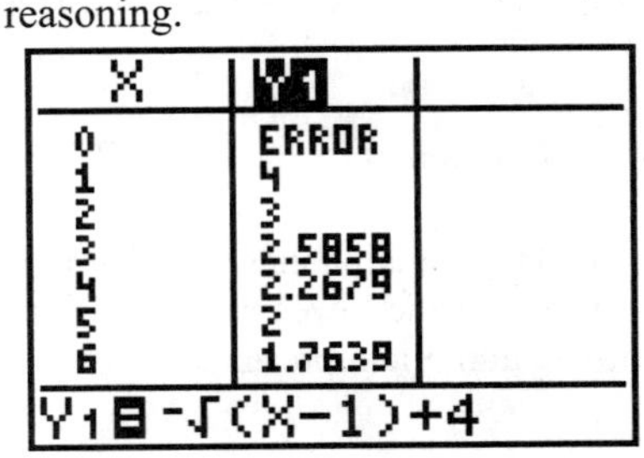

X	Y1	
0	ERROR	
1	4	
2	3	
3	2.5858	
4	2.2679	
5	2	
6	1.7639	

Y1=-√(X-1)+4

Figure 9.1.8

***

Example 4: What is the minimum value of the function $y = 2\sqrt{x+1} - 4$?

Solution: The minimum value of the parent square root function is 0. This occurs when *x* is 0. The graph of the function $y = 2\sqrt{x+1} - 4$ is the same graph as the parent function that has been shifted left 1 and down 4. Since these shifts move every point in the graph, the point (0, 0) has also been moved left 1 and down 4. Thus, the new minimum occurs when *x* is –1 and the minimum is –4. The new point on the graph is (–1, –4). The stretch of 2 does not affect the minimum value of 0 on the parent. The confirmation is shown below on the graph.

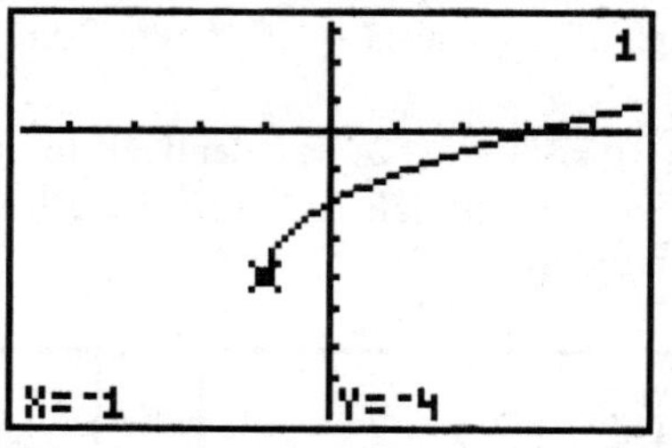

Figure 9.1.9

⁂

Example 5: Find the zero of the function $y = -3\sqrt{x-1.2} + 5.6$ with an error ≤ 0.01.

Solution: Graph the function and use the zero finder as shown in Figure 9.1.10.

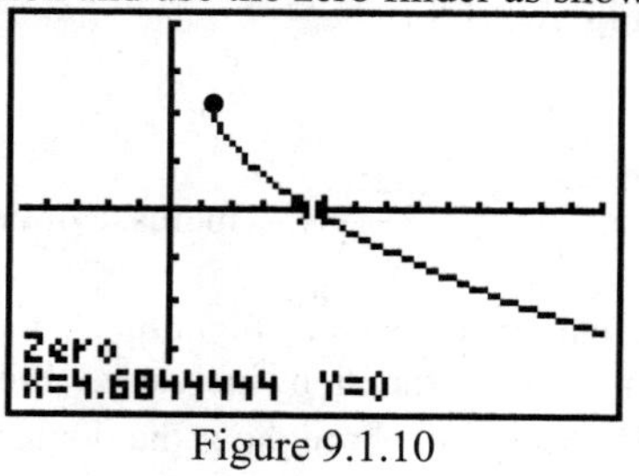

Figure 9.1.10

The zero of the function $y = -3\sqrt{x-1.2} + 5.6$ is 4.684.

⁂

Example 6: Find the zero of the function $f(x) = 4\sqrt{x+2} + 3$.

Solution: There is no zero! The function has a minimum value of 3 and it is increasing throughout the domain. If the lowest value is 3 and it only gets larger in value, there cannot be a zero. The numeric representation of the function confirms this reasoning.

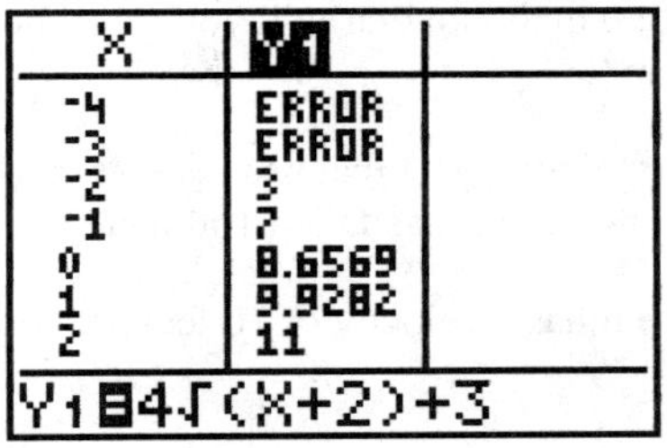

Figure 9.1.11

9.1 EXERCISES.

Many times you can work the exercises by algebraic, numeric, or graphical methods. You can learn about mathematics no matter which method you use. Part of the learning process is for you to decide which method is best for you to use. Please read the text carefully before you do the exercises. Read the **next assigned section** after you finish the assignment below. All numeric answers must have an error ≤ 0.01.

A Quick Look To The Future

-4. Does $\dfrac{\sqrt{3}}{\sqrt{4}}$ equal $\dfrac{\sqrt{3}}{4}$?

-3. What property (from Chapter 1) allows you to add $3x + 5x + 2x$? What about $3\sqrt{x} + 5\sqrt{x} + 2\sqrt{x}$?

-2. Use your calculator to find the value of $5^{\frac{1}{2}}$ and $\sqrt{5}$, $6^{\frac{1}{3}}$ and $\sqrt[3]{6}$, and $2^{\frac{1}{4}}$ and $\sqrt[4]{2}$. What conclusions can you draw from this experience?

-1. Can the square root function of the form $d\sqrt{x+e}+f$ cross the x-axis more than once?

Previous Topics Revisited

1. Solve $\dfrac{5x-4}{x-1} = 4 + \dfrac{7}{x-1}$
2. Solve $\dfrac{2.6}{x+3} + 4.2 = 9$
3. Solve $\dfrac{4x-6}{x-2} = \dfrac{2}{x-2} + 3$
4. Subtract $\dfrac{7x}{x^2+3x-4} - \dfrac{2x}{x+4}$
5. Simplify $\dfrac{\dfrac{3}{x-1} + \dfrac{2}{x+1}}{\dfrac{1}{x^2-1}}$
6. Divide $\dfrac{x^2-x-2}{x^2-2x-3} \div \dfrac{x^2+2x-8}{x^2-3x}$

New Topic Exercises

7. Using the model from the skid mark problem, how fast is a car traveling when the breaks are applied if the skid mark is 200 feet long?

8. If you are driving at a speed of 60 mph, how far will you travel before stopping if you apply the breaks hard enough to lock them?

9. Using the model for the velocity of an object that has been dropped (see the beginning of the section)how fast is a ball traveling after it has fallen 100 feet? 200 feet? 500 feet?

10. How far must a chair fall before it has a velocity of 10 feet per second? 35 feet per second? 100 feet per second.

11. The Texas Instruments graphing calculator TI-81 was made to withstand a drop from a height of 6 feet. How fast is it moving when it hits the floor?

12. The acceleration due to gravity on the moon is 5.47 feet per second per second. Thus, the model for the velocity of an object dropped on the moon can be found by changing the function parameter g to 5.47 in the model $v = \sqrt{2gh}$. It simplifies to $v = 3.31\sqrt{h}$ (Confirm this yourself.) When Neil Armstrong stepped off of the 3 feet ladder on the lunar lander in 1969, what was his velocity when he hit the surface of the moon?

13. If Neil Armstrong's space suit would withstand a maximum velocity of 20 feet per second, from what heights could he have dropped and not damaged the suit?

Are the following functions increasing or decreasing?

14. $y = 1.3\sqrt{x+4} - 7$

15. $y = 1.3\sqrt{x+4}$

16. $f(x) = 1.3\sqrt{x-4} + 7$

17. $V(h) = -2.6\sqrt{h+7} - 9$

18. $f(x) = -2.6\sqrt{x-7}$

19. $y = -2.6\sqrt{x-7} + 9$

20. Under what conditions does the function $f(x) = d\sqrt{x+e} + f$ increase? Decrease?

Find the minimum or maximum value of the following functions; also, find the value of x at the minimum or maximum.

21. $f(x) = 5.6\sqrt{x-3.2} + 11.2$

22. $y = -2\sqrt{x+7} + 11.2$

23. $f(x) = -3\sqrt{x-5} - 11.2$

24. $y = 0.2\sqrt{x+4.4} - 11.2$

25. What are the coordinates of the maximum (or minimum) point on the graph of $y = d\sqrt{x+e} + f$?

Find the zeros of the functions in Exercises 26 - 29.

26. $y = 2.3\sqrt{x+5} - 7.1$

27. $f(x) = -\sqrt{x+3} + 4$

28. $f(x) = 4.5\sqrt{x-2.2} + 0.1$

29. $y = -3.3\sqrt{x+4} - 0.2$

Find the normal domain and range in Exercises 30 - 38.

30. $y = 2\sqrt{x-3} + 2$, $y = 2\sqrt{x-3} + 3$, $y = 2\sqrt{x-3} + 4$, $y = 2\sqrt{x-3} + 7$

31. $f(x) = -2\sqrt{x+3} - 6$, $g(x) = -2\sqrt{x+2} - 6$, $h(x) = -2\sqrt{x+1} - 6$, $i(x) = -2\sqrt{x+(-1)} - 6$

32. $y=-2\sqrt{x+7}-8,\ y=-3\sqrt{x+7}-8,\ y=2\sqrt{x+7}-8,\ y=3\sqrt{x+7}-8$

33. $y=-\sqrt{x}\ ,\ y=\sqrt{-x}\ ,\ y=-\sqrt{-x}$

34. $f(x)=\sqrt{-x+3}-4,\ g(x)=\sqrt{-x+3}-2,\ h(x)=\sqrt{-x+3}-1,\ j(x)=\sqrt{-x+3}+2$

35. $y_1=\sqrt{-x-5}+3,\ y_2=\sqrt{-x-2}+3,\ y_3=\sqrt{-x+2}+3,\ y_4=\sqrt{-x+5}+3$

36. $f(x)=\sqrt{x+3}+\sqrt{x-5}$

37. $g(x)=-\dfrac{1}{2}\sqrt{x+3}\sqrt{x+1}$

38. $f(x)=\sqrt{x+3}+\sqrt{-x+2},\ g(x)=\sqrt{x+3}-\sqrt{-x+2},\ h(x)=\sqrt{x+3}\cdot\sqrt{-x+2},\ k(x)=\dfrac{\sqrt{x+3}}{\sqrt{-x+2}}$

39. How can you change the function $f(x)=\sqrt{x}$ so that it is decreasing instead of increasing?

40. What transformation of the function $f(x)=\sqrt{x}$ will cause the graphical representation of the transformed function to be a reflection of $f(x)=\sqrt{x}$ about the x-axis?

41. How can you change the function $f(x)=\sqrt{x}$ so that it has a minimum value of 7 instead of 0?

42. What transformation of the function $f(x)=\sqrt{x}$ will cause the transformed function to be exactly 7 units higher than the parent $f(x)=\sqrt{x}$?

43. How can you change the function $f(x)=\sqrt{x}$ so that the domain is $[-3, \infty)$ instead of $[0, \infty)$?

44. What transformation of the function $f(x)=\sqrt{x}$ will cause the graphical representation of the transformed function to be 3 units to the left of the parent?

45. How can you change the function $f(x)=\sqrt{x}$ so that it increases 4 times faster?

46. What transformation of the function $f(x)=\sqrt{x}$ will cause the graphical representation of the transformed function to be 4 times as far from the x-axis as the parent?

47. Describe (in English) the transformations of the graph of $f(x)=\sqrt{x}$ that will give the graph of $g(x)=-2\sqrt{x+3}-5$.

48. Find a square root function whose maximum value is 3. There are many answers.

49. Find a square root function whose range is $[2, \infty)$. There are many answers.

50. Find any square root function whose zero is 3.

51. Find any square root function whose domain is $[-6, \infty)$.

52. Find any square root function that is always negative.

53. Find any square root function whose domain is $(-\infty, 5]$.

From Mathematics to English

54. In the function $f(x) = d\sqrt{x+e} + f$, if *d* is zero, what does the graph of the function look like?

55. In the function $y = d\sqrt{x+e} + f$, under what conditions on *d*, *e*, and *f* does the function increase throughout the normal domain?

56. In the function $y = d\sqrt{x+e} + f$, under what conditions on *d*, *e*, and *f* does the function decrease throughout the normal domain?

57. Is it possible to create a function of the form $y = d\sqrt{x+e} + f$ with a domain of all real numbers? Why?

58. After reading this section, make a list of questions that you want to ask your instructor.

59. Will the function parameter 5.5 from the "skid mark problem" increase or decrease if the road is wet? Explain.

60. Describe how a square root function can be negative when you know that square roots are positive.

62. Describe the "zero" of a square root function.

63. Describe how you know a square root function is increasing.

64. What does the statement "the maximum value of the function is 4" mean?

65. Continue in your daily journal and make an entry. In addition to your normal entry on thoughts about the mathematics in this section, list at least two positive comments about what you have learned about this topic.

66. In paragraph format, summarize the material in this section of the text in your daily journal.

67. Describe how your classroom instructor made this topic more understandable and clear.

68. After reading the text and listening to your instructor, what do you not understand about this topic?

Explorations

69. Develop a square root function of the form $y = d\sqrt{x+e} + f$ that meets the criteria specified below.

	Domain	Range	Minimum Value	Maximum Value
a.	$[4, \infty)$		3	
b.	$[-2, \infty)$	$[4, \infty)$		
c.	$[3, \infty)$			2
d.	$[-5, \infty)$		0	
e.	$[2.6, \infty)$	$(-\infty, -5]$		
f.	$(-\infty, 2]$			1

70. Describe the transformation of $y = \sqrt{x}$ that yields the graph of:

 a. $y = \sqrt{2x-6}$

 b. $y = \sqrt{3x-9}$

 c. $y = \sqrt{3x+12}$

 d. $y = \sqrt{2x+8}$

71. Describe what the absolute value transformation does to the graph of $f(x)$:

 a. $f(x) = \sqrt{x}$ Describe how the graph of $\left|\sqrt{x}\right|$ relates to the graph of $f(x)$.

 b. $f(x) = \sqrt{x+2}$ Describe how the graph of $\left|\sqrt{x+2}\right|$ relates to the graph of $f(x)$.

 c. $f(x) = -\sqrt{x}$ Describe how the graph of $\left|-\sqrt{x}\right|$ relates to the graph of $f(x)$.

 d. $f(x) = \sqrt{x} - 4$ Describe how the graph of $\left|\sqrt{x} - 4\right|$ relates to the graph of $f(x)$.

 e. $f(x) = -\sqrt{x} + 4$ Describe how the graph of $\left|-\sqrt{x} + 4\right|$ relates to the graph of $f(x)$.

 f. $f(x) = \sqrt{x+2} - 6$ Describe how the graph of $\left|\sqrt{x+2} - 6\right|$ relates to the graph of $f(x)$.

72. Describe what the squaring transformation does to the graph of $f(x)$:

a. $f(x)=\sqrt{x}$ Describe how the graph of $\left(\sqrt{x}\right)^2$ relates to the graph of $f(x)$.

b. $f(x)=\sqrt{x+2}$ Describe how the graph of $\left(\sqrt{x+2}\right)^2$ relates to the graph of $f(x)$.

c. $f(x)=-\sqrt{x}$ Describe how the graph of $\left(-\sqrt{x}\right)^2$ relates to the graph of $f(x)$.

d. $f(x)=\sqrt{x}-4$ Describe how the graph of $\left(\sqrt{x}-4\right)^2$ relates to the graph of $f(x)$.

e. $f(x)=-\sqrt{x}+4$ Describe how the graph of $\left(-\sqrt{x}+4\right)^2$ relates to the graph of $f(x)$.

f. $f(x)=\sqrt{x+2}-6$ Describe how the graph of $\left(\sqrt{x+2}-6\right)^2$ relates to the graph of $f(x)$.

73. Find the zeros of the following functions. Support your answers with graphs or tables

a. $y=-2\sqrt{x^2-5x-14}+2$

b. $y=\sqrt{(x-2)(x+5)(3x-7)}$

c. $y=-3\sqrt{x+4}-3+2\sqrt{x-2}+9$

d. $f(x)=\sqrt{x+5}\left|x^2+3x-4\right|$

e. $f(x)=3\sqrt{|x+3|-1}-2$

9.2 Properties of Irrational Expressions

What is the Calculator Teaching You?
As you begin to learn about the properties of irrational numbers, consider what the calculator tells you about these properties.

Try more samples like these on your own.

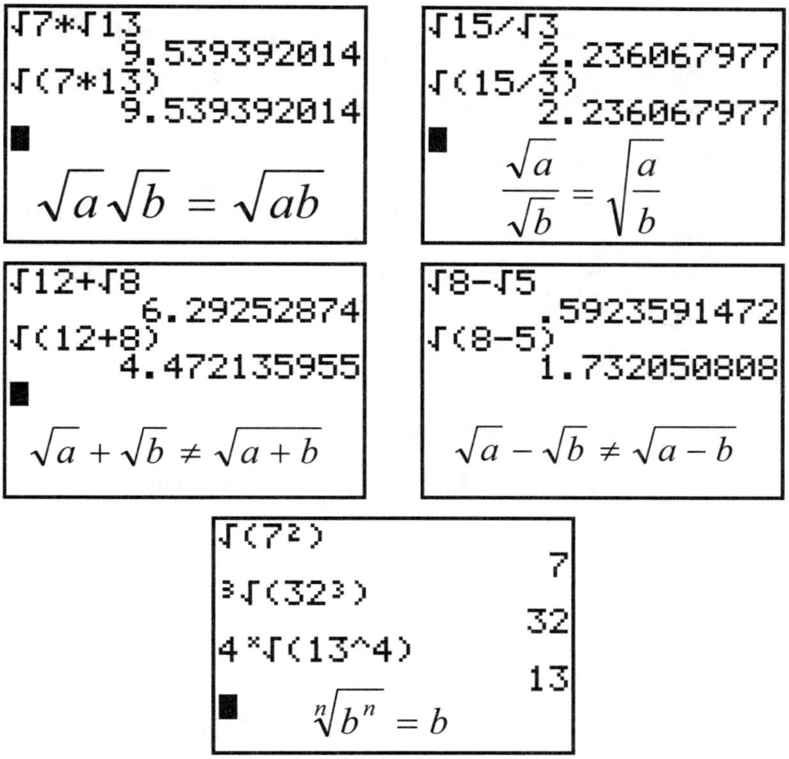

Figure 9.2.1

The square root function studied in the last section is an example of a function that generates irrational numbers. Numbers like $\sqrt{7}$, $\sqrt{5}$, $-\sqrt{3}$, and $\sqrt{2.6}$ are **irrational numbers** because they cannot be written as division of integers. These numbers are generated by the square root function. The number π is another example of an irrational number. Other examples of irrational numbers include cube roots of non-perfect cubes, fourth roots of non-perfect fourth powers, etc. These numbers come from the cube root and fourth root functions $y = \sqrt[3]{x}$ and $y = \sqrt[4]{x}$. Remember, the **cube root** of a number x $\left(\sqrt[3]{x}\right)$ is a number that can be cubed to get x. The **fourth root** of a number x $\left(\sqrt[4]{x}\right)$ is a number that can be raised to the exponent of four to obtain x. The symbol $\sqrt{}$ is called the **radical sign**; thus, some people refer to roots as radicals. In general, $\sqrt[n]{x}$ is a number that can be raised to the nth exponent to get x. The expression $\sqrt[n]{x}$ is read the nth root of x.

Since several functions can generate irrational numbers, they are analyzed in this section and the next. Of course, irrational numbers are real numbers, so all of the properties of real numbers apply to irrational numbers. Because most of the irrational numbers encountered are square roots or cube roots, this section is devoted to these properties. In addition to studying the properties of irrational numbers, simplification of irrational numbers in the form of radicals is also included. Calculators normally evaluate square roots; in this chapter, square roots and cube roots will be simplified. Simplification of radicals (square roots and cube roots) is different than evaluating radicals, as you will see below.

Irrational numbers are created by certain functions.

Please consider the following example that answers the question: Does the product of square roots equal the square root of the product of the radicands?

For example, does $\sqrt{4} \cdot \sqrt{25} = \sqrt{4 \cdot 25}$?

It's known that: $\sqrt{4} = 2$ and $\sqrt{25} = 5$, and 4 times 25 is 100, and $\sqrt{100} = 10$.

Thus, substitute what is known into the problem and verify that it is a true statement.

$$\sqrt{4} \cdot \sqrt{25} = \sqrt{4 \cdot 25}$$

$$2 \cdot 5 = \sqrt{100}$$

$10 = 10$. Yes, it is a true statement.

In general this idea works for all radicals as is stated in the following property.

Product Property

$$\sqrt[n]{a} \cdot \sqrt[n]{b} = \sqrt[n]{ab} \quad \text{and} \quad \sqrt[n]{ab} = \sqrt[n]{a} \cdot \sqrt[n]{b}$$

n is the index of the radical and if it is 2, it is not written.
a and b must be positive when n is even.

The product property allows for the multiplication of any root times any other root as long as the indices are the same. For example, the product property cannot be used to multiply a square root times a cube root. When the indices are the same, the product is found by multiplying the radicands, while not changing the index. By reversing the product property to $\sqrt[n]{ab} = \sqrt[n]{a} \cdot \sqrt[n]{b}$, it is possible to "undo" multiplication. For example, $\sqrt{4 \cdot 5}$ is equivalent to $\sqrt{4} \cdot \sqrt{5}$. This becomes important when simplifying radicals.

Does the same kind of property hold for division? The example below will demonstrate the idea. Does the square root of a quotient simplify to the quotient of the square roots?

For example, does $\sqrt{\dfrac{100}{25}} = \dfrac{\sqrt{100}}{\sqrt{25}}$?

It's known that: $\dfrac{100}{25}$ is 4; thus, $\sqrt{\dfrac{100}{25}} = \sqrt{4} = 2$, and $\sqrt{100} = 10$. $\sqrt{25} = 5$.

Does $\sqrt{\dfrac{100}{25}} = \dfrac{\sqrt{100}}{\sqrt{25}}$?

$2 = \dfrac{10}{5}$ Yes, in general, the square root of a quotient equals the quotient of the square root!

Quotient Property

$$\frac{\sqrt[n]{a}}{\sqrt[n]{b}} = \sqrt[n]{\frac{a}{b}} \quad \text{and} \quad \sqrt[n]{\frac{a}{b}} = \frac{\sqrt[n]{a}}{\sqrt[n]{b}}$$

a and b must be positive when n is even, and $b \neq 0$.

The quotient property allows for division of any root by any other root as long as the indices are the same. When reversed, the root of a fraction can be split into the root of the numerator divided by the root of the denominator. This is useful when simplifying radicals.

One more property is needed to help simplify radicals: the power property. In simplifying radical expressions, an expression like $\sqrt{7^2}$ may be encountered. To simplify it, you know that 7^2 is 49, and $\sqrt{49}$ is 7; thus, $\sqrt{7^2} = 7$. This same process will work for any nth root of a number to the nth exponent.

Power Property

$\sqrt[n]{a^n} = a$ where $a \geq 0$ when the index n, is even.

If $a \in (-\infty, \infty)$ and n is even, the power property becomes $\sqrt[n]{a^n} = |a|$. Please confirm this with your graphing calculator by graphing the function $y = \sqrt{x^2}$, for example! The power property is used to simplify; for example, $\sqrt{3^2}$ simplifies to 3. The number 3 is certainly a simpler looking number than $\sqrt{3^2}$.

With the three properties described above and a few properties of real numbers, radical expressions can be simplified. Some commonly agreed upon general rules identify when a radical expression is said to be simplified. They are listed below.

Rule 1
A square root expression is ***NOT*** simplified if the radicand has any factors that are perfect squares. A cube root expression is ***NOT*** simplified if the radicand has any factors that are perfect cubes, etc.

Rule 2
A radical expression is ***NOT*** simplified if the radicand is a fraction.

Rule 3
A radical expression is ***NOT*** simplified if the denominator contains a radical.

These rules were developed many years before the invention of the calculator, and their primary use was to aid in the process of evaluating radicals. Now that calculators have become available, the rules are no longer needed for evaluation of radicals. One of the reasons the rules are still used is that numbers can easily be compared when in simplified form. Another reason that you are expected to simplify radicals is that addition, subtraction, multiplication, and division of some radicals require simplification before the operation can be done.

Example 1: Simplify $\sqrt{12}$.

Solution: The irrational number $\sqrt{12}$ is not simplified because 12 has a factor that is a perfect square, the number 4. To simplify:

$\sqrt{12}$	given
$\sqrt{4 \cdot 3}$	factor 12 into a perfect square and another number
$\sqrt{4} \cdot \sqrt{3}$	use the product property
$2 \cdot \sqrt{3}$	evaluate the square root of the perfect square

The expression is now simplified because all three rules for simplification are satisfied; that is, 1) there are no factors of 3 that are perfect squares, 2) the radicand is not a fraction, and 3) there is no radical in the denominator. Use your calculator to confirm that $\sqrt{12}$ and $2 \cdot \sqrt{3}$ are the same number.

⁂

Example 2: Simplify $\sqrt{\frac{1}{2}}$.

Solution: By rule 2, the expression is not simplified. Use the quotient rule to convert the expression from one with a fraction under the radical sign to an expression with a radical in the numerator and denominator.

$$\sqrt{\frac{1}{2}} = \frac{\sqrt{1}}{\sqrt{2}}$$

SURE ARE A LOT OF 2'S HERE.

This expression is not simplified because it is in violation of rule 3. To eliminate the radical from the denominator, use the fundamental property of fractions to multiply the numerator and denominator by $\sqrt{2}$. The reason for this is that the denominator becomes $\sqrt{2} \cdot \sqrt{2}$; which is $\sqrt{2^2}$. By the power property, $\sqrt{2^2}$ is 2.

$$\frac{\sqrt{1} \cdot \sqrt{2}}{\sqrt{2} \cdot \sqrt{2}} = \frac{\sqrt{1 \cdot 2}}{\sqrt{2^2}} = \frac{\sqrt{2}}{2}$$

The expression is now simplified because all three rules of simplification are satisfied. Confirm that $\sqrt{\frac{1}{2}}$ is the same number as $\frac{\sqrt{2}}{2}$ with your calculator.

⁂

In considering the above problem, the simplification process makes perfect sense to pre-calculator professionals. Without the aid of a calculator, scientists, mathematicians, and engineers could evaluate a radical like $\sqrt{\frac{1}{2}}$ by simplifying it to $\frac{\sqrt{2}}{2}$. Since they usually memorized the radical $\sqrt{2}$ to be approximately 1.414, all that needed to be done was to divide 1.414 by 2; thus, the value of $\sqrt{\frac{1}{2}}$ is approximately 0.707. As was mentioned earlier in this section, other reasons for simplifying radicals still remain even though our calculator will evaluate $\sqrt{\frac{1}{2}}$ to be approximately 0.707 without simplifying first.

Example 3: Simplify $\sqrt[3]{16}$.

Solution: The cube root of 16 is not simplified because 16 has a factor that is a perfect cube. Factor 16 into a perfect cube factor and another factor; then, use the product property.

$\sqrt[3]{16} = \sqrt[3]{8 \cdot 2} = \sqrt[3]{8} \cdot \sqrt[3]{2}$ Now evaluate the cube root of the perfect cube

$$\sqrt[3]{8} \cdot \sqrt[3]{2} = 2\,\sqrt[3]{2}$$

⁂

Example 4: Simplify $\sqrt{x^3}$.

Solution: The expression $\sqrt{x^3}$ is not simplified because x^3 has a factor that is a perfect square.

$\sqrt{x^3} = \sqrt{x^2 \cdot x}$ by factoring

$= \sqrt{x^2} \cdot \sqrt{x}$ by the product property

$= x\sqrt{x}$ by the power property ($x \geq 0$)

To confirm that, in fact, $\sqrt{x^3}$ does equal $x\sqrt{x}$, the graphing calculator can be used. If you think of $\sqrt{x^3}$ and $x\sqrt{x}$ as functions, then the numerical representations should be identical if the functions are equivalent. Below are the tables.

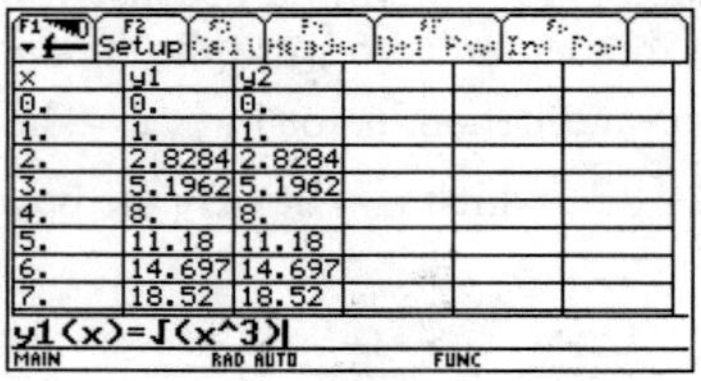

x	y1	y2
0.	0.	0.
1.	1.	1.
2.	2.8284	2.8284
3.	5.1962	5.1962
4.	8.	8.
5.	11.18	11.18
6.	14.697	14.697
7.	18.52	18.52

Figure 9.2.2

Hopefully, you can see that the algebraic work is correct.

⁂

Example 5: Simplify $\sqrt{\frac{2}{3}}$.

Solution:

$\sqrt{\frac{2}{3}} = \frac{\sqrt{2}}{\sqrt{3}}$ by the quotient property

$= \frac{\sqrt{2} \cdot \sqrt{3}}{\sqrt{3} \cdot \sqrt{3}}$ by the fundamental property of fractions

$= \frac{\sqrt{6}}{\sqrt{3^2}}$ by the product property

$= \frac{\sqrt{6}}{3}$ by the power property

⁂

Example 6: Simplify $\sqrt[3]{x^{12}}$.

Solution: x^{12} is a perfect cube. It is $(x^4)^3$. The cube root of x^{12} is x^4 because $x^4 \cdot x^4 \cdot x^4 \cdot = x^{12}$. $\sqrt[3]{x^{12}} = x^4$.

⁂

9.2 EXERCISES

Many times you can work the exercises by algebraic, numeric, or graphical methods. You can learn about mathematics no matter which method you use. Part of the learning process is for you to decide which method is best for you to use. Please read the text carefully before you do the exercises. Read the **next assigned section** after you finish the assignment below. All numeric answers must have an error ≤ 0.01.

A Quick Look To The Future

-4. Do you multiply $(ax + b)$ and $(cx + d)$ by multiplying ax and cx, and then b and d?

-3. Is $3^{\frac{1}{2}}$ between 1 and 2?

-2. If $x = 3$, does $x^2 = 9$?

-1. Do you suppose that the square root function is related to a pendulum?

Previous Topics Revisited

1. What is the domain of the square root function $f(x) = -2\sqrt{x+4.2} - 3.6$?

2. Does the square root function $y = 3\sqrt{x-3} - 3$ increase or decrease on the normal domain?

3. What is the range of $f(x) = \sqrt{x+1.99} - 2.65$?

4. Solve $\dfrac{2}{x+5} + x = \dfrac{4x+22}{x+5}$

5. Solve $\dfrac{x}{x-2} + \dfrac{1}{x+1} = \dfrac{3x}{x^2-x-2}$

6. Simplify $\dfrac{\dfrac{3}{x-1} - \dfrac{5}{x+1}}{\dfrac{4}{x^2-1}}$

New Topic Exercises

Simplify the irrational numbers in Exercises 7 - 22. Confirm that your answer is correct with your calculator.

7. $\sqrt{24}$

8. $\sqrt{50}$

9. $\sqrt{20}$

10. $\sqrt[3]{54}$

11. $\sqrt[3]{72}$

12. $\sqrt[3]{128}$

13. $\dfrac{1}{\sqrt{2}}$

14. $\dfrac{2}{\sqrt{3}}$

15. $\dfrac{1}{\sqrt{3}}$

16. $\dfrac{2}{\sqrt{2}}$

17. $\dfrac{4}{2\sqrt{3}}$

18. $\dfrac{1}{\sqrt{2}} \cdot \dfrac{\sqrt{3}}{2} + \dfrac{1}{\sqrt{2}} \cdot \dfrac{1}{2}$

19. $1 \cdot \dfrac{1}{\sqrt{2}} - 0 \cdot \dfrac{\sqrt{2}}{2}$

20. $\dfrac{\dfrac{1}{\sqrt{3}} + 1}{\sqrt{3}}$

21. $2 \cdot \dfrac{1}{2} \cdot \dfrac{\sqrt{3}}{2}$

22. $\dfrac{\sqrt{3}}{2} \cdot \dfrac{1}{\sqrt{2}}$

Find an equivalent function for each of the following functions. Confirm that your answer is equivalent to the given function with your calculator. ($x \geq 0$)

23. $\sqrt{x^5}$ 24. $\sqrt{x^7x^3}$ 25. $x\sqrt{x^3}$ 26. $\sqrt{x^2}$

27. $\sqrt[3]{x^3}$ 28. $\sqrt[4]{x^4}$ 29. $\sqrt[5]{x^5}$ 30. $-2x+5\sqrt{x^3}$

31. $\sqrt{(x-2)^2}$ 32. $\sqrt[4]{(x+3)^4}$ 33. $\sqrt[5]{(x-1)^5}$

34. The number $\frac{2}{\sqrt{3}}$ simplifies to $\frac{2\sqrt{3}}{3}$, and $\sqrt{3}$ is approximately 1.732.

    Without a calculator, calculate $\frac{2}{\sqrt{3}}$ and $\frac{2\sqrt{3}}{3}$.

35. What symbolic transformation can be applied to the function $y=\sqrt{x}$ to eliminate the radical sign? (Hint: See the explorations in Section 9.1.)

36. Find a square root that simplifies to $5\sqrt{3}$.

37. Find a square root function that simplifies to $x\sqrt{x}+4$.

38. Find values for a and b that make the statement $\sqrt{a^2-b^2}=a-b$ true.

39. Find a value for x that makes the statement $\sqrt{x^2}=x$ not true.

40. Without using technology or algebra, find the solution to the equation $x^2=5$.

From Mathematics to English

41. Does $\sqrt{a+b}=\sqrt{a}+\sqrt{b}$? Why? (Hint: Give a counter-example.)

42. Graph the functions $y=\sqrt{8x^5}$ and $y=2x^2\sqrt{2x}$. Describe the relationship between the two graphs.

43. Explain the statement, "the square root of a product is the product of the square roots."

44. Explain how you might multiply $\sqrt{3}$ and $\sqrt[3]{10}$.

45. Give your definition of an irrational number. Do not use examples.

46. Describe how to simplify $\sqrt{48}$. Do not simplify.

47. After reading this section, make a list of questions that you want to ask your instructor.

48. Continue in your daily journal and make an entry. In addition to your normal entry on thoughts about the mathematics in this section, list at least two positive comments about what you have learned about this topic.

49. In paragraph format, summarize the material in this section of the text in your daily journal.

50. Describe how your classroom instructor made this topic more understandable and clear.

51. After reading the text and listening to your instructor, what do you not understand about this topic?

Explorations

52. Find an irrational number between 2 and 2.01.

53. If $\sqrt{a-b} = \sqrt{a} - \sqrt{b}$ were a true statement, prove that 3 is equal to 1 and prove that 1 is equal to 5. Given these two proofs, why can you conclude that 3 is equal to 5?

54. Are the graphs of the following pairs of functions identical? Explain why or why not?

a. $y = \sqrt{2x}$ and $y = \sqrt{2}\sqrt{x}$

b. $y = \sqrt{\frac{x}{6}}$ and $y = \frac{\sqrt{x}}{\sqrt{6}}$

c. $y = \sqrt{x^2}$ and $y = x$

d. $y = \sqrt{x+5}$ and $y = \sqrt{x} + \sqrt{5}$

e. $y = \sqrt{x-7}$ and $y = \sqrt{x} - \sqrt{7}$

55. Develop a strategy for simplifying the expressions below; then, simplify. Explain your strategy. (x, y, and z are ≥ 0)

a. $\sqrt[3]{x^{10}y^8}$

b. $\sqrt[5]{x^{11}y^{14}z^6}$

c. $\sqrt[6]{128x^9y^{12}}$

d. $\frac{1}{\sqrt[3]{4}}$

e. $\frac{1}{\sqrt[3]{16}}$

f. $\frac{1}{\sqrt[4]{48x^9}}$

g. $\sqrt[5]{\frac{x^{11}}{y^7z^{18}}}$

9.3 Operations with Irrational Expressions

What Else Can You Learn From The Calculator?

As you start to learn how to add, subtract, multiply, and divide irrational numbers, experiment with your calculator, as shown below, to learn how you should perform these operations without the calculator.

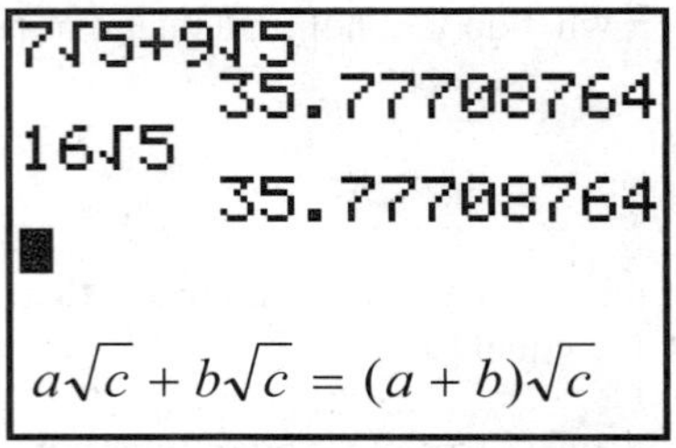

Figure 9.3.1a

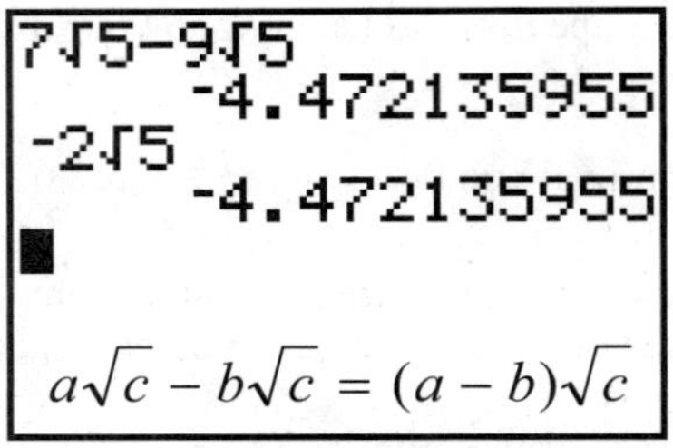

Figure 9.3.1b

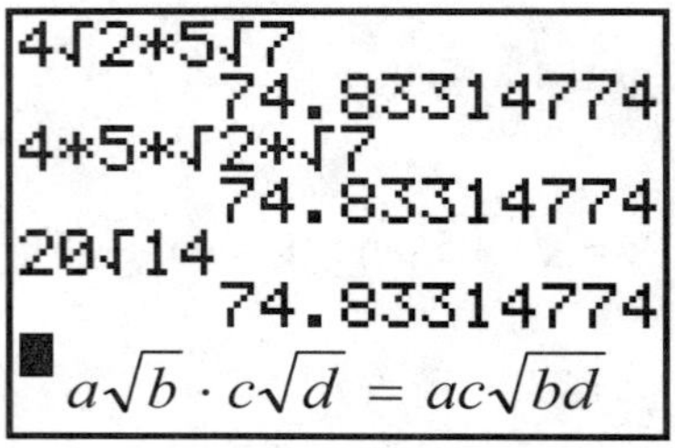

Figure 9.3.1c

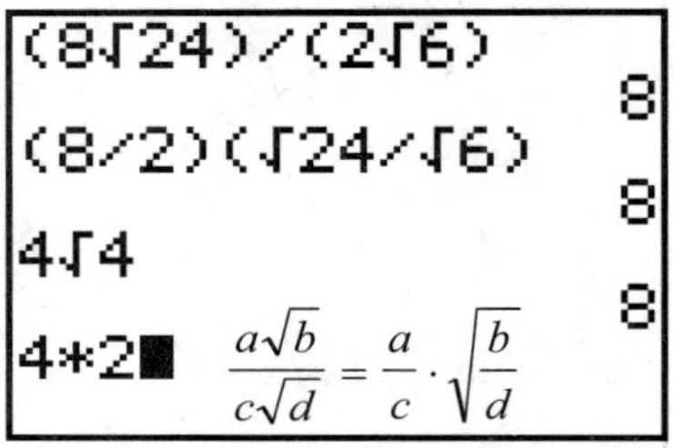

Figure 9.3.1d

Because irrational numbers are real numbers, they are governed by the properties of real numbers. Irrational numbers can be simplified with real number properties as well as the properties studied in the last section. For example, to add $3\sqrt{5}$ and $8\sqrt{5}$, the distributive property is used to rewrite $3\sqrt{5}+8\sqrt{5}$ as $\sqrt{5}(3+8)$. The expression $\sqrt{5}(3+8)$ is equivalent to $\sqrt{5}(11)$, and by the commutative property, $\sqrt{5}\cdot 11$ is commonly written as $11\sqrt{5}$. Thus, the sum of the irrational number $3\sqrt{5}+8\sqrt{5}$ is $11\sqrt{5}$. You can verify this on your calculator by calculating $3\sqrt{5}+8\sqrt{5}$, and then calculate $11\sqrt{5}$. The two answers should be the same. Other examples of simplifying irrational numbers by using the irrational number properties and real number properties follow. The same three rules for simplifying radicals from the last section apply to all radical simplifications.

Example 1: Add $\sqrt{12}+\sqrt{27}$.

Solution: $\sqrt{12}$ and $\sqrt{27}$ are not simplified because the radicands contain factors that are perfect squares.

$\sqrt{12}+\sqrt{27}$	given
$\sqrt{4\cdot 3}+\sqrt{9\cdot 3}$	factorization

$\sqrt{4}\cdot\sqrt{3}+\sqrt{9}\cdot\sqrt{3}$	product property
$2\sqrt{3}+3\sqrt{3}$	evaluation, then do the implied addition
$\sqrt{3}(2+3)$	distributive property
$\sqrt{3}\cdot 5$	addition
$5\sqrt{3}$	commutative property

Since adding $2\sqrt{3}+3\sqrt{3}$ is just like adding $2x + 3x$, addition of similar terms will be done without explanation in the remaining problems.

⁂

Example 2: Simplify $5\sqrt{6}-\sqrt{24}+3\sqrt{54}$.

Solution: The radicands 24 and 54 have factors that are perfect squares; thus, they can be simplified.

$5\sqrt{6}-\sqrt{4\cdot 6}+3\sqrt{9\cdot 6}$	factorization
$5\sqrt{6}-\sqrt{4}\cdot\sqrt{6}+3\sqrt{9}\cdot\sqrt{6}$	product property
$5\sqrt{6}-2\sqrt{6}+3\cdot 3\cdot\sqrt{6}$	square root evaluation
$5\sqrt{6}-2\sqrt{6}+9\sqrt{6}$	multiplication
$12\sqrt{6}$	addition (see problem above)

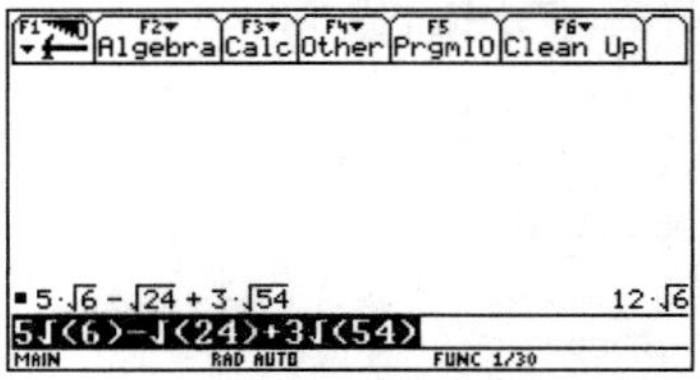

Figure 9.3.2

Figure 9.3.1 shows the arithmetic work is correct.

⁂

Example 3: Multiply $(2\sqrt{5})(\sqrt{15})$.

Solution: All of the radicals are simplified. Perform the implied multiplication.

$(2\sqrt{5})(\sqrt{15})$	given
$2\cdot\sqrt{5}\cdot\sqrt{15}$	associative property of multiplication
$2\cdot(\sqrt{5}\cdot\sqrt{15})$	associative property of multiplication

$2\cdot(\sqrt{5\cdot 15})$	product property
$2\cdot(\sqrt{5\cdot 5\cdot 3})$	factorization
$2\cdot(\sqrt{5^2\cdot 3})$	definition of an exponent
$2\cdot(\sqrt{5^2}\cdot\sqrt{3})$	product property
$2\cdot(5\sqrt{3})$	power property
$10\sqrt{3}$	associative property, multiplication

You should confirm the answer by calculating $(2\sqrt{5})(\sqrt{15})$ and $10\sqrt{3}$. They should have the same value.

⁂

Example 4: Multiply $(\sqrt{3}+\sqrt{2})(4\sqrt{3}-\sqrt{2})$.

Solution: Use the FOIL method to multiply the binomials.

F O I L

$$(\sqrt{3}+\sqrt{2})(4\sqrt{3}-\sqrt{2}) = \sqrt{3}\cdot 4\sqrt{3} - \sqrt{3}\cdot\sqrt{2} + \sqrt{2}\cdot 4\sqrt{3} - \sqrt{2}\cdot\sqrt{2}$$
$$= 4\sqrt{3}\cdot\sqrt{3} - \sqrt{3}\cdot\sqrt{2} + 4\sqrt{2}\cdot\sqrt{3} - \sqrt{2}\cdot\sqrt{2}$$
$$= 4\sqrt{3^2} - \sqrt{6} + 4\sqrt{6} - \sqrt{2^2}$$
$$= 4\cdot 3 - \sqrt{6} + 4\sqrt{6} - 2$$
$$= 12 - \sqrt{6} + 4\sqrt{6} - 2$$
$$= 10 + 3\sqrt{6}$$

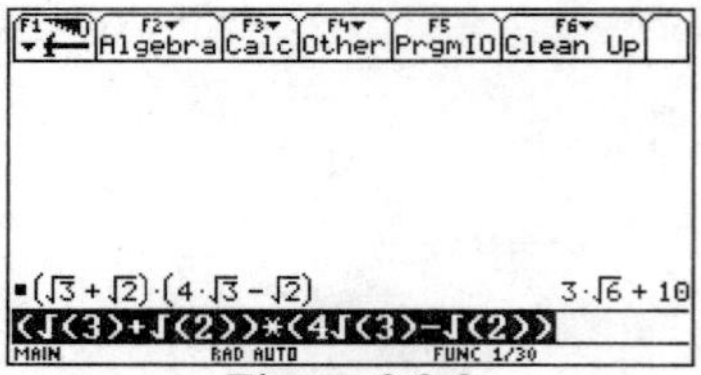

Figure 9.3.3

Why can't 10 be added to $3\sqrt{6}$?

⁂

Example 5: Simplify $(\sqrt{5}-2)(\sqrt{5}+2)$.

Solution: Since the binomials are conjugates, multiply first terms and last terms.

$$(\sqrt{5}-2)(\sqrt{5}+2) = \sqrt{5}\cdot\sqrt{5} - 2\cdot 2$$
$$= \sqrt{5^2} - 4$$
$$= 5 - 4$$
$$= 1$$

⁂

Example 6: Multiply $(\sqrt{3}+\sqrt{7})(\sqrt{3}-\sqrt{7})$.

Solution: The binomials are conjugates; thus, multiply first terms and last terms.

$$(\sqrt{3}+\sqrt{7})(\sqrt{3}-\sqrt{7}) = \sqrt{3}\cdot\sqrt{3}-\sqrt{7}\cdot\sqrt{7}$$
$$= \sqrt{3^2}-\sqrt{7^2}$$
$$= 3 - 7$$
$$= -4$$

The previous two problems demonstrate that when conjugates are multiplied, the product is a rational number. Recall from Chapter Four that $(a+b)(a-b)=a^2-b^2$. By the power property a^2 and b^2 are rational if a and b are square roots. This fact is important when performing division by a binomial. See the examples below.

Example 7: Divide $\dfrac{4}{1+\sqrt{2}}$.

Solution: The expression is not simplified because of the irrational number in the denominator. Multiplying by $\sqrt{2}$ makes the denominator $\sqrt{2}+2$; thus, the expression is still not simplified. As was shown in the previous two examples, multiplying $(1+\sqrt{2})$ by its conjugate $(1-\sqrt{2})$ causes the denominator to become a rational number. To simplify the expression, use the fundamental property of fractions and multiply the top and bottom of the expression by $(1-\sqrt{2})$.

$\dfrac{4}{1+\sqrt{2}}$ given

$\dfrac{4\cdot(1-\sqrt{2})}{(1+\sqrt{2})\cdot(1-\sqrt{2})}$ fundamental property

$\dfrac{4-4\sqrt{2}}{1^2-\sqrt{2^2}}$ multiplication

$\dfrac{4-4\sqrt{2}}{1-2}$ power property

$\dfrac{4-4\sqrt{2}}{-1}$ simplification

OR $-4+4\sqrt{2}$ division

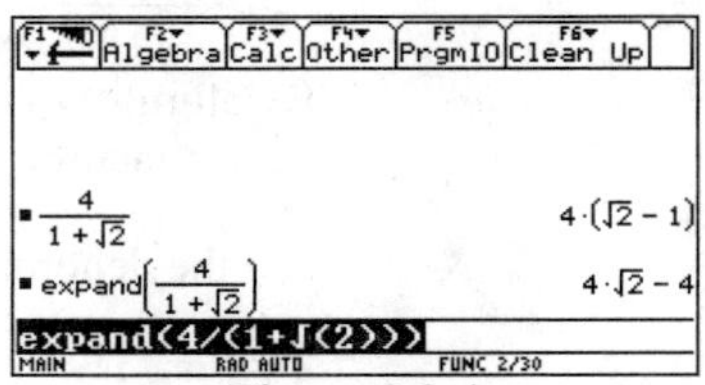

Figure 9.3.4

You may wonder why the directions to Example 7 were to divide when it doesn't look like any "division" has taken place. The division process may not be clear to you; however, do you know a way of proving that $-4+4\sqrt{2}$ is, in fact, the quotient of 4 divided by $1+\sqrt{2}$? Or maybe you should answer the question: How do you prove that 3 is the quotient when 12 is divided by 4?" The proof is that the quotient is a number that can be multiplied by 4 to get 12; 3 is that number. Back to the question, how do you know that $-4+4\sqrt{2}$ is the quotient of 4 divided by $1+\sqrt{2}$? Answer: $-4+4\sqrt{2}$ is the number that can be multiplied by $1+\sqrt{2}$ to get 4.

$$
\begin{aligned}
(-4+4\sqrt{2})(1+\sqrt{2}) &= \\
&= -4 + 4\sqrt{2} - 4\sqrt{2} + 4\sqrt{2}\sqrt{2} \\
&= -4 + \quad 0 \quad + 4\sqrt{4} \\
&= -4 + 8 \\
&= 4
\end{aligned}
$$

Example 8: Divide $\dfrac{2+5\sqrt{3}}{1-\sqrt{3}}$.

Solution: Multiply the numerator and denominator by the conjugate of the denominator.

$$\frac{\left(2+5\sqrt{3}\right)\cdot\left(1+\sqrt{3}\right)}{\left(1-\sqrt{3}\right)\cdot\left(1+\sqrt{3}\right)}$$

$$\frac{\overset{\text{F}}{2\cdot 1}+\overset{\text{O}}{2\cdot\sqrt{3}}+\overset{\text{I}}{1\cdot 5\sqrt{3}}+\overset{\text{L}}{5\sqrt{3}\cdot\sqrt{3}}}{1^2-\sqrt{3^2}}$$

$$\frac{2+2\sqrt{3}+5\sqrt{3}+5\sqrt{3^2}}{1^2-\sqrt{3^2}}$$

$$\frac{2+2\sqrt{3}+5\sqrt{3}+5\cdot 3}{1-3}$$

$$\frac{17+7\sqrt{3}}{-2}$$

By all rules of simplification, the quotient is now in simplified form. The radicand has no factors that are perfect squares; there are no fractions under the radical sign; and the denominator is a rational number. A check of $\dfrac{2+5\sqrt{3}}{1-\sqrt{3}}$ with a quotient of $\dfrac{17+7\sqrt{3}}{-2}$ is shown in Figure 9.3.5 below

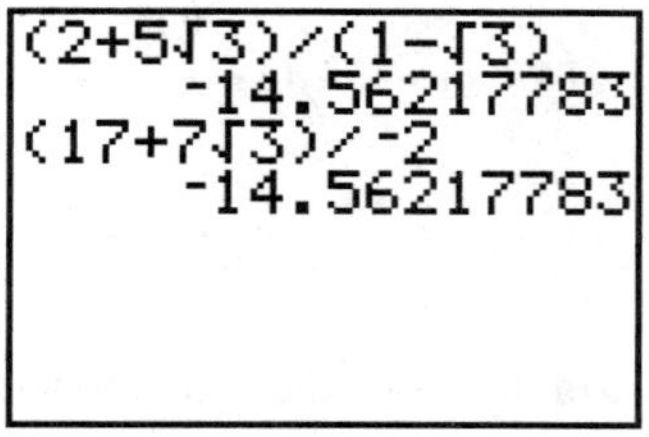

Figure 9.3.5

9.3 EXERCISES

Many times you can work the exercises by algebraic, numeric, or graphical methods. You can learn about mathematics no matter which method you use. Part of the learning process is for you to decide which method is best for you to use. Please read the text carefully before you do the exercises. Read the **next assigned section** after you finish the assignment below. All numeric answers must have an error ≤ 0.01.

A Quick Look To The Future

-4. Is $2^{\frac{1}{2}}$ the same thing as $2 \cdot \frac{1}{2}$?

-3. Is $(x+2)^2$ the same thing as x^2+4?

-2. Do you suppose the square root function is related to the velocity of an object falling?

-1. Can the height of an object thrown straight up be modeled with a linear function? Why?

Previous Topics Revisited

1. Simplify $\frac{1}{\sqrt{8}}$
2. Simplify $\sqrt{12x^5}$
3. Simplify $\sqrt{15} \cdot \sqrt{5}$
4. In the function $f(x) = d\sqrt{x+e}+f$, which parameter is an indication of whether the function is increasing or decreasing?
5. In the function $f(x) = d\sqrt{x+e}+f$, which parameter translates the graph of $y=\sqrt{x}$ horizontally to get the graph of $f(x) = d\sqrt{x+e}+f$?
6. Solve $\frac{3}{x-4.1}+7.2=14.9$

New Topic Exercises

Simplify the expressions in Exercises 7 - 17. Confirm that your answers are correct with your calculator.

7. $\sqrt{7}+5\sqrt{7}-6\sqrt{7}$
8. $\sqrt{20}-\sqrt{45}$
9. $6\sqrt{\frac{1}{2}}+\sqrt{18}$
10. $\sqrt{10} \cdot \sqrt{5}$
11. $2\sqrt{3}(2\sqrt{3}-5\sqrt{12})$
12. $(2\sqrt{3}-7)(4\sqrt{3}+5)$

13. $(2\sqrt{5}+\sqrt{3})(2\sqrt{5}-\sqrt{3})$

14. $(1-\sqrt{3})(1+\sqrt{3})$

15. $\dfrac{\sqrt{7}}{1-\sqrt{7}}$

16. $\dfrac{3}{\sqrt{2}+\sqrt{5}}$

17. $\dfrac{3+\sqrt{2}}{1-\sqrt{2}}$

Simplify the following functions. Confirm that your simplified function is equivalent to the given function with your calculator.

18. $f(x) = 2\sqrt{x} + 5\sqrt{x}$

19. $g(x) = 8\sqrt{x} - 13\sqrt{x} + 7\sqrt{x}$

20. $h(x) = \left(1+\sqrt{x}\right)\left(1-\sqrt{x}\right)$

21. $k(x) = \dfrac{x}{\sqrt{x}-2}$

22. What is the domain of the above function $h(x)$?

23. What is the domain of the above function $k(x)$?

24. Is the above function $h(x)$ increasing or decreasing?

25. What is the zero of the above function $k(x)$?

26. Find two functions whose sum is $3\sqrt{x-1}$.

27. Find two functions whose difference is $7\sqrt[3]{x}$.

28. Find two functions whose product is $4 - x$.

29. Are the functions $\left(3+\sqrt{x+2}\right)\left(3-\sqrt{x+2}\right)$ and $9-(x+2)$ equivalent?

30. What can $\sqrt{3}$ be multiplied by to rationalize it? (Change $\sqrt{3}$ into a rational number.)

31. Rationalize the *numerator* of the fraction: $\dfrac{1+\sqrt{2}}{4}$.

32. What can 4 be multiplied by to change it to an irrational?

33. What is the exact value of $\sqrt{12}\cdot\sqrt{3}$?

34. Rationalize the *numerator* of the fraction: $\dfrac{\sqrt{3}-\sqrt{2}}{6}$.

35. Rationalize the function $(\sqrt{x+h}-\sqrt{h})$.

From Mathematics to English

36. Will multiplying $(2+\sqrt{3})$ times $\sqrt{3}$, rationalize $(2+\sqrt{3})$? Explain.

37. Describe how to divide 7 by $\sqrt{2}$. Do not actually divide.

38. Why do you think $6 + \sqrt{5}$ is an irrational number?

39. Is $\left(\sqrt{3}+\sqrt{2}\right)\left(\sqrt{3}-\sqrt{2}\right)$ an irrational number? Explain.

40. Is $\frac{\sqrt{32}}{\sqrt{2}}$ an irrational number? Explain.

41. Describe how to add $4\sqrt{7}$ and $3\sqrt{28}$. Do not actually add.

42. Give one example of a use of an irrational number.

43. After reading this section, make a list of questions that you want to ask your instructor.

44. Continue in your daily journal and make an entry. In addition to your normal entry on thoughts about the mathematics in this section, list at least two positive comments about what you have learned about this topic.

45. In paragraph format, summarize the material in this section of the text in your daily journal.

46. Describe how your classroom instructor made this topic more understandable and clear.

47. After reading the text and listening to your instructor, what do you not understand about this topic?

Explorations

48. List *3 different expressions* that can be multiplied by the given expression in order to rationalize it. (Change it to a rational number.)

a. $\sqrt{2}$ b. $\sqrt{x}$ c. $\sqrt{a+b}$

d. $\sqrt[3]{2}$ e. $\sqrt[3]{x}$

49. a. Are the graphs of $y=(\sqrt{x}-2)(\sqrt{x}+3)$ and $y=x-6$ the same graph? Why?

b. Are the graphs of $y=(\sqrt{x}-2)(\sqrt{x}+3)$ and $y=\sqrt{x^2}-6$ the same graph? Why?

c. Are the graphs of $y=(\sqrt{x}-2)(\sqrt{x}+3)$ and $y=x+\sqrt{x}-6$ the same graph? Why?

50. Simplify (x, y, and $z \geq 0$)

a. $\sqrt{x^5}$ b. $\sqrt{x^{13}}$ c. $\sqrt{x^7y^8}$

d. $\sqrt{\frac{x^5}{y^3}}$ e. $\sqrt{\frac{x^9y^{11}}{z^4}}$

9.4 Fractional Exponents

Thus far, in the development of mathematical topics in this text, you have learned:

1. Positive integer exponents have been defined as: $b^n = b \cdot b \cdot b\ldots$, n times.
2. Negative integer exponents have been defined as: $b^{-n} = \dfrac{1}{b^n}$, $b \neq 0$.
3. A zero exponent has been defined as: $b^0 = 1$, $b \neq 0$.

The last type of exponent defined in this text is the fractional exponent.

From Section 7.1, the exponential function is defined as $y = b^x$, where $b > 0$ and $b \neq 1$. A quick review of the numeric representation of the function, where b is 2, is shown below .

Table 9.4.1

x	...	−3	−2	−1	0	1	2	3	...
2^x	...	0.125	0.25	0.5	1	2	4	8	...

As shown in Table 9.4.1, you know the meaning of an exponent if it is an integer, but what is the meaning of a number like $2^{\frac{1}{2}}$ or $2^{\frac{3}{4}}$? The demonstration with the calculator in Figure 9.4.1 and the explanation from Section 7.1 should answer this question.

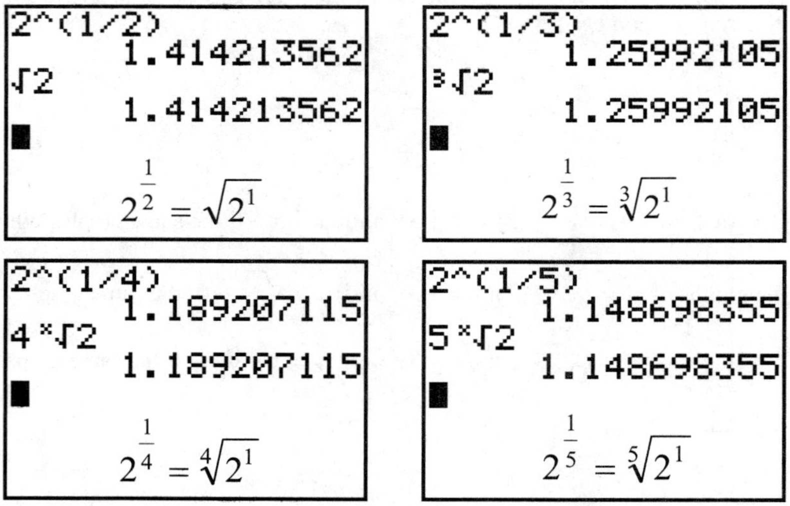

Figure 9.4.1

Hopefully, the pattern established in Figure 9.4.1 is clear. The other question that remains is: What if the exponent is of the form $\dfrac{m}{n}$ where m is not 1? Consider what you already know about the properties of exponents studied in Chapter Seven.

$$b^{\frac{m}{n}} = \left(b^{\frac{1}{n}}\right)^m \text{ by the power property.}$$

That is, something like $2^{\frac{3}{4}}$ is nothing more than $\left(2^{\frac{1}{4}}\right)^3$ or $\sqrt[4]{2}^3$. As you can see below, the exponent 3 can be placed under the radical sign as well as outside it. You should confirm this yourself.

Definition of a Fractional Exponent

$$b^{\frac{n}{m}} = \sqrt[m]{b^n} = \left(\sqrt[m]{b}\right)^n$$

Conversely: $\sqrt[m]{b^n} = \left(\sqrt[m]{b}\right)^n = b^{\frac{n}{m}}$

where n and m are integers and b must not be negative if m is even.

Example 1: What is the meaning of $b^{\frac{1}{2}}$?

Solution: The denominator 2 becomes the index and the numerator 1 becomes the exponent on the base b. It is $\sqrt[2]{b^1}$, $\sqrt[2]{b}^1$, or $\sqrt{b}$ because the index 2 and the exponent 1 are not normally written. An exponent of $\frac{1}{2}$ is the same thing as a $\sqrt{\ }$.

⁂

The definition of a fractional exponent allows the writing of numbers raised to fractional exponents as a radical or it allows the writing of radicals as numbers raised to fractional exponents. Rewriting a radical as a number to an exponent gives information needed to use the calculator to calculate (evaluate) any radical.

Example 2: Evaluate: $\sqrt[3]{16}$.

Solution: Think of $\sqrt[3]{16}$ as $\sqrt[3]{16^1}$ and rewrite $\sqrt[3]{16^1}$ as $16^{\frac{1}{3}}$. You can also use the cube root built into some calculators.

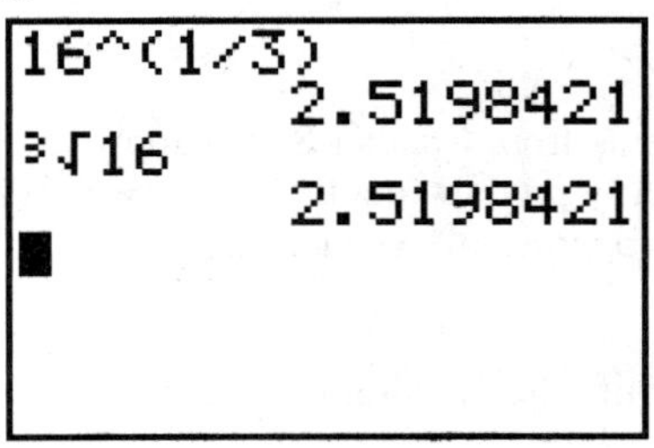

Figure 9.4.2

The value is approximately 2.5198.

⁂

Example 3: Evaluate $\sqrt[5]{17}^{2}$.

Solution: Rewrite $\sqrt[5]{17^2}$ as $17^{\frac{2}{5}}$ and calculate.

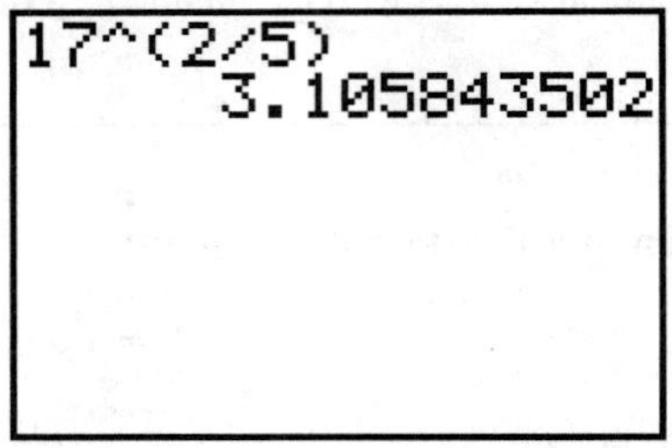

Figure 9.4.3

$\sqrt[5]{17^2}$ is approximately 3.1058.

⁂

Example 4: Evaluate $\sqrt[5]{7}$.

Solution: Think of $\sqrt[5]{7}$ as $\sqrt[5]{7^1}$ and rewrite $\sqrt[5]{7^1}$ as $7^{\frac{1}{5}}$. Use the calculator to find the approximate value. Since $\frac{1}{5}$ is 0.2 in decimal form, you may also use 7 raised to the 0.2 exponent.

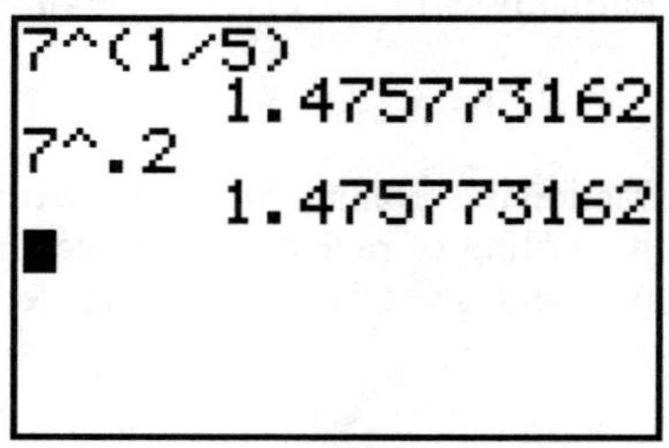

Figure 9.4.4

The approximate value is 1.4758.

⁂

The properties of exponents from Chapter Seven apply to numbers raised to fractional exponents as well as to integer exponents. Any product, quotient, or power of numbers raised to fractional exponents can be simplified by using the properties of exponents.

Example 5: Simplify the function $y = x^{\frac{1}{2}} \cdot x^{\frac{1}{4}}$.

Solution: By the product property, add exponents.

$$y = x^{\frac{1}{2}} \cdot x^{\frac{1}{4}} = x^{\frac{1}{2}+\frac{1}{4}} = x^{\frac{2}{4}+\frac{1}{4}} = x^{\frac{3}{4}}$$

The function simplifies to $x^{\frac{3}{4}}$. If $x^{\frac{3}{4}}$ were now rewritten as a radical, it can be simplified by the rules for simplification of radicals. It is NOT necessary to do this. Don't be shy about using your calculator. It will add ½ and ¼ as shown in Figure 9.4.5.

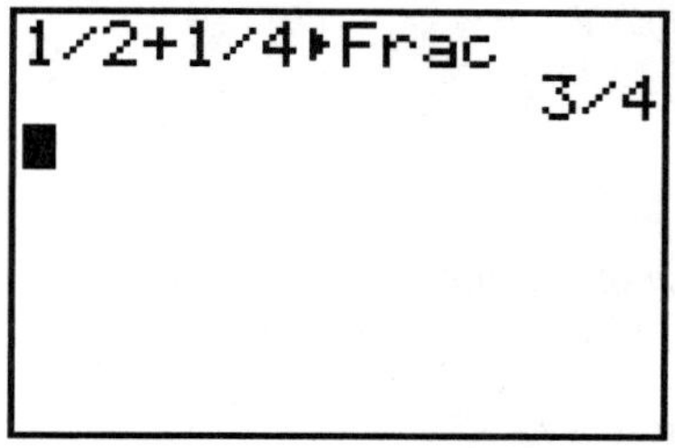

Figure 9.4.5

⁂

Example 6: Simplify the function $f(x) = (x^3)^{\frac{2}{3}}$.

Solution: Use the power property of exponents to simplify.

$$f(x) = (x^3)^{\frac{2}{3}} = x^{3 \cdot \frac{2}{3}} = x^2$$

⁂

Example 7: Rewrite the function $f(x) = \sqrt{2x} - 4\sqrt[3]{x^2}$ in exponential form.

Solution: Think of $\sqrt{2x}$ as $\sqrt{(2x)^1}$ and use the definition of fractional exponents.

$f(x) = \sqrt{2x} - 4\sqrt[3]{x^2}$ is the same thing as $f(x) = (2x)^{\frac{1}{2}} - 4x^{\frac{2}{3}}$.

⁂

9.4 EXERCISES

Many times you can work the exercises by algebraic, numeric, or graphical methods. You can learn about mathematics no matter which method you use. Part of the learning process is for you to decide which method is best for you to use. Please read the text carefully before you do the exercises. Read the **next assigned section** after you finish the assignment below. All numeric answers must have an error ≤ 0.01.

A Quick Look To The Future

-4. If $x = 2$, does $x^3 = 8$?

-3. If two variables are related by a square root function, could they also be related by a quadratic function?

-2. If a quadratic function has a minimum, is the graph rising or falling to the right of the minimum?

-1. If $ab = 0$, what do you know about a or b?

Previous Topics Revisited
Simplify Exercises 1 - 5.

1. $\dfrac{5-\sqrt{3}}{2+\sqrt{3}}$

2. $(3\sqrt{2}+\sqrt{5})(\sqrt{2}-4\sqrt{5})$

3. $(2\sqrt{x}-\sqrt{y})(2\sqrt{x}+\sqrt{y})$

4. $\dfrac{8\sqrt{12}}{2\sqrt{3}}$

5. $(2\sqrt{3})^4$

6. Find the minimum value of the function $2\sqrt{x-1}+4$.

New Topic Exercises
Evaluate the radicals in Exercises 7 - 13.

7. $\sqrt[4]{6}$

8. $\sqrt[9]{2}$

9. $\sqrt[3]{4^2}$

10. $\sqrt[5]{-32}$

11. $\sqrt[3]{125^2}$

12. $\sqrt[4]{3^4}$

13. $\sqrt[6]{12.56^9}$

Simplify the following functions using the properties of exponents. $(x \geq 0)$

14. $y = x^{\frac{1}{2}} \cdot x^{\frac{1}{2}}$

15. $f(x) = \left(x^{\frac{1}{4}}\right)^8$

16. $g(x) = \dfrac{x^{\frac{1}{4}}}{x^{\frac{1}{2}}}$

17. $y = x^{\frac{2}{3}}(x^{\frac{4}{9}} + x^{\frac{1}{3}})$

Simplify the following functions by writing them in exponential form first. $(x \geq 0)$

18. $t(x) = \sqrt[3]{x^2}\left(\sqrt[9]{x^4} + \sqrt[3]{x}\right)$

19. $h(x) = \dfrac{\sqrt[4]{x}}{\sqrt{x}}$

20. $y = \sqrt[4]{x^8}$

21. $y = \sqrt[4]{x}^{\,8}$

22. $y = \sqrt{x}\sqrt{x}$

23. $f(x) = \sqrt[5]{(2x)^{10}}$

24. What is the domain of the function $y = x^{\frac{1}{2}}$?

25. What is the domain of the function $y = x^{\frac{1}{3}}$?

26. What happens on your calculator when you try to evaluate $(-16)^{\frac{1}{4}}$?

27. Express $5^{3\frac{1}{4}}$ in radical form.

28. Evaluate the number $2^{1.35}$.

29. Order the numbers from low to high: $\sqrt[4]{5},\ \sqrt{5},\ \sqrt[3]{5},\ 5^{\frac{2}{3}}$.

30. What is the value of the biggest number in the list $5^{1.2},\ 5^{3.4},\ 5^{2.2},\ 5^{2.1}$?

31. Express the number $\sqrt{\sqrt{\sqrt{3}}}$ in exponential form.

32. What is the domain of $x^{1.5}$?

From Mathematics to English

33. If $n > m$, give a counter-example or use some other argument to show why the statement $b^n > b^m$ is not always a true statement. (b is a number larger than 1)

34. Can your calculator find the value of $(-16)^{\frac{1}{4}}$? If yes, what is it? If no, why not?

35. Describe the 5th root of 20. Do not actually find the 5th root of 20.

36. Describe the number $4^{1.5}$ without actually evaluating it.

37. Give an example where the number $98^{\frac{1}{2}}$ is actually used.

38. After reading this section, make a list of questions that you want to ask your instructor.

39. Continue in your daily journal and make an entry. In addition to your normal entry on thoughts about the mathematics in this section, list at least two positive comments about what you have learned about this topic.

40. In paragraph format, summarize the material in this section of the text in your daily journal.

41. Describe how your classroom instructor made this topic more understandable and clear.

42. After reading the text and listening to your instructor, what do you not understand about this topic?

Explorations

43. Develop a method that would allow radicals with different indices to be multiplied; for example, multiply $\sqrt{x}\cdot\sqrt[3]{x}$.

44. Check your answers to Exercises 14 - 17 by graphing the given function and your answer on the same window. Explain any differences in the graph of the given function and the graph of your answer.

45. Are the graphs of $y = x^{\frac{2}{3}}$ and $y = \left(x^{\frac{1}{3}}\right)^2$ the same graph? Explain.

9.5 Solving Equations Containing the Square Root Function

The Architect and His Computer
An architect has a known position on a piece of property and must find the coordinates of a second position that is a known distance and a known direction from the beginning point. How would his computer calculate the coordinates of the second position?

The programmer of the computer has to look-up two things in this text. She must know that northeast is halfway between east and north. This means she can use the idea of rate of change. If the change in the distance north and the change in the distance east are equal, the direction is northeast.

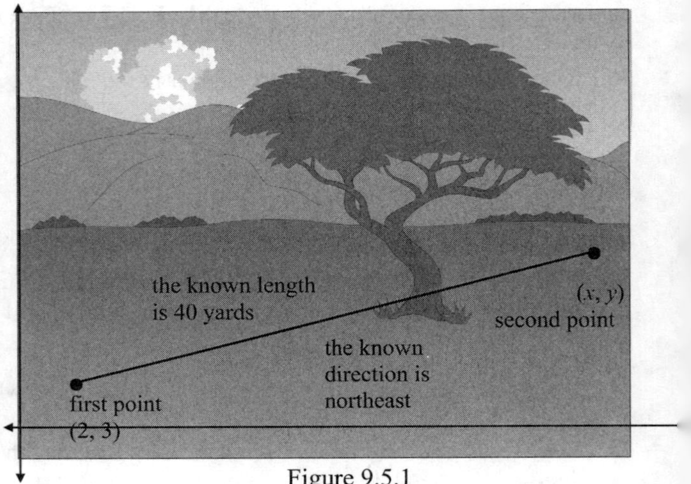

Figure 9.5.1

That is, from the first point, change the vertical position and then the horizontal position by equal amounts to get to the second point. The change in the northerly direction (the difference of the y-coordinates) divided by the change in the easterly direction (the difference of the x-coordinates) is one. So, $\dfrac{y-3}{x-2} = 1$. The second item is found in Section 11.1. The distance from the first point to the second can be expressed as $\sqrt{(x-2)^2 + (y-3)^2}$ (This comes straight from the Pythagorean Theorem.) and this is known to be 40.

So:

1. $\sqrt{(x-2)^2 + (y-3)^2} = 40$ and
2. $\dfrac{y-3}{x-2} = 1$ multiply both sides by $(x-2)$
3. $(y-3) = (x-2)$ and replace $(y-3)$ with $(x-2)$ in 1.
4. $\sqrt{(x-2)^2 + (x-2)^2} = 40$ Solve the equation for x.

This, of course, brings you to the point of the problem. How do you solve equations containing the square root function? Well, you could do like the architect and use a computer shown below in Figure 9.5.2.

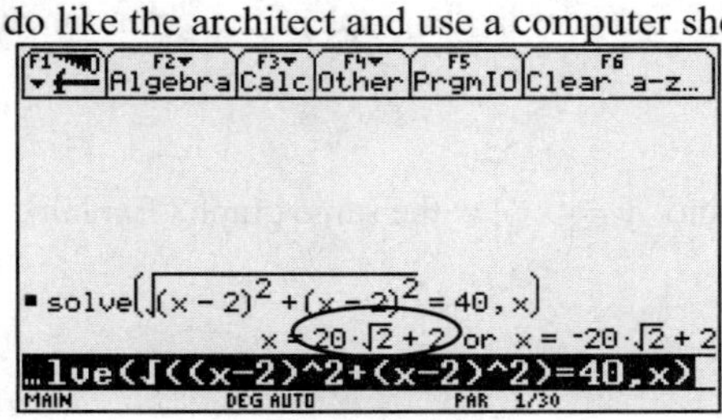

Figure 9.5.2

This shows two values for the x-coordinate in exact form -- not too useful to the architect, but he hasn't learned how to use the computer very well. The second value for x is to the left of the first point and not needed by the architect. The y-coordinate is shown below in Figure 9.5.3.

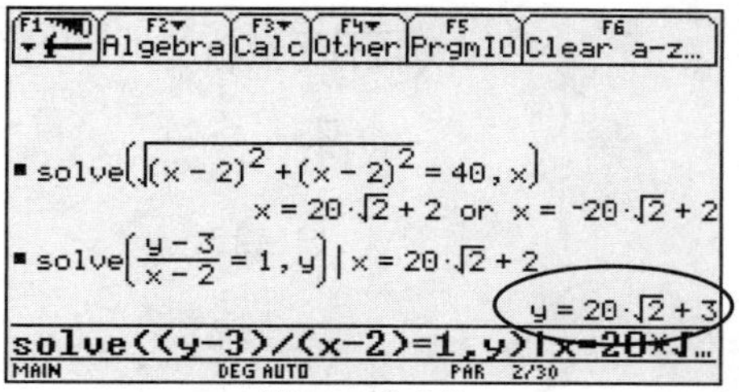

Figure 9.5.3

Below are the answers in approximate form. The second point is at (30.284, 31.284).

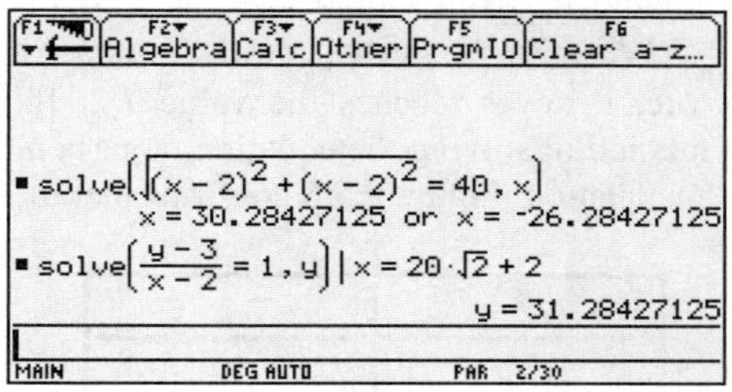

Figure 9.5.4

The four technology based methods used to solve equations in other chapters can also be applied to equations containing the square root function. No new properties are needed. You only need to know the behavior of the square root function presented in Section 9.1. The algebra-based method is studied later and it will require a new property of equality before it can be used.

Example 1: Solve the equation $\sqrt{x-5} = 13$.

Solution: If you use the zeros method, subtract 13 from both sides of the equation and graph the function $y = \sqrt{x-5} - 13$. Find the zero of this function. Because the function is increasing and it has a value of -13 when x is 5, the zero must be quite a bit to the right of 5. Recall that square root functions with a d parameter of 1 increases very slowly. The window of Figure 9.5.5a is $[-10, 100]$ by $[-20, 20]$, and the

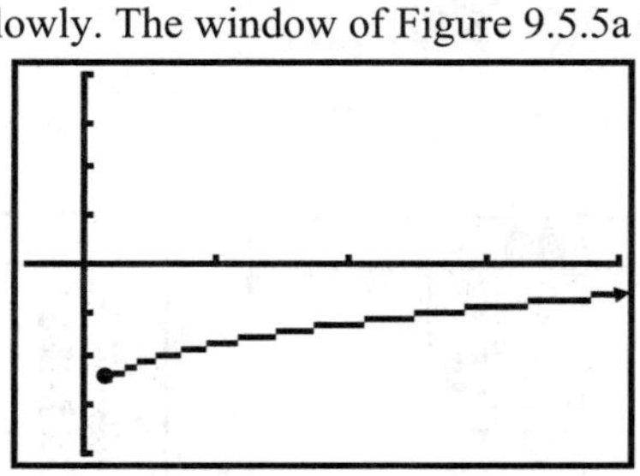
Figure 9.5.5a

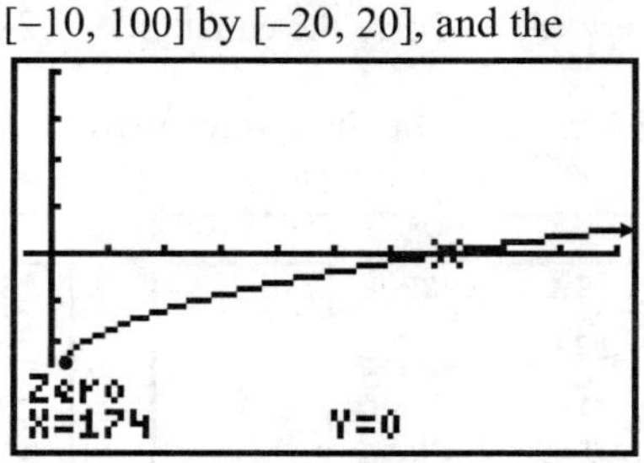

Figure 9.5.5b

window of Figure 9.5.5b is $[-10, 250]$ by $[-20, 20]$. The function has a value of 0 when $x = 174$. A check of the zero shows it to be an exact answer.

Check:

$$\sqrt{x-5} = 13$$
$$\sqrt{174-5} = 13$$
$$\sqrt{169} = 13$$
$$13 = 13$$

⁂

Example 2: Solve the equation $-2\sqrt{x+3.6} = 7.2$.

Solution: If you think graphically, you should be visualizing the graphical representation of the function $y = -2\sqrt{x+3.6}$ as starting on the x-axis when x is -3.6 and decreasing in value; therefore, it never reaches the value 7.2. If you are thinking of using the intersection method of solving the equation, your conclusion has to be that there is no solution to the equation. Figure 9.5.6 confirms that there is no solution.

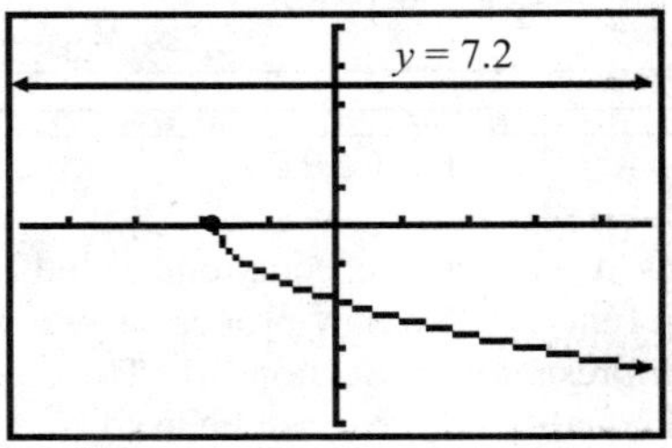

Figure 9.5.6

It doesn't matter what method you use to solve the equation. All methods will fail to find the solution because there is no solution.

⁂

Example 3: Solve the equation $-3.2\sqrt{x+4} = -3.1$.

Solution: The three screens from the TI calculator show the solution to the equation using the numerical method.

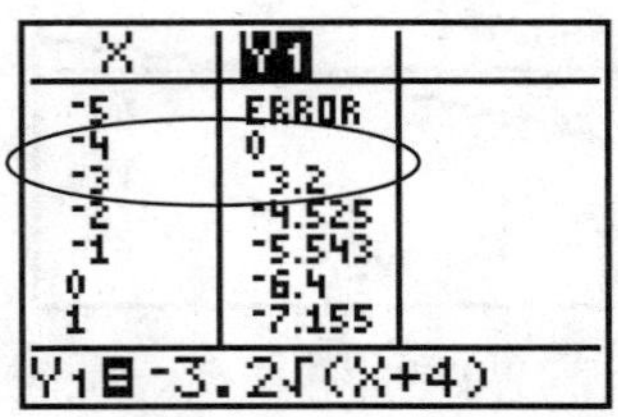

Figure 9.5.7a

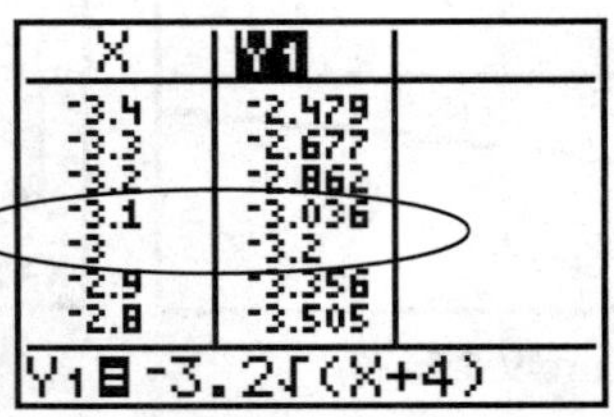

Figure 9.5.7b

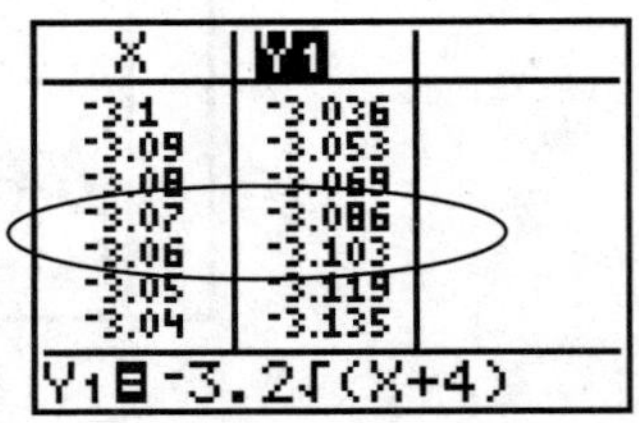

Figure 9.5.7c

Figure 9.5.7c shows that there is a value for x between −3.07 and −3.06 that will cause the function $-3.2\sqrt{x+4}$ to be −3.1. A solution to the equation with an error less than 0.01 is −3.065.

⁂

Since the square root function in Example 4 is different than what was studied in Section 9.1, you may need to experiment with the viewing window to get a complete graph if you solve graphically.

Example 4: Solve the equation $\sqrt{x-2}+\sqrt{2x+3}=5$.

Solution: When the equation is a little more complicated than normal, the zeros method may be the best choice for solving the equation. Subtract 5 from both sides to get a function on the left side of the equation and 0 on the right: $\sqrt{x-2}+\sqrt{2x+3}-5=0$. Graph $y=\sqrt{x-2}+\sqrt{2x+3}-5$ and identify the zeros. Upon inspecting the symbolic form of this function, you will note that the domain is $[2, \infty)$, because if x is less than 2, the function $\sqrt{x-2}$ is not a real number. The graph should start at $x = 2$ and be to the right of 2.

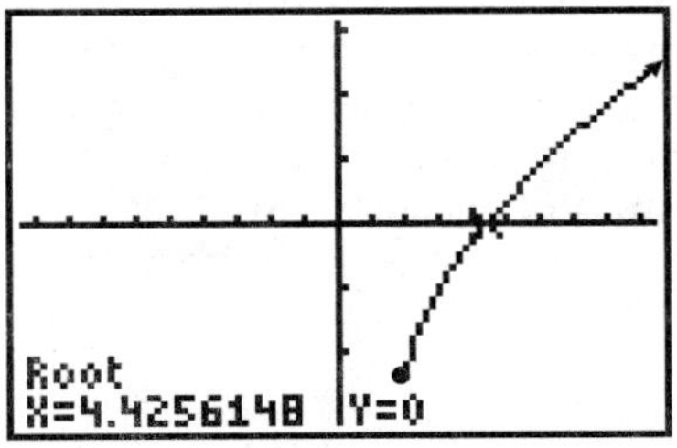

Figure 9.5.8

The function has a zero at 4.4256148... To select a zero with an error ≤ 0.01, you can select the zero to the thousandths place; thus, the solution to the equation is 4.426 with an error ≤ 0.01.

⁂

The analytical method for solving equations containing radicals requires a new property. Earlier properties of equality have allowed for both sides of any equation to be multiplied or divided by a non-zero number, and any number can be added to or subtracted from both sides of any equation without changing the solutions. The property needed to solve equations with radicals does not always guarantee that the solutions remain the same after the property has been used.

Consider the equation	$x = 2$	the solution is 2
Square both sides	$x^2 = 2^2$	the solutions are 2 and −2

The equation $x^2 = 4$ has an extraneous solution. The number −2 is called an **extraneous solution**, an extra solution that was picked up in the process of squaring both sides of the equation. The extraneous solution is discarded because it does not satisfy the original equation. You will note that the original equation ($x = 2$) contains linear functions; however, the equation that has been transformed by the squaring process contains the quadratic function. As you know from Chapter Three, a quadratic function may have none, one, or two zeros and the linear function can only have one zero. You should also note in the example above that the solution to the parent equation was not lost in the transforming process -- just another solution was added.

> **Power Property of Equality**
>
> If $a = b$, then $a^n = b^n$. $\qquad$ $a^n = b^n$ may have extraneous solutions.

An extraneous solution will satisfy the equation $a^n = b^n$, but it will NOT make the original equation a true statement. Since the goal is to solve the original equation, each solution to $a^n = b^n$ must be checked in the equation $a = b$ to make certain it is a solution.

Example 5: Find the exact solution to the equation $\sqrt{x-5} = 13$.

Solution: Since the exact solution is requested, the algebraic method may be the best choice for solving the equation. With a simple square root on one side, if both sides are squared using the power property of equality, the radical sign will drop out. By squaring both sides, the equation is transformed into an equation that can be solved with the remaining properties of equalities. Since the parent equation contains a square root function, which can have none or one zero, and the transformed equation contains a linear function which has one zero, the solution must be checked to make certain it is not extraneous.

Exactly!

$(\sqrt{x-5})^2 = 13^2$ power property of equality

$x - 5 = 169$ power property of radicals

$x = 174$ addition property of equality

The solution, 174, need not be checked because this equation was solved graphically in Example 1, and the number 174 is the exact solution. Instead of doing an analytical check on any problem, you may want to check graphically. The graphical method does NOT cause any extraneous solutions.

Example 6: Find the exact solution to $-2\sqrt{x+3.6} = 7.2$.

Solution: As in Example 5, the algebraic method will guarantee the exact solution. Before squaring both sides, it may be simpler to divide both sides by –2.

$\dfrac{-2\sqrt{x+3.6}}{-2} = \dfrac{7.2}{-2}$ division property of equality

$\sqrt{x+3.6} = -3.6$ simplification

$(\sqrt{x+3.6})^2 = (-3.6)^2$	power property of equality
$x + 3.6 = 12.96$	simplification
$x = 9.36$	subtraction property

The solution, 9.36, must be checked to see if it is extraneous. Because this equation was solved in Problem 2, it is not necessary to check -- it is extraneous. There is *no solution*. You may have known this as soon as you looked at the equation because the function on the left side of the equation is always negative or zero and the right side is a positive number; therefore, they could never be equal. Since the parent equation contains a square root function, which can have none or one zero, and the transformed equation contains a linear function which has one zero, the solution must be checked to make certain it is not extraneous.

⁂

Example 7: Find the exact solution to $x-3=\sqrt{2x+2}$.

Solution:

$x - 3 = \sqrt{2x+2}$	given
$(x-3)^2 = (\sqrt{2x+2})^2$	power property of equality
$x^2 - 6x + 9 = 2x + 2$	simplification
$x^2 - 8x + 7 = 0$	subtraction property of equality
$(x-1)(x-7) = 0$	factorization
$x = 1 \qquad x = 7$	relationship between factors and zeros

Check for extraneous solutions:

$x-3=\sqrt{2x+2}$	$x-3=\sqrt{2x+2}$
$1-3=\sqrt{2\cdot 1+2}$	$7-3=\sqrt{2\cdot 7+2}$
$-2=\sqrt{4}$	$4=\sqrt{16}$
$-2 = 2$ FALSE	$4 = 4$ TRUE
$x \neq 1$	$x = 7$ This is the only solution.

Everyone checks their work.

Everyone checks their work.

Instead of checking analytically, you may find it simpler to check graphically. Figure 9.5.9 shows the graphs of the functions $x-3$ and $\sqrt{2x+2}$. As you can see the intersection method only shows one solution to the equation.

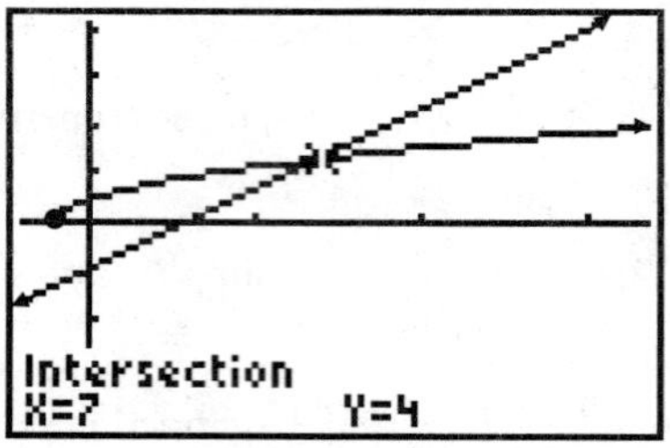

Figure 9.5.9

⁂

If the analytical method is to be used on an equation that contains two radical expressions, the method may work best if there is one radical per side. See the example below.

Example 8: Find the exact solution to $\sqrt{x-3}-\sqrt{x+1}=-2$.

Solution: Add $\sqrt{x+1}$ to both sides before squaring each side.

$$\sqrt{x-3}-\sqrt{x+1}+\sqrt{x+1}=-2+\sqrt{x+1}$$ addition property

$$\sqrt{x-3}=\sqrt{x+1}-2$$ simplification

$$(\sqrt{x-3})^2=(\sqrt{x+1}-2)^2$$ power property

Remember $(a+b)^2 = \quad a^2+2ab+b^2$

$$x-3=x+1-4\sqrt{x+1}+4$$

$$x-3=x+5-4\sqrt{x+1}$$ subtract x and 5

$$-8=-4\sqrt{x+1}$$ divide by –4

$$2=\sqrt{x+1}$$ square both sides

$$2^2=(\sqrt{x+1})^2$$ simplify

$$4=x+1$$

$$x=3$$

check for extraneous solution

$$\sqrt{x-3}-\sqrt{x+1}=-2$$

$$\sqrt{3-3}-\sqrt{3+1}=-2$$

$$\sqrt{0}-\sqrt{4}=-2$$

$$-2=-2$$

The solution is 3. Remember: A graphical check may be a better choice.

⁂

Example 9: Solve $\sqrt{x-3}-\sqrt{x+1}=-2$ again.

Solution: Since the exact solution is not required, you may want to use a graphical method. It too may give the exact solution however. Add 2 to both sides and graph the related function. Graph $y = \sqrt{x-3} - \sqrt{x+1} + 2$ and find values for x where the function has a value of 0.

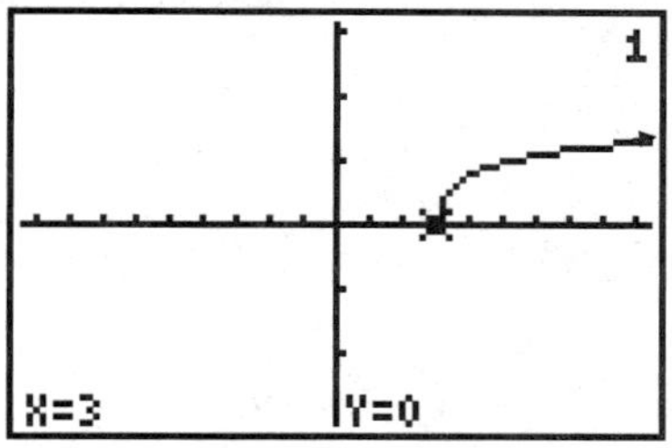

Figure 9.5.10

The solution is exactly 3. Remember, you have five methods for solving an equation. In addition to the three graphical methods and the analytical method, you may also use the numeric method. Below is the numerical method for solving the above equation.

Table 9.5.1

x	3	4	5	6	7	8
$y = \sqrt{x-3} - \sqrt{x+1} + 2$	0	0.76	0.96	1.09	1.17	1.24

The domain is $[3, \infty)$, thus there are no zeros left of 3. The function is increasing; therefore, there are no zeros to the right of 3. The only solution is 3.

⁂

9.5 EXERCISES

Many times you can work the exercises by algebraic, numeric, or graphical methods. You can learn about mathematics no matter which method you use. Part of the learning process is for you to decide which method is best for you to use. Please read the text carefully before you do the exercises. Read the **next assigned section** after you finish the assignment below. All numeric answers must have an error ≤ 0.01.

A Quick Look To The Future

-4. True or false, the square root function has no application to a real-world situation.

-3. Can a quadratic function have just one zero?

-2. If $ax + b = 0$, what is x? If $cx + d = 0$, what is x? If $ex + f = 0$, what is x? If $gx + h = 0$, what is x?

-1. You know that $(x + 2)^2 = x^2 + 4x + 4$. Is $x^2 + 4x + 4$ a perfect square?

Previous Topics Revisited

Simplify the following functions.

1. $y = x^{\frac{1}{3}} \cdot x^{\frac{2}{3}} \cdot x^{\frac{1}{2}}$

2. $f(x) = \dfrac{x^{\frac{2}{5}}}{x^{\frac{-4}{5}}}$

3. $g(x) = \left[\left(x^{\frac{3}{4}}\right)^2\right]^2$

4. $y = (2\sqrt{x} - 3)(4\sqrt{x} + 1)$

5. $f(x) = (\sqrt{x} - 3\sqrt{2})(\sqrt{x} + 3\sqrt{2})$

6. Evaluate the number $\sqrt{9+16}$.

New Topic Exercises

Solve the equations.

7. $2\sqrt{x+3} = 5,\ 2\sqrt{x+3} = 6,\ 2\sqrt{x+3} = 7,\ 2\sqrt{x+3} = 8,\ 2\sqrt{x+3} = 9$

8. $-4\sqrt{x-2} = -7,\ -4\sqrt{x-2} = -8,\ -4\sqrt{x-2} = -9,\ -4\sqrt{x-2} = -10$

9. $-4\sqrt{x-2} = 7$

10. $2\sqrt{x+3} = -5$

11. $\sqrt{x-3.7} + 4.2 = 6.9$

12. $47 = 20\sqrt{x-15} + 23$

13. $0.27 = -0.15\sqrt{x+3.75} + 1.33$

14. $-2\sqrt{x+5} = 0,\ 2\sqrt{x-3} = 0,\ -\frac{1}{3}\sqrt{x+1} = 0,\ 5\sqrt{2x+4} = 0$

15. $\sqrt{2x+8} = 3$

16. $\sqrt{x-3} + \sqrt{x+4} = 6$

17. $x - 1 = \sqrt{x+29}$

18. $x = \sqrt{x+2} - 2$

19. $2 = \sqrt{2x+39} - x$

20. $2\sqrt{x-1} - 3 = -\sqrt{x+2} + 1$

21. In the equation $d\sqrt{x+e} + f = 0$, if d is positive, what values of f cause the equation to have a solution?

22. In the equation $d\sqrt{x+e} + f = 0$, if d is positive, what values of f cause the equation to have no solution?

23. Create an equation containing a square root function that has 2 as the solution.

24. Does the equation $2\sqrt{x+3}\sqrt{x+3} + 4 = 3$ have a solution other than $\varnothing$?

25. Change one of the parameters in the equation $-2\sqrt{x+1} = 5$ so that your new equation has a solution.

From Mathematics to English

26. From a non-graphical perspective, describe how you know that the equation $\sqrt{x+1.1} = -3$ has no solution without ever "solving" the equation.

27. From a graphical perspective, describe how you know that the equation $-2\sqrt{x+1}-4=0$ has no solution without ever "solving" the equation.

28. Describe an advantage of the technical based methods of solving equations containing the square root function over the analytic method.

29. Describe a second method of checking for extraneous solutions -- other than substituting the possible solutions in the original equation.

30. Describe how you would solve the equation $2\sqrt{x-3}+4=6$ using algebra. Do not actually solve the equation.

31. Describe how you would solve the equation $2\sqrt{x-3}+4=6$ using technology. Do not actually solve the equation.

32. Give your definition of an equation.

33. What is your idea of the solution to an equation?

34. After reading this section, make a list of questions that you want to ask your instructor.

35. Continue in your daily journal and make an entry. In addition to your normal entry on thoughts about the mathematics in this section, list at least two positive comments about what you have learned about this topic.

36. In paragraph format, summarize the material in this section of the text in your daily journal.

37. Describe how your classroom instructor made this topic more understandable and clear.

38. After reading the text and listening to your instructor, what do you not understand about this topic?

39. What is the most important idea presented in this section? Why?

Explorations

40. Fill in the chart below with the conditions on *d*, *e*, and *f* that cause the desired results on the equation $d\sqrt{x+e}=f$; use interval notation.

conditions on *d*	conditions on *e*	Conditions on *f*	desired result
			one solution
			no solution

41. Equations containing radicals can get much more complicated than the equations you solved above. However, while the equations can get more complicated, the graphical method for solving the equations remains the same. Solve the equations below. Be sure you have a complete graph before you find the solution.

   a. $x^2 - 6x + 2\sqrt{3x - 4} = |x + 2|$

   b. $3x - \sqrt{x^2 + 2x} = -5$

   c. $\left|4x^2 - 3\sqrt{2x^2 - 6x + 4}\right| = -2$

   d. $-|3x - 5| + \sqrt{x^2 - 4x - 2} - x + 2 = -x^2$

   e. $\frac{1}{2} \cdot 2^{\sqrt{x^2 - x - 12}} = 1$

42. Solve the equation $\sqrt{(x-2)^2 + (x-2)^2} = 40$ using a graphical method.

43. Solve the equation $\frac{y - 3}{x - 2} = 1$ for y when $x = 30.284$.

44. In the opening problem in this section, change the given first point to (4, 5), the given distance to 60 yards and the rate of change (slope) to 2. Find the second point.

45. Explain why the graph used in Exercise 42 is a V.

9.6 The Square Root Function as a Mathematical Model

The models that follow require you to be familiar with the behaviors of the basic square root function, and you should know how to simplify square root expressions. Each model is described and the variables are identified. Each model includes a sample question that has been answered.

Velocity of a Dropped Ball

Physicists have measured the velocity of a ball that has dropped h feet. Table 9.6.1 shows the results of the experiment. (ignoring air resistance)

Table 9.6.1

h	0	1	2	3	4	5	10	15	20	35	50
v	0	8	11.31	13.86	16	17.89	25.30	30.98	35.78	47.33	56.57

The units for v are feet per second, so for example, when the ball has fallen 5 feet, the velocity is about 17.89 feet per second (12.2 miles per hour). The question physicists had to answer was "can this data be modeled with a function -- if yes, what function?" If you are going to answer this question for the physicists, you would want the graphical representation of the data. It is shown below in Figure 9.6.1.

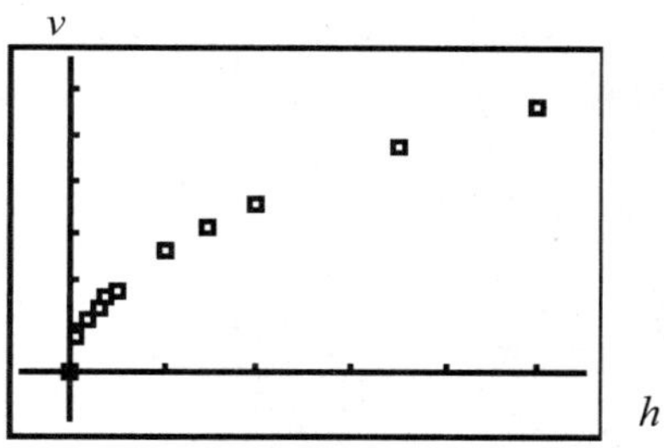

Figure 9.6.1

Right away you may recognize the shape as a square root function. Because the data is increasing and the first data point is at (0, 0), you know that the function parameter d in the function $y = d\sqrt{x+e}+f$ is positive and that e and f are 0. All that remains is to identify d. Study Figure 9.6.2, it shows the graphs of (from bottom to top) $v = 2\sqrt{h}$, $v = 6\sqrt{h}$, $v = 8\sqrt{h}$, and $v = 10\sqrt{h}$.

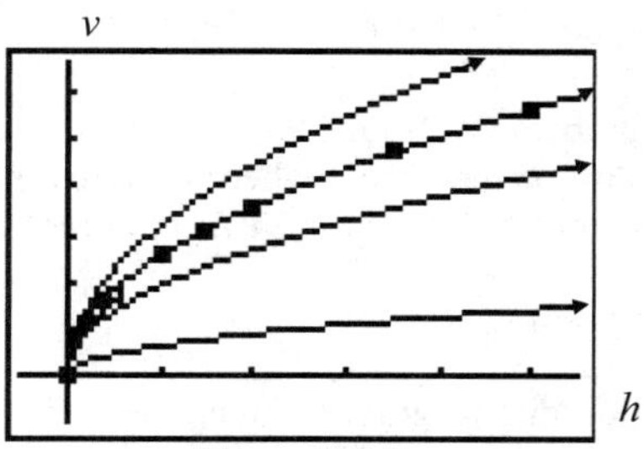

Figure 9.6.2

The function that matches the data is $v = 8\sqrt{h}$. That is, it is possible to predict the velocity of an object in free-fall when you know how far it has fallen -- use the model $v = 8\sqrt{h}$.

Physicists did not have the graphing calculator as a tool to do the above work. They used laws of physics and algebra to develop the model from the data. As you might guess, the model is different on planets with a different acceleration due to gravity. The more massive the planet -- the faster the ball travels because gravity is pulling harder.

The work of the physicists results in the model of this velocity as $v = \sqrt{2gh}$ feet per second when the object has fallen h feet. That is, the velocity function during free-fall is $v = \sqrt{2gh}$, where g is the acceleration due to the earth's gravity, and h is the distance traveled. The acceleration due to gravity is 32 ft/sec^2. When g is replaced with 32, the function becomes $v = \sqrt{2 \cdot 32 \cdot h}$, which simplifies to $v = \sqrt{64 \cdot h}$. This simplifies to $v = 8\sqrt{h}$, where h is in feet and v is in feet per second. The velocity function assumes no air resistance. For example, a ball dropped from a height h above ground level will have a velocity nearly identical to $8\sqrt{h}$ when it strikes the ground. On the other hand, a piece of paper dropped from a height of h will have a velocity that cannot be modeled with the above velocity function because air resistance drastically alters the velocity.

Example 1: How far must a ball fall for it to have a velocity of 100 feet per second? (68 mph)

Solution: Using the function model $v = 8\sqrt{h}$, replace v with 100. This gives an equation that you can solve by any method of your choice. Using algebra might be the best choice.

$$100 = 8\sqrt{h}$$ divide both sides by 8

$$\frac{100}{8} = \sqrt{h}$$ square both sides

$$\left(\frac{100}{8}\right)^2 = h$$ simplify

$$h = 156.25 \text{ feet}$$

⁂

Example 2: A point on the graph of $v = 8\sqrt{h}$ is (9, 24). What do the numbers mean?

Solution: The pair of numbers (9, 24) means that when an object is dropped from a height of 9 feet, the velocity of the object when it strikes the ground is 24 feet per second. Or, after falling 9 feet the velocity is 24 feet per second.

⁂

Example 3: The acceleration due to gravity on Jupiter is 86.8 feet per second per second. How fast is a piece of space junk falling after it has fallen 1000 feet?

Solution: The model $v = \sqrt{2gh}$ becomes $v = \sqrt{2 \cdot 86.8h}$ which simplifies to $v = \sqrt{2 \cdot 86.8}\sqrt{h}$. This is also written as $v = 13.18\sqrt{h}$. To answer the question, replace h with 1000 and evaluate v.

$$v = 13.18\sqrt{h}$$

$$v = 13.18\sqrt{1000}$$

$$v = 416.79 \text{ feet per second}$$

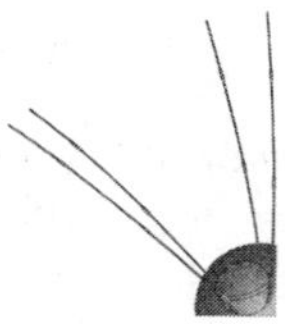

⁂

The Pendulum Swings

A free-swinging weight on the end of a secured string is called a **pendulum**. A picture of a pendulum is shown below in Figure 9.6.3.

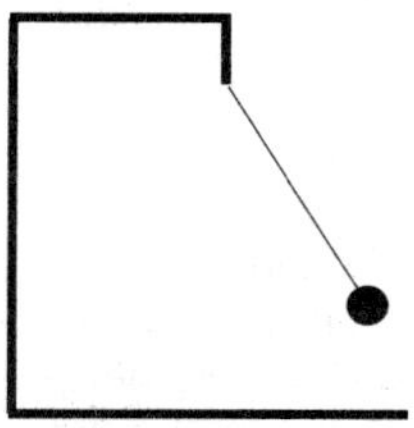

Figure 9.6.3

The **period** of the pendulum is the time it takes for the weight to go through a complete cycle; for example, the time it takes to move from the left-most position back to the left-most position again. Scientists tell you that the period is related to the length of the string by a square root function. The relationship between the period and the length of the pendulum is $t = 2\pi\sqrt{\frac{l}{g}}$, where g is the acceleration due to gravity (32 ft/sec^2), π is approximately 3.14159, l is the length of the string in feet, and t is the period measured in seconds. This function simplifies to: $t = 1.1107\sqrt{l}$ (do this yourself). Like the last model, the pendulum function is the basic square root function stretch by a factor of 1.1107.

Example 4: Is the pendulum function an increasing function?

Solution: YES! Since the function parameter d is a positive number -- it is an increasing function. This means that the longer the pendulum arm, the longer the period.

⁂

In 1851, at the Pantheon in Paris, a French physicist by the name of J. B. L. Foucault used an iron ball on a 200-foot wire as a pendulum to prove that the earth was rotating. This may not seem like a significant event today because everyone "knows" the earth rotates on its axis, but in the middle 1800's, many people still thought the sun revolved around the earth. This simple pendulum experiment shook the foundation of the European religious establishment and redirected the fields of astronomy and physics.

Example 5: How long must the string on a pendulum be for the pendulum to have a period of 6 seconds?

Solution: From the mathematical model that describes pendulum motion, replace t with 6 and solve for l.

$t = 1.1107\sqrt{l}$ replace t with 6

$6 = 1.1107\sqrt{l}$ divide both sides by 1.1107

$5.4020 = \sqrt{l}$ square both sides

$29.2 = l$

If the pendulum has a length of 29.2 feet, the period will be 6 seconds.

⁂

Example 6: Suppose it is the year 2028 and you have just signed on as the world space station resident engineer. The rotating space station simulates gravity and the measured value of the acceleration due to gravity (g) is 15 feet per second per second. Your job is to design a pendulum clock for the space station. How long should the pendulum string be so that it has a period of one second?

Solution: The model for the period of the pendulum becomes $t = 2\pi\sqrt{\frac{l}{15}}$. This simplifies to $t = \frac{2\pi}{\sqrt{15}}\sqrt{l}$, or $t = 1.622\sqrt{l}$. Now replace t with 1 and solve the resulting equation.

$$1 = 1.622\sqrt{l}$$

$$\frac{1}{1.622} = \sqrt{l}$$

$$\left(\frac{1}{1.622}\right)^2 = l$$

$$0.380 = l$$

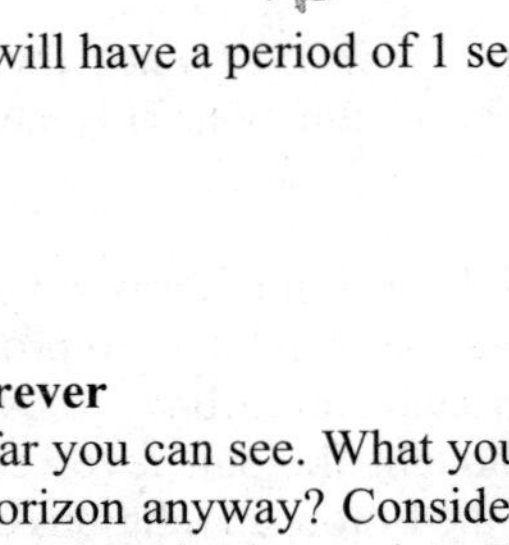

If you make the length 0.38 feet the pendulum will have a period of 1 second.

⁂

On a Clear Day You Can See Forever

If you have ever been in an airplane you may have wondered how far you can see. What you know is that the higher you go the farther you can see. So how far is it to the horizon anyway? Consider the drawing of your height above the Earth and the distance to the horizon in Figure 9.6.4.

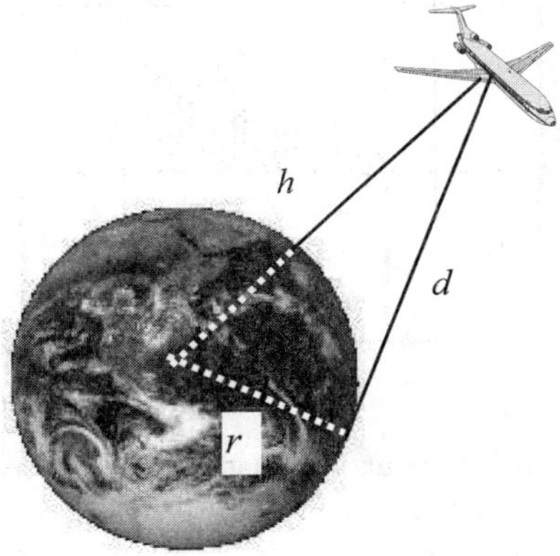

Caution: this drawing is not to scale!

Figure 9.6.4

The triangle is a right triangle and in a right triangle the hypotenuse squared is equal to the sum of the squares of the legs.

That is: $(h+r)^2 = d^2 + r^2$ simplify the left side

$h^2 + 2hr + r^2 = d^2 + r^2$ subtract r^2 from both sides

$h^2 + 2hr = d^2$ take the square root of both sides

$$d = \sqrt{h^2 + 2hr}$$

The distance to the horizon can be modeled by a square root function. The value of r is the radius of the Earth which is about 3977 miles or 20,998,560 feet (about 21 million feet).

Example 7: How far is it to the horizon if you are flying at an altitude of 36,000 feet?

Solution: Miles may be a good choice for units; 36,000 feet is 6.82 miles. Replace h with 6.82 and r with 3977 and calculate d.

$$d = \sqrt{h^2 + 2hr}$$
$$d = \sqrt{6.82^2 + 2 \cdot 6.82 \cdot 3977}$$
$$d = 233.01$$

At an altitude of 36,000 feet, you can see 233.01 miles

⁂

Area Under $\sqrt{x}$

The last model of the square root function is an area function. The area in question is a region bounded by the x-axis on the bottom, the graph of $y = \sqrt{x}$ on top, and the vertical line through a, where $a > 0$. The figure below shows the region.

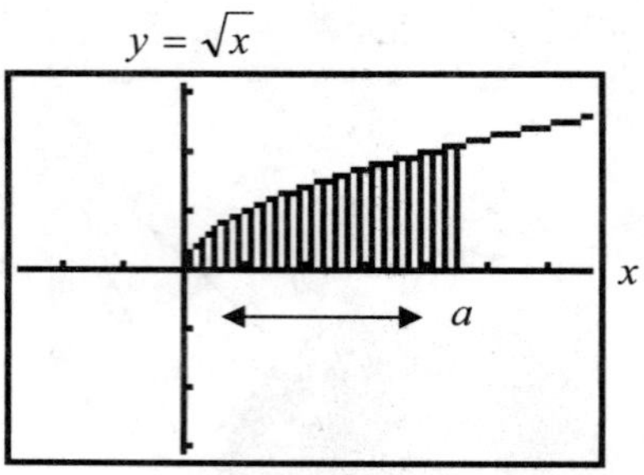

Figure 9.6.5

Although you may have already learned about formulas (functions) used to calculate areas of geometric shapes, the shape generated by the function $y = \sqrt{x}$ cannot be found from formulas typically taught at earlier mathematical levels. The area of the square root region is described by the area function $A = \frac{2}{3}a\sqrt{a}$. The function A represents the area of the geometric shape, and the function variable a represents where the vertical line is drawn.

Example 8: Find the area bounded by the x-axis, $\sqrt{x}$, and the vertical line at $x = 4$.

Solution: Using the area function in symbolic form, the area is:

$A = \frac{2}{3} \cdot 4 \cdot \sqrt{4} = \frac{2}{3} \cdot 4 \cdot 2 = \frac{2}{3} \cdot 8 = \frac{16}{3}$ square units. A picture of the area is shown below in Figure 9.6.6

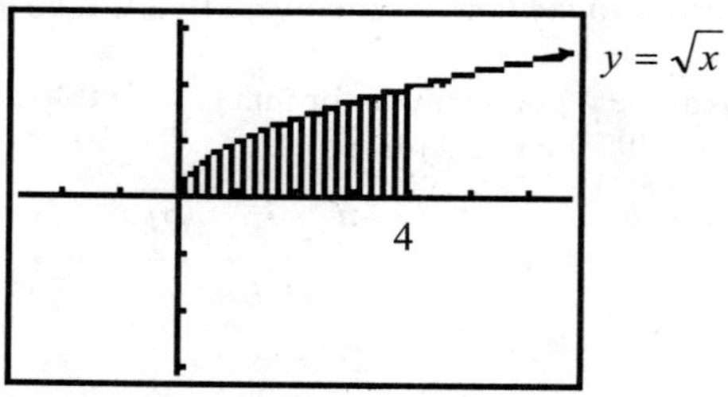

Figure 9.6.6

⁂

Example 9: Graph the area function $A = \frac{2}{3}a\sqrt{a}$.

Solution: Since the graphing calculator may not be able to use the variable a or the function name A, graph the function $y = \frac{2}{3}x\sqrt{x}$.

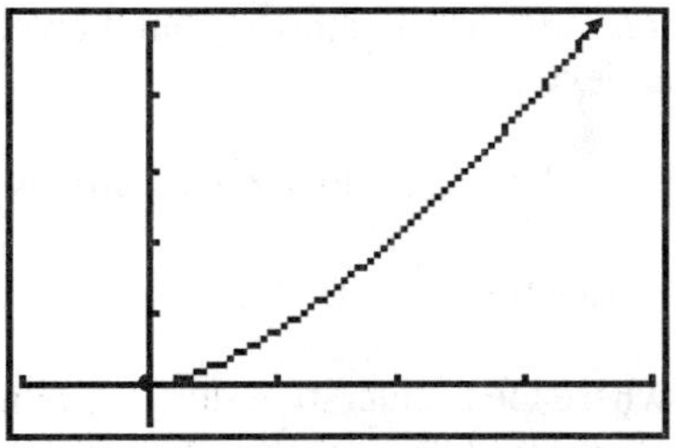

An appropriate window might be [–5, 20] by [–5, 50].

Figure 9.6.7

The function behaves the way you might expect: the function is positive because it represents an area, which is a positive number. Also note the area function increases. You expect this as well, because as the region gets larger the area should get larger.

⁂

9.6 EXERCISES

Many times you can work the exercises by algebraic, numeric, or graphical methods. You can learn about mathematics no matter which method you use. Part of the learning process is for you to decide which method is best for you to use. Please read the text carefully before you do the exercises. Read the **next assigned section** after you finish the assignment below. All numeric answers must have an error ≤ 0.01.

A Quick Look To The Future

-4. If a quadratic function has a minimum of 1, does it have any zeros?

-3. If $ab = 1$, find three pairs of numbers for a and b that make the statement true.

-2. If $x^2 + 4x = 12$, can you add 4 to both sides of the equation?

-1. What is the "formula" for solving every equation of the form $ax + b = 0$?

Previous Topics Revisited

1. Solve $2\sqrt{x-3}+5=8$

2. Solve $-2.55\sqrt{x+1.2}-4.1=0$

3. Solve $x-\sqrt{2x+20}=2$

4. Simplify $x^{\frac{3}{5}} \cdot x^{\frac{-4}{5}} \cdot x^{\frac{2}{10}}$

5. Simplify $\left\{\left[x^{\frac{3}{4}}\right]^{-\frac{2}{3}}\right\}^2$

6. Simplify $3\sqrt{12}-5\sqrt{27}$

New Topic Exercises

Velocity of a Dropped Object Exercises

7. Find the velocity of a water balloon as it strikes the ground when it is dropped from a height of 100 feet.

8. An object is dropped from a height of 120 feet above ground level. Find the velocity of the object when it is 100 feet above ground level.

9. An object should be dropped from what height to attain a velocity of 32 feet per second?

10. What is the range of the velocity function $v = 8\sqrt{h}$?

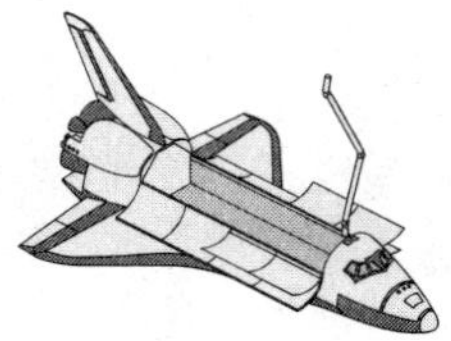

11. If you are on the space shuttle where the acceleration due to gravity might be $\frac{1}{100}$ the value of g on Earth, find the velocity of a wrench accidentally dropped after falling 6 feet.

12. Is the velocity of the wrench in Exercise 11 equal to $\frac{1}{100}$ of the velocity of the wrench had it fallen 6 feet near the surface of the Earth?

13. How far must a ball fall near the surface of the moon to have a velocity of 100 feet per second? The value of g on the moon is 5.47? How far must it fall on Earth to have the same velocity?

14. The velocity function for free-fall, $v = 8\sqrt{h}$, was derived from $v = \sqrt{2gh}$ where g is 32 ft/sec^2 and is used in the English system of measurement. In the metric system, $g = 9.8$ m/sec^2 and gives rise to the function $v = \sqrt{2 \times 9.8 \times h}$; simplify it. (Remember: 9.8 is a fraction.)

15. A point on the graph of the free-fall velocity model is (100, 80). What is the meaning of these numbers?

16. Rework Example 3 when on earth.

Pendulum Exercises

17. What is the period of a pendulum with an arm of 500 feet?

18. How long must a pendulum arm be to have a period of an hour? (1 hour is 3600 seconds)

19. The period of a pendulum is dependent on the function parameter g as you can see in the model $t = 2\pi\sqrt{\frac{l}{g}}$. If you are on the space shuttle in Exercise 11, what is the period of a pendulum with an arm of 6 feet? What is the period of the same pendulum on Earth?

20. A point on the graph of the Earth-based pendulum model is (0.5, 0.7828). What is the meaning of these numbers?

21. What are the domain and range of the pendulum function model? Is the model increasing or decreasing?

22. What is the period of the Foucault pendulum?

23. A pendulum of what length would cause a period of 1 second?

On a Clear Day Exercises

24. How far is the horizon from an airplane flying at a height of 45,000 feet?

25. Suppose you are flying over Cleveland and you think you see the Ohio River 200 miles away on the horizon. How high are you?

26. If you are landing a space vehicle on the moon and your altimeter reads 200,000 feet, how far is it to the horizon? The moon has a radius of 1080 miles.

Area Under $\sqrt{X}$ Exercises

27. Find the area of the region bounded by the x-axis on the bottom, $y = \sqrt{x}$ on the top, and the line $x = 25$.

28. Does the graph of the area function $A = \frac{2}{3}a\sqrt{a}$ intercept the x-axis? If yes, where?

29. A point on the graph of $A = \frac{2}{3}a\sqrt{a}$ is (81, 486), what do these numbers mean?

30. Find the area of the region bounded by the x-axis on the bottom, the function $y = \sqrt{x}$ on the top, the line $x = 4$ on the left, and the line $x = 16$ on the right.

31. Find the domain and range of the area function $A = \frac{2}{3}a\sqrt{a}$.

32. What value of a will cause the area bounded by the x-axis, $y = \sqrt{x}$, and the line $x = a$ to equal 100 square units?

From Mathematics to English

33. The velocity function $v = 8\sqrt{h}$ is a mathematical model for the behavior of an object falling near the surface of the earth, ignoring air resistance. Explain "ignoring air resistance."

34. Explain why you would expect the function $A = \frac{2}{3}a\sqrt{a}$ to never be negative.

35. Explain "the distance to the horizon."

36. What is your real-world interpretation of the zero of the pendulum model?

37. After reading this section, make a list of questions that you want to ask your instructor.

38. Continue in your daily journal and make an entry. In addition to your normal entry on thoughts about the mathematics in this section, list at least two positive comments about what you have learned about this topic.

39. In paragraph format, summarize the material in this section of the text in your daily journal.

40. Describe how your classroom instructor made this topic more understandable and clear.

41. After reading the text and listening to your instructor, what do you not understand about this topic?

42. What is the most important idea presented in this section? Why?

Explorations

43. Make a list of square root functions that you have used in other classes. Explain what the variables represent, and identify the domain and range of each. Draw a complete graph of each function.

44. Develop a square root application for modeling the length of the diagonal of a square of side x. Using the model, find the length of the diagonal of a square with the length of a side of 4 meters, 26 meters, 32 meters, and 72 meters. Find the length of the sides of a square whose diagonal is 10 meters, 20 meters, 32 meters, and $8\sqrt{2}$ meters.

45. Consider the isosceles triangle with equal legs of 4 inches, a base of x inches, and a height of h.

The height is perpendicular to the base and splits the isosceles triangle into two identical right triangles.

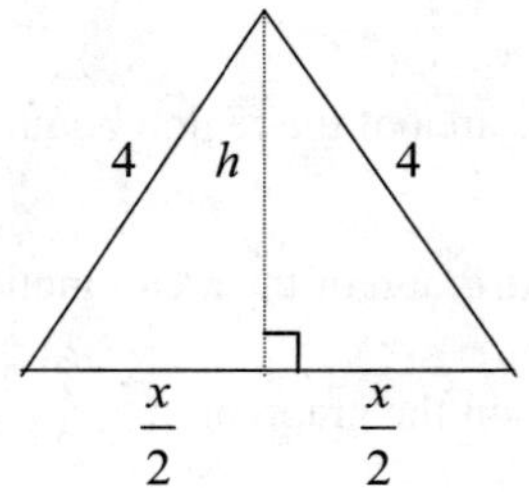

Using the Pythagorean Theorem and the "formula" for the area of a triangle, show that the relationship between the area A and the base x is a square root function.

46. Find the area of the isosceles triangle when the base is 5, 10, 15, 20, 25, 30, and 35 inches.

CHAPTER NINE TEST

1. Find the normal domain and range of $y = -2\sqrt{x-3} + 7$.

2. Is the function $f(x) = 0.3\sqrt{x+1} - 4$ increasing or decreasing on the normal domain?

3. What is the maximum value of the function $y = -5\sqrt{x-2} + 4$?

4. What is the zero of $f(x) = -5\sqrt{x-2} + 4$?

5. In the function $y = d\sqrt{x+e} + f$, describe all values of d that cause the function to be decreasing.

6. Find any square root function whose minimum value is 47.

7. Simplify $\dfrac{5}{\sqrt{5}}$

8. Simplify $\sqrt{75}$

9. Multiply and simplify $(\sqrt{3} - 2)(\sqrt{12} + 1)$

10. Divide and simplify $\dfrac{1}{1+\sqrt{2}}$

11. Evaluate $\sqrt[5]{20}$

12. Simplify $(x^{-6})^{\frac{1}{2}}$

13. Does the equation $-3\sqrt{x+4} - 2 = 0$ have a solution? If so, what is it?

14. Solve $2.1\sqrt{x-3.6} = 4.9$

15. Solve $x + 2 = \sqrt{x-5}$

16. If $d < 0$ and $d\sqrt{x+e} + f = 0$, what are all values of f that cause the equation to have a solution?

17. Using the mathematical model $A = \dfrac{2}{3}a\sqrt{a}$ to calculate the area bounded by $y = \sqrt{x}$, the x-axis, and the vertical line $x = a$, what value of a will cause the area to be one square unit?

18. With the mathematical model $v = 8\sqrt{h}$, how high would you have to drop a ball from to have it reach a velocity of 50 feet per second just as it strikes the ground?

19. Using the mathematical model $t = 1.1107\sqrt{l}$, what is the period of a pendulum with an arm length of three feet?

20. Using the mathematical model for the distance to the horizon $d = \sqrt{h^2 + 2hr}$, where the radius of the Earth is 3977 miles, how high must you be to see the horizon at 50 miles?

CHAPTER TEN
ADVANCED ANALYSIS OF THE QUADRATIC FUNCTION

10.0 Introduction

I Want to Maximize My Profits, But How?

When pricing merchandise for retail sales, a manufacturer cannot make the suggested selling price too low because she/he must cover the cost of manufacturing the product in order to make a profit. Likewise she/he cannot make the suggested retail price too high because people will not purchase a product when priced too high. This means that the manufacturer must use historical data and mathematics to help decide on a selling price that will maximize the profit for the item being sold. Below is the numeric representation of the daily profit earned by selling graphing calculators at x dollars each. The goal of the manufacturer is to develop a symbolic representation of a function that describes the daily profit from selling the calculator at x dollars each; that is, she/he must develop a mathematical model of the profit function. It may require an arithmetic combination of functions studied in this chapter or previous chapters to develop the symbolic representation of the function. The manufacturer may take advantage of his or her knowledge of the relationship between the behavior of a function and the parameters of the symbolic representation of the function. She/he may also use transformations like a stretch, shrink, reflection, horizontal translation, and/or vertical translation.

x	25	37	50	63	75	88	100	115
P	–500	8,200	16,500	19,000	19,000	16,000	8,100	1,000

The thinking of the manufacturer may require her/him to ask many questions. You will be able to help the manufacturer after you finish this chapter.

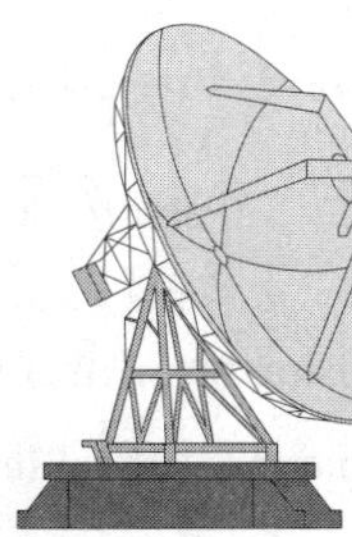

The Quadratic Function in Action

Have you ever wondered why the TV satellite dishes are so flat? Why are they shaped like they are, and what does this have to do with the quadratic function? As you learned in Chapter Three, the graph of the quadratic function is a parabola. The parabola has an interesting characteristic that can be used to collect radio and TV radiation (and other radiation). If parallel radio waves strike the surface of a paraboloid (If you spin a parabola on its axis, the three-dimensional shape generated is a **paraboloid**.), they are all reflected to a point called the **focus of the paraboloid**. Radio (TV) waves coming from a communications satellite are nearly parallel; therefore, they bounce off the sides and move to the focus. At the focus, they are collected and sent to the TV set.

Where is the focus? Recalling your work from Chapter Three, you should remember that the function parameters in the symbolic representation of the quadratic function tell you about the behavior of the graphical representation. You learned that in the function $y = \frac{1}{8}(x-3)^2 + 4$, the number $\frac{1}{8}$ tells you that the parabola opens up. The numbers –3 and 4 tell you that the vertex of the parabola is (3, 4), etc. So where is the focus? The number $\frac{1}{8}$ will also give you the location of the focus. If you think of the standard form of a quadratic function as $y = d(x+e)^2 + f$, then the focus of the parabola (or paraboloid) is located inside the parabola at a distance from the vertex of $\frac{1}{4d}$. In the sample parabola $y = \frac{1}{8}(x-3)^2 + 4$, the focus is $\frac{1}{4 \cdot \frac{1}{8}}$, or $\frac{1}{\frac{1}{2}}$, or 2 units from the vertex. The parabola $y = 6x^2$ $\left(y = 6(x+0)^2 + 0\right)$ has a focus $\frac{1}{4 \cdot 6}$, or $\frac{1}{24}$ unit from the vertex. From your experiences in Chapters Two & Three, you know that the graph of $y = \frac{1}{8}x^2$ is very flat and the graph of $y = 6x^2$ is very skinny. If the units of measurement are in feet, then the distance from the vertex to the focus in $y = 6x^2$ is $\frac{1}{24}$ of a foot (0.5 inch) and is very impractical for a satellite dish. The radio receiver put at the focus is relatively large and the parabola is so skinny that the receiver will block many of the incoming waves. If the focus is 2 feet from the vertex and the paraboloid is very flat as in $y = \frac{1}{8}x^2$, the receiver only covers a small area of the paraboloid and is at a usable distance from the dish -- a much better choice.

If you want to know more about other properties of the parabola, spend some time in your library. The parabola has many interesting applications that you may want to discover on your own.

You have now finished your study of the linear, absolute value, exponential, rational, and square root functions. The investigation of the quadratic function continues in Chapter Ten. You will investigate characteristics such as the domain, range, increasing, decreasing, transformations, zeros, and maximum or minimum. Once you are familiar with the basic behavior of the quadratic function, you will see how the zeros of the function are used to solve quadratic equations. Finally, you will study an application of the quadratic function. The quadratic function is a good model of projectile motion. When an object is thrown or shot, the description of its position, velocity, height, acceleration, and time in flight is called **projectile motion.**

In all previous material studied in this text, which includes all of the representations of a function, you have only considered real values for the function variable, function parameters, and real values of the function. In this chapter, you will be introduced to another type of number called a **complex number**. Complex numbers are found as the zeros of some quadratic functions and roots to some quadratic equations.

10.1 The Quadratic Function

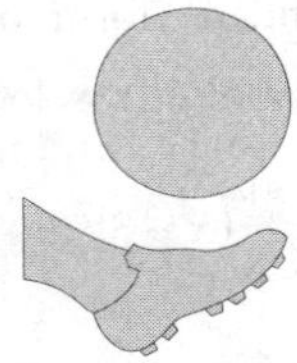

Kick That Ball One More Time

Imagine that you kick a ball straight up. You know what happens to the height of the ball. The height increases for awhile. It reaches a maximum height, and this is followed by the height decreasing until it has a zero height. What is so significant about this event is that a quadratic function can model this same behavior! To help understand the behaviors of the mathematical function, you can make a connection between the behavior of the height of the ball and the behavior of the function modeling the height of the ball. They are the same. Physicists will tell you that if the ball is thrown upward at an initial velocity of 30 feet per second, it leaves your hand when it is 6 feet above the ground, and you are on Earth, the model of the height of the ball is nearly $s = -16t^2 + 30t + 6$. The function parameter -16 is there because you are on the Earth. Maybe you can guess where the other two parameters come from.

Making the connection between the height of the ball and the function modeling the height of the ball will even help you set a graphing window on your calculator. For example, you know that the entire event only takes a few seconds; thus, set *xmax* at 3 or 4 seconds. Further, negative time is not a part of the event; therefore, you may want to set *xmin* at 0. You might estimate that the ball will reach a maximum height of 35 to 40 feet; thus, set *ymax* at around 40. Since the *x*-axis is zero height and the ball cannot go below zero height, you know to set *ymin* at 0. Actually, it makes the graph look a little nicer if you set the window a bit larger than the suggested values. It is nice to see the ground (the *x*-axis) when trace is on.

If you aren't quite certain when the function $s = -16t^2 + 30t + 6$ is increasing, think about when the ball is increasing. Figure 10.1.1 shows that the ball reaches a maximum height after 0.9375 second; thus,

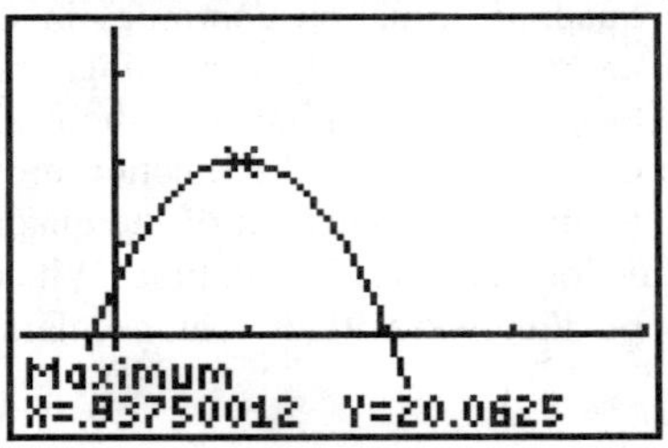

Figure 10.1.1

the function must be increasing until time in flight is 0.9375 seconds. On the problem domain this is when time is on the interval (0, .9375). On the normal domain, the function is increasing on $(-\infty, .9375)$.

Try making connections between the real-world events and function behaviors as you proceed through the section. For example, finding the zeros of the function $s = -16t^2 + 30t + 6$ is nothing more than finding when the ball is on the *x*-axis (the ground). Thinking about mathematics in the context of a real-world problem will help you understand the mathematics.

You were first introduced to the behavior of the quadratic function in Chapters Two and Three. In Chapter Four, the quadratic function was used to help factor trinomials. In this chapter, the investigation of the function continues.

> The **Quadratic Function** in symbolic form is
>
> $y = ax^2 + bx + c$ where a, b, and c are real numbers and $a \neq 0$.

If a has a value of 0, the function simplifies to the linear function; thus, it is no longer quadratic. You may remember from Chapters Two and Three that the shape of the graph is called a **parabola**.

A Look at Behaviors Using Transformations

The investigation of the quadratic function starts by checking to see if the notion of transformations applies. Remember, the graphs of the transformed functions do not move, but it "appears" that the graph of the basic function has been moved, flipped, or stretched. The graph of any function is simply the plot of all points (x, y), where x represents each element in the domain, and y represents the corresponding value of the function.

The basic quadratic function is $y = x^2$ or $f(x) = x^2$. As presented in Chapter Three, it is a parabola with a vertex at (0, 0) and has a minimum value of 0 at the vertex. The range is the interval $[0, \infty)$. The graph is shown below in Figure 10.1.2 along with the graphs of $y = x^2 + 4$ and $y = x^2 - 6$ on the window [–4.7, 4.7] by [–10, 15].

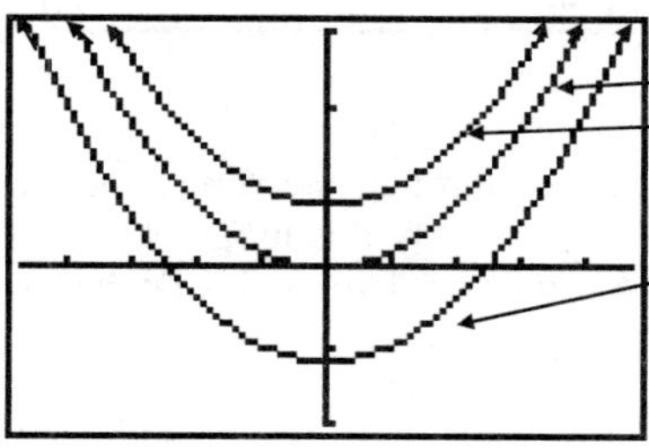

Figure 10.1.2

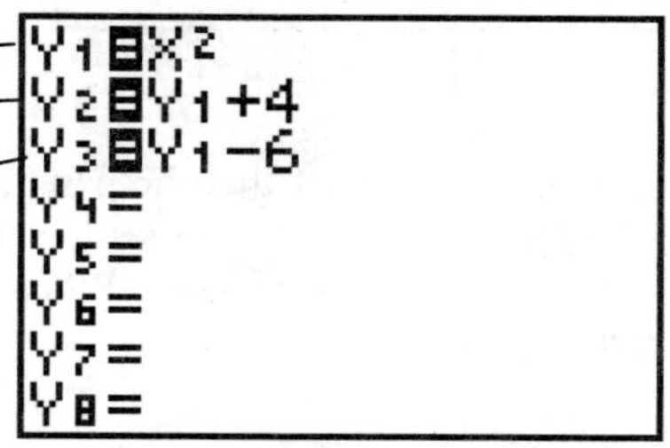

Figure 10.1.3

The above view (or any view) does not clearly show that for any value of x, Y_2 is always exactly 4 units above Y_1. Likewise, Y_3 is exactly 6 units below Y_1 for every x in the domain. However, if you use trace to jump from graph to graph, you will see that for any x you choose the function values (Y) differ by 4 or –6 from the basic function. Thus, you may conclude the idea of a vertical shift applies to the basic quadratic function. The graph of $y = x^2 + 4$ appears to be the graph of $y = x^2$, but just shifted up by 4 units. The graph of $y = x^2 - 6$ appears to be the graph of $y = x^2$, but shifted down by 6 units.

When x is replaced with "$x + e$" in the basic parent function, does the graph shift left or right? Recall from Section 3.6 that this process is called the composition of $y = x^2$ with $x + e$. Since the calculator accepts function notation, the composition is shown in Figure 10.1.4. To find out what happens, consider the graphs of $y = (x + 3)^2$ and $y = (x - 5)^2$ below on the window [–9.4, 9.4] by [–10, 10].

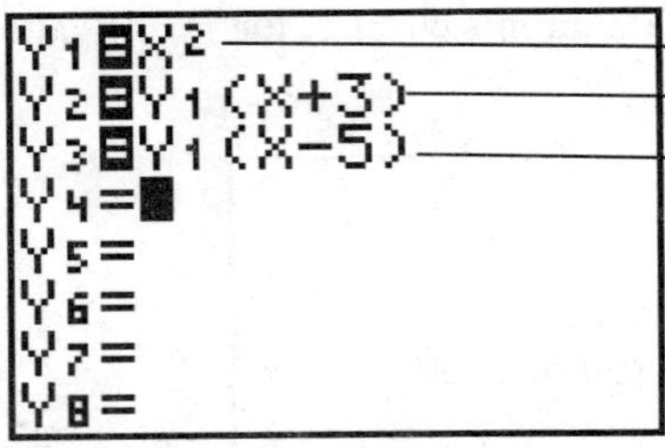

Figure 10.1.4

Review Sections 4.1 and 4.6 if you have forgotten function notation shown in Figure 10.1.4.

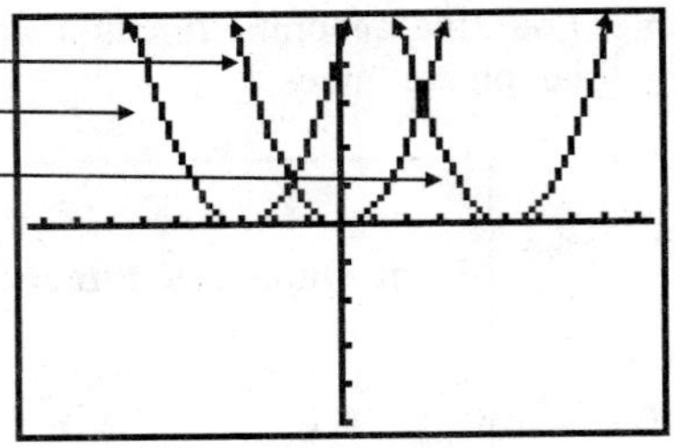

Figure 10.1.5

As can be seen, the graphs are the shifted graphs of $y = x^2$. As has happened with other functions studied, the graph of $y = (x + 3)^2$ is the graph of $y = x^2$ shifted to the LEFT by 3 units. The graph of $y = (x - 5)^2$ is the graph of $y = x^2$ shifted RIGHT by 5 units.

With the functions previously studied, when the parent function is multiplied by a negative number, the graph is inverted. In the function $y = -4x^2$, the parent function has been multiplied by -4. This should invert the parent as well as stretch it by a factor of 4. Figure 10.1.6 is the graph of $y = -4x^2$.

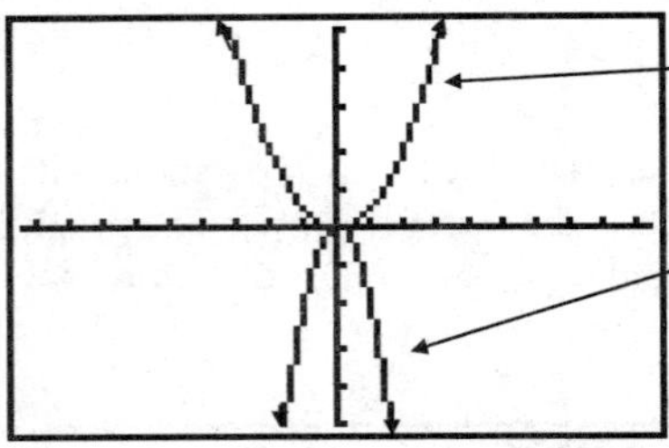

Figure 10.1.6

Figure 10.1.7

The graph demonstrates that the parent does reflect about a horizontal line through the vertex and each point on the new graph is 4 times lower than the graph of $y = -x^2$. You can verify this by graphing both functions on the same window. When trace is used, the function $-4x^2$ will be 4 times lower than the function $-x^2$.

Example 1: Graph $f(x) = -2(x + 4)^2 - 5$.

Solution: The graph is stretched by a factor of 2 and it opens down instead of up because transformations apply to the quadratic function. To get a quick sketch, make the graph of x^2 a little skinnier and put the vertex at $(-4, -5)$. You know the vertex of the parent quadratic function is $(0, 0)$, and since every point on the graph of the transformed function has been shifted left 4 and down 5, the new vertex is $(-4, -5)$.

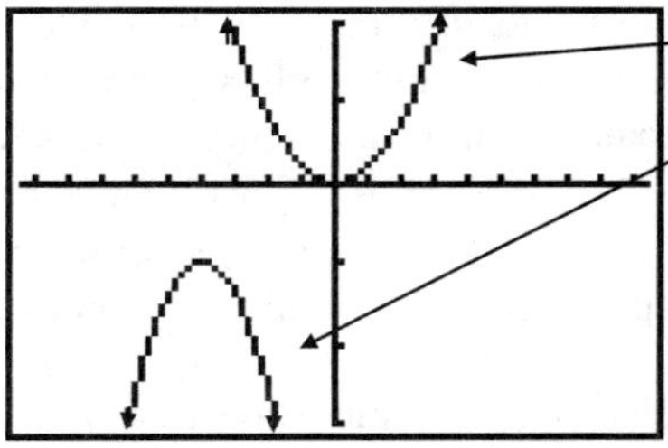

Figure 10.1.8

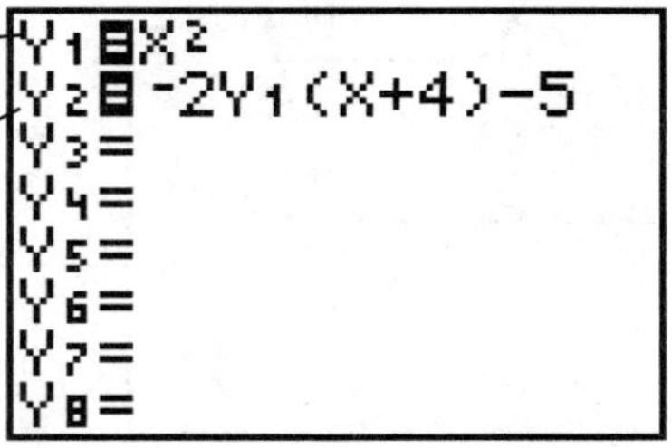

Figure 10.1.9

⁂

Example 2: Graph the function $y = x^2 - 4x + 4$.

Solution: The calculator can be used to graph the function; however, if the function were in the form of the quadratic functions in Chapter Three, standard form, it would be easy to graph without the calculator. The function can easily be put in standard form by factoring the trinomial $x^2 - 4x + 4$. The function becomes $y = (x - 2)^2$. This is the graph of the parent function ($y = x^2$) shifted to the right by 2 units. Since standard form is $y = d(x + e)^2 + f$, you also recall from the study of function parameters in Chapter Three that since d is positive, the graph opens up and $(-e, f)$ are the coordinates of the minimum point on the graph. Draw a graph that opens up, is the same shape as the parent, and has as coordinates of the minimum (2, 0).

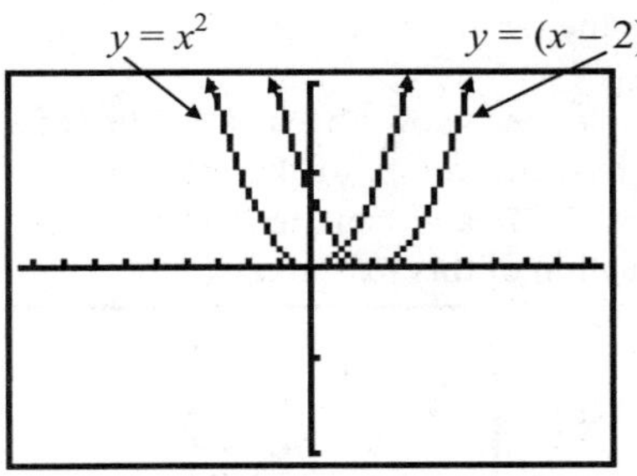

Figure 10.1.10

⁂

The range of a function is the set of all y-coordinates for every x in the domain. The range of any quadratic function can be found if the **vertex** (the minimum or maximum point) of the graph is known. If the vertex of the quadratic function has a y-coordinate of f and the graph opens up, every other y-coordinate is larger than f. This makes the range $[f, \infty)$. The range includes all y-coordinates, including the y-coordinate of the vertex as well as all y-coordinates larger than the y-coordinate of the vertex. See Figure 10.1.11.

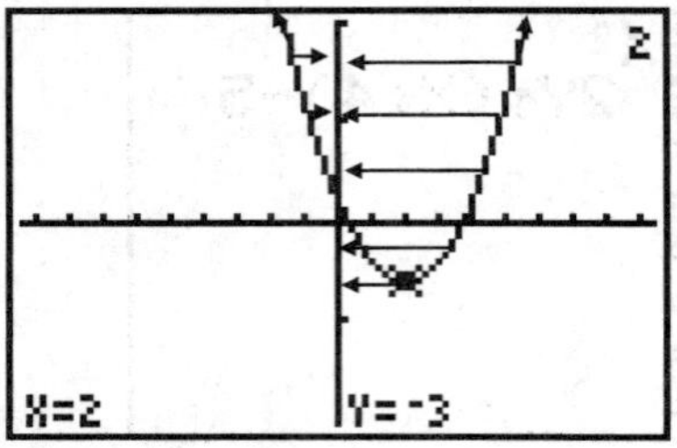

Figure 10.1.11

In the example $y = (x - 2)^2 - 3$, the smallest value of the function is –3 and every other point on the graph has y-values larger than –3; thus, the range is $[-3, \infty)$. In general, the range of $y = d(x+e)^2 + f$ is $[f, \infty)$ provided d is positive. If the vertex of the quadratic function has a y-coordinate of f and the graph opens down, every other y-coordinate on the graph would have to be less than f. This makes the range $(-\infty, f]$. The value of f belongs to the range because f is a y-coordinate -- the y-coordinate of the vertex. See Figure 10.1.12 below where the range is $(-\infty, -3]$.

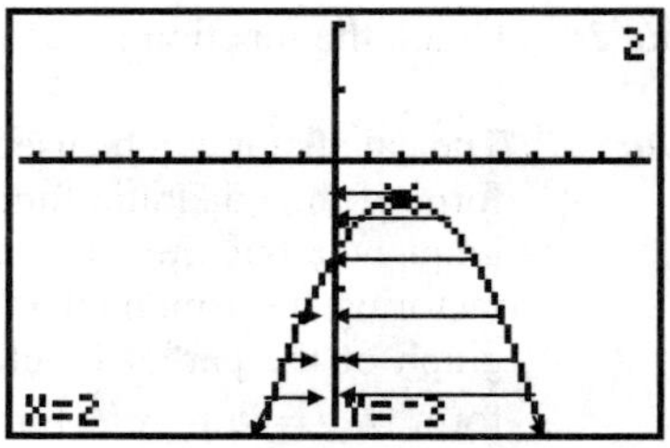

Figure 10.1.12

Example 3: What is the range of the function $y = -(x + 3.16)^2 + 2.77$?

Solution: The graph opens down because of the reflection factor. The shifts are left 3.16 and up 2.77. Thus, the vertex of the graph has a y-coordinate of 2.77. The range is all real numbers 2.77 and smaller. The range is $(-\infty, 2.77]$. See Figure 10.1.13 for confirmation of this reasoning.

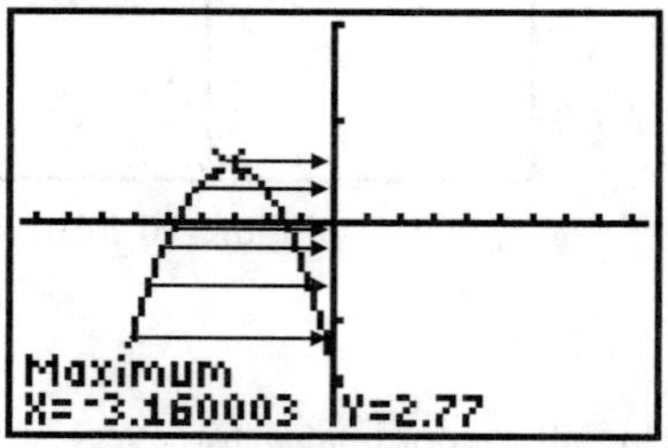

Figure 10.1.13

⁂

Example 4: Find the range of the function $y = x^2 - 6x + 9$.

Solution: The function is not in standard form. The calculator can be used, but it is easily put in standard form. The right side is a perfect square trinomial and factors into $(x - 3)^2$. The function is the parent quadratic function shifted right by 3 units; thus, the vertex is at (3, 0) and the graph opens up. The range of the function is $[0, \infty)$.

⁂

To find the range of the quadratic function analytically, the function should be in standard form. Standard form requires a function has to be in the form $y = d(x + e)^2 + f$; that is, the quadratic function must have a binomial squared times the stretch factor plus the vertical shift. In the examples above, the perfect square trinomial was given, and all that was necessary was to factor the trinomial. To make a perfect square trinomial from a trinomial $x^2 + bx + c$, consider the following examples:

binomial^2		product
$(x - 2)^2$	=	$x^2 - 4x + 4$
$(x + 7)^2$	=	$x^2 + 14x + 49$
$(x - 8)^2$	=	$x^2 - 16x + 64$.

The coefficient of the linear term (the x term) in the <u>trinomial</u> is twice (2 times) the constant in the <u>binomial</u>. The constant in the <u>trinomial</u> is the constant of the <u>binomial</u> squared. If this reasoning is reversed, the coefficient of the linear term of the <u>trinomial</u> can be halved (divided by 2) to get the constant in the <u>binomial</u>.

Square that sucker!

For example:

given: $x^2 + 6x$ — half of 6 squared (9) will give the constant in the trinomial that will make it a perfect square trinomial

$x^2 + 6x + 9$ factors into $(x + 3)^2$

given: $x^2 - 12x$ — half of -12 squared (36) will make the polynomial a perfect square

$x^2 - 12x + 36$ factors into $(x - 6)^2$

Example 5: Put the function $y = x^2 + 2x + 9$ in standard form.

Solution: The polynomial $x^2 + 2x$ can be made a perfect square trinomial by adding 1 (½ of 2, squared). Since you know that adding 1 is a vertical shift of up 1, you must move the graph back down 1 by subtracting 1.

$$y = x^2 + 2x \qquad + 9$$

$$y = x^2 + 2x + 1 \quad + 9 - 1$$

$$y = (x + 1)^2 + 8 \text{ is standard form.}$$

⁂

Example 6: Put the function $f(x) = x^2 - 5x + 3$ in standard form and find the range.

Solution: The polynomial $x^2 - 5x$ can be made a perfect square trinomial by adding $\frac{1}{2}$ of (-5) squared. The calculator keystrokes are:

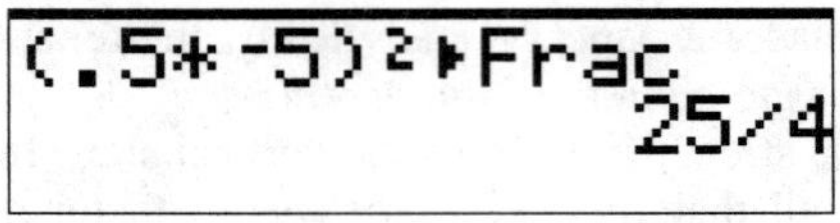

Since this moves the function up $\frac{25}{4}$ you must also subtract $\frac{25}{4}$ from the function.

$$y = x^2 - 5x \qquad\qquad + 3$$

$$y = x^2 - 5x + \frac{25}{4} \qquad + 3 - \frac{25}{4}$$

The first three terms form a perfect square trinomial. Factor them.

$$y = \left(x - \frac{5}{2}\right)^2 + \frac{12}{4} - \frac{25}{4}$$

$$y = \left(x - \frac{5}{2}\right)^2 - \frac{13}{4}$$

Since the graph opens up ($d = 1$) and the minimum is $-\frac{13}{4}$ ($f = -\frac{13}{4}$), the range is $\left[-\frac{13}{4}, \infty\right)$. Confirm this with your calculator using the minimum finder.

⁂

Example 7: Put the function $y = 3x^2 + 12x$ in standard form and find when it is increasing.

Solution: The quadratic term ($3x^2$) must have a coefficient of 1 for the above technique to work. Factor the coefficient of x^2 before completing the polynomial into a perfect square.

$$y = 3x^2 + 12x$$

$$y = 3(x^2 + 4x \quad)$$ complete the square inside the parentheses

$$y = 3(x^2 + 4x + 4) - 12$$ (–12 is needed to compensate for the + 3 times 4 inside the parentheses)

$$y = 3(x + 2)^2 - 12$$

The function opens up and the vertex is at $(-2, -12)$. This means that the function is decreasing until x is -2, then it starts to increase. The function is increasing when $x \in (-2, \infty)$. Confirm this on your calculator.

⁂

Example 8: Find the minimum value of the function $y = 2x^2 - 5x + 3$ analytically.

Solution: The function must be in standard form if you are going to use algebra.

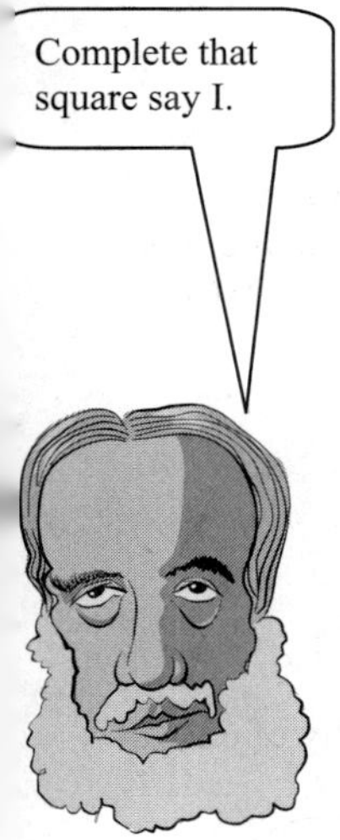

factor 2 from $2x^2 - 5x$: $\quad y = 2\left(x^2 - \frac{5}{2}x + \quad\right) + 3$

complete the square: $\quad y = 2\left(x^2 - \frac{5}{2}x + \frac{25}{16}\right) + 3 - \frac{25}{8}$

$\left(\text{the } -\frac{25}{8} \text{ compensates for } 2 \times \frac{25}{16}\right)$

factor the trinomial and simplify: $\quad y = 2\left(x - \frac{5}{4}\right)^2 + \frac{24}{8} - \frac{25}{8}$

simplify $\quad y = 2\left(x - \frac{5}{4}\right)^2 - \frac{1}{8}$

The stretch factor is positive; thus, the graph opens up. The vertex is at $\left(\frac{5}{4}, -\frac{1}{8}\right)$; therefore, the minimum is $-\frac{1}{8}$.

Comment: The calculator keystrokes for taking $\left(\frac{1}{2} \text{ of } \frac{5}{2}\right)^2$ and $3 - \frac{25}{8}$ are:

⁂

Example 9: When is $y = -x^2 + 4x - 3$ increasing? When is it decreasing? What is the minimum or maximum? What is the range?

Solution: The coefficient of x^2 tells you that the graph opens down; thus, there is a maximum. The vertex is needed to figure when it is increasing or decreasing. It is also needed to find the maximum and the range. All this work can be done with the maximum finder on the calculator or you can do it analytically as shown below.

$$y = -(x^2 - 4x \quad) \quad - 3$$

$$y = -(x^2 - 4x + 4) \quad - 3 \; + \; 4$$

$$y = -(x - 2)^2 + 1$$

The vertex is (2, 1) and the graph opens down. Thus, the function is increasing until x is 2, that is, when $x \in (-\infty, 2)$. The function is decreasing when $x \in (2, \infty)$. Since the maximum value of the function is 1, the range is $(-\infty, 1]$. Confirm all answers with your calculator.

⁂

10.1 EXERCISES

Many times you can work the exercises by algebraic, numeric, or graphical methods. You can learn about mathematics no matter which method you use. Part of the learning process is for you to decide which method is best for you to use. Please read the text carefully before you do the exercises. Read the **next assigned section** after you finish the assignment below. All numeric answers must have an error ≤ 0.01.

A Quick Look To The Future

-4. What value of x makes $x + 2$ zero?

-3. Is $x^2 + 6x + 9$ a perfect square?

-2. What "formula" will solve any equation of the form $dx + e = f$?

-1. Can a quadratic function model the distance traveled by a car moving at a constant velocity of 65 miles per hour?

Previous Topics Revisited

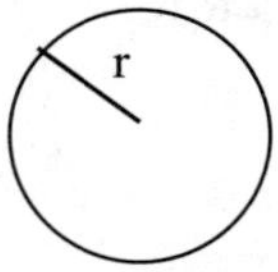

1. The radius of a circle can be found by the function $r = \dfrac{\sqrt{A}}{\sqrt{\pi}}$. This simplifies to $r = 0.5642\sqrt{A}$. Is the function increasing? What is the domain of the function? Find A when $r = 10$ inches.

2. Imagine a coiled spring hanging from the ceiling, with a mass attached to the spring. If the mass is pulled down a small amount and released, it will bounce up and down and have a period of $T = \dfrac{2\pi}{\sqrt{k}}\sqrt{m}$. The period is T seconds for a mass of m. The value of k is the spring constant and is different for each spring. Suppose the spring constant of a certain spring is 0.01. The function then simplifies to $T = 20\pi\sqrt{m}$, or simply $T = 62.83\sqrt{m}$. If a mass of 7 grams is set in motion, what is its period? What is the period for a 14 gram mass?

3. In the above situation, what mass (in grams) can be hung from the spring to maintain a period of 2 seconds?

4. Solve $-\sqrt{2x-6} + 3 = 0$

5. Solve $\sqrt{x+4} + \sqrt{4-x} = 3$

6. Simplify $\left[\left(256^{\frac{1}{2}}\right)^{\frac{1}{2}}\right]^{\frac{1}{2}}$

New Topic Exercises

7. Where is the focus of the parabola $y = 3(x-2)^2 + 5$? (See "The Quadratic Function in Action.")

8. Where is the focus of the parabola $y = 0.01x^2 + 3x - 4$?

9. What is the symbolic representation of any parabola whose focus is 2 units from the vertex?

10. If the focus of a parabola is at (3, 6) and the vertex is at (3, 8), what is the symbolic representation of the parabola?

11. The symbolic representation of the model of the height of a ball thrown straight upward on the planet Saturn is approximately $h = -18.6t^2 + 40t + 6$; provided it is thrown at an initial velocity of 40 feet per second and it leaves your hand when it is 6 feet above the surface. What is the model if the ball leaves your hand at a velocity of 60 feet per second and it is 5 feet from the surface when it leaves your hand? (Note: The surface is a dense gas -- not solid like the Earth. It may be difficult to stand on.)

12. What horizontal and vertical shifts of $y = x^2$ will give the graph of $y = (x + 3)^2 - 4$?

13. What horizontal and vertical shifts of $y = -3x^2$ will give the graph of $y = -3(x - 4)^2 + 2$?

14. Is $y = -5(x + 2)^2 - 3$ the shifted graph of $y = x^2$ or $y = -5x^2$? Why?

15. Where is the vertex of the graph of $y = -1.2(x - 3.6)^2 + 4.7$?

16. Does the graph of $y = 3(x - 4)^2 + 2$ have a zero? (Recall Section 2.5.)

Put the following in standard form and find when each function is increasing and when it is decreasing.

17. $y = x^2 + 18x - 81$

18. $y = -x^2 - 40x + 2$

19. $y = x^2 - 5x + 3$

20. $y = 3x^2 - 12x + 1$

21. $y = 2x^2 + 7x - 5$

Find the range and minimum or maximum.

22. $y = 3(x + 7)^2 - 5$

23. $y = -1.5(x - 2.1)^2 + 4.7$

24. $y = x^2 + 6x - 4.2$

25. $y = -x^2 - 9x + 14$

26. $y = 5x^2 + 2(x - 1)$

27. Find the zeros of $f(x) = 2x^2 + 3x - 5$.

28. Find when $f(x) = 2x^2 + 3x - 5$ is negative.

29. Find when $f(x) = 2x^2 + 3x - 5$ is positive.

30. What is the average rate of change of the function $f(x) = 2x^2 + 3x - 5$ as x changes from 1 to 2?

31. Find the zeros of $f(x) = 2x^2 - 3x - 5$.

32. Find when $f(x) = 2x^2 - 3x - 5$ is negative.

33. Find when $f(x) = 2x^2 - 3x - 5$ is positive.

34. What is the average rate of change of the function $f(x) = 2x^2 - 3x - 5$ as x changes from 1 to 2?

35. Does the graph of $y = 10x^2$ look "wider" or "narrower" than $y = x^2$?

36. Does the graph of $y = 0.1x^2$ look "wider" or "narrower" than $y = x^2$?

37. Does $y = 2 - x^2$ open up or down? Explain your answer.

38. What viewing window makes $y = 0.000001x^2$ "look" like a parabola? There are many answers.

39. What viewing window makes $y = 5000000x^2$ "look" like a parabola? There are many answers.

40. Give an example of a quadratic function whose zeros are 2 and π.

41. Find any quadratic function with a range of $[-3, \infty)$.

42. Find any quadratic function with a range of $(-\infty, -3]$.

43. Find any quadratic function with a maximum of 7.

44. Find any quadratic function with a minimum of 7.

45. Find any quadratic function that changes from increasing to decreasing at -2.

46. Find any quadratic function that changes from decreasing to increasing at -2.

47. Find any quadratic function that has a range of $(-\infty, 3]$ and changes from increasing to decreasing at 4.

48. If a point on the graph of $f(x)$ is $(-3, 5)$, what are the coordinates of the transformed point on the graph of $f(x) + 2$?

49. If a point on the graph of $f(x)$ is $(4, -3)$, what are the coordinates of the transformed point on the graph of $f(x) - 2$?

50. If a point on the graph of $f(x)$ is $(5, 1)$, what are the coordinates of the transformed point on the graph of $4f(x)$?

51. If a point on the graph of $f(x)$ is $(5, 1)$, what are the coordinates of the transformed point on the graph of $-4f(x)$?

From Mathematics to English

52. Describe the difference between a vertical shift of 7 and a stretch of 7.

53. What is your definition of a quadratic function?

54. Why doesn't every quadratic function have real zeros?

55. Describe a data relationship that can be modeled with a quadratic function.

56. Explain why the graph of every quadratic function has a y-intercept.

57. Explain why the graphs of quadratic functions may or may not have an x-intercept.

58. After reading this section, make a list of questions that you want to ask your instructor.

59. Continue in your daily journal and make an entry. In addition to your normal entry on thoughts about the mathematics in this section, list at least two positive comments about what you have learned about this topic.

60. In paragraph format, summarize the material in this section of the text in your daily journal.

61. Describe how your classroom instructor made this topic more understandable and clear.

62. After reading the text and listening to your instructor, what do you not understand about this topic?

63. Describe the symmetry of the graph of $y = d(x + e)^2 + f$.

Explorations

64. Find a symbolic representation of any parabola that opens to the right or left.
Graph it on the graphing calculator.

65. Graph $y = x^2 + bx + 3$, where $b = 1$, then 2, then 3, and finally 4, all on the same window.

Are all the graphs **congruent** (identical shape and size)? Why?

Why do all of the above graphs cross the y-axis at 3?

In the function $y = 1x^2 + bx + 3$, describe what b does to the graph of the function, and what b does to the zeros of the function.

66. Describe the horizontal shift of $y = x^2 + 3$ for each of the following graphs:

a. $y = (x - 2)^2 + 3$

b. $y = (3x - 6)^2 + 3$

c. $y = (5x - 10)^2 + 3$

d. $y = (6 - 3x)^2 + 3$

67. Describe what the absolute value transformation does to each graph below.

a. Describe the graph of $y = |-x^2|$ as compared to the graph of $y = -x^2$.

b. Describe the graph of $y = |-x^2 - 5|$ as compared to the graph of $y = -x^2 - 5$.

c. Describe the graph of $y = |-(x-4)^2|$ as compared to the graph of $y = -(x-4)^2$.

d. Describe the graph of $y = \left|\frac{1}{2}(x+2)^2 - 3\right|$ as compared to the graph of $y = \frac{1}{2}(x+2)^2 - 3$.

In general, describe what the absolute value transformation does to the graph of every quadratic function.

68. Describe what the squaring transformation does to the graph of each function below.

a. Describe the graph of $y = (-x^2)^2$ as compared to the graph of $y = -x^2$.

b. Describe the graph of $y = (x^2 - 5)^2$ as compared to the graph of $y = x^2 - 5$.

c. Describe the graph of $y = (-(x-4)^2)^2$ as compared to the graph of $y = -(x-4)^2$.

d. Describe the graph of $y = \left[\frac{1}{2}(x+2)^2 - 3\right]^2$ as compared to the graph of $y = \frac{1}{2}(x+2)^2 - 3$.

In general, describe what the squaring transformation does to the graph of every quadratic function.

10.2 Solving Quadratic Equations of the Form $(ax + b)(cx + d) = 0$

When is the Ball on the Ground?

Continuing with the ball problem started in the last section -- when does the ball strike the ground? If you toss a ball straight upward with an initial velocity of 16 feet per second and it leaves your hand when it is 5 feet above the ground, the mathematical model of the height of the ball is $h = -16t^2 + 16t + 5$ (ignoring air resistance). You want to know when the height h is 0. Since h is $-16t^2 + 16t + 5$, you can express the problem as the statement $-16t^2 + 16t + 5 = 0$. You recognize this as an equation. If you review Section 4.5, you know that if you can factor the function $-16t^2 + 16t + 5$, you also know when it is zero. That is, $-16t^2 + 16t + 5$ can be written as $-(16t^2 - 16t - 5)$ or $-(4t - 5)(4t + 1)$. Back to the question -- you want to know when $-(4t - 5)(4t + 1)$ equals 0. Section 4.5 tells you that the function is zero when t is $\frac{5}{4}$ or $-\frac{1}{4}$. Finally, the ball hits the ground at 1.25 seconds $\left(1.25 = \frac{5}{4}\right)$ after it leaves your hand.

The real number solutions to any quadratic equation can be found graphically. Recall that the real solutions to the *quadratic equation* $ax^2 + bx + c = 0$ are the zeros of the related *quadratic function* $y = ax^2 + bx + c$. That is, the zeros of the quadratic function are values for x that cause the function to be zero. Any function has a value of zero on the x-axis. Thus, the real solutions to a quadratic equation are the x-intercepts of the related quadratic function.

The particular interest in this section of the text is; however, solving equations of the form $(ax + b)(cx + d) = 0$. Consider the zeros of the function $y = (1x + 4)(1x - 3)$ in the graph below.

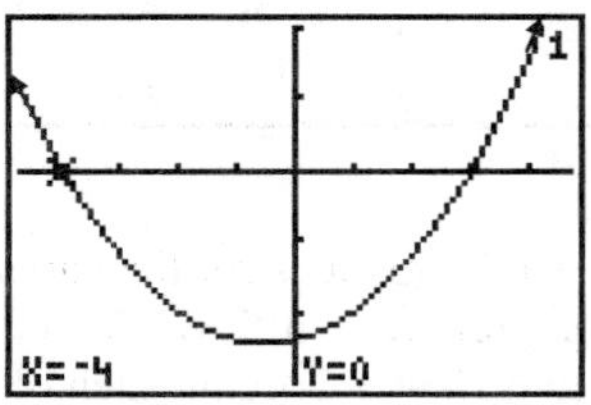

Figure 10.2.1a

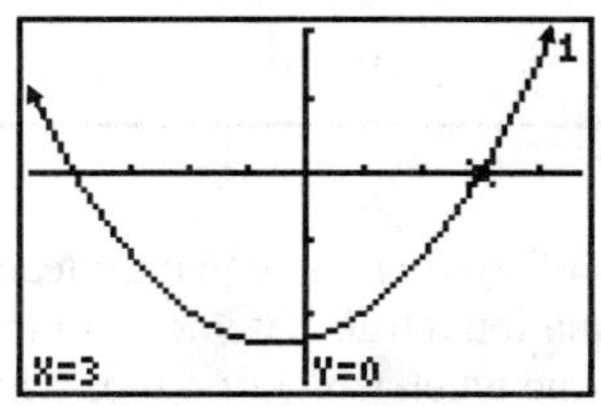

Figure 10.2.1b

The zeros of the function $y = (1x + 4)(1x - 3)$ and the solutions to the equation $(1x + 4)(1x - 3) = 0$ are -4 and 3, as shown above in Figure 10.2.1. The zeros of the function $y = (1x - 4)(1x + 3)$ and the solutions to the equation $(1x - 4)(1x + 3) = 0$ are 4 and -3. They are displayed in Figure 10.2.2 below.

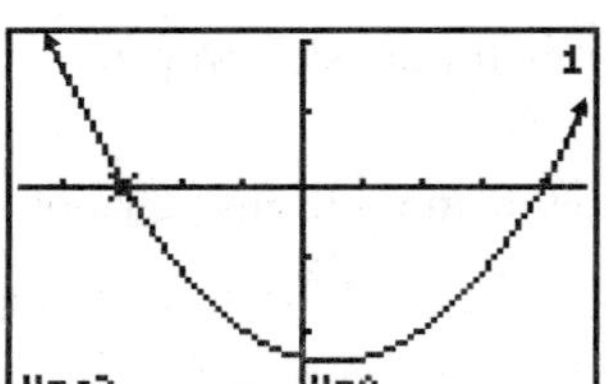

Figure 10.2.2a

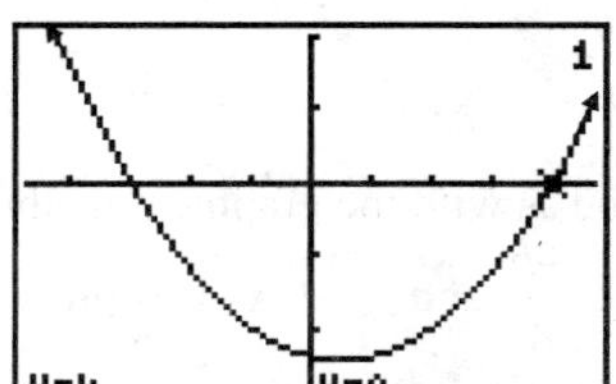

Figure 10.2.2b

The solutions to the equation $(2x + 5)(1x - 6) = 0$ are displayed below as the zeros of the function $y = (2x + 5)(1x - 6)$. (Negative 2.5 is $-\frac{5}{2}$).

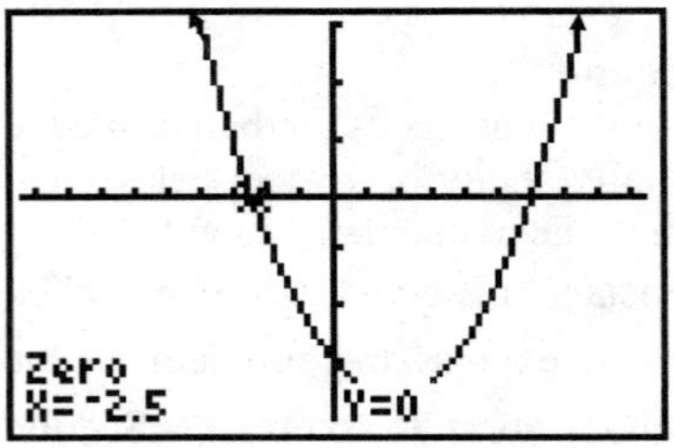

Figure 10.2.3a

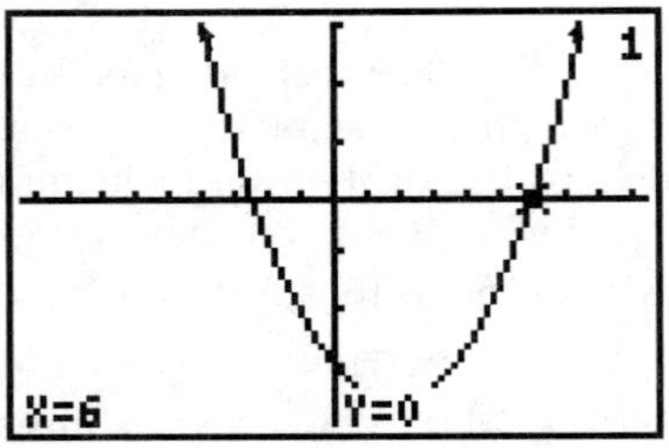

Figure 10.2.3b

As noted in the above examples, there is nothing new in the way the graphical method is used to solve quadratic equations. Next you will look at an analytical method for solving quadratic equations.

As was developed in Section 4.5 and again demonstrated above, the solutions to an equation of the form $(ax + b)(cx + d) = 0$ are $\frac{-b}{a}$ and $\frac{-d}{c}$. This conclusion is generalized below.

Zero Property of Equality

> If $(ax + b)(cx + d) = 0$, then the solutions to the equation are $\frac{-b}{a}$ and $\frac{-d}{c}$.

Proof:

The process of solving any equation leads to finding values for the variable that make the equation a true statement. In the equation $(ax + b)(cx + d) = 0$, the equation is true when $(ax + b) = 0$ or when $(cx + d) = 0$. Solving each of these equations yields the following results:

$ax + b = 0 \qquad cx + d = 0$

$ax = -b \qquad cx = -d \qquad$ by the subtraction property

$x = \frac{-b}{a} \qquad x = \frac{-d}{c} \qquad$ by the division property

The results are the same as with the graphical method; the solutions to <u>any equation</u> of the form $(ax + b)(cx + d) = 0$ are $\frac{-b}{a}$ and $\frac{-d}{c}$! As was presented in Section 4.5, the zeros of <u>any function</u> of the form $y = (ax + b)(cx + d)$ are $\frac{-b}{a}$ and $\frac{-d}{c}$!

Example 1: Solve the equation $(3x + 7)(2x - 9) = 0$

Solution: Algebraically, the solutions are $\frac{-7}{3}$ and $\frac{-(-9)}{2}$. The solution set is $\left\{\frac{-7}{3}, \frac{9}{2}\right\}$.

⁂

Example 2: Solve the equation $3x^2 + 7x - 15 = 5$ analytically, and check graphically.

Solution: To solve analytically, the one side must be zero and the other side must be in factored form.

$3x^2 + 7x - 15 = 5$ given
$3x^2 + 7x - 20 = 0$ subtraction property of equality
$(3x - 5)(x + 4) = 0$ factorization

$x = \frac{5}{3}$ and $x = -4$ zero property of equality

To check graphically with the zeros method, graph the function $y = 3x^2 + 7x - 20$ and locate the zeros.

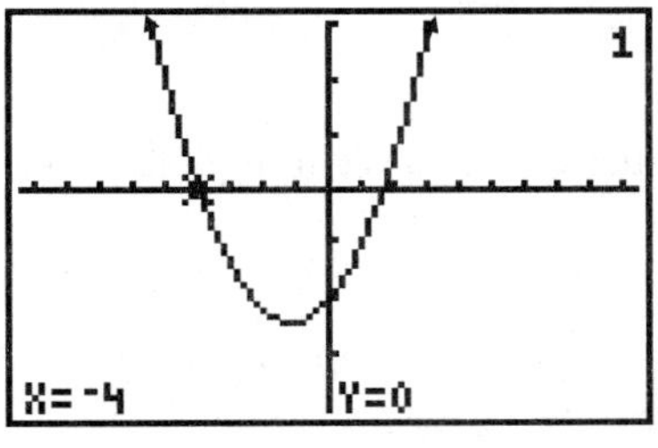

Figure 10.2.4a

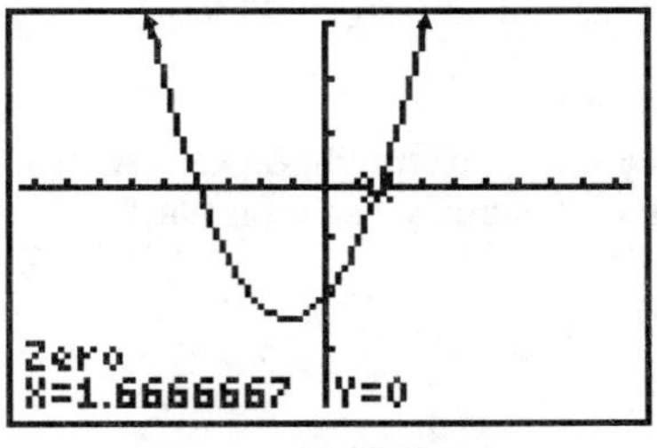

Figure 10.2.4b

⁂

You now have the first of three non-technology based methods for solving quadratic equations -- the factoring method. A word of caution however, this method does not work if the function in the equation does not factor. You will find that many quadratic equations cannot be solved with this method. As you will see in Section 10.4, there are also quadratic equations that cannot be solved by the technology based graphical and numerical methods either.

10.2 EXERCISES

Many times you can work the exercises by algebraic, numeric, or graphical methods. You can learn about mathematics no matter which method you use. Part of the learning process is for you to decide which method is best for you to use. Please read the text carefully before you do the exercises. Read the **next assigned section** after you finish the assignment below. All numeric answers must have an error ≤ 0.01.

A Quick Look To The Future

-4. Is the left side of the equation $x^2 - 8x + 16 = 3$ a perfect square?

-3. Do you suppose there is a "formula" for solving the equation $ax^2 + bx + c = 0$?

-2. Could the quadratic function be used to model the world population of humans?

-1. What number is half-way between 2 and 12?

Previous Topics Revisited

1. What is the range of the function $y = x^2 - 5x + 4$?

2. Put $y = 2x^2 + 14x - 6$ in standard form. What are the coordinates of the vertex of the parabola?

3. What translation of the graph of $y = x^2$ will yield the graph of $y = (x - 5)^2 + 2$?

Heron's formula for the area of any triangle with sides a, b, and c is $A = \sqrt{s(s-a)(s-b)(s-c)}$, where $s = \frac{1}{2}(a+b+c)$.

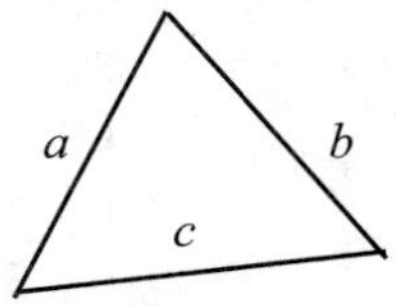

4. Simplify the area expression when $a = 4$, $b = 6$, and $c = 8$.

5. Simplify the area expression when $a = 6$, $b = 10$, and $c = 12$.

6. Solve the equation $x - \sqrt{6x+37} = -5$.

New Topic Exercises

Find when the ball thrown straight upward with an initial velocity of v_0 from h_0 feet above the ground, strikes the ground. Assume you are on Earth.

7. $v_0 = 2$, $h_0 = 5$
8. $v_0 = 43$, $h_0 = 15$
9. $v_0 = 62$, $h_0 = 8$
10. $v_0 = 78$, $h_0 = 10$
11. Using the data from Exercise 7, find when the ball is 2 feet above the ground.
12. Using the data from Exercise 9, find when the ball is 54 feet above the ground.
13. Using the data from Exercise 10, find when the ball is 45 feet above the ground.

Solve the equations.

14. $(2x - 5)(3x + 4) = 0$
15. $(15x - 13)\left(\frac{1}{2}x + 8\right) = 0$
16. $(x + 1)(x - 3) = -4$
17. $(x - 2)(x - 5) = -2$
18. $\left(3x - \frac{1}{4}\right)\left(2x + \frac{1}{2}\right) = 0$
19. $\left(\frac{2}{3}x + 1\right)\left(7x + \frac{2}{5}\right) = 0$
20. $x^2 - 8x = -15$
21. $x^2 + 2x + 5 = 0$
22. $(x - 2)(x + 3)(2x - 5)(x + 4) = 0$
23. $4(x + 1)(x - 3)\left(x + \frac{1}{2}\right) = 0$
24. $x^2 - 16 = 0$
25. $(x + 3)^2 = 0$
26. $2(x + 6)^2 = 0$
27. $(x + 1)^2 = 4$
28. Find an equation that has –2, 4, –3, and 1 as solutions.

29. Find a quadratic equation that has only one rational solution. How many times does the related function intercept the x-axis?

30. Find a quadratic equation that has only one irrational solution (irrational constants and coefficients are acceptable). Does the graph of the related function intercept the x-axis? If so, how many times?

31. Write an equation in general form ($ax^2 + bx + c = 0$) that has solutions of π and $\sqrt{3}$.

32. Find four equations that have $\frac{2}{3}$ and $-\frac{1}{5}$ as solutions.

From Mathematics to English

33. Find any quadratic equation that has no real solution. Does the graph of the related quadratic function have any real zeros? Why or why not?

34. Find a quadratic equation that has no real solution. Does the graph of the related function intercept the x-axis? Why?

35. How many solutions does the equation $(x - 1)(x - 1)(x - 1)(x - 1) = 0$ have? Explain your answer.

36. Describe how can you solve a quadratic equation that is not factorable?

37. After reading this section, make a list of questions that you want to ask your instructor.

38. Continue in your daily journal and make an entry. In addition to your normal entry on thoughts about the mathematics in this section, list at least two positive comments about what you have learned about this topic.

39. In paragraph format, summarize the material in this section of the text in your daily journal.

40. Describe how your classroom instructor made this topic more understandable and clear.

41. After reading the text and listening to your instructor, what do you not understand about this topic?

Explorations

42. Make a list of five quadratic equations of the form $ax^2 + bx + c = 0$, where the numbers a, b, and c are chosen at random. How many of the above equations have rational solutions? Let a, b, and c be integers between 1 and 10, and list another five equations. How many of these have rational solutions? What conclusions can you draw from this exercise?

43. Describe the behavior of the graph of each function near the zeros of the function.
    a. $y = (x - 1)^1 (x + 3)^2$
    b. $y = (x - 1)^2 (x + 3)^1$
    c. $y = (x - 1)^2 (x + 3)^2$
    d. $y = (x - 1)^2 (x + 3)^3$
    e. $y = (x - 1)^3 (x + 3)^2$

    Find the symbolic representation of any function with a graph that passes through the x-axis at -2, 3, and $\frac{3}{5}$, and only touches (is tangent to) the x-axis at 5.

10.3 Solving Quadratic Equations by the Completing-the-Square Method

A Look Back at the Soda Pop Can Problem

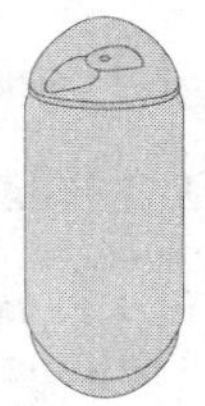

Section 3.2 starts with a model of the volume of the soda in a pop can. You may recall that the volume of the soda increases as the temperature rises above 4° C (This is 39.2° F.), and the volume increases as the temperature falls below 4° C. This is certainly an unusual behavior. Further, the model for the volume is $V = 0.003(T-4)^2 + 354.8824$ when the volume is measured in milliliters. A packaging engineer must be able to answer questions like "At what temperature is the volume 354.9 ml?" What she is asking is: "What is the solution to the equation $0.003(T-4)^2 + 354.8824 = 354.9$?" Not many engineers would ever try to solve this equation by the factoring method. It may be factorable but no one wants to try because of the decimals involved. Below is the simplification of the equation -- just to see what it looks like.

$0.003(T-4)^2 + 354.8824 = 354.9$	square the binomial
$0.003\left(T^2 - 8T + 16\right) + 354.8824 = 354.9$	multiply both sides be 10,000
$30\left(T^2 - 8T + 16\right) + 3548824 = 3549000$	simplify the left side
$30T^2 - 240T + 480 + 3548824 = 3549000$	simplify the left side again
$30T^2 - 240T + 3549304 = 3549000$	subtract 3549000 from both sides
$30T^2 - 240T + 304 = 0$	divide both sides by 2
$15T^2 - 120T + 152 = 0$	The quadratic function on the left is not factorable. The factoring method cannot be used.

The point of the argument is that most equations that come in the context of a real-world problem cannot be solved by the factoring method. What does the engineer do? Before the computer and sophisticated calculator were invented, she would use the method presented in this section or the next. Below is the development of the completing-the-square method of solving the equation. The process of "completing the square" is a technique used in Section 10.1 to put the quadratic function in standard form. Here, it will be used to help solve a quadratic equation.

The equation $x^2 = 4$ can be solved by factoring.

$x^2 - 4 = 0$	subtraction property of equality
$(x-2)(x+2) = 0$	factoring
$x = 2 \quad \mid \quad x = -2$	zero property of equality

The equation $(x-3)^2 = 16$ can be solved by factoring.

$(x-3)^2 - 16 = 0$	subtraction property of equality
$[(x-3)+4][(x-3)-4] = 0$	factoring

$(x-3)+4=0$	$(x-3)-4=0$	zero property of equality
$x-3+4=0$	$x-3-4=0$	associative property
$x=-1$	$x=7$	simplification

OR

$(x-3)^2-16=0$	subtraction property of equality
$x^2-6x+9-16=0$	simplification
$x^2-6x-7=0$	simplification
$(x-7)(x+1)=0$	factoring
$x=7 \quad x=-1$	zero property of equality

While $x^2 = 4$ and $(x - 3)^2 = 16$ can be solved by factoring, another method will also work. This new method involves taking the square root of both sides of the equation. For example, from the equation $x^2 = 4$, taking the square root of both sides appears to yield only one solution, as shown below in the series of "equivalent" equations.

$$x^2 = 4$$
$$\sqrt{x^2} = \sqrt{4}$$
$$x = 2.$$

From above, the factoring method for solving $x^2 = 4$ gave two solutions, 2 and –2. Checking both 2 and –2 in the original equation confirms that both are solutions. This leads to the conclusion that there must be an error in taking the square root of both sides! The $\sqrt{4}$ is definitely 2 but, the $\sqrt{x^2}$ is not x. The square root of x^2 is x only if $x \geq 0$; for example,

$$\sqrt{3^2} = 3 \qquad \text{because } 3^2 = 9 \text{ and } \sqrt{9} = 3.$$

The square root of $(-2)^2$ is NOT –2, that is,

$$\sqrt{(-2)^2} \neq -2 \quad \text{because } \sqrt{(-2)^2} = \sqrt{2^2} = 2.$$

The square root of a number is zero or larger. A mathematical function that makes a number always be zero or larger is the *absolute value function*. This leads to the conclusion:

Square Root of x^2

$$\sqrt{x^2} = |x|$$

Square Root Property of Equality

$$\text{If } x^2 = a, \text{ then } |x| = \sqrt{a}$$

You can solve the equation $x^2 = 4$ by taking the square root of both sides.

$x^2 = 4$	given
$\|x\| = \sqrt{4}$	square root property of equality
$\|x\| = 2$	simplification
$x = 2$ or $x = -2$	absolute value (See Section 6.3.)

Check: $|2| = 2$ and $|-2| = 2$.

The short-cut notation used to denote positive *and* negative 2 is ± 2. Positive and negative 5 are written ± 5. When the symbol is written between two numbers it means plus and minus.

To solve $(x - 3)^2 = 16$, take the square root of both sides.

$(x - 3)^2 = 16$	given
$\|x - 3\| = 4$	by the square root property of equality
$x - 3 = 4$ or $x - 3 = -4$	absolute value (See Section 6.3.)
$x - 3 = 4 \quad x - 3 = -4$	add 3 to both sides
$x = 7 \quad x = -1$	

The statement $|x| = 2$ is true when $x = \pm 2$. The statement $|x - 3| = 4$ is true when $x - 3 = \pm 4$. It appears from these examples that instead of replacing $\sqrt{x^2}$ with $|x|$, the square root of the constant can be replaced with ± in front of the $\sqrt{\ }$. A common short-cut used when taking the square root of both sides of an equation is:

Square Root Property (Abbreviated Notation)

$$\text{If } x^2 = a, \text{ then } x = \pm\sqrt{a}$$

Example 1: Find the exact solutions to $x^2 = 49$.

Solution: Use the abbreviated version of the square root property of equality.

$$x^2 = 49$$

$$\sqrt{x^2} = \pm\sqrt{49}$$

$x = \pm 7$

Example 2: Find the exact solutions to $(x - 5)^2 = 9$.

Solution: Use the abbreviated version of the square root property of equality.

$$(x - 5)^2 = 9$$

$$\sqrt{(x-5)^2} = \pm\sqrt{9}$$

$$x - 5 = \pm 3 \qquad \text{add 5 to both sides}$$

$$x = 5 \pm 3$$

$$x = 5 + 3 \text{ and } x = 5 - 3$$

$$x = 8, x = 2$$

Now that the square root property has been developed, solving a quadratic equation by completing-the-square follows.

Example 3: Find the exact solutions to $x^2 + 6x - 2 = 0$.

Solution: The equation cannot be solved by factoring. The following method is called **completing-the-square**.

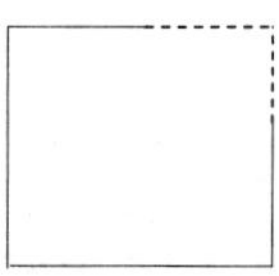

$x^2 + 6x - 2 = 0$	given
$x^2 + 6x \quad = 2$	addition property
$x^2 + 6x + 9 = 2 + 9$	complete the square by adding 9 to both sides of the equation
$(x + 3)^2 = 11$	factor the perfect square
$x + 3 = \pm\sqrt{11}$	square root property
$x = -3 \pm\sqrt{11}$	subtraction property

$$x = -3 + \sqrt{11} \text{ and } -3 - \sqrt{11}$$

The exact solutions are $-3 + \sqrt{11}$ and $-3 - \sqrt{11}$. The graphical based methods for solving this equation will not give the exact solution, but an approximate solution. You are strongly encouraged to solve this equation graphically to confirm the answers.

Example 4: Find the exact solutions to $x^2 - 3x - 5 = 0$.

Solution: The equation cannot be solved by factoring. Solving by the graphical method will not necessarily yield exact solutions. The completing-the-square method will work.

$x^2 - 3x - 5 = 0$	given
$x^2 - 3x \quad = 5$	addition property
$x^2 - 3x + \frac{9}{4} = 5 + \frac{9}{4}$	complete the square
$\left(x - \frac{3}{2}\right)^2 = \frac{29}{4}$	factor and simplify
$x - \frac{3}{2} = \pm\frac{\sqrt{29}}{2}$	square root property
$x = \frac{3}{2} \pm \frac{\sqrt{29}}{2}$	addition property

$$x = \frac{3}{2} + \frac{\sqrt{29}}{2} \text{ and } x = \frac{3}{2} - \frac{\sqrt{29}}{2}$$

Since both fractions have the same denominator, add fractions.

$$x = \frac{3 + \sqrt{29}}{2} \text{ and } x = \frac{3 - \sqrt{29}}{2}$$

Please confirm that the answers are correct by solving the equation by a technology-based method.

Example 5: Find the exact solutions to $2x^2 + 8x - 1 = 0$.

Solution: Solve by completing the square.

$$2x^2 + 8x - 1 = 0$$

$$2x^2 + 8x \quad = 1$$

$$2(x^2 + 4x \quad) = 1$$

Add 4 inside parentheses, but it is being multiplied by 2, so add 8 to the right side.

$$2(x^2 + 4x + 4) = 1 + 8$$

$$2(x + 2)^2 = 9$$

$$(x + 2)^2 = \frac{9}{2}$$

$$x+2=\pm\frac{3}{\sqrt{2}} \qquad \left(\frac{3}{\sqrt{2}}=\frac{3\sqrt{2}}{2}\right)$$

$$x=-2\pm\frac{3\sqrt{2}}{2}$$

$$x=\frac{-4}{2}\pm\frac{3\sqrt{2}}{2}$$

$$x=\frac{-4\pm3\sqrt{2}}{2}$$

$$x=\frac{-4+3\sqrt{2}}{2} \text{ and } \frac{-4-3\sqrt{2}}{2}$$

⁂

Example 6: Solve the pop can equation $15T^2-120T+152=0$.

Solution: As you know, much of the work has already been done. Proceed with the method.

$15T^2-120T+152=0$ — subtract 152 from both sides

$15T^2-120T = -152$ — divide both sides by 15

$T^2-8T = -\frac{152}{15}$ — complete the square by adding 16 to both sides

$T^2-8T+16 = 16-\frac{152}{15}$ — factor the left and simplify the right side (Use your calculator!)

$(T-4)^2 = \frac{88}{15}$ — use the square root property

$T-4 = \pm\sqrt{\frac{88}{15}}$ — add 4 to both sides

$$T=4\pm\sqrt{\frac{88}{15}}$$

As you can see, there are two temperatures that will cause the pop to expand to 354.9 *ml*.

⁂

The previous examples pose an unusual circumstance. What is the solution to the equation if the number under the square root symbol is a negative number? What you know is that the square root of a negative number is not real; thus, the solutions to the equation will not be real. If the solutions are not real, what are they? In the next section of the text you will look at the square root of a negative number. Once the square root of a negative number has been defined, equations containing square roots of negative numbers can be solved.

10.3 EXERCISES

Many times you can work the exercises by algebraic, numeric, or graphical methods. You can learn about mathematics no matter which method you use. Part of the learning process is for you to decide which method is best for you to use. Please read the text carefully before you do the exercises. Read the **next assigned section** after you finish the assignment below. All numeric answers must have an error ≤ 0.01.

A Quick Look To The Future

-4. Does $\dfrac{1+3}{2}$ equal $1+\dfrac{3}{2}$?

-3. When downloading software from an Internet site, you quite often see that the data is being transferred to your computer at a constant rate. Could you use a quadratic function to model the time remaining until the software it totally downloaded?

-2. Can you find the length of a line segment without knowing where it starts and where it stops?

-1. How many sides does a triangle have?

Previous Topics Revisited

1. Solve $(2x - 5)(3x + 7) = 0$.
2. Solve $(2x + 3)(3x - 1)(3 + x)(5x - 7) = 0$.
3. Solve $3x^2 - 10x = 8$.
4. What are the coordinates of the minimum point for the function $y = x^2 - 5x + 2$?
5. When is $y = -2(x + 1)^2 - 3$ increasing? Decreasing?
6. The distance d from a point h kilometers above the earth's surface to the horizon is $\sqrt{12800h + h^2}$. In function form, $d = \sqrt{12800h + h^2}$. Graph the function on the interval [0, 20] for h. What is the distance to the horizon if you are in an airplane flying at an altitude of 33,000 feet? (33,000 feet ≈ 10.06 *km*)

New Topic Exercises

Solve the following equations analytically and check them graphically.

7. $(x - 2)^2 = 121$
8. $(2x - 5)^2 = 1$
9. $(x + 6)^2 - 16 = 0$
10. $x^2 + 4x - 3 = 0$
11. $x^2 - 10x = -25$
12. $x^2 + 5x - 1 = 0$
13. $x^2 - 3x + 2 = 0$
14. $2x^2 + 6x - 1 = 0$
15. $2x^2 - 5x + 1 = 0$
16. $4x^2 - 20x + 25 = 0$
17. $9x^2 + 48x = -64$

From Mathematics to English

18. Describe the major differences between the answers to equations that can be solved by factoring and those quadratic equations that cannot be solved by factoring?

19. What does completing-the-square mean to you?

20. Describe how to solve the equation $x^2 + 5x - 3 = 0$ by completing-the-square method. Do not actually solve the equation.

21. List the methods you can use to solve the equation $x^2 + 5x - 3 = 0$.

22. Using the completing-the-square method for solving the quadratic equation $-2(x^2 + 6x + \quad) = 5$, 9 is added inside the parentheses on the left side of the equation. To satisfy the addition property of equality, what is added to the right side of the equation? Why?

23. If you developed a "cube root property" similar to the square root property shown in this section, will a "±" symbol precede the cube root on the right side of $x^3 = 27$ when it is solved? Why?

24. In reference to your above cube root property, will the left side of the equation be $|x|$ when using your cube root property? Why?

25. Describe the differences between putting the quadratic function in standard form and solving a quadratic equation by the completing-the-square method.

26. Describe how to solve $0.2x^2 - 0.3x - 1.5 = 0$. Do not actually solve the equation.

27. Find the real solutions to $x^4 = 16$. What property is used to solve the equation?

28. After reading this section, make a list of questions that you want to ask your instructor.

29. Continue in your daily journal and make an entry. In addition to your normal entry on thoughts about the mathematics in this section, list at least two positive comments about what you have learned about this topic.

30. In paragraph format, summarize the material in this section of the text in your daily journal.

31. Describe how your classroom instructor made this topic more understandable and clear.

32. After reading the text and listening to your instructor, what do you not understand about this topic?

Explorations

33. Develop an algorithm (a series of steps) that can be used to solve a fourth degree equation of the form $ax^4 + bx^2 + c = 0$. Assume the left side of the equation cannot be factored.

34. What conclusions can you draw about the zeros of a quadratic function that cannot be factored? What about one that can be factored?

10.4 Solving Quadratic Equations with the Quadratic Formula

In this section, you will learn how to solve quadratic equations by the last algebraic method. You will also learn to solve equations that do not have real numbers for solutions; the solutions are called **complex numbers**. All of the exercises and sample problems in previous sections have been designed to avoid non-real solutions. In this section, non-real solutions to quadratic equations are addressed. Before quadratic equations are solved with the quadratic formula, however, a brief study of complex numbers is necessary.

Consider the equation $x^2 = -1$. A solution to the equation is a number that must be squared to get the result of -1. There is no real number that can be squared to equal -1. Then what can be squared to obtain -1? This question is much like asking a beginning fifth grader the question, "What is 7 – 12?" The fifth grader has never heard of negative integers. To the fifth grader, there is no answer until the teacher explains the negative integers. Once the student understands what a negative number is, the problem 7 – 12 has a solution. It is entirely possible that you have never heard of a number that can be squared to get -1. Whatever the case, here is the definition of a number that can be squared to get -1.

> **Definition of *i***
>
> ***i*** is a number that can be squared to get -1. That is, $i^2 = -1$.

Now the equation $x^2 = -1$ can be solved. Use the abbreviated square root property from Section 10.3 and take the square root of both sides.

$$x^2 = -1$$

$$x = \pm\sqrt{-1}$$

$$x = \pm i$$

The problem now becomes how to solve an equation like $x^2 = -4$. You know that there is no real number that you can square to get -4. Does this mean that the equation has no solution? No. It means that you need to learn about another kind of number -- the imaginary number.

> **Definition of an Imaginary Number**
>
> ***bi*** is an imaginary number where b is real.

The equation $x^2 = -4$ can now be solved. Use the abbreviated square root property from Section 10.3.

$$x^2 = -4$$

$$x = \pm\sqrt{-4}$$

$$x = \pm\sqrt{4}\sqrt{-1}$$

$$x = \pm 2i$$

The last problem is to solve an equation like $(x + 3)^2 = -9$. It has answers that are not imaginary numbers nor are the answers real numbers. The square root property is applied below.

$$(x + 2)^2 = -9$$

$$x + 2 = \pm\sqrt{-9}$$

$$x + 2 = \pm\sqrt{9}\sqrt{-1}$$

$$x + 2 = \pm 3i \qquad \text{subtract 2 from both sides}$$

$$x = -2 \pm 3i$$

If you are going to solve quadratic equations, you will need to use this new number.

Definition of a Complex Number

> ***a* + *bi*** is a complex number, where a and b are real numbers.
> a is called the **real part** and b is called the **imaginary part**.

In order to solve equations with complex solutions, a brief discussion of complex numbers follows. The operations of addition or subtraction of complex numbers is very similar to adding or subtracting polynomials, as you will see in the examples. Multiplication of complex numbers is much like multiplying binomials. Division is similar to dividing a real number by an irrational number in binomial form. The rules that follow can be used to perform the operations, but you may want to consider using methods developed in the examples.

Addition of Complex Numbers

$$(a + bi) + (c + di) = (a + c) + (b + d)i$$

(Add the real parts and the imaginary parts.)

Quickie Example: $(-2 + 5i) + (6 + 3i)$ is $(-2 + 6) + (5 + 3)i$, which simplifies to $4 + 8i$.

Subtraction of Complex Numbers

$(a + bi) - (c + di) = (a - c) + (b - d)i$
(Subtract the real parts and the imaginary parts.)

Quickie Example: $(-5 + 7i) - (-3 - 4i)$ is $(-5 - (-3)) + (7 - (-4))i$, which simplifies to $-2 + 11i$.

Multiplication of Complex Numbers

$(a + bi)(c + di) = (ac - bd) + (ad + bc)i$

Quickie Example: $(1 - 2i)(-3 + i)$ is $(1(-3) - (-2)1) + (1(1) + (-2)(-3))i$, which simplifies to $-1 + 7i$.

Division of Complex Numbers

$$\frac{a+bi}{c+di} = \frac{(ac+bd)+(bc-ad)i}{c^2+d^2}$$

Quickie Example: $\frac{-2+3i}{1+i}$ is $\frac{\left(-2(1)+3(1)\right)+\left(3(1)-(-2)1\right)i}{1^2+1^2}$ which simplifies to $\frac{1+5i}{2}$ or $0.5 + 2.5i$.

(You will find a better way soon.)

Example 1: Add $2 + 4i$ and $-3 + 5i$.

Solution: $(2 + 4i) + (-3 + 5i) = (2 + (-3)) + (4 + 5)i$ add similar to $(2 + 4x) + (-3 + 5x)$

$= -1 + 9i$

⁂

Example 2: Subtract $(-2 + 5i) - (3 - 7i)$.

Solution: $(-2 + 5i) - (3 - 7i) = (-2 - 3) + (5 - (-7))i$ subtract similar to $(-2 + 5x) - (3 - 7x)$

$= -5 + 12i$

⁂

The rules for multiplication and division are a little messy and another suggested method for multiplying and dividing is shown below.

Example 3: Multiply $(-2 + 4i)(3 + 5i)$.

Solution: Instead of using the multiplication rule, treat the product just as if the binomials $(-2 + 4x)(3 + 5x)$ are being multiplied, except replace i^2 with -1.

$(-2 + 4i)(3 + 5i)$given

F O I L

$(-2)(3) + (-2)(5i) + 3(4i) + (4i)(5i)$

$-6 \quad - \quad 10i + 12i + 20i^2$

$-6 \quad - \quad 10i + 12i + 20(-1)$

$-6 - 20 - 10i + 12i$

$-26 + 2i$

If the multiplication rule is used, the product is:

$(-2 + 4i)(3 + 5i) = [(-2)(3) - (4)(5)] + [(-2)(5) + (4)(3)]i$

$(-6 - 20) + (-10 + 12)i$

$-26 + 2i$

⁂

Since the FOIL method works just as well as the multiplication rule, you may choose either method. A similar situation happens with the quotient of complex numbers.

Example 4: Divide: $\dfrac{2+3i}{1-i}$.

Solution: Instead of using the quotient rule, divide the complex numbers using the same method used to divide irrational binomials.

$\dfrac{2+3i}{1-i}$ given

$\dfrac{(2+3i)\mathbf{(1+i)}}{(1-i)\mathbf{(1+i)}}$ multiply by the conjugate of the denominator

$\dfrac{2\cdot 1+2i+3i+3i^2}{1^2-i^2}$ FOIL

$$\frac{2+2i+3i+3(-1)}{1-(-1)} \qquad (i^2=-1)$$

$$\frac{-1+5i}{2} \text{ or } \frac{-1}{2}+\frac{5i}{2} \text{ or } -0.5+2.5i$$

⁂

If complex numbers are raised to an exponent of 3 or more, or if 3 or more complex numbers are multiplied, the number i to an exponent larger than 2 will be in the answer. You should be able to simplify i to any exponent to either 1, –1, i, or $-i$. Below are a few examples of i raised to an exponent.

Example 5: Raise i to an exponent of 3.

Solution: i^3 is the same as $i \cdot i^2$, and $i^2 = -1$, so i^3 is $i \cdot (-1)$ or $-i$.

i^3 simplifies to $-i$.

⁂

Example 6: Simplify i^4.

Solution: i^4 is the same as $i^2 \cdot i^2$, and since i^2 is –1, i^4 is (–1)(–1). $i^4 = 1$.

⁂

Example 7: Simplify i^{27}.

Solution:

$$i^{27} = (i^4)^6 \cdot i^3$$

$= (1)^6 \cdot i^3$ from earlier work $i^4 = 1$

$= 1 \cdot i^3,$ from earlier work, $i^3 = -1$

$= 1 \cdot (-i)$

$= -i$

⁂

Example 8: Simplify i^{82}.

Solution:

$i^{82} = (i^4)^{20} \cdot i^2$ from earlier work $i^4 = 1$

$= (1)^{20} \cdot (-1)$

$= 1 \cdot (-1)$

$= -1$

⁂

Example 9: Simplify i^{101}.

Solution:

$i^{101} = (i^4)^{25} \cdot i$ from earlier work $i^4 = 1$

$= (1)^{25} \cdot i$

$= 1 \cdot i$

$= i$

⁂

While you may certainly perform arithmetic operations on complex numbers with pencil and paper, you may also want to use a calculator like the TI-84. The example below shows that complex number arithmetic is just as simple as real number arithmetic for the calculator.

Example 10: Add, subtract, multiply and divide $(2 - i)$ and $(1 + 3i)$.

Solution:

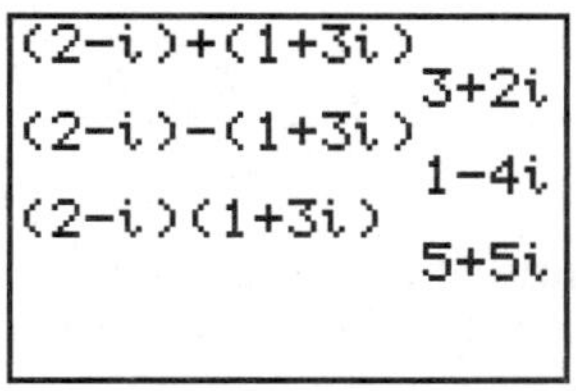

Figure 10.4.1a

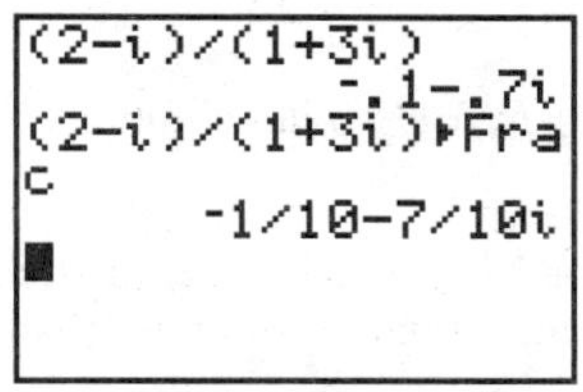

Figure 10.4.1b

⁂

Performing an operation on complex numbers or raising i to any exponent is useful in a course in college algebra or in several courses in technical mathematics. A more thorough investigation of complex numbers is covered in several other courses. In this text, complex numbers are not used a great deal, but it is necessary to introduce them because some of the equations solved below have complex answers.

Now back to the goal of this section: to solve a quadratic equation with the quadratic formula. To develop the "quadratic formula," the equation $ax^2 + bx + c = 0$ will be solved by the completing-the-square method. That is, the above equation will be solved for x. The solution for x will contain the known parameters of a, b, and c.

$$ax^2 + bx + c = 0 \qquad \text{given}$$

$$ax^2 + bx = -c \qquad \text{subtract } c \text{ from both sides}$$

$$x^2 + \frac{b}{a}x = \frac{-c}{a} \qquad \text{divide by both sides by } a$$

$$x^2 + \frac{b}{a}x + \frac{b^2}{4a^2} = \frac{-c}{a} + \frac{b^2}{4a^2} \qquad \text{complete-the-square}$$

$$\left(x + \frac{b}{2a}\right)^2 = \frac{b^2 - 4ac}{4a^2} \qquad \text{factor the left/simplify the right}$$

$$\sqrt{\left(x+\frac{b}{2a}\right)^2}=\pm\sqrt{\frac{b^2-4ac}{4a^2}}$$ take the square root of both sides

$$x+\frac{b}{2a}=\pm\frac{\sqrt{b^2-4ac}}{2a}$$ simplify

$$x=-\frac{b}{2a}\pm\frac{\sqrt{b^2-4ac}}{2a}$$ subtract $\frac{b}{2a}$ from both sides

$$x=\frac{-b\pm\sqrt{b^2-4ac}}{2a}$$ combine the fractions

The equation $ax^2 + bx + c = 0$ has solutions of: $x=\frac{-b\pm\sqrt{b^2-4ac}}{2a}$

The equation solved for x is called the **quadratic formula**. If the equation $ax^2 + bx + c = 0$ is solved on a calculator like the Voyage 200, the solution doesn't quite look the same as the above, but it is the same.

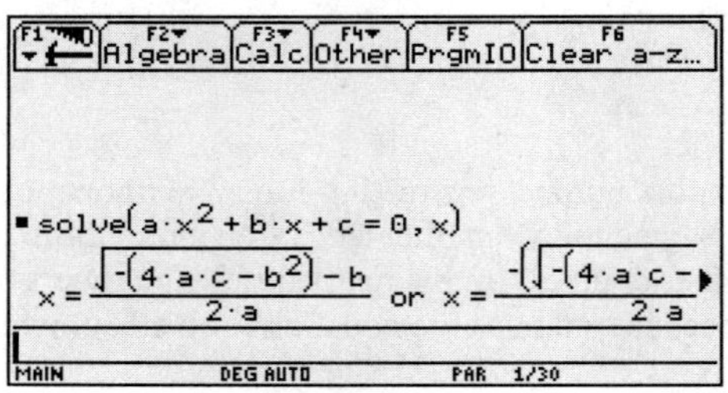

Figure 10.4.2

Every quadratic equation ($ax^2 + bx + c = 0$) can be solved with the quadratic formula, whether it can be factored or not, and whether it has real solutions or not.

Example 11: Find the exact solutions to the equation $x^2 + 3x - 1 = 0$.

Solution: If the solutions are real, a graphical method can be used; however, the values for x may not be exact. The completing-the-square method or the quadratic formula can be used to get the exact solutions. Factoring cannot be used because the trinomial does not factor with rational numbers. The values of a, b, and c are needed to use the quadratic formula. The values are identified as follows: a is the coefficient of x^2, b is the coefficient of x, and c is the constant term, provided the equation is in the form of $ax^2 + bx + c = 0$. In the above equation, $a = 1$, $b = 3$, and $c = -1$. The solution is found by replacing a, b, and c in quadratic formula with their values.

$$x=\frac{-b\pm\sqrt{b^2-4ac}}{2a}$$

$$x = \frac{-3 \pm \sqrt{3^2 - 4 \cdot 1(-1)}}{2 \cdot 1}$$ the expression must be simplified

$$x = \frac{-3 \pm \sqrt{9+4}}{2}$$

$$x = \frac{-3 \pm \sqrt{13}}{2}$$

The exact solutions to the equation $x^2 + 3x - 1 = 0$ are: $\dfrac{-3+\sqrt{13}}{2}$ and $\dfrac{-3-\sqrt{13}}{2}$.

⁂

Example 12: Solve the equation $2x^2 - 3x + 5 = 0$.

Solution: The trinomial does not factor; thus, it cannot be solved by the factoring method. The completing-the-square method will work, as will the quadratic formula. The graphical method will not work because the graph does not have any real zeros. The quadratic formula is used below. The value of $a = 2$, $b = -3$, and $c = 5$.

$$x = \frac{-b \pm \sqrt{b^2 - 4ac}}{2a}$$

$$x = \frac{-(-3) \pm \sqrt{(-3)^2 - 4 \cdot 2 \cdot 5}}{2 \cdot 2}$$ this expression must be simplified

$$x = \frac{3 \pm \sqrt{9 - 40}}{4}$$

$$x = \frac{3 \pm \sqrt{-31}}{4}$$ $\sqrt{-31} = \sqrt{-1} \cdot \sqrt{31} = i\sqrt{31}$

The exact solutions to the equation $2x^2 - 3x + 5 = 0$ are: $\dfrac{3+i\sqrt{31}}{4}$ and $\dfrac{3-i\sqrt{31}}{4}$

⁂

Example 13: Solve $2x^2 + 5x = -5(3 - x)$.

Solution: The equation must have 0 on one side so that values of a, b, and c can be determined. $2x^2 + 5x = -5(3 - x)$

$$2x^2 + 5x = -15 + 5x$$

$$2x^2 = -15$$

$$2x^2 + 15 = 0$$

The values of *a*, *b*, and *c* are 2, 0, and 15. The quadratic formula is used below; the square root property can also be used.

$$x = \frac{-b \pm \sqrt{b^2 - 4ac}}{2a}$$

$$x = \frac{0 \pm \sqrt{0^2 - 4 \cdot 2 \cdot 15}}{2 \cdot 2}$$

$$x = \frac{\pm\sqrt{-120}}{4} \qquad \sqrt{-120} = \sqrt{4} \cdot \sqrt{-1} \cdot \sqrt{30} = 2i\sqrt{30}$$

The exact solutions to the equation are: $x = \dfrac{\pm 2i\sqrt{30}}{4}$ or $\dfrac{\pm i\sqrt{30}}{2}$

⁂

10.4 EXERCISES

Many times you can work the exercises by algebraic, numeric, or graphical methods. You can learn about mathematics no matter which method you use. Part of the learning process is for you to decide which method is best for you to use. Please read the text carefully before you do the exercises. Read the **next assigned section** after you finish the assignment below. All numeric answers must have an error ≤ 0.01.

A Quick Look To The Future

-4. The shape of the water at a drinking fountain seems to be parabolic; does this suggest that a quadratic function could model the behavior of the water? True? False?

-3. What number is halfway between *a* and *b*?

-2. If you sum the lengths of any two sides of a triangle, how is this sum related to the third side?

-1. What four-sided polygon shape has opposite sides that are parallel?

Previous Topics Revisited

1. Solve by using the completing-the-square method $2x^2 + 4x + 5 = 0$.

2. Solve by using the completing-the-square method $3x^2 - 10 = 0$.

3. Solve by using the completing-the-square method $x^2 - 3x + 2 = 0$.

4. Solve $2x^2 - 5x - 3 = 0$ by the factoring method.

5. Solve $9x^2 - 16 = 0$ by the factoring method.

6. Put the quadratic function $y = x^2 - 6x + 3$ in standard form and determine the minimum value of the function, the range, when the function in increasing, and when it is decreasing.

New Topic Exercises

Do the indicated operation below.

7. $(-3 + 5i) + (-2 - 6i)$
8. $(5 - 6i) - (-2 + 7i)$
9. $(1 - 3i)(-2 + 4i)$
10. $\frac{4}{1+i}$
11. $(1.2 - 6.5i) + (-1.2 + 6.5i)$
12. $3i - 5i - 2i + 4i$
13. $(2i)^5$
14. $\frac{1+i}{2-i}$
15. $\frac{(1+i)(1-i)}{(1+i)}$
16. i^{145}
17. i^{-13}

Solve the equations below for real or complex values.

18. $x^2 - x + 1 = 0$
19. $x^2 = -25$
20. $(x - 3)^2 = -9$
21. $(x + 2)^2 = 25$
22. $x^2 = 9x$
23. $0.3x^2 - 2x = 12$
24. $5x^2 - 7x + 4 = 0$
25. $100x^2 - 500x = 1000$

The solutions to the quadratic equation $ax^2 + bx + c = 0$ are: $\frac{-b \pm \sqrt{b^2 - 4ac}}{2a}$, where a, b, and c are real numbers, $a \neq 0$.

26. If $b^2 - 4ac$ is a negative number, what kind of number is x?
27. If $b^2 - 4ac$ is a perfect square, what kind of number is x?
28. If $b^2 - 4ac$ is zero, how many solutions does the equation have? Are they real or complex?
29. Is it possible for a quadratic equation to have one and only one complex solution? Why? If yes, give an example.
30. Is it possible for a quadratic equation to have one and only one rational solution? Why? If yes, give an example.
31. Is it possible for a quadratic equation to have one and only one irrational solution? Why? If yes, give an example.

From Mathematics to English

32. Can you give an example of a quadratic equation that has 3 answers? Why?

33. Describe how you can tell if a quadratic equation has real solutions or complex solutions.

34. Describe how to solve the equation $3x^2 + 5x + 4 = 0$. Do not actually solve it.

35. What is your own definition of a quadratic equation?

36. List five different ways of solving the equation $2x^2 - 6x - 3 = 17$.

37. List as many methods as possible you can use to solve the equation $x^2 + 4 = 0$.

38. After reading this section, make a list of questions that you want to ask your instructor.

39. Continue in your daily journal and make an entry. In addition to your normal entry on thoughts about the mathematics in this section, list at least two positive comments about what you have learned about this topic.

40. In paragraph format, summarize the material in this section of the text in your daily journal.

41. Describe how your classroom instructor made this topic more understandable and clear.

42. After reading the text and listening to your instructor, what do you not understand about this topic?

Explorations

43. Fill in the table:

	Coordinates of the Vertex	Solutions to $y = 0$
$y = (x - 2)^2 + 3$		
$y = (x + 5)^2 + 4$		
$y = (x - 4)^2 + 3$		
$y = (x + 3)^2 + 2$		
$y = x^2 = 4x + 7$		
$y = x^2 + 6x + 10$		
$y = x^2 - 10x + 25$		

What is the relationship between the coordinates of the vertex and the complex zeros of the quadratic function? $y = x^2 + bx + c$?

44. Develop a method by which the quadratic formula can be used to solve the following equations. The equations are said to be in "quadratic form." Quadratic form means the exponent in the first term is twice the exponent in the second term. Check the real solutions with a graphing calculator.

a. $2x^4 - 6x^2 - 20 = 0$

b. $x^4 - 3x^2 - 10 = 0$

c. $5x - 2x^{\frac{1}{2}} - 3 = 0$

d. $x^{\frac{2}{3}} - 6x^{\frac{1}{3}} + 7 = 0$

e. $2(x - 1)^6 + 5(x - 1)^3 + 3 = 0$

10.5 The Quadratic Function as a Mathematical Model

When an object near the surface of the earth is dropped, the force of gravity pulls it toward earth. In the early 1600s, Galileo Galilei discovered the distance an object travels in free-fall is a quadratic relationship with the length of time traveled. The mathematical model he developed for free-fall motion is $s = \frac{1}{2}gt^2$, where g is the acceleration due to gravity, t is the time of free-fall, and s is the distance traveled during free fall.

In the metric system of measurement, the constant number g is –9.8 meters per second squared. In the English system it is –32 feet per second squared. On other planets, the constant g is different. Even on the planet Earth, the number g varies slightly according to the distance from the center of the Earth. For example g is slightly smaller in absolute value on a mountain than at sea level. The difference is small and if ignored, will not harm calculations and graphical work. For all practical purposes, it will be considered as a constant for the planet Earth.

The quadratic function $s = \frac{1}{2}gt^2$ models free-fall motion because it can be used to accurately predict the distance traveled when t is known, or it will accurately predict t when the distance s is known. The conditions that apply to this function are: 1) the function describes free-fall motion near the surface of the Earth; 2) the function can be applied to free-fall situations where the initial velocity of the object is zero (it is dropped, not thrown); and 3) the initial height of the object is considered to be at "ground zero". The initial conditions for free-fall motion apply when time t is zero. If the initial conditions are not satisfied as above, a different form of the function is needed.

Example 1: Make a numeric representation for the function $s = \frac{1}{2}gt^2$ and find the distance (s) traveled by a rock in free-fall after 1, 2, 3, 4, 5, and 6 seconds.

Solution: Using g = – 32, the function becomes $s = -16t^2$. Since the calculator may not accept the variables s and t, use the function $y = -16x^2$.

X	Y1
0	0
1	-16
2	-64
3	-144
4	-256
5	-400
6	-576

Y1 = -16X²

Figure 10.5.1

As you can see in the table, the rock will fall, for example, 144 feet in 3 seconds.

The coordinates of each point (t, s) tell you the time and distance traveled. It is a little strange to see distance expressed as a negative number. The problem can be better understood if the x-axis (now the time-axis) is imagined as being ground level (ground zero) and the object is dropped into a hole. It then makes a little sense to think of numbers above the x-axis as positive distances and numbers below the x-axis as negative distances. To solve this problem, the x-axis (*time*-axis) can be drawn any place, even above ground level. This will allow a visualization of a dropped object other than into a hole.

Example 2: How long does it take for a dropped ball to travel 64 feet, ignoring wind resistance?

Solution: The problem can be solved using Figure 10.5.1. In 2 seconds, the ball will fall 64 feet. The point (2,–64) is on the graph; thus, when time is 2 seconds, the distance traveled is 64 feet. If you want to answer the question using algebra, use the mathematical model $s = \frac{1}{2}gt^2$, replace s with –64, replace g with –32, and solve the equation for t.

$-64 = \frac{1}{2}(-32)t^2$ simplify

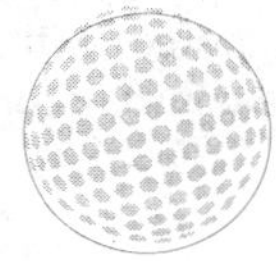

$-64 = -16t^2$ divide by –16

$4 = t^2$ take the square root

$\pm 2 = t$

Disregard the –2 because a time of –2 implies 2 seconds before the ball was dropped. It takes 2 seconds for a ball to fall 64 feet.

Now consider the case where actual ground level is the time-axis and the object is dropped from a distance s_0 above ground level. From Section 10.1 and from many other discussions about functions, you know how to give any graph a vertical translation. If the graph is moved up by an amount s_0, the mathematical model is now more realistic. By adding s_0 to the function $s = \frac{1}{2}gt^2$, the function becomes $s = \frac{1}{2}gt^2 + s_0$ where s_0 is the height above ground level from which an object is dropped. The graph has been translated up by an amount of s_0. Ground level is now thought of as the time-axis. When the object is above ground level, the s-coordinate will be a positive number, and the object can be thought of as below ground level, when s is negative. The s-coordinate of any point on the graph measures the distance above or below ground level (the t-axis).

Example 3: A water balloon is dropped from a height of 43 feet. How long does it take to strike the ground?

Solution: To solve the problem graphically, graph the function $y = -16x^2 + 43$ and identify the x-coordinate (time) when the y-coordinate (distance coordinate) is zero. That is, since y measures the distance above or below ground level, y is zero when the balloon is on the ground.

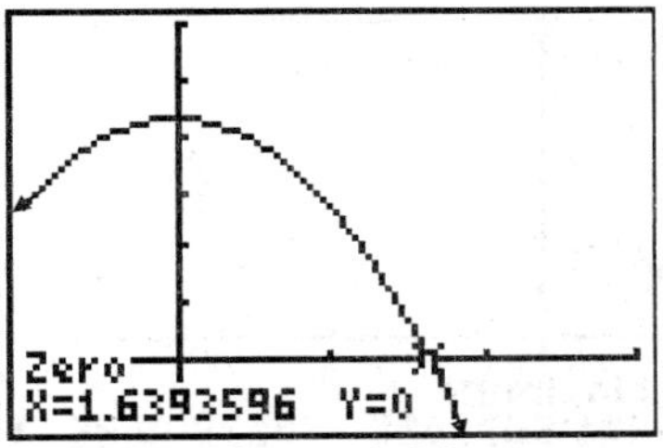

Figure 10.5.2

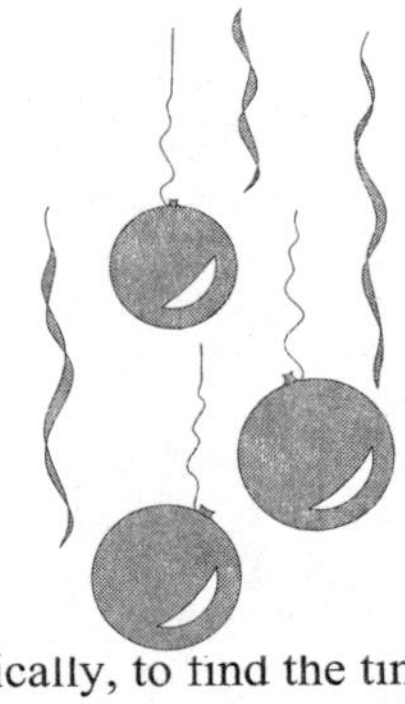

The graph crosses when x is 1.639 with an error < 0.01. Analytically, to find the time it takes a water balloon to drop 43 feet, replace y with 0 and solve for x.

$$0 = -16x^2 + 43$$

$$-43 = -16x^2$$

$$2.6875 = x^2$$

$$1.639 = x$$

⁂

One more dimension will now be added to the free-fall model. What if the object is not dropped, but is thrown or shot straight up or straight down? If an object is not just dropped -- but thrown -- it is called **projectile motion**. The new initial condition on the free-fall problem is that the object is given an initial boost or velocity. The new mathematical model is NOT the previous model plus the velocity. From the model $s = \frac{1}{2}gt^2 + s_0$, s represents a distance, $\frac{1}{2}gt^2$ represents a distance and s_0 also represents a distance. If the initial velocity is symbolized with v_0, the function $s = \frac{1}{2}gt^2 + s_0 + \mathbf{v_0}$ does not make sense. That is, adding a velocity to a distance does not yield a distance. You may recall from other courses the formula: *distance* = *rate* × *time*. This formula allows for multiplication of the initial velocity (rate) times time (t) and the product is a distance. If the distance v_0t is added to the existing mathematical model, the new model for projectile motion with an initial position s_0 and initial upward velocity v_0 is: $s = \frac{1}{2}gt^2 + v_0t + s_0$. The new model for projectile motion with an initial position s_0 and initial downward velocity v_0 is: $s = \frac{1}{2}gt^2 - v_0t + s_0$. Each of the terms in the model represents a distance.

Example 4: From a height of 100 feet, an apple is thrown straight up with an initial velocity of 50 feet per second. What is the maximum height of the apple above ground level?

Solution: If the function $s = -16t^2 + 50t + 100$ is graphed, you know it is a parabola with a maximum point. The coordinates of the point (t, s) at the maximum point tell the time it takes to reach the maximum distance above ground as well as the maximum distance above ground level.

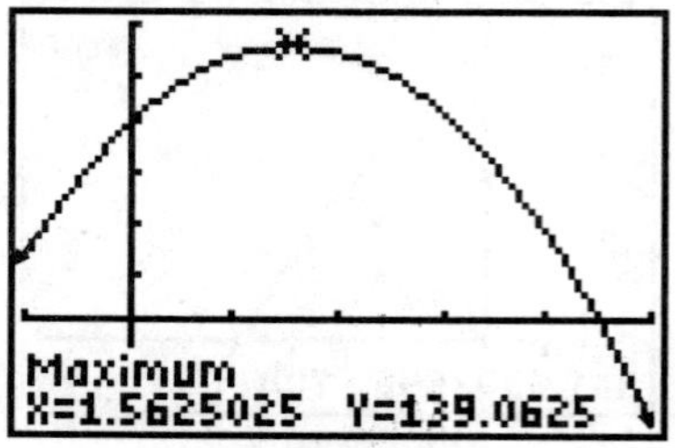

Figure 10.5.3

The maximum distance is 139.06 feet above ground level. The problem can be solved for the exact solution analytically. The model must be put in standard form.

$$s = -16t^2 + 50t \qquad + 100$$

$$s = -16\left(t^2 - \frac{50}{16}t \qquad \right) + 100$$

$$s = -16\left(t^2 - \frac{25}{8}t \qquad \right) + 100$$

$$s = -16\left(t^2 - \frac{25}{8}t + \frac{625}{256}\right) + 100 + \frac{625}{16}$$

$$s = -16\left(t - \frac{25}{16}\right)^2 + \frac{2225}{16}$$

$$s = -16\left(t - \frac{25}{16}\right)^2 + 139.0625$$

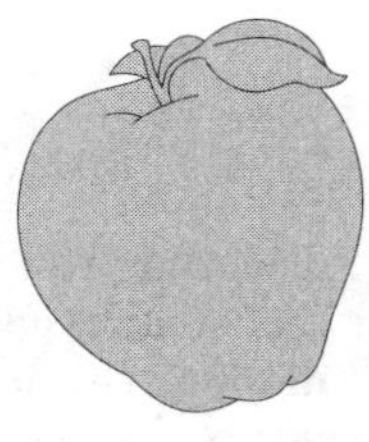

The maximum height of the apple is 139.0625 feet, and it happens 1 and $\frac{9}{16}$ seconds after it is thrown.

⁂

The quadratic functions used to model free-fall motion and projectile motion are certainly all related. They were all derived from observations of motion made by Galileo well over 300 years ago. These same functions may be studied at a higher level in college algebra as well as in a first course in calculus and they are studied in most beginning physics courses.

10.5 EXERCISES

Many times you can work the exercises by algebraic, numeric, or graphical methods. You can learn about mathematics no matter which method you use. Part of the learning process is for you to decide which method is best for you to use. Please read the text carefully before you do the exercises. Read the **next assigned section** after you finish the assignment below. All numeric answers must have an error ≤ 0.01.

A Quick Look To The Future

-4. What is the distance between the numbers a and b on the real number line?

-3. Does a triangle have three or four angles?

-2. Do parallel lines ever touch?

-1. Draw a circle on your calculator screen.

Previous Topics Revisited

1. Simplify $\dfrac{(-2+3i)(-1-5i)}{(-1-5i)}$

2. Find the complex zeros of $f(x) = 2x^2 + 5x + 7$ by the quadratic formula.

3. Find the complex roots of $3x^2 - 7x + 5 = 0$ by the quadratic formula.

4. Find the real roots of $(x - 5)^2 = 15$ by the square root property.

5. Find the real roots of $0 = x^2 + 6x - 1$ by using the completing-the-square method.

6. Solve $(x + 2)^2 - 9 = 0$ by factoring.

New Topic Exercises

Use the function $s = \dfrac{1}{2}gt^2$ or $s = \dfrac{1}{2}gt^2 + s_0$ to solve the following problems; s is measured in feet and t is measured in seconds.

7. How far does an object fall in 6 seconds?

8. How long does it take for an object to fall 200 feet?

9. The acceleration due to gravity (g) on the moon is -5.47 feet per second^2 ($g = -5.47$). How long does it take an object on the moon to fall 100 feet? How long does it take the same object to fall 100 feet on Earth?

10. The acceleration due to gravity on the planet Jupiter is -86.8 feet per second^2 ($g = -86.8$). How far does an object in free-fall travel in 10 seconds? How far does the same object fall on Earth in 10 seconds?

11.

Planet	Value of g
Mercury	–12.9
Venus	–28.9
Mars	–12.9
Jupiter	–86.8

Graph the functions $s = \frac{1}{2}gt^2$ for all of the planets listed. Put all graphs on the same screen. Approximately how far will a stone fall in 3 seconds on each of the planets?

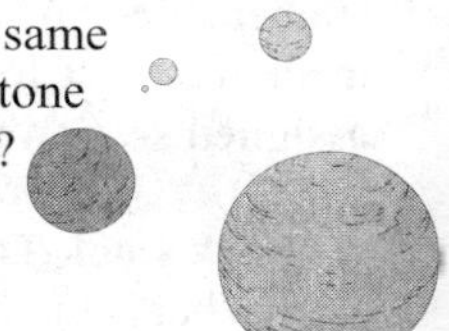

12. Does a 3-pound box of chocolates fall faster than a 1-pound box?

13. A rock is dropped from an initial height of 1 mile; how long does it take to reach a point 600 feet above ground level? (1 mile = 5280 feet)

14. If a ball is dropped from a height of 86 feet, approximately how long does it take to fall the first 43 feet? The second 43 feet? What does this imply about the average velocity (speed) during each portion of the fall?

15. If a stone is dropped from a plane flying in a horizontal pattern, what is the shape of the path of the stone? Why?

In the next set of exercises, use the function $s = \frac{1}{2}gt^2 \pm v_0t + s_0$.

16. A small caliber gun is fired upward with an initial velocity of 1,000 feet per second from a height of 10 feet above ground level; how long does it take the bullet to strike the ground – assuming no air resistance?

17. In the above exercise, what is the maximum height the bullet reaches as measured from ground level? How long does it take?

18. If a water balloon is thrown straight up from an initial position of 5 feet with an initial velocity of 50 feet per second, how long will it take for the balloon to strike the 6 foot 4 inch thrower on the top of his head – assuming no air resistance?

19. What initial velocity will cause a stone thrown upward to reach a height of 50 feet in 2 seconds? Assume the stone is thrown from a position 6 feet above ground level, and assume no air resistance.

20. Will a ball thrown upward from ground level with an initial velocity of 100 feet per second reach a height of 157 feet – assuming no air resistance?

21. A ball is thrown downward with an initial velocity of 30 feet per second from a height of 200 feet. How long does it take to reach the ground – assuming no air resistance?

22. If you throw your shoe downward at an initial velocity of 3 feet per second from a height of 8 feet, when will it hit the floor – assuming no air resistance?

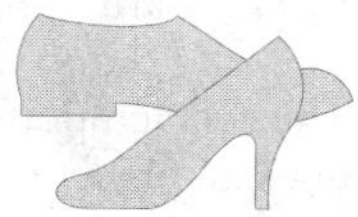

23. At the base of the clouds on the planet Saturn the acceleration due to gravity is –37.1 feet per second per second. If you are floating along in a balloon 200 feet above what looks like the surface of Saturn and you throw a bolt downward at 10 feet per second, when will it strike the surface?

24. If you are piloting a small military jet and find it necessary to use the ejection seat, how high above the plane will you reach if the ejection seat has an initial velocity of 100 feet per second?

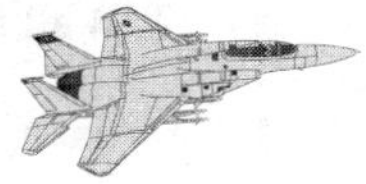

From Mathematics to English

25. Write a short essay about Galileo Galilei.

26. Describe the velocity of a ball as it travels after being thrown upward.

27. Would you weigh more or less on the moon. Explain.

28. After reading this section, make a list of questions that you want to ask your instructor.

29. Continue in your daily journal and make an entry. In addition to your normal entry on thoughts about the mathematics in this section, list at least two positive comments about what you have learned about this topic.

30. In paragraph format, summarize the material in this section of the text in your daily journal.

31. Describe how your classroom instructor made this topic more understandable and clear.

32. After reading the text and listening to your instructor, what do you not understand about this topic?

Explorations

33. The value of g can be found experimentally with a pendulum. With a pendulum of length l and period t, the acceleration due to gravity can be found with the mathematical model: $g = \frac{4\pi^2 l}{t^2}$. Calculate g. Record all of your data used to find g. Perform the experiment 5 times each for string lengths of 2, 3, 4, 5, and 6 feet.

34. Find the maximum height of a ball thrown upward from ground level with an initial velocity of 100 feet per second on the moon, Earth, Mercury, Venus, Mars, and Jupiter. See Exercises 9 and 11 for values of g.

CHAPTER TEN TEST

1. List all transformations of the graph of $y = x^2$ that yield the graph of $y = -2(x + 3)^2 - 1$.
2. What are the coordinates of the vertex of $y = -2(x + 3)^2 - 1$?
3. What are the coordinates of the vertex of $y = x^2 - 3x + 2$?
4. What is the range of the function $y = -2(x + 3)^2 - 1$?
5. Using interval notation, describe values of x when the function $y = -2(x + 3)^2 - 1$ is decreasing.
6. List any quadratic function with zeros of –4 and 7.
7. Solve the equation $(2x + 1)(x - 4) = 0$.
8. Solve the equation $(x - 2)^2 = 9$.
9. Solve the equation $6x^2 - 18x + 11 = 0$.
10. Find $(2 + 7i) + (-3 + 4i)$ and $(-2 + 3i)(-2 - 3i)$
11. Find $(-3 + 2i) - (4 - 5i)$ and $\dfrac{2i}{(1+i)}$
12. Does $x^2 + 5x + 7 = 0$ have real or complex solutions?
13. Find the complex solutions to $x^2 - 4x + 13 = 0$
14. Find all solutions to $0.6x^2 - 3.7x - 5.1 = 0$.
15. Give an example of a quadratic equation with one and only one solution.
16. Find the exact solutions to $x^2 = \pi$.
17. How far does a dropped object fall in 16 seconds if it is dropped while on the Earth, the mathematical model used to describe free-fall motion is $s = -16t^2$?
18. The mathematical model for projectile motion with an initial upward velocity is $s = -16t^2 + v_0t + s_0$. An object thrown upward at 50 feet per second from 4 feet above ground level will return to ground level in how many seconds?
19. What is the maximum height of the object thrown in Question 18?
20. An object is thrown downward at an initial velocity of 100 feet per second from a height of 200 feet. How long does it take to reach ground level? $(s = -16t^2 - v_0t + s_0)$

CHAPTER ELEVEN
BASIC GEOMETRY

11.0 Introduction

Chapter Eleven integrates analytical geometry with the study of traditional Euclidean geometry. Analytical geometry is the study of geometry on the coordinate plane. The graphing calculator is used to solve minimum distance problems in Section One, and the graphs of circles are developed in Section Four. This theme is used somewhat throughout Chapter Eleven; however, the last three sections emphasize the characteristics and properties of the triangle, parallelogram, and circle.

The characteristics and properties of the triangle, parallelogram, and circle are introduced with no formal proofs as might be found in a geometric course. Since geometric terms are necessary for a review of geometric properties, each section includes identification of the components that make-up or are related to the geometric figures studied. You will learn about angle bisectors, perpendicular bisectors, tangents, secants, isosceles triangles, etc.

Geometry in Action

Consider Figure 11.0.1. If the construction worker knows her height and measures her shadow and the shadow of the tree, she can find the height of the tree using geometry. The triangles superimposed on top of the drawings are similar triangles because the angle of elevation of the sun is the same in both triangles. The construction worker need only solve the equation, $\frac{6}{3} = \frac{x}{9}$ with x as the height of the tree.

Figure 11.0.1

11.1 The Distance and Midpoint Formulas

Imagine two points drawn on a flat piece of paper like shown below in Figure 11.1.1; the distance between the points can be found by putting a ruler's scale on the points and subtracting the numbers adjacent to the points – just as a carpenter would do. That is, if the points fall at 0 and 7 on the ruler, the distance between the points is $7-0$, or 7. If the points fall at 11 and 4, the distance between the points is $11-4$, or 7.

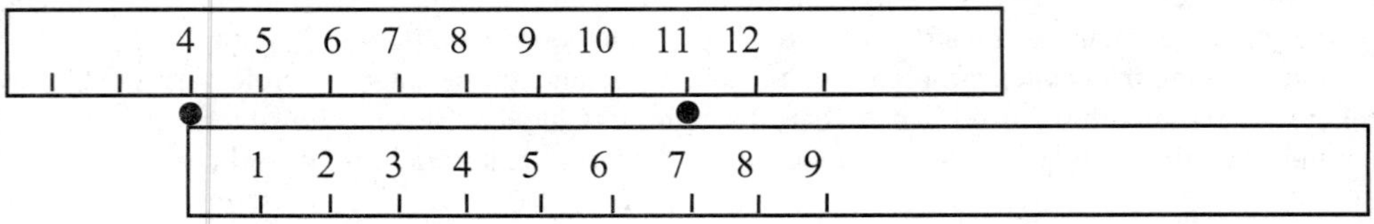

Figure 11.1.1

These numbers (4 and 11 or 0 and 7) can be thought of as the coordinates of the two points. They are, in effect, in a one-dimensional space. If the points are in two dimensional space -- on the *xy*-coordinate plane -- the calculation for finding the distance between two points is a little more complicated – but not much.

To develop a method for finding the distance between points on a plane, start with two points on a horizontal line or a vertical line. Consider the points $(8,3)$ and $(-5,3)$; they are on a horizontal line and the distance between the points can be found in a similar fashion to using a ruler.

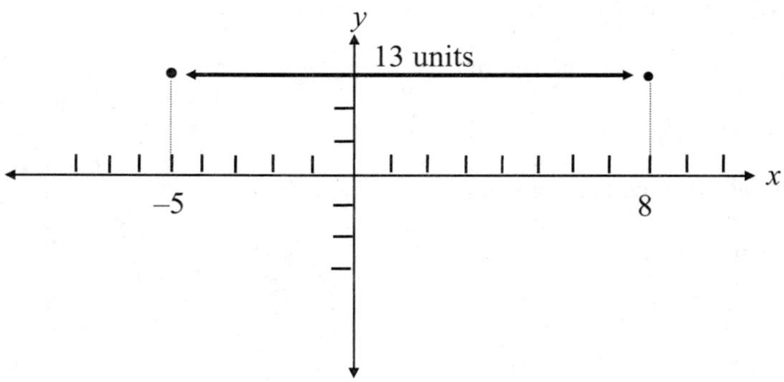

Figure 11.1.2

The *x*-axis behaves like the ruler; if the *x*-coordinates are subtracted, the distance between the points is known. Obviously, the number –5 is not found on a ruler; however, the process of finding the distance between the points is the same -- subtract the *x*-coordinates. The distance between the above points is $8-(-5)$, or 13. There is a minor problem with the algorithm "subtract the *x*-coordinates." If the coordinates are reversed, the distance is $(-5)-8=-13$. The question must be asked, "Which answer is correct, 13 or –13?" Positive numbers are normally used to represent distances. If it is agreed to express the distance as a positive number, there is a simple mathematical function you are familiar with that will always generate a positive distance; it is the absolute value function. If the absolute value of the difference of the x-coordinates is taken, the distance will be positive. To calculate the distance between the points $(8, 3)$ and $(-5, 3)$, do either of the calculations $|8-(-5)|$ or $|-5-8|$. Both calculations give the length of 13.

> In general, if the x-coordinates of two points are x_1 and x_2, the horizontal distance between the points is $|x_2 - x_1|$.

The distance between two points on a vertical line can be found in a similar fashion. When points are on a vertical line, the y-axis behaves as the ruler.

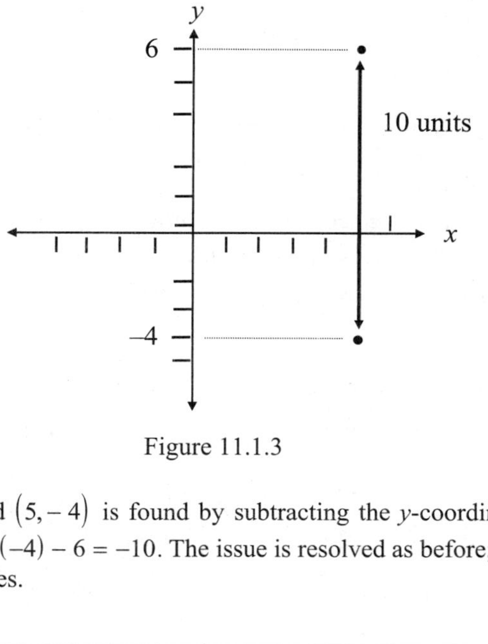

Figure 11.1.3

The distance between $(5, 6)$ and $(5, -4)$ is found by subtracting the y-coordinates. Again, there can be two answers; $6-(-4)=10$ and $(-4)-6=-10$. The issue is resolved as before; take the absolute value of the difference of the y-coordinates.

> In general, for two points with y-coordinates y_1 and y_2, the vertical distance between the points is $|y_2 - y_1|$.

The Distance Formula

Finally, you are ready to find the distance between any two points in the coordinate plane. Consider the points $P_1(x_1, y_1)$ and $P_2(x_2, y_2)$ as shown in Figure 11.1.4.

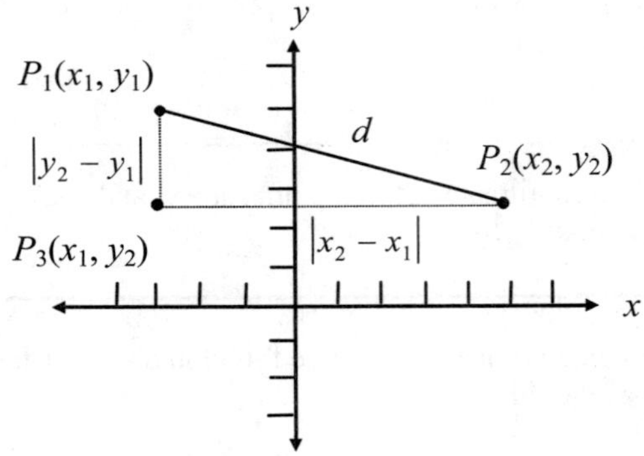

Figure 11.1.4

The extra lines are drawn to aid in finding the distance between the points. The third point P_3 is on the same vertical line as is P_1, so it has the same x-coordinate as P_1. The point P_3 is on the same horizontal line as P_2, so it has the same y-coordinate as P_2. As shown earlier, the vertical line segment has a length of $|y_2 - y_1|$ and the horizontal line segment has a length of $|x_2 - x_1|$. When a horizontal and vertical line meet, they form a right angle (90°). This makes the triangle $P_1P_2P_3$ a right triangle. In any right triangle the hypotenuse squared equals the sum of the squares of the legs; that is, $d^2 = (P_1P_3)^2 + (P_2P_3)^2$. In the case above:

$d^2 = \left(\lvert x_2 - x_1\rvert\right)^2 + \left(\lvert y_2 - y_1\rvert\right)^2$	squaring a number or its absolute value gives the same answer, so drop the absolute value
$d^2 = \left(x_2 - x_1\right)^2 + \left(y_2 - y_1\right)^2$	take the square root of both sides
$d = \sqrt{\left(x_2 - x_1\right)^2 + \left(y_2 - y_1\right)^2}$	

This is the formula used to calculate the distance between any two points in the xy-plane. It is commonly called the "distance formula". It tells you how to calculate the distance between two points. You may recognize the expression $x_2 - x_1$ as the change in the x-coordinates (Δx). You may also recognize $y_2 - y_1$ as the change in the y-coordinates (Δy). If Δx and Δy represent the amount of change in the x and y-coordinates respectively from $P_1(x_1, y_1)$ to the point $P_2(x_2, y_2)$, then the distance between points is $d = \sqrt{(\Delta x)^2 + (\Delta y)^2}$. If you think of the distance formula in this way, you may be able to do much of the calculation in your head.

THE DISTANCE FORMULA:

The distance between points (x_1, y_1) and (x_2, y_2) is

$$d = \sqrt{(\Delta x)^2 + (\Delta y)^2}\ .$$

Example 1: Find the length of the line segment joining $(-2,3)$ and $(-5,-1)$.

Solution: Use the distance formula to find the length of the line segment. It does not make any difference which point is thought of as being P_1 or P_2. Choose P_1 to be $(-2,3)$ and P_2 is $(-5,-1)$. $d = \sqrt{(\Delta x)^2 + (\Delta y)^2}$. The x changes from -2 to -5, or a decrease of $3(\Delta x = -3)$. The y changes from 3 to -1, or a decrease of $4(\Delta y = -4)$.

$d = \sqrt{(-3)^2 + (-4)^2}$

$d = \sqrt{9+16}$

$d = \sqrt{25} = 5$ The distance between $(-2,3)$ and $(-5,-1)$ is 5.

⁂

Example 2: Find the distance between $(-4,3)$ and $(-4,-7)$.

Solution: The distance formula does not need to be used because the points are on a vertical line. To find the distance between the points, find the absolute value of the difference of the y-coordinates. $d = |-7\ -3| = |-10| = 10$.

⁂

Example 3: Find the distance between $(3,-6)$ and $(4,\ 5)$.

Solution: Let $P_1 = (3,-6)$ and $P_2 = (4,5)$, and evaluate the distance formula.

$d = \sqrt{(\Delta x)^2 + (\Delta y)^2}$ The x changes from 3 to 4, or an increase of $1(\Delta x = 1)$.

The y changes from -6 to 5, or an increase of 11 $(\Delta y = 11)$.

$d = \sqrt{1^2 + 11^2}$

$d = \sqrt{1+121}$

$d = \sqrt{122}$

⁂

Example 4: Find an expression representing the distance between the point (0, 0) and the line described by the linear function $y = x + 3$.

Solution: Every point on the linear function $y = x + 3$ can be represented by the ordered pair of numbers $(x, x+3)$. Use the distance formula, where $P_1 = (0,0)$ and $P_2 = (x, x+3)$.

$$d = \sqrt{(x_2 - x_1)^2 + (y_2 - y_1)^2}$$

$$d = \sqrt{(x-0)^2 + (x+3-0)^2}$$

$$d = \sqrt{x^2 + (x+3)^2}$$

$$d = \sqrt{x^2 + x^2 + 6x + 9}$$

$d = \sqrt{2x^2 + 6x + 9}$. This distance function can be used to find the distance between any point on $y = x + 3$ $\{(x, x+3)\}$ and the origin (0, 0). For example, when $x = 2$, the point on $y = x + 3$ is (2, 5) and the distance from that point to (0, 0) is found by letting $x = 2$ in the function $d = \sqrt{2x^2 + 6x + 9}$. $d = \sqrt{2 \cdot 2^2 + 6 \cdot 2 + 9}$, or $d = \sqrt{29} = 5.39$.
When $x = -3$, the point on $y = x + 3$ is $(-3, 0)$; the distance between the point and (0, 0) can be found by letting $x = -3$ in the above function,

$$d = \sqrt{2(-3)^2 + 6(-3) + 9} = \sqrt{9} = 3$$

⁂

Example 5: Graph the distance function $d = \sqrt{2x^2 + 6x + 9}$ from the previous problem, and give an interpretation of the graph.

Solution: The function can be graphed on the calculator as shown below.

$$d = \sqrt{2x^2 + 6x + 9}$$

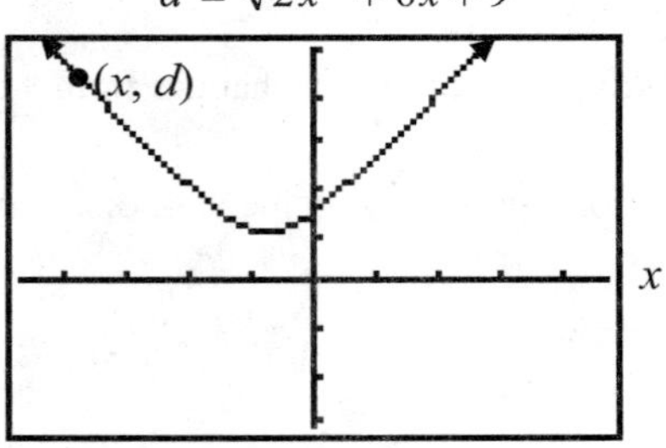

Figure 11.1 5

The coordinates of all points on the graph above are (x, d). The x-coordinate on the above graph comes from the x-coordinate of any point on the graph of $y = x + 3$. The y-coordinate on the above graph represents the distance from the point (0, 0) to a point on the line $y = x + 3$. For example, the point $(-4, 4.123106)$ tells you the distance between the point (0, 0) and the point $(-4, -1)$ on the line $y = x + 3$ is 4.123106.

⁂

Example 6: Find the minimum distance between the point (0, 0) and the line $y = x + 3$.

Solution: The graph in Figure 11.1.5 is a graph of the distance between (0, 0) and $y = x + 3$. The minimum distance is at the minimum point on the distance function. The minimum value of the function is shown below.

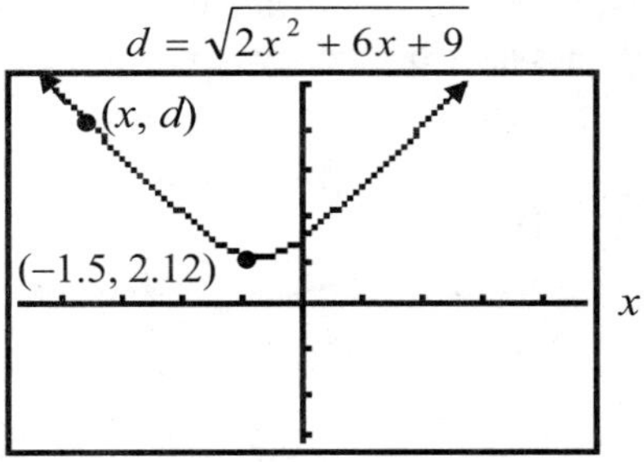

Figure 11.1.6

The minimum point on the distance function is $(-1.50, 2.12)$. This means the shortest distance between (0, 0) and any point on $y = x + 3$ is 2.12. That is, the distance between (0, 0) and $(-1.5, 1.5)$ is 2.12. When $x = -1.5$, the y-coordinate of the point on $x + 3$ is 1.5. No other point on $y = x + 3$ has a shorter distance to (0, 0).

⁂

The Midpoint Formula

The second topic in this section is the midpoint formula. You just learned to find the length of a line segment; now, you will learn how to find the coordinates of the point in the middle of a line segment. Below is the line segment joining the points $P_1(x_1, y_1)$ and $P_2(x_2, y_2)$. The point (x_m, y_m) is the point halfway between P_1 and P_2 it is called the **midpoint**.

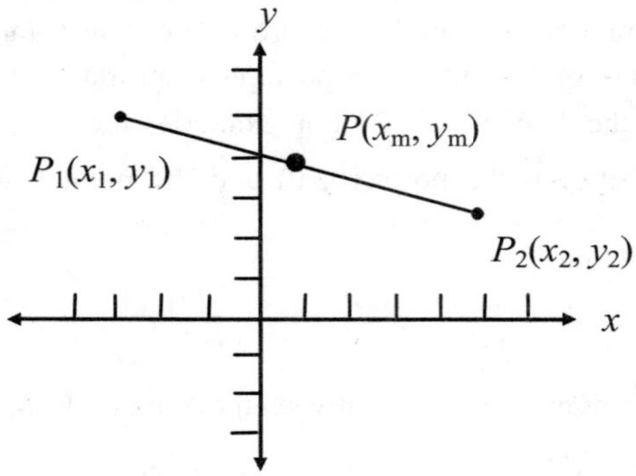

Figure 11.1.7

The goal is to find a way of calculating x_m and y_m when the coordinates of P_1 and P_2 are given. In Figure 11.1.8 extra horizontal and vertical lines have been drawn to help explain how to find the midpoint.

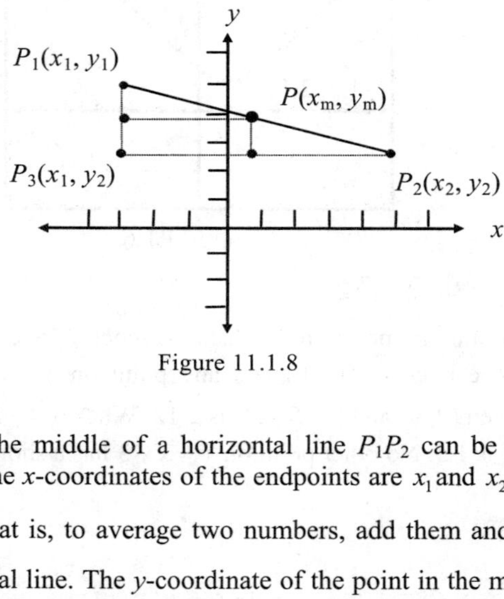

Figure 11.1.8

The x-coordinate of the point in the middle of a horizontal line P_1P_2 can be found by averaging the x-coordinates of the endpoints. If the x-coordinates of the endpoints are x_1 and x_2, the x-coordinate of the point in the middle is $\frac{x_1 + x_2}{2}$. That is, to average two numbers, add them and divide by 2. The same calculation can be done for a vertical line. The y-coordinate of the point in the middle of the vertical line P_1P_3 is the average of the y-coordinates of the endpoints. If the endpoints of a vertical line have y-coordinates of y_1 and y_2, the y-coordinate of the point in the middle is the average, $\frac{y_1 + y_2}{2}$. Without proof, as shown in Figure 11.1.9 below, the midpoint of the line joining P_1 and P_2 has the same x-coordinate as the midpoint of the line joining P_3 and P_2. The midpoint of the line joining P_1 and P_2 has the same y-coordinate as the midpoint of the vertical line P_1P_3.

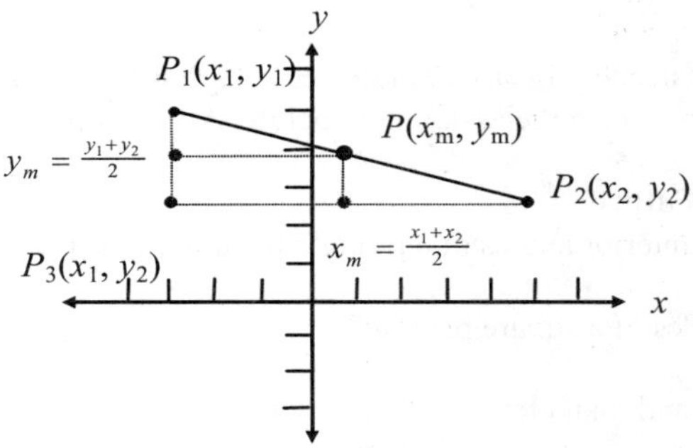

Figure 11.1.9

> THE MIDPOINT FORMULA: The midpoint of the line joining
>
> P_1 and P_2 is $\quad (x_m, y_m) = \left(\dfrac{x_1 + x_2}{2}, \dfrac{y_1 + y_2}{2}\right)$

Example 7: Find the midpoint of the line segment joining (2, 5) and $(6, -3)$.

Solution: Use the midpoint formula: $x_m = \dfrac{2+6}{2} = \dfrac{8}{2} = 4$

$$y_m = \frac{5+(-3)}{2} = \frac{2}{2} = 1$$

The midpoint of the line segment is (4, 1).

⁂

Example 8: Find the midpoint of the line segment joining $(-2, 6)$ and $(5, -1)$.

Solution: From the midpoint formula: $x_m = \dfrac{-2+5}{2} = \dfrac{-3}{2}$

$$y_m = \frac{6+(-1)}{2} = \frac{5}{2}$$

The midpoint of the line segment is $\left(\dfrac{-3}{2}, \dfrac{5}{2}\right)$.

⁂

11.1 EXERCISES

Use a graphing utility as needed. Please read the text carefully before you do the exercises. Read the **next assigned section** after you finish the assignment below. All numeric answers must have an error ≤ 0.01.

A Quick Look To The Future

-4. Can the sum of the interior angles of a triangle be larger that 180°?

-3. Are the opposite sides of a square parallel?

-2. How are the radius and diameter of a circle related?

-1. What units do you use to measure the size of an angle?

Previous Topics Revisited

1. Fireworks are shot straight up from ground level at an initial velocity 500 feet per second; 5 seconds later they explode, how high up is the explosion?

2. In the above problem, do the fireworks explode at the maximum height of the trajectory?

3. What approximate initial velocity should be used to have the fireworks reach 2100 feet 5 seconds after being shot from ground level?

4. Solve $3x^2 - 4x + 2 = 0$ for complex roots - use the quadratic formula.

5. Solve $-2x^2 + 5x - 1 = 0$ by the quadratic formula.

6. Solve $x^2 + 4x - 6 = 0$ by completing-the-square.

New Topic Exercises

Find the length and midpoint of the line segment joined by the given points.

7. $(-3,-6)$ and $(-5,-8)$

8. $(4,2)$ and $(3,8)$

9. $(-3,4)$ and $(5,-2)$

10. $(5,-2)$ and $(4,3)$

11. $(0,-1)$ and $(9,13)$

12. $(-42,100)$ and $(-16,0)$

13. $(0,0)$ and (a,b)

14. Find the distance between (0, 0) and $(x, x-2)$.

15. Find the minimum distance between (0, 0) and $y = x - 2$.

16. Find the distance between (1, 3) and $(x, x-3)$

17. Find the minimum distance between (1, 3) and $y = x - 3$

18. What is the distance from LA to NYC?

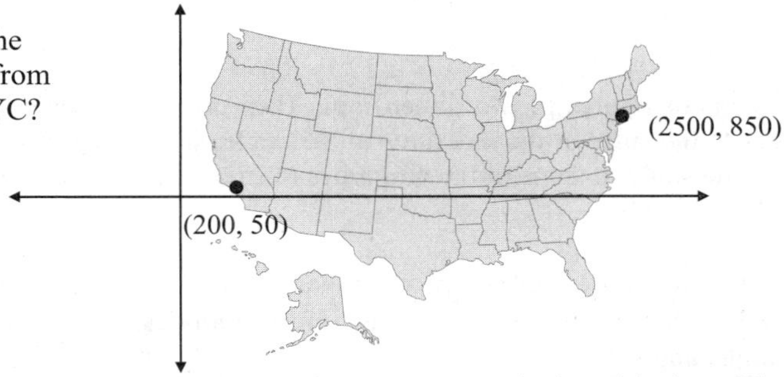

19. If (2, 3) is an endpoint of a line segment and (4, 6) is the midpoint of the segment, where is the other endpoint?

20. Find a formula for the distance between the origin and any point (x, y).

From Mathematics to English

21. What is a midpoint of a line segment?

22. Describe how to find the length of the line joining (–3, 4) and (2, 1). Do not actually find the length.

23. What is the difference between a line and a line segment?

24. After reading this section, make a list of questions that you want to ask your instructor.

25. Continue in your daily journal and make an entry. In addition to your normal entry on thoughts about the mathematics in this section, list at least two positive comments about what you have learned about this topic.

26. In paragraph format, summarize the material in this section of the text in your daily journal.

27. Describe how your classroom instructor made this topic more understandable and clear.

28. After reading the text and listening to your instructor, what do you not understand about this topic?

Explorations

29. Develop an argument *verifying* that $\left(\frac{x_1+x_2}{2}, \frac{y_1+y_2}{2}\right)$ is the point in the middle of the line segment joining (x_1, y_1) and (x_2, y_2).

30. Below are the coordinates of vertices of four right triangles. If the midpoints of each side are joined by line segments, are these new triangles right triangles? Why? What conclusions can you draw about all right triangles?
    a. (4, 0), (0, 6), and (0, 0)
    b. $(-2,8)$, (4, 2), and $(0,-2)$
    c. $(-4,-6)$, (8, 6), and $(-2,16)$
    d. $(1,-5)$, $(7,-1)$, and $(-3,14)$

Make-up the vertices of your own right triangle, and test your conclusion made above.

11.2 Triangles

Triangles are three-sided polygons or figures in two dimensions. They come in many shapes and sizes and are identified by the lengths of the sides or the measures of the angles in the triangle. In this section, the names of triangles as well as the study of some of the properties of triangles will be studied.

Angles that measure from 0° to 90° are **acute angles**. An angle that measures 90° is called a **right angle**, and angles whose measures are between 90° and 180° are called **obtuse angles**. Finally, an angle whose measure is 180° is called a **straight angle**.

The names of triangles as related to the angles in the triangles are:

ACUTE TRIANGLE	**all angles are acute**
RIGHT TRIANGLE	**one angle measures 90°**
OBTUSE TRIANGLE	**one angle is obtuse**

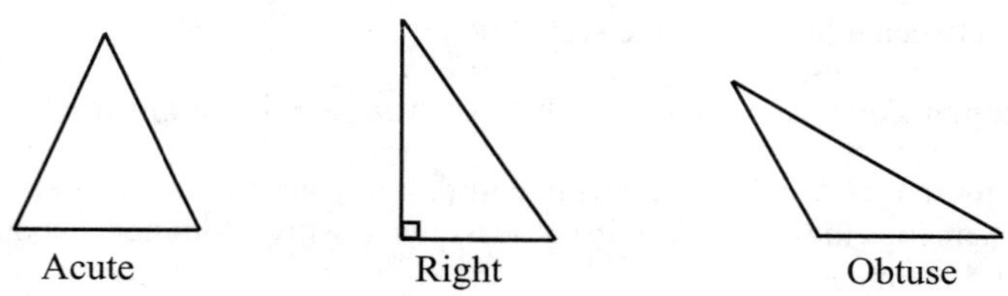

Figure 11.2.1

The names of triangles as related to sides are:

EQUILATERAL TRIANGLE	**all sides have the same length**
ISOSCELES TRIANGLE	**two sides have the same length**

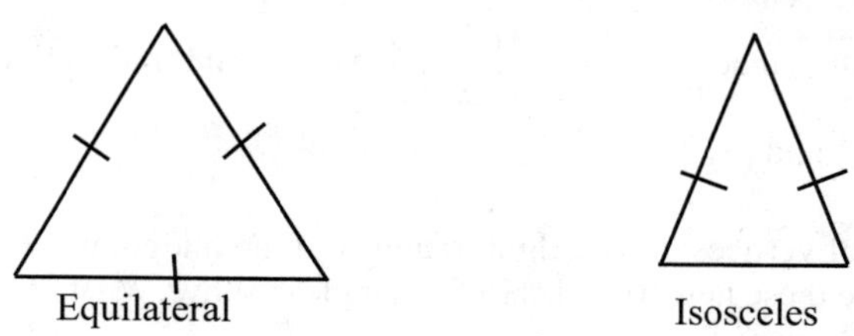

Figure. 11.2.2

The angles of a triangle are normally identified with capital letters and the sides opposite each angle are labeled with the same letter except in the lower case. In a right triangle, the side opposite the right angle (the hypotenuse) might commonly be labeled c, and the right angle C.

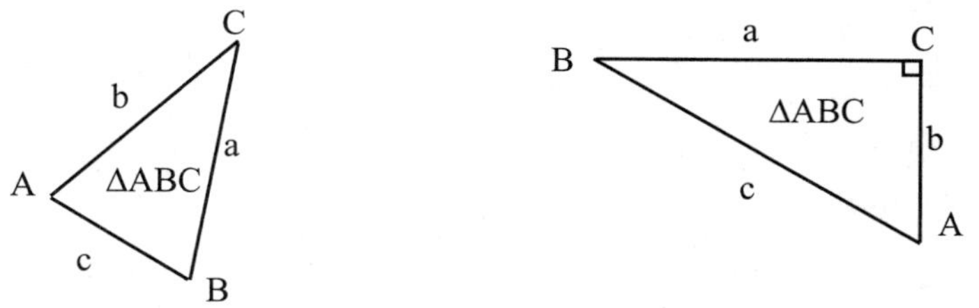

Figure 11.2.3

Two Important Properties
You may remember two relationships about triangles. The sum of the measures of the interior angles of any triangle is 180°; in triangle ABC, ∠A + ∠B + ∠C = 180°. The second is the Pythagorean Theorem; in right triangle ABC, $c^2 = a^2 + b^2$; the hypotenuse squared equals the sum of the squares of the legs. Remember, the Pythagorean Theorem is true for right triangles only and the first relationship is true for all triangles in a plane.

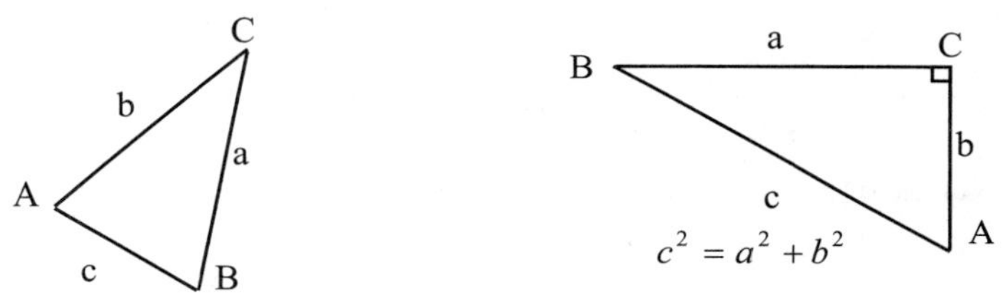

$\angle A + \angle B + \angle C = 180°$

Figure 11.2.4

Example 1: In triangle ABC, ∠A = 50°, ∠C = 120°, find ∠B.

Solution: From ∠A + ∠B + ∠C = 180°, replace ∠A and ∠C with 50° and 120° respectively.

$$50 + \angle B + 120 = 180$$

$$\angle B + 170 = 180$$

$$\angle B = 10°$$

⁂

Example 2: In right triangle ABC, $a = 5$, $b = 12$, find c.

Solution: From the Pythagorean Theorem, solve for c.

$$c^2 = a^2 + b^2$$

$$c^2 = 5^2 + 12^2$$

$$c^2 = 25 + 144$$

$$c^2 = 169$$

$$c = 13$$

⁂

Example 3: In an equilateral triangle, not only are all sides the same length, but the angles are the same size as well. What are they?

Solution: Since all angles are the same, call them x, this means in the statement

$\angle A + \angle B + \angle C = 180°$, replace all angles with x and solve.

$$\angle A + \angle B + \angle C = 180°$$

$$x + x + x = 180$$

$$3x = 180$$

$$x = 60°$$

Each angle of an equilateral triangle is 60°.

⁂

Isosceles Properties

The next set of properties relates to the isosceles triangle. In an isosceles triangle, the side not equal to either of the other two sides is called the base. The angle opposite the base is the **vertex angle** and the angles opposite the equal sides are called the **base angles**.

Isosceles Properties

I. A line drawn from the vertex perpendicular to the base, bisects the base.

II. The perpendicular bisector of the base in an isosceles triangle bisects the vertex angle.

The properties tell you the line drawn from the vertex and perpendicular to the base bisects the base and bisects the vertex angle. **Bisects** means to divide into two equal parts. Because the line is perpendicular to the base, it forms right angles with the base; thus, the perpendicular bisector divides the isosceles triangle into two right triangles of the same size.

Example 4: Show that the base angles of an isosceles triangle are equal in size.

Solution: When a perpendicular bisector is drawn, it divides the isosceles triangle into two right triangles and it bisects the vertex angle of the isosceles; thus, angles C_1 and C_2 are equal. Since the perpendicular bisector forms right angles at the base, each of the angles D_1 and D_2 are equal. You can conclude from this argument that the third angles, $\angle A$ and $\angle B$ are also equal because the sum of the angles in each right triangle must equal 180°. That is, the first two respective angles in each triangle are equal; therefore, the third angles must be equal or the 180° sum property would be violated.

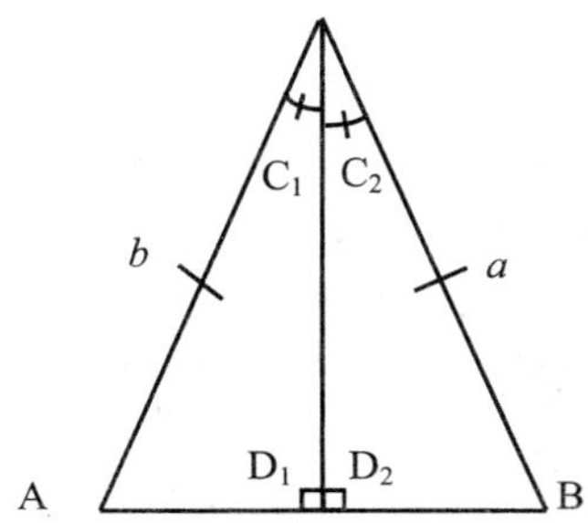

Figure 11.2.5

⁂

The Altitude
A line drawn from any vertex of a triangle perpendicular to the opposite side is called an **altitude** or **height** of the triangle. In some cases, the altitude must be drawn to a line containing the side opposite the vertex.

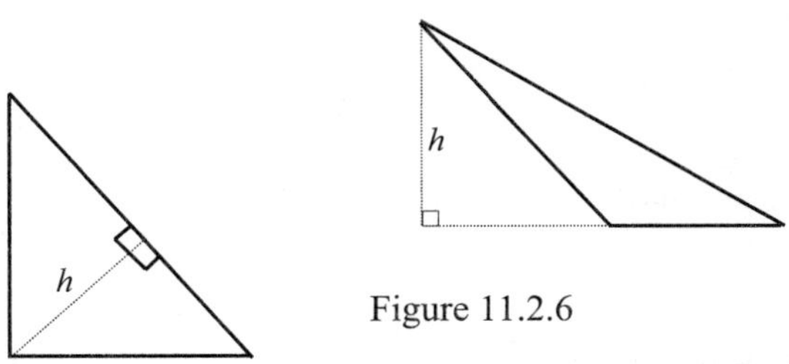

Figure 11.2.6

Example 5: Find the length of the altitude of an equilateral triangle whose sides are 2 units long.

Solution: Draw an altitude h to one side. Because the triangle is isosceles, it bisects the side and divides the equilateral into two right triangles. In either of the right triangles, the hypotenuse is 2 and the shortest leg is 1. Use the Pythagorean Theorem to find the third side (altitude) of the right triangle.

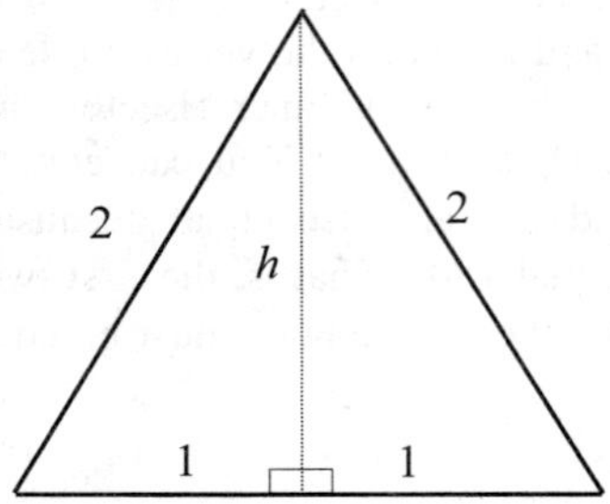

Figure 11.2.7

From the right triangle, $2^2 = 1^2 + h^2$

$$4 = 1 + h^2$$
$$3 = h^2$$
$$\sqrt{3} = h$$

In an equilateral triangle with sides 2, the altitude is $\sqrt{3}$ units long. Each of the right triangles has acute angles of 30° opposite the side of length 1, and 60° opposite the side of length $\sqrt{3}$.

⁂

The Right Isosceles

A triangle that is right and isosceles is called a **right isosceles triangle**. The right angle is the vertex angle and the base is the hypotenuse.

Example 6: What are the acute angles of a right isosceles triangle?

Solution: Since the acute angles are the base angles of an isosceles triangle, they are equal in size. Identify them as x. From the 180° property: $x + x + 90° = 180°$

$$2x + 90° = 180°$$

$$2x = 90°$$

$$x = 45°$$

Each of the acute angles of a right isosceles is 45°.

⁂

Similar Triangles

Triangles that 'look' similar are shown in Figure 11.2.8.

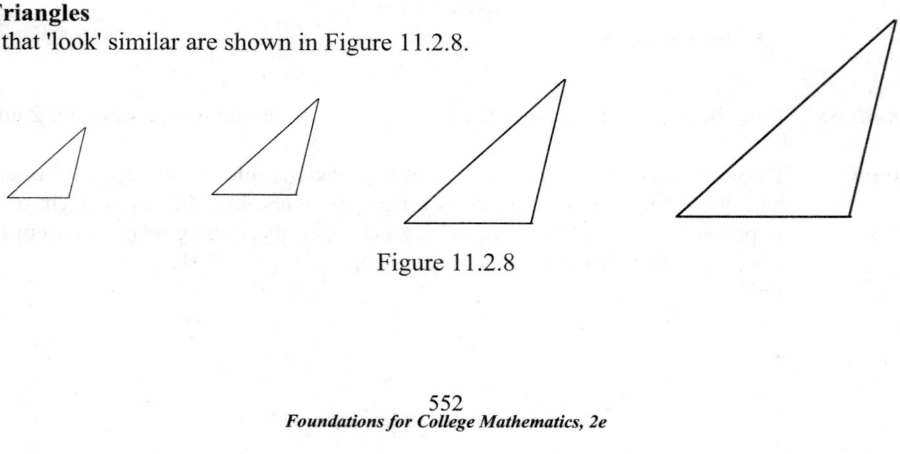

Figure 11.2.8

While the above triangles may 'look' similar to you, a formal interpretation of what it means for polygons to be similar is needed. Triangles (and polygons) with corresponding angles equal are called **similar triangles** (similar polygons), and in similar triangles (polygons) corresponding sides are in proportion as shown below.

Similar Triangles

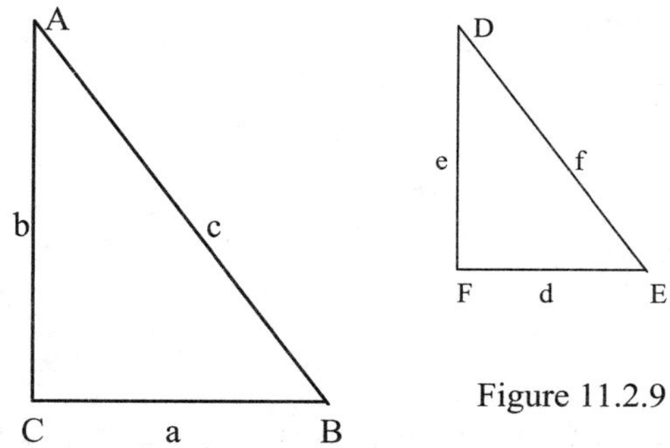

If

$\angle A = \angle D$
$\angle B = \angle E$
$\angle C = \angle F$

then

$$\frac{a}{d} = \frac{b}{e} = \frac{c}{f}$$

Figure 11.2.9

Example 7: In similar triangles ABC and A' B' C', $a = 5, b = 7$, and $a' = 9$. Find side b'.

Solution: Similar triangles have corresponding sides in proportion;

thus $\frac{a}{b} = \frac{a'}{b'}$

$\frac{5}{7} = \frac{9}{b'}$ multiply both sides by $7b'$

$5b' = 63$ divide by 5

$b' = 12.6$

⁂

Perimeter and Area

The discussion of triangles ends by looking at the area and perimeter of any triangle.

> The perimeter of any polygon is the sum of the lengths of the sides that make up the polygon.

Triangle ABC has a perimeter of $P = a + b + c$. The perimeter is measured in linear units; that is, units like feet, inches, centimeters, meters, yards, etc, because perimeter is a length or distance and these units measure length.

Example 8: The perimeter of a triangle is 17 feet, side $a = 5$ feet, and side $b = 8$ feet, find the third side.

Solution: From the formula $P = a + b + c$, find the third side.

$$P = a + b + c$$

$$17 = 5 + 8 + c$$

$$17 = 13 + c$$

$$4 = c$$

The third side is 4 feet long.

⁂

The most commonly used formula for the area of a triangle requires that the length of a side (usually called the base) be known and the length of the height (altitude to the known side) must also be known.

> The Area of any Triangle is: $A = \frac{1}{2}bh$, where b is any side and h is the height (altitude) to that side.

The area is measured in square units; for example, square inches $\left(\text{in}^2\right)$, square feet$\left(\text{ft}^2\right)$, square centimeters $\left(\text{cm}^2\right)$, etc.

Example 9: Find the area of an equilateral triangle whose sides are 2 cm.

Solution: All sides are 2 cm, so the base can be 2 cm. From Example 5, the height is $\sqrt{3}$.
The area is: $A = \frac{1}{2} \cdot 2 \cdot \sqrt{3}$. The area is $\sqrt{3}\ \text{cm}^2$.

⁂

> Another formula for the area of a triangle is Heron's formula:
> $A = \sqrt{s(s - a)(s - b)(s - c)}$, where $s = \frac{1}{2}(a + b + c)$ and a, b, and c are the lengths of the sides of the triangle, where s is called the **semi-perimeter**.

Example 10: Find the area of an equilateral triangle with sides of 2.

Solution: $s = \frac{1}{2}(2+2+2) = \frac{1}{2}(6) = 3$

$$A = \sqrt{3(3-2)(3-2)(3-2)} = \sqrt{3 \times 1 \times 1 \times 1} = \sqrt{3}\ .$$

11.2 EXERCISES

Use a calculator as needed. Please read the text carefully before you do the exercises. Read the **next assigned section** after you finish the assignment below. All numeric answers must have an error ≤ 0.01.

A Quick Look To The Future

-4. What does the word "parallelogram" mean to you?

-3. What is the distance around a circle called?

-2. What is a unit of angular measure that is smaller that a degree?

-1. If the sides of a triangle are a, b, and c, how many possible ratios (fractions) can be formed from these three numbers?

Previous Topics Revisited

1. Find the length and midpoint of the line joining $(-2,-6)$ and $(3,-4)$.

2. Find the distance between the point $(-1,1)$ and $(x, x+1)$.

3. Find the minimum distance between the point $(-1,1)$ and the line $y = x+1$.

4. A water balloon is thrown downward with an initial velocity of 10 feet per second. How long does it take to hit the ground 50 feet below?

5. A water balloon is thrown upward with an initial velocity of 10 feet per second. How long does it take to hit the ground 50 feet below?

6. Find the complex roots to the equation $3x^2 - 4x + 4 = 0$.

New Topic Exercises

7. Identify the triangles relative to angles and sides.

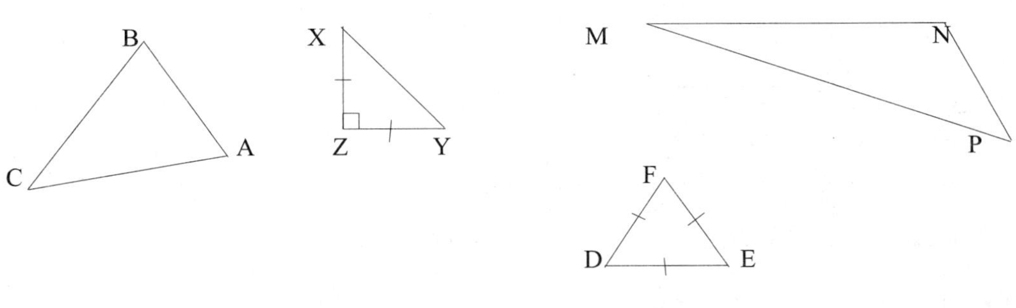

8. A triangle has vertices in the coordinate plane at (0, 0) (0, 4) and (6, 2). What kind of triangle is it? (list all possible identifications)

9. A triangle has vertices in the coordinate plane at (0, 0) (0, 4) and (4, 0). What kind of triangle is it? (list all possible identifications)

10. The vertex of an isosceles triangle is 100°, what are the base angles?

11. The right triangle ABC has legs of 4 and 6; how long is the hypotenuse?

12. Find the lengths of an altitude in each of the equilateral triangles whose sides are 3, 4, and 5. Express the lengths in simplified form.

13. Triangles ABC and ADE are similar; $b = 5$, $c = 8$, $d = 12$, find side e.

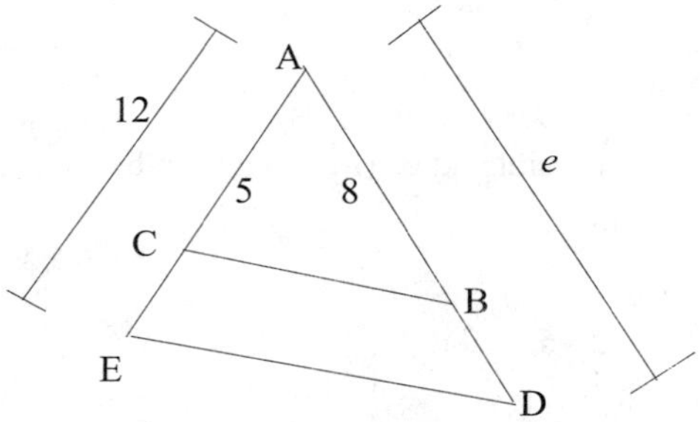

14. The right triangles ABC and CDE are similar; $b = 4$, $a = 3$, $e = 8$, find the length of DE.

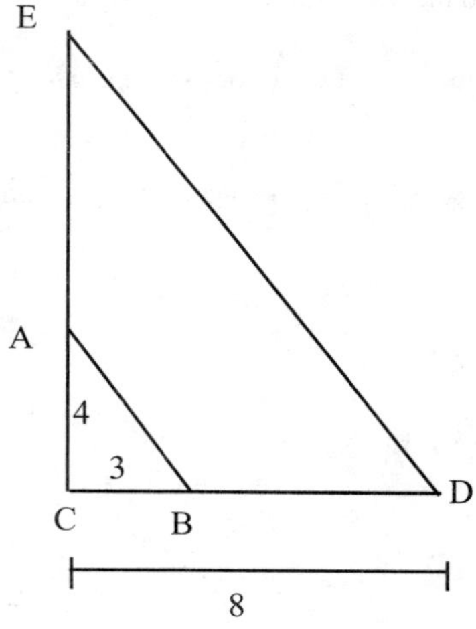

15. A right triangle has sides of 5, 12, and 13. Find the area and perimeter of the triangle.

16. A triangle has sides of 6, 8, and 10. Find the area and perimeter of the triangle.

17. Find the area and perimeter of a triangle whose vertices are at (0, 0). (6, 0), and (0, –8).

18. Find the area and perimeter of a triangle whose vertices are at (2, 2), (–2, 2), and (0, 0).

19. Find the area and perimeter of a triangle whose vertices are at (3, 0), (6, 4), and (–2, –2).

From Mathematics to English

20. Can a triangle drawn in a plane contain two obtuse angles? Why?

21. Explain why the sum of the measures of the interior angles of a rectangle is 360°.

22. Can a right triangle also be an obtuse triangle? Why?

23. Define, using your own terminology, a. Triangle b. Similar triangles c. Area of a triangle

24. Describe how to find the area of the triangle whose vertices are at (–1, 3), (4, 5), and (2, –3). Do not actually find the area.

25. Describe how to find the perimeter of the triangle whose vertices are at (–1, 3), (4, 5), and (2, –3). Do not actually find the perimeter.

26. Why can a right triangle never be equilateral?

27. After reading this section, make a list of questions that you want to ask your instructor.

28. Continue in your daily journal and make an entry. In addition to your normal entry on thoughts about the mathematics in this section, list at least two positive comments about what you have learned about this topic.

29. In paragraph format, summarize the material in this section of the text in your daily journal.

30. Describe how your classroom instructor made this topic more understandable and clear.

31. After reading the text and listening to your instructor, what do you not understand about this topic?

Explorations

32. Draw an irregular shaped polygon with six sides. Using only a ruler and Heron's formula, find the area of the polygon.

33. What kind of triangle is formed by joining the points in succession? Justify your answer.

    a. $(0,0)$, $(-6,0)$, and $(-3,7)$
    b. $(-4,4)$, $(6,4)$, and $(6,12)$
    c. $(-1,-1)$, $(3,-5)$, and $(7,-1)$
    d. $(-5,5)$, $(3,3)$, and $(-3,-3)$

34. What kind of triangle is formed by the intersection of the three lines below?

    $y=\sqrt{3}\cdot x+6,\ \ y=-\sqrt{3}\cdot x+6,$ and $y=0x-6$.

11.3 Parallelograms

A **parallelogram** is a four-sided polygon (quadrilateral) with opposite sides parallel. This includes the more familiar quadrilaterals -- the rectangle and square. The parallelogram gets its name from the fact that opposite sides of the quadrilateral are parallel.

Properties
In addition to opposite sides being parallel, you probably have noted that:

> **Properties of a Parallelogram:**
>
> opposite sides are equal in length
>
> angles at opposite corners are equal in measure

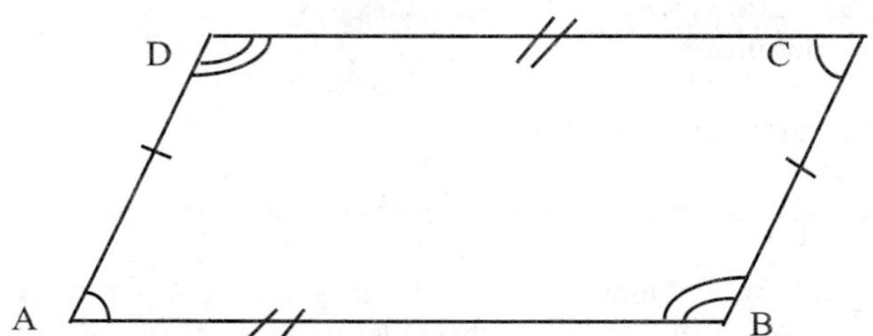

Figure 11.3.1

The rectangle is a parallelogram with right angles at the corners and a square is a rectangle with equal sides. While these descriptions point out some of the obvious characteristics of the rectangle and square, below are a few more:

> **Properties of the Square and Rectangle**
>
> The diagonals of a rectangle are equal in length.
>
> The diagonals of a square are perpendicular.

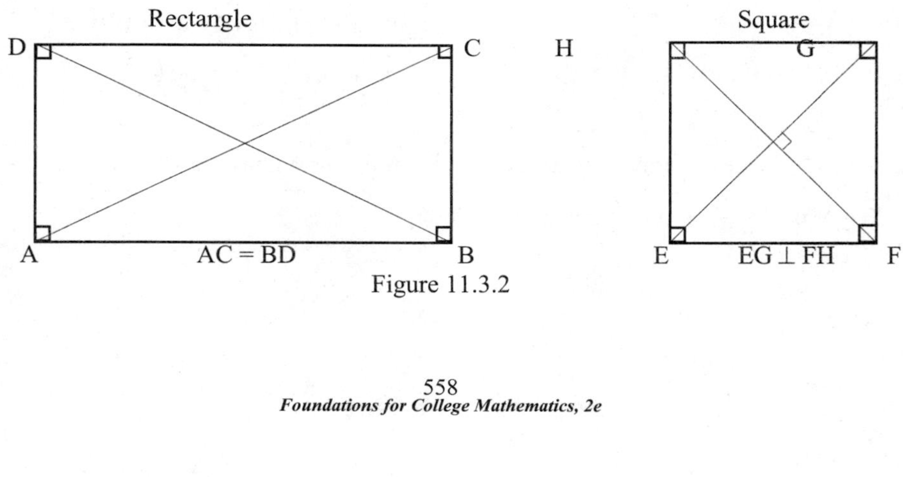

Figure 11.3.2

Perimeter and Area

The idea of a perimeter is that of the "distance around a polygon." If adjacent sides of a parallelogram are labeled *1* and *w*, for length and width, the perimeter *P* is $1 + w + 1 + w$, or $P = 2(1 + w)$.

The perimeter of a parallelogram is $P = 2(1 + w)$

Below, parallelogram ABCD is divided into two triangles, ABD and BCD, by a diagonal. The triangles ABD and BCD are identical because opposite sides of a parallelogram equal, and the diagonal of the parallelogram is common to both triangles; thus, the triangles are identical (**congruent**).

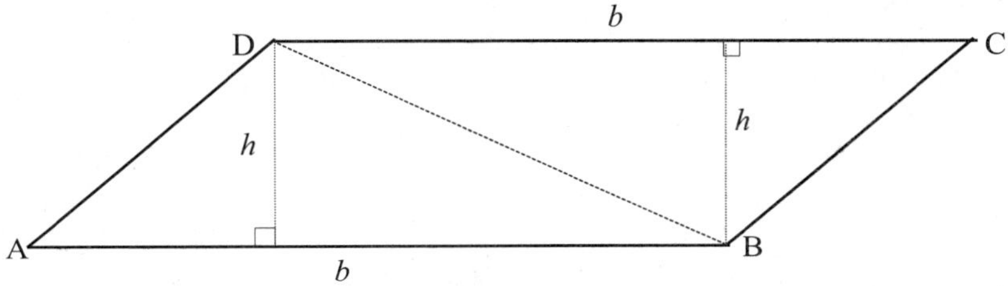

Figure 11.3.3

The height of the triangles are marked as *h*, and the side the heights are drawn to are labeled as *b*. Remember that the side and height must form a right angle. The area of the parallelogram is the sum of the areas of the triangles. That is: $A = \frac{1}{2}bh + \frac{1}{2}bh = 1bh$.

The area of a parallelogram is $A = bh$

Example 1: What is the formula for the area of a rectangle?

Solution: The area of any *parallelogram* is the product of a side times the height drawn to that side. In the rectangle below, *1* is a side and *w* is the height to that side; thus, the area is *1w*. Area of a rectangle is $A = l \cdot w$, where *1* is any side and *w* is the adjacent side.

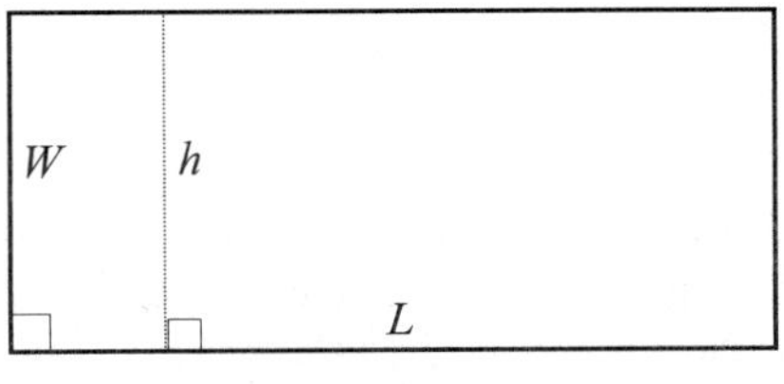

Figure 11.3.4

⁂

Example 2: How long is the diagonal of a square whose side is 10 meters.

Solution: The diagonal forms a right triangle and is the hypotenuse. The sides of the square are the legs of the triangle.

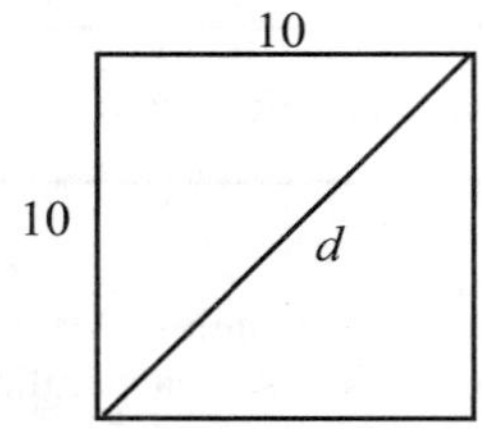

Figure 11.3.5

From the Pythagorean Theorem: $d^2 = 10^2 + 10^2$

$$d^2 = 100 + 100$$

$$d^2 = 200$$

$$d = \sqrt{200} = 10\sqrt{2} = 14.142$$

⁂

Example 3: Is the polygon formed by connecting the given points in succession with line segments a parallelogram? Points: $A(-3,2)$, $B(5,6)$, $C(7,9)$, $D(-1,4)$

Solution: It is helpful to plot the points, but the visualization does not prove it is a parallelogram.

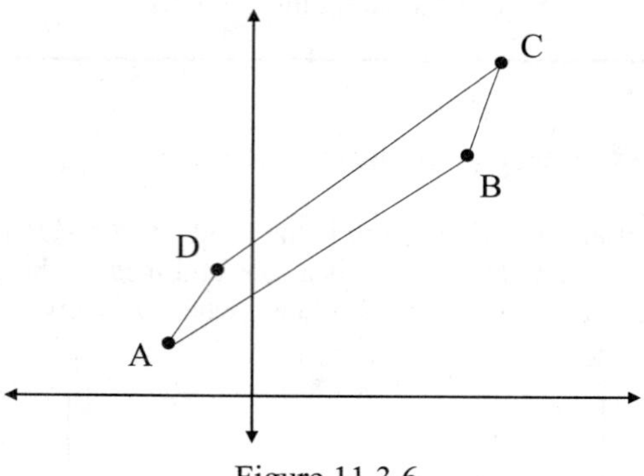

Figure 11.3.6

The definition requires that opposite sides be parallel. From Chapter Five, you learned parallel lines have the same slopes. To prove the quadrilateral is a parallelogram, show that opposite sides have the same slopes.

$$\text{Slope of AB} = \frac{4}{8} = \frac{1}{2}$$

$$\text{Slope of CD} = \frac{-5}{(-8)} = \frac{5}{8}$$

NO! The quadrilateral is not a parallelogram; because, the opposite sides AB and CD do not have equal slopes.

⁂

11.3 EXERCISES

Please read the text carefully before you do the exercises. Read the **next assigned section** after you finish the assignment below. All numeric answers must have an error ≤ 0.01. Use the calculator as needed.

A Quick Look To The Future

-4. Is the area different than the circumference of a circle?

-3. Is there more than one set of units used to measure the size of an angle?

-2. A line drawn from the origin of the *xy*-plane into quadrant I contains the point (3, 4). How far is that point from the origin?

-1. If the legs of a right triangle are known, how do you find the hypotenuse?

Previous Topics Revisited

1. The base angles of an isosceles triangle are each 20°, find the vertex angle.

2. Find the altitude to the base of an isosceles triangle, if the base is 10 inches and the equal sides are 15 inches.

3. Triangles ABC and ADE are similar, side AB = 5, AD = 12, AE = 15, find the length of side AC.

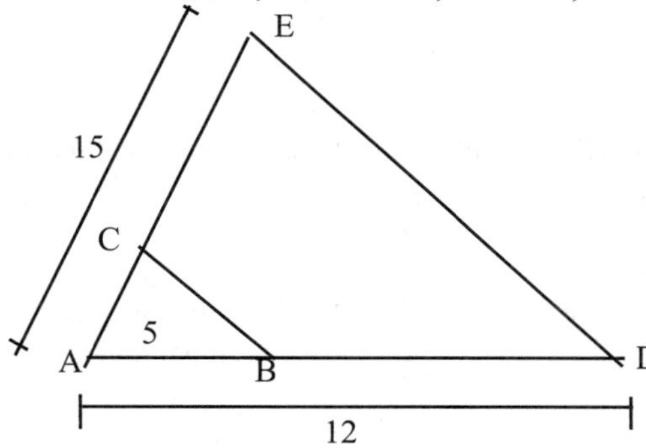

4. For the line segment whose endpoints are $(-2,5)$ and $(-3,-7)$, find the length and midpoint.

5. Find the distance between $(-2,-2)$ and $(x, x+2)$; also find the minimum distance between $(-2,-2)$ and the line $y = x + 2$.

6. An object is thrown upward from ground level at an initial velocity of 100 feet per second; what is its maximum height?

New Topic Exercises

7. Find ∠ACB and ∠CAB in parallelogram ABCD

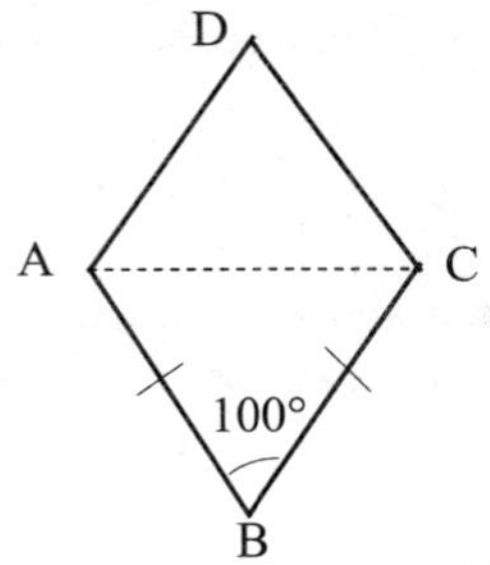

8. Find x in parallelogram ABCD.

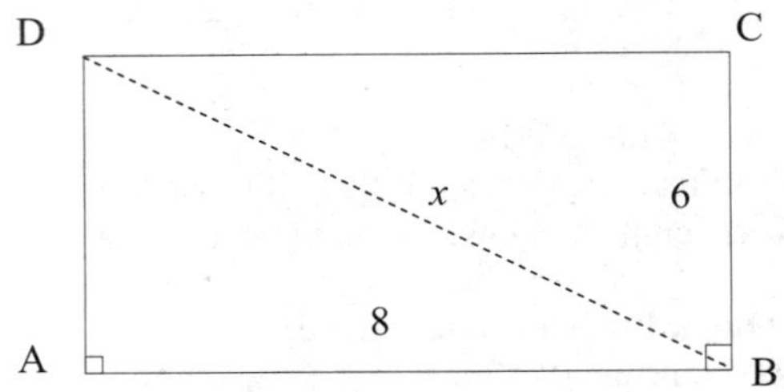

9. The points $A(-5,2)$, $B(3,-4)$, $C(4,-7)$, and $D(?,?)$ when connected in succession, form a parallelogram. Find point D.

10. Find the lengths of the diagonals of the three squares with sides of 6, 7, and 8. Express the diagonals in simplified radical form.

11. Find θ in parallelogram ABCD.

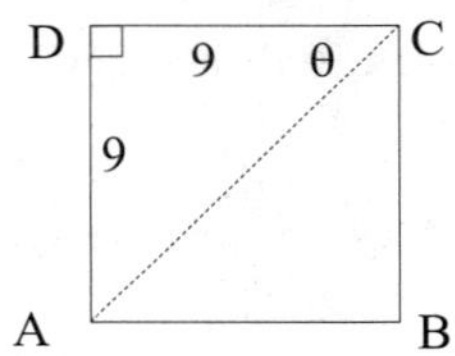

12. If the points $A(-1,-4)$, $B(3,-2)$, $C(1,1)$, and $D(-3,-1)$ are joined in succession to form a quadrilateral, is it a parallelogram?

13. Find the perimeter and the area of the figure.

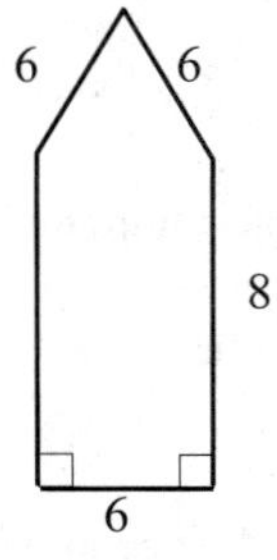

14. Find the perimeter and the area of the 4 × 8 parallelogram.

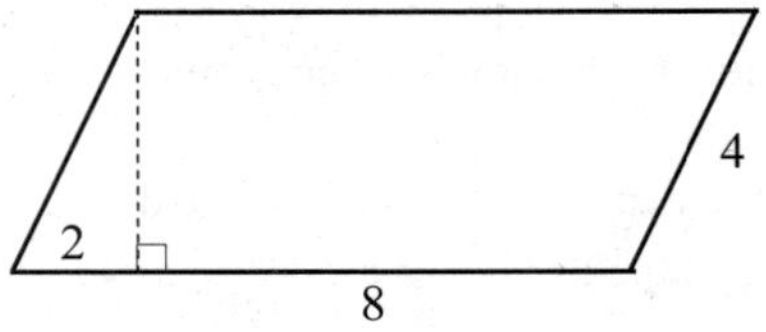

15. Find A in the parallelogram.

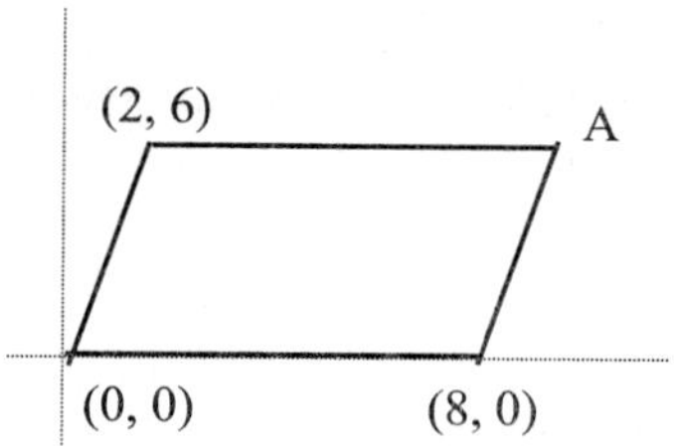

16. Find the area by two different methods. Show your work. All angles are 90°.

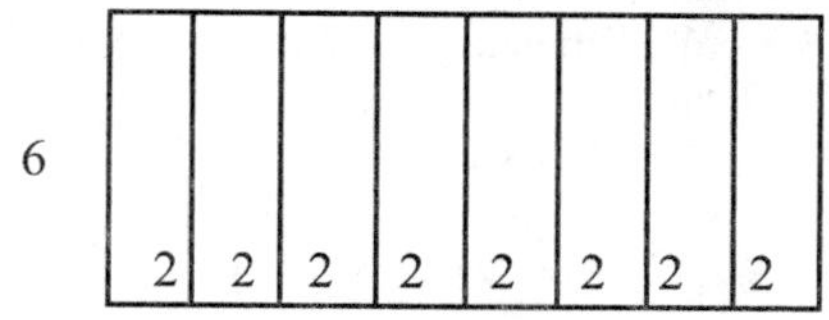

17. Find the area. All angles are 90°.

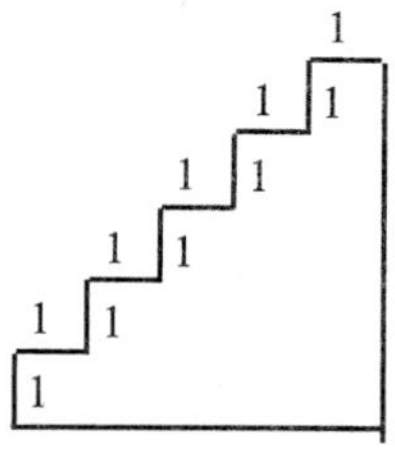

18. The midpoints of the sides of the 6 × 10 rectangle are connected. Find the area of the shaded region and the un-shaded region.

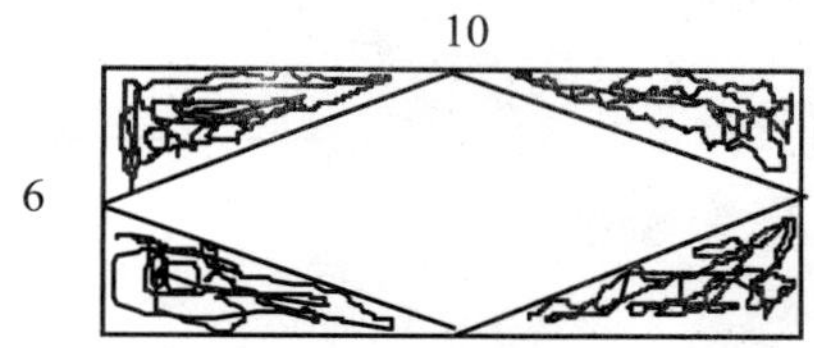

From Mathematics to English

19. Describe how to find the area of the figure drawn below. Do not find the area. All line segments are 4 inches. The middle figure is a parallelogram.

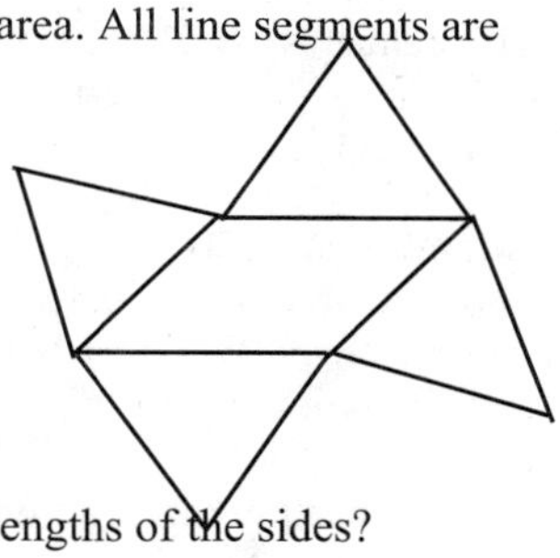

20. Make a list of everything you know about the rectangle.

21. Explain why opposite interior angles in a parallelogram are the same size.

22. What is the difference between a rectangle and a square -- other than the lengths of the sides?

23. After reading this section, make a list of questions that you want to ask your instructor.

24. Continue in your daily journal and make an entry. In addition to your normal entry on thoughts about the mathematics in this section, list at least two positive comments about what you have learned about this topic.

25. In paragraph format, summarize the material in this section of the text in your daily journal.

26. Describe how your classroom instructor made this topic more understandable and clear.

27. After reading the text and listening to your instructor, what do you not understand about this topic?

Explorations

28. Draw a complex geometric shape consisting of triangles and parallelograms. Using only a ruler, find the approximate area of the region.

29. If a gallon of paint covers 500 square feet, will 1 gallon cover the walls and ceiling of a room 10 feet wide by 12 feet long and 8 feet high? Will it cover the walls of your classroom?

11.4 Circles

A **circle** is the set of all points (x, y) that are at a given distance from a given point. The given (known) distance r is called the **radius** of the circle and the given (known) point (h, k) is called the **center** of the circle.

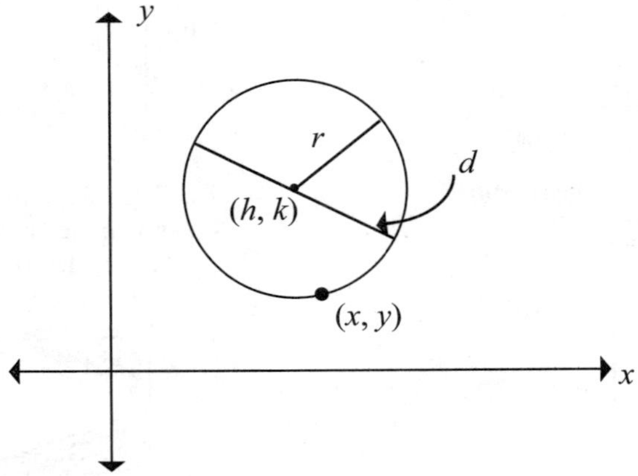

Figure 11.4.1

The center is NOT part of the circle. The radius is NOT part of the circle. The diameter - except the endpoints - is NOT part of the circle. The circle is the set of points (x, y) at the given distance from the given point.

Equation of a Circle
The equation of the circle is not a function because for most x's in the domain there can be two values for y in the range; that is, it fails the vertical line test. The relationship between the center (h, k), the radius r, and the points (x, y) that make-up the circle is found in the distance formula. The "distance" from the center
(h, k) to each point (x, y) on the circle is r. The distance formula is used to express the distance between (h, k) and (x, y), and this distance is equal to r.

$\sqrt{(x-h)^2+(y-k)^2}=r$ square both sides

$(x-h)^2+(y-k)^2=r^2$ This is the standard form of the equation of a circle.

> The equation of a circle with center (h, k), radius r, and whose points are represented by (x, y) is $(x-h)^2+(y-k)^2=r^2$.

Example 1: The circle whose center is at the origin and whose radius is 1 is called the **unit circle**; find the equation of the unit circle.

Solution: The center is given to be (0, 0) and r is given to be 1. With (h, k) being (0, 0) and r as 1, replace the letters h, k, and r with the given values in the above equation.

$$(x-h)^2+(y-k)^2=r^2$$
$$(x-0)^2+(y-0)^2=1^2$$
$$x^2+y^2=1$$

⁂

Example 2: Find the equation of the circle whose center is $(-3,5)$ and whose radius is 2.

Solution: The given information is h, k, and r; (h, k) is $(-3,5)$ and r is 2. Substitute these values in the equation of the circle above.

$$(x-h)^2+(y-k)^2=r^2$$
$$(x-(-3))^2+(y-5)^2=2^2$$
$$(x+3)^2+(y-5)^2=4$$

⁂

Example 3: What is the equation of a circle whose endpoints of a diameter are (–3, 6) and (5, –8).

Solution: While a visualization of the problem won't solve the problem, it will be helpful in understanding the problem. Below is a picture of the circle.

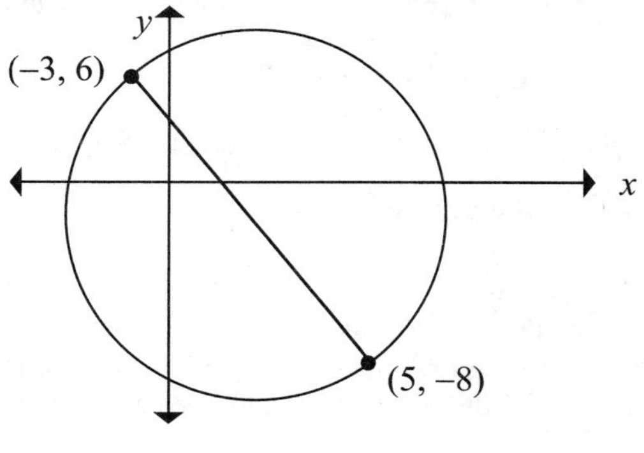

Figure 11.4.2

No matter what is given, to find the equation of the circle, the center and radius are needed. The center is the midpoint of the diameter; thus, the center is:

$$\left(\frac{-3+5}{2},\frac{6+(-8)}{2}\right)=(1,-1)=(h,k)$$

The radius is the distance from the center to the circle. The center is $(1,-1)$ and a given point on the circle is $(-3,6)$; thus the radius is:

$$r=\sqrt{(1-(-3))^2+(-1-6)^2}$$

$$r=\sqrt{4^2+(-7)^2}$$

$$r=\sqrt{16+49}$$

$$r=\sqrt{65}$$

The equation of the circle is:

$$(x-h)^2+(y-k)^2=r^2$$

$$(x-1)^2+(y-(-1))^2=\sqrt{65}^2$$

$$(x-1)^2+(y+1)^2=65$$

⁂

The graphing calculator can be used to graph the equation of the circle even though it is not a function. It is not quite as simple as graphing a function, but nearly. As you know, to graph any function, the y must be by itself on one side of the equal sign. To graph $(x-1)^2+(y+1)^2=65$, isolate y.

from	$(x-1)^2+(y+1)^2=65$	subtract $(x-1)^2$
	$(y+1)^2=\pm\sqrt{65-(x-1)^2}$	take the square root
	$y+1=\pm\sqrt{65-(x-1)^2}$	subtract 1
	$y=-1\pm\sqrt{65-(x-1)^2}$	

Graph $y=-1+\sqrt{65-(x-1)^2}$ and $y=-1-\sqrt{65-(x-1)^2}$ on the same screen to get the graph of $(x-1)^2+(y+1)^2=65$. (Suggestion: use a decimal window.)

Depending on the viewing window, the graph of any circle may look more like an ellipse than a circle. Also, depending on the window, the graph may not be a complete circle. It may be a lot simpler to graph the equation by hand. If the center and radius are known, plot the center and draw a circle with the known radius.

General Form of the Equation of a Circle

Another way of looking at the equation of a circle is to take the above equation and perform the following operations.

From the equation of the circle $\qquad (x-h)^2+(y-k)^2=r^2$

square the binomials $$x^2 - 2hx + h^2 + y^2 - 2ky + k^2 = r^2$$

or $$x^2 + y^2 - 2hx - 2ky + \left(h^2 + k^2 - r^2\right) = 0\,.$$

By looking at the equation of the circle in this form - called general form - there is a $1x^2$ term, a $1y^2$ term, an x and a y term, and a constant. Many times the equation of the circle is given in general form. When given in general form, how can you graph the circle? Or, how can you find the center or radius?

The answer to these questions can be found by taking the given equation in general form and putting it in standard form. When it is in standard form, the center and radius are immediately known and this information can be used to graph the equation. All of the tools needed to put the equation in standard form were used earlier when you found the vertex of the parabola algebraically, and when the completing-the-square method of solving the quadratic equation was used. That is, complete-the-square on the x terms and on the y terms while moving constants to the other side of the equation results in the standard form of the equation.

Example 4: Find the center and radius of the circle $x^2 + y^2 + 4x - 6y = 3$.

Solution: Group the x's and group the y's and complete-the-square.

$$x^2 + 4x \quad + y^2 - 6y \quad = 3$$

$$x^2 + 4x + 4 + y^2 - 6y + 9 = 3 + 4 + 9$$

$$(x+2)^2 + (y-3)^2 = 16$$

The center is $(-2, 3)$ and the radius is 4.

⁂

Example 5: Graph $x^2 + y^2 - 5x + 8y - 3 = 0$.

Solution: Put the equation in standard form; then, use the center and radius to graph the equation.

$$x^2 - 5x \quad + y^2 + 8y \quad = 3$$

$$x^2 - 5x + \frac{25}{4} + y^2 + 8y + 16 = 3 + \frac{25}{4} + 16$$

$$\left(x - \frac{5}{2}\right)^2 \quad + (y+4)^2 \quad = \frac{12}{4} + \frac{25}{4} + \frac{64}{4}$$

$$\left(x - \frac{5}{2}\right)^2 + (y+4)^2 = \frac{101}{4}$$

Draw a circle with radius $\frac{\sqrt{101}}{2} \approx 5$, and center at $\left(\frac{5}{2}, -4\right)$.

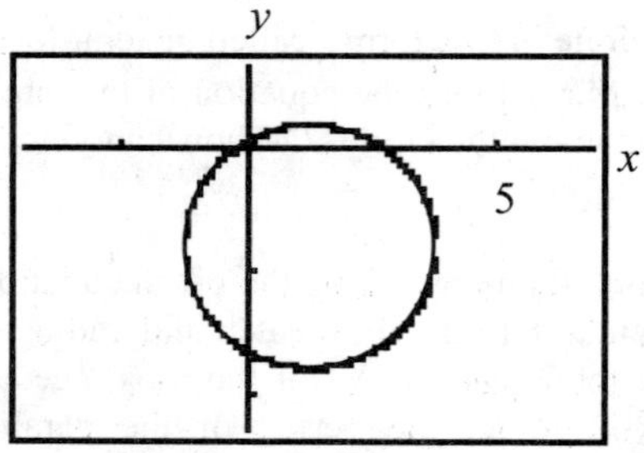

Figure 11.4.3

⁂

Circumference and Area of a Circle
The distance around a circle is usually not called the perimeter, but the **circumference** of the circle. The distance around the circle is found by the formula, $C = 2\pi r$ or $C = \pi d$, where d is the diameter and r is the radius. The area of the circle is $A = \pi r^2$. As before, circumference is measured in linear units and area is measured in square units.

Example 6: Find the area and circumference of the circle $(x-3)^2 + (y+5)^2 = 16$.

Solution: The center is not used for the area nor circumference. The only parameter needed is the radius. Since the radius squared is 16, the radius is 4. From $C = 2\pi r$, $C = 2 \times \pi \times 4$, or $C = 8\pi$. If you want an approximate circumference evaluate 8π. $C \approx 8 \times 3.14159 = 25.13272$. The area is $\pi 4^2$. $A = 16\pi$. In approximate form, the area is 50.2655.

⁂

Example 7: Find the circumference and area of the circle: $(x+3.7)^2 + (y-2.5)^2 = 27$

Solution: All that is needed is the radius; it is $\sqrt{27}$. The circumference is, $C = 2 \cdot \pi \cdot \sqrt{27} = 32.65$ units. The area is, $A = \pi \cdot \sqrt{27}^2 = 84.82$ square units.

⁂

Secant and Tangent Lines
While most people are familiar with the circumference and area of a circle, the relationship between the circle and the line may not be as commonly known. When a line intersects a circle, they can have two points in common as shown below.

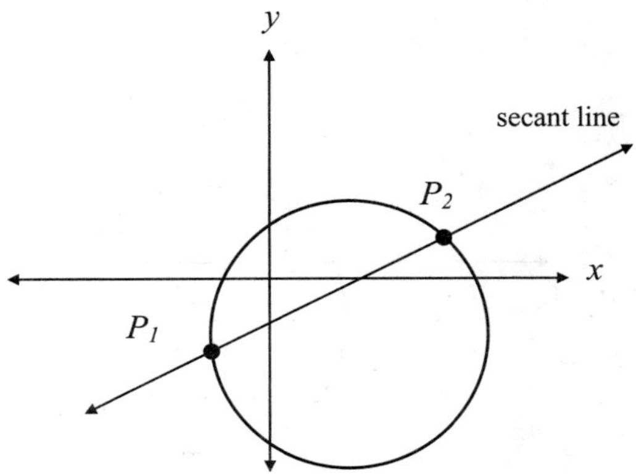

Figure 11.4.4

When the line intersects the circle twice, it is called a **secant line**. When the line intersects the circle once; that is, the line and the circle have one point in common, the line is called a **tangent line**. A tangent line is shown below.

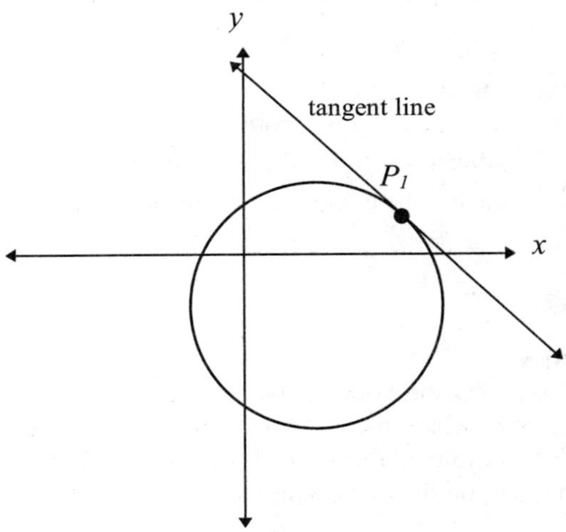

Figure 11.4.5

When a radius is drawn to the point of contact between the circle and the tangent line, the tangent and the radius meet at a right angle.

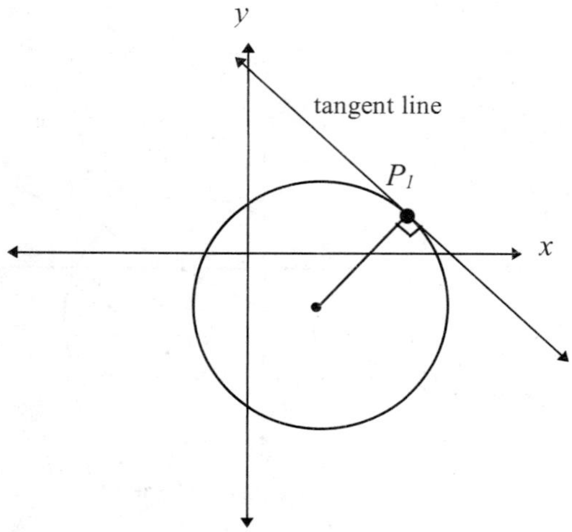

Figure 11.4.6

Example 8: Is the line $y = 5$ a secant line, a tangent line, or neither, to the circle $(x+3)^2 + (y-2)^2 = 9$?

Solution: The line $y = 5$ is a horizontal line through 5. The circle has a center at $(-3, 2)$ with a radius of 3. Since the center is up 2 from the x-axis and the radius is 3, the line just touches the circle once 3 units above the center at $(-3, 5)$. The line is a tangent line.

⁂

11.4 EXERCISES

Many times you can work the exercises by algebraic, numeric, or graphical methods. You can learn about mathematics no matter which method you use. Part of the learning process is for you to decide which method is best for you to use. Please read the text carefully before you do the exercises. Read the **next assigned section** after you finish the assignment below. All numeric answers must have an error ≤ 0.01.

A Quick Look To The Future

-4. What is the circumference of a circle of radius 1?

-3. The following points are on a line drawn from the origin: (3, 4), (6, 8), and (9, 12). Find the ratio of the x-coordinate of each point to the distance the point is from the origin.

-2. How many triangles can be drawn with interior angles of 30°, 60°, and 90°?

-1. Imagine riding a Ferris wheel with a large rubber band tied to your seat and to the ground right next to the wheel. As you go round and round with passing time, can you think of a function you have studied that models the length of the rubber band?

Previous Topics Revisited

1. Why is a square a parallelogram?

2. Are the diagonals of a rectangle perpendicular to one another? Why?

3. What is the area of a parallelogram when one side is 8 inches in length and the height to the 8 inch side is 4 inches?

4. The base of an isosceles triangle is 4 meters and the equal sides are each 10 meters; what is the area of the triangle?

5. Find the area of an equilateral triangle with sides of 8 inches. Use the formula $A = \frac{1}{2}bh$ first; then use Heron's formula to calculate the area a second time.

6. What is the center of a circle if the endpoints of a diameter are (–5, 3) and (7, –4)?

New Topic Exercises

Find the center, radius, circumference, and area of each circle.

7. $(x+3)^2 + (y-2)^2 = 9$

8. $(x-1.2)^2 + (y-3.7)^2 = 2.5$

9. $x^2 + (y-1)^2 = 5$

10. $(x+5)^2 + y^2 = 1$

11. $x^2 + y^2 = 100$

12. $x^2 + y^2 - 10x + 16y = 1$

13. $x^2 + y^2 + 8x - 2y - 3 = 0$

14. $x^2 + y^2 - 5x + 4y + 2 = 0$

15. $x^2 + y^2 + 3x + 7y = \frac{1}{2}$

Find the equation of the circles described in Exercises 16 - 20. Graph the equation.

16. Center $(-2, 2)$ and radius $\sqrt{5}$.

17. Center $(-3, -5)$ and radius $\frac{2}{3}$.

18. Endpoints of a diameter $(5, -2)$ and $(3, 8)$.

19. Center at $(3, 5)$ and tangent to the x-axis.

20. Center at $(3, 5)$ and tangent to the line $x = 0$.

21. Is $y = x + 2$ a tangent, secant, or neither to $(x+3)^2 + (y-1)^2 = 1$?

22. Is $y = -3x + 5$ a tangent, secant, or neither to $x^2 + y^2 + 10x + 20y + 25 = 0$.

23. Find a secant line and a tangent line to the circle $(x-5)^2+(y-1)^2=9$

24. Find a secant line and a tangent line to the circle $x^2+y^2-4x+8y-5=0$.

25. Find the circumference and area of $(x+5)^2+(y-1)^2=6$.

26. What are the dimensions of a square and a circle that have the same area?

From Mathematics to English

27. Is $(x+2)^2-(y-3)^2=4$ the equation of a circle? Why?

28. Is $(x+2)^2+(y-3)^2=-4$ the equation of a circle? Why?

29. Is $\frac{(x+2)^2}{4}+\frac{(y-3)^2}{4}=1$ the equation of a circle? Why?

30. After reading this section, make a list of questions that you want to ask your instructor.

31. Continue in your daily journal and make an entry. In addition to your normal entry on thoughts about the mathematics in this section, list at least two positive comments about what you have learned about this topic.

32. In paragraph format, summarize the material in this section of the text in your daily journal.

33. Describe how your classroom instructor made this topic more understandable and clear.

34. After reading the text and listening to your instructor, what do you not understand about this topic?

Explorations

35. $1(x-3)^2+1(y-4)^2=12$ is the equation of a circle. Analyze the graph of $a(x-3)^2+b(y-4)^2=12$, where a and b are positive factors of 12. What is the graph when $a=b$? What is the graph when $a \neq b$? What is the difference in the graphs when $a>b$ compared to $a<b$?

36. Develop possible equations for the "Olympic Circles".

37. Develop possible equations for a "bulls eye".

CHAPTER ELEVEN TEST

1. Find the length and midpoint of the line segment formed by the points $(-5, 4)$ and $(-1, -6)$.

2. Find the distance between $(1, 1)$ and $(x, x + 2)$; simplify as much as possible.

3. Find the minimum distance from the point $(1, 1)$ to the line $y = x + 2$.

4. If $(-1, 2)$ is the endpoint of a line segment and $(2, 6)$ is the midpoint of the segment, find the other endpoint of the line segment.

5. Can a triangle drawn in a plane have two right angles? Why?

6. What type of triangle is formed by joining the vertices at $(0, 2)$, $(3, 0)$, and $(0, 0)$?

7. If the vertex of an isosceles triangle is 10°, what are the base angles?

8. If the base of a right isosceles triangle is 4 inches, how long are the sides?

9. How long is the altitude of an equilateral triangle with sides of 6 feet?

10. Triangles ABC and A′B′C′ are similar; if $a = 2$, $b = 6$, and $a' = 1$, find b'.

11. Find the area of a triangle with sides 5, 12, and 13.

12. A rectangle has adjacent sides of 9 inches and 12 inches, how long is the diagonal?

13. Do the points $(0, 0)$, $(5, 3)$, $(5, 7)$, and $(0, 3)$ when joined in succession, form a parallelogram?

14. What is the measure of an angle formed by a diagonal and any side of a square?

15. What is the area of a parallelogram formed by joining the points $(0, 0)$, $(6, 0)$, $(8, 3)$, and $(2, 3)$ in succession, ending the figure with the starting point?

16. What is the radius and the center of the circle $(x - 2)^2 + (y + 4)^2 = 36$?

17. What is the radius and center of the circle $x^2 + y^2 + 6x - 8y - 11 = 0$?

18. What is the symbolic representation of the circle with a center of $(-1, 2)$ and a *diameter* of 8?

19. Is the line $y = x$ a secant, tangent, or neither to the circle $(x - 6)^2 + (y - 3)^2 = 4$?

20. Find the circumference and area of circle $(x - 3.2)^2 + (y + 7.1)^2 = 16$.

CHAPTER TWELVE
BASIC TRIGONOMETRY

12.0 Introduction

When I Grow Up, I Want to be Six Feet Tall

Below is the numeric representation of the height of a typical human over an 18-year growth cycle. A pediatrician typically monitors the growth of a child checking for particular medical abnormalities that are reflected in an unusual growth pattern. Medical personnel can develop a symbolic representation of the function that describes the height of the human over the 18-year period. They may use arithmetic combinations of functions studied in this chapter or previous chapters to develop the symbolic representation of the model. They may also take advantage of their knowledge of the relationship between the behavior of a function and the function parameters to develop the symbolic representation of the model.

Table 12.0.1

t	0	0.5	1	1.5	2	2.5	3	3.5	4	4.5	5	5.5	6	6.5	7	7.5	8	8.5	9
H	1.9	2.1	2.3	2.4	2.5	2.6	2.6	2.6	2.6	2.7	2.9	3.1	3.3	3.5	3.7	3.8	3.9	4	4

T	9.5	10	10.5	11	11.5	12	12.5	13	13.5	14	14.5	15	15.5	16	16.5	17	18
H	4	4	4.1	4.3	4.4	4.7	4.9	5.1	5.2	5.3	5.3	5.4	5.4	5.4	5.5	5.6	6

Researchers may ask and must answer many questions about the growth pattern. By the time you finish Chapter Twelve, you should also be able to answer many questions.

You learned of the properties of triangles in Chapter Eleven. In this chapter, you will learn of the relationships between the acute angles of a right triangle and the sides of the triangle. The sine relationship is developed from a functional view to show that trigonometric relationships are functions just like all the relationships studied in Chapters Two through Ten; however, the relationships are further studied in ratios and equations.

When mathematicians and scientists use angles in their work, they quite frequently use radians instead of degrees to measure the size of an angle. Thus, in Section 12.1, methods are developed for converting between measures of degrees and measures of radians.

In Section 12.2, the sine function is developed as a data relationship. This is followed by the definitions of cosine and tangent functions. The functions are then used to find trigonometric ratios for the special angles. These are crucial to any further study of trigonometry and are required to apply trigonometry to situations you may already be familiar with. Please study these definitions carefully and pay special attention to the homework exercises.

You will get a chance to use the trigonometric definitions in Sections 12.3 and 12.4. In Section 12.3, you are given two parts of a right triangle and asked to use the definitions and your calculator to find all of the remaining parts of the triangle. In Section 12.4, the definitions are applied to a variety of situations that you may have already encountered in everyday life.

Trigonometry in Action
Civil engineers use right triangle trigonometry nearly on a daily basis. One of the jobs of the civil engineer is to survey land. If you sell or refinance your house, the financial institutions involved will normally require a survey of the property. A survey will tell you the length and direction of each side of the property. The property below poses a minor problem for the surveyor. How do you measure the length of the side containing a pond or many trees?

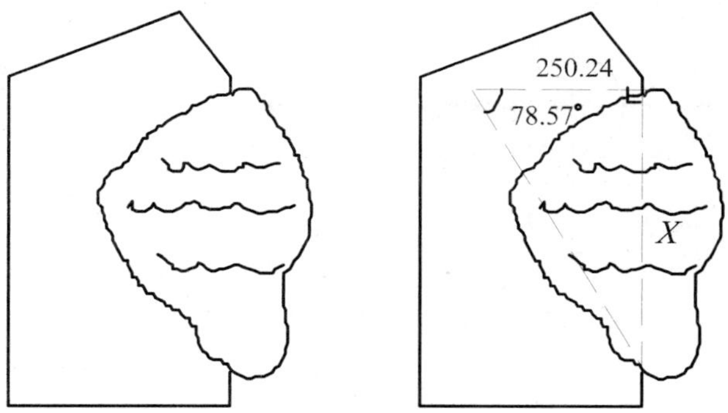

Figure 12.0.1

The right triangle superimposed on the right diagram contains an angle and a side that the surveyors can measure. Once measured, they use the definitions found later in this chapter to calculate the length of the pond. This length is added to the remainder of the side to get the length of the side containing to pond. The mathematics is shown below.

$$\tan 78.57° = \frac{x}{250.24} \quad \text{use the calculator to find tan 78.57°}$$

$$4.9461 = \frac{x}{250.24} \quad \text{multiply both sides times 250.24}$$

$$1237.71 = x$$

The distance across the pond is 1237.71 feet. Now the entire length of the side can be found.

12.1 Conversions between Degrees and Radians

Many people learned how to measure angles using the degree system of measurement. It was developed by mathematicians well over 2500 years ago and it is still used in many places in the world. Over 200 years ago, mathematicians and scientists developed the radian system of measurement. Today, several scientific fields do not use degrees and some engineering fields use degrees. Thus, the problem is that many beginning mathematics students are familiar with degrees, and radians are used. The material in this section is designed to start the transition from using degrees to using radians.

When a half-line, called a **ray**, is rotated counter-clockwise about its endpoint by one complete revolution, the measure of the angle formed is defined as 360°. Imagine the ray as a copy of the positive x-axis.

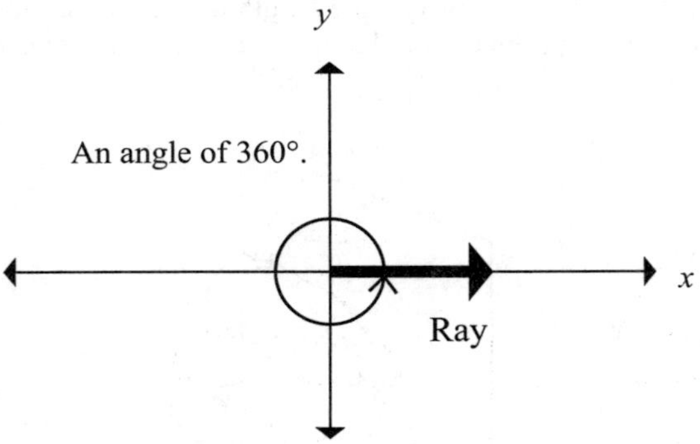

Figure 12.1.1

From this basic definition, you can see below, why a straight angle measures 180°; because, it is $\frac{1}{2}$ of a revolution, or $\frac{1}{2}$ of 360°. Likewise, the right angle is 90° because it is $\frac{1}{4}$ of 360°. By similar reasoning, an angle of 60° must be $\frac{1}{6}$ of a revolution, or $\frac{1}{6}$ of 360°.

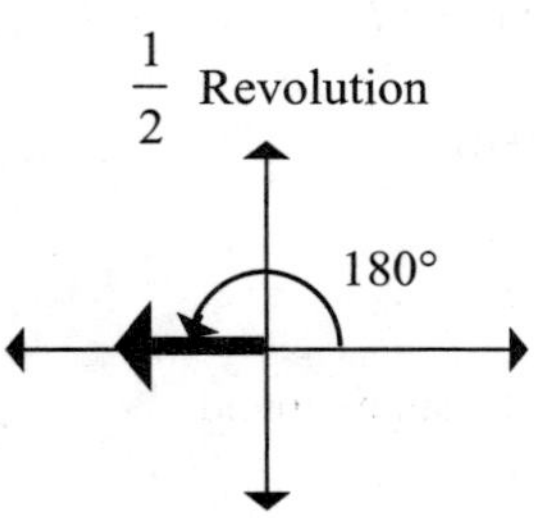

Figure 12.1.2a

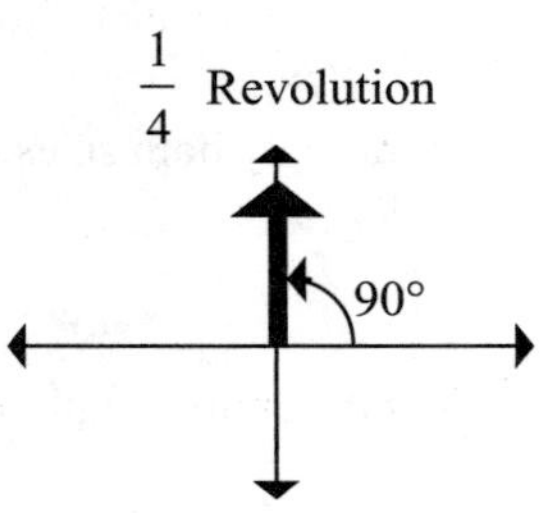

Figure 12.1.2b

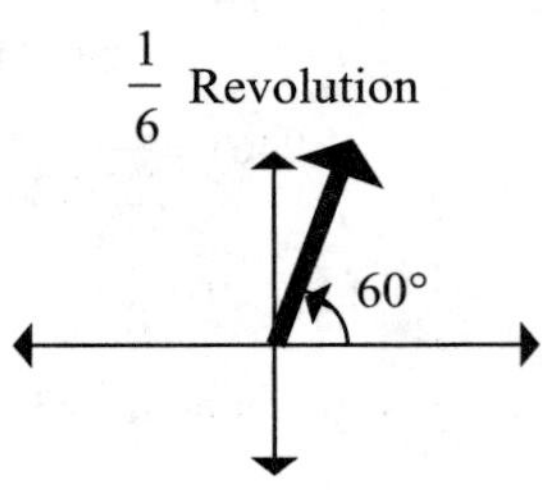

Figure 12.1.2c

About 300 years ago, astronomers needed to measure angles with more accuracy than to the nearest degree. As a result of this need, the angle with a measure of 1° was split into 60 parts. Each part is an angle with a measure of 1′, read "one minute." An angle with a measure of one minute is $\frac{1}{60}$ of $\frac{1}{360}$ of a revolution. Another way of thinking of this relationship is $1° = 60'$, or $1' = \frac{1}{60}°$. Further refinements in the degree system of measurements were made by dividing the minute into 60 parts. The size of each of these parts is called 1", read "one second." An angle with a measure of one second is $\frac{1}{60}$ of $\frac{1}{60}$ of $\frac{1}{360}$ of a revolution. Another way of thinking of this relationship is $1'' = \frac{1}{60}'$, and $1'' = \frac{1}{3600}°$.

In the radian system of measurement, the angle of one counter-clockwise revolution of the ray is **2π radians**.

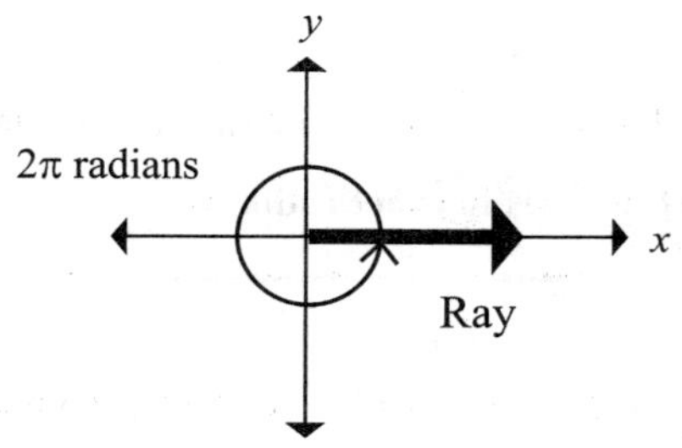

Figure 12.1.3

It may be helpful to think of the 2π as the circumference of a circle of radius 1 unit. When using radians, no units of measure are attached. While you might think the units should be in inches, feet, meters, or some other linear unit because the 2π comes from a circumference -- in the coordinate plane, length is measured with real numbers. For example, the distance between the points (3, 0) and (0, –4) is 5; not 5 inches, not 5 meters, just 5. Thus, the distance around a circle of radius 1 is 2π.

From comparing the two known angles 360° and 2π, both being equal to 1 revolution, a method of converting from one system to the other exists. The basic conversion formula is

$$360° = 2\pi\,.$$

If both sides of the equation are divided by 2, the basic conversion formula becomes

180° = π radians.

From this basic conversion formula, formulas can be developed to convert between degrees and radians.

$180° = \pi$	Given (basic conversion).
$1° = 0.01745$	Derived by dividing both sides of the basic conversion formula by 180 and replace π with 3.14159. **Used to convert degrees to decimal radians.**
$57.3° = 1$	Derived by dividing both sides of the basic conversion by 3.14159. **Used to convert decimal radians to degrees.**
$\frac{180°}{\pi} = 1$	Derived by dividing both sides of the basic conversion by π. **Used to convert exact radians to degrees.**
$\frac{\pi}{180°} = 1$	Derived by dividing both sides of the basic conversion by 180°. **Used to convert degrees to exact radians.**

Conversions between degrees and radians can be done with the above conversion formulas as demonstrated below in the sample problems. *When an angle is measured in radians, it may contain the symbol π or it may be in decimal form.* Many mathematicians and scientists use radian angle measures containing π, these are in exact form. When radians are expressed in decimal form without the π symbol, the measure is radians in decimal form and is approximate in value. This form is quite often used by some calculators.

Example 1: Convert 15° to decimal radians.

Solution: The formula $1° = 0.01745$, indicates that 1 degree is equivalent to 0.01745 radian. There are 15°; thus, multiply 15° times 0.01745. $15° = 15 \cdot (0.01745) = 0.26175$ radian.

Alternate Solution: Since degrees and radians are related linearly, proportions can be used to do all conversions. Below is an example.

$$\frac{x \text{ radians}}{15°} = \frac{3.14159 \text{ radians}}{180°}$$

$$x = \frac{3.14159 \text{ radians}}{180°} \cdot 15°$$

$$x = 0.2618 \text{ radian}$$

⁂

Example 2: Convert 30° to radians containing π.

Solution: From π = 180°, $\frac{\pi}{180°} = 1$; thus, when 30° is multiplied by 1 in the form of $\frac{\pi}{180°}$, the degree units cancel and 30° is converted to π radians.

$$30° = 30° \cdot \left(\frac{\pi}{180°}\right) = \frac{30\pi}{180} = \frac{\pi}{6} \text{ radian.}$$

Alternative Solution: Use a proportion.

$$\frac{x \text{ radians}}{30°} = \frac{\pi \text{ radians}}{180°}$$

$$x = \frac{\pi \text{ radians}}{180°} \cdot 30°$$

$$x = \frac{\pi}{6} \text{ radian}$$

Since radians are real numbers, there is no need to write the word radian after the measure of the angle in radians.

⁂

Example 3: Convert $\frac{5\pi}{6}$ to degrees.

Solution: Since 180° = π, replace π with 180°. $\frac{5\pi}{6} = 5 \cdot \left(\frac{180°}{6}\right) = 5 \cdot (30°) = 150°$

Alternative Solution: Use a proportion.

$$\frac{x^\circ}{\frac{5\pi}{6}} = \frac{180^\circ}{\pi}$$

$$x = \frac{180^\circ}{\pi} \cdot \frac{5\pi}{6} = 150^\circ$$

⁂

Example 4: Convert 2.65 radians to degrees.

Solution: 1 radian equals 57.3°; thus 2.65 radians is 2.65(57.3°), or 151.845° ≈ 152°

Alternative Solution: Use the following proportion:

$$\frac{x^\circ}{2.65} = \frac{180^\circ}{3.14159}$$

$$x = \frac{180^\circ}{3.14159} \cdot 2.65 = 151.834^\circ \approx 152^\circ$$

⁂

Example 5: Convert 45° to radians containing π (exact radians.)

Solution: Multiply 45° times 1 in the form of $\frac{\pi}{180^\circ}$.

$$45^\circ \cdot \left(\frac{\pi}{180^\circ}\right) = \frac{45\pi}{180} = \frac{\pi}{4}$$

Alternative Solution: Use the following proportion:

$$\frac{x}{45^\circ} = \frac{\pi}{180^\circ}$$

$$x = \frac{\pi}{180^\circ} \cdot 45^\circ = \frac{\pi}{4}$$

⁂

Example 6: Convert $\frac{3\pi}{5}$ to decimal radians.

Solution: Replace π with 3.14159 and simplify.

$$\frac{3\pi}{5} = \frac{3 \cdot (3.14159)}{5} = 1.88495 \approx 1.88$$

⁂

As was mentioned at the beginning of this section, the angle of one revolution is measured in a counter-clockwise fashion. When angles are measured in a counter-clockwise fashion, they are assigned a positive measure. This is just like numbers to the right of the origin on the number line are assigned positive values. If angles are measured in a clockwise fashion, they are assigned a negative value; like numbers to the left of the origin on the number line are assigned negative values. Like real numbers, angles can represent numbers in the interval $(-\infty,\infty)$. Below are several examples of angles you may not find in triangles and parallelograms, but are used elsewhere.

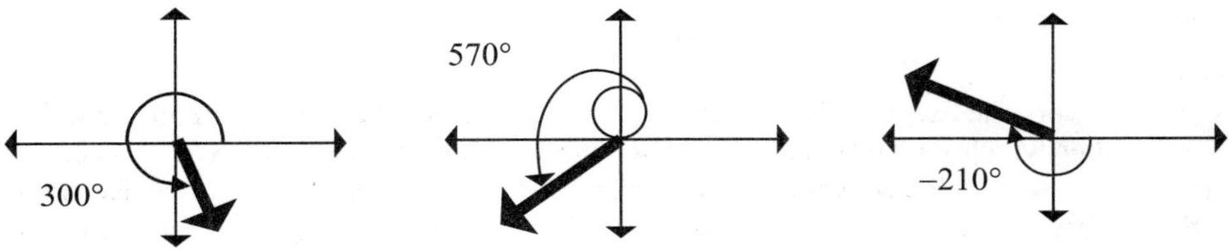

Figure 12.1.4

Normally, all of the angles are drawn with the initial side on the positive x-axis and the terminal side can be anywhere in the plane; this is called **standard position**. When angles have the same terminal side, they are called **coterminal angles**. Below are some examples of coterminal angles.

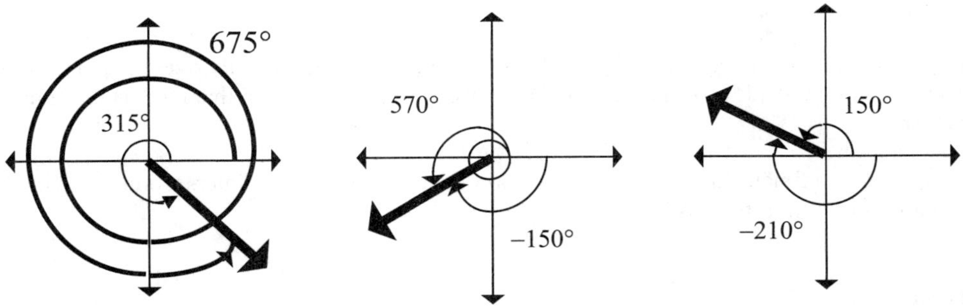

Example 7: Find several coterminal angles to 50°.

Solution: The given angle is $\frac{50}{360}$ of one revolution; thus, another angle that is coterminal will be 1 and $\frac{50}{360}$ revolutions, or 360° + 50°, which is **410°**. Another angle coterminal to the given angle is 2 and $\frac{50}{360}$ revolutions, or 720° + 50°, which is **770°**. This process can continue indefinitely. On the other hand, a coterminal angle can be found by starting at $\frac{50}{360}$ revolution and adding 1 clockwise revolution. A coterminal angle is $\frac{50}{360}+(-1)$ revolutions, or $\frac{50}{360}-1$, or $50° - 360° = \mathbf{-310°}$.

This same idea can be repeated; for example,

$\frac{50}{360}+(-2)$ revolutions or, $\frac{50}{360}-2$ revolutions or, $50° - 720° = \mathbf{-670°}$.

***

12.1 EXERCISES

Many times you can work the exercises by algebraic, numeric, or graphical methods. You can learn about mathematics no matter which method you use. Part of the learning process is for you to decide which method is best for you to use. Please read the text carefully before you do the exercises. Read the **next assigned section** after you finish the assignment below. All numeric answers must have an error ≤ 0.01.

A Quick Look To The Future

-4. The following points are on a line drawn from the origin: (3, 4), (6, 8), and (9, 12). Find the ratio of the y-coordinate of each point to the distance the point is from the origin.

-3. If you know the three sides of a right triangle are 3, 4, and 5 inches, can mathematics be used to find the angles?

-2. Imagine a pole standing in an empty field and the sun comes up, goes directly over the pole and then sets. The pole casts a shadow on a clear day. Do you know of any function that models the length of the shadow?

-1. If the graphs of two functions intersect, are the coordinates of the point of intersection in the numeric representations of both functions?

Previous Topics Revisited

1. Find the area and circumference of the circle $(x-2)^2 + y^2 = 9$.
2. What is the equation of a circle with a center at (–3, 5) and a diameter of 3?
3. Find the center, radius, and any two points on the circle $x^2 + y^2 - 12x - 4y - 9 = 0$.
4. Find the diagonal of the square with a side of 2.
5. Verify that the two diagonals of a rectangle are equal in length by calculating the lengths of each diagonal in a rectangle with length of l and width of w.
6. The height of an equilateral triangle is $\sqrt{3}$, find the length of the sides.

New Topic Exercises

7. Fill in the table; round degrees to a whole number and express the radian measures containing π.

DEGREES	RADIANS
30°	
	$\frac{\pi}{4}$
60°	
	$\frac{\pi}{2}$
120°	
	$\frac{3\pi}{4}$
–150°	
	$-\pi$
–210°	
	$\frac{-5\pi}{4}$

8. Fill in the table, round to hundredths.

DEGREES	DECIMAL RADIANS
12°	
	6.28
–200°	
	–1
0.2°	
	100

9. Find one positive and one negative coterminal angle to the given angle.

ANGLE	POSITIVE COTERMINAL	NEGATIVE COTERMINAL
240°		
–60°		
500°		
π		
$\frac{7\pi}{4}$		

10. Draw an angle of *approximately* 2000°. (No protractor needed.)

11. Draw an angle of *approximately* –1000°. (No protractor needed.)

12. Find the coordinates of two points on the terminal side of each angle drawn in standard position.

    a. 90° b. π c. 270° d. 2π

13. Find one point on the terminal side of each angle in standard position, and calculate the length of the line from the point to the origin.

    a. 180° b. –270° c. $-\pi$

From Mathematics to English

14. Describe a degree.

15. How do you know if two angles are coterminal?

16. What is 1 radian?

17. Write a paragraph explaining why you think the radian system of measurements was developed.

18. After reading this section, make a list of questions that you want to ask your instructor.

19. Continue in your daily journal and make an entry. In addition to your normal entry on thoughts about the mathematics in this section, list at least two positive comments about what you have learned about this topic.

20. In paragraph format, summarize the material in this section of the text in your daily journal.

21. Describe how your classroom instructor made this topic more understandable and clear.

22. After reading the text and listening to your instructor, what do you not understand about this topic?

Explorations

23. a. When an angle terminates in one of the four quadrants, a positive ***acute angle*** is formed by the terminal ray and the negative x-axis or positive x-axis - which ever is the closest to the terminal ray. This angle is called the **reference angle** to the original angle. Develop a method by which the reference angle can be found for an angle in each quadrant. That is, for each quadrant, describe in English or mathematics how to find the reference angle.

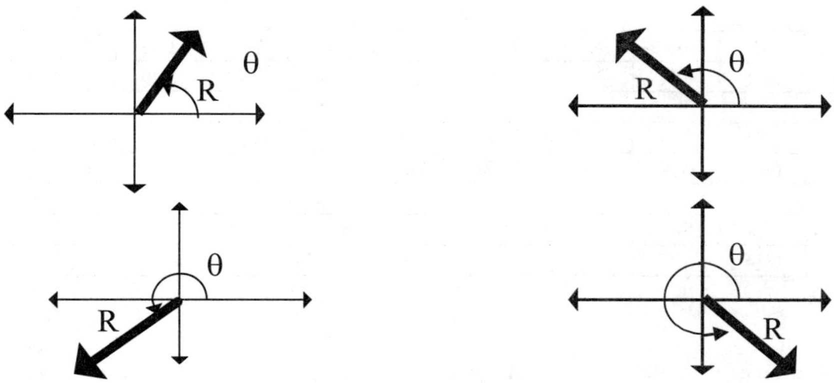

b. Find the reference angle for the following angles:

i. 120°, 135°, 150°, 210°, 225°, 240°, 300°, 315°, 330°, 30°, 45°, 60°.

ii. –30°, –45°, –60°, –120°, –135°, –150°, –210°, –225°, –240°, –300°, –315°, –330°.

iii. 170°, –190°, 530°, –550°, 890°, –910°.

24. Given the following similar triangles, find the ratio of sides a to b, a to c, and b to c in each of the three triangles Express the ratios in fraction and decimal forms.

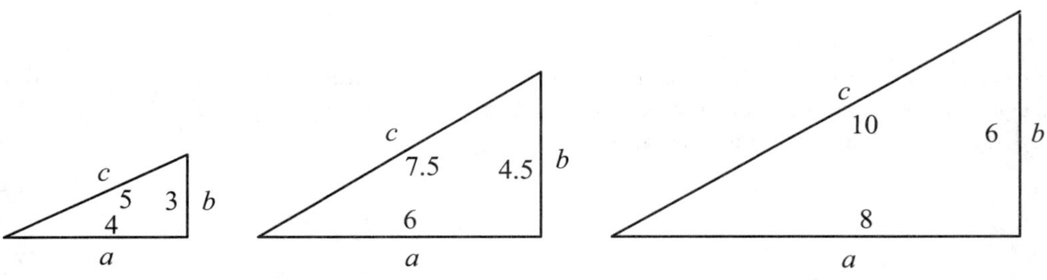

	a to b	a to c	b to c
small triangle			
medium triangle			
large triangle			

12.2 Trigonometric Definitions

Sun, Your Pattern of Behavior Keeps Repeating
Below is the data relationship that shows the day of the year (for almost two years) as the independent variable and the number of minutes of sunlight on that day as taken in Florida.

Table 12.2.1

t	1	21	41	61	81	101	121	141	161	181	201	221
S	564	589	621	681	734	786	835	874	897	899	877	839

t	241	261	281	301	321	341	361	381	401	421	441	461
S	791	740	688	629	595	567	561	564	589	621	681	734

t	481	501	521	541	561	581	601	621	641	661	681	701
S	786	835	874	897	899	877	839	791	740	688	629	595

The graphical representation of the data is below in Figure 12.2.1. This, of course, doesn't look like any

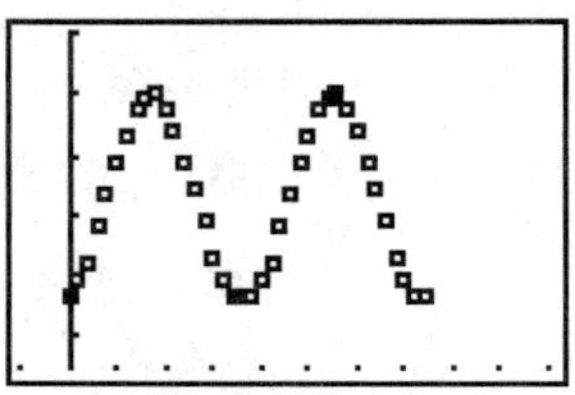

Figure 12.2.1

of the functions studied thus far. It appears that you will need to study a new function if you are going to predict the Sun's behavior!

Consider Figure 12.2.2 below, it is a visualization of *any* angle with a measure of θ (theta); it is referred to as angle θ and is in standard position. The point (x, y) can be *any point* on the terminal side of the angle and r is the distance from the point to the origin, where $r > 0$.

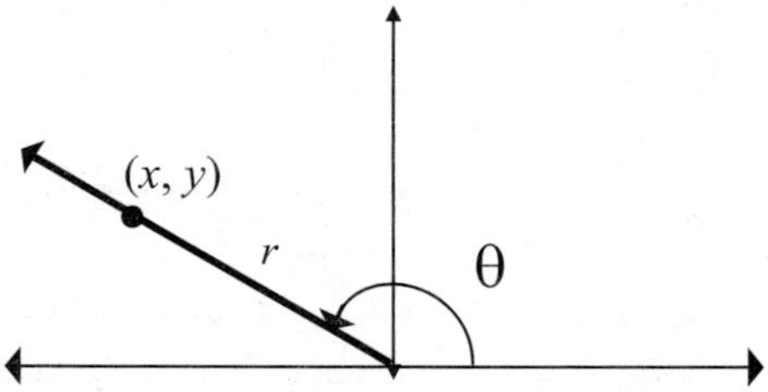

Figure 12.2.2

To begin the investigation of trigonometry, the ratio of y to r (the fraction $\frac{y}{r}$) for every angle θ is analyzed. The point (x, y) is on the terminal side of θ and r is the distance from the point (x, y) to the origin. *To make the calculations easy*, a value for r of 1 is used. That is, below is a data relationship between the angle θ and the y-coordinate of the point on the terminal ray of θ one unit from the origin. The data pair is the set of numbers (θ, y). The data can be collected by making a drawing similar to that shown in Figure 12.2.3.

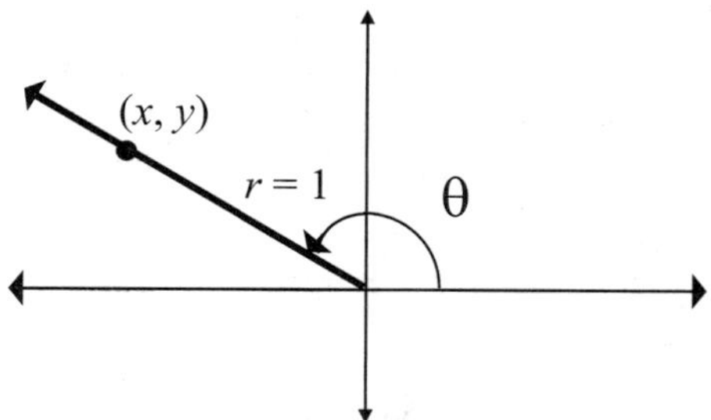

Figure 12.2.3

Use a protractor to measure the angle and a ruler to measure the y-coordinate. You may want to try this experiment yourself. Below in Table 12.2.2 is the collected data. Theta is in degrees.

Table 12.2.2

θ	−45	−30	−15	0	15	30	45	60	75	90	105	120	135	150	165
y	−.71	−.5	−.26	0	.26	.5	.71	.87	.97	1	.97	.87	.71	.5	.26

θ	180	195	210	225	240	255	270	285	300	315	330	345	360	375	390	405
y	0	−.26	−.5	−.71	−.87	−.97	−1	−.97	−.87	−.71	−.5	−.26	0	.26	.5	.71

The domain of the data relationship does not have to start at −45 and stop at 405. Angles can be any real number from −∞ to ∞. As you know by now, looking at the numeric representation of a data relationship doesn't tell you much of what you may want to know about the behavior of the relationship. Looking at the graphical representation of the relationship may help identify the data type. The graphical representation of the data is shown below in Figure 12.2.4.

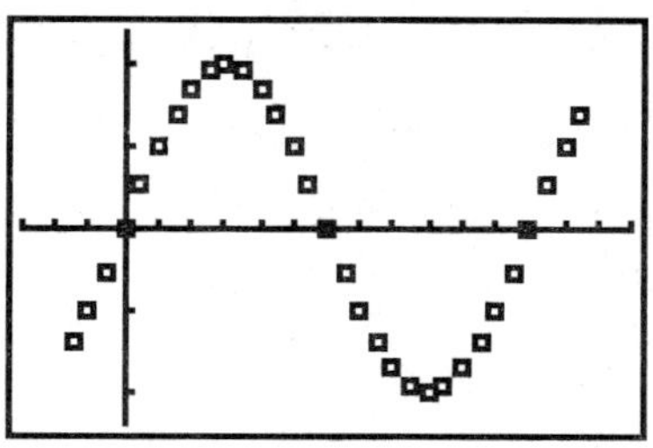

Figure 12.2.4

If you review all of the functions studied, you will not find one that behaves like this data relationship. You might argue that it looks like two quadratic functions. If this is so, how will you connect the quadratic functions to get this shape? Further, you must remember that the data does not stop at –45° and 405°. If you try plotting more data, you will discover that the relationship between θ and the y-coordinate is not even close to being two quadratic functions. So if the shape is not like anything studied, what is it?

The data relationship is a new kind of function. You may note in the table above, the values for $\frac{y}{r}$ are on the interval [–1, 1]. This is true for all angles because the y-coordinate of a point on the terminal side of θ is always less than or equal to the distance from the point to the origin. The conclusion is that the range of the new function is [–1, 1]. The new function has a domain of $(-\infty, \infty)$ and a range of [–1, 1]. There is a mathematical function that behaves just like the data. This particular function is named the sine function. Recall that many of the functions already studied have names; for example, $(x, mx+b)$ is called the linear function and is commonly written $y = mx + b$. The notation used to describe the new function is a little different than studied earlier. The mathematical notation for the sine function is below.

> **Definition of the Sine Function**
>
> $\sin\theta = \frac{y}{r}$, where θ is any angle on $(-\infty, \infty)$, (x, y) is a point on the terminal side of θ, and r is the distance from (x, y) to (0, 0).

This notation is used even though it is different than normal $f(x)$ or y notation. When you see a statement like $\sin\ 45° = 0.707$, it is describing the pair of numbers (45°,0.707) -- a point on the graph. In the statement $\sin\ 90° = 1$, the pair of numbers (90°,1) is being described. This notation will be modified later to make it look like functions studied in Chapters Three through Ten.

A calculator can calculate the sine ratio for any angle entered. The calculator will work in degrees or radians; thus, make certain the calculator is in the correct mode before calculating a sine ratio.

Example 1: Find the sine of 72°.

Solution: Select degrees from the mode menu. Then use the keystrokes:

SIN (72) <ENTER>

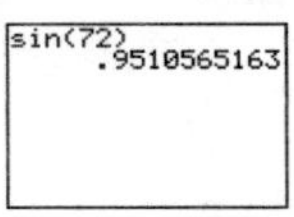

The calculator should then display 0.9510565163. This is the ratio $\frac{y}{r}$ for the angle of 72°. That is, the sine graph contains the point (72, 0.951).

⁂

Example 2: Find the sine of $\frac{5\pi}{6}$.

Solution: Select radians from the mode menu. Then use the keystrokes:

SIN (5 π ÷ 6) <ENTER>

The calculator should display 0.5. This is the ratio $\frac{y}{r}$ for the angle of $\frac{5\pi}{6}$.

The point $\left(\frac{5\pi}{6}, .5\right)$ is on the graph of the sine function.

⁂

As you may have guessed, not only will the calculator find the sine ratio for any angle, it will also plot all points (θ, sin θ) for the domain and range specified. The calculator requires a function in the form $y = f(x)$; to enter the sine function on the calculator enter sin x. The calculator understands that in the function $y = \sin x$; x is the angle and y is the sine of the angle. These are not to be confused with the point (x, y) on the terminal side of θ.

Example 3: Graph the function (θ, sin θ).

Solution: Select a mode - degree mode. Enter $y = \sin x$. Set the viewing window to [–720, 720] by [–2, 2]. Figure 12.2.5 is the graph of the sine function and the original data relationship.

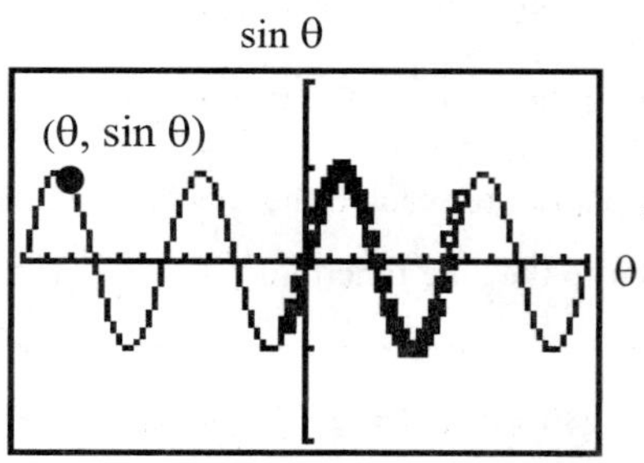

Figure 12.2.5

⁂

Example 4: Is the graph of $y = 3 \cdot \sin x$, the graph of $y = \sin x$ stretched by a factor of 3?

Solution: Yes, the y-coordinate of every point on the original graph is sin x and the y-coordinate of the new graph is 3 times the y-coordinate of the original graph $3 \cdot \sin x$. The graphs below on the window [–360, 360] by [–4, 4] confirm this conclusion. Each vertical tic mark is 1.

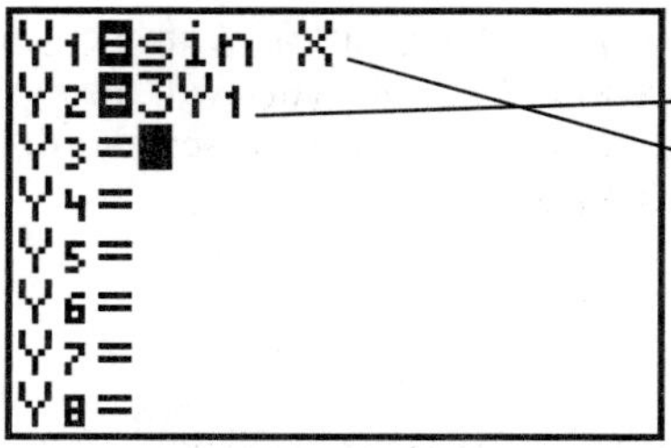

Figure 12.2.6a

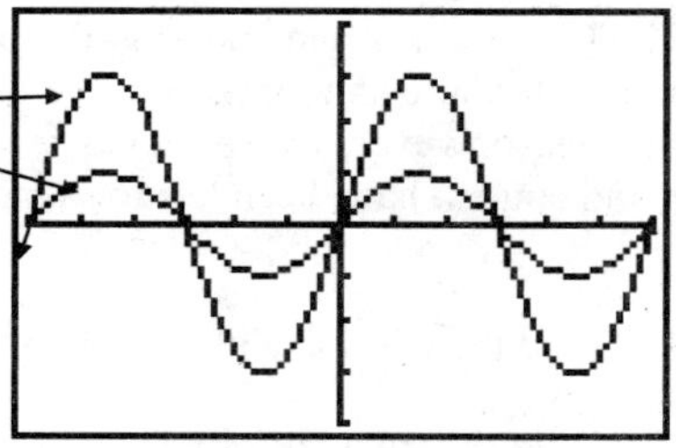

Figure 12.2.6b

⁂

The investigation of the graph of the sine function can continue in great detail; however, this analysis will be left for another course. Relationships between x, y, and r in Figure 12.2.7 are studied here, where the sine ratio is just one of the relationships.

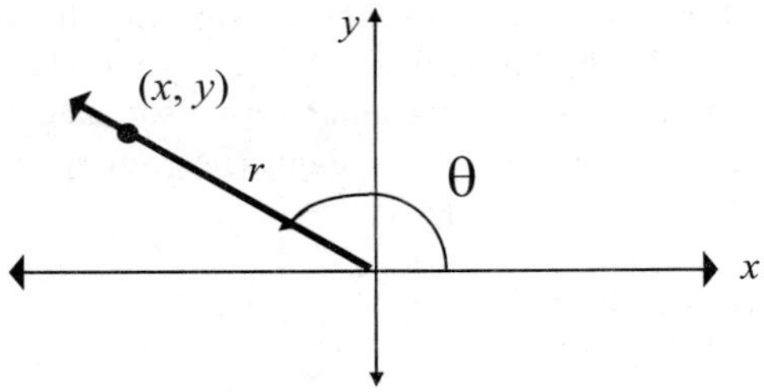

Figure 12.2 7

In addition to analyzing the ratio $\frac{y}{r}$, arguments similar to the above discussion can be made for the ratios $\frac{x}{r}$ and $\frac{y}{x}$. If θ is paired with each of these ratios, the resulting pairs of numbers form a function. That is, $(\theta, \frac{x}{r})$ and $(\theta, \frac{y}{x})$ are functions; because, for each θ there is one and only one $\frac{x}{r}$ and $\frac{y}{x}$. The notation for describing these functions is similar to the sine function.

> **Definition of the Cosine and Tangent Functions**
>
> $\cos\theta = \frac{x}{r}$ $\qquad\qquad$ $\tan\theta = \frac{y}{x}$ Where θ is any angle on $(-\infty, \infty)$, for the cosine ratio, (x, y) is any point on the terminal side of θ and r is the distance from (x, y) to $(0, 0)$. For the tangent function, $x \neq 0$; that is, θ cannot be on the y-axis.

The statement $\cos 90° = 0$ is another way of writing the ordered pair of numbers $(90°, 0)$; 0 is the ratio $\frac{x}{r}$ in Figure 12.2.7. The statement $\tan\pi = 0$ is the ordered pair $(\pi, 0)$. Just as with the sine function, the calculator will calculate the cosine ratio and the tangent ratio. The calculator will also graph the cosine and tangent functions; however, except for a few questions in the exercise set, the discussion of the graphs of cosine and tangent have been left for a more advance course.

Example 5: Find the cos 30° with the calculator.

Solution: With the calculator in degree mode, enter the following keystrokes:

cos (30) <ENTER>

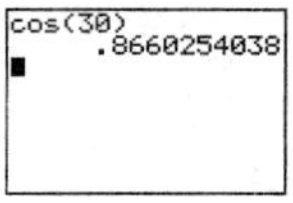

The display should read 0.8660254038; if it doesn't perhaps you did not select degree mode. The number 0.8660254038 is the ratio $\frac{x}{r}$ for an angle of 30°.

⁂

Example 6: Find the tan $\frac{\pi}{2}$ with the calculator.

Solution: Select radian mode and enter the keystrokes: tan (π ÷ 2) <ENTER>
The display should be an error message. The angle $\frac{\pi}{2}$ is on the positive y-axis; on the y-axis, the value of any x-coordinate is 0. The ratio $\frac{y}{x}$ is undefined. The calculator cannot find undefined numbers; the tan $\frac{\pi}{2}$ is undefined.

⁂

Special Angles
In Section 11.2, two triangles of special interest were developed. These triangles are the right isosceles with legs of 1 and the triangle developed from the equilateral triangle with sides of 2; they are displayed below.

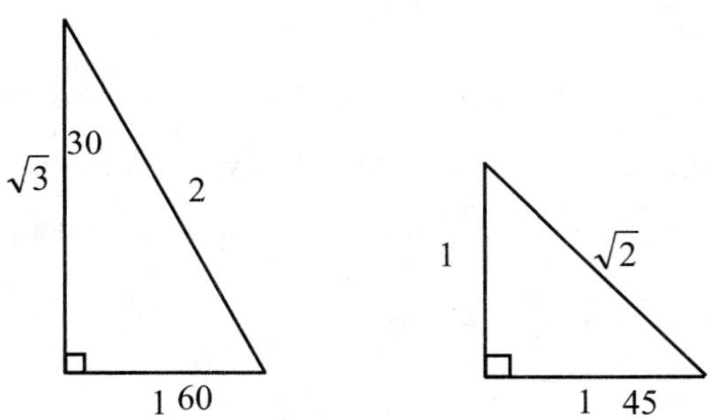

$$60° = \frac{\pi}{3}$$

$$45° = \frac{\pi}{4}$$

$$30° = \frac{\pi}{6}$$

Figure 12.2.8

From these triangles and the definitions of sine, cosine, and tangent, many trigonometric ratios can be calculated without a calculator. The calculator usually gives approximate values for the trigonometric ratios and the algebraic method will give exact values – as will CAS calculators like the TI-89 or Voyage 200.

Example 7: Find the exact value of cos 150°.

Solution: The angle formed by the terminal side of 150° and the negative x-axis is 30°. This angle is called the **reference angle**. What is significant about the reference angle is that it can be used to find a point on the terminal side of 150°. The 30-60-90 triangle has been superimposed on the angle of 150° so that the 30° reference angle and the 30° angle in the triangle match. Further, the 30-60-90 has been placed so that one leg is on the x-axis. The hypotenuse must lie on the terminal side of the given angle of 150°.

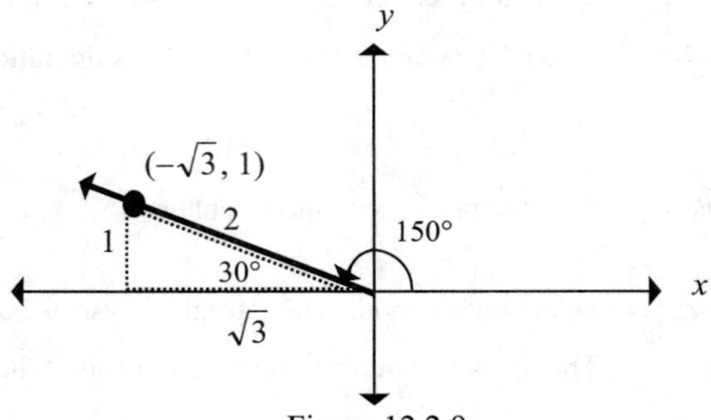

Figure 12.2.9

The point on the terminal side of 150° is $\left(-\sqrt{3},1\right)$ and the distance from $\left(-\sqrt{3},1\right)$ to the origin is 2. The point (x, y) and r are now known. The exact value of cos 150° is $\frac{-\sqrt{3}}{2}$ because the definition of $\cos\theta = \frac{x}{r}$.

⁂

Example 8: Find the exact value of $\tan\frac{\pi}{3}$.

Solution: Superimpose the 30-60-90 $\left(\frac{\pi}{6},\frac{\pi}{3},\frac{\pi}{2}\right)$ triangle on top of the $\frac{\pi}{3}$ angle so that the angles match and one side of the triangle is on the x-axis. The hypotenuse must lie on the terminal side of the given angle of $\frac{\pi}{3}$. The required point on the terminal side is $\left(1,\sqrt{3}\right)$, and r is 2. The exact value of $\tan\frac{\pi}{3}$ is $\frac{\sqrt{3}}{1}$, or $\sqrt{3}$.

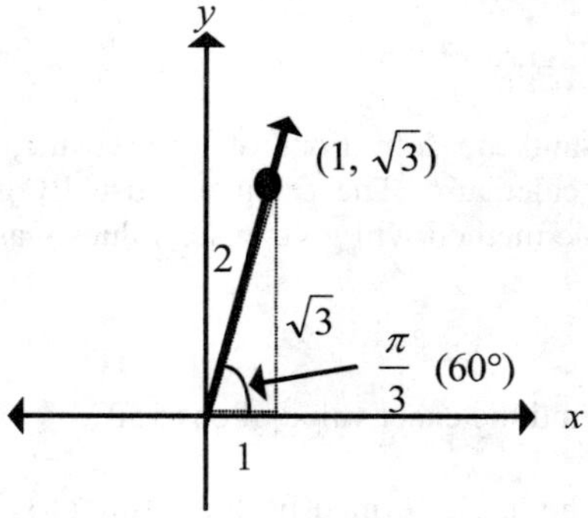

Figure 12.2.10

⁂

Example 9: Find the exact values for the sine, cosine, and tangent ratios of the angle of $\frac{\pi}{2}$ radians.

Solution: One point on the terminal side is (0, 1) because the terminal side of $\frac{\pi}{2}$ lies on the positive y-axis. The distance from (0, 0) to (0, 1) is 1; thus, $x = 0$, $y = 1$, and $r = 1$.

$$\sin\frac{\pi}{2} = \frac{y}{r} = \frac{1}{1} = 1$$

$$\cos\frac{\pi}{2} = \frac{x}{r} = \frac{0}{1} = 0$$

$$\tan\frac{\pi}{2} = \frac{y}{x} = \frac{1}{0} = \text{undefined}$$

The calculator can also be used to find the ratios. With the calculator in radian mode, enter the keystrokes:
sin (π ÷ 2) <ENTER>
cos (π ÷ 2) <ENTER>
tan (π ÷ 2) <ENTER> (display shows an error message)

⁂

Example 10: Find the sine, cosine, and tangent of 225°.

Solution: Superimpose the right isosceles on the 225° angle so that one side of the triangle is on the x-axis and the reference angle (45°) of 225° matches the 45° angle in the triangle. The hypotenuse must lie on the terminal side of the given angle of 225°. A point on the terminal side of 225° is (–1, –1) and the distance from the point to the origin is $\sqrt{2}$. From this information, the exact values of sin 225°, cos 225°, and tan 225° are found.

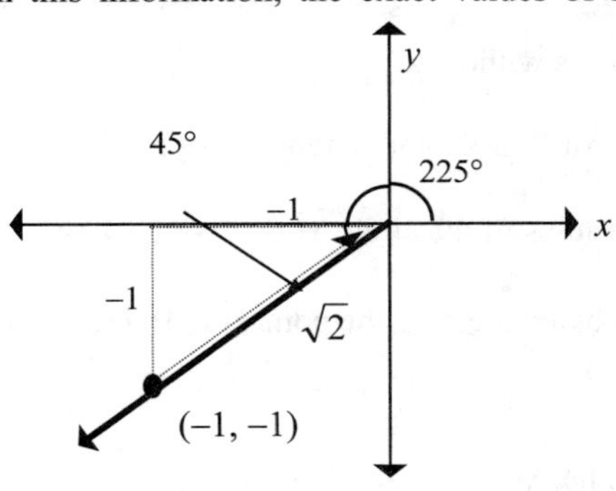

Figure 12.2.11

$$\sin 225° = \frac{-1}{\sqrt{2}}, \text{ which simplifies to } \frac{-\sqrt{2}}{2}$$

$$\cos\ 225° = \frac{-1}{\sqrt{2}}, \text{ which simplifies to } \frac{-\sqrt{2}}{2}$$

$$\tan\ 225° = \frac{-1}{(-1)}\ , \text{ which simplifies to } 1$$

⁂

12.2 EXERCISES

Many times you can work the exercises by algebraic, numeric, or graphical methods. You can learn about mathematics no matter which method you use. Part of the learning process is for you to decide which method is best for you to use. Please read the text carefully before you do the exercises. Read the **next assigned section** after you finish the assignment below. All numeric answers must have an error ≤ 0.01.

A Quick Look To The Future

-4. Do you think a carpenter who knows one side and one acute angle of a right triangle, can find the other angle?

-3. Do you think the time-height relationship of a person from age 0 to about 18 is linear?

-2. Given to linear functions, do their graphs always intersect?

-1. If $2x + 1 = 3$, and $5x - 3 = 2$, does $2x + 1\ +\ 5x - 3 = 3\ +\ 2$?

Previous Topics Revisited

1. Convert $\frac{7\pi}{4}$ to degrees.

2. Convert 2.45 radians to degrees.

3. Convert 450° to radians without π.

4. What is the equation of a circle with endpoints of a diameter of (–3, 4) and (–5, –10)?

5. Find the center and radius of the circle $x^2 + y^2 - 5x + 12y - 1 = 0$.

6. Is the figure formed by connecting the points (4, 4), (3, –2), (–2, –2), and (–1, 3) in succession, a parallelogram? Why?

New Topic Exercises

7. Fill in the table with EXACT values.

θ	cos θ	sin θ	tan θ
45°			
120°			
180°			
210°			
270°			
315°			

8. Fill in the table. You may use a calculator.

θ	$\cos\theta$	$\sin\theta$	$\tan\theta$
$\frac{\pi}{6}$			
$\frac{\pi}{2}$			
$\frac{2\pi}{3}$			
$\frac{5\pi}{4}$			
$\frac{11\pi}{6}$			
2π			

9. What is the distance from the point (0, 0) to the point (x, y)?

Identify the distance found in Exercise 9 as r and solve Exercises 10 - 12.

10. If $x = 4$ and $y = -3$, find r.

11. If $y = 0$ and $r = 7$, find x. (Hint: There are 2 answers.)

12. If $r = 6$ and $x = -5$, find y. (Hint: There are 2 answers.)

13. If $\sin\theta = 0$, what is the y-coordinate of any point on the terminal side of θ?

14. If $\cos\theta = 0$, what is the x-coordinate of any point on the terminal side of θ?

15. If $\tan\theta = 1$, what is the relationship between x and y?

16. Graph the sine function on the viewing window [−8, 8] by [−2, 2]. Use radian mode. Find the zero of the sine function that is between 3 and 4.

17. List five positive and five negative angles whose tangent is undefined.

From Mathematics to English

18. Graph the tangent function on the window $[-\pi, \pi]$ by [−10, 10] using radian mode. Describe the behavior of the function near $\frac{-\pi}{2}$ and $\frac{\pi}{2}$.

19. Describe a situation, based on your experiences, that behaves "like" the sine function.

20. Why is the sine function never undefined?

21. Why is the cosine function never undefined?

22. What conditions cause the sine function to be zero?

23. What conditions cause the tangent function to be zero?

24. If θ is in quadrant I, explain why the sine of θ is on the interval (0, 1).

25. Write your own definition of the sine function.

26. Describe how to find the exact value of the sine of 315°.

27. How many zeros does the sine function have? How far apart are they?

28. After reading this section, make a list of questions that you want to ask your instructor.

29. Continue in your daily journal and make an entry. In addition to your normal entry on thoughts about the mathematics in this section, list at least two positive comments about what you have learned about this topic.

30. In paragraph format, summarize the material in this section of the text in your daily journal.

31. Describe how your classroom instructor made this topic more understandable and clear.

32. After reading the text and listening to your instructor, what do you not understand about this topic?

33. What is the most important idea presented in this section? Why?

Explorations

34. Using the table in Exercise 7, for each angle, square the value of the sine, square the value of the cosine. Add the squares together. What is the sum? Is it always the same number no matter what θ is used? Develop an argument that would convince classmates that $(\sin\theta)^2 + (\cos\theta)^2$ is always the same number as found above. Find one more relationship between sine, cosine, and tangent.

35. You have solved a variety of equations - both algebraically and graphically. Equations many times involve trigonometric functions as well as linear, quadratic, absolute value, square root, exponential, and rational functions. While an algebraic method for solving the equations below is beyond the scope of this text, or impossible, the graphical method is the same method you have used since solving the equations in Chapter Five. Be sure your calculator is in radian mode and solve the equations below:

    a. $2^{4\cos x} = 5\sqrt{x}$

    b. $|x-3| - 2\sin x = -1$

    c. $\sqrt{5\cos(2x)} + 3.6 = |x|$

    d. $3^{\cos x} \cdot |\sin x| - x = \sqrt{(x+2)(x-1)}$

    e. $(\sin x)^2 + (\cos x)^2 = 5 - 3x - x^2$

    f. $\dfrac{\sin x}{\cos x} + \sqrt{x^2 - x - 12} = \tan x$

    g. $|\sin x| + 2|\cos x| + \left|\dfrac{1}{2}x\right| = 2$

12.3 Solving Right Triangles

Based on Figure 12.3.1 below, the definitions of the trigonometric functions of sine, cosine, and tangent for ANY angle in the appropriate domain are:

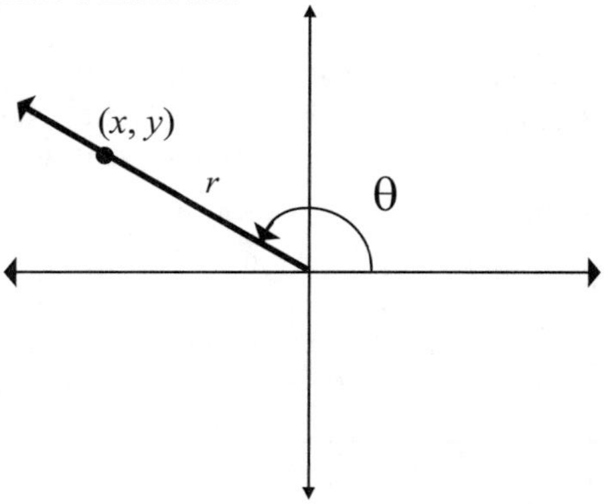

Figure 12.3.1

Definition of Sine, Cosine, and Tangent Functions

$$\sin\theta = \frac{y}{r} \qquad \cos\theta = \frac{x}{r} \qquad \tan\theta = \frac{y}{x}.$$

For the purpose of applying the above definitions to a right triangle, ***the domain is restricted on each of the above trigonometric functions to be the interval* (0°, 90°) *in degrees or*** $\left(0, \frac{\pi}{2}\right)$ ***in radians***. By restricting the domain, the reference angle is always equal to θ. Since θ has now been restricted to be an acute angle, a right triangle can always be superimposed on *top of the angle* θ*;* with the hypotenuse on the terminal side of θ and one side on the positive x-axis - see Figure 12.3.2. This is like earlier exercises where a right triangle was superimposed on top of the reference angle to θ; however now, θ and the reference angle are the same.

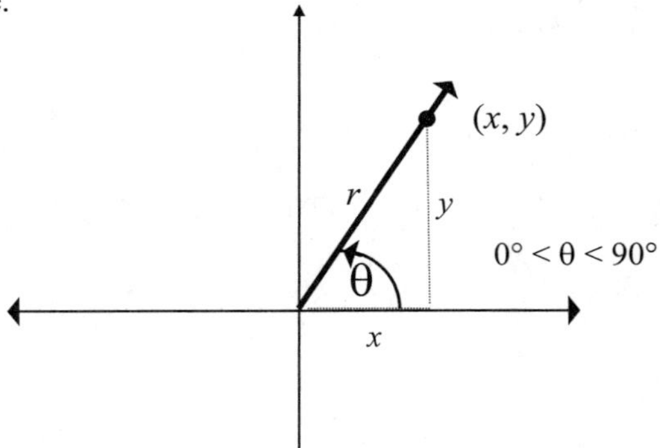

Figure 12.3.2

While in quadrant I, the length of the horizontal side of the triangle is identical to the x-coordinate of the point on the terminal side of θ, the length of the vertical side of the triangle is identical to the y-coordinate of the point on the terminal side of θ, and the length of the hypotenuse is identical to the distance from the point on the terminal side to the origin.

The next step in developing new trigonometric definitions related to the right triangle is to re-label the sides of the right triangle. The side "opposite" the angle θ is "opp", the hypotenuse is labeled "hyp", and the third side of the triangle looks like it is "adjacent" to the angle θ so it is labeled "adj".

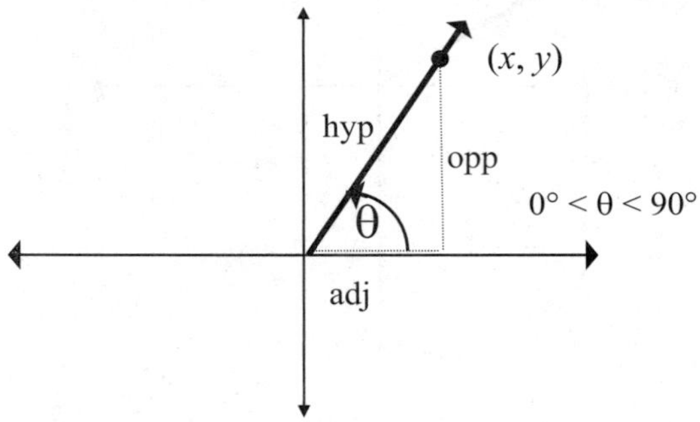

Figure 12.3.3

Alternative Definitions of Sine, Cosine, and Tangent

$$\sin\theta = \frac{y}{r} = \frac{\text{opp}}{\text{hyp}} \qquad \cos\theta = \frac{x}{r} = \frac{\text{adj}}{\text{hyp}} \qquad \tan\theta = \frac{y}{x} = \frac{\text{opp}}{\text{adj}}.$$

where θ is an acute angle of the right triangle.

Finally, if the rectangular coordinate system is erased, the triangle is free to move or rotate to any position. The definitions are no longer dependent on the coordinates of a point on the terminal side of θ, they are only dependent on the sides of the right triangle.

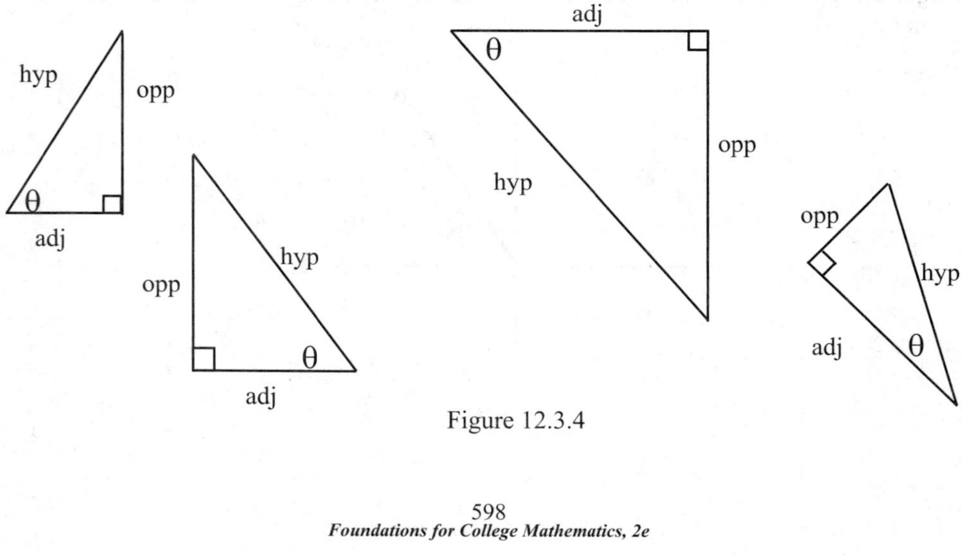

Figure 12.3.4

As shown in the right triangles above:

$$\sin\theta = \frac{\text{opp}}{\text{hyp}}$$

$$\cos\theta = \frac{\text{adj}}{\text{hyp}}$$

$$\tan\theta = \frac{\text{opp}}{\text{adj}}.$$

The angle θ can be either of the two acute angles of the right triangle. The figure below demonstrates that whichever angle is chosen to be labeled θ, the only things that change are the actual side that is opposite θ and the side that is adjacent to θ.

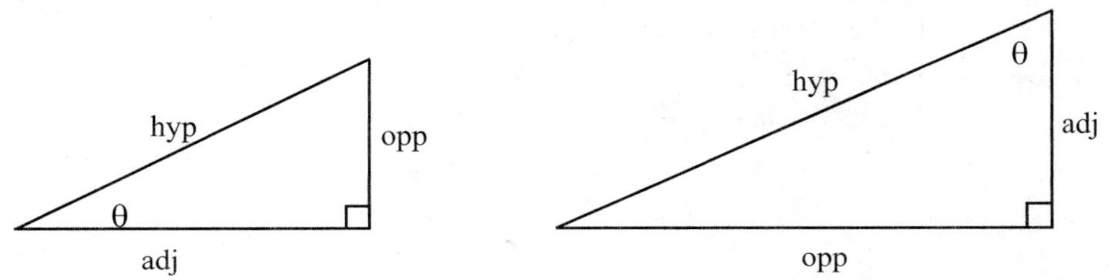

Figure 12.3.5

Since the angle θ is no longer on the coordinate plane, but a part of a right triangle, the acute angles can be identified by the labels in the triangle.

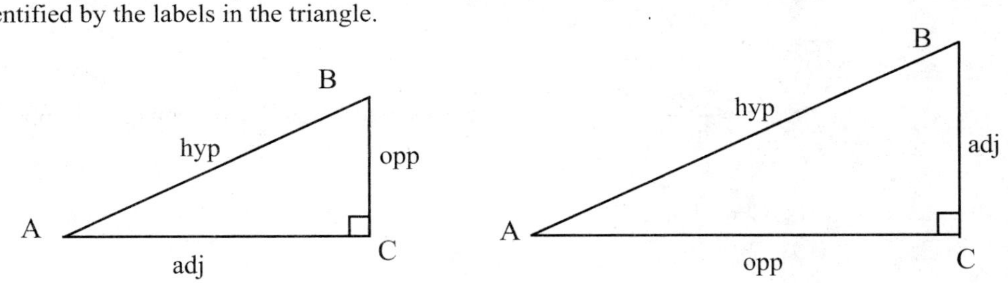

Figure 12.3.6

$$\sin A = \frac{\text{opp}}{\text{hyp}} \qquad \sin B = \frac{\text{opp}}{\text{hyp}}$$

$$\cos A = \frac{\text{adj}}{\text{hyp}} \qquad \cos B = \frac{\text{adj}}{\text{hyp}}$$

$$\tan A = \frac{\text{opp}}{\text{adj}} \qquad \tan B = \frac{\text{opp}}{\text{adj}}$$

In a statement like $\sin A = \frac{\text{opp}}{\text{hyp}}$, you may want to think of this definition as a formula containing three variables -- sin *A*, opp, and hyp. If two of the three variables have known values, the value of the third can be found. In a statement like $\tan A = \frac{\text{opp}}{\text{adj}}$, you may want to think of this definition as a formula containing three variables -- tan *A*, opp, and adj. If two of the three variables have known values, the value of the third can be found.

Example 1: In a right triangle, find the side opposite an angle of 20°, if the hypotenuse is 12 inches long.

Solution: The information in the problem is an angle, the hypotenuse, and the side opposite the angle. The right triangle definition that includes an angle, the side opposite the angle, and the hypotenuse is the sine definition.

$$\sin\theta = \frac{\text{opp}}{\text{hyp}}$$

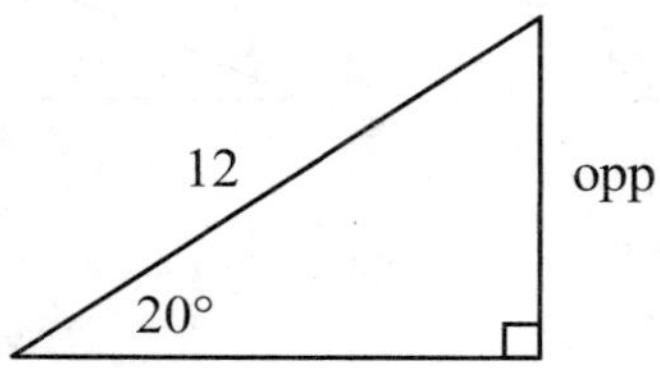

Figure 12.3.7

$$\sin 20° = \frac{\text{opp}}{\text{hyp}}$$ evaluate the sin 20° and solve the equation for opp

$$0.342 = \frac{\text{opp}}{12}$$

$$12 \cdot (0.342) = \left(\frac{\text{opp}}{12}\right) \cdot 12$$

$$4.104 = \text{opp}$$ The side opposite 20° is 4.1 in.

⁂

While you know the calculator will find the sine, cosine, or tangent ratio, it will also find the angle if the ratio is known. The calculator technique is demonstrated in the problem below.

Example 2: In right triangle ABC, side $a = 8$ and side $b = 15$, find angle *A*.

Solution: The desired angle is opposite side *a*, side *b* is adjacent to the angle needed.

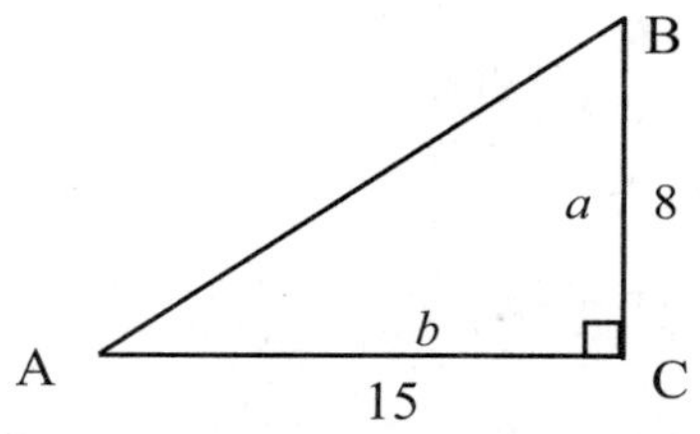

Figure 12.3.8

From the definitions,

$$\tan A = \frac{\text{opp}}{\text{adj}}$$

$$\tan A = \frac{8}{15}$$

$$\tan A = 0.5333$$

The problem is to find angle A when the tangent of A is a known number. If the angle's measure is to be in degrees, calculator must be in degree mode. To find the angle, enter something like the following keystrokes:

2nd TAN (.5333) <ENTER>

If the mode is set on degrees, the display should read 28.071. That is, angle A is 28.071°.

⁂

To find the acute angle whose trig ratio is known, use something like the following keystrokes: (set mode to degrees)

$\sin A = 0.2953$ → 2nd SIN .2953) <ENTER> ← $A = 17.18°$

$\cos A = 0.8491$ → 2nd COS .8491) <ENTER> ← $A = 31.89°$

$\tan A = 14.29$ → 2nd TAN 14.29) <ENTER> ← $A = 86.00°$

When the keystrokes 2nd and sin are entered, the display may look like $\sin^{-1}$. This notation does not mean that the sine has been raised to the −1 exponent. For now, ignore the notation, it will be discussed in a later course or you may ask your instructor for an interpretation.

For now,

you may think of the notation $\sin^{-1}x$ as meaning the angle whose sine is x.

You may think of the notation $\cos^{-1}x$ as meaning the angle whose cosine is x.

You may think of the notation $\tan^{-1}x$ as meaning the angle whose tangent is x.

Example 3: In right triangle ABC, $a = 13$ and $c = 15$, find angle A in radians.

Solution: The information includes an angle (A), the side opposite the angle (a), and the hypotenuse (c). The definition needed is

$$\sin A = \frac{\text{opp}}{\text{hyp}}$$

$$\sin A = \frac{13}{15}$$

$\sin A = 0.8667$ with the calculator in radian mode, enter:

2nd SIN .8667 <ENTER>

The angle A is 1.049 radians.

⁂

Using the right triangle trigonometric definitions and the Pythagorean theorem, if any two parts of the triangle are known, except the two acute angles, finding all of the remaining parts is called solving the triangle.

Example 4: Solve the right triangle, if $B = 25°$ and $b = 4$.

Solution: The unknown parts are a, c, and A.

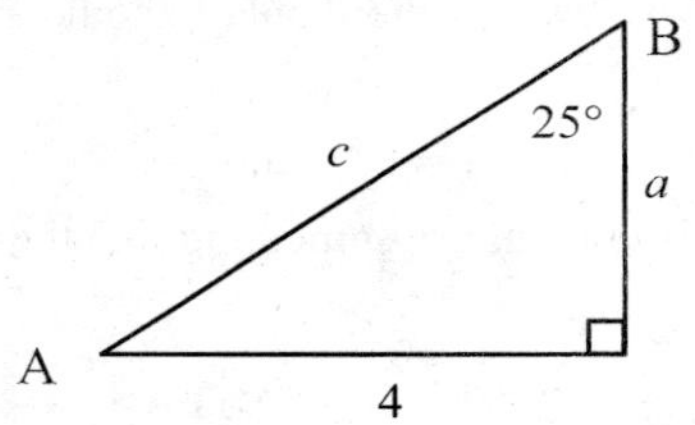

Figure 12.3.9

It does not make any difference which part is found first. The simplest to find is angle A because $A + B = 90°$, so $A + 25° = 90°$ or $A = 65°$. To find side a, the angle B is known and the side opposite B is known. Side a is adjacent to angle B; that is:

$$\tan B = \frac{\text{opp}}{\text{adj}}$$

$$\tan 25° = \frac{4}{a}$$

$$0.4663 = \frac{4}{a}$$

$$0.4663a = 4$$

$$a = 8.58$$

The only remaining part is the hypotenuse, from the Pythagorean theorem:

$$c^2 = a^2 + b^2$$

$$c^2 = (8.58)^2 + 4^2$$

$$c^2 = 89.62$$

$$c = 9.47$$

The solution to the right triangle is: $a = 8.58$, $c = 9.47$, and $A = 65°$.

⁂

12.3 EXERCISES

Many times you can work the exercises by algebraic, numeric, or graphical methods. You can learn about mathematics no matter which method you use. Part of the learning process is for you to decide which method is best for you to use. Please read the text carefully before you do the exercises. Read the **next assigned section** after you finish the assignment below. All numeric answers must have an error ≤ 0.01.

A Quick Look To The Future

-4. List one thing in your life that is predictably cyclic.

-3. At most, how many times can the graph of a quadratic function intersect with the graph of a square root function?

-2. Suppose that $2x + 5 = 9$ and $x = y + 1$, does $2(y + 1) + 5 = 9$?

-1. Suppose that $\begin{pmatrix} 2 & 3 \\ -1 & 4 \end{pmatrix}$ is a short-cut way of writing the expression $\begin{matrix} 2x + 3y \\ -x + 4y \end{matrix}$, what does $\begin{pmatrix} -2 & 1 \\ 3 & -5 \end{pmatrix}$ mean?

Previous Topics Revisited

1. What is the exact value of sin 330°?
2. What angle has a sine ratio and cosine ratio that are equal in value? There are many answers when using the original definition from the last section.
3. The point $(x,-3)$ is on the terminal side of θ, $r = 5$, find x.
4. Convert 210° to π radians.
5. Convert $\frac{7\pi}{4}$ to radians without π.
6. Is $x^2 - y^2 + 4x - 2y + 4 = 0$ the equation of a circle? Why?

New Topic Exercises

Find θ, if θ is on the interval (0°, 90°).

7. $\sin \theta = 0.111$
8. $\cos \theta = 0.111$

9. $\tan \theta = 0.111$

In right triangle ABC, find the specified part.

10. $a = 4.7$, $c = 9.2$, find B.

11. $B = 89°$, $b = 100$, find a.

12. $c = 12$, $A = 19°$, find b.

Solve the right triangles.

13. $a = 5$ and $b = 2$.

14. $A = 30°$ and $a = 4$.

15. $B = 70°$ and $c = 10$.

From Mathematics to English

16. Explain why the trigonometric definitions from this section do not apply to non-right triangles.

17. For the right triangle definitions, why does the domain not include 0° and 90°?

18. Can the right triangle with the two acute angle as given information be solved? Why or why not?

19. Is it possible for a right triangle to have a hypotenuse of 10 inches and a leg of 12 inches? Why?

20. Describe how you identify opposite, adjacent, and hypotenuse.

21. After reading this section, make a list of questions that you want to ask your instructor.

22. Continue in your daily journal and make an entry. In addition to your normal entry on thoughts about the mathematics in this section, list at least two positive comments about what you have learned about this topic.

23. In paragraph format, summarize the material in this section of the text in your daily journal.

24. Describe how your classroom instructor made this topic more understandable and clear.

25. After reading the text and listening to your instructor, what do you not understand about this topic?

Explorations

26. Use the definitions from this section as the basis for an argument to explain why:

 a. $\tan \theta = \dfrac{\sin \theta}{\cos \theta}$

 b. In right triangle ABC, $\sin A = \cos B$ and $\cos A = \sin B$.

27. If triangle ABC is *NOT* a right triangle, and an altitude is drawn to side b. If side b is 8 inches long, side c is 10 inches long, and angle A is 50°, find the area of the triangle.

28. In right triangle ABC, using the Pythagorean Theorem and the definitions from this section, show that for either acute angle A or B of triangle ABC, $(\sin A)^2 + (\cos A)^2 = 1$.

12.4 Trigonometric Functions as Mathematical Models

The situations that follow all involve right triangles. The first thing you must do to solve a problem associated with the situation is to identify the parts of the right triangle given in the problem; and secondly, you must determine which part of the triangle you are being asked to find. From this point, you can use the definition of cosine, sine, or tangent to obtain an equation that can be solved for the unknown quantity. In a few cases, the unknown quantity may be needed to find a second unknown quantity. The final step is to interpret the numeric answer and make it fit the situation. For example, the numeric answer to a problem may be the size of an angle; thus, the units of degrees or radians must be attached. If the numeric value of an answer is the length of a side of a triangle, the proper units of length must be added.

One situation involving right triangle trigonometry involves angles called angles of elevation and angles of depression. These angles are used because they can be measured with instruments like a transit, clinometer, or a sextant. Simply stated, an **angle of elevation** is an angle formed by a horizontal ray and an intersecting ray *above* the horizontal. An **angle of depression** is an angle formed by a horizontal ray and an intersecting ray *below* the horizontal. See Figure 12.4.1.

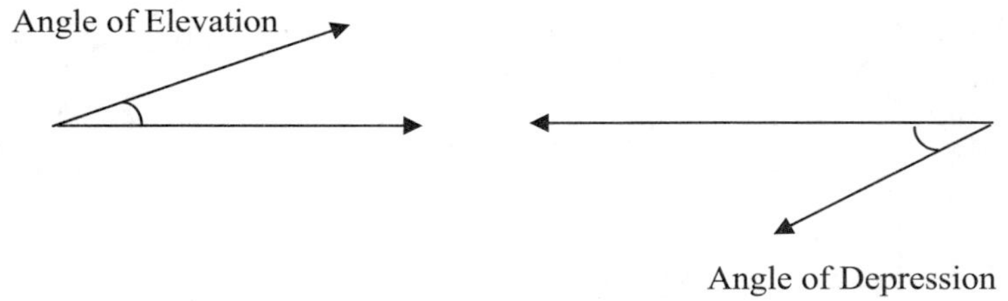

Figure 12.4.1

Example 1: The angle of elevation of the top of a sequoia tree is 21°. It is measured at a distance of 800 feet from the base of the tree. How high is the tree?

Solution: As shown in Figure 12.4.2, an angle and the side adjacent to the angle are known. The side opposite the known angle is the part of the right triangle that must be found.

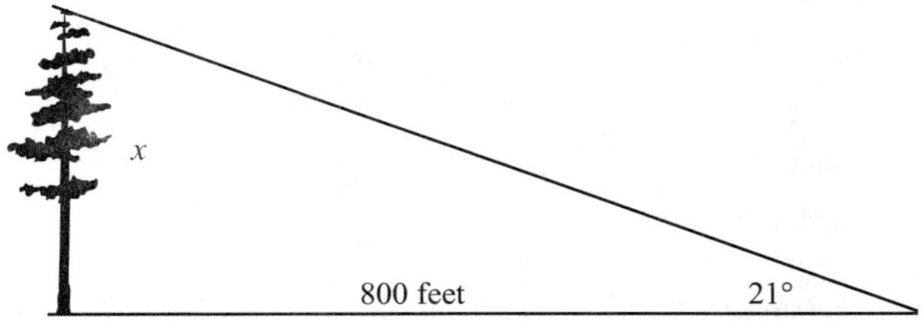

Figure 12.4.2

Use the definition of tangent because it contains an angle, the side adjacent to the angle, and the side opposite the angle.

$$\tan\theta = \frac{\text{opp}}{\text{adj}}$$

$$\tan 21° = \frac{x}{800}$$

Multiply both sides by 800 $\quad 800 \cdot \tan 21° = x.$

Calculate in degree mode $\quad$ 800 TAN 21) <ENTER>

The height of the sequoia tree is 307 feet.

⁂

Example 2: A bank of solar collectors is arranged as shown below. What must be the distance between collectors to avoid the shadow of one collector falling on the surface of another collector, when the angle of elevation of the sun is 87° and the vertical height of the collectors is 8 feet. (Drawing not to scale.)

Solution:

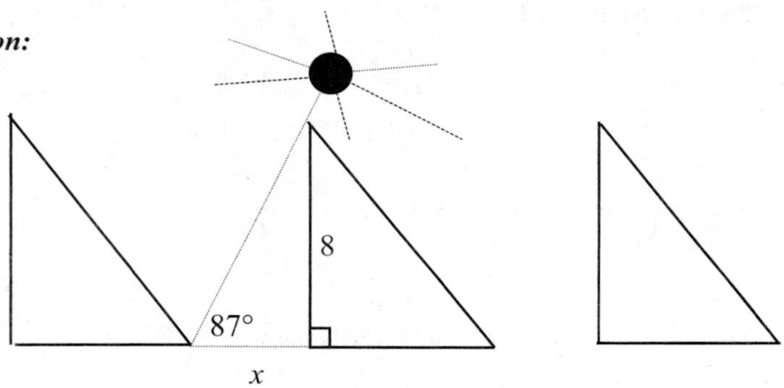

Figure 12.4.3

The right triangle used to solve this problem has an angle and the side opposite the angle as given information. The side adjacent to the given angle must be found. With information like an angle, the opposite side, and the adjacent side, the tangent definition should be used.

$$\tan\theta = \frac{\text{opp}}{\text{adj}}$$

$$\tan 87° = \frac{8}{x}$$

multiply by x $\quad x \cdot \tan 87° = 8$

divide by tan 87° $\quad x = \dfrac{8}{\tan 87°}$

keystrokes: $\quad$ 8 ÷ TAN 87) <ENTER> (Use degree mode.)

The distance between solar collectors is 0.42 feet or 5 inches.

⁂

Example 3: If the distance down a mountain slope is approximately 15,250 feet and the average angle of elevation of the mountain is 8.5°, how high is the mountain?

Solution: As can be seen in Figure 12.4.4, the given information includes an angle and the hypotenuse. The part required is the side opposite the angle. An angle, the side opposite the angle, and the hypotenuse imply the use of the sine definition.

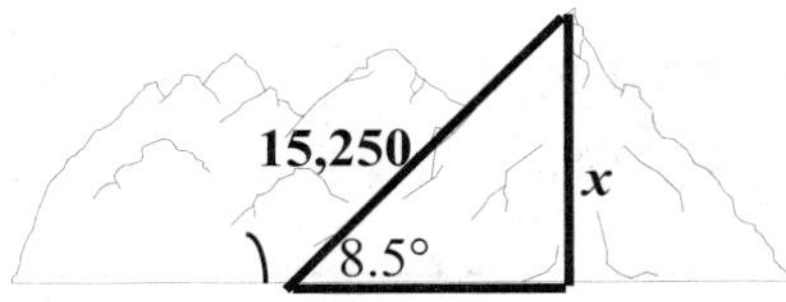

Figure 12.4.4

$$\sin\theta = \frac{\text{opp}}{\text{hyp}}$$

$$\sin 8.5° = \frac{x}{15250}$$

multiply by 15250 $\qquad 15250 \cdot \sin 8.5° = x$

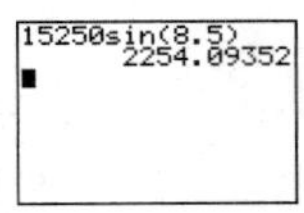

calculate $\qquad$ 15250 SIN 8.5) <ENTER>

The mountain is approximately 2,254 feet high.

Example 4: What angle does the graph of $y = \left(\frac{2}{3}\right)x$ make with the x-axis?

Solution: From Figure 12.4.5, note the right triangle. The point (3, 2), for example, is on the graph.

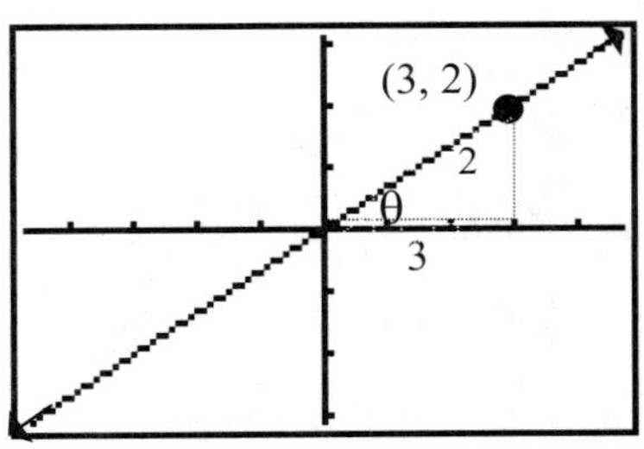

Figure 12.4.5

The sides opposite θ and the side adjacent to θ are known. An angle, the opposite, and the adjacent imply the use of the tangent.

$$\tan\theta = \frac{\text{opp}}{\text{adj}}$$

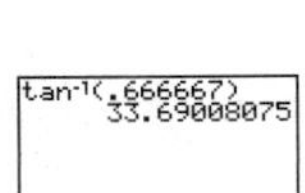

$$\tan\theta = \frac{2}{3}$$

$$\tan\theta = .666667$$

Use the keystrokes: 2nd TAN .666667 <ENTER>. The angle the graph makes with the x-axis is 33.7°.

Example 5: What angle does the graph of $y = \left(\frac{2}{3}\right)x + 4$ make with the x-axis?

Solution: As shown in Figure 12.4.6, the graph forms a right triangle with the x and y-axes.

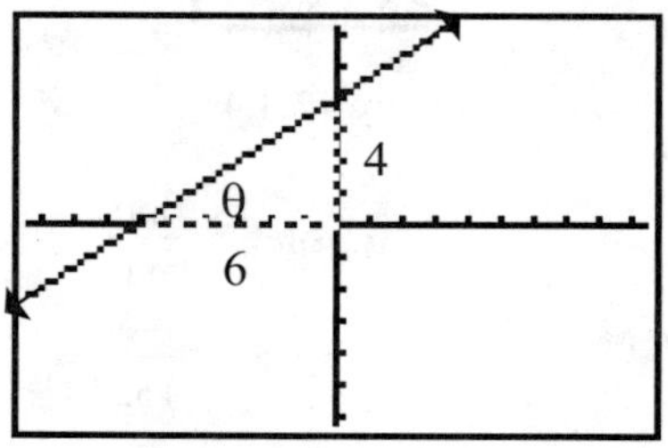

Figure 12.4.6

The side opposite the angle θ is the y-intercept and the side adjacent to θ can be found from the x-intercept. The y-intercept is 4. To find the x-intercept, replace y with 0 and solve for x.

$$0 = \left(\frac{2}{3}\right)x + 4 \qquad \text{subtract 4}$$

$$-4 = \left(\frac{2}{3}\right)x \qquad \text{multiply by } \frac{3}{2}$$

$$-6 = x$$

The **triangle** has the side opposite as 4 and the side adjacent as 6. The definition of tangent should be used to find θ.

$$\tan\theta = \frac{4}{6}$$

$$\tan\theta = .666667$$

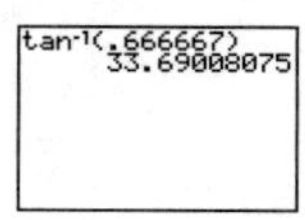

keystrokes: 2nd TAN .666667) <ENTER>

The angle formed by the graph and the x-axis is 33.7°.

12.4 EXERCISES

Many times you can work the exercises by algebraic, numeric, or graphical methods. You can learn about mathematics no matter which method you use. Part of the learning process is for you to decide which method is best for you to use. Please read the text carefully before you do the exercises. Read the **next assigned section** after you finish the assignment below. All numeric answers must have an error ≤ 0.01.

A Quick Look To The Future

-4. How would you enter $3x + 5y = 6$ on the function screen of your calculator?

-3. If $a = b$ and $c = d$, does $a + c = b + d$?

-2. Do you think it is possible for x to represent something other than a real number?

-1. Does an electrical engineer use mathematics?

Previous Topics Revisited

1. Solve the right triangle ABC with side $b = 4{,}000{,}000$ and $\angle A = 45°$.

2. Solve the right triangle ABC with $\angle B = 1°$ and the hypotenuse, $c = 0.001$.

3. Solve the right triangle ABC with $a = 9$ and $b = 13$.

4. If the sine of θ is $\frac{1}{2}$, what is the cosine of θ?

5. If θ is in quadrant II and the distance from the origin to a point on the terminal side of θ is 13, find the x-coordinate of the point if the y-coordinate is 5.

6. Convert $\frac{6\pi}{5}$ to degrees.

New Topic Exercises

7. In making a teepee, what length poles must be used if the height in the middle of the teepee is to be 12 feet and the angle made between level ground and the sides of the teepee is 70°? What is the area of the circular floor space?

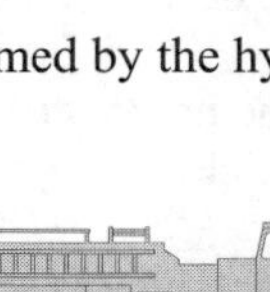

8. What angle does the roof of a house make with the horizontal if the pitch of the roof is 8 feet up for every 12 feet in?

9. What is the area of a right triangle with a hypotenuse of 12 and an angle formed by the hypotenuse and a leg is 25°?

10. A fire truck has a 100 foot ladder. How high will the ladder reach if it is raised to an angle of elevation of 15°? 30°? 45°? 60°? 75°?

11. A spotlight is directed vertically upward toward the cloud ceiling. From a distance of 200 feet away from the spotlight, a ground observer notes the angle of elevation of the ceiling is 76°. What is the cloud ceiling?

12. From an observation point 40 feet above sea level, a small boat makes an angle of depression of 5°. How far is the boat from the base of the observation point?

13. Suppose each city block is 1000 feet north/south by 800 feet east/west. You walk north one block then east one block and you repeat this path four more times. What is the direct distance from where you started to where you end? What is the angle made by the direct distance and an east/west line?

14. The space shuttle wants to land by approaching the landing site at an angle of 15° with the horizontal. If its approach altitude is 15,000 feet, what ground distance from the landing strip must it start its descent?

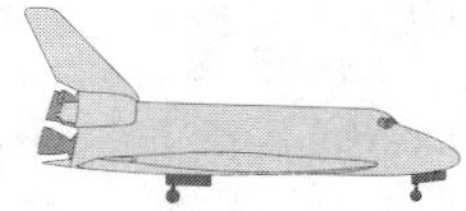

15. When the space shuttle lifts off, it makes an angle of 80° with the horizontal. When it has flown 100,000 feet, what is its elevation?

16. An 80 foot tree stands 60 feet ground feet from a power line that is 50 feet above ground level. When the tree is cut down, it accidentally falls toward the power line. Does it hit the power line?

17. Suppose you live 20,000 feet from the airport. Planes taking off must ascend at an angle of 35°. How high are the planes when they are directly above your house?

18. Find the angle made by the graph and the x-axis for:

a. $y = \left(\dfrac{3}{4}\right)x$ b. $y = \left(\dfrac{5}{8}\right)x$ c. $y = \left(\dfrac{2}{7}\right)x$

d. $y = \left(\dfrac{3}{4}\right)x + 5$ e. $y = \left(\dfrac{5}{8}\right)x - 7$ f. $y = \left(\dfrac{2}{7}\right)x + 1$

19. Find the angle formed by the graph and the x-axis for:

a. $y = \dfrac{1}{2}x$ b. $y = \dfrac{1}{2}x + 3$ c. $y = \dfrac{1}{2}x + 5$

d. $y = \dfrac{1}{2}x - 2$ e. $y = \dfrac{1}{2}x - 6$

From Mathematics to English

20. Describe a method you can use to find the height of the tree in Exercise 16.

21. Can the definitions of cosine, sine, and tangent be used on a non-right triangle? Why?

22. Describe how you can *calculate* the angle of elevation of the sun with nothing but a meter-stick (yard-stick).

23. After reading this section, make a list of questions that you want to ask your instructor.

24. Continue in your daily journal and make an entry. In addition to your normal entry on thoughts about the mathematics in this section, list at least two positive comments about what you have learned about this topic.

25. In paragraph format, summarize the material in this section of the text in your daily journal.

26. Describe how your classroom instructor made this topic more understandable and clear.

27. After reading the text and listening to your instructor, what do you not understand about this topic?

Explorations

28. Make-up at least 3 application problems that can be solved using right triangle trigonometry.

29. Solve the equations graphically. Describe what you did to solve the equations. Check your work algebraically.

a. $\sin 42^\circ = \dfrac{5}{x}$ b. $\cos 19^\circ = \dfrac{x}{9}$ c. $\tan x^\circ = \dfrac{3}{4}$

30. In your room, find two lines that meet to form a non-right angle. Using trigonometry and a ruler, calculate the size of the angle. Describe the lines and how you found the size of the angle.

31. The equations below may be difficult or impossible to solve analytically; try to solve them graphically or numerically. Be sure your calculator is in radian mode.

a. $3\sin(x-1) = -x$ b. $2\cos(4x) = 2x - 1$

c. $2\sin(5x+1) = 3\sqrt{x-2}$ d. $3\sin x = \cos(6x) + 1 - 3x$

CHAPTER TWELVE TEST

1. Convert 135° to π radians and pure radians.
2. Convert $\frac{7\pi}{4}$ radians to degrees.
3. Convert 4.29 radians to degrees.
4. List one positive and one negative coterminal angle to 30°.
5. Find any point on the terminal side of an angle with a measure of –90°.
6. What is the exact value of tan 60°?
7. What is the approximate value of $\sin\left(\frac{9\pi}{4}\right)$?
8. If the point (–5, 12) is on the terminal side of angle θ, find the distance from the point to the origin (r).
9. If tan θ = 0, list any possible angle θ.
10. If tan θ = 0, what is the y-coordinate of any point on the terminal side of θ?
11. If θ is in the interval (0°, 90°), find θ when sin θ = 0.875.
12. In triangle ABC, side $a = 3$ and side $b = 4$, find cos B.
13. In triangle ABC, side $a = 36$ and $\angle A = 20°$, find side b.
14. In triangle ABC, side $b = 7$ and side $c = 14$, find $\angle A$.
15. Solve the right triangle ABC when $a = 4$ and $\angle A = 30°$, that is, find:

    $b =$ $c =$ $\angle B =$

16. What angle does the roof of a building make with the horizontal if the pitch of the roof is 16 feet up for every 12 feet in?
17. How high does a 60-foot fire truck ladder reach when the angle of elevation of the ladder is 65°?
18. If the angle of elevation of the flight path of a plane in take-off is 20°, how high is the plane after it has traveled 30 miles in the air?
19. What is the angle made by the graph of $y = \left(\frac{4}{5}\right)x$ and the x-axis?
20. What is the angle made by the graph of $y = \left(\frac{4}{5}\right)x + 2$ and the x-axis?

CHAPTER THIRTEEN
SYSTEMS OF EQUATIONS AND INEQUALITIES

13.0 Introduction to Systems of Equations

A **system of equations or inequalities** is a series of equations or inequalities describing a particular object or situation. While the examples below only contain 2 equations or inequalities in 2 variables, engineers may write as many as 100 or more equations in describing an object or situation being designed or analyzed. Beginning electronics students may need to write a series of 4 or 5 equations describing a simple electronic circuit being analyzed. For the sake of understanding, systems with 2 equations or 2 inequalities are presented here, and systems with 3 equations follow. Once you can solve a 2×2 system, the methods used can be applied to a 100×100 system (for example).

In this chapter, systems will be solved by a variety of methods. Each method has merits and drawbacks. Some methods can be processed on the calculator and some require an algebraic approach. The concentration is on the method of solving the system. The last section of this chapter contains application problems that can be solved by using systems of equations to describe the problem.

Systems in Action

To earn extra money while in college, during football games you sell hot dogs at \$8 each and soft drinks at \$3.75 each. You are supplied with a total of 200 dogs and drinks total, and your goal is to earn at least \$850 from the sale of the drinks and dogs. A series of inequalities can describe the situation.

Think of x as symbolizing the *number* of hot dogs sold.
Think of y as symbolizing the *number* of drinks sold.

The number of items you will sell is $x + y$ and it must be 200 or under; that is: $x + y \leq 200$

Your income must be at least \$850; that is $8x + 3.75y \geq 850$

The inequalities form a system of inequalities.

$$\begin{aligned} x + y &\leq 200 \\ 8x + 3.75y &\geq 850 \end{aligned}$$

Solving the system will tell you how you can reach your financial goal.

As another simple example, consider the coins in your pocket. Suppose you have 12 coins, all nickels and dimes. The value of the coins is \$1.00. Based on the known information about the coins, a series of two equations can be written describing the coins.

Think of x as symbolizing the *number* of nickels.

Think of y as symbolizing the *number* of dimes.

Since you have 12 coins, then $x + y = 12$.

The value of 4 nickels is 5×4 cents. The value of 6 nickels is 5×6 cents. Thus, the value of x nickels is $5x$ cents. The value of 3 dimes is 10×3 cents. The value of 7 dimes is 10×7 cents. Therefore, the value of y dimes is $10y$ cents.

Since the value of the coins is 100 cents, then $5x + 10y = 100$.

The equations:

$$\begin{aligned} x + y &= 12 \\ 5x + 10y &= 100 \end{aligned}$$

form a system of equations. These equations describe the coins in your pocket. Solving the system will tell you the number of coins.

13.1 Solving Systems Graphically

Solving a system means finding values for the variables that make all of the equations or inequalities true statements. The graphical method of solving systems allows you to "see" the solution. When an equation written as a function is graphed, you see in the viewing window all of the values (x, y) that make the equation a true statement. If two graphs are displayed on the same appropriate viewing window, all of the values (x, y) that make both equations true can be seen at the same time. Because the solution is values for x and y that make both equations true, the solution, if there is a solution, is the point(s) of intersection of the graphs. What you should recognize is that the graphical method for solving systems of equations is the same as the intersection method for solving a single equation. For example, in Chapter Six you solved an equation like $x + 35 = 1.25x + 30$. If you used the intersection method, you were solving a system like in Example 1 below.

Example 1: TV repair service company A charges \$35 plus \$1 per minute of labor. TV repair service company B charges \$30 plus \$1.25 per minute of labor. The equations that describe this situation are $C = 1x + 35$ and $C = 1.25x + 30$. For what length of time do the repair companies charge the same amount? To find out, solve the system.

Solution: Graph both equations on the same window and use the intersection finder.

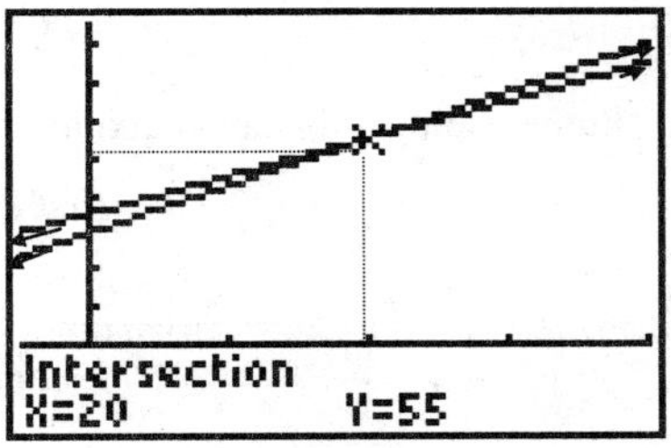

Figure 13.1.1

At 20 minutes, each service repair company charges \$55. That is, the point $(x, C) = (20, 55)$ is the only point the functions have in common. Thus, the solution to the system is $(20, 55)$.

⁂

Example 2: Solve the system graphically.

$$\begin{aligned} 2x - 5y &= -19 \\ -3x + 4y &= 0. \end{aligned}$$

Solution: The graphical approach requires that each equation be in function form.

$$y = \frac{2}{5}x + \frac{19}{5} \qquad \text{first equation}$$

$$y = \frac{3}{4}x \qquad \text{second equation}$$

Graph each function and identify the point of intersection.

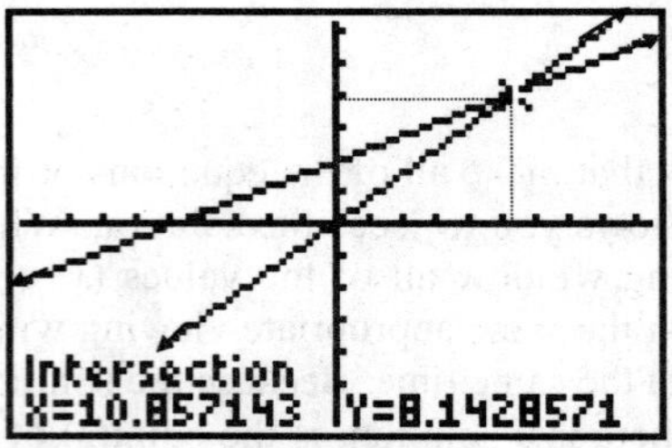

Figure 13.1.2a

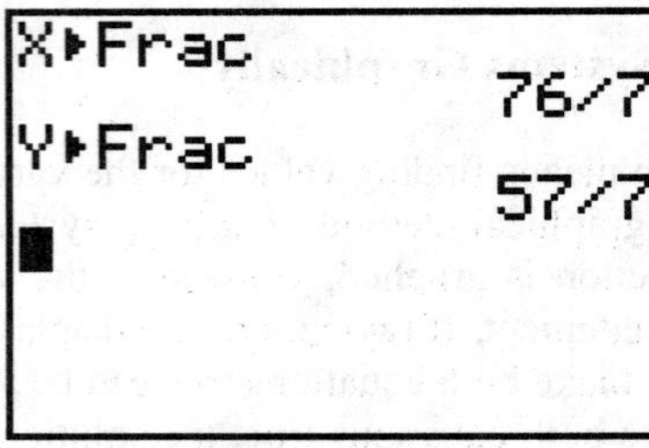

Figure 13.1.2b

The solution is (10.86, 8.14). If you prefer exact answers, the solution is $\left(\frac{76}{7}, \frac{57}{7}\right)$.

When x is $\frac{76}{7}$, y is $\frac{57}{7}$ in both equations.

⁂

Example 3: Solve the system graphically.

$$-4x + 3y = -1$$
$$12x - 9y = 11.$$

Solution: To graph the equations, isolate y in each equation.

$$y = \frac{4}{3}x - \frac{1}{3} \qquad \text{first equation}$$

$$y = \frac{4}{3}x - \frac{11}{9} \qquad \text{second equation}$$

There is no solution. The graphs have the same slope and different y-intercepts; thus, the graphs are parallel.

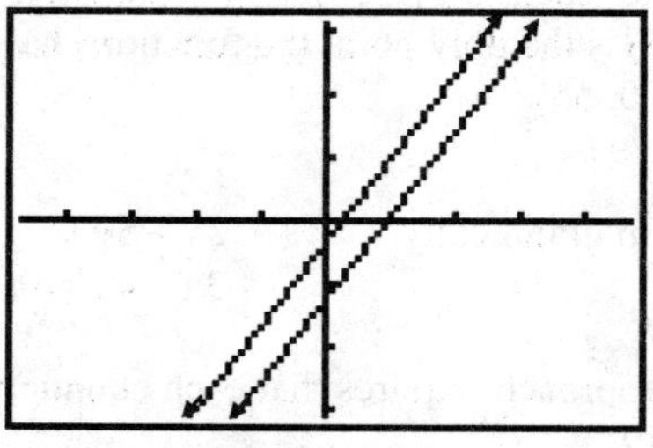
Figure 13.1.3

⁂

Example 4: Solve the system graphically.

$$-2x + 5y = -3$$
$$8x - 20y = 12.$$

Solution: The equations must be in slope-intercept form.

$$y = \frac{2}{5}x - \frac{3}{5} \qquad \text{first equation}$$

$$y = \frac{2}{5}x - \frac{3}{5} \qquad \text{second equation}$$

The equations have the same graph! Any solution to one is a solution to the other. There are an infinite number of solutions.

⁂

Systems of Inequalities

Before systems of inequalities are solved, a brief discussion of how inequalities are graphed is needed. Below is the graph of ALL points where the *y*-coordinate EQUALS $\frac{1}{2}$ of the *x*-coordinate minus 3; that is, it is the graph of $y = \frac{1}{2}x - 3$.

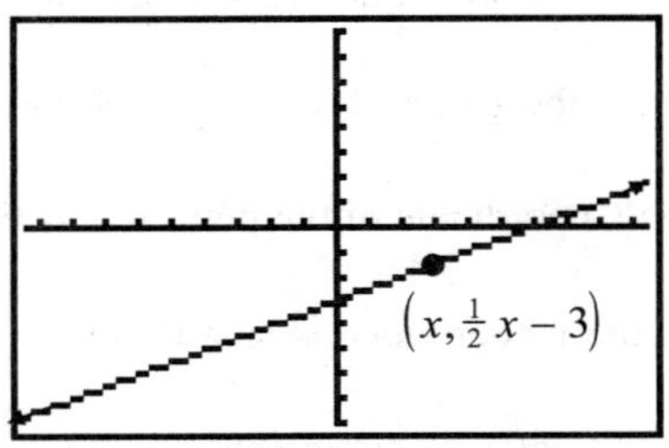

Figure 13.1.4

The above graph shows all points where the *y*-coordinate EQUALS $\frac{1}{2}$ of the *x*-coordinate minus 3; *every other* point in the coordinate plane has a *y*-coordinate that DOES NOT EQUAL $\frac{1}{2}$ of the *x*-coordinate minus 3. The graph below shows all of these points. Note that the line $y = \frac{1}{2}x - 3$ is a dashed line indicating it is not part of the graph of $y \neq \frac{1}{2}x - 3$. Further, note that the graph of $y \neq \frac{1}{2}x - 3$ is not a function; it is a relation.

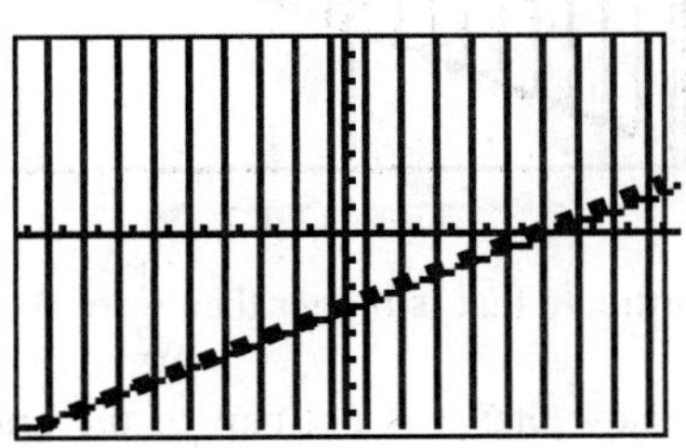

Figure 13.1.5

Below is the graph of all points where the y-coordinate is LESS THAN $\frac{1}{2}$ of the x-coordinate minus 3; that is, the graph of $y < \frac{1}{2}x - 3$.

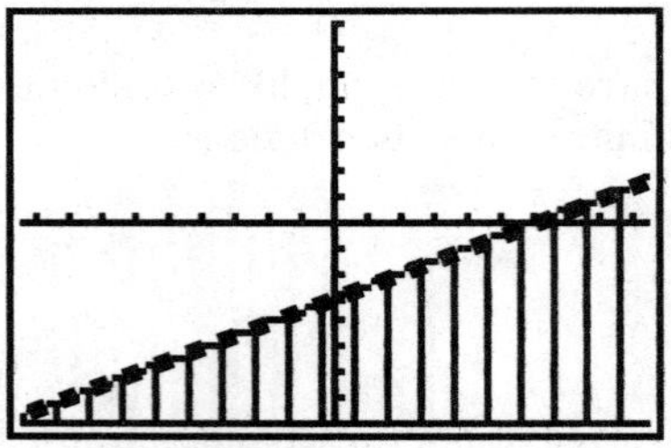

Below!
Below!

Figure 13.1.6

The boundary of the half plane is a dashed line; because, it is not part of the set of points that make $y < \frac{1}{2}x - 3$ a true statement. All points in Figure 13.1.6 that are shaded make $y < \frac{1}{2}x - 3$ a true statement; that is, the shaded region is the graph of $y < \frac{1}{2}x - 3$ and the solution to the inequality. For every point in the shaded region, the y-coordinate of the point is less than $\frac{1}{2}$ of the x-coordinate minus 3. *For example,* at the point (4, –15), the y-coordinate is smaller than $\frac{1}{2} \cdot 4 - 3$; all points (x, y) *below* the point (4, –1) (or any point on the graph) have a y-coordinate less than $\frac{1}{2}$ the x-coordinate (4) minus 3.

Finally, consider all points where the y-coordinate is larger than $\frac{1}{2}$ of the x-coordinate minus 3: the graph of $y > \frac{1}{2}x - 3$.

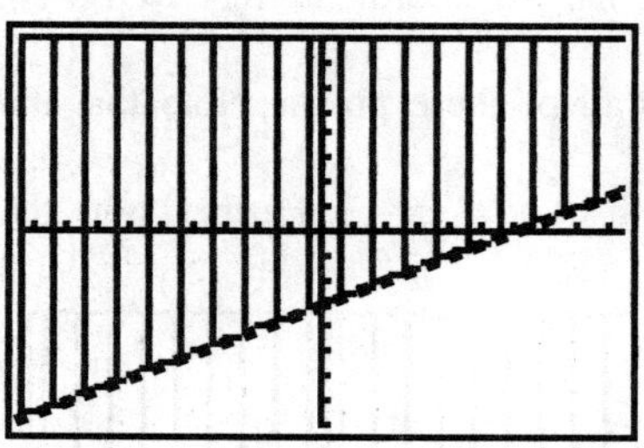

Greater Than.

Figure 13.1.7

Each point on the graph has a y-coordinate that is bigger than $\frac{1}{2}$ of the corresponding x-coordinate minus 3. The boundary line is dashed to show that it is not part of the graph of $y > \frac{1}{2}x - 3$. The graph of $y \geq \frac{1}{2}x - 3$ has a boundary line that is solid; this indicates it is part of the graph.

Example 5: Graph $y \leq -2x+5$.

Solution: Graph the **solid** boundary line $y = -2x+5$ and shade in all points where the y-coordinate is $\leq -2x+5$ (below the boundary line). The boundary line is a solid line because the graph is to show all points where the y-coordinate is less than OR EQUAL TO -2 times the x-coordinate plus 5.

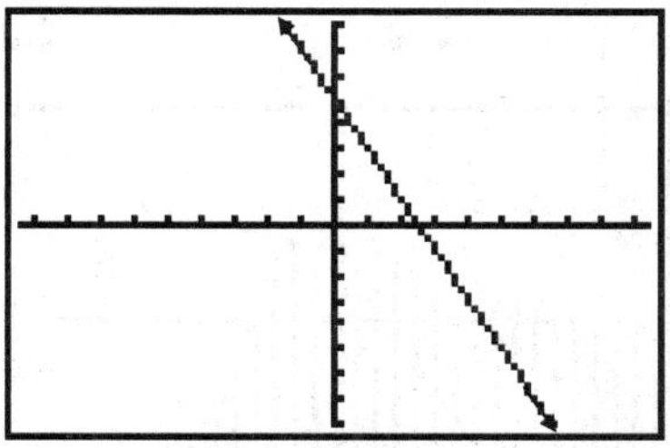

Figure 13.1.8a

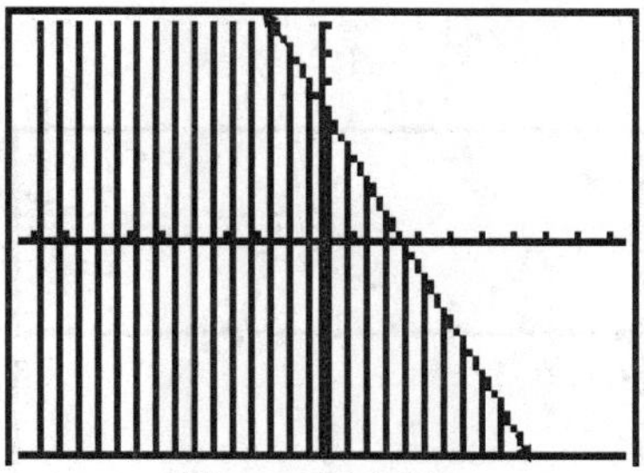

Figure 13.1.8b

⁂

Example 6: Graph $3x-5y < 15$.

Solution: Isolate the y on the left side of the inequality.

$-5y < -3x+15$ multiply by $\frac{-1}{5}$

$y > \frac{3}{5}x-3$ less than changes to greater than

Graph the **dashed** boundary line $y=\frac{3}{5}x-3$ and shade in **above** the line.

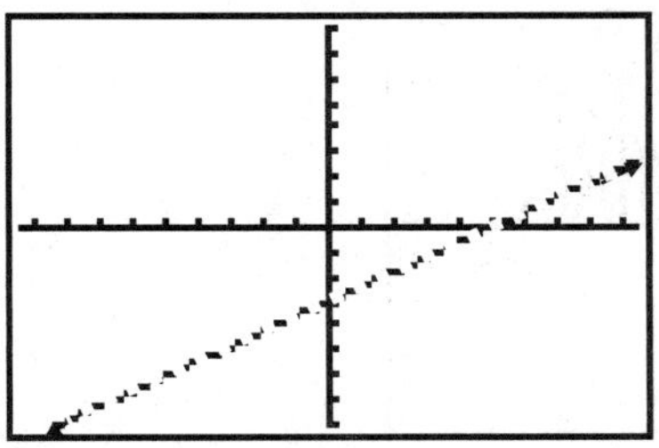

Figure 13.1.9a

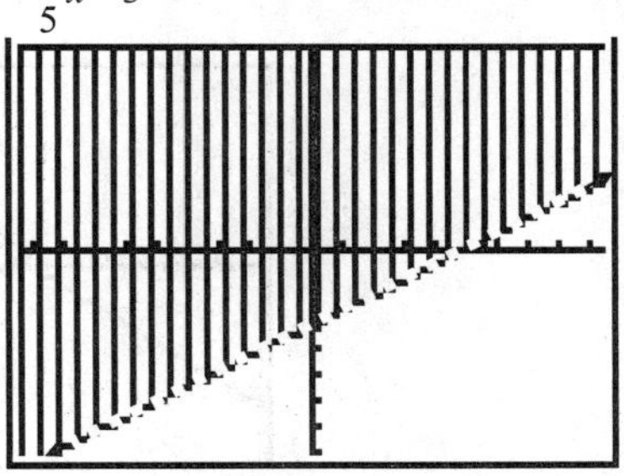

Figure 13.1.9b

⁂

The solution to a system of equations or a system of inequalities is all points the graphs have in common. In the case of the system of equations, you have seen that the graphs normally have one point in common. In the case of a system of inequalities each graph is a half plane; thus, you will find points in the solution where the half planes intersect. You will find an infinite number of solutions.

Example 7: Solve the system

$$y < -3x+4$$
$$y < \frac{1}{4}x+1.$$

Solution: Graph each inequality; the points where the graphs intersect is the solution to the system. The solution is displayed graphically; however, you can also describe the solution set in English as "all points that are below $y = -3x+4$ *and* below $y = \frac{1}{4}x+1$."

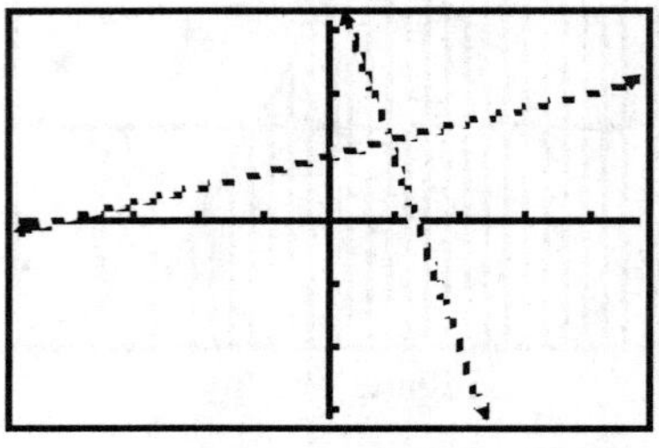

Figure 13.1.10a

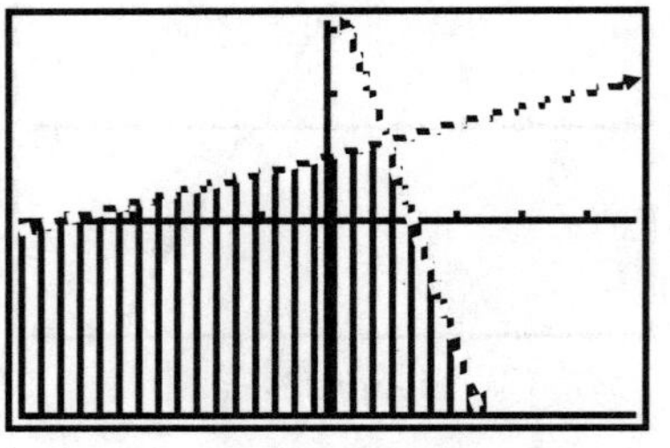

Figure 13.1.10b

⁂

Example 8: Solve the system

$$y \le -2x+6$$
$$y \le x+1$$
$$y \ge 0$$
$$x \ge 0.$$

Solution: Graph all of the inequalities and shade in the region of the intersection of all graphs.

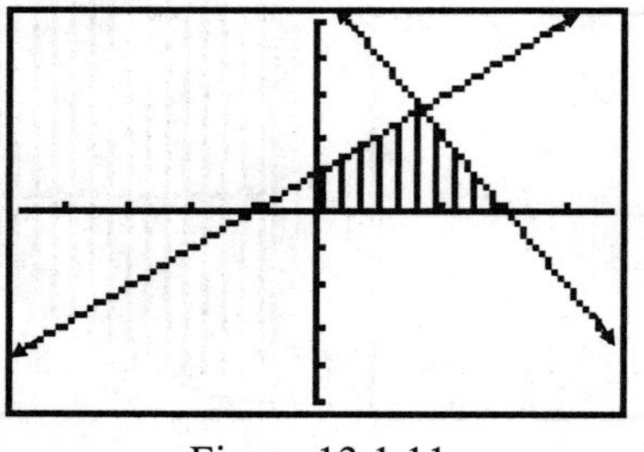

Figure 13.1.11

For every point shaded, the y-coordinate is $\le -2x+6$, AND

the y-coordinate is $\le x+1$, AND

the y-coordinate is ≥ 0, AND

the x-coordinate is ≥ 0.

⁂

Example 9: Solve the system

$$2x - 5y \le 0 \quad (1)$$
$$-3x + 2y \ge -8 \quad (2).$$

Solution: Isolate the y on the left and graph the inequalities; the intersection of the graphs is the solution to the system.

$$-5y \le -2x \qquad \text{from (1)}$$

$$y \ge \frac{2}{5}x \qquad \text{from (1)}$$

$$2y \ge 3x - 8 \qquad \text{from (2)}$$

$$y \ge \frac{3}{2}x - 4 \qquad \text{from (2)}$$

Graph $y \ge \frac{2}{5}x$ and $y \ge \frac{3}{2}x - 4$.

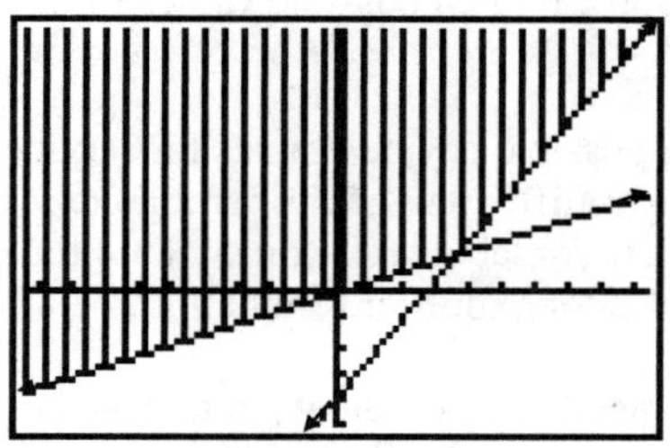

AND.

Figure 13.1.12

The solution is shown graphically.

13.1 EXERCISES

Many times you can work the exercises by algebraic, numeric, or graphical methods. You can learn about mathematics no matter which method you use. Part of the learning process is for you to decide which method is best for you to use. Please read the text carefully before you do the exercises. Read the **next assigned section** after you finish the assignment below. All numeric answers must have an error ≤ 0.01.

A Quick Look To The Future

-4. What types of polynomials have a sum that is zero?

-3. Just suppose that you call the variable **A** $\begin{pmatrix} -1 & 2 \\ 8 & 1 \end{pmatrix}$, and the variable **B** $\begin{pmatrix} 4 & 1 \\ 3 & -2 \end{pmatrix}$. Propose what you think **A** + **B** might be.

-2. Propose how you would mathematically write "The length of a rectangle is 2 feet longer than the width, and the area is 14."

-1. What exponent on 2 will cause the expression to be 128?

Previous Topics Revisited

1. A pole makes a right angle with the ground and casts a 12-foot shadow. The angle of elevation of the sun is 80°. How tall is the pole?

2. The elevation at the top of a ski slope is 4500 feet above sea level. The elevation at the bottom of the slope is 3200 feet above sea level. The mountain makes an angle of 22° with the horizontal. How long is the trip down the ski slope?

3. When garbage is dumped in a pile, the height of the pile is normally half the distance across the base of the garbage pile and the pile is symmetric to the vertical line through the center. If the base is 1000 feet across, what is the angle formed by the ground and the pile of garbage?

4. Solve the right triangle whose legs are 7000 feet and 13,000 feet.

5. Solve the right triangle whose leg is .005 mile with an angle of 72° opposite the leg.

6. If the sin θ is 0.5 and $0° \le \theta \le 360°$, find all possible values for θ.

New Topic Exercises

7. Ms. Exhaust is considering two possible payment plans for service charges to customers of her muffler shop. One is \$50 for any muffler plus \$0.75 per minute for labor. Plan two is to charge \$20 for any muffler plus \$1 per minute for labor. How much labor time will cause the charges to be the same whether she uses plan 1 or 2 for a muffler installation?

8. Printer *X* holds 500 sheets of paper and can print at a rate of 40 pages per minute. Printer *Y* holds 750 sheets of paper and can print at a rate of 60 pages per minute. If both printers are started at the same time, how long will it take for the paper trays to contain the same number of sheets of paper?

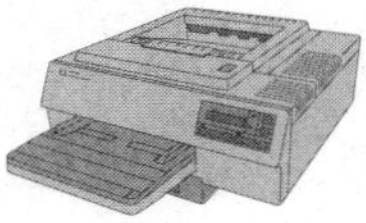

Solve the systems.

9. $y = -5x + 3$
   $y = 2x - 4$

10. $y = \frac{2}{3}x - 1$
    $y = \frac{-1}{5}x + 3$

11. $3x - 7y = 16$
    $-2x + y = 7$

12. $-2x + 3y = -5$
    $6x - 9y = 2$

13. $-3x + y = -2$
    $4x - 3y = 5$

14. $y = 2.5x - 3.1$
    $y = -.1x + 5.6$

15. $y = -5.23x + 7.22$
    $y = 4.75$

16. $-5x + 2y = -1$
    $15x - 6y = 3$

Solve the system by showing the solution set on a coordinate system.

17. $y \geq -3x + 9$
    $y \leq x - 4$

18. $y > 4x - 5$
    $y > -\frac{1}{2}x + 3$

19. $3x - 2y \leq -1$
    $x + 4y \geq -12$

20. $-2x + 4y > -2$
    $3x - 5y < 20$

21. $y \leq \frac{1}{2}x + 2$
    $y \leq -2x + 6$
    $y > 0$
    $x > 0$

22. $y \geq 3x + 2$
    $y \geq -2x + 10$

23. $y \leq \frac{1}{4}x + 1$
    $y \leq -2x + 12$
    $y \geq 0$
    $x \geq 0$

24. Solve the coin problem in the introduction to this chapter.

25. Solve the hot dog and drink problem posed in the introduction to this chapter.

26. Give three different combinations of numbers of dogs and drinks sold that satisfy the conditions of the dog and drink problem.

27. Give an example of a system of linear equations that has no solution.

28. Give an example of a system of linear inequalities that has no solution, if possible.

29. What are the coordinates of the four corners of the region that is the solution to Exercise 21?

30. What are the coordinates of the four corners of the region that is the solution to Exercise 23?

From Mathematics to English

31. What is your interpretation of the solution to any system of equations?

32. Describe a "system of equations".

33. Describe how to solve the system of equations $2x + y = 3$ and $x - y = 0$. Do not actually solve the system.

34. How do you know that the system $2x + y = 3$ and $-4x - 2y = -9$ has no solution?

35. How do you know that the system $2x + y = 3$ and $-4x - 2y = -6$ has an infinite number of answers?

36. Create a system of equations to describe a situation. Tell what each equation describes about the situation.

37. After reading this section, make a list of questions that you want to ask your instructor.

38. Continue in your daily journal and make an entry. In addition to your normal entry on thoughts about the mathematics in this section, list at least two positive comments about what you have learned about this topic.

39. In paragraph format, summarize the material in this section of the text in your daily journal.

40. Describe how your classroom instructor made this topic more understandable and clear.

41. After reading the text and listening to your instructor, what do you not understand about this topic?

Explorations

42. Given the system of equations $a_1x + b_1y = c_1$

$$a_2x + b_2y = c_2,$$

what is the relationship between a_1 and a_2, b_1 and b_2, and c_1 and c_2;

if the graphs are 1) parallel
2) the same line?

43. Find a system of linear inequalities with a solution that is a parallelogram formed by connecting the points (3, 2), (–2, 2), (–3, –2), (2, –2), and (3, 2) in succession.

44. This section includes systems of equations that are called linear systems. Below is a series of non-linear systems; solve the systems.

a. $y = -|x - 3| + 5$
$x + 2y = 13$

b. $y = -|x - 3| + 5$
$x + 2y = 11$

c. $y = (x + 2)^2 - 3$
$y = 2^{x+2} - 4$

13.2 Solving Systems by the Addition and Substitution Methods

What Happens If You Add Two Equations?

Below are the graphs of the system of equations

$$y = 2x + 7$$
$$y = -x + 1.$$

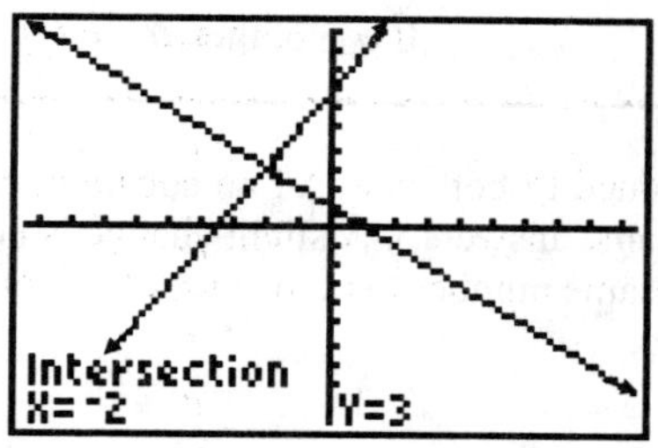

Figure 13.2.1

Figure 13.2.1 also shows the solution to the system to be (–2, 3). When you are curious, sometimes you do things just to see what happens. Like what happens if you add the equations by adding corresponding sides. That is, do it as shown below.

$$\begin{array}{r} y = 2x + 7 \\ \underline{y = -x + 1} \\ 2y = x + 8 \end{array}$$

There doesn't seem to be anything exciting going on thus far. Try graphing the sum on the same window as the first two equations. To graph, isolate y first by multiplying both sides by ½. The new equation becomes $y = \frac{1}{2}x + 4$. Figure 13.2.2 shows the graphs of:

$$y = 2x + 7$$
$$y = -x + 1 \text{ and } y = \frac{1}{2}x + 4.$$

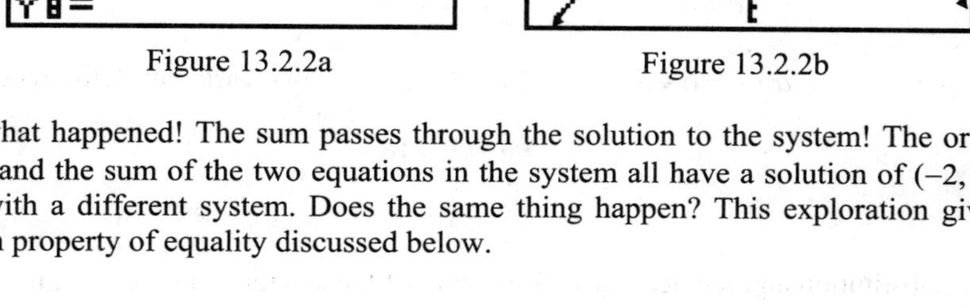

Figure 13.2.2a

Figure 13.2.2b

Question answered.

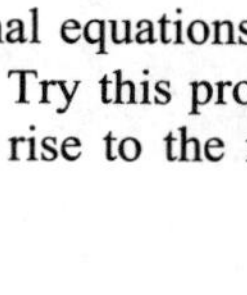

Look what happened! The sum passes through the solution to the system! The original equations in the system and the sum of the two equations in the system all have a solution of (–2, 3). Try this procedure again with a different system. Does the same thing happen? This exploration gives rise to the revised addition property of equality discussed below.

The property used to solve a system of equations by the analytical method called the addition method is the addition property of equality. In Chapters One and Six you used the property to solve equations and to isolate numbers or variables on one side of the equal sign in an equation or formula.

> **Addition Property** as used until now is:
>
> If $a = b$, then $a + c = b + c$.

That is, the same number may be added to both sides of an equation, and the new equation is equivalent to the original equation (It has the same answers.). A slight change is needed to use this property to solve systems. Not only may you add the same number to both sides of an equation, but you may also add equal numbers to both sides of an equation.

> **Revised Addition Property** If $a = b$ AND $c = d$, then $a + c = b + d$.

Because the addition property produces equivalent equations, this means that if (x, y) is the common solution to the system $a = b$ and $c = d$, it is also a common solution to the equation $a + c = b + d$, see Figure 13.2.2. Keep this idea in mind as you study the examples below.

All of the other properties of equality and the substitution property are also needed to solve systems analytically. The general approach used to solve a system is to apply the addition property to eliminate a variable from an equation. This produces an equation with one variable, which can normally be solved. Once one of the variables is known, the substitution property can be used to find the second variable. See the example problem below.

Example 1: Solve the system $3x - 2y = 10$ and $x + 2y = 6$.

Solution: If you think of $3x - 2y = 10$ as the equation $a = b$, and think of the equation $x + 2y = 6$ as $c = d$, when the revised addition property is used, the y variable is eliminated.

$$\begin{array}{ccl} a = b & 3x - 2y = 10 & \\ \underline{c = d} & \underline{x + 2y = 6} & \\ a + c = b + d & 4x + 0 = 16 & \text{one variable is eliminated} \\ & 4x = 16 & \\ & x = 4 & \end{array}$$

Use substitution and replace x with 4 in either of the equations in the system.

$$\text{From} \quad 3x - 2y = 10 \quad \text{replace } x \text{ with } 4.$$

$$3 \cdot 4 - 2y = 10$$

$$12 - 2y = 10$$

$$-2y = -2$$

$$y = 1$$

The solution is (4, 1). In the world of business, science, and engineering, it is a common practice to confirm that your work is correct. You can confirm that your work is correct by solving the system using another method. The equations must be put in function form. The equation $3x - 2y = 10$ becomes $y = \frac{3}{2}x - 5$, and the equation $x + 2y = 6$ becomes $y = -\frac{1}{2}x + 3$. Below is the confirmation that the solution (4, 1) is correct.

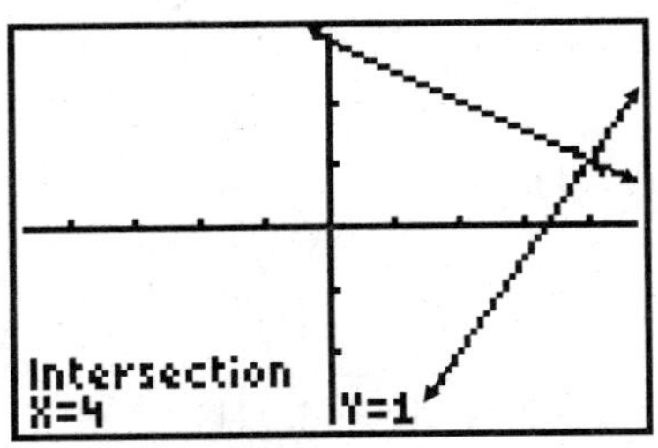

Figure 13.2.3

In Example 1, when the addition property is used, the *y*-variable is eliminated because the coefficients of *y* are opposites. When opposites are added, the sum is always zero. If the coefficients of either variable are not opposites, they can be made opposites by using the multiplication property of equality. Example 2 demonstrates how to make coefficients of either variable be opposites.

Example 2: Solve the system $2x - 3y = 9$
$5x - 4y = 19$.

Solution: Decide which variable you want to eliminate. If you want to eliminate the *x* variable, the coefficients should be opposites. To make the coefficients of *x* opposites, the first equation can be multiplied by −5 and the second by 2. There are many other choices as well, but this choice will work fine.

$$-5[2x - 3y] = -5[9]$$

$$2[5x - 4y] = 2[19]$$

These simplify to: $-10x+15y=-45$

$$10x-8y=38$$

add

$$\begin{array}{r} -10x+15y=-45 \\ \underline{10x-8y=38} \\ 7y=-7 \\ y=-1 \end{array}$$

by substitution in the first equation

$$\begin{aligned} 2x-3y&=9 \\ 2x-3(-1)&=9 \\ 2x+3&=9 \\ 2x&=6 \\ x&=3 \end{aligned}$$

The solution is (3, –1).

Example 3: Solve the system

$$\begin{aligned} -2x+5y&=-1 \quad (1) \\ 6x-15y&=4. \quad (2) \end{aligned}$$

Solution: The x can be eliminated by multiplying the first equation by 3.

$$3[-2x+5y]=3[-1] \qquad \text{from (1)}$$

$$6x-15y=4 \qquad (2)$$

the system now becomes:

$$-6x+15y=-3 \qquad \text{from (1)}$$

$$6x-15y=4 \qquad (2)$$

add:

$$\begin{array}{rl} -6x+15y=-3 & \text{from (1)} \\ \underline{6x-15y=4} & (2) \\ 0=1 & \end{array}$$

Both variables are eliminated and a false statement is left! What is the solution to the system? What does the false statement mean? To answer these questions, the graphs of the equations should help. To graph the equations, isolate y.

from (1) $\quad -2x + 5y = -1$

$$5y = 2x - 1$$

$$y = \frac{2}{5}x - \frac{1}{5}$$

from (2) $\quad 6x - 15y = 4$

$$-15y = -6x + 4$$

$$y = \frac{6}{15}x - \frac{4}{15}$$

$$y = \frac{2}{5}x - \frac{4}{15}$$

The graphs have the same slope and different y-intercepts! The graphs are parallel. There is no solution. When both variables are eliminated and a false statement remains, the system has no solution.

⁂

A second analytical method for solving systems of equations is called the **substitution method**. It works because at the intersection of two graphs (the solution) the values for both x and y from either equation are the same. Because both the x's and the y's from each equation are the same at the solution, the x or y variable from one of the equations can be substituted in the other equation. For example, from the equation $y = 2x - 3$, y in the other equation in the system can be replaced with $2x - 3$. This will produce an equation with one variable, which can normally be solved using methods from Chapter Six.

Example 4: Solve the system

$$2x - y = 4 \quad (1)$$
$$-3x + 2y = -6. \quad (2)$$

Solution: From (1), isolate y.

$$2x - y = 4$$

$$-y = -2x + 4$$

$$y = \mathbf{2x - 4} \quad (3)$$

Using substitution, replace y in (2) with the y from (3).

$$-3x + 2y = -6$$

$$-3x + 2(\mathbf{2x - 4}) = -6 \quad \text{solve for } x$$

$$-3x + 4x - 8 = -6$$

$$x - 8 = -6$$

$$x = 2$$

Using substitution in any of the above equations containing both x and y, replace x with 2 and solve for y.

$$y = 2x - 4 \text{ from (3)}$$

$$y = 2 \cdot 2 - 4$$

$$y = 4 - 4$$

$$y = 0$$

The solution is (2, 0). Below is the confirmation.

Figure 13.2.4a

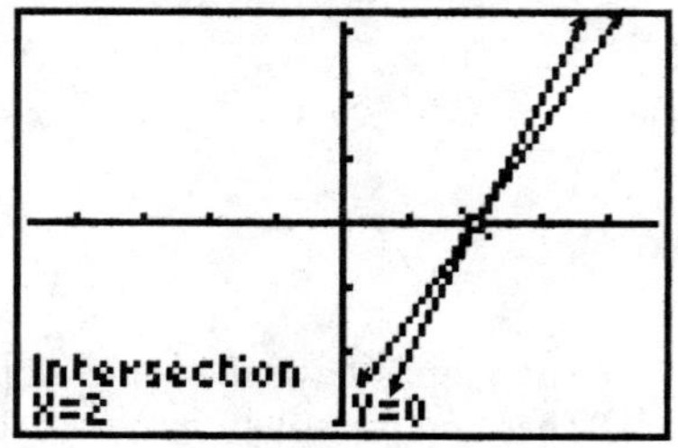

Figure 13.2.4b

Example 5: Solve the system

$$-3x + 5y = -7 \quad (1)$$
$$x - 4y = 2. \quad (2)$$

Solution: From (2), isolate x.

$$x - 4y = 2$$

$$x = \mathbf{4y + 2} \quad (3)$$

Using substitution, replace x in (1) with the x from (3).

$$-3x + 5y = -7$$

$$-3(\mathbf{4y + 2}) + 5y = -7 \qquad \text{solve for } y$$

$$-12y - 6 + 5y = -7$$

$$-7y - 6 = -7$$

$$-7y = -1$$

$$y = \frac{1}{7}$$

Solve for y.

Use substitution in any of the above equations; replace y with $\frac{1}{7}$.

$$x = 4y + 2 \qquad \text{from (3)}$$

$$x = 4\left(\frac{1}{7}\right) + 2$$

Below are the keystrokes for x.

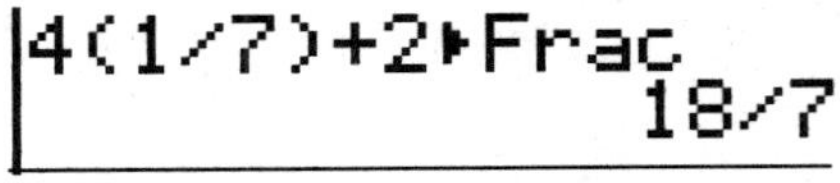

The solution is $\left(\frac{18}{7}, \frac{1}{7}\right)$.

The confirmation is:

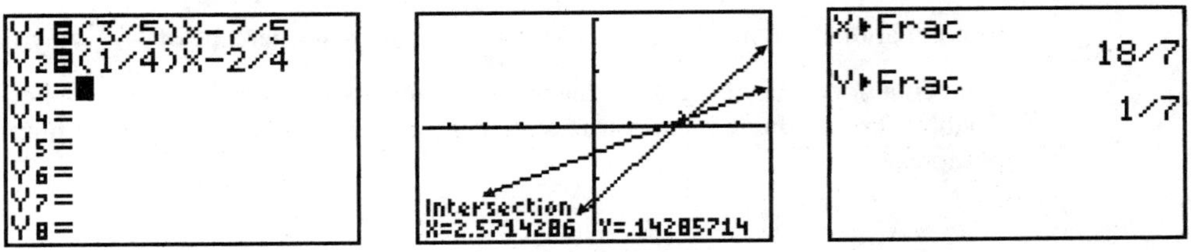

Figure 13.2.5

⁂

Example 6: Solve the system

$$-2x + 5y = -3 \quad (1)$$
$$6x - 15y = 9. \quad (2)$$

Solution: To use substitution, isolate a variable in one of the equations.

From equation (1)

$$-2x + 5y = -3$$
$$5y = 2x - 3$$
$$y = \frac{2}{5}x - \frac{3}{5} \quad (3)$$

Using substitution, replace y in (2) with y from (3).

$$6x - 15y = 9 \quad (2)$$
$$6x - 15\left(\frac{2}{5}x - \frac{3}{5}\right) = 9$$
$$6x - 6x + 9 = 9$$
$$9 = 9$$

It's true. It's true.

Both variables have been eliminated and a true statement is left! What is the solution to the system? What does the true statement mean? To answer these questions, the graphs of the equations should help. To graph the equations, isolate y.

$$y = \frac{2}{5}x - \frac{3}{5} \quad \text{from (3)}$$

$$6x - 15y = 9 \quad \text{from (2)}$$
$$-15y = -6x + 9$$

$$y = \frac{6}{15}x - \frac{9}{15}$$

$$y = \frac{2}{5}x - \frac{3}{5} \qquad (4)$$

Both equations (3) and (4) have a slope of $\frac{2}{5}$ and a y-intercept of $\frac{-3}{5}$; they are the same equation! Anything that makes (1) true also makes (2) true. When both variables are eliminated and a true statement is left, there are an infinite number of solutions.

⁂

13.2 EXERCISES

Many times you can work the exercises by algebraic, numeric, or graphical methods. You can learn about mathematics no matter which method you use. Part of the learning process is for you to decide which method is best for you to use. Please read the text carefully before you do the exercises. Read the **next assigned section** after you finish the assignment below. All numeric answers must have an error ≤ 0.01.

A Quick Look To The Future

-4. So far, you have studied three ways of solving systems of equations. Are there more ways?

-3. Is there a unique solution to the problem, "I have a total of 17 dimes and nickels, how many of each do I have?"

-2. Fill in the blank space: $3^{\text{-----}} = 81$, $4^{\text{-----}} = 64$, $2^{\text{-----}} = 512$

-1. What is $b^x \cdot b^y$?

Previous Topics Revisited

Solve the next three systems graphically.

1. $y = \frac{-2}{3}x + 4$
   $y = \frac{5}{2}x - 6$

2. $y = -2.6x + 15.7$
   $y = 5.1x - {}^{-}17.1$

3. $y \le -x + 5$
   $y \le 2x + 1$
   $y \ge 0$
   $x \ge 0$

4. A granary is a building used to store grain. A window in the granary is 16 feet above ground level. How long must a grain elevator be to reach the window if the maximum angle on the elevator is 45°?

5. The cord attached to a kite is 500 feet long, and it is fully extended. The angle the kite string makes with the ground is 75°. Assuming the string is straight, how high is the kite?

6. Solve the right triangle ABC if $a = 6.3$ million kilometers and $\angle B = 0.5°$.

New Topic Exercises

Solve by an analytical method. Confirm by another method.

7. $\begin{aligned} y &= 2x - 7 \\ y &= -3x + 13 \end{aligned}$

8. $\begin{aligned} 5x - 2y &= -13 \\ -2x + y &= 6 \end{aligned}$

9. $\begin{aligned} 7x + 2y &= -3 \\ -14x - 4y &= 8 \end{aligned}$

10. $\begin{aligned} -3x + 2y &= 11 \\ 2x - 5y &= 0 \end{aligned}$

11. $\begin{aligned} 2.5x - 6.0y &= 4.3 \\ -3.0x - 1.5y &= -2.9 \end{aligned}$

12. $\begin{aligned} -4x &= 20 \\ 2x + 3y &= 2 \end{aligned}$

13. $\begin{aligned} 1000x + 3500y &= -8250 \\ 10x - 500y &= 0 \end{aligned}$

14. $\begin{aligned} -3x + 5y &= -1 \\ 9x - 15y &= 3 \end{aligned}$

15. How many solutions does the following system have? There is no need to solve the system. $\begin{aligned} y &= -2x + 5 \\ y &= -2x - 4 \end{aligned}$

16. How many solutions does the following system have? There is no need to solve the system. $\begin{aligned} y &= \frac{1}{4}x - 10 \\ y &= \frac{5}{20}x - 10 \end{aligned}$

17. Give an example of ANY system that has 2 solutions.

18. Create a system of equations that has (–2, 5) as the solution.

19. Create a system of equations that has (0, –4) as the solution.

20. Create a system of equations that has $\left(\frac{1}{5}, \frac{3}{10}\right)$ as the solution.

21. Create a system of equations that has no solution.

22. Create a system of equations that has an infinite number of solutions.

From Mathematics to English

23. The first equation of a system is $2x - 5y = 3$. Find a second equation that will make the system have no solution. Explain how to get your answer.

24. The first equation of a system is $2x - 5y = 3$. Find a second equation that will cause the system to have an infinite number of solutions. Explain your answer.

25. When solving a system analytically, explain why the system has no solution if both variables drop out, leaving a false statement.

26. Can a linear system of equations have exactly 2 solutions? Why?

27. Without solving the system, explain how you know that (−6, 10) is a solution to

$$\begin{aligned} x + y &= 4 \\ x - y &= -16. \end{aligned}$$

28. What is a system of equations?

29. Describe the substitution method for solving a system of equations.

30. Describe the addition method for solving a system of equations.

31. After reading this section, make a list of questions that you want to ask your instructor.

32. Continue in your daily journal and make an entry. In addition to your normal entry on thoughts about the mathematics in this section, list at least two positive comments about what you have learned about this topic.

33. In paragraph format, summarize the material in this section of the text in your daily journal.

34. Describe how your classroom instructor made this topic more understandable and clear.

35. After reading the text and listening to your instructor, what do you not understand about this topic?

Explorations

36. Solve the system below by addition and again by substitution. Simplify your answers. Compare solutions.

$$\begin{aligned} a_1X + b_1Y &= c_1 \\ a_2X + b_2Y &= c_2 \end{aligned}$$

37. Apply the addition or substitution method and solve the systems.

a. $$\begin{aligned} 2x - y + z &= 2 \\ x + 2y - z &= 2 \\ x - y - 2z &= -2 \end{aligned}$$

b. $$\begin{aligned} 3x - 2y + 4z &= -1 \\ x + 5y - 3z &= -6 \\ -2x + y + 2z &= 1 \end{aligned}$$

38. Traditional analytical (algebraic) methods may or may not help to solve the following systems of equations. You may want to try graphical methods

a. $$\begin{aligned} y &= x^2 - 2x + 3 \\ y &= -\sqrt{x-1} + 3 \end{aligned}$$

b. $$\begin{aligned} -2x + y &= 3 \\ y &= 2^{x-1} + 2 \end{aligned}$$

c. $$\begin{aligned} 2xy &= 4x - 1 \\ y &= -2\sqrt{x+2} + 2 \end{aligned}$$

d. $$\begin{aligned} y &= 2x^2 + 6x - 7 \\ 2x - y &= 9 \end{aligned}$$

13.3 Solving Systems Using Cramer's Rule and Matrices

It Seems like We Are Always Looking for a Quicker Way

Consider the system of equations
$$\begin{aligned} 3x - 5y &= 7 \\ -2x + 7y &= -6. \end{aligned}$$

What if you agreed to <u>ALWAYS</u> write the equations in any linear system with the x term first and the y term second and the constant on the right side? This means that the "stuff" on the left side of the equations can, in fact,

be written as $\begin{bmatrix} 3 & -5 \\ -2 & 7 \end{bmatrix}$ and you understand that this notation

really refers to the expressions $\begin{bmatrix} 3x & -5y \\ -2x & 7y \end{bmatrix}$. In this section you will explore what you can do with the system written in matrix form.

The fourth method used to solve linear systems of equations requires a brief introduction to a concept called a **matrix**, which is a series of real numbers arranged in rows and columns. The rows and columns have brackets around them as shown in the following matrix:

$$\begin{bmatrix} 3 & -5 \\ -2 & 7 \end{bmatrix}.$$

The entire group of numbers is a matrix. As in algebra, it can be identified with a variable; for example, it might be called matrix $\boldsymbol{A}$. The above matrix becomes:

$$\boldsymbol{A} = \begin{bmatrix} 3 & -5 \\ -2 & 7 \end{bmatrix}$$

Since matrix $\boldsymbol{A}$ has 2 rows and 2 columns, it is referred to as a 2 by 2 matrix; this is commonly written 2×2; that is, the dimension of the matrix is 2×2. The first row contains the elements 3 and -5, and the second row consists of the elements -2 and 7. The elements are normally identified by their respective row AND column. For example, the element -2 is in the second row and first column, and is commonly identified as a_{21}. This is read as "a sub two one", not "a sub twenty-one." It is the element in the second row and first column of matrix $\boldsymbol{A}$. The element in the first row and second column is identified as a_{12}. This is read as "a sub one two." Just as -2 and 7 form a row of the above matrix, 3 and 7 form the main diagonal, -5 and -2 form a diagonal as well.

A further discussion of matrices is delayed until later in this section, except for finding the determinant of a matrix below. When the number of rows in a matrix equals the number of columns, it is called a **square matrix**. Associated with the square matrix is a number called the determinant of the matrix. It is an operation performed on the elements of a matrix to obtain a single real number from the matrix that can be used to help solve systems of equations. When the operation of finding the determinant of a square matrix is desired, vertical bars on either side of the matrix are used. For example,

$|A| = \begin{vmatrix} 3 & -5 \\ -2 & 7 \end{vmatrix}$. The vertical bars imply the value of the determinant of matrix $\boldsymbol{A}$.

> Determinant of a **2 × 2** Matrix is $\begin{vmatrix} a & b \\ c & d \end{vmatrix} = ad - bc$.

The determinant of a 2×2 matrix is the product of the elements on the main diagonal *minus* the product of the elements on the other diagonal. The determinant of matrix $\boldsymbol{A}$ above is $3 \cdot 7 - (-5) \cdot (-2)$, or 11. The determinant of a 2×2 matrix will be used to solve a system of equations.

Example 1: Evaluate $\begin{vmatrix} -2 & 3 \\ 4 & 7 \end{vmatrix}$.

LOOKS EASY.

Solution: Multiply the elements on the main diagonal, and then subtract the product of the elements on the other diagonal.

$$(-2) \cdot 7 - 3 \cdot 4 = -14 - 12 = -26$$

⁂

Example 2: Evaluate $\begin{vmatrix} 4 & -7 \\ 5 & -1 \end{vmatrix}$.

Solution: Take the product of the elements on the main diagonal *minus* the product of the elements on the other diagonal.

$$4 \cdot (-1) - (-7) \cdot 5 = -4 - (-35) = -4 + 35 = 31. \begin{vmatrix} 4 & -7 \\ 5 & -1 \end{vmatrix} = 31.$$

⁂

You may be wondering what all this has to do with solving a system of equations. The problem under consideration is how to solve a system using Cramer's rule. This is developed below.

Below is the solution to the system $\begin{aligned} a_1x + b_1y &= c_1 \quad (1) \\ a_2x + b_2y &= c_2 \quad (2) \end{aligned}$ using addition.

multiply (1) by b_2	$b_2(a_1x + b_1y) = b_2c_1$	(3)
multiply (2) by $-b_1$	$-b_1(a_2x + b_2y) = b_1c_2$	(4)

simplify $\quad a_1b_2x + b_1b_2y = b_2c_1 \quad$ from (3)

$\quad -a_2b_1x - b_1b_2y = b_1c_2 \quad$ from (4)

add
$$\begin{array}{r} a_1b_2x + b_1b_2y = b_2c_1 \\ -a_2b_1x - b_1b_2y = b_1c_2 \\ \hline \end{array}$$

simplify $\quad (a_1b_2 - a_2b_1)x + 0 \ = b_2c_1 - b_1c_2$

simplify $\quad (a_1b_2 - a_2b_1)x = b_2c_1 - b_1c_2$

divide by $a_1b_2 - a_2b_1 \quad \dfrac{(a_1b_2 - a_2b_1)x}{a_1b_2 - a_2b_1} = \dfrac{b_2c_1 - b_1c_2}{a_1b_2 - a_2b_1}$

$$x = \frac{b_2c_1 - b_1c_2}{a_1b_2 - a_2b_1}$$

Using the addition method to eliminate the x terms from both equations yields a similar result for the variable y.

$$y = \frac{a_1c_2 - a_2c_1}{a_1b_2 - a_2b_1}$$

The solution to the system $\begin{array}{l} a_1x + b_1y = c_1 \\ a_2x + b_2y = c_2 \end{array}$ is: $x = \dfrac{b_2c_1 - b_1c_2}{a_1b_2 - a_2b_1} \qquad y = \dfrac{a_1c_2 - a_2c_1}{a_1b_2 - a_2b_1}$

Now, to tie the two topics together, consider the following determinants.

$$|D| = \begin{vmatrix} a_1 & b_1 \\ a_2 & b_2 \end{vmatrix} = a_1b_2 - a_2b_1 \qquad |D_x| = \begin{vmatrix} c_1 & b_1 \\ c_2 & b_2 \end{vmatrix} = b_2c_1 - b_1c_2 \qquad |D_y| = \begin{vmatrix} a_1 & c_1 \\ a_2 & c_2 \end{vmatrix} = a_1c_2 - a_2c_1$$

When you compare the numerator of x with $|D_x|$, they are the same! When you compare the numerator of y with $|D_y|$, they are the same! The denominator of both x and y is the determinant $|D|$! In other words, from the algebraic solution for x and y shown above, you now have a formula to solve any linear system.

The solution to the system $\begin{array}{l} a_1x + b_1y = c_1 \\ a_2x + b_2y = c_2 \end{array}$ is: $x = \dfrac{b_2c_1 - b_1c_2}{a_1b_2 - a_2b_1} \qquad y = \dfrac{a_1c_2 - a_2c_1}{a_1b_2 - a_2b_1}$.

The solution to the system $\begin{aligned} a_1x + b_1y = c_1 \\ a_2x + b_2y = c_2 \end{aligned}$ is: $x = \dfrac{|D_x|}{|D|}$ $\quad y = \dfrac{|D_y|}{|D|}$

One method for remembering $|D|, |D_x|$, and $|D_y|$ is to visualize the system of equations where the x's and y's have been erased and all that remains are the coefficients on the left and the constants on the right.

$$\begin{aligned} a_1x + b_1y &= c_1 \\ a_2x + b_2y &= c_2 \end{aligned} \qquad \begin{matrix} a_1 & b_1 & c_1 \\ a_2 & b_2 & c_2 \end{matrix}$$

then $|D| = \begin{vmatrix} a_1 & b_1 \\ a_2 & b_2 \end{vmatrix}$

Determinant $|D|$ consists of the coefficients on the left sides of the equations. When the coefficients of x are replaced with the constants c_1 and c_2, determinant $|D|$ becomes $|D_x|$.

$$|D_x| = \begin{vmatrix} c_1 & b_1 \\ c_2 & b_2 \end{vmatrix}$$

When the coefficients of y in $\boldsymbol{D}$ are replaced with the constants c_1 and c_2, determinant $|D|$ becomes $|D_y|$.

$$|D_y| = \begin{vmatrix} a_1 & c_1 \\ a_2 & c_2 \end{vmatrix}$$

Cramer's Rule

The solution to the system $\begin{aligned} a_1x + b_1y = c_1 \\ a_2x + b_2y = c_2 \end{aligned}$

is: $x = \dfrac{|D_x|}{|D|}$ $\quad y = \dfrac{|D_y|}{|D|}$ provided $|D| \neq 0$.

Example 3: Solve the system using Cramer's Rule.

$$\begin{aligned} 2x - 5y &= -4 \\ -3x + 4y &= 7 \end{aligned}$$

Solution: Calculate $|D|$, $|D_x|$, and $|D_y|$ and use Cramer's Rule to calculate x and y.

$$|D| = \begin{vmatrix} 2 & -5 \\ -3 & 4 \end{vmatrix} = 2 \cdot 4 - (-5) \cdot (-3) = 8 - 15 = -7.$$

$$\left|D_x\right| = \begin{vmatrix} -4 & -5 \\ 7 & 4 \end{vmatrix} = (-4) \cdot 4 - (-5) \cdot 7 = -16 - (-35) = 19.$$

$$\left|D_y\right| = \begin{vmatrix} 2 & -4 \\ -3 & 7 \end{vmatrix} = 2 \cdot 7 - (-4) \cdot (-3) = 14 - 12 = 2$$

$$x = \frac{|D_x|}{|D|} = \frac{19}{-7} \qquad y = \frac{|D_y|}{|D|} = \frac{2}{-7}$$

$x = -\frac{19}{7}$, $y = -\frac{2}{7}$. The solution is $\left(\frac{-19}{7}, \frac{-2}{7}\right)$.

Your calculator can also evaluate a determinant. Below are the TI-84 keystrokes for the solution to the system.

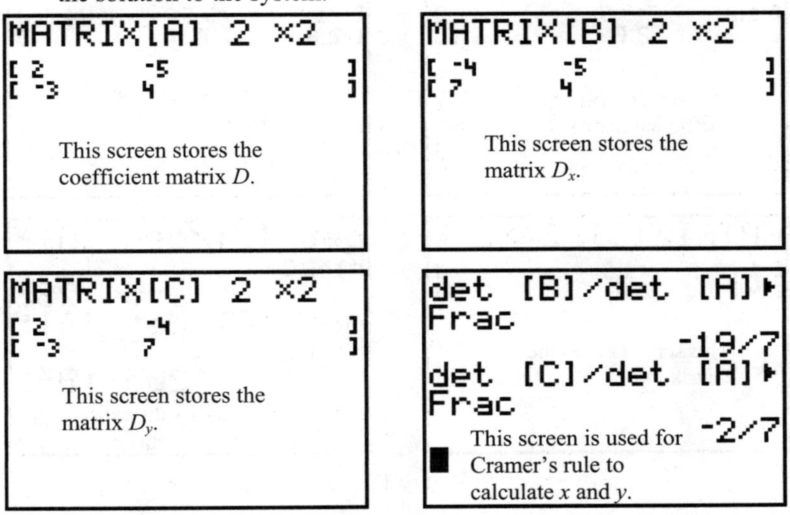

Figure 13.3.1

⁂

Example 4: Solve the system with Cramer's Rule.

$$1.2x - 3.5y = \mathbf{-1.9}$$
$$4.7x + 2.8y = \mathbf{2.6}$$

Solution:

$$|D| = \begin{vmatrix} 1.2 & -3.5 \\ 4.7 & 2.8 \end{vmatrix} = (1.2)(2.8) - (-3.5)(4.7) = 19.81$$

$$|D_x| = \begin{vmatrix} \mathbf{-1.9} & -3.5 \\ \mathbf{2.6} & 2.8 \end{vmatrix} = (-1.9)(2.8) - (-3.5)(2.6) = 3.78$$

$$|D_y| = \begin{vmatrix} 1.2 & \mathbf{-1.9} \\ 4.7 & \mathbf{2.6} \end{vmatrix} = (1.2)(2.6) - (-1.9)(4.7) = 12.05$$

$$x = \frac{|D_x|}{|D|} = \frac{3.78}{19.81} = 0.19 \qquad y = \frac{|D_y|}{|D|} = \frac{12.05}{19.81} = 0.61$$

The solution is (0.19, 0.61) rounded to hundredths. On a problem like this, the "matrix" key on your calculator can be used to simplify the calculations; it can find the value of a determinant in a few simple keystrokes. Below are the TI-84 keystrokes. The exact answers are shown instead of the approximate answers as above.

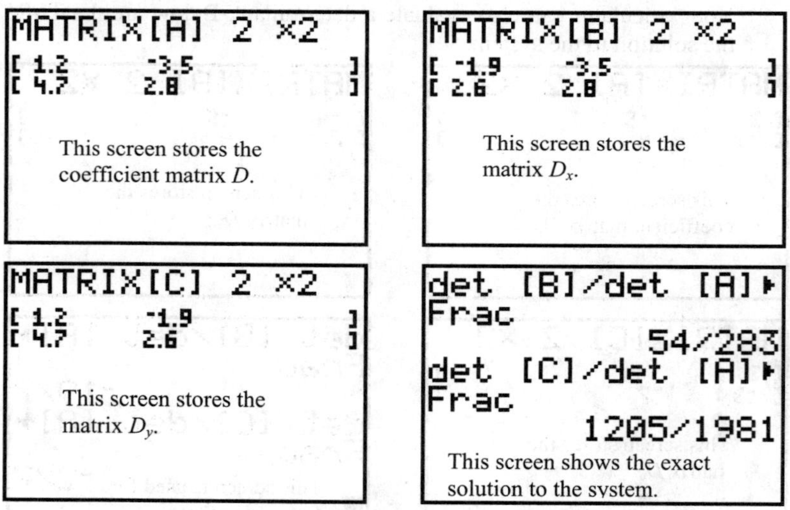

Figure 12.3.2

⁂

Example 5: Solve the system with Cramer's Rule.

$$\begin{aligned} -2x + 5y &= -3 \\ 6x - 15y &= 5 \end{aligned}$$

Solution:

$$|D| = \begin{vmatrix} -2 & 5 \\ 6 & -15 \end{vmatrix} = (-2)(-15) - (6)(5) = 0$$

Cramer's Rule requires that $D \neq 0$; thus, it does not apply. A check of the graphs reveals they have the same slopes and different y-intercepts; therefore, there is no solution.

⁂

A Matrix Method for Solving Systems of Equations

Thus far you know that you can find the determinant of a matrix. You may be wondering what else can be done with matrices and will it have any application to systems of equations. If you recall your preschool days, you may remember that you learned about numbers. You didn't learn what you can do with numbers until you were in school. This is your current situation with matrices. Just like with numbers, do you suppose you can add matrices? Can you subtract matrices? The answer is yes and it is quite simple.

Consider the matrices $\boldsymbol{A}$ and $\boldsymbol{B}$, $\boldsymbol{A} = \begin{bmatrix} -4 & 3 \\ 5 & 0 \end{bmatrix}$ and $\boldsymbol{B} = \begin{bmatrix} 7 & -2 \\ 2 & 4 \end{bmatrix}$. These two matrices have been entered on the calculator and the screen below shows the sum to be the matrix $\begin{bmatrix} 3 & 1 \\ 7 & 4 \end{bmatrix}$.

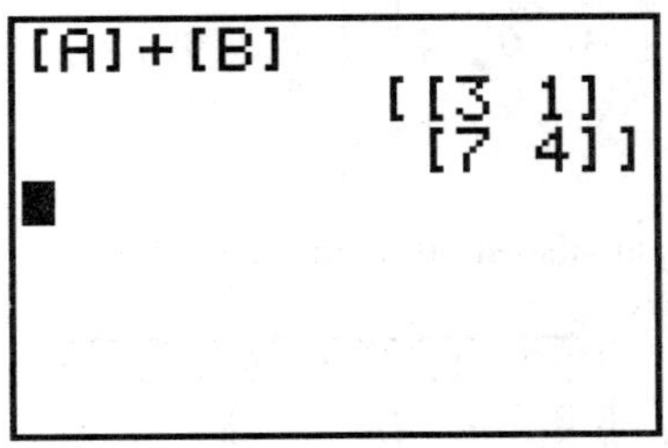

Figure 13.3.3

Can you figure out an algorithm for adding matrices based on this example? If not, try a little exploration on your calculator by doing several addition problems. Do you see that the matrices must be the same dimension and that the sum is found by adding corresponding elements of the matrices? Below is the difference of matrix $\boldsymbol{A}$ and $\boldsymbol{B}$.

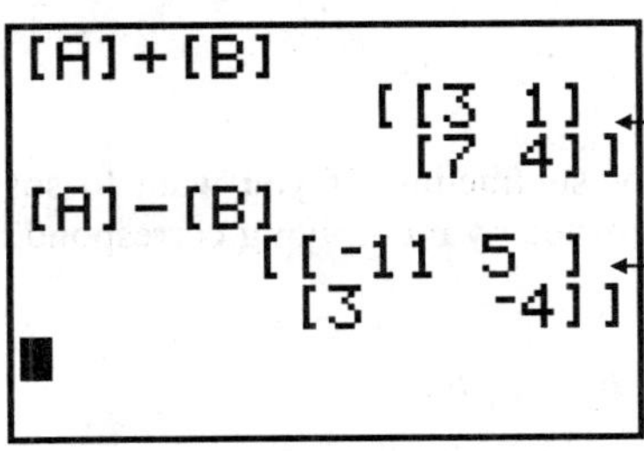

$$\begin{bmatrix} -4 & 3 \\ 5 & 0 \end{bmatrix} + \begin{bmatrix} 7 & -2 \\ 2 & 4 \end{bmatrix} = \begin{bmatrix} 3 & 1 \\ 7 & 4 \end{bmatrix}$$

$$\begin{bmatrix} -4 & 3 \\ 5 & 0 \end{bmatrix} - \begin{bmatrix} 7 & -2 \\ 2 & 4 \end{bmatrix} = \begin{bmatrix} -11 & 5 \\ 3 & -4 \end{bmatrix}$$

Figure 13.3.4

Can you figure out an algorithm for subtracting matrices based on this example? If not, try a little exploration on your calculator by doing several subtraction problems. Do you see that the matrices must be the same dimension and that the difference is found by subtracting corresponding elements of the matrices?

Example 6: Find the sum and difference of $\boldsymbol{A}=\begin{bmatrix}-1 & 4 & 0\\ 2 & -5 & 1\\ 7 & -3 & 6\end{bmatrix}$ and $\boldsymbol{B}=\begin{bmatrix}8 & -2 & 1\\ 0 & 2 & 5\\ -3 & 0 & 4\end{bmatrix}$.

Solution: To add matrices, add corresponding elements.

add

$$\begin{bmatrix}-1 & 4 & 0\\ 2 & -5 & 1\\ 7 & -3 & 6\end{bmatrix}+\begin{bmatrix}8 & -2 & 1\\ 0 & 2 & 5\\ -3 & 0 & 4\end{bmatrix}=\begin{bmatrix}7 & 2 & 1\\ 2 & -3 & 6\\ 4 & -3 & 10\end{bmatrix}$$

To subtract matrices, subtract corresponding elements.

$$\begin{bmatrix}-1 & 4 & 0\\ 2 & -5 & 1\\ 7 & -3 & 6\end{bmatrix}-\begin{bmatrix}8 & -2 & 1\\ 0 & 2 & 5\\ -3 & 0 & 4\end{bmatrix}=\begin{bmatrix}-9 & 6 & -1\\ 2 & -7 & -4\\ 10 & -3 & 2\end{bmatrix}$$

subtract

The calculator can also subtract matrices. Figure 13.3.5 shows confirmation of the answer above.

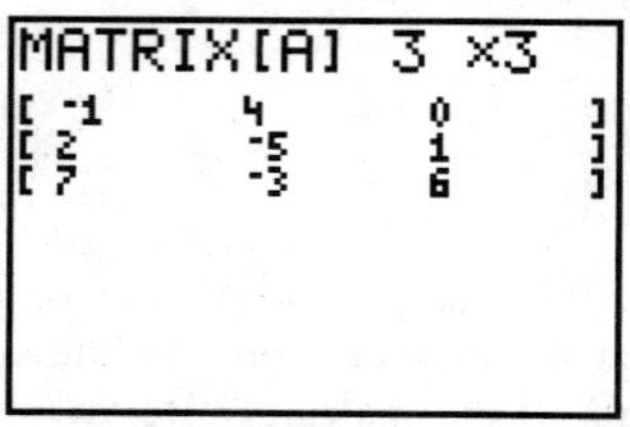

Figure 13.3.5a

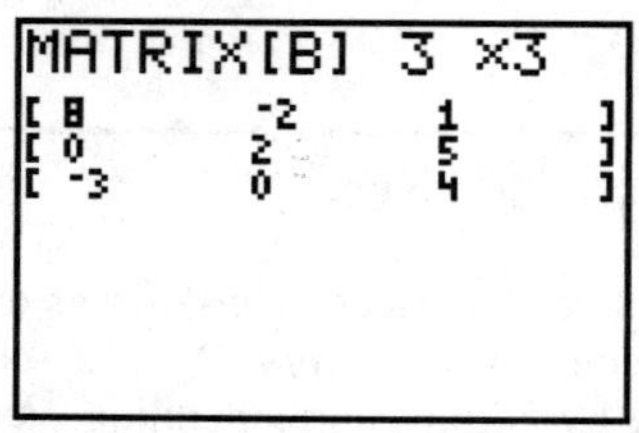

Figure 13.3.5b

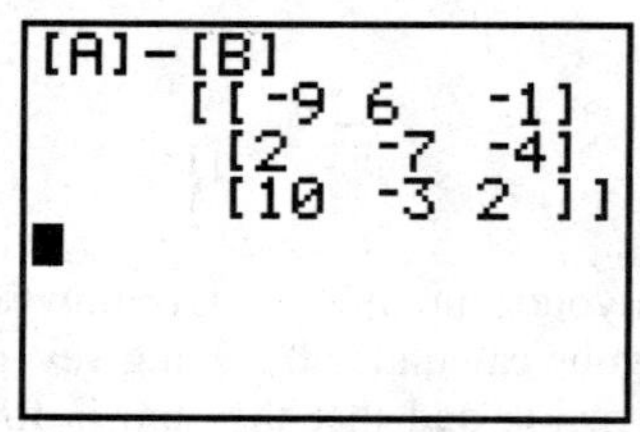

Figure 13.3.5c

⁂

Multiplication of matrices is a little more involved than addition or subtraction. If you want to solve systems of equations with matrices, multiplication cannot be accomplished by multiplying corresponding elements. Please study the example of multiplication below.

$$\begin{bmatrix}a_{11} & a_{12}\\ a_{21} & a_{22}\end{bmatrix}\times\begin{bmatrix}b_{11} & b_{12}\\ b_{21} & b_{22}\end{bmatrix}=\begin{bmatrix}a_{11}b_{11}+a_{12}b_{21} & a_{11}b_{12}+a_{12}b_{22}\\ a_{21}b_{11}+a_{22}b_{21} & a_{21}b_{12}+a_{22}b_{22}\end{bmatrix}$$

The element in the <u>1st row and 1st column</u> of the product comes from

$$\begin{array}{c} a_{11} \times b_{11} \\ + \\ a_{12} \times b_{21} \end{array}$$

. It is like the <u>1st row</u> has been rotated 90° clockwise and the elements are multiplied by the corresponding elements of the <u>1st column</u>. By similar reasoning, to get the element of the product in the <u>1st row and 2nd column</u>, rotate the <u>1st row</u> by 90° and multiply the elements by the corresponding elements of the <u>2nd column</u>, as shown below.

$$\begin{array}{c} a_{11} \times b_{12} \\ + \\ a_{12} \times b_{22} \end{array} \qquad \text{(element in the 1st row and 2nd column)}$$

To get the complete product, continue this process until finished.

As you might guess, the calculator can multiply matrices using the technique described above.

Example 7: Multiply $\begin{bmatrix} -2 & 3 \\ 5 & -8 \end{bmatrix} \times \begin{bmatrix} 5 & 9 \\ 1 & 4 \end{bmatrix}$.

Solution: Enter each matrix on the calculator and multiply.

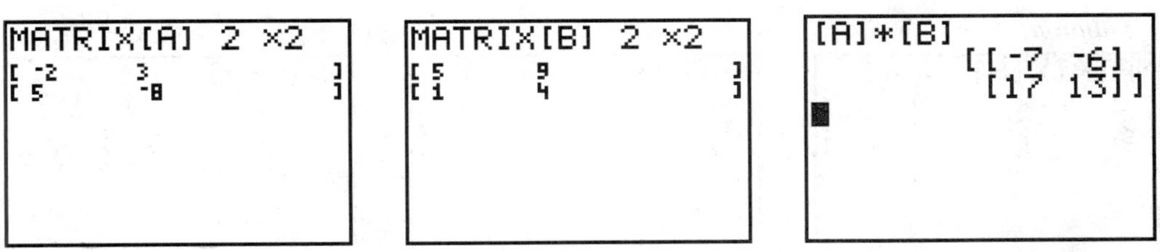

Figure 13.3.6

The product is $\begin{bmatrix} -7 & -6 \\ 17 & 13 \end{bmatrix}$.

⁂

The interesting features of multiplication are that multiplication is <u>not</u> commutative (You should confirm this on your calculator.), and the matrices being multiplied do not have to be the same dimension. For example, you can multiply a 2 × 2 matrix and a 2 × 1 matrix. The number of columns in the first (left) matrix must equal the number of rows in the second (right) matrix. The product will have the same number of rows as the first (left) matrix and it will have the same number of columns as the second (right) matrix. Thus, for example, you can multiply a 2 × 2 matrix by a 2 × 1 matrix because the number of columns in the left matrix (2) equals the number of rows in the right matrix (2). The product will have the same number of rows as the left (2) and the same number of columns as the right (1). Likewise, you can multiply a 3 × 3 matrix times a 3 × 1 matrix.

Example 8: Multiply $\begin{bmatrix} -3 & 5 \\ 0 & 2 \end{bmatrix} \times \begin{bmatrix} 1 \\ 0 \end{bmatrix}$.

Solution:

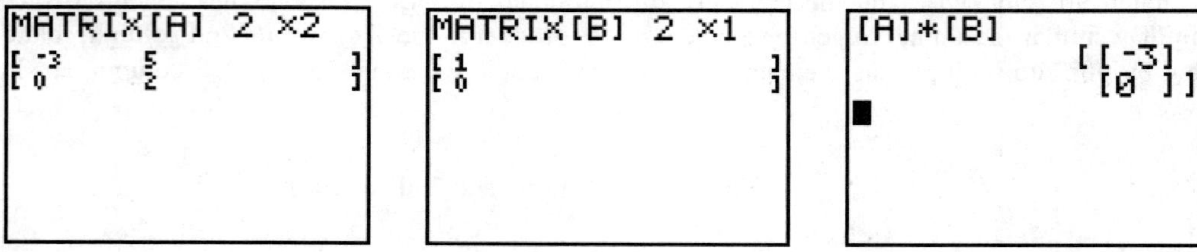

Figure 13.3.7

The product is $\begin{bmatrix} -3 \\ 0 \end{bmatrix}$.   Note: the times sign is not needed on the calculator.

⁂

Example 9: Multiply $\begin{bmatrix} -1 & 4 & 0 \\ 3 & -1 & 8 \\ 3 & 5 & 7 \end{bmatrix} \times \begin{bmatrix} 4 \\ 1 \\ -1 \end{bmatrix}$.

Solution:

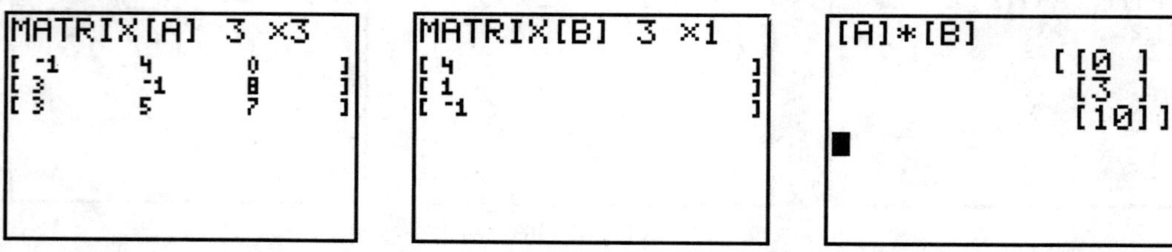

Figure 13.3.8

The product is $\begin{bmatrix} 0 \\ 3 \\ 10 \end{bmatrix}$.   Note: the times sign is not needed.

⁂

Where all of this multiplication is headed is solving systems using matrices. Any linear system of equations can be described by a product of matrices. The system can be 2 × 2, 3 × 3, 4 × 4, etc.

Please consider the system of equations $\begin{matrix} 2x + 5y = 11 \\ -x + 2y = -1 \end{matrix}$

and the matrix statement

$$\begin{bmatrix} 2 & 5 \\ -1 & 2 \end{bmatrix} \times \begin{bmatrix} x \\ y \end{bmatrix} = \begin{bmatrix} 11 \\ -1 \end{bmatrix}.$$

The product of the matrices on the left side of the equation is

$$\begin{bmatrix} 2x+5y \\ -x+2y \end{bmatrix}.$$

This leaves the statement

$$\begin{bmatrix} 2x+5y \\ -x+2y \end{bmatrix} = \begin{bmatrix} 11 \\ -1 \end{bmatrix}.$$

When two matrices are equal, corresponding elements must be equal; thus, this statement means

$$\begin{aligned} 2x+5y &= 11 \\ -x+2y &= -1. \end{aligned}$$

The conclusion to this reasoning is that $\begin{aligned} 2x+5y &= 11 \\ -x+2y &= -1 \end{aligned}$ is the same thing as $\begin{bmatrix} 2 & 5 \\ -1 & 2 \end{bmatrix} \times \begin{bmatrix} x \\ y \end{bmatrix} = \begin{bmatrix} 11 \\ -1 \end{bmatrix}$!

If symbols are used to represent the matrices, this system can be represented as the equation

$$\underbrace{\begin{bmatrix} 2 & 5 \\ -1 & 2 \end{bmatrix}}_{\boldsymbol{A}} \times \underbrace{\begin{bmatrix} x \\ y \end{bmatrix}}_{\boldsymbol{X}} = \underbrace{\begin{bmatrix} 11 \\ -1 \end{bmatrix}}_{\boldsymbol{B}}$$

$$\boldsymbol{A\,X} = \boldsymbol{B}.$$

If you can solve this equation for $\boldsymbol{X}$, you have a "formula" that will solve any linear system. To solve this equation for $\boldsymbol{X}$, simply multiply both sides of the equation by the inverse of $\boldsymbol{A}$. This is the exact same method you may use to solve the equation $2x = 6$. Just multiply both sides times the inverse (reciprocal) of 2, and the equation is solved. That is:

$2x = 6$	$\boldsymbol{A\,X} = \boldsymbol{B}$
$2^{-1} \times 2x = 2^{-1} \times 6$	$\boldsymbol{A}^{-1}\boldsymbol{A\,X} = \boldsymbol{A}^{-1}\boldsymbol{B}$
$x = 3$	$\boldsymbol{X} = \boldsymbol{A}^{-1}\boldsymbol{B}.$

The solution to the system $\begin{aligned} 2x+5y &= 11 \\ -x+2y &= -1 \end{aligned}$ is $\begin{bmatrix} x \\ y \end{bmatrix} = \begin{bmatrix} 2 & 5 \\ -1 & 2 \end{bmatrix}^{-1} \times \begin{bmatrix} 11 \\ -1 \end{bmatrix}$. The calculator has an inverse key. It is the key labeled $\boxed{x^{-1}}$. The solution to the system is

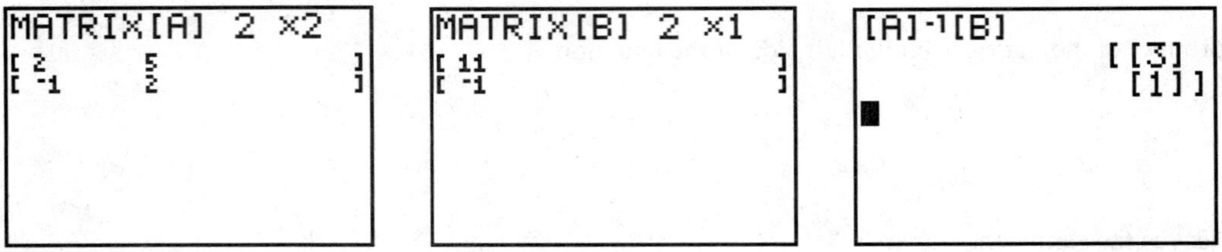

Figure 13.3.9

That is, $\begin{bmatrix} x \\ y \end{bmatrix} = \begin{bmatrix} 3 \\ 1 \end{bmatrix}$. The solution is the point (3, 1).

Example 10: Solve the system $\begin{matrix} 4x - 7y = 9 \\ -6x + 4y = -7 \end{matrix}$.

Solution: The solution is $\begin{bmatrix} x \\ y \end{bmatrix} = \begin{bmatrix} 4 & -7 \\ -6 & 4 \end{bmatrix}^{-1} \begin{bmatrix} 9 \\ -7 \end{bmatrix}$. That is $\begin{bmatrix} x \\ y \end{bmatrix} = \begin{bmatrix} \frac{1}{2} \\ -1 \end{bmatrix}$.

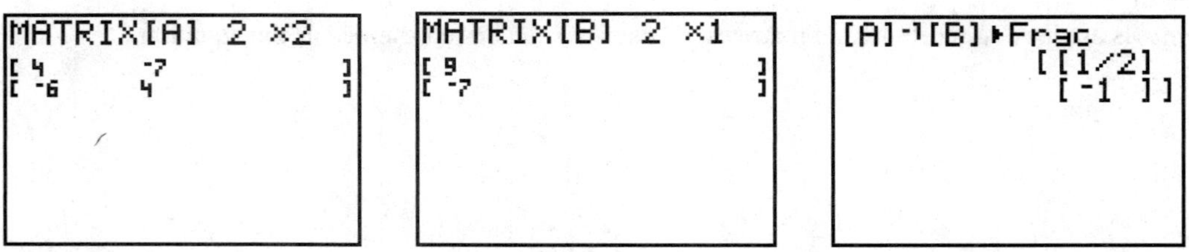

Figure 13.3.10

The solution to the system is the point (½, –1).

***

As was discussed at the beginning of this chapter, systems can definitely have more than two variables. For example, consider the following situation:

A purse contains 28 coins, they are all nickels, dimes, and quarters.

There are twice as many nickels as dimes. The value of the coins is 425 cents.

If x represents the number of nickels, y represents the number of dimes, and z represents the number of quarters, then:

there are 28 coins	$x + y + z = 28$
twice as many nickels as dimes	$2y = x$
the value is 425	$5x + 10y + 25z = 425$

When the second equation is rewritten, the system is:

$$\begin{aligned} x + \quad y + \quad z &= 28 \\ -x + \ 2y + \ 0z &= \ \ 0 \\ 5x + 10y + 25z &= 425. \end{aligned}$$

You can use the addition method or the substitution method to solve this system. A solution is a value for x, y, and z that makes each equation a true statement. The graphs of the three equations cannot be done on the graphing calculator. Graphing an equation with three variables is beyond the scope of this text; thus, the graphing method is not an option. Cramer's Rule can be applied to this problem and systems of higher order as well; however, by far the simplest method is the matrix method.

Example 11: Solve the above coin problem.

Rather simple.

Solution: The solution is $\begin{bmatrix} x \\ y \\ z \end{bmatrix} = \begin{bmatrix} 1 & 1 & 1 \\ -1 & 2 & 0 \\ 5 & 10 & 25 \end{bmatrix}^{-1} \begin{bmatrix} 28 \\ 0 \\ 425 \end{bmatrix}$

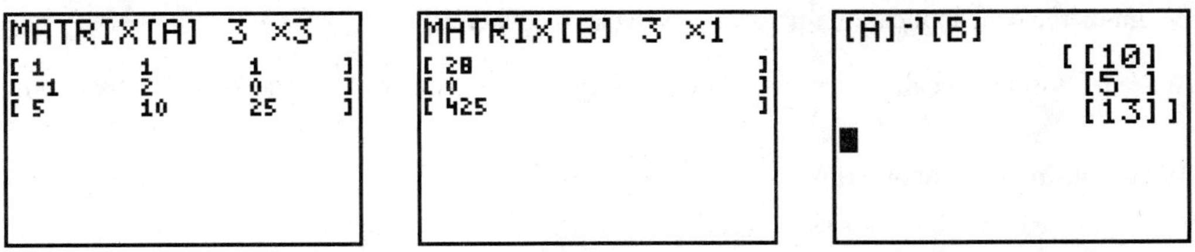

Figure 13.3.11

The solution is $\begin{bmatrix} x \\ y \\ z \end{bmatrix} = \begin{bmatrix} 10 \\ 5 \\ 13 \end{bmatrix}$; that is, the solution is 10 nickels, 5 dimes, and 13 quarters.

⁂

Example 12: Solve the system

$$\begin{aligned} 2x - 3y + 2z &= -21 \\ -3x + 5y - 4z &= \ \ 34 \\ 4x - 7y + 5z &= -47. \end{aligned}$$

Solution: The solution is $\begin{bmatrix} x \\ y \\ x \end{bmatrix} = \begin{bmatrix} 2 & -3 & 2 \\ -3 & 5 & -4 \\ 4 & -7 & 5 \end{bmatrix}^{-1} \begin{bmatrix} -21 \\ 34 \\ -47 \end{bmatrix}$.

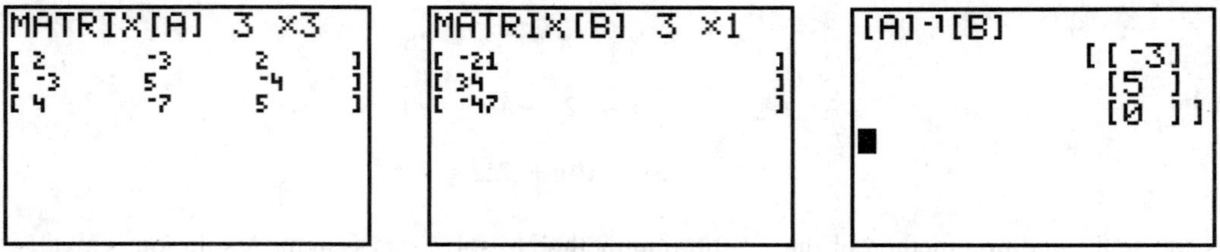

Figure 13.3.12

The solution is $\begin{bmatrix} x \\ y \\ z \end{bmatrix} = \begin{bmatrix} -3 \\ 5 \\ 0 \end{bmatrix}$; that is, the solution is (–3 ,5 ,0).

13.3 EXERCISES

Many times you can work the exercises by algebraic, numeric, or graphical methods. You can learn about mathematics no matter which method you use. Part of the learning process is for you to decide which method is best for you to use. Please read the text carefully before you do the exercises. Read the **next assigned section** after you finish the assignment below. All numeric answers must have an error ≤ 0.01.

A Quick Look To The Future

-4. Might there be an application for solving systems of equations?

-3. Write the following numbers as another number to an exponent: 16, 125, 10000, 1024, 1000000.

-2. What is a simpler way of writing $\frac{b^x}{b^y}$?

-1. Solve the equation $100000 = 10^?$.

Previous Topics Revisited

1. Solve by the addition method $\begin{aligned} 3x + \ \ y &= 0 \\ x + \frac{1}{3}y &= 0 \end{aligned}$

2. Solve by the substitution method $\begin{aligned} -x + 5y &= -1 \\ 3x + 2y &= 0 \end{aligned}$

3. Solve by the substitution method $\begin{aligned} y &= -x^2 + 1 \\ y &= x^2 - 1 \end{aligned}$

4. Solve by the graphical method $\begin{aligned} -2x + y &= -1 \\ 10x - 5y &= -3 \end{aligned}$

5. Solve by the graphical method $\begin{aligned} y &= -x^2 + 1 \\ y &= x^2 - 1 \end{aligned}$

6. How long must the house overhang be so that the sun does not enter through the window when the angle of elevation of the sun is 81°?

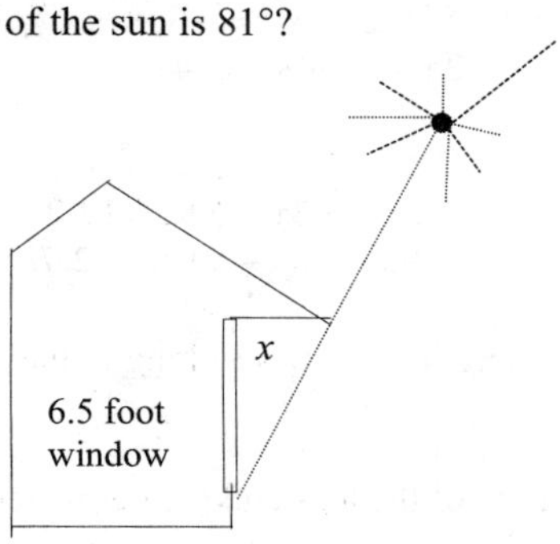

New Topic Exercises

Evaluate the determinants in Exercises 7 - 11. Simplify the matrices in Exercises 12 - 16.

7. $\begin{vmatrix} -4 & 5 \\ 2 & -3 \end{vmatrix}$

8. $\begin{vmatrix} -3 & -5 \\ 3 & 5 \end{vmatrix}$

9. $\begin{vmatrix} -7 & 2 \\ 21 & 6 \end{vmatrix}$

10. $\begin{vmatrix} 0 & 0 \\ 3 & 2 \end{vmatrix}$

11. $\begin{vmatrix} -2 & 3 & -1 \\ 5 & 2 & 10 \\ 3 & -2 & 4 \end{vmatrix}$

12. $\begin{bmatrix} -2 & 1 \\ 8 & 3 \end{bmatrix} + \begin{bmatrix} 5 & -1 \\ 0 & 4 \end{bmatrix}$

13. $\begin{bmatrix} -5 \\ 2 \\ -3 \end{bmatrix} - \begin{bmatrix} 3 \\ 6 \\ -1 \end{bmatrix} + \begin{bmatrix} 0 \\ 1 \\ -3 \end{bmatrix}$

14. $\begin{bmatrix} 2 & 1 \\ 0 & 1 \end{bmatrix} \times \begin{bmatrix} 1 & 5 \\ -1 & 3 \end{bmatrix}$

15. $\begin{bmatrix} -1 & 2 \\ -4 & 5 \end{bmatrix} \times \begin{bmatrix} 5 \\ 1 \end{bmatrix}$

16. $\begin{bmatrix} 1 & 0 & 2 \\ 0 & 3 & 0 \\ -1 & 2 & 0 \end{bmatrix} \times \begin{bmatrix} -1 \\ 8 \\ 1 \end{bmatrix}$

Solve the following systems using Cramer's Rule.

17. $\begin{aligned} 2x - 5y &= -3 \\ -3x + 4y &= 7 \end{aligned}$

18. $\begin{aligned} 1.5x + 2.7y &= 1.9 \\ 3.2x - 4.1y &= 2.1 \end{aligned}$

19. $\begin{aligned} 2x - 6y &= 12 \\ -x + 3y &= -15 \end{aligned}$

Solve using matrices.

20. $\begin{aligned} 250x - 525y &= -1250 \\ 24x + 874y &= 926 \end{aligned}$

21. $\begin{aligned} -7x + 3y &= -8 \\ 2y - 5x &= 2 \end{aligned}$

22. $\begin{aligned} y &= \frac{2}{3}x - 5 \\ y &= \frac{-4}{5}x - 2 \end{aligned}$

23. $\begin{aligned} 3y &= 4x - 7 \\ 12x &= 9y + 21 \end{aligned}$

24. $\begin{aligned} x+y+z&=4 \\ x-y+z&=2 \\ x-y-z&=-6 \end{aligned}$

25. $\begin{aligned} 3x-5y+2z&=4 \\ -5x+2y-4z&=-6 \\ 2x-7y+8z&=-4 \end{aligned}$

26. $\begin{aligned} 7x-2y-3z&=25 \\ -4x\qquad +5z&=-2 \\ 6y-7z&=-44 \end{aligned}$

27. $\begin{aligned} 3.6x-5.1y+2.3z&=-4.5 \\ -2.2x+4.3y-6.8z&=12.2 \\ -1.9x+1.8y+5.9z&=-2.7 \end{aligned}$

28. The perimeter of a rectangle is 16 feet and the length is 3 times as long as the width, what are the dimensions of the rectangle?

29. The base of an isosceles triangle is half the length of the legs and the perimeter is 10 inches. How long are the legs and base?

30. If the value of $\begin{vmatrix} a & b \\ c & d \end{vmatrix}$ is W, what is the value of $\begin{vmatrix} 2a & 2b \\ c & d \end{vmatrix}$?

31. If the value of $\begin{vmatrix} a & b \\ c & d \end{vmatrix}$ is W, what is the value of $\begin{vmatrix} 2a & b \\ 2c & d \end{vmatrix}$?

32. If the value of $\begin{vmatrix} a & b \\ c & d \end{vmatrix}$ is W, what is the value of $\begin{vmatrix} 2a & 2b \\ 2c & 2d \end{vmatrix}$?

33. Does $\begin{vmatrix} a & b \\ c & d \end{vmatrix} = \begin{vmatrix} c & d \\ a & b \end{vmatrix}$? (The rows were interchanged.)

34. Does $\begin{vmatrix} a & b \\ c & d \end{vmatrix} = \begin{vmatrix} b & a \\ d & c \end{vmatrix}$? (The columns were interchanged.)

From Mathematics to English

35. Multiply the matrix $\boldsymbol{I} = \begin{bmatrix} 1 & 0 \\ 0 & 1 \end{bmatrix}$ times several 2×2 matrices of your choice. Describe the product of $\boldsymbol{I}$ and any other 2×2 matrix.

36. For $\boldsymbol{A} = \begin{bmatrix} -2 & 3 \\ 4 & 1 \end{bmatrix}$, find $\boldsymbol{A}^{-1}$ with your calculator and then multiply $\boldsymbol{A}$ times $\boldsymbol{A}^{-1}$. Repeat this experiment with several matrices of your choosing. Describe each product.

37. Describe how to solve the system $\begin{aligned} 2x-5y&=1 \\ -3x+2y&=-1 \end{aligned}$ using matrices. Do not actually solve the system.

38. After reading this section, make a list of questions that you want to ask your instructor.

39. Continue in your daily journal and make an entry. In addition to your normal entry on thoughts about the mathematics in this section, list at least two positive comments about what you have learned about this topic.

40. In paragraph format, summarize the material in this section of the text in your daily journal.

41. Describe how your classroom instructor made this topic more understandable and clear.

42. After reading the text and listening to your instructor, what do you not understand about this topic?

Explorations

43. What conditions on $|D|$, $|D_x|$, and $|D_y|$ cause any 2×2 system of equations to have exactly:

a) one solution

b) in infinite number of solutions

c) no solution?

44. Solve the system. (Use whatever method you think best.)

$$\begin{aligned}
u + v + w + x + y \quad &= 0 \\
v \quad + x \quad + z &= 0 \\
2u - 5v - w + x + 4y - z &= 0 \\
-3u \quad + 2w + 2x + 3y + z &= 0 \\
4v - w \quad - y + 2z &= 0 \\
3u - 3v + w + x - 2y - 2z &= 0
\end{aligned}$$

45. Now that you have learned five different methods for solving systems of <u>linear</u> equations, try to solve the following <u>non-linear</u> systems of equations. Your calculator must be in radian mode.

a. $y = \cos x + x^2 - 2$

$y = \sqrt{x + \sin x}$

b. $y = \sqrt{8 \cos x} + x$

$y = \sqrt{(x-3)(x+1)}$

c. $y = \sqrt{(x-2)(x+4)(x-5)} - \sqrt{9 \sin(2x)}$

$(x-3)^2 + (y+2)^2 = 16$

13.4 Modeling with Systems of Equations

You will recall that a series of equations mathematically describing a situation or object is called a system of equations. To solve a particular problem with a system of equations, you must convert the information from the given English form to mathematical form. From the given information, you should assign a variable to each of the requested unknown quantities in the situation. Using these variables, write a series of equations describing the situation. When all of the equations have been written about the situation, solve the system. Once the system has been solved, the values for the variables should be interpreted as to their meaning in the problem.

Example 1: Your food budget for pizza is \$48. You want 5 pizzas, some with everything and a few with just pepperoni. Pepperoni pizzas are \$8 each, and pizzas with everything cost \$12 each. How many of each can be purchased for exactly \$48?

Solution: Let x symbolize the *number* of pepperoni pizzas purchased.
Let y symbolize the *number* of everything pizzas purchased.

Since you want 5 pizzas: $x + y = 5.$
Since you can spend \$48: $8x + 12y = 48.$
Below is the matrix method for solving the system.

$$\begin{bmatrix} x \\ y \end{bmatrix} = \begin{bmatrix} 1 & 1 \\ 8 & 12 \end{bmatrix}^{-1} \begin{bmatrix} 5 \\ 48 \end{bmatrix}$$

$$\begin{bmatrix} x \\ y \end{bmatrix} = \begin{bmatrix} 3 \\ 2 \end{bmatrix}$$

You should buy 3 pepperoni pizzas and 2 with everything pizzas.

Example 2: An isosceles triangle has a perimeter of 32 feet. The base is 4 feet less than the lengths of the equal sides. How long are the equal sides and how long is the base?

Solution: Let x represent the length of each of the equal sides.
Let y represent the length of the base.

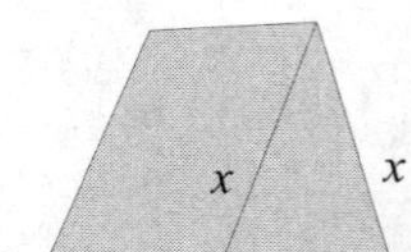

Since the perimeter is 32 feet: $x + x + y = 32$ or $2x + y = 32.$
If 4 feet is added to the length of the base, it will equal the length of each of the equivalent sides: $y + 4 = x.$

The system is $2x + y = 32$ (1)

$x = y + 4,$ (2) Substitution may be the simplest method.

Replace x in $2x + y = 32$ with $y + 4$ from (2):

$$2(y + 4) + y = 32$$

$$2y+8+\ y=32$$
$$3y+8=32$$
$$3y=24$$
$$y=8.$$

From (2), replace y with 8:

$$x=8+4$$
$$x=12$$

Each of the equal sides is 12 feet long and the base is 8 feet long.

⁂

Example 3: A box contains 100 worms. It only has red and green worms, and the red worms outnumber the green by 4 to 1. How many worms of each color are in the box?

Solution: Let x represent the number of red worms and y the number of green worms.
$x + y = 100$ (1) since there are 100 worms
$x = 4y$ (2) since the reds outnumber the greens by 4 to 1
Replace x in (1) with $4y$ and solve by substitution.

$$4y+y=100$$
$$5y=100$$
$$y=20.$$

If $y = 20$ and $x = 4y$, then $x = 4\cdot 20 = 80$.

There are 80 red worms and 20 green worms.

⁂

Consider the following situations:

a) The level of the drug Imipramine in the blood of a patient rises at a constant rate (for example 60 nanograms per week) until the patient is at the prescribed level of 180 nanograms per *ml* of blood, and then the rate of change remains at 0% until the patient is taken off the drug at a constant rate (for example 90 nanograms per *ml* per week).

b) For users of electricity who also use a heat pump, many power companies charge, for example, $0.08 per kWh used for the first 1000 kWh and $0.06 for the next 1000 and finally, $0.05 for any consumption over 2000 kWh.

c) Sliding scale commissions are business examples of multi-constant rates of change and are used to provide incentive for higher employee productivity. For example, a business may pay the sales staff 3% commission on sales from $0 to $10,000, 5% on sales from $10,001 to $15,000, and 8% commission on sales $15,001 and over.

d) The pay scale for piece-work by a telemarketing company is $0.35 per call for the first 300 calls in the week, $0.42 per call for the next 200 calls, and $0.65 per call for any call over 500 calls.

e) A San Diego cab company charges a fare of \$2.75 entrance fee plus \$1.80 per $\frac{1}{5}$ mile for the first mile, \$1.50 per $\frac{1}{4}$ mile for the next mile, and \$1.00 per mile for lengths longer than 2 miles.

f) A new long-distance phone company has the following rate schedule:
12¢ per minute for the first 15 minutes
9¢ per minute for the next 10 minutes, and
6¢ per minute for any time over 25 minutes.

g) The 1994 (or any year with slightly different rates) United States federal income tax form 1040 schedule (for single filers) had rates of taxation of 15% on the first \$22,750 of taxable income, 28% on the next \$32,350, 31% on the next \$59,900 of taxable income, etc. That is, the taxable income brackets are at \$22,750, \$55,100, and \$115,000. (There were two more brackets that will be ignored for the sake of brevity.)

h) A commercial airline flight from Reno, Nevada to St. Louis, Missouri has an on-board computer that directs it to ascend at a constant rate (after initial take-off) of 1100 feet per minute for 30 minutes (until it reaches a cruising altitude of 33,000 feet). It then levels off until it is 60 minutes into the flight. At 60 minutes into the flight, it has burned off enough fuel to ascend to 37,000 feet at a rate of 400 feet per minute. This takes 10 minutes. The plane remains at 37,000 feet until 175 minutes of flight time when it descends at a rate of 1233 feet per minute and then lands in St. Louis.

All of these situations are identical in structure. Since the rates of change are constant on each of the subsets of the problem domain described in each situation, the model for each of these situations is linear on each of these subsets. If you go back and look at Chapter Four, Section 6 and Chapter Six, Section 3, you will find functions that have linear graphs on subsets of the domain – just like the situations described above.

As a quick reminder, notice what behavior occurs when absolute value functions of the form $f(x) = d|x+e| + f$ are added:

$y1(x) = 2|x+5| + |x-2| - 15$

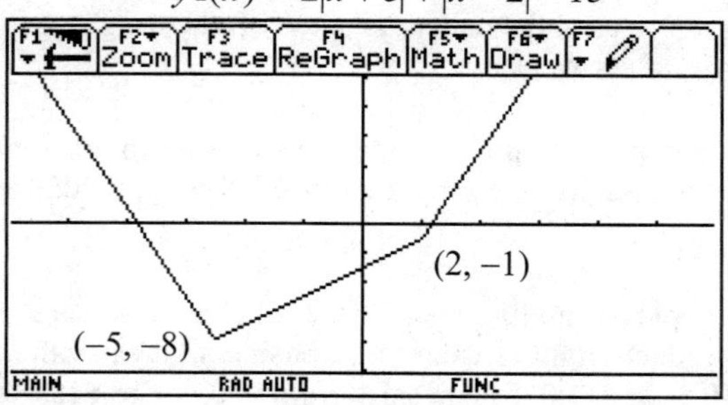

Figure 13.4.1

$y2(x) = -3|x+2| + 2|x-3|$

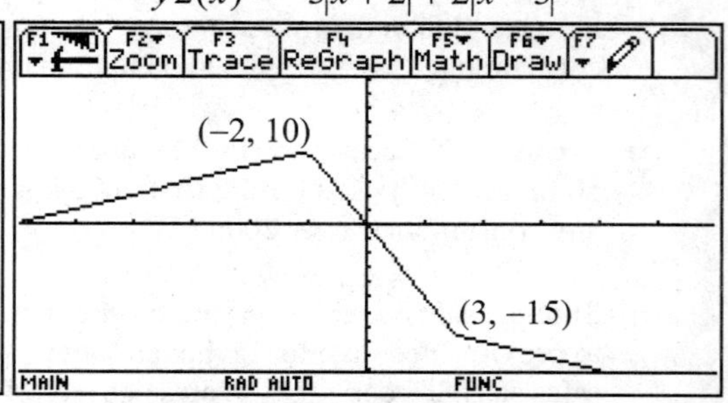

Figure 13.4.2

While the graphical representations of these functions are not proof that the functions are linear on each piece of the domain, using a numeric representation and calculating the rate of change from point-to-point is a convincing argument for most people. Proof lies in the algebraic simplification as shown for $y1(x)$ in Figure 13.4.3. As you can see, by using a Voyage 200 to simplify the $y1(x)$ function on each piece of the domain, it simplifies to a linear function. Functions that behave like linear functions on pieces of the domain were developed in Sections 4.6 and 6.3. Studying the examples above should convince you that the corners of the graph of a sum of absolute value functions like $f(x) = a|x + e_1| + b|x + e_2| + c$ are at $-e_1$ and $-e_2$. To help develop a method for finding the model for situations a) through h) without the aid of the Voyage 200, please consider the examples below.

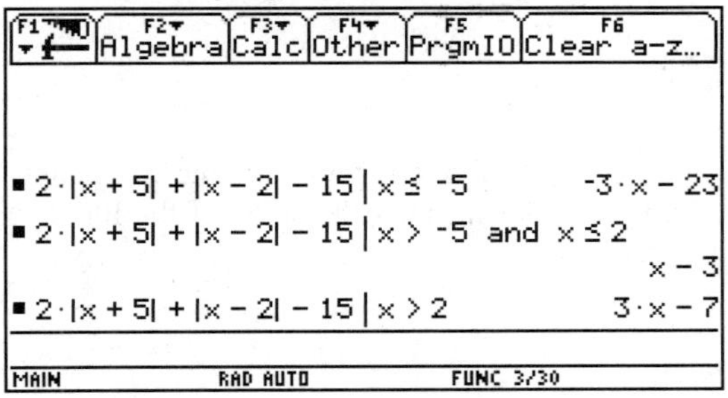

Figure 13.4.3

The Income Tax Example

This example demonstrates how to model the 1994 1040 tax rate schedule for a single taxpayer, with a sum of absolute value functions. The tax form gives the following information:

\$0	\$22,750	pay 15%
\$22,750	\$55,100	pay \$3,412.50 + 28% of excess over \$22,750
\$55,100	∞	pay \$12,470.5 + 31% of excess over \$55,100.

The tax schedule suggests the corners of the graphical representation of the model are at 0, 22750, and 55100. There are three rates of change; thus, the model is the sum of three absolute value functions $T = a|I - 0| + b|I - 22750| + c|I - 55100| + f$, where T is taxes owed, I is taxable income, and a, b, c, and f are function parameters. On each tax bracket, the individual absolute value functions simplify to:

	$a\|I\|$	$b\|I - 22750\|$	$c\|I - 55100\|$	(See the next two
$I \in [0, 22750]$	aI	$-b(I - 22750)$	$-c(I - 55100)$	examples for more
$I \in [22750, 55100]$	aI	$b(I - 22750)$	$-c(I - 55100)$	simplifications of
$I \in [55100, \infty)$	aI	$b(I - 22750)$	$c(I - 55100)$	absolute value functions.)

Since the tax bracket constants above have no effect on the slope for each interval, this table can be simplified to show only the slopes of each bracket – the coefficients of I.

$I \in [0, 22750]$	aI	$-bI$	$-cI$
$I \in [22750, 55100]$	aI	bI	$-cI$
$I \in [55100, \infty)$	aI	bI	cI .

Finally, form 1040 gives the tax rate for each interval, and each row above shows this rate of change (slope) for each interval also. Just factor out the variable I and the coefficient is the slope of the line for each interval. That is, the slopes (rates of change) on each interval are:

$I \in [0, 22750]$	a	+	$-b$	+	$-c$
$I \in [22750, 55100]$	a	+	b	+	$-c$
$I \in [55100, \infty)$	a	+	b	+	c

Set them equal to the given tax rates of change of 15%, 28%, and 31% and solve the system for a, b, and c. Figure 13.4.4 shows the solution.

$$a - b - c = 0.15$$
$$a + b - c = 0.28$$
$$a + b + c = 0.31$$

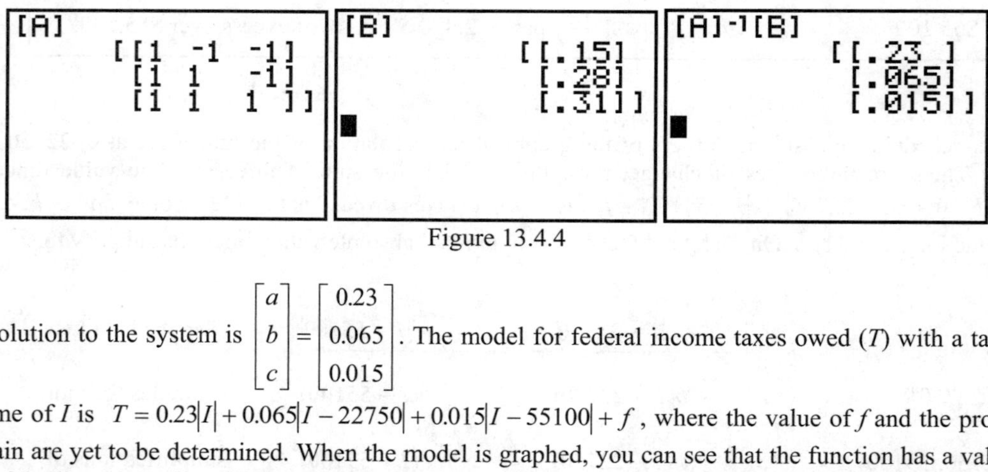

Figure 13.4.4

the solution to the system is $\begin{bmatrix} a \\ b \\ c \end{bmatrix} = \begin{bmatrix} 0.23 \\ 0.065 \\ 0.015 \end{bmatrix}$. The model for federal income taxes owed (T) with a taxable income of I is $T = 0.23|I| + 0.065|I - 22750| + 0.015|I - 55100| + f$, where the value of f and the problem domain are yet to be determined. When the model is graphed, you can see that the function has a value of 2305.25 when I is 0; thus, if –2305.25 is added to the model, the vertical shift will make the taxes start at \$0 for \$0 in taxable income. A reasonable problem domain might be [0, 115000] and the domain-controlling function $0\sqrt{-I(I - 115000)}$ can be added to restrict the domain of the model. The final model is $T = 0.23|I| + 0.065|I - 22750| + 0.015|I - 55100| - 2305.25 + 0\sqrt{-I(I - 115000)}$. The graph of the model in Figure 13.4.5 is shown in two windows, [–5000, 56000] by [–5000, 10000] and [–25000, 120000] by [–5000, 35000] so that each branch can be seen.

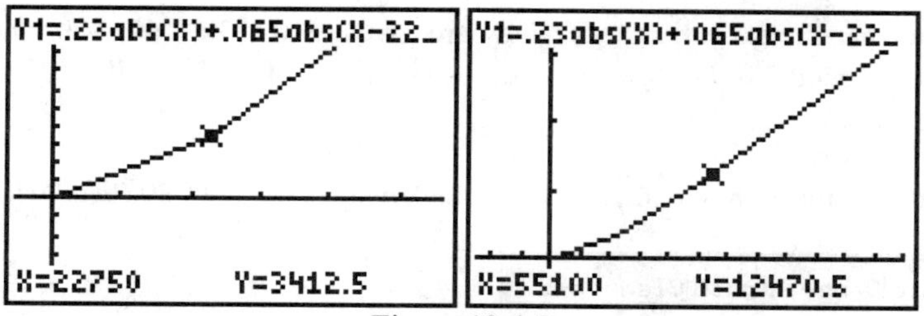

Figure 13.4.5

So, knowing how the corners are connected to the function parameters, knowing that other parameters control the rates of change on each branch, and observing that there is a connection between the number of corners and number of rates of change, you are ready to find a model. Perhaps listing the algorithm will help your understanding.

The Algorithm for Finding the Model.

1. Recognize that the structure of the model for all of the above situations is $M = a|x+e_1| + b|x+e_2| + \ldots\ d|x+e_n| + f$, where there are n rates of change given in the situation.
2. Find the corners (e_1, e_2, ...) [corners are given in the situation]
3. Simplify the model for each rate interval using the TI-92 **(bear with this for a bit yet)**
4. Set the coefficients of x equal to each rate of change. [rates of change are given in the situation]
5. Solve the system.
6. Find f using a geometric transformation of a vertical shift.

Airplane Flight Height:

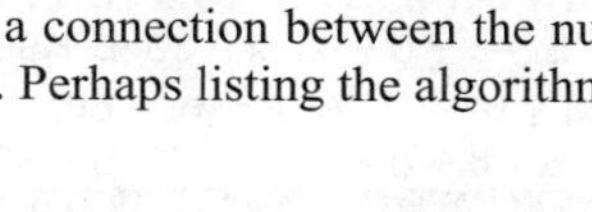

The on-board computer on a commercial airline flight from Reno, Nevada to St. Louis, Missouri commands the plane to ascend at a constant rate (after initial take-off) of 1100 feet per minute for 30 minutes (until it reaches a cruising altitude of 33,000 feet). It then levels off until it is 60 minutes into the flight. At 60 minutes into the flight, it has burned off enough fuel to ascend to 37,000 feet at a rate of 400 feet per minute. This takes 10 minutes. The plane remains at 37,000 feet until 175 minutes of flight time when it descends at a rate of 1233 feet per minute and then lands in St. Louis.

1. Recognize that the structure of the model is $h(x) = a|x+e_1| + b|x+e_2| + c|x+e_3| + d|x+e_4| + e|x+e_5| + f$, where there are 5 rates of change.

2. Find the corner parameters (e_1, e_2, …). They are given as 0, –30, –60, –70, and –175. Thus, the initial attempt at a model is $h(x) = a|x| + b|x - 30| + c|x - 60| + d|x - 70| + e|x - 175| + f$, where x is time in minutes.

3. Simplify the model for each rate interval using the Voyage 200 or TI-89 **(bear with this for a bit yet)**

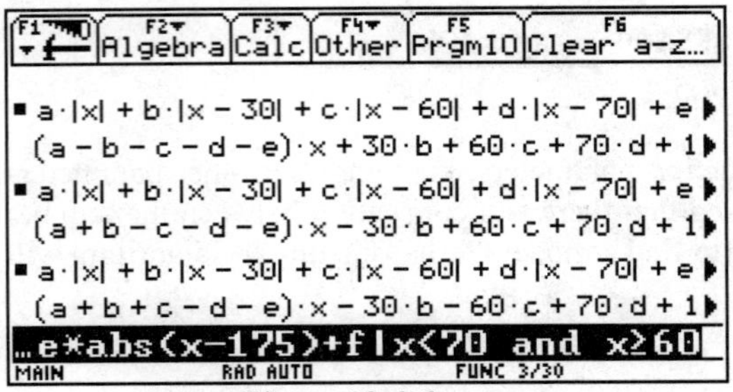

Conditions

$x \geq 0$ and $x < 30$

$x \geq 30$ and $x < 60$

$x \geq 60$ and $x < 70$

Figure 13.4.6

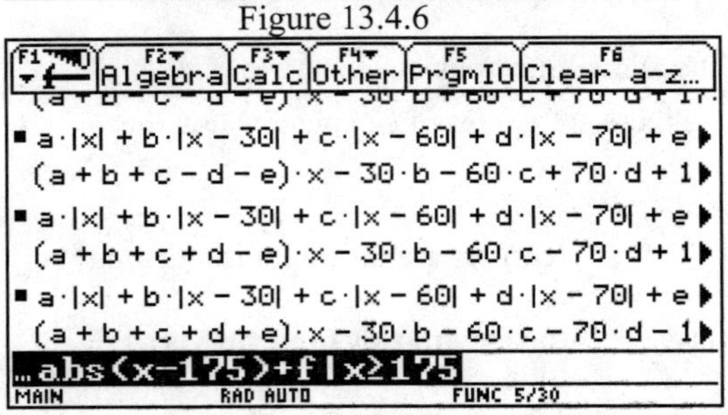

Conditions

$x \geq 70$ and $x < 175$

$x \geq 175$

Figure 13.4.7

4. Set the coefficients of x equal to each rate of change; that is, each piece of the model simplifies to a linear function as shown on the Voyage 200 screens. Thus, the coefficient of x is the rate of change for each piece.

	Conditions
$a - b - c - d - e = 1100$	$x \geq 0$ and $x < 30$
$a + b - c - d - e = 0$	$x \geq 30$ and $x < 60$
$a + b + c - d - e = 400$	$x \geq 60$ and $x < 70$
$a + b + c + d - e = 0$	$x \geq 70$ and $x < 175$
$a + b + c + d + e = -1233$	$x \geq 175$

5. Solve the system.

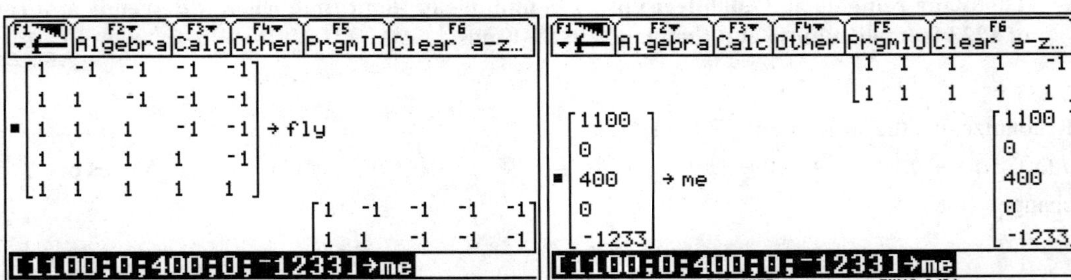

Figure 13.4.8

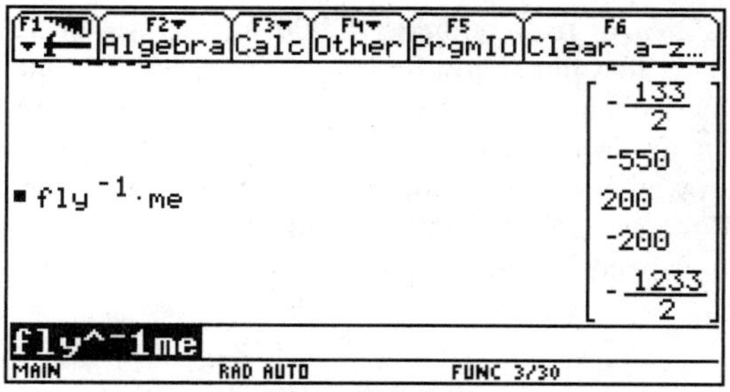

Figure 13.4. 9

The model for the height control function is:

$$h(x) = -\frac{133}{2}|x| - 550|x-30| + 200|x-60| - 200|x-70| - \frac{1233}{2}|x-175| + f.$$

Find f using a geometric transformation of a vertical shift.

When x is 0, for example, the function is -126387.5. Add 126387.5 as the parameter f.

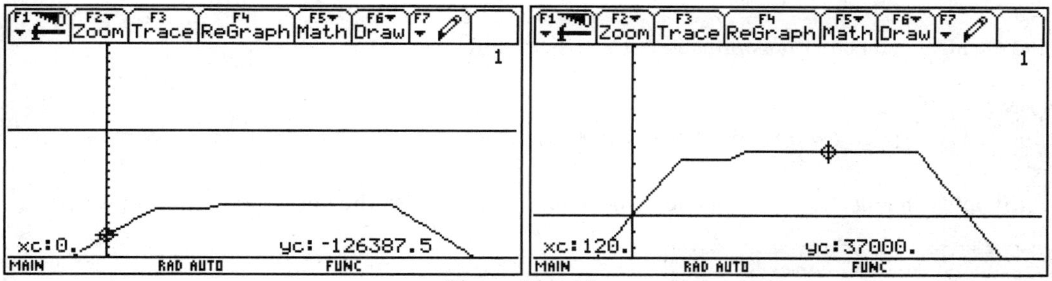

Figure 13.4.10

The final model needed by the computer is

$$h(x) = -\frac{133}{2}|x| - 550|x-30| + 200|x-60| - 200|x-70| - \frac{1233}{2}|x-175| + 126387.5.$$

1996 Income Tax:

The 1996 U. S. individual income tax rate schedule for filers who are married and filing jointly or are qualifying widow(er) with dependent children is published as follows:

Taxable income		Pay +	% on excess over lower bracket amount
$0	**$40,100**	**$0**	**15%**
$40,100	**$96,900**	**$6,015**	**28%**
$96,900	**$147,700**	**$21,919**	**31%**
$147,700	**$263,750**	**$37,667**	**36%**
$263,750	**and over**	**$79,445**	**39.6%**

The column numbers labeled as "Pay +" come from the previous tax bracket rate times the difference in the previous taxable income bracket numbers, plus previous taxes, plus current bracket taxes. For example, if you have taxable income in the 28% bracket, the tax to be paid is $40{,}100 \times 0.15$ ($6,015) plus

28% of the excess over \$40,100. If you have taxable income in the 31% bracket, the taxes owed is \$6015 (15% of the first \$40,100) plus 28% of the amount (\$96,900 – \$40,100), or \$21,919 plus 31% of income over \$96,900. What this means is that the single function relating taxable income to taxes owed can be thought of as continuous. Further, since the rate on each bracket is constant, this means that each piece of the function that models the taxes owed in linear. Put this information together and we find that the function that models the taxes owed is a sum of absolute value functions. The corners of the function are given as 0, 40100, 96900, 147700, and 263750. The rates of change (slopes) are given as 0.15, 0.28, 0.31, 0.36, and 0.396. This is all that is needed to find a single function that models taxes owed. The algorithm for finding the model is:

1. The model is $T = a|x+e_1| + b|x+e_2| + \ldots d|x+e_n| + f$, where there are n rates of change given.
2. Find the corners (e_1, e_2, …) [corners are known]
3. Simplify the model for each rate interval using a CAS calculator **(bear with this for a bit yet)**
4. Set the coefficients of x equal to each rate of change. [rates of change are known]
5. Solve the system.
6. Find f using a geometric transformation of a vertical shift.

Steps 1 & 2. The tax model is:

$$T = a|x-0| + b|x-40100| + c|x-96900| + d|x-147700| + e|x-263750| + f$$

Step 3. Simplify the function on each known interval: **(bear with the development for a bit yet)**

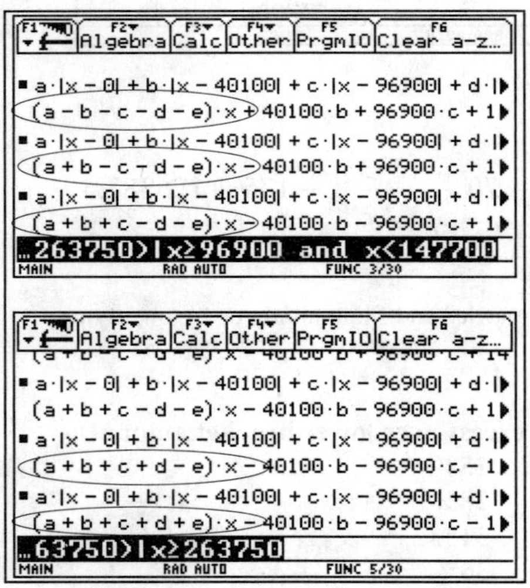

$x \ge 0$ and $x < 40100$

$x \ge 40100$ and $x < 96900$

$x \ge 96900$ and $x < 147700$

$x \ge 147700$ and $x < 263750$

$x \ge 263750$

Figure 13.4.11

Of course, for each bracket, the function simplifies to a linear function. The rate of change (slope) is circled above. These are known values – the tax rate for each bracket. That is: (Steps 4 and 5)

$$a-b-c-d-e=0.15$$
$$a+b-c-d-e=0.28$$
$$a+b+c-d-e=0.31$$
$$a+b+c+d-e=0.36$$
$$a+b+c+d+e=0.396$$

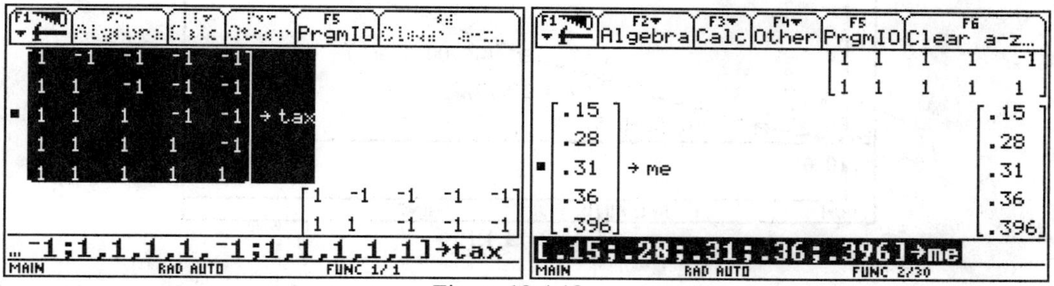

Figure 13.4.12

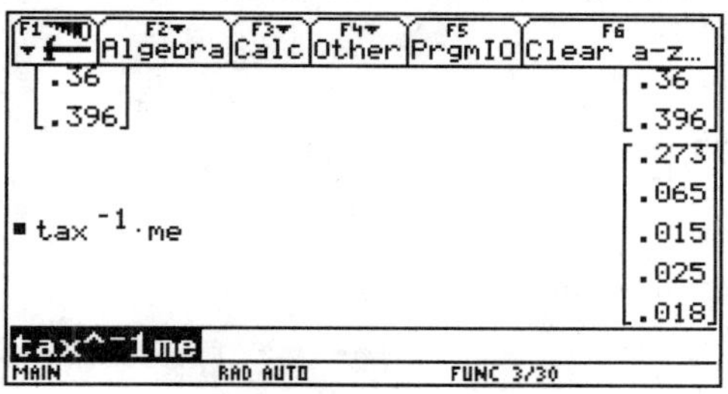

Figure 13.4.13

The solution to the system is $\begin{bmatrix} a \\ b \\ c \\ d \\ e \end{bmatrix} = \begin{bmatrix} 0.273 \\ 0.065 \\ 0.015 \\ 0.025 \\ 0.018 \end{bmatrix}$ and the model of the taxes owed becomes

$$T = 0.273|x-0| + 0.065|x-40100| + 0.015|x-96900| + 0.025|x-147700| + 0.018|x-263750| + f\ .$$

When this function is graphed you will notice that it has a value of 12500 when $x = 0$; thus, if a vertical shift of -12500 is given, the function is finally correct. (Step 6)

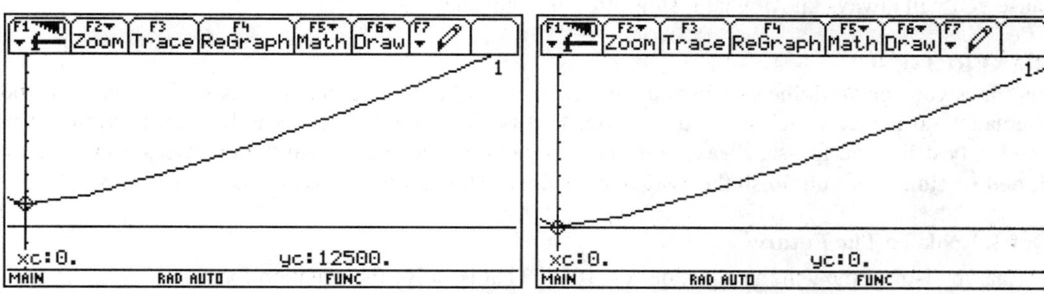

Figure 13.4.14

The final model is
$$T = 0.273|x-0| + 0.065|x-40100| + 0.015|x-96900| + 0.025|x-147700| + 0.018|x-263750| - 12500.$$

To confirm that the model is correct, shown below are several data points that are known to be correct. You are encouraged to try any taxable income.

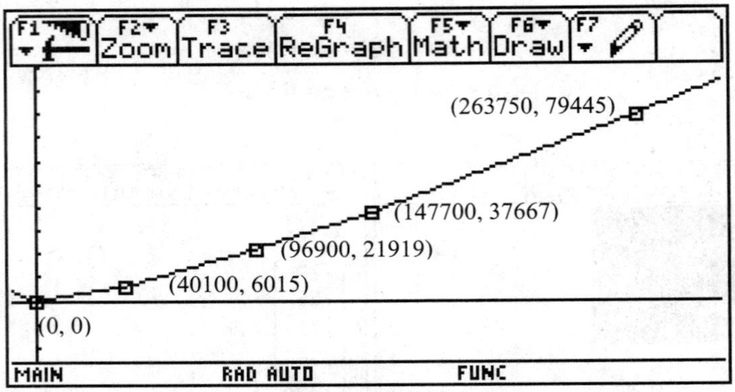

Figure 13.4.15

You may not have a Voyage 200 or TI-89 to simplify the sum of the absolute value functions like in the three examples above; however, this is not necessary. Did you notice that in every example the slopes for each interval always simplify the same way? For example, if there are three rates of change, the slopes on each interval became:

$$\begin{bmatrix} a & -b & -c \\ a & b & -c \\ a & b & c \end{bmatrix} = \begin{bmatrix} m_1 \\ m_2 \\ m_3 \end{bmatrix}.$$

In a system with four rates of change, the slopes on each interval became:

$$\begin{bmatrix} a & -b & -c & -d \\ a & b & -c & -d \\ a & b & c & -d \\ a & b & c & d \end{bmatrix} = \begin{bmatrix} m_1 \\ m_2 \\ m_3 \\ m_4 \end{bmatrix}.$$

This is always the case; thus, you don't need to simplify the absolute value sum function on each interval because you will always know what it simplifies to in advance.

13.4 EXERCISES

Many times you can work the exercises by algebraic, numeric, or graphical methods. You can learn about mathematics no matter which method you use. Part of the learning process is for you to decide which method is best for you to use. Please read the text carefully before you do the exercises. Read the **next assigned section** after you finish the assignment below. All numeric answers must have an error ≤ 0.01.

A Quick Look To The Future

-4. Can you isolate the y in the equation $x = 10^y$? (That is, solve the equation for y?)

-3. Simplify $\left(2^x\right)^y$.

-2. What is the solution to the equation $10^x = 100000$?

-1. Have you ever heard of the Richter Scale?

Previous Topics Revisited

1. Evaluate the determinate $\begin{vmatrix} -2.1 & -3.7 \\ 5.9 & 4.8 \end{vmatrix}$

2. Solve the system using Cramer's Rule $\begin{aligned} 2x &= 3y - 7 \\ y &= 6 - 4x \end{aligned}$

3. Solve the system using matrices. $\begin{aligned} 2x - y + 3z &= 5 \\ x + 3y - 2z &= -1 \\ 4x - 5y - z &= -17 \end{aligned}$

4. Solve by addition $\begin{aligned} 3x - 5y &= 0 \\ -2x + 3y &= 4 \end{aligned}$

5. Solve by substitution $\begin{aligned} 4x - 7y &= -12 \\ 3x + y &= 5 \end{aligned}$

6. Solve graphically $\begin{aligned} y &= \frac{-2}{3}x - 5 \\ x - y &= 4 \end{aligned}$

Use the method of your choice on the following problems.

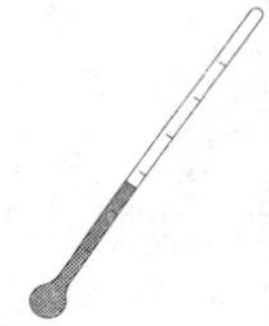

7. The relationship between the Centigrade and Fahrenheit temperature scales is $C = \frac{5}{9}(F - 32)$. What temperature on the Centigrade scale is equivalent to one half that on the Fahrenheit scale?

8. What temperature on the Fahrenheit scale is 3 times the equivalent temperature on the Centigrade scale?

9. You have \$3.55 in nickels and quarters. If you have 39 coins total, how many are nickels and how many are quarters?

10. A bag contains a mixture of 28 pieces of candy weighing 72 ounces. The big pieces weigh 3 ounces each and the small pieces weigh 2 ounces each. How many of each type are in the bag?

11. A rectangle with a perimeter of 22 inches has a length of 2 more than twice the width. Find the length and width.

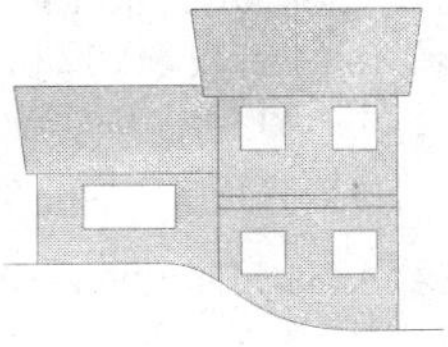

12. The electric company offers a discount for electricity used by a heat pump. They charge 5 cents per kilowatt hour for use by the heat pump and 7 cents per kilowatt hour for all other use. For 2200 kilowatt hours used, the bill is \$134. Find the number of kilowatt hours used by the heat pump and for all other use.

13. The circumference of a circle is 10.708 inches greater than the diameter. Find the circumference and the diameter.

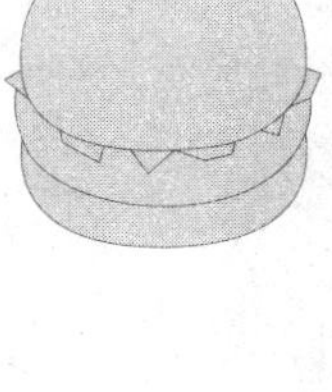

14. Hamburgers sell for \$1.25 each and fries for \$.90 each. You can only spend \$20.80. If you want to buy 4 more fries than burgers, how many of each can you buy?

15. If you can work 10 word problems (some easy and some hard) in 82 minutes, exactly how many easy and how many hard problems can you work if easy word problems take 7 minutes each and hard problems take 10 minutes each?

16. Your teacher has a habit of scratching his nose and wiping chalk from his hands. Since you already know the classroom material because you read the text, you count the total number of the above activities at 42. Your friend tells you that the number of nose scratchings is 8 more than the number of chalk wipings. How many times did your teacher scratch and how many times did he wipe?

17. The difference between the second and third angles of a triangle is 5°. The difference between the second and first is 40°. How many degrees are in each of the three angles of the triangle?

18. Duane, a retired pilot, decided to raise pine trees with all of his extra time. His wife figured that with the amount of land owned he could grow 1000 trees. The county extension agent suggested that he plant 100 more spruce than Austrian pine. The Christmas Tree Society recommended that the number of white pine be two and a half times the number of Austrian pine planted. If Duane uses all of the recommendations, how many of each type of tree should he plant?

19. The level of the drug Imipramine in the blood of a patient rises at a constant rate of 60 nanograms per week until the patient is at the prescribed level of 180 nanograms per *ml* of blood in three weeks, and then the rate of change remains at 0% for 20 weeks until the patient is taken off the drug at a constant rate of 90 nanograms per *ml* per week for two weeks. Find the model for the level of Imipramine in the patient's blood.

20. For users of electricity who also use a heat pump, many power companies charge, for example, \$0.08 per kWh used for the first 1000 kWh and \$0.06 for the next 1000 and finally, \$0.05 for any consumption over 2000 kWh. Find the model that the power company uses to calculate customer's monthly charges.

21. Sliding scale commissions are business examples of multi-constant rates of change and are used to provide incentive for higher employee productivity. For example, a business may pay the sales staff 3% commission on sales from \$0 to \$10,000, 5% on sales from \$10,001 to \$15,000, and 8% commission on sales \$15,001 and over. Find the model the company's computer uses to calculate monthly gross wages.

22. The pay scale for piece-work by a telemarketing company is \$0.35 per call for the first 300 calls in the week, \$0.42 per call for the next 200 calls, and \$0.65 per call for any call over 500 calls. Find the model the district supervisor uses to calculate monthly gross wages for employees.

23. A San Diego cab company charges a fare of \$2.75 entrance fee plus \$1.80 per $\frac{1}{5}$ mile for the first mile, \$1.50 per $\frac{1}{4}$ mile for the next mile, and \$1.00 per mile for lengths longer than 2 miles. Find the model the cabby's meter uses to calculate cab fare.

24. A new long-distance phone company has the following rate schedule:
    12¢ per minute for the first 15 minutes
    9¢ per minute for the next 10 minutes, and
    6¢ per minute for any time over 25 minutes. Find the model the company's computer uses to calculate the charges per call.

From Mathematics to English

25. Make up three different application problems that can be solved by using systems of equations.

26. What method did you use to solve Exercise 18? Why?

27. What is your favorite method of solving linear systems of equations? Why?

28. Why should you know more than one method for solving systems of equations?

29. Describe how you know whether a system of equations has a solution.

30. After reading this section, make a list of questions that you want to ask your instructor.

31. Continue in your daily journal and make an entry. In addition to your normal entry on thoughts about the mathematics in this section, list at least two positive comments about what you have learned about this topic.

32. In paragraph format, summarize the material in this section of the text in your daily journal.

33. Describe how your classroom instructor made this topic more understandable and clear.

34. After reading the text and listening to your instructor, what do you not understand about this topic?

Explorations

35. Below are the equations of a circle and an ellipse. Do they intersect? If so, where?

$$(x-5)^2+(y-4)^2=4$$
$$\frac{(x-1)^2}{38}+\frac{(y-4)^2}{9}=1$$

36. Solve the following system.

$$\begin{aligned} 2v+w-x+y-z&=4\\ v+w+x+y+z&=13\\ v-2w+3x-7y+3z&=-12\\ 5w-y&=0\\ 5v+4w&=x+y+z \end{aligned}$$

CHAPTER THIRTEEN TEST

1. Solve the system graphically (Show your work.)
$$y = 2x + 5$$
$$y = -3x - 2$$

2. Solve the system graphically (Show your work.)
$$3x - 2y = -22$$
$$-x + y = 11$$

3. Solve the system by showing the solution on a coordinate plane.
$$y > 2x - 5$$
$$y < -x + 3$$

4. Does the system
$$2x - 9y = -4$$
$$-10x + 43y = 16$$
have a solution? If no, why? If yes, why?

5. Solve by *addition* (Show your work.)
$$5x - 7y = 8$$
$$-2x + y = -5$$

6. Solve by *substitution* (Show your work.)
$$-3x + 5y = -11$$
$$2x - y = 5$$

7. Solve by *Cramer's Rule* (Show your work.)
$$3x - 5.7y = 2.1$$
$$9.6x + 4.2y = -0.6$$

8. Solve by *Matrices* (Show your work.)
$$-5x + 3y + 9z = 4$$
$$x - 6y - 8z = 2$$
$$3x + y - z = 0$$

9. Buckeye Central High School (BC) has one more class period per day than does Big Walnut High School (BW). If BC periods are 38 minutes and BW 44 minutes, and the total time in school for the schools together is 612 minutes per day, how many periods does each school have per day?

10. A rectangle with a length 7 inches less than twice its width has an area of 72 square inches. Find the length and width.

11. In preparing for the wedding of Jennifer and Bill, Nancy decided that she would need 12 pounds of mixed nuts for the reception. The caterer recommended that she buy 3 times as many pounds of roasted pecans as brazil nuts. The father of the bride demanded that there be twice as many pounds of filbert nuts as brazil nuts. How many pounds of each kind of nut should Nancy buy?

CHAPTER FOURTEEN
INTRODUCTION TO THE ANALYSIS OF THE LOGARITHMIC FUNCTION

14.0 Introduction

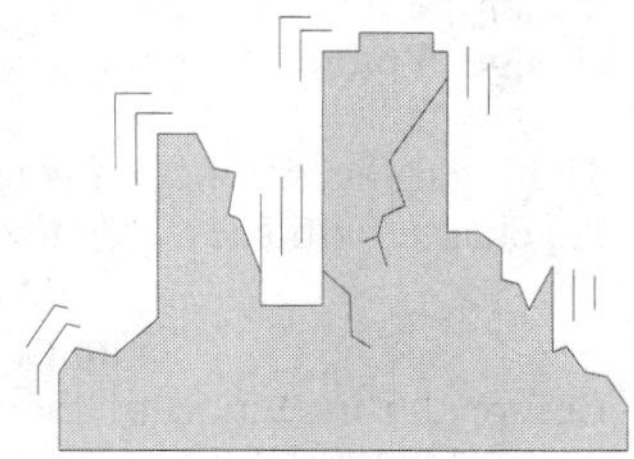

The Earthquake Situation

Below is a table showing the Richter Scale value for earthquakes of known intensity. The intensity (*I*) is expressed in terms of how many times more intense than an earthquake that can barely be detected by humans. For example, an earthquake of intensity 5000 means 5000 times more intense than one that can barely be discerned by humans -- one of minimal intensity.

Table 14.0.1

I	1	1 000	50 000	100 000	1 000 000	10 000 000	50 000 000	80 000 000
R	0	3	4.70	5	6	7	7.70	7.90

Hopefully, you can see that the data relationship is a function. The question is, what function? To get an idea, perhaps if you graph the data relationship you will recognize the shape as a known function. Below is the graphical representation of the data as a scatter plot and as data connected. Is the shape of the data

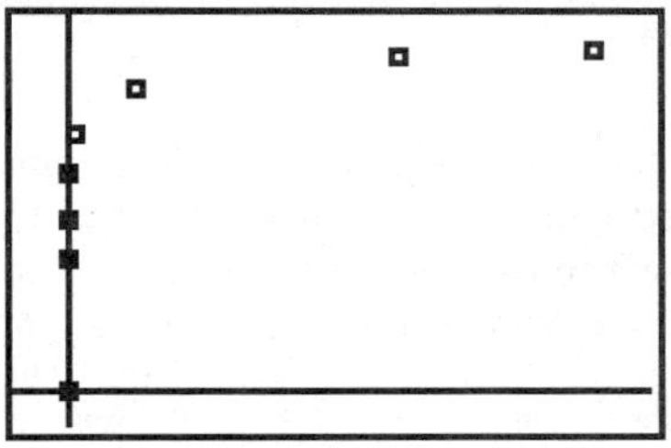

Figure 14.0.1a

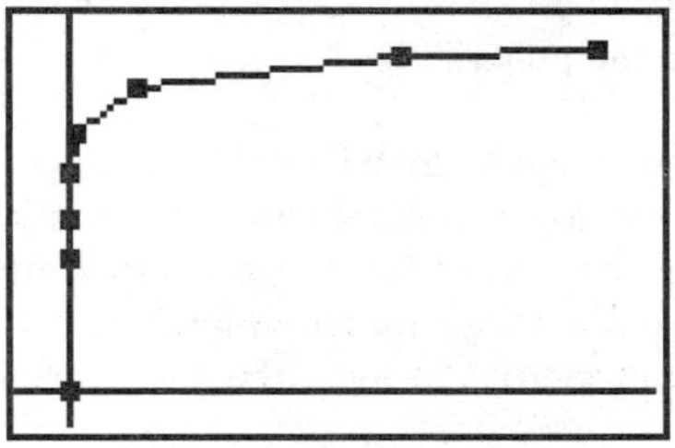

Figure 14.0.1b

relationship like any function studied thus far? Is the data relationship linear? Absolute value? Quadratic? Exponential? Rational? Square root? Sine or cosine? While you may suspect that it looks like one piece of a rational function, there is no horizontal asymptote. The data relationship is a new kind of function. It is called a logarithmic function -- the focus of study in this chapter. Richter's model for the number that is an indication of the intensity of an earthquake is $R = \log I$, where *I* is the ratio of the intensity of the measured quake to that of one of minimal intensity.

Logarithmic functions are closely related to the exponential functions developed in Chapter Seven. You may need to review the behavior of the exponential function as well as the properties of exponential expressions before continuing in this chapter. This material is needed to develop the logarithmic function as well as the properties of logarithms.

You already know the calculator can raise a positive number to any exponent. In this chapter, you will reverse this process, and you will learn to find the exponent on any positive number that yields any other desired positive number. For example, the exponent on 10 that yields 126.59 is 2.1023994. That is, $10^{2.1023994} = 126.59$. The exponent number 2.1023994 is called the common logarithm of 126.59. These numbers (exponents) are used throughout this chapter.

In Chapter Seven, you solved exponential equations by using technology based methods; logarithms will allow you to solve exponential equations using an analytical method. In addition to solving exponential equations, the properties of logarithms will also give you an analytic method for solving a limited number of equations containing logarithms. As always, the technology based methods for solving equations will allow you to solve logarithmic equations that are a little more complicated than the equations solved with the properties of logarithms. The technology based methods for solving these new equations will not change. You may use the numeric, trace, intersection, or zeros methods to solve logarithmic equations.

The logarithmic function is used in many scientific fields. The applications that are included at the end of the chapter will show you that even a simple logarithmic function can be used in a significant way.

The Logarithmic Function in Action

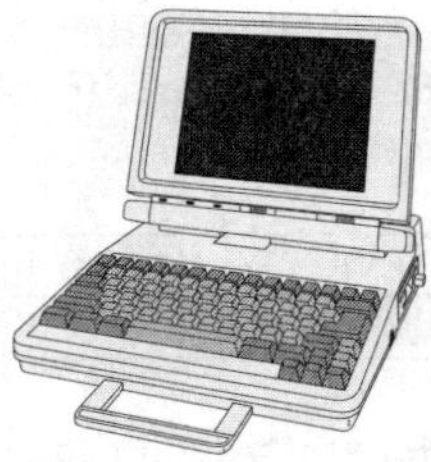

Before the invention of the hand-held calculator, the desktop calculator, or the computer, how did practicing engineers, scientists, or business persons calculate products, quotients, square roots, or powers? Of course, without technology they used pencil and paper and a lot of time. Suppose you have to calculate 4260^4 as an astronomer might have to do. Try it without a calculator. Multiply 4260 times 4260 times 4260 times 4260. If you think this is a long problem, think about what the business person had to do to find a savings account balance. Part of the calculation looks like 1.006666667^{60}, a calculation that no one ever really would want to calculate. Do you know how to find $\sqrt[3]{783.255}$ or $\sqrt{283}$? While you really can calculate 1.006666667^{60} if you had the time, many of us no longer know how to solve problems like $\sqrt[3]{783.255}$ or $\sqrt{283}$ using paper and pencil.

How did professional people do calculations similar to those above? With logarithmic tables (the numeric representation of the logarithmic function) complicated calculations can be reduced to a four to five minute process. In the section on the properties of logarithms, you will see properties that will make nearly impossible calculations rather simple. Below is an example; however, you may not understand it until you study the properties of logarithms.

Find 1.006666667^{60}: let $N = 1.006666667^{60}$

take the log of both sides $\log N = \log 1.006666667^{60}$

use the power property $\log N = 60 \cdot \log 1.006666667$

find the log $\log N = 60(0.0028856884)$

multiply $\log N = 0.173141304$

find the number whose log is 0.173141304 $N = 1.48984574$ (1.006666667^{60} is 1.48984574)

This five-step process gives the answer and is a lot faster than multiplying 1.006666667 times itself 60 times. You may want to check the answer and smile when you realize how fast you can do the problem with a calculator.

14.1 The Logarithmic Function

The basic exponential function studied in Chapter Seven is $y = b^x$, where b is greater than 0 and $\neq 1$. You may recall the base is positive to avoid non-real values for the function. The base is not 0 because it simplifies to a constant function and is undefined when $x = 0$. You may want to review Section 7.1 before proceeding.

Remember the exponential function!

A function related to the exponential function is the logarithmic function, which is derived by interchanging the variables x and y in the basic (parent) exponential function. This is certainly an unusual endeavor because the variable x has always been the variable in the function and the variable y has always represented the value of the function. To interchange the variables doesn't seem to make mathematical sense. However, on closer inspection, you will find it is an acceptable move to make.

The parent exponential function has one and only one function value for every x in the domain; if it didn't, it would not be a function. Recalling the graph of the basic (parent) function, it is also true that for every value of the function there is one and only one corresponding value of x. Thus, interchanging the variables in the exponential function causes the new statement to satisfy the definition of a function as well! From $y = b^x$, interchanging the variables yields $x = b^y$. This process is a part of **finding the inverse** of a function.

> **Definition of the Logarithmic Function**
>
> $x = b^y$ is the logarithmic function in exponential form where $b > 0$ and $\neq 1$, and y is called the logarithm of x base b.

By interchanging x and y, the domain of the exponential function (real numbers) becomes the range of the logarithmic function and the range of the exponential function $y = b^x$ becomes the domain of the logarithmic function $x = b^y$. The equation appears strange because nearly all other functions studied have had x as the variable in the function and y as the function. Well, this is no different. The variable y still represents the logarithmic function and x is the variable in the function. What is strange is that y is not isolated on one side of the equal sign with an expression containing x on the other side of the equal sign. This is not totally unusual. The linear function was expressed like this in Chapters Twelve and Thirteen. For example, the linear function $y = -2x + 3$ was expressed as $2x + y = 3$. You learned how to express $2x + y = 3$ in standard function form, and the logarithmic (log) function $x = b^y$ can also be put in standard function form. Before this is developed any further, a study of the function as it stands follows.

Example 1: When $x = 81$ and $b = 3$, find y. In new terminology, this is finding the logarithm of 81 base 3.

Solution: From the statement $x = b^y$, replace x with 81 and b with 3 and find y.

$$x = b^y$$

$$81 = 3^y$$

What exponent on 3 will give the number 81? By trial and error, $3^2 = 9$, $3^3 = 27$, $3^4 = 81$. Since 4 is the exponent on 3 that results in 81, 4 is the logarithm of 81, base 3.

⁂

Example 2: When $x = 100$ and $b = 10$, find y. In the new terminology, this example shows how to find the logarithm of 100 base 10.

Solution: From the statement $x = b^y$, replace x with 100 and b with 10 and find y.

$$x = b^y$$

$$100 = 10^y$$

What exponent on 10 will give 100? Of course, 10^2 is 100; thus, the logarithm of 100 base 10 is 2.

⁂

Example 3: Develop a numeric representation of the logarithmic function $x = 2^y$.

Solution: The simplest approach is to replace y with any number in the range and calculate x. The normal range is any real number. From $x = 2^y$, create the table.

Table 14.1.1

x	.125	0.25	0.5	1	2	4	8	16	32	64
y	–3	–2	–1	0	1	2	3	4	5	6

Notice that the above domain contains only positive numbers. Further, no matter what base 2 is raised to, the value of x will always be positive.

Upon inspection of the values in the numeric representation, the function values are *increasing* as x gets larger.

⁂

Example 4: Graph the logarithmic function $x = 2^y$.

Solution: From the numeric representation of the function above, plot the points and connect the points with the suggested geometric shape. The window below is [–5, 80] by [–8, 8].

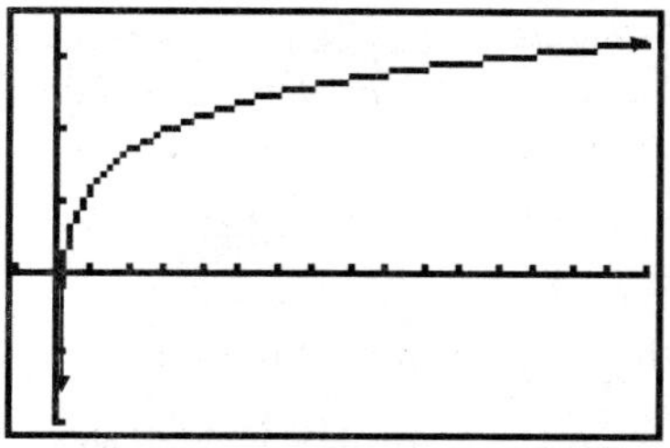

Figure 14.1.1

⁂

Example 5: Graph the logarithmic function $x = \left(\frac{1}{2}\right)^y$.

Solution: Make a numeric representation from the symbolic representation and plot the points in the table. As in the previous table, choose values for y and calculate corresponding values for x.

Table 14.1.2

x	.0625	0.125	0.25	0.5	1	2	4	8	16	32
y	4	3	2	1	0	–1	–2	–3	–4	–5

Plot the points and connect them with a smooth curve as shown below on the window [–5, 40] by [–6, 6].

Figure 14.1.2

Recall that the exponential function with a base less than 1 is decreasing over the normal domain. Likewise, the logarithmic function with a base less than 1 is decreasing (*see above*). Similarly, the exponential function with a base larger than 1 is increasing over the normal domain, like the logarithmic function with a base of 2 in Example 3. Another observation about the graph of the logarithmic function is that it has a vertical asymptote at the beginning of its domain (note the graphs above).

⁂

Most applications of logarithmic functions use a base that is larger than 1. In this text, most of the logarithmic functions studied will also contain a base that is larger than 1.

You may be wondering how the function $x = b^y$ can be put in standard function form. To isolate *y*, a new notation is needed. *From the definition above, the number y is called the logarithm of x base b.* The new notation is the symbolic form of this English statement.

> **Standard Form of the Logarithmic Function**
>
> From $x = b^y$, $y = \log_b x$. Read "*y* is the log of *x*, base *b*."
>
> Log is an abbreviation of logarithm.
>
> $y = \log_b x$ is logarithmic form of the logarithmic function $x = b^y$.

You still want to think of the statement $y = \log_b x$ as "y is the exponent on b that yields x."

Example 5: Find $\log_5 125$.

Solution: Find an exponent on 5 that will yield a value of 125. Since 5^3 is 125, 3 is the logarithm of 125 base 5. That is, $\log_5 125 = 3$.

⁂

The calculator has the base 10 logarithmic function built-in. When the base is 10, the base in the logarithmic form of the function is NOT written. That is, $y = \log_{10} x$ is written $y = \log\ x$. The calculator key "log" is the logarithmic function, base 10. This key is used to find the base 10 logarithm of any number in the domain of the function $y = \log\ x$. Logarithms with a base of 10 are called **common logarithms**.

Example 6: Graph the parent (basic) logarithmic function $y = \log\ x$.

Solution: On a viewing window such as $[0,500]$ by $[-6,6]$, let the calculator graph the function. You should expect it to look like the graph of the logarithmic function graphed earlier without the calculator.

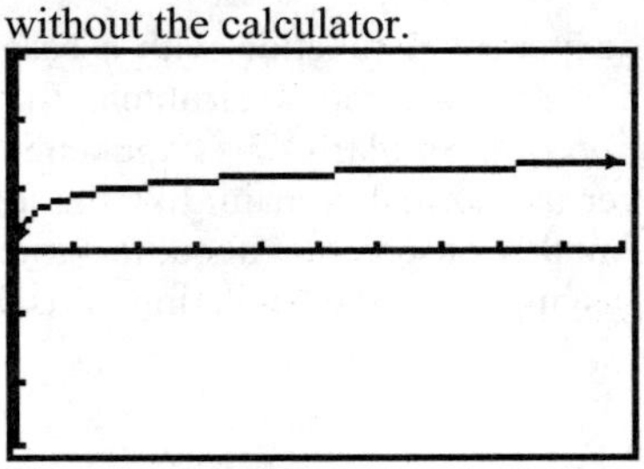

Figure 14.1.3

Note the graph is increasing on its domain, and at the beginning of the domain is a vertical asymptote. As you can see from the graph, the log function does not rise very fast for large *x*; that is, the rate of change is quite small for large *x*.

⁂

Transformations of the Logarithmic Function

Do the concepts of horizontal and vertical translations apply to the logarithmic function? Does multiplying the log function by -1 reflect the graph of $y = \log\ x$ about the x-axis? Does multiplying by a real number stretch or shrink the parent graph? To answer these questions, several log functions are graphed below to demonstrate these characteristics. You are encouraged to explore these ideas on your own.

Below are the graphs of $y = \log\ (x)$ and $y = \log(x) + 3$ and $y = \log(x) - 5$. The parentheses around x are to emphasize that a number is being added or subtracted to the parent function, not to the x.

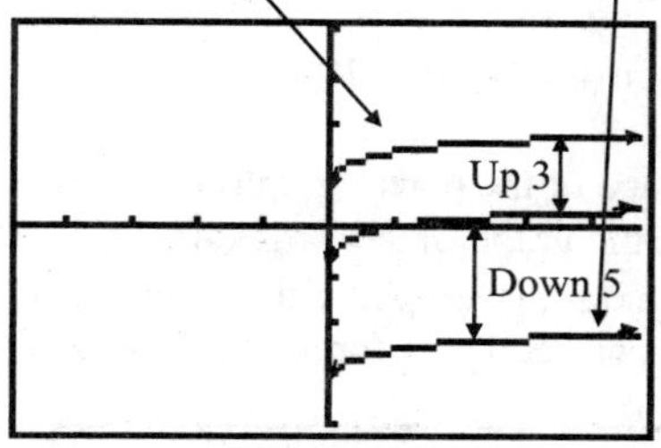

Figure 14.1.4

As you might have suspected, the idea of vertical translation does apply. Check this with your trace key. Jump between functions and note that $y = \log(x)$ and $y = \log\ (x) + 3$ are exactly 3 units apart for every x value you try.

To see if horizontal translations apply, below are the graphs of $y = \log(x)$ and $y = \log(x+3)$ and $y = \log(x-5)$.

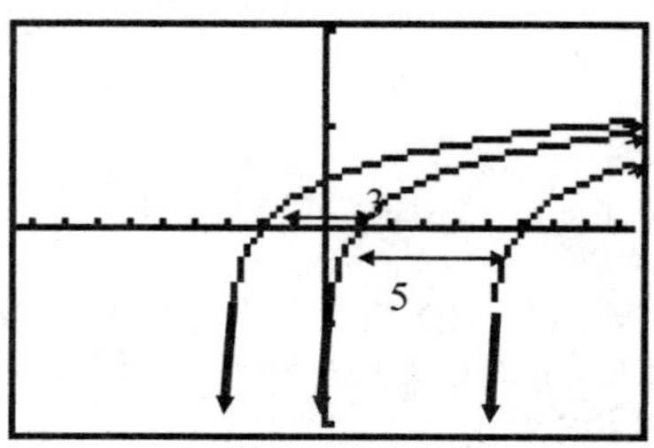

Figure 14.1.5

Hopefully, it is demonstrated in Figure 14.1.5 that the concept of horizontal translation does apply. That is, when x is replaced with $x + 3$, the graph of the parent log function is translated left 3 units. When x is replaced with $x - 5$, the graph of the parent log function is translated right by 5 units. This includes the vertical asymptote; the new vertical asymptote has been moved right 5 units. This graph also shows the domain is $(5, \infty)$.

The graphs of $y = 4\log(x)$ and $y = -\log(x)$ below demonstrate that stretches and reflections also apply. Check these with your calculator and the trace key.

$y = 4\log(x)$

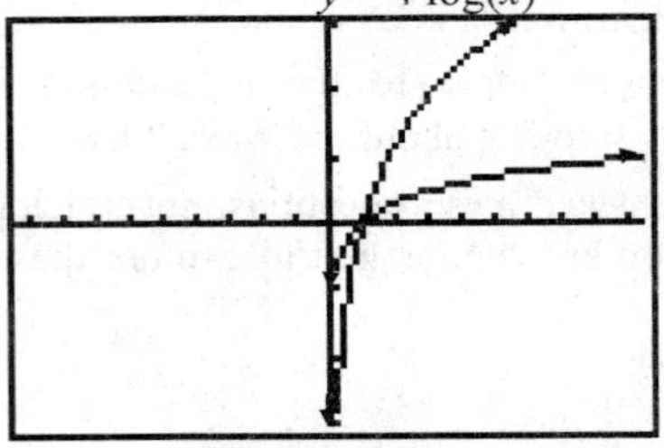

Figure 14.1.6a

$y = -\log(x)$

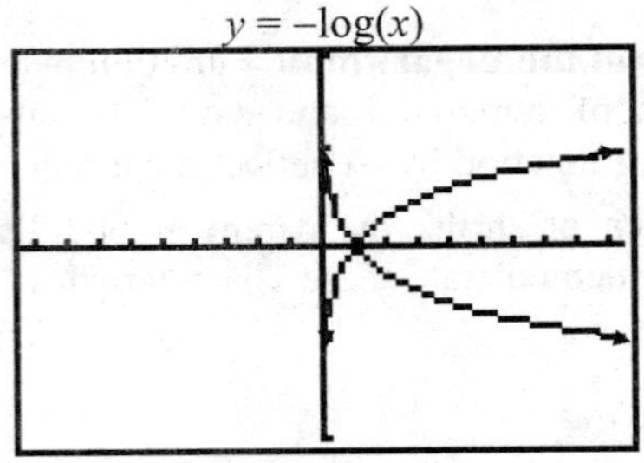

Figure 14.1.6b

Example 7: Graph the function $y = \log(x+2)+5$.

Solution: If the general shape of the parent graph is known, move it to the left by 2 units and up by 5 units to get the graph of $y = \log(x+2)+5$. Drawing a vertical asymptote at –2 may be used as an aid in graphing the function. The calculator can be used to verify that the graph is correct. Don't forget, the base on the function is 10.

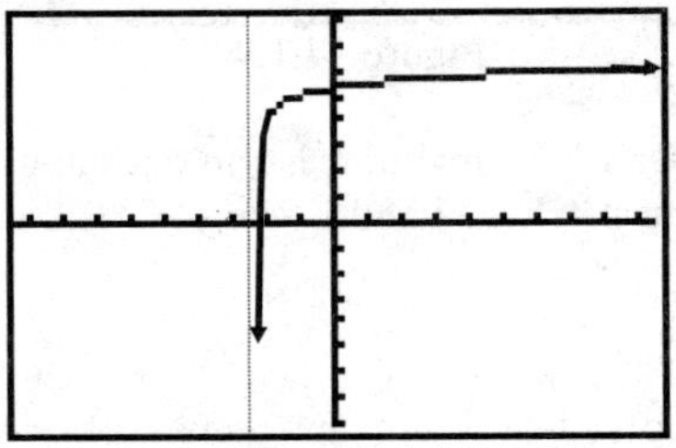

Figure 14.1.7

Example 8: Find the common log of 126.59.

Solution: Enter log 126.59 <ENTER> on the calculator to find the exponent on 10 that yields 126.59. The log of 126.59 is 2.1023994; that is, 10 raised to the exponent of 2.1023994 is 126.59.

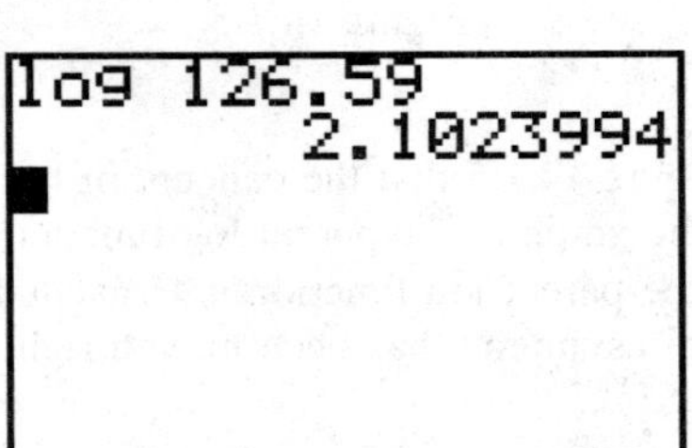

Figure 14.1.8

In addition to using the calculator to find common logarithms, it is also used to do the reverse process -- to find an **antilogarithm**. This means the common logarithm is known and the number whose logarithm is known is needed. If this process is called for, it can be written in the following way:

If $\log x = 1.362$, what is x?

However, it has been a common practice for many years to write it as:

antilog 1.362

Read as "the antilog of 1.362."

The notation "antilog 1.362" is the number whose logarithm is 1.362. As with common logarithms, the base is not written. The notation "antilog 1.362" means "antilog_{10} 1.362."

Example 9: Find antilog 1.362.

Solution: Since a common logarithm is the exponent on 10 that yields the desired number, raise 10 to the exponent of 1.362. This can be done by entering the keystrokes: 10 ^ 1.362 <ENTER>. It can also be done by entering the keystrokes:

2nd log 1.362 <ENTER>. The antilog 1.362 = 23.01441817, or

$10^{1.362} = 23.01441817$.

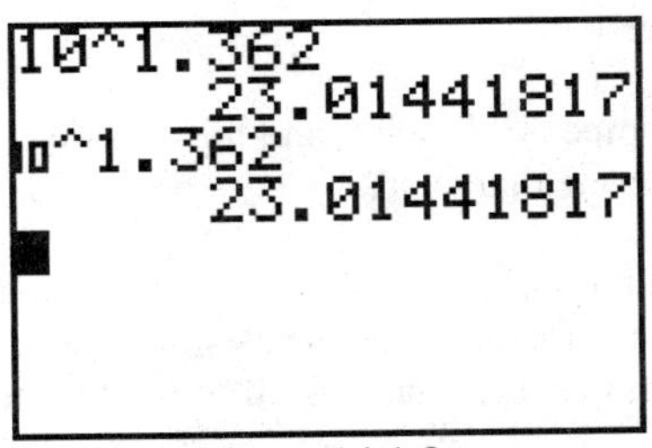

Figure 14.1.9

Example 10: If $\log x = -1.894$, find x.

Solution: The problem asks for the antilog –1.894. Enter the keystrokes:

2nd log (–) 1.894 <ENTER>. The value of x is 0.0127643881.

That is, $10^{-1.894}$ is 0.0127643881.

Figure 14.1.10

14.1 EXERCISES

Many times you can work the exercises by algebraic, numeric, or graphical methods. You can learn about mathematics no matter which method you use. Part of the learning process is for you to decide which method is best for you to use. Please read the text carefully before you do the exercises. Read the **next assigned section** after you finish the assignment below. All numeric answers must have an error ≤ 0.01.

A Quick Look To The Future

-3. Find log (5×13) and find log 5 + log 13.

-2. Find b, if $\log_b x = 1$.

-1. Do you know what a decibel is?

Previous Topics Revisited

1. At a social gathering of 60 people, the ratio of women to men is 3 to 2. Find the number of females present.

2. The inside circumference of a pipe is 33 inches and the outside circumference is 37.714. What is the thickness of the pipe and the inside diameter?

3. When Jupiter and Mars are on the same side of the sun and in line with it, the distance between them is 341,770,000 miles. When they are on opposite sides of the sun and in line with it, the distance between them is 624,856,000 miles. How far is each from the sun?

4. Solve the system.
$$\begin{aligned} 2x - 5y &= -18 \\ -3x + 4y &= 20 \end{aligned}$$

5. Solve the system.
$$\begin{aligned} x + y + z &= -3 \\ 3x - y &= -2 \\ 4x - 3z &= -1 \end{aligned}$$

6. Solve by any analytical method. Check graphically.
$$\begin{aligned} -2x + 7y &= -8 \\ 3x + y &= 12 \end{aligned}$$

New Topic Exercises

7. If an earthquake measures 100,000 times as intense as one you can barely feel, what is the Richter Scale measurement for this quake? (See the Earthquake Situation.)

8. If an earthquake measures 100,000,000 times as intense as one you can barely feel, what is the Richter Scale measurement for this quake?

9. If an earthquake measures 4.5 on the Richter Scale, how much more intense is it than a quake that can barely be felt?

10. If an earthquake measures 8.5 on the Richter Scale, how much more intense is it than a quake that can barely be felt?

Calculate the following logarithms.

11. $\log_4 1,\ \log_4 4,\ \log_4 16,\ \log_4 64,\ \log_4 256,$ and $\log_4 1024$

12. $\log_5 \frac{1}{25},\ \log_5 \frac{1}{5},\ \log_5 1,\ \log_5 5,\ \log_5 25,$ and $\log_5 625$

13. $\log_{\frac{1}{2}} \frac{1}{4},\ \log_{\frac{1}{2}} \frac{1}{2},\ \log_{\frac{1}{2}} 1,\ \log_{\frac{1}{2}} 2,\ \log_{\frac{1}{2}} 4,\ \log_{\frac{1}{2}} 8,$ and $\log_{\frac{1}{2}} 16$

14. $\log_{\frac{1}{4}} 1,\ \log_{\frac{1}{4}} 4,\ \log_{\frac{1}{4}} 16,\ \log_{\frac{1}{4}} 64,$ and $\log_{\frac{1}{4}} 256$

15. $\log_2 \frac{1}{8},\ \log_2 \frac{1}{4},\ \log_2 \frac{1}{2},\ \log_2 1,\ \log_2 2,\ \log_2 4,\ \log_2 8,\ \log_2 16,\ \log_2 32,$ and $\log_2 128$

16. $\log_3 \frac{1}{9},\ \log_3 \frac{1}{3},\ \log_3 1,\ \log_3 3,\ \log_3 9,\ \log_3 27,$ and $\log_3 243$

17. log 0.01, log 0.1, log 1, log 10, log 100, log 1000, log 10000, and log 100000

18. log 50 + log 10 and log 500

19. log 20 + log 30 and log 600

20. log 3 + log 4 and log 12

21. log 5 + log 6 and log 30

22. log 7 + log 8 and log 56

Find the antilogarithms in Exercises 23 - 26.

23. antilog 1, antilog 10, antilog 100, antilog 1000, antilog 10000, and antilog 100000

24. antilog –4, antilog –3, antilog –2, and antilog –1

25. antilog 3.45, antilog 34.5, antilog 345, antilog 3450, antilog 34500, and antilog 345000

26. If $\log x = 2.915$, find x.

27. Is the function $y = \log_3(x+2)$ increasing or decreasing on its domain?

28. Is the function $y = \log_5(x+2)$ increasing or decreasing on its domain?

29. Is the function $y = \log_{\frac{1}{2}}(x+2)$ increasing or decreasing on its domain?

30. Is the function $y = \log_{\frac{1}{4}}(x+2)$ increasing or decreasing on its domain?

31. What is the domain of $y = \log(x+4) - 3$?

32. Where is the vertical asymptote for the graph of $y = \log(x+4) - 3$?

33. What is the domain of $y = \log(x-5) + 2$?

34. Where is the vertical asymptote for the graph of $y = \log(x-5) + 2$?

35. What is the maximum value of the function $y = 2 \log(x + 5) - 3$?

36. What is the minimum value of the function $y = -3 \log(x - 1) + 4$?

37. Find the zeros of the functions
$y = \log_2 x$, $y = \log_3 x$, $y = \log_4 x$, $y = \log_5 x$, $y = \log_6 x$, and $y = \log_{13} x$.
(Technology may be a poor choice for trying to answer this question.)

38. Find the zeros of the functions $y = \log_{\frac{1}{2}} x$, $y = \log_{\frac{1}{3}} x$, $y = \log_{\frac{1}{4}} x$, and $y = \log_{\frac{2}{13}} x$.

39. Is the logarithmic function $y = \log x$ increasing faster near the beginning of the normal domain or near the end of the normal domain?

40. When are the functions in Exercise 37 negative?

Describe the transformations of the graph of $y = \log x$ that will result in the graph of the functions in Exercises 41 - 44.

41. $y = 2 \cdot \log(x+6) + 4$

42. $y = \log(x-3) + 2$

43. $y = \frac{1}{2}\log(x-5) + 1$

44. $y = -4\log(x+2) - 5$

45. What is the domain of $y = \log(-x)$?

46. Create any logarithmic function with a domain of $(3, \infty)$.

47. Create any logarithmic function that is decreasing.

48. Create any logarithmic function that is decreasing and has a vertical asymptote at -1.

49. Create any logarithmic function that has a zero when x is 3.

50. Create any logarithmic function that is decreasing and has a zero at −5.

51. Create any logarithmic function that passes through the point (3, 2).

52. What is the range of every logarithmic function studied in this section?

53. Does the graph of the logarithmic function $y = d\log(x+e)+f$ have a horizontal asymptote? If yes, where? (*d, e,* and *f* are real numbers and $d \neq 0$)

54. In the function $y = \log\, x$, what values of x make y a negative number?

55. In the function $y = \log\, x$, what values of x make y a positive number?

From Mathematics to English

56. Explain why the base of the logarithmic function cannot be a negative number.

57. In the function in Exercise 53, can $(x+e)$ be a negative number? Why?

58. In the function in Exercise 53, can y be a negative number? Why?

59. What happens when you try to find log (-5) on your calculator? Explain why you think the calculator does what it does.

60. After reading this section, make a list of questions that you want to ask your instructor.

61. Continue in your daily journal and make an entry. In addition to your normal entry on thoughts about the mathematics in this section, list at least two positive comments about what you have learned about this topic.

62. In paragraph format, summarize the material in this section of the text in your daily journal.

63. Describe how your classroom instructor made this topic more understandable and clear.

64. After reading the text and listening to your instructor, what do you not understand about this topic?

Explorations

65. Explain in detail how the graphs of $y = \log(x-e)$ and $y = \log(e-x)$ are related. Assume e is a positive constant.

66. Graph $y = 10^x$, $y = \log_{10} x$, and $y = x$ on the same window. Describe the relationship between the graphs of $y = 10^x$ and $y = \log_{10} x$.

67. a. Find $\log 2 + \log 3$ and find the logarithm of the product $\log (2\times3)$.

b. Find $\log 4 + \log 5$ and find the logarithm of the product $\log (4\times5)$.

c. Find $\log 6 + \log 7$ and find the logarithm of the product $\log (6\times7)$.

d. Find $\log 8 + \log 9$ and find the logarithm of the product $\log (8\times9)$.

e. Find $\log 42 + \log 93$ and find the logarithm of the product $\log (42\times93)$.

What conjecture can you make about the logarithm $\log g + \log h$.

68. a. Find $\log 2 - \log 3$ and find the logarithm of the quotient $\log (2/3)$.

b. Find $\log 4 - \log 5$ and find the logarithm of the quotient $\log (4/5)$.

c. Find $\log 6 - \log 7$ and find the logarithm of the quotient $\log (6/7)$.

d. Find $\log 8 - \log 9$ and find the logarithm of the quotient $\log (8/9)$.

e. Find $\log 47 - \log 83$ and find the logarithm of the quotient $\log (47/83)$.

What conjecture can you make about the logarithm $\log g - \log h$.

69. a. Find $2 \log 3$ and find $\log (3^2)$.

b. Find $4 \log 5$ and find $\log (5^4)$.

c. Find $6 \log 7$ and find $\log (7^6)$.

d. Find $8 \log 9$ and find $\log (9^8)$.

e. Find $13 \log 21$ and find $\log (21^{13})$.

What conjecture can you make about the logarithm $\log (g^h)$.

14.2 Properties of Logarithms

The logarithmic function is commonly expressed two ways: when the logarithmic function is written as $x = b^y$, it is in exponential form, and when it is written as $y = \log_b x$, it is in logarithmic form. As you will discover in the next section, simply converting the function from one form to the other will help to explain the properties of logarithms, and it will help solve simple logarithmic or exponential equations.

Any positive number can be written as a base to an exponent. For example, the number 1000 can be written as 10^3. The number 5.3 can be written as $10^{0.72427588696}$. The number 0.72427588696 is the common logarithm of 5.3. Recall that a logarithm of a number is an exponent on a base that equals the given number. The calculator can be used to find common logarithms. Thus, by this reasoning, *any positive number* can be written as 10 raised to the exponent that is the logarithm of the positive number. Of course, any base can be used, but the calculator can find logarithms base 10. Following is a development of the product property of logarithms. You may have already figured out the property if you worked the explorations in Section 14.1.

The positive number N can be written as b^P. The positive number M can be written as b^Q.

That is: $\quad N = b^P \qquad M = b^Q \qquad (1)$

Rewrite these statements in logarithmic form.

$$P = \log_b N \qquad Q = \log_b M \qquad (2)$$

Consider the product of two numbers $N \cdot M$:

$$N \cdot M = b^P \cdot b^Q \qquad \text{by substitution}$$

$$N \cdot M = b^{P+Q} \qquad \text{by the product property of exponents}$$

Rewrite this statement in logarithmic form.

$P + Q = \log_b N \cdot M$ and now use substitution from (2)

$$\log_b N + \log_b M = \log_b N \cdot M$$

The Product Property of Logarithms

$$\log_b N \cdot M = \log_b N + \log_b M$$

The logarithm of a product of numbers is the same as the *sum* of the logarithms of the numbers.

Below is a confirmation of the product property. The numeric representation of the logarithm of the *product* of 3 and x and the numeric representation of the *sum* of the logarithm of 3 plus the logarithm of x are shown in Figure 14.2.1.

	log(3x)	log 3 + log x
X	Y1	Y2
0	ERROR	ERROR
1	.47712	.47712
2	.77815	.77815
3	.95424	.95424
4	1.0792	1.0792
5	1.1761	1.1761
6	1.2553	1.2553

X=0

Figure 14.2.1

The numeric representations are equivalent. The conclusion is that the logarithm of the product $3x$ is equivalent to the sum of the logarithms of the factors 3 and x. That is, the log $(3x)$ is equivalent to $\log 3 + \log x$. Below is a second confirmation of the product property using the *product* $\log(4(x-1))$.

	log(4(x – 1))	log 4 + log (x – 1)
X	Y1	Y2
0	ERROR	ERROR
1	ERROR	ERROR
2	.60206	.60206
3	.90309	.90309
4	1.0792	1.0792
5	1.2041	1.2041
6	1.301	1.301

X=0

Figure 14.2.2

Example 1: Does the log of 100000 equal the log of 100 plus the log of 1000?

Solution: Since $100000 = 100 \cdot 1000$

$$\log 100000 = \log 100 \cdot 1000?$$

$$\log 100000 = \log 100 + \log 1000 \text{ by the product property}$$

$$5 = 2 + 3 \qquad \text{evaluate the logs}$$

Yes, the log of 100000 does equal the log of 100 plus the log of 1000.

Example 2: Express $\log 8 + \log n$ as a single logarithm.

Solution: By the product property, $\log 8 + \log n = \log(8n)$.

Below is the development of the quotient property of logarithms. The development is very similar to the product property.

The positive number N can be written as b^P. The positive number M can be written as b^Q.

That is: $N = b^P \qquad M = b^Q \qquad (1)$

Rewrite these statements in logarithmic form.

$$P = \log_b N \qquad Q = \log_b M \qquad (2)$$

Consider the quotient of the two numbers $\frac{N}{M}$:

$$\frac{N}{M} = \frac{b^P}{b^Q} \qquad \text{by substitution}$$

$$\frac{N}{M} = b^{P-Q} \qquad \text{by the quotient property of exponents}$$

Rewrite this statement in logarithmic form.

$$P - Q = \log_b \frac{N}{M} \qquad \text{and now use substitution from (2)}$$

$$\log_b N - \log_b M = \log_b \frac{N}{M}$$

The Quotient Property of Logarithms

$$\log_b \frac{N}{M} = \log_b N - \log_b M$$

The logarithm of a quotient of numbers is the same as the *difference* of the log of the numerator minus the log of the denominator.

Below are two confirmations of the quotient property: the numeric representation of the logarithm of the *quotient* of 3 divided by x, and the numeric representation of the *difference* of the logarithm of 3 minus the logarithm of x.

$\log\left(\frac{3}{x}\right)$ $\log 3 - \log x$

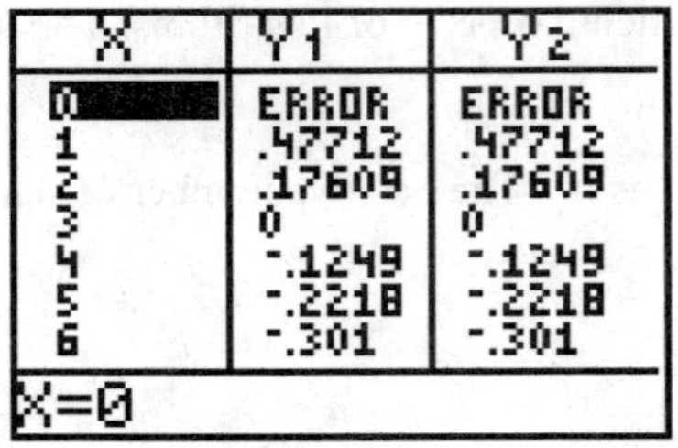

X	Y_1	Y_2
0	ERROR	ERROR
1	.47712	.47712
2	.17609	.17609
3	0	0
4	-.1249	-.1249
5	-.2218	-.2218
6	-.301	-.301

X=0

Figure 14.2.3

Below is a second confirmation of the quotient property using the *quotient* function $\log \frac{3}{(x+2)}$ and the *difference* function $\log 3 - \log(x+2)$.

$\log\left(\frac{3}{x+2}\right)$

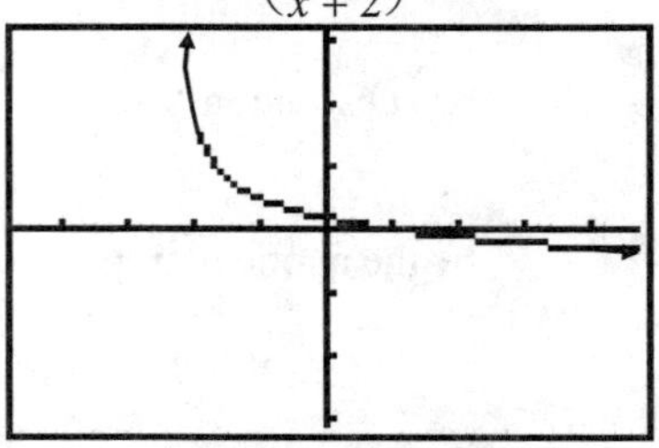

Figure 14.2.4a

$\log 3 - \log(x+2)$

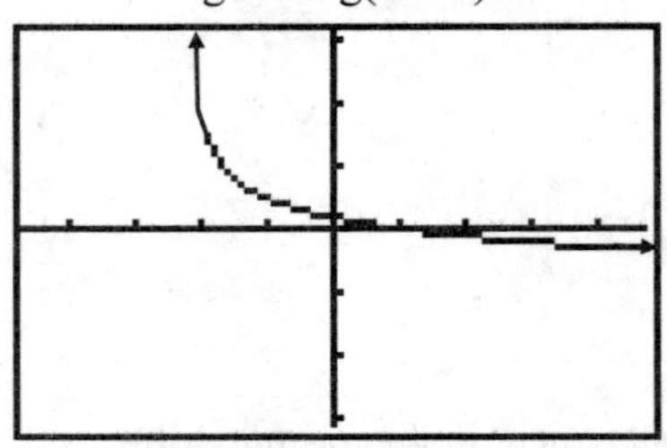

Figure 14.2.4b

Example 3: Express $\log 2x - \log 17$ as a single logarithm.

Solution: By the quotient property, the logarithm of a difference is equivalent to the logarithm of a quotient, $\log 2x - \log 17 = \log \frac{2x}{17}$. (Caution, just like reducing fractions may cause the domain to change, the properties of logarithms may also cause the domain to change. See Example 6.)

⁂

Example 4: Express $\log .01$ as the difference of two logarithms.

Solution: Since $\log .01 = \log \frac{1}{100}$ and $\log \frac{1}{100} = \log 1 - \log 100$, then, $\log .01 = \log 1 - \log 100$.

⁂

Example 5: Express $\log(x+2) - \log x + \log 5$ as a single logarithm.

Solution: The statement $\log(x+2) - \log x$ matches the condition on the quotient property; thus, $\log(x+2) - \log x = \log \dfrac{(x+2)}{x}$.

The statement $\log \dfrac{(x+2)}{x} + \log 5$ matches the condition on the product property; therefore, $\log \dfrac{(x+2)}{x} + \log 5 = \log 5\dfrac{(x+2)}{x}$.

$\log(x+2) - \log x + \log 5 = \log \dfrac{5(x+2)}{x}$. (Caution, see below.)

⁂

Example 6: Verify that $\log(x+2) - \log x + \log 5$ and $\log \dfrac{5(x+2)}{x}$ are not always equivalent functions.

Solution: Make numeric representations of the function $y = \log(x+2) - \log x + \log 5$ and the function $y = \log \dfrac{5(x+2)}{x}$. If the tables are identical, the functions are equivalent. You can see in Figure 14.2.5 that the tables are not the same. The difference in the

$\log(x+2) - \log x + \log 5$ ↘ $\log\left(\dfrac{5(x+2)}{x}\right)$ ↓

X	Y1	Y2
-4	ERROR	.39794
-3	ERROR	.22185
-2	ERROR	ERROR
-1	ERROR	ERROR
0	ERROR	ERROR
1	1.1761	1.1761
2	1	1

X= -4

Figure 14.2.5

tables is explained by the fact that the function $y = \log(x+2) - \log x + \log 5$ has a normal domain of $(0,\infty)$ and the function $y = \log \dfrac{5(x+2)}{x}$ has a normal domain of $(-\infty,-2) \cup (0,\infty)$. That is, since the log of a negative number is not a real number, x

must be greater than zero in the original function, but in the simplified function, when x is less than -2, the function $\frac{5(x+2)}{x}$ is positive; thus, the log function exists left of -2. The domain includes $(-\infty,-2)$, and the original function does not include $(-\infty,-2)$ in its domain. *If the domain of the simplified function is restricted to that of the original function, the functions are equivalent!* The conclusion is that the $\log(x+2) - \log x + \log 5$ is equivalent to $\log \frac{5(x+2)}{x}$ on the domain of the original function. The graphs below verify this reasoning.

$\log(x+2) - \log x + \log 5$

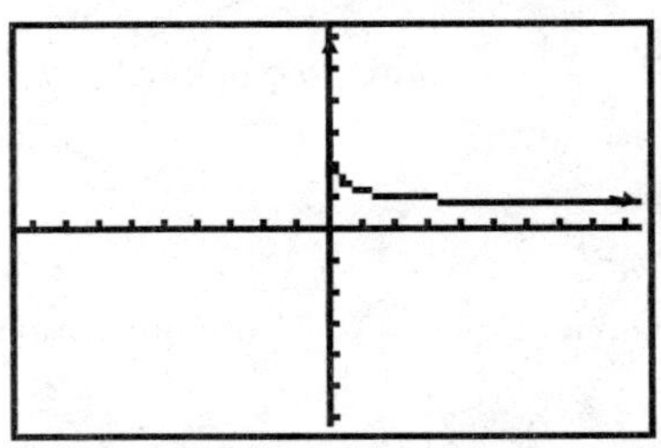

Figure 14.2.6a

$\log\left(\frac{5(x+2)}{x}\right)$

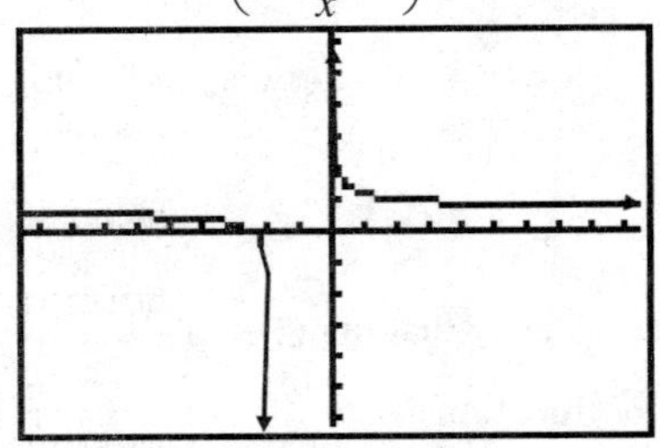

Figure 14.2.6b

⁂

The power property of logarithms can be developed using a method similar to the methods used for the product and quotient properties. A development of the power property follows.

The number N can be written as b^P.	$N = b^P$.	(1)
In logarithmic form it is	$P = \log_b N$.	(2)

Consider the number N^Q:

$N^Q = (b^P)^Q$	by substitution from (1)
$N^Q = b^{PQ}$	by the power property of exponents

Rewrite this statement in logarithmic form, treating PQ as the logarithm (the exponent).

$$PQ = \log_b N^Q \qquad \text{and now use substitution from (2)}$$

$$Q \cdot \log_b N = \log_b N^Q \qquad \text{interchange sides}$$

$$\log_b N^Q = Q \cdot \log_b N$$

The Power Property of Logarithms

$$\log_b N^Q = Q \log_b N$$

The logarithm of a number to an exponent is the same as the exponent times the log of the number.

Below is visual confirmation of the power property. The graph of the logarithm of the *power* function x^3 and the graph of the 3 *times* the logarithm of the linear function x are shown on the same window.

$\log(x^3)$ $\qquad$ $3\log x$

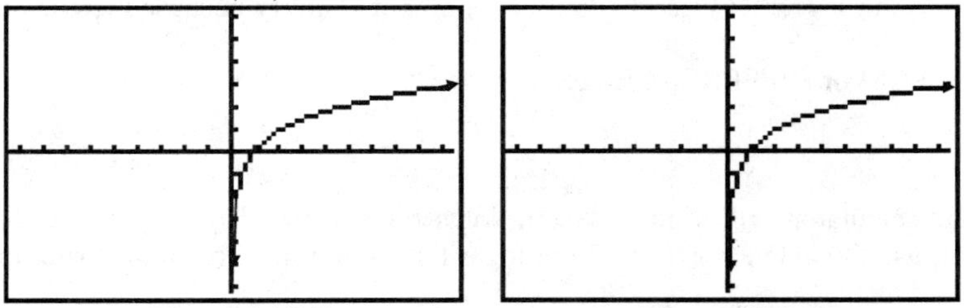

Figure 14.2.7a $\qquad$ Figure 14.2.7b

Since both graphs do not show when graphed at the same time, they have been split into two separate windows. If you try the above graphs on your calculator, you will see only one graph because the log x^3 is equivalent to $3\log x$! You can verify that both graphs are on your calculator by using trace and switching back and forth between functions. Since the graphs are identical, the values of the functions will remain the same.

Example 7: Find the value of the log 1000^4.

Solution: By the power property

$$\log 1000^4 = 4 \log 1000$$
$$= 4 \cdot 3 \qquad (10^3 = 1000)$$
$$= 12$$

⁂

Example 8: Express as a single logarithm: $2\log x + \log(x-3) - \log 6$.

Solution: Use the power property first $\log x^2 + \log(x-3) - \log\ 6$

Use the product property $\log x^2(x-3) - \log\ 6$

Use the quotient property $\log\ \dfrac{x^2(x-3)}{6}$

⁂

Example 9: Expand the expression $\log\ \dfrac{(x+4)^2(x-3)}{x^2}$.

Solution: Start with the product and the quotient properties:

$$\log(x+4)^2 + \log(x-3) - \log x^2$$

Now use the power property:

$$2\log(x+4) + \log(x-3) - 2\log\ x$$

Note: $\log(x+4)$ and $\log(x-3)$ do not simplify because there is no SUM or DIFFERENCE property.

⁂

As you know, the calculator will find common logarithms (base 10). Using the next property of logarithms, you can use the calculator to find the logarithm of any positive number to any positive base except 1.

Suppose the logarithm of a number N base a is called L. That is,

$$\log_a N = L \qquad (1)$$

Rewrite this in exponential form. $a^L = N \qquad (2)$

Since the numbers a^L and N are equal, the logarithms of these numbers must also be equal. The base can be any base, so choose base b.

$$\log_b a^L = \log_b N$$

Use the power property on the left side. $L \cdot \log_b a = \log_b N$

Divide both sides by $\log_b a$. $L = \dfrac{\log_b N}{\log_b a}$

Use substitution on L from (1) $\qquad \log_a N = \dfrac{\log_b N}{\log_b a}$

Change of Base Property

$$\log_a N = \frac{\log_b N}{\log_b a}$$

The log of a number N base a is the same as the log of the number N to any other base divided by the log of a to the other base. Since the calculator can find base 10 logs, the property can also be written as

$$\log_a N = \frac{\log N}{\log a}.$$

Below is visual confirmation of the change of base property. The graphs of the $\log_2 x$ and $\dfrac{\log x}{\log 2}$ are shown on the same window.

$\log_2 x$

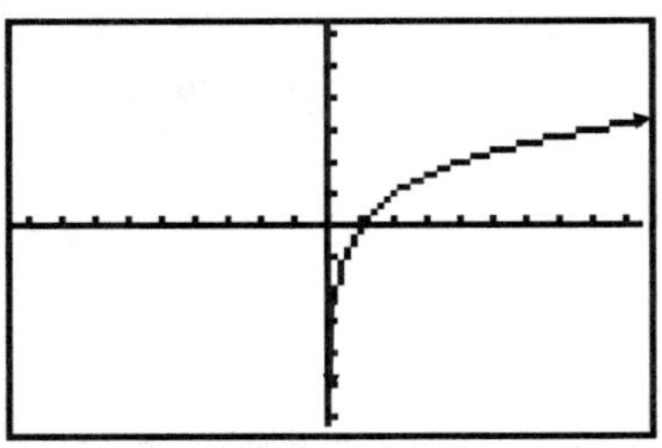

Figure 14.2.8a

$\dfrac{\log x}{\log 2}$

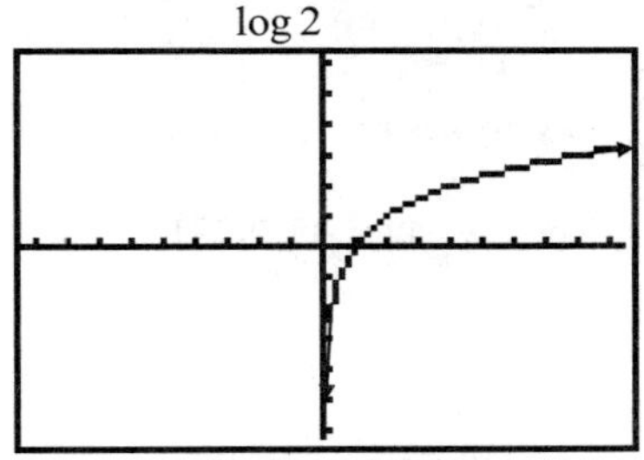

Figure 14.2.8b

Since both graphs do not show when graphed at the same time, they have been split into two separate windows. If you try the above graphs on your calculator, you will see only one graph because the $\log_2 x$ is equivalent to $\dfrac{\log x}{\log 2}$. You can verify that both graphs are the same by using trace and switching back and forth between functions for any value of x you choose. Since the graphs are identical, the values of the functions will remain the same. Using the table feature of the two functions will provide a more convincing argument.

Example 10: Calculate $\log_{17} 24$.

Solution: Estimation is helpful in knowing whether calculator errors are made. Since $17^1 = 17$ and $17^2 = 289$, the desired log must be between 1 and 2 because 24 is between 17 and 289. From the change of base property, $\log_{17} 24 = \dfrac{\log\ 24}{\log\ 17}$, or

1.121714. The keystrokes are log 24 ÷ log 17 < ENTER >

Example 11: Calculate $\log_3 25$.

Solution: Since $3^2 = 9$ and $3^3 = 27$, the log should be between 2 and 3. The change of base property allows for the calculation of $\log_3 25$. The keystrokes are: log 25 ÷ log 3 < ENTER > The $\log_3 25$ is approximately 2.929947.

Example 12: Calculate $\log_{\frac{1}{2}} 42$.

Solution: Since $\left(\dfrac{1}{2}\right)^{-5} = 32$ and $\left(\dfrac{1}{2}\right)^{-6} = 64$, the log should be between –5 and –6. The

keystrokes are log 42 ÷ log .5 < ENTER >

The $\log_{\frac{1}{2}} 42 = -5.392317$.

14.2 EXERCISES

Many times you can work the exercises by algebraic, numeric, or graphical methods. You can learn about mathematics no matter which method you use. Part of the learning process is for you to decide which method is best for you to use. Please read the text carefully before you do the exercises. Read the **next assigned section** after you finish the assignment below. All numeric answers must have an error ≤ 0.01.

A Quick Look To The Future

-2. Convert $\log_3 x = 5$ to exponential form.

-1. Do you suppose that the logarithmic function has any application in the real world?

Previous Topics Revisited

1. Where is the vertical asymptote for the graph of $y = 2 \cdot \log(x+3) - 5$?

2. Is the graph of $y = -3 \cdot \log_{\frac{1}{4}}(x-2) + 4$ increasing or decreasing over the normal domain?

3. What translations of the graph of $y = 2 \cdot \log\ x$ will give the graph of $y = 2 \cdot \log(x-7) - 35$?

4. You bought 50 – CD's and 25 - $3\frac{1}{2}$ inch diskettes for \$45. The CD's cost twice as much as the $3\frac{1}{2}$ inch diskettes. What is the price per disk for each type of discs?

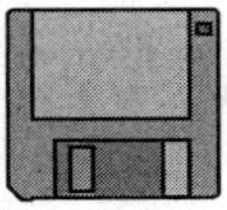

5. The perimeter of a rectangle is 28 meters and the area is 24 square meters. Find the length and width of the rectangle.

6. Find the value of the determinate: $\begin{vmatrix} 0.2 & 1.5 \\ -1.6 & 3.1 \end{vmatrix}$

New Topic Exercises

Convert to logarithmic form.

7. $1024 = 2^{10}$

8. $5^{x+1} = 17$

9. $x^7 = 19$

10. $3^{x-2} \cdot 3^{x+5} = 81$

11. $2 \cdot 6^{x+1} = 42$

Convert to exponential form.

12. $\log x = 14$

13. $3 \cdot \log x = 5$

14. $\log_2(x+3) + \log_2(x-5) = 42$

15. $\log_5(x+1) = \log_5(x-3) + 4$

16. $2\log(x+6) = 14$

17. $3\log_8(x-1) + 2\log_8(x+1) = 6$

Write the following expressions as a single logarithmic expression. Specify a domain where the given expression and the single logarithmic expression are equal.

18. $\log 7 + 2\log(x+4)$

19. $3\log(x-2) - \log(x+1)$

20. $2\log x - 4\log(x+3) + \log(x-1)$

21. $\log(2x-5) - 3\log(x+4)$

22. $\frac{1}{2}\log(x+7) + 2\log(x+1) - 3\log(x-4)$

23. $2\log(x+3) + 1$

Expand the following logarithmic expressions.

24. $\log\frac{(x+4)^2}{(x-5)}$

25. $\log\frac{x^2}{(x-1)^4}$

26. $\log\frac{(x+2)(x-1)^2}{(x+4)}$

27. $\log\frac{x^2(x+3)(x-4)}{(x+4)(x-3)}$

Calculate the logarithms in Exercises 28 - 33.

28. $\log_{25} 165$

29. $\log_2 1000$

30. $\log_{3.6} 225$

31. $\log_{0.3} 15$

32. $\log_{1.5} 12.7$

33. $\log_{12} 250$

34. Can your calculator find $\log_2(-15)$? Why?

35. What is the domain of the expression $\log(x+3)+\log(x-1)$?

36. Are the graphs of $\log(x-1)+\log(x+2)$ and $\log(x-1)(x+2)$ the same graph? Why?

37. What is the domain of $\log(2-x)$?

From Mathematics to English

38. Does $\log_2(x+3)-\log(x-1)$ simplify? Why?

39. Does $\dfrac{\log(x+2)}{\log(x+4)}$ simplify to $\log(x+2)-\log(x+4)$? Why?

40. Does $\log(x+5)\cdot\log(x-3)$ simplify to $\log(x+5)(x-3)$? Why?

41. What is the base in the logarithmic expression $\log x$?

42. Research and explain the LN key on your calculator.

43. Explain any difficulties you had understanding logarithm material.

44. After reading this section, make a list of questions that you want to ask your instructor.

45. Continue in your daily journal and make an entry. In addition to your normal entry on thoughts about the mathematics in this section, list at least two positive comments about what you have learned about this topic.

46. In paragraph format, summarize the material in this section of the text in your daily journal.

47. Describe how your classroom instructor made this topic more understandable and clear.

48. After reading the text and listening to your instructor, what do you not understand about this topic?

49. Give as many reasons as you can why $\log_1 25$ does not make sense.

Explorations

50. Develop a method for finding 12.72^4 using logarithms. (Hint: see properties.)

51. Develop a method for finding $\sqrt[5]{28}$ using logarithms. (Hint: see properties.)

52. Graph $y=2^x$, $y=\log_2 x$, and $y=x$ on the same window -- observe the graphs.
Next, graph $y=5^x$, $y=\log_5 x$, and $y=x$ on the same window -- observe the graphs.
Next, graph $y=7.3^x$, $y=\log_{7.3} x$, and $y=x$ on the same window -- observe the graphs.

Based on your observations above, what do you propose as the relationship between the graphs of $y=b^x$ and $y=\log_b x$?

14.3 Solving Logarithmic and Exponential Equations

The Problem with People

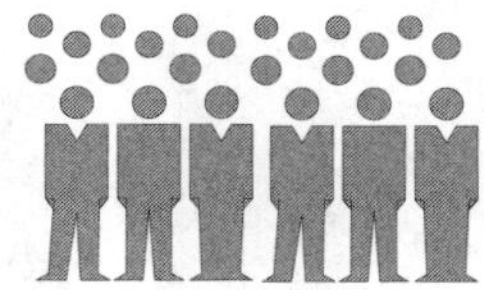

The function that models the number of people on earth developed in Chapter Seven is $P = P_0(1+r)^t$. Suppose you want to know how long it will take for 6.5 billion people on Earth in 2006 to become 10 billion at an average growth rate of 2% per year. The population model becomes the equation $10 = 6.5(1+.02)^t$. You solved equations like this in Chapter Seven by using one of the technology-based methods. Now, you can solve it analytically using the properties of logarithms from Section 14.2.

$10 = 6.5(1+.02)^t$	given
$10 = 6.5 \cdot 1.02^t$	simplification
$\dfrac{10}{6.5} = 1.02^t$	division property of equality
$\log\left(\dfrac{10}{6.5}\right) = \log\left(1.02^t\right)$	logs of equal numbers are equal
$\log 10 - \log 6.5 = t \log 1.02$	log division and power properties
$\dfrac{\log 10 - \log 6.5}{\log 1.02} = t$	division property of equality
$21.75 = t$	simplification

It will take 21.75 years for the 2006 population of 6.5 billion people to become 10 billion at an average growth rate of 2% per year.

The graphical method for solving equations will certainly work for solving logarithmic equations. You now know the nature of the graph of the logarithmic function $y = d\log_b(x+e)+f$; thus, you expect the logarithmic equation $d\log_b(x+e)+f=0$ to have one real solution, or the logarithmic function one zero. If the function related to the logarithmic equation gets more complicated than the above function, you may have to experiment with the graphing calculator to make certain you have a complete graph of the function.

The analytic method for solving exponential and logarithmic equations will make use of the properties from the last section. The examples below will demonstrate how the logarithmic properties are used to solve logarithmic and exponential equations.

Example 1: Find the exact solution to the equation $2^{x+3} = 15$.

Solution: Since the exact solution is desired, an analytical method may be called for. To solve the equation, one analytical method is to rewrite the equation in logarithmic form.

$2^{x+3} = 15$	given
$\log_2 15 = x + 3$	rewrite in log form and subtract 3 from both sides
$\log_2 15 - 3 = x$	the exact solution is $\log_2 15 - 3$

An approximate solution can be found with the following keystrokes:

log 15 ÷ log 2 − 3 < ENTER >

It is approximately 0.9068906. To verify that the answer is correct, graph the exponential function $y = 2^{x+3} - 15$ and find the zero.

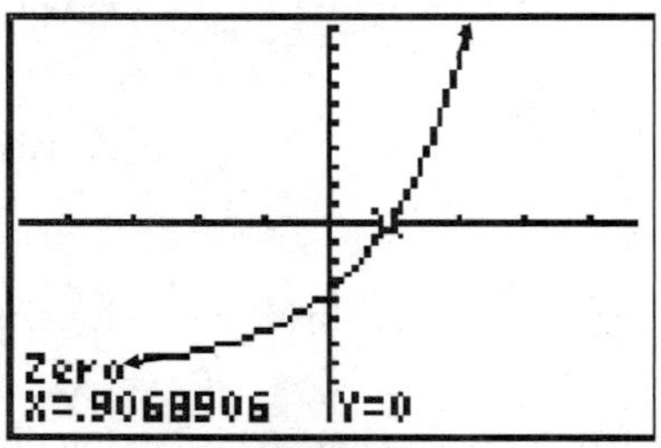

Figure 14.3.1

***

Example 2: Solve the equation $4\log(x+5) - 1 = 4.6$ with an error < 0.01.

Solution: Since an exact solution is not called for, below is the zeros method. Subtract 4.6 from both sides of the equation and graph the related function $y = 4\log(x+5) - 5.6$. The zero of the function is the solution to the equation.

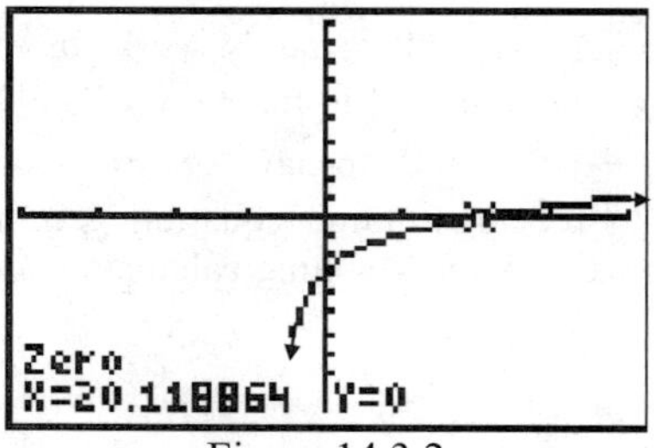

Figure 14.3.2

The solution to the equation is 20.119 with an error < 0.01.

***

Example 3: Find the exact solution to the equation $\log(x-3) + 5 = 7$.

Solution: Subtract 5 from both sides and then rewrite the equation in exponential form.

$$\log(x-3)=2$$
$$10^2 = x-3$$
$$100 = x-3$$
$$103 = x$$

⁂

Example 4: Find the exact solution to $\log_2(x+2)+\log_2(x-1)=2$.

Solution: Use the product property on the left side to yield:

$$\log_2(x+2)(x-1)=2$$

Rewrite the equation in exponential form:

$$2^2=(x+2)(x-1)$$

Square 2 and simplify the right side:

$$4=x^2+x-2$$

The equation is quadratic; solve by factoring:

$$0=x^2+x-6$$
$$0=(x+3)(x-2)$$

When x is –3 and 2, the product is zero; thus, the solutions to the equation are –3 and 2. However, the domain of the function $\log_2(x+2)+\log_2(x-1)$ is $(-2, \infty)$. This means that –3 cannot be a solution. The only solution is 2. Below is confirmation using the numerical method. Like the graphical method, it will not cause extraneous solutions because none of the log simplification properties are used, but the answers may be approximate.

Table 14.3.1

x	–1	0	1	2	3	4	5	6
$\log_2(x+2)+\log_2(x-1)-2$	-	-	-	0	1.32	2.17	2.81	3.32

The zero is 2 and the function is not real on the interval $(-\infty, 1]$.

⁂

Example 5: Solve the equation $\log_2(x+2)+\log_2(x-1)=2$ with an error < 0.01.

Solution: The exact solution is not called for, below is the intersection method for solving equations. Graph the related functions $y=\log_2(x+2)+\log_2(x-1)$ and $y=2$. To enter $\log_2(x+2)$ and $\log_2(x-1)$ on the calculator use the change of base property. The keystrokes are:

log (x + 2) ÷ log 2 + log (x − 1) ÷ log 2

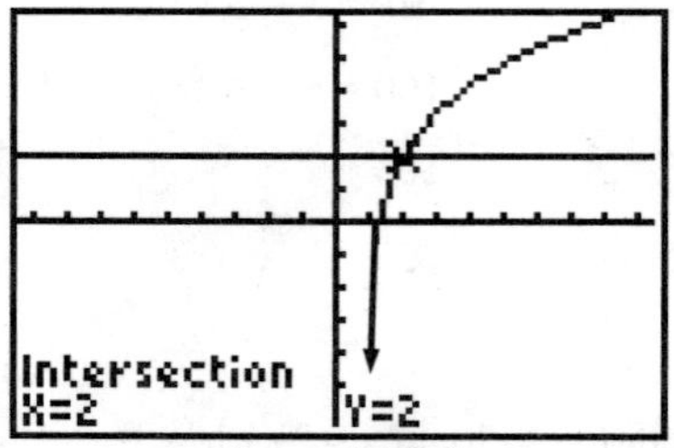

Figure 14.3.3

When x is 2, the functions are equal. The solution is 2 with an error < 0.01.

⁂

In the next problem, a single logarithm can be obtained on each side of the equation. When this is possible, the argument can be used that if the logarithms of two numbers are equal, the numbers must also be equal. This leads to an equation without logarithms. When the equation does not contain logarithms, conventional analytical methods can be used to solve the equation.

Example 6: Find the exact solution to the equation $\log(x+6) = \log(x-1) + \log(x+2)$.

Solution: Simplify the right side using the product property.

$$\log(x+6) = \log(x-1)(x+2)$$

(Note: the domain has now changed from $(1,\infty)$ to $(1,\infty)\cup(-6,-2)$. Any solution in the interval $(-6,-2)$ will be extraneous.)

Continuing, if the logarithms of two numbers are equal, the numbers must be equal.

$$x+6=(x-1)(x+2)$$

solve analytically

$$x+6=x^2+x-2$$

$$0=x^2-8$$

$$8=x^2$$

$$\pm\sqrt{8}=x$$

OR

$$x=\pm 2\sqrt{2}$$

If log a = log b, then a = b.

The apparent solution $-2\sqrt{2}$ cannot possibly be a solution because it is not in the domain of the function on the right side of the ***original*** equation. Since the graphing calculator will only plot points in the domain of the function being graphed, solving the original equation graphically will only show the correct solution, $2\sqrt{2}$. Below is the graph of the related logarithmic function used in the zeros method, $y=\log(x+6)-\log(x-1)-\log(x+2)$.

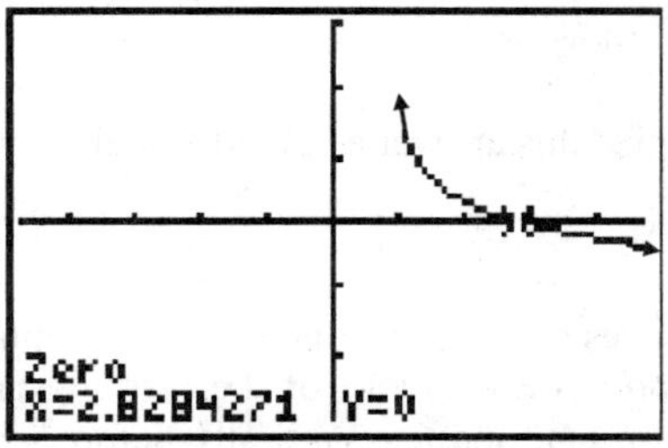

Figure 14.3.4

The solution to the equation is 2.828 with an error < 0.01.

⁂

Example 7: Find the exact solution to $3 \cdot 5^{x-2} = 18$.

Solution: To solve the equation analytically, divide both sides by 3.

$5^{x-2} = 6$ rewrite in log form

$x - 2 = \log_5 6$ add 2 to both sides

$x = 2 + \log_5 6$

The exact solution is $2 + \log_5 6$. The keystrokes for an approximate solution are:

2 + log 6 ÷ log 5. The approximate solution is 3.11328. Of course, the approximate solution can be found graphically or numerically.

⁂

A second analytic method can be used to solve exponential equations. It is based on the idea that if two numbers are equal, their logarithms are equal. See the next problem.

Example 8: Find the exact solution to the equation $5^{x-2} = 6$.

Solution: Given that the number 5^{x-2} and the number 6 are equal, the log of 5^{x-2} must equal the log of 6. That is:

If a = b, then log a = log b.

$\log 5^{x-2} = \log 6$ use the power property

$(x-2) \cdot \log 5 = \log 6$ divide by log 5

$x - 2 = \dfrac{\log 6}{\log 5}$ add 2 to both sides

$$x = \frac{\log\ 6}{\log\ 5} + 2$$

You should recognize this answer as the identical answer to the previous problem.

14.3 EXERCISES

Many times you can work the exercises by algebraic, numeric, or graphical methods. You can learn about mathematics no matter which method you use. Part of the learning process is for you to decide which method is best for you to use. Please read the text carefully before you do the exercises. Read the **next assigned section** after you finish the assignment below. All numeric answers must have an error ≤ 0.01.

A Quick Look To The Future

-1. Is an Earthquake that measures 6 on the Richter Scale twice as destructive as one that measures 3?

Previous Topics Revisited

1. Find $\log_{12} 125$.

2. Express $2\log(x+3) - 5\log(x-7)$ as a single logarithm and specify a domain that makes both expressions equal.

3. Expand $\log\frac{3(x+5)}{(x-2)^2}$ into three logarithms.

4. Where is the vertical asymptote for the graph of $y = 2\log(x+4) - 7$?

5. Is the graph of $y = 4\log_{\frac{1}{4}}(x-3) + 5$ increasing or decreasing throughout the normal domain?

6. In the part-time job you have at the CD and tape store, you sell old cassette tapes at \$4.99 each and new releases at \$8.99 each. After a good day, you notice that receipts for the day are \$991.71 and the count of tapes sold is 129. How many oldies and how many new releases did you sell?

New Topic Exercises

Solve the following equations by the method of your choice.

7 Using the human population growth model from Chapter Seven $\left(P = P_0(1+r)^t\right)$, find how long it will take for a city with a population of 20,000 to grow to 50,000 at an average growth rate of 5% per year. Repeat this problem with growth rates of 4%, 3%, 2%, and 1%.

8. Using the E. coli population growth model from Chapter Seven $\left(P = P_0 2^{\frac{t}{20}}\right)$, find how long it takes for 10000 E. coli bacteria to become 1000000 under unrestricted growth conditions.

9. Using the money growth model from Chapter Seven $\left(P = P_0(1+r)^t\right)$, find how long it will take for \$500 to become \$5000 at an annual interest rate of 8%.

10. $5^x = 72$

11. $3.3^x = 17$

12. $4^x = 5^{x+1}$

13. $7^{2x-1} = 1$

14. $1.5^{x+3} = 4$

15. $7.3^{4x-2} = 7.3^{x+2}$

16. $6.7^{x+3} = 24$

17. $9^{3x-5} = -4$

18. $2^{3x-1} \cdot 2^{x+1} = 19$

19. $4^{x+2} = 5^{2x-1}$

20. $\log(x+3) = -2$

21. $\log_3(x-2) = 4$

22. $\log(x+1) - \log(x-3) = \log 5$

23. $\frac{1}{2}\log(x-4) = \log(2x-3)$

24. $\log_2(x+5) + \log_2 x = \log_2 24$

25. $\log\left(x^2 - 2x + 2\right) = 1$

26. $\log_5(x-1) + \log_5(x+3) = \log_5(x-1)$

27. $\log_4(x^2 + 3x - 1) = 2$

28. Create a logarithmic equation that has a solution of 5.

29. Create a logarithmic equation that has a solution of –3.

30. Create a logarithmic equation that has a solution of –4 and has a related logarithmic function that is decreasing.

31. Create a logarithmic equation that has solutions of –1 and 3.

From Mathematics to English

32. Explain how it is possible for a logarithmic equation to have a negative number for a solution.

33. As best you can, explain why the analytic method for solving logarithmic equations gives rise to extraneous solutions.

34. Is the graphical method for solving logarithmic equations better than an analytic method? Why?

35. Describe two analytic methods for solving the equation $6^x = 5$.

36. How do you decide which method to use when solving a logarithmic equation?

37. After reading this section, make a list of questions that you want to ask your instructor.

38. Continue in your daily journal and make an entry. In addition to your normal entry on thoughts about the mathematics in this section, list at least two positive comments about what you have learned about this topic.

39. In paragraph format, summarize the material in this section of the text in your daily journal.

40. Describe how your classroom instructor made this topic more understandable and clear.

41. After reading the text and listening to your instructor, what do you not understand about this topic?

Explorations

42. Develop a strategy for solving the equation $(3x-7)^4 = 7$ using logarithms. Show the solution and write the reason for each step.

43. While the algebraic method for solving equations may be used on Exercises 7 - 27 above, it cannot be used on the equations below. Use the graphical method to solve the following equations.

a. $\log(\sin(3x)+2) = \sqrt{x+5}$

b. $\cos(\log(x-1)) + |x+4| = 7$

c. $\log\left(x^2 - 3x - 10\right) - \log(\sin x + 3) = 2^{\log x}$

d. $\cos x + \log\left|x^2 + 2x - 3\right| = 5\sin x - 5$

e. $\log(\sin x) - |3x+2| = \sqrt{x+4}$

f. $-\log\left(x^2 + 5x - 14\right) + \log(\sin x + 2) = 16\left|5 \cos \frac{x}{3}\right|$

14.4 The Logarithmic Function as a Mathematical Model

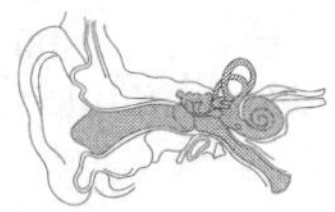

The logarithmic function is used as a model in this chapter for three different physical situations that many of you have experienced. Each of the topics is seemingly totally different from the other topics, yet they are all modeled by nearly the same logarithmic function. This function was developed by the German psycho physicist G. T. Fechner in 1860. In essence, his law states that a human sensory response is proportional to the common logarithm of a physical stimulus. In mathematical terms, it is $R = k \log S$; where R is a measure of the human response, k is the constant of proportionality that depends on the stimulus, and S is a measure of the intensity of the stimulus. The human sensory response is a reaction to light, movement, sound, and the like. The stimuli in this section refers to the light from a star, the movement of the earth from an earthquake, and sounds from any source.

The first application is the apparent brightness of the stars and planets. Astronomers call the apparent brightness, the **magnitude** of the star. The second application is the Richter scale for measuring the magnitude of an earthquake. The third logarithmic model will relate how humans interpret the loudness of a sound of a given intensity level.

Magnitudes of the Stars

Over 2500 years ago (600 BC), the astronomer Ptolemy grouped the visible stars into six categories according to their brightness. The first group contained stars that were brightest to the naked eye and the sixth group contained the faintest. Over 2400 years later (in the mid-1800s), astronomers formalized the brightness of the stars and planets by identifying the brightest stars as having a magnitude of 1 and the faintest group as having a magnitude 6. The telescope had been invented and they could now see stars fainter than before; thus, they extended the chart to include stars fainter than could be seen with the naked eye. In measuring the brightness of stars at each magnitude level, they discovered that a star of magnitude 1 is 2.512 or 2.512^1 times as bright as a star of magnitude 2. Further, a first magnitude star is 6.3 or 2.512^2 times as bright as a star of magnitude 3. The data astronomers measured is displayed in Table 14.4.1.

Table 14.4.1

x	x times as bright	m magnitude
$2.512^0 =$	1	1
$2.512^1 =$	2.512	2
$2.512^2 =$	6.3	3
$2.512^3 =$	15.84	4
$2.512^4 =$	39.8	5
$2.512^5 =$	100.0	6
$2.512^6 =$	251.2	7

A first magnitude star's brightness as compared to other stars.

For example, a 1st magnitude star is 6.3 times as bright as a 3rd magnitude star. A 1st magnitude star is 100 times brighter than a 6th magnitude star.

Below is a plot of the above numeric representation of the relative brightness of a first magnitude star compared to stars of other magnitudes. The second graph is the same data superimposed on the graph of the logarithmic function $m = 1 + \log_{2.512} x$.

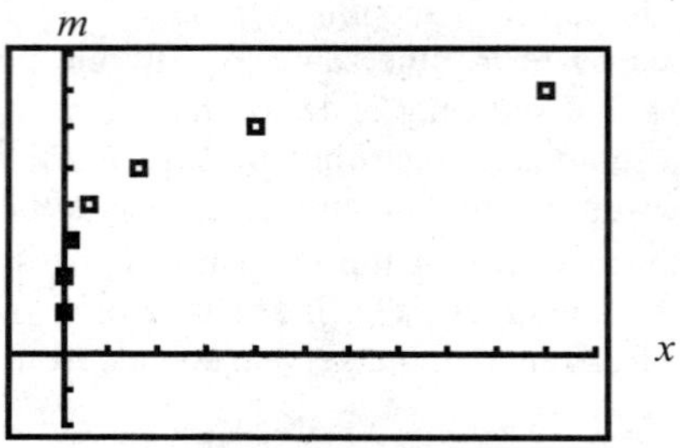

Figure 14.4.1a

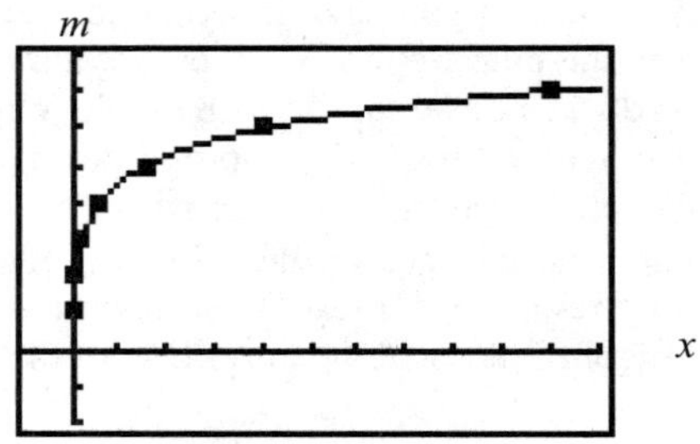

Figure 14.4.1b

The data is a match. The apparent magnitude of a star can be modeled with a logarithmic function. If the brightness (x) of a star relative to a first magnitude star is known, the magnitude can be calculated. Also, if the magnitude (m) of a star is known, you can calculate how much brighter a first magnitude star is compared to the star of known magnitude.

Example 1: How much brighter is a star of magnitude 1 than a star of magnitude 9?

Solution: Using the model $m = 1 + \log_{2.512} x$, replace m with 9 to produce the equation $9 = 1 + \log_{2.512} x$. Use any method of your choice to solve the equation. Below is the analytical method with a check using the intersection method.

$9 = 1 + \log_{2.512} x$	given
$8 = \log_{2.512} x$	1 has been subtracted from both sides
$x = 2.512^8$	the equation has been rewritten in exponential form
$x = 1585.47$	A 1st magnitude star is 1585.47 times as bright as a 9th magnitude star.

The check shows the graph of the function $m = 9$ and $m = 1 + \log_{2.512} x$ on the window[–200, 2500] by [–2, 12]

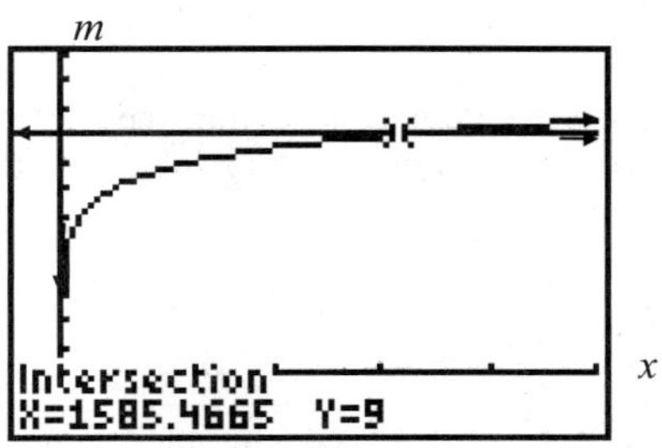

Figure 14.4.2

The magnitude model is 9 when a 1st magnitude star is 1585.47 times as bright as a 9th magnitude star.

Example 2: What is the magnitude of a star that is 500 times fainter than a star of magnitude 1?

Solution: If a star is 500 times fainter than a first magnitude star, then a first magnitude star is 500 times brighter. Use the log function $m = 1 + \log_{2.512} x$, where x is 500.

$m = 1 + \log_{2.512} 500$ The calculator keystrokes are

$m = 7.7$ 1 + log 500 ÷ log 2.512 <ENTER>

Please study the graph of the star magnitude model, $m = 1 + \log_{2.512} x$ on the window $[-1,5]$ by $[-5,5]$:

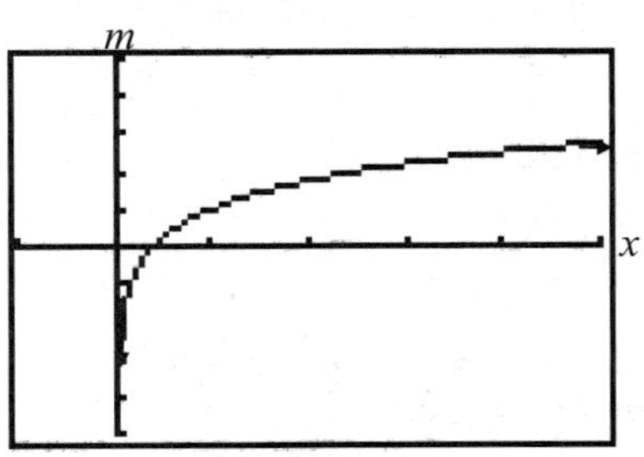

Figure 14.4.3

What you should notice is there are values of m that are negative! This is not surprising because the range of the logarithmic functions studied in Section One is $(-\infty,\infty)$; this certainly includes negative numbers.

Example 3: What does it mean if a magnitude is negative?

Solution: Since the brightest stars were given a magnitude of 1, a magnitude less than 1 must mean brighter than a first magnitude star. Many things are brighter than a first magnitude star, such as Venus, Jupiter, and the Moon.

The model for the magnitude of a star contains a logarithm base 2.512. While this is OK, it would be more convenient if the base were 10. In Section 2 you learned how to change the base to any positive number. Below is the derivation of the model in base 10.

$m = 1 + \log_{2.512} x$ given

$m = 1 + \dfrac{\log x}{\log 2.512}$ by the change of base formula

$m = 1 + \dfrac{\log x}{0.4000}$ evaluation of the log 2.512

$m = 1 + 2.5 \log x$ division be 0.4 equals multiplication by 2.5

The model is now a little easier to work with.

The mathematical model $m = 1 + 2.5 \log x$ can be generalized to model the difference in magnitudes between any two stars. That is, if m and n are magnitudes of two stars and $\frac{x_m}{x_n}$ measures the relative brightness of one star compared to the other, then $m - n = 2.5 \log\left(\frac{x_m}{x_n}\right)$. The model $m - n = 2.5 \log \frac{x_m}{x_n}$ is the model for relative magnitude between stars of brightness x_m and x_n.

Example 4: What is the difference in magnitude between two stars when one star is 1000 times as bright as the other?

Solution: The difference in magnitude is $m - n$, this difference is $m - n = 2.5 \log 1000$.
The keystrokes are: 2.5 log 1000 < ENTER >
$m - n = 7.5$

Thus, if one star is of magnitude 8, the other star would have a magnitude of either $8 + 7.5$ or $8 - 7.5$, (15.5 or 0.5).

⁂

Example 5: The magnitude of the planet Venus is –4, how much brighter than a first magnitude star is Venus?

Solution: The difference in magnitude is $1 - (-4)$, or 5.

Algebraically, from $5 = 2.5 \log\left(\frac{x_m}{x_n}\right)$, solve for $\frac{x_m}{x_n}$ by dividing both sides by 2.5

$$2 = \log\left(\frac{x_m}{x_n}\right) \qquad \text{rewrite in exponential form}$$

$$\frac{x_m}{x_n} = 10^2$$

$$\frac{x_m}{x_n} = 100$$

Venus is 100 times brighter than a first magnitude star.

Example 6: Many stars (the sun not included) vary in brightness over time. If a variable star rises 7 magnitudes between minimum and maximum brightness, how much does the brightness increase?

Solution: From the function $m - n = 2.5\log\left(\frac{x_m}{x_n}\right)$, the difference in magnitude is known and the relative brightness $\frac{x_m}{x_n}$ from minimum to maximum is desired. Solving the problem algebraically, replace $m - n$ with 7 and solve for $\frac{x_m}{x_n}$.

$$7 = 2.5\log\left(\frac{x_m}{x_n}\right) \qquad \text{divide by 2.5}$$

$$2.8 = \log\left(\frac{x_m}{x_n}\right) \qquad \text{rewrite in exponential form}$$

$\frac{x_m}{x_n} = 10^{2.8}$, $\quad \frac{x_m}{x_n} = 631$ The variable star is 631 times as bright at maximum as compared to its minimum brightness.

Richter Scale of Earthquake Intensity

The Earth's crust is made up of approximately 20 rigid chunks called **plates**. As the plates bump into one another, one plate will pass under the other, causing heaving and shifting of the upper plate; this is called an **earthquake**. The surface of the earth upon which buildings and roads are built moves and shakes as the plate below is shoved. Earthquakes happen at a worldwide rate of about 1,000,000 per year. As a point of reference, the energy required to create a slight movement in a plate is approximately equivalent to the energy released from exploding 10,000 of the first atomic bombs.

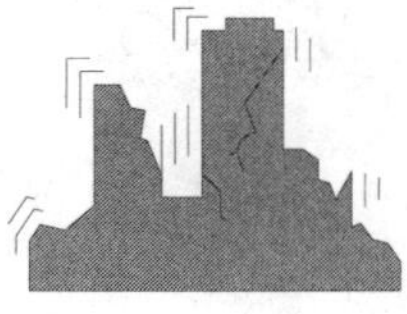

In 1556 an earthquake in central China killed over 800,000 people. In China alone, over 1.6 million people have died from earthquakes. Early in the 18th century, the current site of Tokyo was hit by an earthquake that killed 200,000 people. The San Francisco earthquake of 1906 killed around 3,000 people.

With such a significant event affecting the lives of millions of people, the scientific community has responded by analyzing earthquakes. Seismologists are getting closer to being able to predict when and where a major earthquake will happen. As part of the study of earthquakes, an American seismologist, Charles F. Richter, developed a method by which the intensity of an earthquake can be described, called the **Richter scale**.

How strong is an earthquake that measures 4 on the Richter scale? Is it twice as strong as one that measures 2? Is it four times as strong as an earthquake that measures 1 on the Richter scale? Does the intensity increase in direct proportion to the amount of slippage between the plates that make up the earth's crust? Is there a mathematical model that can describe the intensity? Hopefully, the mathematics that follows will give you a better understanding of how the intensity of an earthquake is measured.

As with the relationship between the magnitude and apparent brightness of a star, the relationship between the magnitude of an earthquake and its intensity is also logarithmic. Further, the magnitude of an earthquake is related to the logarithm of the ratio of the intensity of the earthquake to the intensity of an earthquake of minimal intensity. If I_0 is the intensity of an earthquake that can just be felt and I is the intensity of the earthquake being measured, the magnitude (R) of the earthquake on the Richter scale is:

$$R = \log\left(\frac{I}{I_0}\right)$$

Example 7: What is the magnitude of an earthquake that is 4000 times as intense as I_0? That is, $I = 4000I_0$

Solution: The magnitude is: $R = \log\left(\frac{4000I_0}{I_0}\right)$

$$R = \log\ 4000$$

$$R = 3.6$$

An earthquake that is 4000 times intense as one that can just be felt measures 3.6 on the Richter scale.

Example 8: What do the coordinates of each point on the graph $R = \log\left(\frac{I}{I_0}\right)$ mean?

Solution: The coordinates are $\left(\frac{I}{I_0}, R\right)$. The first coordinate is intensity of an earthquake as compared to an earthquake of minimal intensity. The second coordinate is the measure of the magnitude of the intensity using the Richter scale.

⁂

Example 9: The strongest earthquake on record measured 8.9 on the Richter scale. How much stronger was this earthquake compared to one that can just be felt?

Solution: Using the intersection method, graph the function $y = \log(x)$ and y = 8.9, find a value for x when $y = 8.9$. The window is [0, 2 000 000 000] by [6, 12].

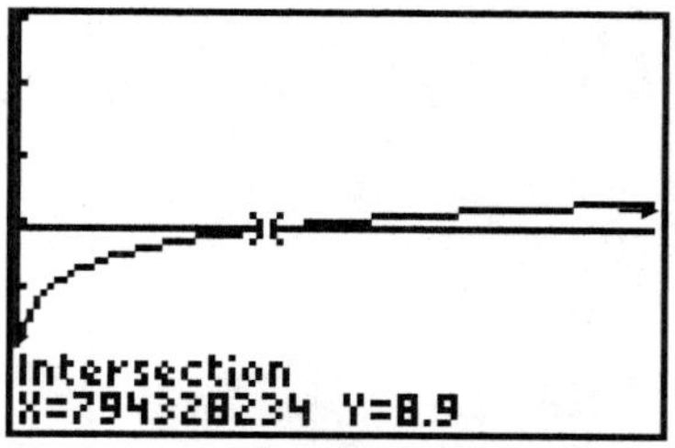

Figure 14.4.4

Since the point (794 328 234, 8.9) is on the graph, an earthquake that measures 8.9 on the Richter scale is approximately 794 million times stronger than one of minimal strength. Using an algebraic approach to check the work, replace R with 8.9 and solve the related equation for $\frac{I}{I_0}$.

$$8.9 = \log\left(\frac{I}{I_0}\right)$$ rewrite in exponential form

$$\frac{I}{I_0} = 10^{8.9}$$

$$\frac{I}{I_0} = 794{,}328{,}234$$ multiply both sides by I_0

$$I = 794{,}328{,}234 I_0$$

The intensity of an earthquake measuring 8.9 on the Richter scale is 794 million times stronger than one of minimal strength.

⁂

The log function that models the magnitude of an earthquake can be used in the same way the magnitude/brightness function of a star was used. That is, if you *think* of R as being the difference in magnitude between an earthquake of intensity I and any earthquake of intensity I_0, the mathematical models are basically the same.

Example 10: How much stronger is an earthquake of magnitude 4 as compared to one of magnitude 2?

Solution: The difference in magnitude is $4-2$, or 2. From the function $R = \log\left(\frac{I}{I_0}\right)$, let $R = 2$ and solve the resulting equation for $\frac{I}{I_0}$.

$$2 = \log\left(\frac{I}{I_0}\right) \quad \text{rewrite in exponential form}$$

$$\frac{I}{I_0} = 10^2$$

$$\frac{I}{I_0} = 100 \quad \text{multiply by } I_0$$

$$I = 100I_0$$

An earthquake of magnitude 4 is 100 times as strong as one of magnitude 2.
It is NOT twice as strong.

⁂

Sound Intensity

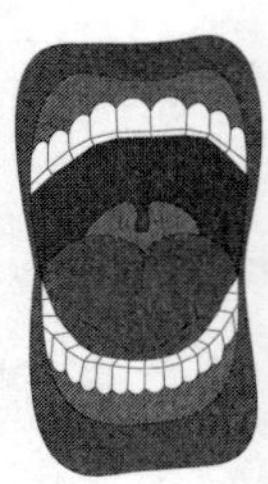

It may be helpful to think of sound waves as being like the waves made by throwing a stone in still water. However, unlike the waves made on water, sound waves emanate in three dimensions. The ripples made in the water can be in a variety of frequencies (cycles per second), just as sound is in a multitude of frequencies. On average, the human ear can detect frequencies as low as 16 cycles per second and as high as 18,000 cycles per second. A high-pitched sound has a high frequency. Many frequencies are either too high for your ears to hear or too low (ultrasound). For example, a nuclear explosion generates a very intense low-frequency sound wave that can be detected with instruments after it has traveled around the earth. A dog whistle has a frequency above the normal human hearing range. As the wave (water or sound) moves outward from its source, the intensity of the wave lessens. The waves have crests and depressions. *The difference in height between the crest and the depression can be thought of as the measure of the intensity.* The intensity of a sound wave is usually measured in decibels. (The *bel* comes from the inventor of the telephone -- Alexander Graham Bell.) The faintest sounds an ear can detect cause the ear drum to move less than one billionth of an inch. The loudest sounds the human ear can tolerate without pain are a trillion (1,000,000,000,000) times more intense than the faintest sound it can detect.

The intensity of sound is modeled with a logarithmic function. As you will see, it is strikingly similar to the model for the magnitude of stars and the model used for the Richter scale. As before, it is a human sensory response to a physical stimulus. The mathematical model that measures the magnitude of a sound of intensity I is:

$$\alpha = 10\log\left(\frac{I}{I_0}\right)$$

where α (alpha) is in decibels, and I_0 is usually a minimal sound intensity.

Think of I_0 as a reference intensity level; for example, the softest sound humans can hear (on average).

Example 11: Normal conversation at one meter measures about 60 decibels. How much more intense is conversation than minimal sound?

Solution: The problem can be solved two ways. Algebraically, replace α with 60 and solve for I.

$$60 = 10\log\left(\frac{I}{I_0}\right) \quad \text{divide by 10}$$

$$6 = \log\left(\frac{I}{I_0}\right) \quad \text{rewrite in exponential form}$$

$$\frac{I}{I_0} = 10^6 \quad \text{multiply by } I_0$$

$$I = 1000000I_0$$

Normal conversation is 1 million times as intense as the faintest sound humans can hear.

Graphically, the problem can be solved by graphing the function $\alpha = 10\log\left(\frac{I}{I_0}\right)$ as $y = 10\log(x)$ and finding x when $y = 60$. The intersection method is shown below. The graphs of $y = 60$ and $y = 10 \log x$ are on the window [0, 2 000 000] by [–10, 80].

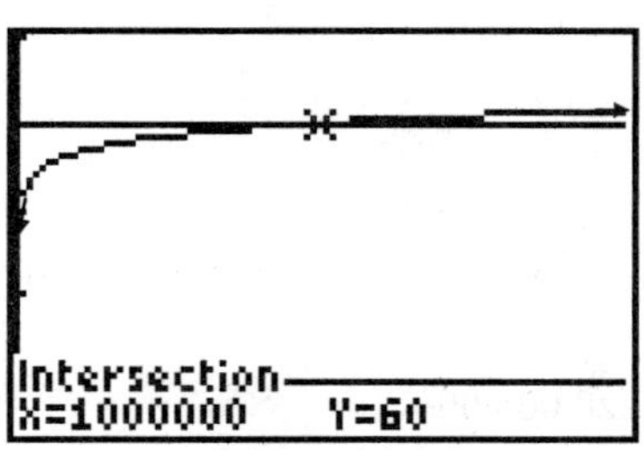

Figure 14.4.5

When $y = 60$, x is approximately 1,000,000. That is, a sound of 60 decibels is 1,000,000 times as intense as minimal sound.

⁂

Example 12: How loud is a sound whose intensity is 1 billion (1E9) times as intense as a minimal sound?

Solution: Replace $\frac{I}{I_0}$ with 1E9 and calculate α.

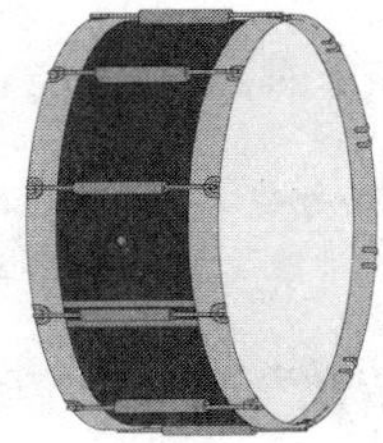

$$\alpha = 10\log(1\text{E}9)$$

$$\alpha = 10 \times 9$$

$$\alpha = 90 \text{ decibels}$$

⁂

People who work or play in a noisy environment may not only suffer large elevations in minimal sound levels, but also suffer permanent loss of hearing if exposed to daily loud noise levels. A more complicated mathematical model can predict the amount of this hearing loss.

Example 13: Humans experience pain when the magnitude of sound is about 125 decibels. How much more intense is this sound compared to minimal sound?

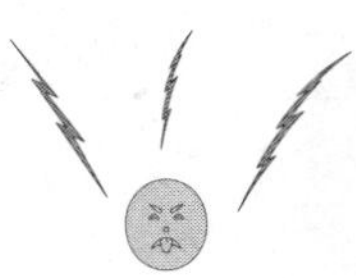

Solution: Using the algebraic method, replace α with 125 and solve for I.

$$125 = 10\log\left(\frac{I}{I_0}\right) \quad \text{divide by 10}$$

$$12.5 = \log\left(\frac{I}{I_0}\right) \quad \text{rewrite in exponential form}$$

$$\frac{I}{I_0} = 10^{12.5}$$

$$\frac{I}{I_0} = 3{,}162{,}228{,}000{,}000 \quad \text{multiply by } I_0$$

$$I = 3{,}162{,}228{,}000{,}000\, I_0$$

This is approximately 3.2 trillion times as intense as minimal sound.

⁂

As in all of the above mathematical models, α also measures the *difference* in magnitude between sounds of ANY intensity levels, I and I_0.

Example 14: How much more intense is a sound that measures 140 decibels as compared to a sound of 125 decibels?

Solution: $\alpha = 140 - 125 = 15$. Solve $\alpha = 10\log\left(\frac{I}{I_0}\right)$ for $\frac{I}{I_0}$ when α is 15.

$$15 = 10\log\left(\frac{I}{I_0}\right)$$

$$1.5 = \log\left(\frac{I}{I_0}\right)$$

$$\frac{I}{I_0} = 10^{1.5}$$

$$\frac{I}{I_0} = 31.6$$

A sound measuring 140 decibels is 31.6 times as intense as a sound measuring 125 decibels. Does this mean if humans feel pain at 125 decibels, at 140 decibels the pain is 31.6 times as severe?

⁂

14.4 EXERCISES

Many times you can work the exercises by algebraic, numeric, or graphical methods. You can learn about mathematics no matter which method you use. Part of the learning process is for you to decide which method is best for you to use. Please read the text carefully before you do the exercises. Read the **next assigned section** after you finish the assignment below. All numeric answers must have an error ≤ 0.01.

A Quick Look To The Future

-1. Is there a future?

Previous Topics Revisited

Solve the equations in Exercises 1 - 3.

1. $-\log(x-5)+7=3$
2. $3\log_2(x+4.3) = -1.23$
3. $\log(x)+\log(x-3)=\log(10)$
4. Expand the expression: $\log\frac{(x-1)^2(x)}{(x+3)}$
5. Calculate: $\log_{12} 48$

6. Find the zero of the function $y = -\log(x-5)+4$

Star Magnitude Exercises

7. A nova is an exploding star. A nova called Persei increased in brightness by a factor of 25,000 between February 20 and February 22, 1901. How many magnitudes did it rise?

8. The universe consists of many different types of stars. One type is a variable star, which may vary in brightness on a regular basis. The variable star Chi Cygni ranges in brightness from a magnitude of 13 to a magnitude 5 on a 400-day cycle. How much does its brightness increase during the cycle?

9. The full moon has a magnitude of –12.5. How much brighter is the full moon than a first magnitude star? (Hint: from the function $m-n = 2.5\log\left(\frac{x_m}{x_n}\right)$, let $m=1$ and $n=-12.5$.)

10. The sun has a magnitude of –26.7. How much brighter is the sun than a first magnitude star?

11. Give an interpretation of the coordinates of points on the graph of the function $m = 1+\log_{2.512}(x)$.

Earthquake Exercises

12. How much stronger is an earthquake of magnitude 5 than one of minimal intensity?

13. How much stronger is an earthquake of magnitude 8 than one of magnitude 3?

14. The 1906 San Francisco earthquake was 200,000,000 (200 million) times as intense as an earthquake of minimal intensity. What was its measure on the Richter scale?

15. The Mexico City earthquake of 1985 was 126,000,000 (126 million) times as intense as an earthquake of minimal intensity. What was its Richter scale magnitude?

16. How much stronger was the San Francisco earthquake compared to the Mexico City earthquake?

17. What is the difference in intensity between earthquakes of magnitude 5 and 8.5?

Sound Exercises

18. While casual conversation measures about 60 decibels, noise that will cause immediate deafness measures about 150 decibels. What is the difference in intensity between 60 and 150 decibels?

19. The sound from an Atlas rocket launch at 100 meters measures 200 decibels. How much more intense is 200 decibels compared to the threshold of hearing? Write the answer using standard notation as opposed to scientific or exponential notation.

20. The sound level in the middle of a 75-piece orchestra is 3.16227766 trillion (3,162,277,660,000) times as intense as minimal sound. What is the decibel level?

21. Is the sound intensity at conversational level (60 decibels) half the intensity of the sound made by an aircraft takeoff at 120 decibels?

22. The rustling of leaves has a sound that measures 10 decibels. How much less intense is this sound compared to normal conversation at 60 decibels?

23. The average noise from a rock band is 100 billion times as intense as the threshold of hearing. What is the decibel magnitude of this noise?

From Mathematics to English

24. Based on an understanding of Fechner's mathematical model of the relationship between a stimulus outside the human body and the body's response to that stimulus, which of the senses of sight, touch, or hearing is most sensitive? Explain why.

25. Considering that the magnitude of stars was based on observations made by Ptolemy over 2500 years ago, if you were to go outside on a clear dark night, could you see a sixth magnitude star? Explain why or why not.

26. Describe the conditions after an earthquake with a magnitude of 12 hit your city.

27. Research the relationship between the magnitude of sound and the loss of hearing. Write a brief report.

28. After reading this section, make a list of questions that you want to ask your instructor.

29. Continue in your daily journal and make an entry. In addition to your normal entry on thoughts about the mathematics in this section, list at least two positive comments about what you have learned about this topic.

30. In paragraph format, summarize the material in this section of the text in your daily journal.

31. Describe how your classroom instructor made this topic more understandable and clear.

32. After reading the text and listening to your instructor, what do you not understand about this topic?

33. What is the most important idea presented in this section? Why?

Explorations

34. Invent a mathematical model that relates the strength of any taste to the "magnitude" of the taste. List 10 foods along with probable magnitudes of taste of the food.

CHAPTER FOURTEEN TEST

1. Find $\log_8 64$
2. Find log 1,000,000
3. Find antilog 3
4. Is the function $y = \log_4(x-2) + 3$ increasing or decreasing on its domain?
5. What is the domain of $y = \log_4(x-2) + 3$?
6. List the transformations of the graph of $y = \log x$ that yield the graph of $y = -3\ \log(x+2) - 9$.
7. Convert $3 \cdot 7^{x+1} = 12$ to logarithmic form.
8. Convert $\log x + \log(x-2) - \log(x+1) = 7$ to exponential form.
9. Expand the expression $\log \dfrac{xy^2}{z}$
10. Find $\log_5 63$
11. Solve $2^{x+1} = 5^{x-1}$
12. Solve $3.9^{x-5} = 9.8$
13. Solve $\log_5(x-3) = 2$
14. Solve $\log(x-2) - \log x = -0.2614$
15. Solve $\log_2\left(x^2 + 5x - 6\right) = 3$
16. A star is 1,000,000 times as faint as a first magnitude star; using the mathematical model $m = 1 + \log_{2.512} x$, find the magnitude of the star.
17. What is the intensity of an earthquake as compared to one of minimal intensity, if it measures 6 on the Richter Scale? (hint: $R = \log\left(\dfrac{I}{I_0}\right)$)
18. How much stronger is an earthquake of magnitude 6 as compared to an earthquake of magnitude 3?
19. Using the mathematical model $\alpha = 10\log\left(\dfrac{I}{I_0}\right)$, what is the decibel level of the noisiest spot at Niagara Falls, if the sound intensity is 3 billion times as intense as a minimal sound?
20. How much more intense is the threshold of pain, which measures 140 decibels, than normal conversation at 70 decibels?

Answers to Selected Odd Exercises in

Foundations for College Mathematics, 2e

CHAPTER ONE, NUMBERS

1.1 Properties of Numbers, Equality, and Inequality .... Page 15

1. −11	3. −13	5. 2
7. Assoc. property of addition	9. Assoc. property of addition	11. Multiplicative identity
13. Mult. Ident., Commutative property of addition, Distributive property	15. Subtraction property of equality	17. Subtraction property of inequality
19. Multiplication property of inequality	33. If $b = 1$	35. $c > 0$
51. $2 \le x$	53. $car \le 40$ mph	55. $c \le 0$
57. Examples of correct responses: $2 - 5 \neq 5 - 2$ $3 - (5 - 7) \neq (3 - 5) - 7$ $\frac{6}{2} \neq \frac{2}{6}$ $2(3 \times 5) \neq 2 \times 3 \times 2 \times 5$ $0 - 7 \neq 7$ $\sqrt{3^2 + 4^2} \neq \sqrt{3^2} + \sqrt{4^2}$ $\lvert 2 + (-3) \rvert \neq \lvert 2 \rvert + \lvert -3 \rvert$		

1.2 Data Analysis .... Page 23

1. Distributive	5. 5	7. mean = 63922 median =33700 maximum = 291600 minimum = 400 range = 291200
9. Weaker	11. They are all 18.	13. The median is near the center. The other measures of central tendency are not.

1.3 Describing Sets of Numbers With Interval Notation .... Page 34

1. $\frac{x_1 + x_2 + x_3 + \ldots x_n}{n}$	3. mean = 30.2 median = 31 maximum = 38 minimum = 15 range = 23	5. Distributive property
7. −5 0 	9 −3 	11. −1 3 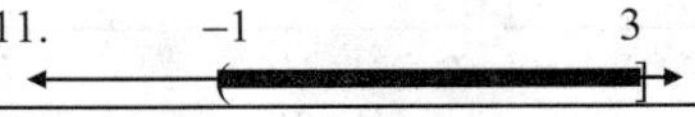
13. 0 	15. $(-\infty, 2000)$	17. $(0, \infty)$
19. $[0, \infty)$	21. $(-4, 3) \cup (5, 7)$	23. $\varnothing$
25. $(-\infty, 3)$	27. $(0, \infty)$	29. $(-\infty, 0) \cup (5, \infty)$
31. $(-4, -1)$	33. $[-1, 1]$	

1.4 Big and Little Numbers .... Page 43

1 −3 2 	3. $(-1, 4)$	7. 2.997925×10^{10}
9. 4.24×10^5	11. 9.11×10^{-28}	13. 2×10^{-6}

15. 1.7×10^{-4} %	17. 0.004 inch	19. 6.4
21. 3.7×10^{11} pounds	23. $998,640,000,000	25. 19.1%
27. 0.000 000 41 667%	29. 95,710,000 people	31. The sun is 332831 times more massive than the earth.

CHAPTER TWO, REPRESENTATION AND BEHAVIOR OF FUNCTIONS

2.1 Data Relationships Represented Numerically and Graphically ....Page 56

1. Distributive Property	3. $7 = \dfrac{5+6-4}{(4-3)}$	5. 5
7. It looks exponential. It is increasing from 1901 to 2000 except from 1920 to 1930 when it decreased.	9. The relationship looks linear or maybe quadratic; it is increasing when *time* $\in$ (1960, 2000)	11. Linear; increasing when *time* $\in$ (1980, 2003).
13. It looks like an up-side down V and is increasing when *time* $\in$ (1910, 1950) and is decreasing when *time* $\in$ (1950, 2000)	15. No particular shape, but somewhat linear for most. It decreases when *time* $\in$ (1, 19) and decreases from 19 to 27	17. The shape looks exponential and is increasing for all time in the data set.
19. No particular shape and increasing/decreasing seem to be meaningless	21. The relationship seems parabolic in shape. The volumn is decreasing when *Temp* $\in$ (−10, 4) and increasing when *Temp* $\in$ (4, 20).	23. Linear. Increasing for all *time* in the problem.
25. Exponential in shape, and increasing for all *time* in the problem.	27. Nearly linear; and increasing for all *Decade's* in the problem.	29. Parabolic, and decreasing for all *d* in the situation.

2.2 Data Relationships Represented Symbolically...Page 69

1. 3.55×10^{-4}	3. 2,901,750,000,000	5. All real numbers from −4 to 3, including −4.
7. $S = 1c + 100$	9. $MC = 0.11Kwh + 9.5$	11. $H = 4000 \times \left(\dfrac{1}{2}\right)^t$
13. Yes	15. Yes	17. Yes
19. $x \in (-\infty, \infty)$	21. $x \in (-\infty, \infty)$	23. $x \in (-\infty, -5) \cap (-5, \infty)$
25. $x \in (-\infty, -5) \cup (-5, \infty)$	27. $x \in [-3, \infty)$	29. $x \in [-5, \infty)$
31. $x \in [5, \infty)$	33. $x \in [5, \infty)$	35. $x \in [-4, 2) \cup (2, \infty)$
37. $x \in (-\infty, -5) \cup (-5, -2) \cup (-2, 1) \cup (1, \infty)$	39. $x \in (-3, \infty)$	

2.3 Geometric Behaviors of Data Relationships...Page 85

1. No	3. $x \in (-\infty, 0]$	5. 17,250,000
7d. normal domain is $x \in (-\infty, \infty)$. range $-x^2 + 4$ is $(-\infty, 4]$. $-x^2 + 4$ is increasing when $x \in (-\infty, 0)$. $-x^2 + 4$ is decreasing when $x \in (0, \infty)$.	the maximum value of $-x^2 + 4$ is 4. there is no minimum value of $-x^2 + 4$. $-x^2 + 4$ changes on an average of −3 units when x changes from 1 to 2.	$-x^2 + 4$ is zero when $x = \pm 2$. $-x^2 + 4$ is negative when $x \in (-\infty, -2) \cup (2, \infty)$. $-x^2 + 4$ is positive when $x \in (-2, 2)$.

9. It is decreasing when $x \neq 2$.	11. ½, increasing	13. $\frac{4}{3}$, increasing
15. $\frac{7}{2}$, increasing	17. $\frac{3}{5}$, increasing	19. $-\frac{3}{5}$, decreasing

2.4 Functions Represented Graphically..Page 95

1. $-\lvert x-3\rvert$ is zero when $x = 3$. $-\lvert x-3\rvert$ is never positive. $-\lvert x-3\rvert$ is negative when $x \in (-\infty, 3) \cup (3, \infty)$ $-\lvert x-3\rvert$ changes at a rate of -1 when $x > 3$.	$-\lvert x-3\rvert$ changes at a rate of 1 when $x < 3$. $-\lvert x-3\rvert$ is increasing when $x \in (-\infty, 3)$. $-\lvert x-3\rvert$ is decreasing when $x \in (3, \infty)$.	The maximum value of $-\lvert x-3\rvert$ is 0. The domain of $-\lvert x-3\rvert$ is $(-\infty, \infty)$. The range of $-\lvert x-3\rvert$ is $(-\infty, 0)$
5. $x \in [-3, 1) \cup (1, \infty)$	7. They are all linear.	9. They are all V's.
11. They all look like half of a parabola on its side.	13. They are all parabolas.	15. 22 units, 4 units, 2 units, 4 units, and 22 units.
17. 0	19. 0 and 0	21. $(-2, 5)$
39. $x^2 + 4$	41. $[-2, 6]$	

CHAPTER THREE: COMMON BEHAVIORS OF FUNCTIONS

3.1 An Introduction to the Analysis of the Linear Function $dx + e$ ....Page 109

1. minimum of –1, no maximum	3. minimum of 0, no maximum	5. Never.
7. 4, 5, 6, 7, 8, 9, 10, 11	9. $\frac{1}{2}, \frac{1}{3}, \frac{1}{4}, \frac{1}{5}, \frac{1}{6}$	11. 1, 2, 3, 4, 5, 6
13. 1, –1, 2, –2, 3, –3, 4, –4	15. $\frac{1}{3}, \frac{2}{3}, \frac{3}{3}, \frac{4}{3}, \frac{5}{3}, \frac{6}{3}, \frac{7}{3}, \frac{8}{3}, \frac{9}{3}$	17. 0, 0, 0, 0, 0, 0
19. For Exercise 11, when $x \in$ $(-\infty, 1)$; $(-\infty, 2)$; $(-\infty, 3)$; $(-\infty, 4)$; $(-\infty, 5)$; $(-\infty, 6)$.	And for Exercise 13, when $x \in$ $(1, \infty)$; $(-1, \infty)$; $(2, \infty)$; $(-2, \infty)$; $(3, \infty)$; $(-3, \infty)$; $(4, \infty)$; $(-4, \infty)$.	21. All are increasing.
23. All are decreasing.	25. Increasing. Decreasing. Increasing. Decreasing.	27. All are neither increasing nor decreasing.
51. $-4x + 200$	53. $-30x + 43$	55. 120 000 000 000 – 50 000x represents the number of brain cells left after x days have passed.
57. 200 – 1.4x represents the weight of a person after x days have passed after the start of the diet.		

3.2 An Introduction to the Analysis of the Quadratic Function $d(x+e)^2 + f$ ....Page 123

1. It rises.	3. –2	5. They are both decreasing.
7. –3 & 3; –4 & 2; –5 & 1; –6 & 0; –7 & –1.	9. –2 & 2; –1 & 3; 0 & 4; 1 & 5; 2 & 6.	11. –2.525 & 7.525; 1.863 & 5.637; –9.197 & 6.197; no real zeros.
13. $(-\infty, -2) \cup (2, \infty)$; $(-\infty, -1) \cup (3, \infty)$; $(-\infty, 0) \cup (4, \infty)$; $(-\infty, 1) \cup (5, \infty)$; $(-\infty, 2) \cup (6, \infty)$.	15. Each function is decreasing when $x \in (-4, \infty)$; $(-3, \infty)$; $(-2, \infty)$; $(-1, \infty)$; $(0, \infty)$.	17. Each function is decreasing when $x \in (-\infty, 1.25)$; $(-0.1667, \infty)$; $(-\infty, -5)$; $(10, \infty)$.

19. Each function has a minimum of 5.	21. Each function has a domain of $(-\infty, \infty)$. The ranges are: $(-\infty, 16]$; $[16, \infty)$; $[-23, \infty)$; $(-\infty, -15]$.	23. –48; –9; –4.25; –0.81
25. 102 feet; 7.9 seconds into the flight; 3.9 seconds into the flight; 79.2 feet; 43.6 feet per second, about 40 feet per second; at release; yes, at about 3.9 seconds.	49. x^2	

3.3 An Introduction to the Analysis of the Absolute Value Function $d\lvert x+e\rvert+f$ ....Page 133

1. –3	3. –22.25 & 2.25	5. –27
7. The minimum temperature is 7° at 9 PM. The temperature is decreasing until 9 PM and then increases after 9 PM. It is decreasing at 2 ° per hour, then increases at 2° per hour. The temperature is never 0 or negative. It is positive throughout the Clipper.	9. All have minimums of: –4, –1, 1, 5, 56, and –4.	11. –7 & –1, –6 & 0, –5 & 1, –4 & 2, –3 & 3, –2 & 4, –1 & 3.
13. –12 & 4, –6.5 & 8.5, no zeros, no zeros, –4 & 2, –7.142 & –0.858	15. $[-7, \infty)$, $[-7, \infty)$, $[-7, \infty)$, $(-\infty, -7]$, $(-\infty, -7]$, $(-\infty,-7]$	17. The first five functions are Dec. when $x \in (-\infty, -2)$ Inc. when $x \in (-2, \infty)$ Dec. when $x \in (-\infty, 2)$ Inc. when $x \in (2, \infty)$
19. Pos. when $x \in (-\infty, -7) \cup (-1, \infty)$ Neg. when $x \in (-7, -1)$. Pos. when $x \in (-\infty, -6) \cup (0, \infty)$ Neg. when $x \in (-6, 0)$.	Pos. when $x \in (-\infty, -5) \cup (1, \infty)$ Neg. when $x \in (-5, 1)$. Pos. when $x \in (-\infty, -4) \cup (2, \infty)$ Neg. when $x \in (-4, 2)$.	Pos. when $x \in (-\infty, -3) \cup (3, \infty)$ Neg. when $x \in (-3, 3)$. Pos. when $x \in (-\infty, -2) \cup (4, \infty)$ Neg. when $x \in (-2, 4)$. Pos. when $x \in (-\infty, -1) \cup (3, \infty)$ Neg. when $x \in (-1, 3)$.
21. Neg. when $x \in (-\infty, -12) \cup (4, \infty)$ Pos. when $x \in (-12, 4)$. Pos. when $x \in (-\infty, -6.5) \cup (8.5, \infty)$ Neg. when $x \in (-6.5, 8.5)$.	Always positive. Always negative. Neg. when $x \in (-\infty, -4) \cup (2, \infty)$ Pos. when $x \in (-4, 2)$.	Pos. when $x \in (-\infty, -7.14) \cup (-0.858, \infty)$ Neg. when $x \in (-7.14, -0.858)$.
23. From left to right on the graph, the rates are: $\frac{1}{4}, -\frac{1}{4}$; $\frac{1}{2}, -\frac{1}{2}$; 1,–1; 2,–2; 5,–5	49. $\lvert x-7\rvert$	

3.4 An Introduction to the Analysis of the Square Root Function $d\sqrt{x+e}+f$ 143

1. 35	3. $[6.2, \infty)$	5. 3
7. $x \in [-5, \infty)$; $[-3, \infty)$; $[-1, \infty)$; $[0, \infty)$; $[2, \infty)$; $[4, \infty)$.	9. The functions have values of: $(-\infty, 5]$; $(-\infty, 4]$; $(-\infty, 3]$; $(-\infty, 2]$; $(-\infty, 1]$; $(-\infty, 0]$.	11. The functions have maximums of: 5; 4; 3; 2; 1; and 0.

13. All functions in # 9 are decreasing on their domains and all functions in # 10 are increasing on their domains.	15. −4; −2; −1; 1; 3; and 6.	17. −2.56; 574; 29.56; 6080.
19. 17.9 feet per second; 27.7 feet per second.	21. 8; 1.21	47. $\sqrt{x-4}$

3.5 An Introduction to the Analysis of the Exponential Function $d \cdot b^{x+e} + f$ 150

1. There is no zero.	3. $x \in [-1, \infty)$	5. Inc. when $x \in (-\infty, -2)$ and decreasing when $x \in (-2, \infty)$.
7. $y = 10000 \times 2^{\frac{x}{20}}$	9. 2049 (sometime in February)	11. Min is 6.6 billion and the max is 13.2 billion. The range is [6.6, 13.2].
13. $2^{x-1} + 3 \to 3$	15. 1	17. Same
19. Increasing	21. There is no zero. The function only approaches zero.	35. $T = T_0 \times 2^x$, where T_0 is the initial thickness, and x is the number of cuts.

CHAPTER FOUR, FUNCTIONS: NOTATION AND OPERATIONS

4.1 Definition of a Function, Again .... Page 165

1. 2	3. $x \in [-4, \infty)$	5. $[3, \infty)$
7. No, For the first number 1 there are two second numbers.	9. Yes, for each first number there is only one second number.	11. Yes, for each x there is only one number related it.
13. No, for each x there are two related values.	15. Yes, it passes the vertical line test.	17. No, it fails the vertical line test.
19. $f(-1) = -7; f(-3) = 1; f(4) = 78$	21. 42	23. $g(1) = 0; g(-1) = 0;$ $g(2000) = 3{,}999{,}999$
25. −9	27. $q(0) = 5; q(4) = -1.828;$ $q(-4) = 15; q(1) = 2.764;$ $q(\pi) = -1.9556$	29. 2
31. $f(0) = 9; f(5) = 97$	33. $m(z)$, z	37. 5
53. $f(x) = x^2 - 15$, where x can be any variable.	55. $p(x) = \sqrt{x+4}$, where x can be any variable.	57. $f(5) = 9$

4.2 Addition and Subtraction of Polynomial Functions .... Page 176

1. 2	3. 4	5. No, for many x's there are two values for $\pm\sqrt{x^2 - 15}$.
7. $-x^2 - 7x + 5$	9. $4x^3 - 8x^2 - 14x + 18$	11. $4x^3 - 5x^2 - 6x + 11$
13. no	15. no	17. $x \in (-\infty, \infty)$
23. 20, 18, 625, yes, no	37. $(f+g)(x)$	39. $(f-g)(x)$

4.3 Multiplication of Polynomial Functions .... Page 188

1. $-2x^2 + 2x + 2$	3. $x^2 - 5x - 2$	5. 10498
7. $-20x^{17}$	9. x^6	11. $-6x^5 + 15x^4 - 12x^3 + 24x^2$
13. $-6x^5 + 4x^{18} - 10x^{33}$	15. $x^4 - 4x^3 + 4x^2 - 4x + 3$	17. $2x^3 - 7x^2 + 11x - 4$

19. $9x^2 - 16$	21. $25x^2 - 81$	23. $x^6 - 25$
25. $3x^2 + 17^x 10$	27. $8a^2 - 10a - 3$	29. $-10x^2 + 19x - 6$
31. $6x^2 - 31x + 40$	33. $3x^4 + 13x^2 - 10$	35. $x^2 - 18x + 81$
37. $4x^2 + 4x + 1$	39. $9x^2 + 72x + 144$	41. $8x^3 + 27$
43. $x^3 - 216$	65. $(x + 32)(2x - 4)$	67. $\frac{4}{3}x^2 + 7$

4.4 Factoring: Common Factors, Grouping, and Difference of Squares..Page 199

1. $4x^4 - 13x^2 + 9$	3. 5th degree	5. 3rd degree
7. $16(x^2 - 2x + 3)$	9. $ab(x^3 - x^2 + x - 2)$	11. $(2x - 1)(5x + 2)$
13. $(3x - 4)(4x - 3)$	15. $(x + 8)(x - 8)$	17. $(x^2 + 1)(x - 1)(x - 1)$
19. $(4x + 9)(4x - 9)$	21. $-3(x^2 - \frac{1}{3}x - \frac{2}{3})$	23. $-4(x^2 - \frac{6}{4}x + \frac{5}{4})$
25. 2, –5, 7, –1	27. $\frac{5}{2}, \frac{-7}{3}$	31. 2, $(x - 2)$
51. $x^2 - 4 = (x + 2)(x - 2)$	53. $63 = 7 \times 9$	

4.5 Factoring the Trinomial..Page 210

1. $8x^2(2x^2 - 4x + 1)$	3. $(3x^2 - 11)(3x^2 + 11)$	5. $6x^4 - 18x^3 + 15x^2$
7. $(x - 3)(x + 1)$	9. $(x - 12)(x - 3)$	11. Prime
13. $(7x + 1)(x + 1)$	15. Prime	17. $(5x - 1)(2x + 1)$
19. $-1, \frac{1}{3}$; $-1, \frac{1}{3}$; $-1, \frac{1}{3}$; $-1, \frac{1}{3}$	21. Sample: 2 and 5	23. Sample: 1 and 1
39. $(x + 3)(x - 7)$		

4.6 Function Operations from a Graphical Perspective..Page 218

1. $(3x + 4)(6x - 7)$	3. Sample: $(3x - 2)(x + 8)$	7. Inc. when $x \in (-3, \infty)$ Dec. when $x \in (-\infty, -3)$ Minimum value of 5 Ave. rate = 1 Pos. when $x \in (-\infty, \infty)$ Neg. when $x \in \varnothing$ No zeros Domain $(-\infty, \infty)$ Range $[5, \infty)$
9. Inc. when $x \in (-2, -1) \cup (1, \infty)$ Dec. when $x \in (-1, 1)$ Minimum value of 0 Ave. rate = 2 Pos. when $x \in (-2, 1) \cup (1, \infty)$ Neg. when $x \in \varnothing$ Zeros of –2 and 1 Domain $[-2, \infty)$ Range $[0, \infty)$	11. $[0, \infty)$	13. $[0, \infty)$
15. $[-2, \infty)$	17. No, the rate of change isn't constant.	19. Yes.

CHAPTER FIVE, ADVANCED ANALYSIS OF THE LINEAR FUNCTION

5.1 Rate of Change, Initial Conditions, and the Zero -- Slope and Intercepts of the Linear Function Page 233

1. −1, 3	3. 0.3643	5. 9(4x + 5)(4x − 5)
7. $\Delta y = 5$, $\Delta x = -3$	9. 1/−1 or −1	11. −2/2 or −1
13. 0	15. 2/3	17. 0
19. 2 and −79	21. −50	23. 28.5
25. Increasing.	27. Neither, it is constant.	29. Yes. 2
31. 0.1908	35. 75%	37. Below.
59. $f(x) = 4x - 37$		

5.2 Slope-Intercept Method of Graphing .... Page 240

1. −14/10 or −7/5	3. Increasing.	5. −1
17. *k*	33. $y(x) = 7x + 47$	35. $S = f(t)$

5.3 Point-Slope Form: $y = m(x - x_1) + y_1$ .... Page 249

5. Decreasing.	7. $y = \frac{90}{131}(x - 950) + 600$	9. $y = x - 2 + 0\sqrt{-(x-2)(x-6)}$
11. $f(x) = -0.3x + 17$	13. For example: $y = \frac{5}{7}(x - 0.6) + 0.6$	15. $y = -x + 14$
17. $y = -65(x + 4) + 2$	19. 4	21. Yes. $P(d) = 0.000346d$
23. a. 2 pieces per minute b. rate or slope c. 6000 pieces d. an initial condition e. $S = 6000 - 2t$, where *t* is in seconds.	f. 5760; 5040 g. 240; 960 h. 3000 seconds i. 3000 seconds	25. $y = m(x + 3) - 5$, where $m \in (-\infty, \infty)$
27. There are many answers. Two samples are: $y = x$ and $y = -x + 6$	29. There are many answers. A sample is $y = -\frac{2}{5}x$.	41. $e = g$. No.
49. $A = \frac{2}{3}S - 4$		

5.4 The Linear Function as a Mathematical Model .... Page 258

1. One possible form of the answer is: $y = 30x - 60$	3. No.	7. a. Yes, increasing. b. 0, 6, $W = 6H$ c. 15 billion d. 6 billion e. It seems the world reached it around 1990.
9. a. 0.004, 0.4% b. $P = 0.004t + 0.8 + 0\sqrt{-t(t-100)}$ c. 1.6 billion. 2.8 billion. The model should not be linear. d. [0.8, 1.2] e. −2.2 billion f. In the year calendar year 1667.	11. a. 8.3333 b. 10.1 c. 833.33% d. 1010% e. Not quite. f. 1965, 196, 9.22, $I = 9.22(t - 1965) + 196$ g. 611 h. About the year 1998.	13. a. No, 1/12. b. The initial condition. c. $W = 1873 - \frac{1}{12}t$ d. $t \in [0, 22476]$ e. $W = 1873 - \frac{1}{12}t + 0\sqrt{-t(t-22476)}$ f. Yes. g. 973 h. $W = 935 - \frac{1}{12}t$ i. $W = 1873 - \frac{1}{15}t$ j. $t \in [0, 11220]$

CHAPTER SIX, EQUATIONS AND INEQUALITIES

6.1 Solving Equations Containing the Linear Function ....Page 278

1. $A = 55n$	3. 182	5. $y = -\frac{2}{3}x - 2$
7. 1.8 years; 3.4 years; 5.5 years	9. 125 cars	11. They will never be equal.
13. Decreasing. At about 57.14 minutes. At 71.42 minutes. At 37.04 minutes.	15. About 14.79 hours. 19.72 hours. \$621.18.	17. 0.448
19. $\frac{2}{\pi}$	21. 8.842	23. 142,500
25. 86.237	27. 14.13	29. 48
53. $3x = 52$	55. $7x = 4 - x$	57. a. –0.147 c. 0.441

6.2 Solving Inequalities Containing the Linear Function ....Page 288

1. –3.5	3. –6013.1	5. American = 91980 lbs. Japanese = 57816 lbs.
7. [0, 8.438); [0,16.25); 18.75 days.	9. $6\frac{2}{3}$ miles; AAA because its rate schedule gives a larger fare.	11. $(721, \infty)$
13. $x \in (-\infty, 11]$	15. $x \in (-\infty, -2)$	17. $x \in \varnothing$
19. $x\, \infty\, (-\infty, 3.99]$	21. $x \in (-\infty, -73.776]$	23. $x \in (-\infty, -0.52]$
45. $2w + 2l < 74$	47. $x \leq 20$	51a. [–1.359, 7.359]

6.3 Solving Inequalities and Equations Containing the Absolute Value Function ....Page 303

1. $x \geq -\frac{23}{7}$	3. $x \leq 0.823$	5. 0.823
7. $t \in \left(22\frac{2}{9}, 37\frac{7}{9}\right)$; $t \in \left[0, 11\frac{1}{9}\right) \cup \left(48\frac{8}{9}, 60\right]$	9. 48 hours. 24 hours. $t \in [0, 15) \cup (33, 48]$	11. $x \in (-\infty, -9] \cup [1, \infty)$ $x \in (-\infty, -8] \cup [2, \infty)$ $x \in (-\infty, -7] \cup [3, \infty)$ $x \in (-\infty, -6] \cup [4, \infty)$
13. $x \in \left[-\frac{167}{66}, -\frac{91}{66}\right]$. Or $x \in [-2.53, -1.38]$	15. $x \in [-6, 1]$	17. $\frac{12}{5}$ or $\frac{18}{5}$
19. $x \in \varnothing$	21. $x \in \varnothing$	23. –1.111 or $-\frac{11}{9}$
25. $x \in (-\infty, 0.25] \cup [2.5, \infty)$	27. $x \in (-\infty, -0.887] \cup [7.322, \infty)$	29. $x \in (-\infty, -6) \cup \{1\} \cup [4, \infty)$
31. ? $\in (-\infty, 0]$	51. $\lvert x \rvert - 3 \geq 4$ or it can be interpreted as $\lvert x - 3 \rvert \geq 4$.	53. $\lvert x \rvert > 0$

6.4 Formulas and Direct Variation ....Page 314

1. $x \in (-\infty, 1.52] \cup [3.08, \infty)$	3. $x \in (0, 11.98)$	5. $x \in [1.619, \infty)$
7. 58.67°; 129 cpm; 171 cpm; 150 cpm; 107 cpm	9. 212°F; 0°C; 15.56°C; 22.22°C; 37.78°C.	11. a. 1410 b. $t \in (2010, \infty)$; $t \in (2032.22, \infty)$; $t \in (2043.33, \infty)$; $t \in (2076.67, \infty)$.
13. $r = \dfrac{C}{2\pi}$; 0.318; 0.637; 0.955; 1.910; 2.865; 3.820	15. $b = \dfrac{2A}{h} - B$; 6; 11; 16; 21; 28.5	17. $h = \dfrac{2A}{b}$; 16.667; 10.526; 6.667; 3.077; 2.222

19. $F=\frac{9}{5}C+32$; –4; 14; 32; 86; 104; 212.	21. \$2.04	23. 567 seconds
25. \$1.50	27. 5.642 in; 7.979 in; 12.616 in; 17.841 ft; 21.851 mi. 12.566 ft^2; 113.097 m^2; 78.54 mi^2	29. 64.004 ft; 256.016 ft; 1024.062 ft; 2304.14 ft; 14400.876 ft; 0.791 sec; 10 sec; 1.225 sec; 1.768 sec; 3.535 sec

CHAPTER SEVEN, ADVANCED ANALYSIS OF THE EXPONENTIAL FUNCTION

1. $m=\frac{y-y_1}{x-x_1}$	3. 900%	5. $x\in(-\infty, 1.9]\cup[5.1,\infty)$
7. f controls where the graph crosses the y-axis, or how high or low the whole shape is relative to the x-axis.	9. d seems to control how sharp the corner is when the graph starts its rapid rise.	11. The single cancer cell has grown to 0.0094 *cc*.
13. 45 months	15. 1991	17. 1.281×10^{124}
19. Decreasing	21. Increasing	23. Decreasing
25. –2.834; pos: $x\in(-2.834,\infty)$; neg: $x\in(-\infty,-2.834)$; 4.778	27. No zero; pos: $x\in(-\infty,\infty)$; neg: $x\in\varnothing$; 1	29. –0.842; pos: $x\in(-0.842,\infty)$; neg: $x\in(-\infty,-0.842)$; 12.732
31. range: $(-3,\infty)$; $y=-3$	33. range: $(-11,\infty)$; $y=-11$	41. A horizontal shift of left 3 and a vertical shift of down 1.
43. A reflection about the x-axis, a stretch of a factor of 2, and a horizontal shift of left 3.	45. $y=2^{x-1}$	47. $y=\left(\frac{1}{2}\right)^{x-1}$
53. a. Vertical shift of up 3 b. Horizontal shift of right 3	c. Horizontal shift of left 2 and a vertical shift of down 7	d. A reflection about the x-axis, a horizontal shift of right 5, and a vertical shift of up 2

1. Increasing.	3. 0.25	5. 2001
7. $f(x)=(x-2)^7$	9. $h(x)=2^{3x+6}$	11. $y=2\cdot3^{x-9}+7$
13. $y=\frac{1}{36}x^{-20}$	15. $g(x)=\frac{1}{4x^6}$	17. $y=x^{-3000}$
19. $Y_1(x)=-4x^{-7}$	21. $f(x)=2^{x^2+x-1}$	23. $f(x)=-\frac{1}{8}\cdot5^{-3x^2-6}$
25. $h(x)=2^{6x}$	27. $j(x)=3^{2x-3}-5$	51. x^3
53. $\frac{4}{3}\pi r^3$	55. $(x-3)^4$	

1. No, they have different numeric representations.	3. 2^{-3x+24}	5. Sample: $y=0.2^x-5$
7. \$15,325.49	9. 17.42%	11. 1973
13. 22 hours; 17.5 hours	15. $x=\frac{7}{3}$	17. $x=-\frac{5}{2}$

19. $x \in (-\infty, -2.229)$	21. $x = -1.152$	23. $x \in \{-3, -1, 3, 5\}$
25. $x \in \{-2.553, 3.117\}$	39. $5^{x-1} = \left(\frac{2}{3}\right)^{x+1} - 2$	

7.4 The Exponential Function as a Mathematical Model..............Page 365

1. $x \in \{-0.996, 4.188\}$	3. $x = \frac{17}{3}$	5. 2^{6x^2-8x+6}
7. 66.44 minutes	9. 20 minutes	11. about 169
13. about 114 years	15. 7.11; 7.77; 8.49; 9.29; 10; 11.1 (billions)	17. $700,649,232.20
19. $11,507.73	21. $509,501.59	23. Better.
25. 14.87% or larger		

CHAPTER EIGHT, ANALYSIS OF THE RATIONAL FUNCTION

8.1 Rational Functions..............Page 381

1. 76 E. coli	3. 0.0244%	5. –3
7. Change the 300 to 250.	9. 75	11. With the strength in % form: $S = 100 \cdot \frac{12}{12+2x}$
13. 50%	15. –1.885; –1.344	17. $x \in (-\infty, -50)$
19. Horz: –7 Vert: 0 Dom: $(-\infty, 0) \cup (0, \infty)$ Rng: $(-\infty, -) \cup (-7, \infty)$	21. Horz: –4 Vert: 4 Dom: $(-\infty, 4) \cup (4, \infty)$ Rng: $(-\infty, -4) \cup (-4, \infty)$	23. Horz: –18 Vert:–14 Dom: $(-\infty, -14) \cup (-14, \infty)$ Rng: $(-\infty, -18) \cup (-18, \infty)$
25. Horz: 3 Vert: 2 Dom: $(-\infty, 2) \cup (2, \infty)$ Rng: $(-\infty, 3) \cup (3, \infty)$	27. –0.0074	29. –2
31. Inc: $(-\infty, -3) \cup (-3, \infty)$; never decreasing	33. Inc: $(-\infty, -3) \cup (-3, \infty)$; never decreasing	35. y is zero when $x = \frac{451}{105}$; the y-intercept is about –4.876
37. The zero is about 2.276; the y-intercept is about –1.843	39. A vertical shift of up 4.	41. A stretch of 4.
43. A reflection about the x-axis, a stretch of 2, a horizontal shift of left 3, and a vertical shift of down 6.	53. From the left, $-\infty$; from the right, ∞.	55. From the left, $-\infty$; from the right, ∞.
57. $d \in (0, \infty)$	59. Horz: f; Range: $(-\infty, f) \in (f, \infty)$	

8.2 The Fundamental Property of Rational Functions..............Page 392

1. $y = -5$	3. For example: $f(x) = \frac{1}{x+3} + 4$	5. 49.83 months.
7. $y = \frac{x-2}{2}$, when x isn't –2	9. $f(x) = \frac{2}{x^2 - 7x + 13}$, always equal.	11. $f(x) = \frac{2x+5}{4x-3}$ when x isn't 1/3.
13. $g(x) = -\frac{3x+2}{2}$, when x isn't 7/5.	15. $g(x) = -1$, when x isn't 3.562 or –5.616 – approximately	17. At $x = -3$.

8.3 Multiplication and Division of Rational Functions...Page 399

1. $\frac{2x+1}{x+4}$	3. $\frac{-1}{3x+2}$	5. The line $y = 5$.
7. $(fg)(x) = \frac{x-3}{x}$	9. $(fg)(x) = \frac{x}{x+1}$	11. $(fg)(x) = -\frac{2x^2+11x+12}{x+2}$
13. $\left(\frac{f}{g}\right)(x) = \frac{2x^2+7x+5}{x^2+3x-4}$	15. $\left(\frac{f}{g}\right)(x) = \frac{2x+4}{3x+2}$	17. Unsimplified: all real numbers except $x \neq -2$, 3/2, and 5/2. Simplified: $x \neq -2$
19. Division		

8.4 Addition and Subtraction of Rational Functions and Simplification of Complex Functions.........Page 412

1. $(fg)(x) = \frac{x-1}{3x+2}$	3. $\left(\frac{f}{g}\right)(x) = \frac{x^2-4}{2x^2-x-3}$	5. $f(x) = -x-2$
7. $(f+g)(x) = \frac{8x-2}{x^2-1}$, $\Re$ except −1 and 1.	9. $(f+g)(x) = \frac{3x^2+9x+7}{x^2-2x-15}$, $\Re$ except −3 and 5	11. $(f-g)(x) = \frac{5x^2-4x-5}{4x^2-9}$, $\Re$ except ±3/2.
13. $(f-g)(x) = \frac{9x^2-13x+4}{9x^2-1}$, $\Re$ except ±1/3	15. $\frac{5}{3}$	17. $f(x) = \frac{1}{4x-6}$, $\Re$ except ±2 and 3/2

8.5 Solving Equations and Inequalities Containing the Rational Function and Inverse Variation .....Page 426

1. $\frac{7x-2}{x^2-4}$	3. $\frac{3x^2-9x-5}{5x^2-15x+2}$	5. $\frac{x^2-2x-3}{x^2-5x+4}$
7. 10.77 cups	9. 25%	11. $5\frac{5}{9}$
13. 1.54% oil	15. 1.62 ounces	17. The 128.
19. −7.315	21. $x \in (-\infty, -2) \cup [4, \infty)$	23. 1.47
25. $x \in (-2.423, -2) \cup (1.548, 2)$	27. −9/4	29. −7/12
31. No real solution.	33. $R = \frac{0.10816}{d^2}$, 1.731 Ω, Impossible	35. ? = 1

CHAPTER NINE, ADVANCED ANALYSIS OF THE SQUARE ROOT FUNCTION

9.1 The Square Root Function..Page 439

1. 7	3. $x \in \varnothing$	5. $5x + 1$
7. about 78 mph	9. about 80 f/s; 113 f/s; 179 f/s	11. about 19.6 f/s
13. about 6.25 feet	15. increasing	17. decreasing
19. decreasing	21. minimum of 11.2 when x is 3.2	23. maximum of −11.2 when x is 5
25. $(-e, f)$	27. 13	29. no zeros
31. dom: $x \in [-3, \infty)$ rng: $f(x) \in (-\infty, -6]$; dom: $x \in [-2, \infty)$ rng: $g(x) \in (-\infty, -6]$; dom: $x \in [-1, \infty)$ rng: $h(x) \in (-\infty, -6]$; dom: $x \in [1, \infty)$ rng: $i(x) \in (-\infty, -6]$.	33. dom: $x \in [0, \infty)$ rng: $y \in (-\infty, 0]$; dom: $x \in (-\infty, 0]$ rng: $y \in [0, \infty)]$; dom: $x \in (-\infty, 0]$ rng: $y \in (-\infty, 0]$.	35. dom: $x \in (-\infty, -5]$ rng: $y_1 \in [3, \infty)$; dom: $x \in (-\infty, -2]$ rng: $y_2 \in [3, \infty)$; dom: $x \in (-\infty, 2]$ rng: $y_3 \in [3, \infty)$; dom: $x \in (-\infty, 5]$ rng: $y_4 \in [3, \infty)$.

37. dom: $x \in [-1, \infty)$ rng: $g(x) \in)-\infty, 0]$.	39. For example, multiply it by –1.	41. For example, add 7 to $f(x)$.
43. For example, replace x with $x + 3$.	45. Multiply it by 4.	

9.2 Properties of Irrational Expressions ....Page 451

1. $x \in [-4.2, \infty)$	3. $f(x) \in [-2.65, \infty)$	5. $x \in \varnothing$
7. $2\sqrt{6}$	9. $2\sqrt{5}$	11. $2\sqrt[3]{9}$
13. $\frac{\sqrt{2}}{2}$	15. $\frac{\sqrt{3}}{3}$	17. $\frac{2\sqrt{3}}{3}$
19. $\frac{\sqrt{2}}{2}$	21. $\frac{\sqrt{3}}{2}$	23. $x^2\sqrt{x}$
25. $x^2\sqrt{x}$	27. x	29. x
31. $x - 2$	33. $x - 1$	35. The squaring transformation.

9.3 Operations with Irrational Expressions ....Page 459

1. $\frac{\sqrt{2}}{4}$	3. $5\sqrt{3}$	5. Parameter e.
7. 0	9. $6\sqrt{2}$	11. –48
13. 17	15. $\frac{-\sqrt{7}-7}{6}$	17. $-4\sqrt{2}-5$
19. $g(x) = 2\sqrt{x}$	21. $k(x) = \frac{x\sqrt{x}+2x}{x-4}$	23. $x \in [0, 4) \cup (4, \infty)$
25. 0	29. Only on the domain $[-2, \infty)$.	31. $\frac{-1}{4-4\sqrt{2}}$ or $\frac{1}{-4+4\sqrt{2}}$
33. 6	35. $\left(\sqrt{x+h}-\sqrt{h}\right)\left(\sqrt{x+h}+\sqrt{h}\right) = x$	

9.4 Fractional Exponents ....Page 465

1. $3\sqrt{3}+7$	3. $4x - y$ where x & $y \in [0, \infty)$	5. 144
7. 1.565	9. 2.52	11. 25
13. 44.513	15. $f(x) = x^2$	17. $y = x^{\frac{10}{9}} + x$
19. $h(x) = x^{-\frac{1}{4}}$ or $h(x) = \frac{1}{\sqrt[4]{x}}$	21. $y = x^2$	23. $f(x) = 4x^2$
25. $x \in (-\infty, \infty)$	27. $\sqrt[4]{5^{13}}$	29. $\sqrt[4]{5}, \sqrt[3]{5}, \sqrt{5}, 5^{\frac{2}{3}}$
31. $3^{\frac{1}{8}}$		

9.5 Solving Equations Containing the Square Root Function ....Page 475

1. $y = x^{\frac{3}{2}}$	3. $g(x) = x^3$, if $x \geq 0$	5. $f(x) = x - 18$, if $x \in [0, \infty)$
7. 3.25; 6; 9.25; 13; 17.25	9. $x \in \varnothing$	11. 10.99
13. 46.188	15. 0.5	17. 7
19. 5	21. $f \in (-\infty, 0]$	

9.6 The Square Root Function as a Mathematical Model..Page 485

1. $\frac{21}{4}$	3. 8	5. x
7. 80 f/s	9. 16 feet	11. 1.96 f/s
13. 914.08 feet; 156.25 feet	15. After falling 100 feet, the velocity is 80 f/s.	17. 24.836 feet
19. 27.21 sec; 2.721 sec	21. The normal domain and range are (0, ∞), and the model is increasing.	23. 0.811 feet, or about 9.7 inches.
25. 5.026 miles or 26,537 feet	27. 83 and 1/3 square units.	29. 81 is the vertical line and 486 is the area under $y = \sqrt{x}$, above the x-axis, and to the left of the vertical line $x = 25$.
31. The domain and range are both [0,∞).		

CHAPTER TEN, ADVANCED ANALYSIS OF THE QUADRATIC FUNCTION

10.1 The Quadratic Function..Page 500

1. Yes; [0, ∞); 314.15	3. 0.001 grams, or 1 milligram	5. −3.968 & 3.968
7. $\left(2, 5\tfrac{1}{12}\right)$	9. $y = \pm\frac{1}{8}(x+e)^2 + f$	11. $h = -18.6t^2 + 60t + 5$
13. Horz. of right 4, and a vert. of up 2.	15. At (3.6, 4.7).	17. $y = (x+9)^2 - 162$ Inc: $x \in (-9, \infty)$ Dec: $x \in (-\infty, -9)$
19. $y = \left(x - \frac{5}{2}\right)^2 - \frac{13}{4}$ Inc: $x \in \left(\frac{5}{2}, \infty\right)$ Dec: $x \in \left(-\infty, \frac{5}{2}\right)$	21. $y = 2\left(x + \frac{7}{4}\right)^2 - \frac{89}{8}$ Inc: $x \in \left(-\frac{7}{4}, \infty\right)$ Dec: $x \in \left(-\infty, -\frac{7}{4}\right)$	23. Range = (−∞, 4.7] Maximum = 4.7
25. Range = $\left(-\infty, \frac{137}{4}\right]$ Maximum = $\frac{137}{4}$	27. −2.5 & 1	29. $f(x)$ is positive when $x \in (-\infty, -2.5) \cup (1, \infty)$
31. −1 & 2.5	33. $f(x)$ is positive when $x \in (-\infty, -1) \cup (2.5, \infty)$	35. narrower
37. Down because the coefficient of x^2 is negative.	49. (4,−5)	51. (5,−4)

10.2 Solving Quadratic Equations of the Form $(ax + b)(cx + d) = 0$..Page 507

1. [−9/4, ∞)	3. Right 5 and up 2.	5. $8\sqrt{14}$
7. 0.625 second	9. 4 seconds	11. 0.5 second
13. At 0.5 second and at 4.375 seconds.	15. $\frac{13}{15}$ and -16	17. 3 and 4
19. $-\frac{3}{2}$ and $-\frac{2}{35}$	21. No real solutions.	23. −1, −½, and 3
25. −3	27. −3 and 1	

10.3 Solving Quadratic Equations by the Completing-the-Square Method .... Page 516

1. $\frac{5}{2}$ and $-\frac{7}{3}$	3. $-\frac{2}{3}$ and 4	5. Increasing when $x \in (-\infty, -1)$. Decreasing when $x \in (-1, \infty)$.
7. –9 and 13	9. –10 and –2	11. 5
13. 1 and 2	15. $\frac{5 \pm \sqrt{17}}{4}$	17. –8/3

10.4 Solving Quadratic Equations with the Quadratic Formula .... Page 526

1. No real solutions. The complex solution is $\frac{-2 \pm i\sqrt{6}}{2}$	3. 1 & 2	5. $\pm\frac{4}{3}$
7. $-5 - i$	9. $10 + 10i$	11. 0, or $0 + 0i$
13. $32i$	15. $1 - i$	17. $-i$
19. $\pm 5i$	21. –7, 3	23. $\frac{10 \pm 2\sqrt{115}}{3}$
25. $\frac{5 \pm \sqrt{65}}{2}$	27. rational	

10.5 The Quadratic Function as a Mathematical Model .... Page 533

1. $-2 + 3i$	3. $\frac{7 \pm i\sqrt{11}}{6}$	5. $-3 \pm \sqrt{10}$
7. 576 feet	9. 6.05 sec; 2.5 seconds	11. Mercury -- 58.05 feet Venus -- 130.05 Mars -- 58.05 Jupiter -- 1290.6
13. 17.1 seconds	15. Parabolic, because gravitation is quadratic.	17. A maximum of 15,635 feet at 31.25 seconds.
19. 54 f/s	21. 2.72 seconds	23. 3.025 seconds

CHAPTER ELEVEN, BASIC GEOMETRY

11.1 The Distance and Midpoint Formulas .... Page 546

1. 2100 feet high	3. 500 f/s	5. $\frac{5 \pm \sqrt{17}}{4}$
7. $2\sqrt{2}$; (–4, –7)	9. 10; (1, 1)	11. $\sqrt{277}$
13. $\sqrt{a^2 + b^2}$; $\left(\frac{a}{2}, \frac{b}{2}\right)$	15. 1.414 or $\sqrt{2}$	17. 3.536
19. (6, 9)		

11.2 Triangles .... Page 555

1. $\sqrt{29}$; (0.5, –5)	3. 0.707 units	5. 1.483 seconds
7. ΔABC is acute. ΔXYZ is right isosceles. ΔDEF is acute equilateral. ΔMNP is obtuse.	9. Right and isosceles.	11. $2\sqrt{13}$
13. 19.2	15. $A = 30$ sq. units; $P = 30$ units.	17. $A = 24$ sq. units; $P = 24$ units.

19. $A = 7$ sq. units; $P = 15 + \sqrt{29}$ units.		

11.3 Parallelograms..Page 561

1. 140°	3. 6.25 units	5. $d = \sqrt{(x+2)^2 + (x+4)^2}$; 1.414 units
7. They are both 40°.	9. (–4, –1)	11. 45°
13. $P = 34$; $A = 48 + 9\sqrt{3}$	15. (10, 6)	17. 15 sq. units

11.4 Circles..Page 570

1. Because opposites are parallel.	3. 32 sq. in.	5. $16\sqrt{3}$
7. $C = (-3, 2)$; $r = 3$ units; $c = 6\pi$ units; $A = 9\pi$ sq. units.	9. $C = (0, 1)$; $r = \sqrt{5}$ units; $c = 2\pi\sqrt{5}$ units; $A = 5\pi$ sq. units.	11. $C = (0, 0)$; $r = 10$ units; $c = 20\pi$ units; A = 100π sq. units.
13. $C = (-4, 1)$; $r = 2\sqrt{5}$ units; $c = 4\pi\sqrt{5}$ units; $A = 20\pi$ sq. units.	15. $C = \left(-\frac{3}{2}, -\frac{7}{2}\right)$; $r = \sqrt{15}$ units; $c = 2\pi\sqrt{15}$ units; A = 15π sq. units.	17. $(x+3)^2 + (y+5)^2 = \frac{4}{9}$
19. $(x-3)^2 + (y-5)^2 = 25$	21. Neither.	25. $c = 2\pi\sqrt{6}$ units; $A = 6\pi$ sq. units

CHAPTER TWELVE, BASIC TRIGONOMETRY

12.1 Conversions Between Degrees and Radians..582

1. $A = 9\pi$ sq. units; $C = 6\pi$ units	3. $C = (6, 2)$; $r = 7$ units; sample points (–1, 2) and (6, –5)	5. In both cases, the length is $\sqrt{l^2 + w^2}$
7. (by row) $\frac{\pi}{6}$, 45°, $\frac{\pi}{3}$, 90°, $\frac{2\pi}{3}$, 135°, $-\frac{5\pi}{6}$, –180°, $-\frac{7\pi}{6}$, –225°	9. (sample answers by row) 1320°, –120°, 300°, –420°, 860°, –220°, 3π, $-\pi$, $\frac{15\pi}{4}$, $-\frac{\pi}{4}$	11. 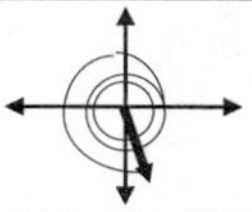
13. (sample points) a. (–1, 0); 1 b. (0, 1); 1 c. (–2, 0); 2		

12.2 Trigonometric Definitions..Page 594

1. 315°	3. $\frac{5\pi}{2}$	5. $\left(\frac{5}{2}, -6\right)$; $\frac{\sqrt{173}}{2}$
7. 45°; $\frac{\sqrt{2}}{2}$, $\frac{\sqrt{2}}{2}$, 1 120°; $-\frac{1}{2}$, $\frac{\sqrt{3}}{2}$, $-\sqrt{3}$ 180°; –1, 0, 0 210°; $-\frac{\sqrt{3}}{2}$, $-\frac{1}{2}$, $\frac{\sqrt{3}}{3}$ 270°; 0, –1, und. 315°; $\frac{\sqrt{2}}{2}$, $-\frac{\sqrt{2}}{2}$, –1	9. $\sqrt{x^2 + y^2}$	11. 7 & –7

13. 0	15. $x = y$	17. Samples: 90°, 270°, 450°, 630°, 810°, –90°, –270°, –450°, –630°, –810°

12.3 Solving Right Triangles .... Page 603

1. –1/2	3. –4 or 4	5. 5.498
7. 6.373°	9. 6.334°	11. 1.746
13. $c = \sqrt{29}$; ∠A = 68.2°; ∠B = 21.8°	15. ∠A = 20°; a = 3.42; b = 9.40	

12.4 Trigonometric Functions as Mathematical Models .... Page 609

1. ∠B = 45°; a = 4,000,000; $c = 2000\sqrt{2}$	3. $c = 5\sqrt{10}$; ∠A = 34.70°; ∠B = 55.30°	5. –12
7. 59.9 ft^2	9. 27.58 sq. ft	11. 802 feet
13. d = 6403.12 ft; ∠ = 51.34°	15. 98,481 ft	17. 14,004 ft
19. All lines make an ∠ of 26.57°.		

CHAPTER THIRTEEN, SYSTEMS OF EQUATIONS AND INEQUALITIES

13.1 Solving Systems Graphically .... Page 621

1. 68.06 ft	3. 45°	5. ∠ = 18°; hyp = 0.00526 mile; other leg = 0.00162 mile
7. 120 minutes (2 hours)	9. $(x, y) = (1, -2)$	11. $(x, y) = (-5.91, -4.82)$
13. $(x, y) = (0.2, -1.4)$	15. $(x, y) = (0.472, 4.75)$	17. (each tic mark = 2) 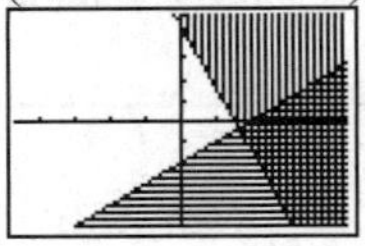
19. (each tic mark = 1) 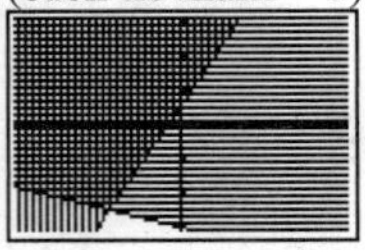	21. (each tic mark = 1) 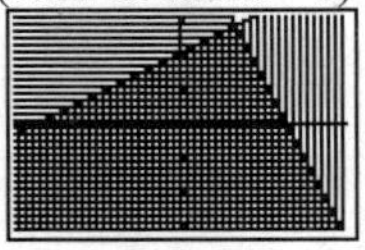 use only points above the x-axis and to the right of the y-axis.	23. (each tic mark = 1) 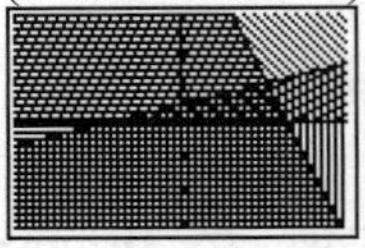 use only points above the x-axis and to the right of the y-axis.

13.2 Solving Systems by the Addition and Substitution Methods .... Page 632

1. $(x, y) = \left(\frac{60}{19}, \frac{36}{19}\right)$	3. (each tic mark = 1) 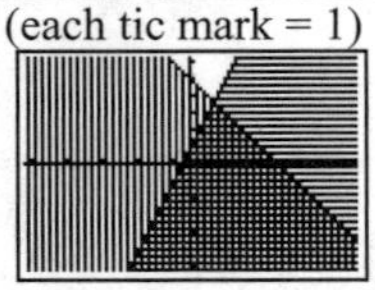 use only points above the x-axis and to the right of the y-axis.	5. 483 ft

7. $(x, y) = (4, 1)$	9. No solution.	11. $(x, y) = (1.097, -0.260)$
13. $(x, y) = (-7.71, -0.15)$	15. None.	

13.3 Solving Systems Using Cramer's Rule and Matrices

1. An infinite number of solutions. A dependent system.	3. $(x, y) = (-1, 0)$ and $(1, 0)$	5. $(x, y) = (-1, 0)$ and $(1, 0)$
7. 2	9. −84	11. −10
13. $\begin{bmatrix} -8 \\ -3 \\ -5 \end{bmatrix}$	15. $\begin{bmatrix} -3 \\ -15 \end{bmatrix}$	17. $(x, y) = (-3.286, -0.714)$
19. No solution. An inconsistent system.	21. $(x, y) = (1.172, 0.069)$	23. An infinite number of solutions. A dependent system.
25. $(x, y, z) = (2, 0, -1)$	27. $(x, y, z) = (13.98, 11.045, 0.671)$	29. base = 2, leg = 4
31. $2W$	33. No.	

13.4 Modeling with Systems of Equations

1. 11.75	3. $(x, y, z) = (-1, 2, 3)$	5. $(x, y) = (0.92, 2.24)$
7. 160°C = 320°F	9. 31 nickels, 8 quarters	11. $w = 3, l = 8$
13. $d = 5, c = 5\pi$	15. 6 easy, 4 hard	17. $f = 35°, s = 75, t = 70$
19. $L = -15\lvert t\rvert - 30\lvert t-3\rvert - 45\lvert t-23\rvert + 1125$	21. $S = 0.055\lvert t\rvert + 0.01\lvert t-10000\rvert + 0.015\lvert t-15000\rvert - 325$	23. $F = 5\lvert d\rvert - 1.5\lvert d-1\rvert - 2.5\lvert d-2\rvert + 9.25$

CHAPTER FOURTEEN, INTRODUCTION TO THE ANALYSIS OF THE LOGARITHMIC FUNCTION

14.1 The Logarithmic Function

1. 24 males & 36 females	3. Jupiter is 483,313,000 miles Mars is 141,543,000 miles	5. $(x, y, z) = (-1, -1, -1)$
7. 5 on the Richter scale	9. 31,623 times as intense	11. 0; 1; 2; 3; 4; 5
13. 2; 1; 0; −1; −2, −3; −4	15. −3; −2; −1; 0; 1; 2; 3; 4; 5; 7	17. −2; −1; 0; 1; 2; 3; 4; 5
19. both are 2.77815	21. both are 1.47712	23. 10; 10^{10}; 10^{100}; 10^{1000}; 10^{10000}; 10^{100000}
25. 2818.38 or $10^{3.45}$; 3.162278×10^{34} or $10^{34.5}$; 10^{345}; 10^{3450}; 10^{34500}; 10^{345000}	27. Increasing	29. Decreasing
31. $x \in (-4, \infty)$	33. $x \in (5, \infty)$	35. There is no maximum.
37. 1; 1; 1; 1; 1; 1	39. Beginning	41. A stretch of 2, horizontal shift of left 6, and a vertical shift of up 4.
43. A shrink of ½, a horizontal shift of right 5, and a vertical shift of up 1.	45. $x \in (-\infty, 0)$	

14.2 Properties of Logarithms

1. $x = -3$	3. A horizontal shift of right 7 and a vertical shift of down 35.	5. length = 12, width = 2
7. $10 = \log_2 1024$	9. $7 = \log_x 19$	11. $x + 1 = \log_6 21$

13. $10^5 = x^3$ or $10^{\frac{5}{3}} = x$	15. $5^4 = \frac{x+1}{x-3}$	17. $8^6 = (x-1)^3(x+1)^2$
19. $\log\frac{(x-2)^3}{(x+1)}$ where $x \in (2, \infty)$	21. $\log\frac{(2x-5)}{(x+4)^3}$ where $x \in (2.5, \infty)$	23. $\log 10(x+3)^2$ where $x \in (-3, \infty)$
25. $2\log x - 4\log(x-1)$	27. $2\log x + \log(x+3) + \log(x-4) - \log(x+4) - \log(x-3)$	29. 9.96578
31. −2.24926	33. 2.221999	35. $x \in (1, \infty)$
37. $x \in (-\infty, 2)$		

14.3 Solving Logarithmic and Exponential Equations ....Page 698

1. 1.94306	3. $\log 3 + \log(x+5) - 2\log(x-2)$	5. Decreasing
7. 18.78; 23.36; 31; 46.27; 92.09 years	9. 29.92 years	11. 2.37
13. 0.5	15. $\frac{4}{3}$	17. $x \in \varnothing$
19. 2.39	21. 83	23. $x \in \varnothing$
25. −2, 4	27. −5.89, 2.89	

14.4 The Logarithmic Function as a Mathematical Model....................Page 711

1. 10005	3. 5	5. 1.55789
7. 12	9. 251,189 times as bright	11. x is the relative intensity of one star compared to another and m is the measure of magnitude.
13. 100,000 times stronger	15. 8.1 on the Richter	17. 3162 times as intense.
19. 100,000,000,000,000,000,000	21. No.	23. 140 decibels

